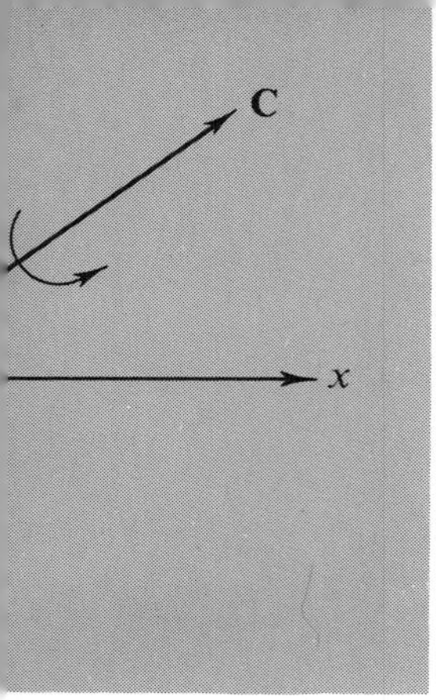

MATHEMATICAL METHODS FOR PHYSICISTS

GEORGE ARFKEN

Miami University
Oxford, Ohio

SECOND EDITION

ACADEMIC PRESS New York San Francisco London
A Subsidiary of Harcourt Brace Jovanovich, Publishers

ACADEMIC PRESS, INC.
111 Fifth Avenue, New York, New York 10003

United Kingdom Edition published by
ACADEMIC PRESS, INC. (LONDON) LTD.
24/28 Oval Road, London NW1

Library of Congress Catalog Card Number: 73-119611

PRINTED IN THE UNITED STATES OF AMERICA

TO CAROLYN

CONTENTS

Chapter 2. COORDINATE SYSTEMS

Chapter 3. TENSOR ANALYSIS

Chapter 4. DETERMINANTS, MATRICES, AND GROUP THEORY

Chapter 5. INFINITE SERIES

Chapter 6. FUNCTIONS OF A COMPLEX VARIABLE I. ANALYTIC PROPERTIES, CONFORMAL MAPPING

Chapter 7. FUNCTIONS OF A COMPLEX VARIABLE II. CALCULUS OF RESIDUES

Chapter 8. SECOND-ORDER DIFFERENTIAL EQUATIONS

Chapter 17. CALCULUS OF VARIATIONS

PREFACE TO THE SECOND EDITION

This second edition of *Mathematical Methods for Physicists* incorporates a number of changes, additions, and improvements made on the basis of experience with the first edition and the helpful suggestions of a number of people. Major revisions have been made in the sections on complex variables, Dirac delta function, and Green's functions. New sections have been included on oblique coordinates, Fourier-Bessel series, and angular momentum ladder operators. The major addition is a series of sections on group theory. While these could have been presented as a separate group theory chapter, there seemed to be several advantages to include them in Chapter 4, Matrices. Since the group theory is developed in terms of matrices the arrangement seems a reasonable one.

PREFACE TO THE FIRST EDITION

Mathematical Methods for Physicists is based upon two courses in mathematics for physicists given by the author over the past eighteen years, one at the junior level and one at the beginning graduate level. This book is intended to provide the student with the mathematics he needs for advanced undergraduate and beginning graduate study in physical science and to develop a strong background for those who will continue into the mathematics of advanced theoretical physics. A mastery of calculus and a willingness to build on this mathematical foundation are assumed.

This text has been organized with two basic principles in view. First, it has been written in a form that it is hoped will encourage independent study. There are frequent cross references but no fixed, rigid page-by-page or chapter-by-chapter sequence is demanded.

The reader will see that mathematics as a language is beautiful and elegant. Unfortunately, elegance all too often means elegance for the expert and obscurity for the beginner. While still attempting to point out the intrinsic beauty of mathematics, elegance has occasionally been reluctantly but deliberately sacrificed in the hope of achieving greater flexibility and greater clarity *for the student*.

Mathematical rigor has been treated in a similar spirit. It is not stressed to the point of becoming a mental block to the use of mathematics. Limitations are explained, however, and warnings given against blind, uncomprehending application of mathematical relations.

The second basic principle has been to emphasize and re-emphasize physical examples in the text and in the exercises to help motivate the student, to illustrate the relevance of mathematics to his science and engineering.

This principle has also played a decisive role in the selection and development of material. The subject of differential equations, foı example, is no longer a series of trick solutions of abstract, relatively meaningless puzzles but the solutions and general properties of the differential equations the student will most frequently encounter in a description of our real physical world.

ACKNOWLEDGMENTS

Any work of this sort necessarily represents the influence and help of many people. I acknowledge gratefully my debt to the professors who taught me physics and mathematics and who instilled in me a love of these disciplines. Among these men, Professors G. Breit, H. Margenau, and E. J. Miles deserve special mention. I thank my colleagues, particularly Professors D. C. Kelly and P. A. Macklin, for their assistance and helpful criticism. A special note of thanks is due to my students over the past eighteen years for their criticisms and reactions which have played a major part in shaping this book, Mr. J. Clow and Mr. S. Orfanides were particularly helpful. A special acknowledgment is owed Mrs. Juanita Killough for so patiently and conscientiously typing this manuscript.

INTRODUCTION

Many of the physical examples used to illustrate the applications of mathematics are taken from the fields of electromagnetic theory and quantum mechanics. For convenience the main equations are listed below and the symbols identified. References in these fields are also given.

Electromagnetic theory

MAXWELL'S EQUATIONS (MKS UNITS—VACUUM)

$$\nabla \cdot \mathbf{D} = \rho \qquad \nabla \times \mathbf{E} = -\frac{\partial \mathbf{B}}{\partial t}$$

$$\nabla \cdot \mathbf{B} = 0 \qquad \nabla \times \mathbf{H} = \frac{\partial \mathbf{D}}{\partial t} + \mathbf{J}$$

Here **E** is the electric field defined in terms of force on a static charge and **B** the magnetic induction defined in terms of force on a moving charge. The related fields **D** and **H** are given (in vacuum) by

$$\mathbf{D} = \varepsilon_0 \mathbf{E} \qquad \text{and} \qquad \mathbf{B} = \mu_0 \mathbf{H}$$

The quantity ρ represents free charge density while **J** is the corresponding current. For additional details see: J. M. Marion, *Classical Electromagnetic Radiation*, New York: Academic Press (1965); W. K. H. Panofsky and M. Phillips, *Classical Electricity and Magnetism*, Reading, Mass.: Addison-Wesley (1955); J. D. Jackson, *Classical Electrodynamics*, New York: Wiley (1962).

Note that Marion and Jackson prefer Gaussian units. A glance at the last two texts and the great demands they make upon the student's mathematical competence should provide considerable motivation for the study of this book.

Quantum Mechanics

SCHRÖDINGER WAVE EQUATION (TIME INDEPENDENT)

$$-\frac{\hbar^2}{2m}\nabla^2\psi + V\psi = E\psi$$

ψ is the (unknown) wave function. The potential energy, often a function of position, is denoted by V while E is the total energy of the system. The mass of the particle being described by ψ is m. $\hbar$ is Planck's constant h divided by 2π. Among the extremely large number of beginning or intermediate texts we might note: A. Messiah, *Quantum Mechanics* (2 vols), New York; Wiley (1961): R. H. Dicke and J. P. Wittke, *Introduction to Quantum Mechanics*, Reading Mass.: Addison-Wesley (1960); E. Merzbacher, *Quantum Mechanics* (Second Edition), New York: Wiley (1970).

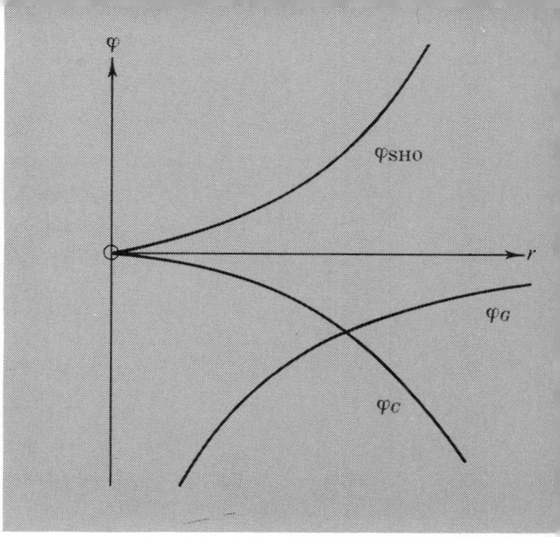

CHAPTER 1

VECTOR ANALYSIS

1.1 Definitions, Elementary Approach

In science and engineering we frequently encounter quantities which have magnitude and magnitude only: mass, time, and temperature. These we label scalar quantities. In contrast, many interesting physical quantities have magnitude and, in addition, an associated direction. This second group includes displacement, velocity, acceleration, force, momentum, and angular momentum. Quantities with magnitude and direction are labeled vector quantities. Usually, in elementary treatments, a vector is defined as a quantity having magnitude and direction. To distinguish vectors from scalars we identify vector quantities with boldface type, that is, **V**.

As an historical sidelight, it is interesting to note that the vector quantities listed are all taken from mechanics but that vector analysis was not used in the development of mechanics and, indeed, had not been created. The need for vector analysis became apparent only with the development of Maxwell's electromagnetic theory and in appreciation of the inherent vector nature of quantities such as electric field and magnetic field.

Our vector may be conveniently represented by an arrow with length proportional to the magnitude. The direction of the arrow gives the direction of the vector, the positive sense of direction being indicated by the point. In this representation vector addition

$$\mathbf{C} = \mathbf{A} + \mathbf{B} \tag{1.1}$$

consists in placing the rear end of vector **B** at the point of vector **A**. Vector **C** is then represented by an arrow drawn from the rear of **A** to the point of **B**. This procedure the triangle law of addition, assigns meaning to Eq. 1.1 and is illustrated in Fig. 1.1.

By completing the parallelogram we see that

$$\mathbf{C} = \mathbf{A} + \mathbf{B} = \mathbf{B} + \mathbf{A}, \tag{1.2}$$

as shown in Fig. 1.2. Similarly

$$\mathbf{D} = \mathbf{A} - \mathbf{B}$$

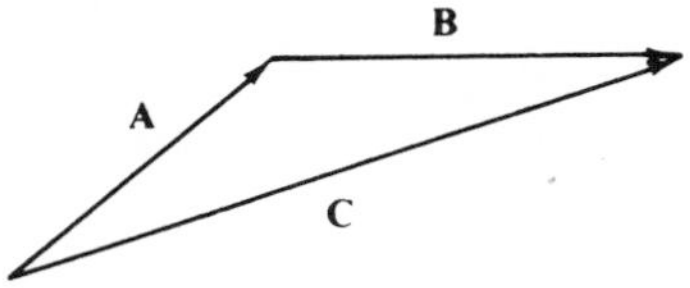

FIG. 1.1 Triangle law of vector addition

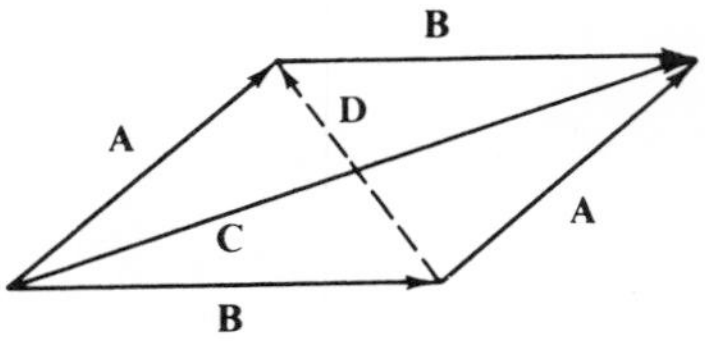

FIG. 1.2 Parallelogram law of vector addition

A direct physical example of this parallelogram addition law is provided by a weight suspended by two cords. If the junction point (O in Fig. 1.3) is in equilibrium, the vector sum of the two forces $\mathbf{F}_1$ and $\mathbf{F}_2$ must just cancel the downward force of gravity, $\mathbf{F}_3$. Here the parallelogram addition law is subject to immediate experimental verification.[1]

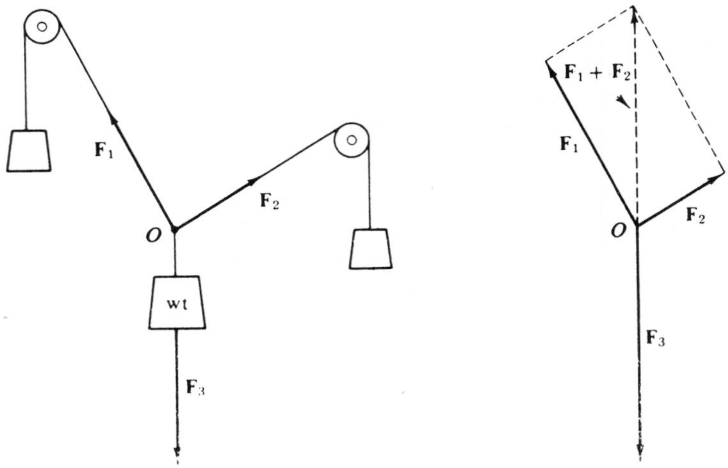

FIG. 1.3 Equilibrium of forces. $\mathbf{F}_1 + \mathbf{F}_2 = -\mathbf{F}_3$

[1] Strictly speaking the parallelogram addition was introduced as a definition. Experiments show that if we assume that the forces are vector quantities and we combine them by parallelogram addition the equilibrium condition of zero resultant force is satisfied.

Note that the vectors are treated as geometrical objects which are independent of any coordinate system. Indeed, we have not yet introduced a coordinate system. This concept of independence of a preferred coordinate system is developed in considerable detail in the next section.

The representation of vector **A** by an arrow suggests a second possibility. Arrow **A** (Fig. 1.4), starting from the origin,[1] terminates at the point (x_1, y_1, z_1). Thus, if we agree that the vector is to start at the origin, the positive end may be specified by giving the cartesian coordinates (x_1, y_1, z_1) of the arrow head.

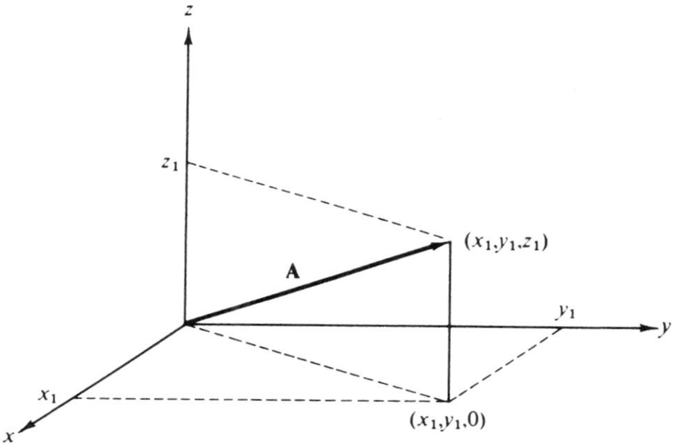

FIG. 1.4 Cartesian components

Although **A** could have represented any vector quantity (momentum, electric field, etc.,), one particularly important vector quantity, the displacement from the origin to the point (x_1, y_1, z_1), is denoted by the special symbol **r**. We then have a choice of referring to the displacement either as the vector **r** or as the collection (x_1, y_1, z_1), the coordinates of its end point.

$$\mathbf{r} = (x_1, y_1, z_1). \tag{1.3}$$

Using r for the magnitude of vector **r**, Fig. 1.5 shows that the end-point coordinates and the magnitude are related by

$$x_1 = r \cos \alpha, \qquad y_1 = r \cos \beta, \qquad z_1 = r \cos \gamma. \tag{1.4}$$

Cos α, cos β, and cos γ are called the *direction cosines*, α being the angle between the given vector and the positive x-axis, and so on. One further bit of vocabulary: the quantities x_1, y_1, and z_1 are known as the (cartesian) *components* of **r** or the *projections* of **r**.

If we proceed in the same manner, any vector **A** may be resolved into its components (or projected onto the coordinate axes) to yield

$$A_x = A \cos \alpha, \tag{1.5}$$

[1] The reader will see that we could start from any point in our cartesian reference frame, We choose the origin for simplicity.

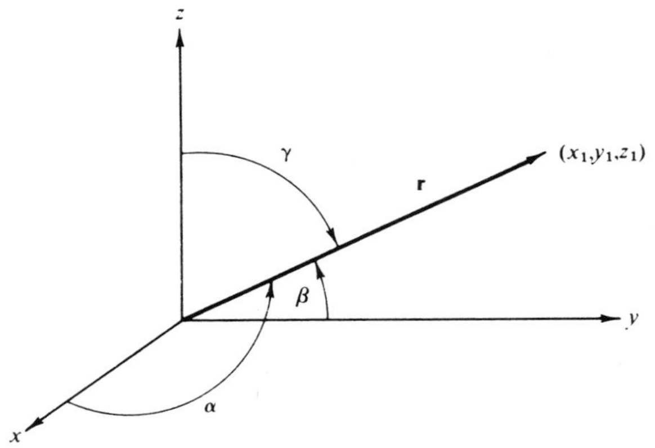

FIG. 1.5 Direction cosines

in which α is the angle between **A** and the positive x-axis. Again we may choose to refer to the vector as a single quantity **A** or to its components (A_x, A_y, A_z). Note that the subscript x in A_x denotes the x component and not a dependence on the variable x. A_x may be a function of x, y, and z as $A_x(x, y, z)$.

At this stage it is convenient to introduce unit vectors along each of the coordinate axes. Let **i** be a vector of unit magnitude pointing in the positive x-direction, **j**, a vector of unit magnitude in the positive y-direction, and **k**, a vector of unit magnitude in the positive z-direction. Then $\mathbf{i}A_x$ is a vector with magnitude equal to A_x and in the positive x-direction. By vector addition

$$\mathbf{A} = \mathbf{i}A_x + \mathbf{j}A_y + \mathbf{k}A_z, \tag{1.6}$$

which states that a vector equals the vector sum of its components. Note that if **A** vanishes all of its components must vanish individually; that is, if

$$\mathbf{A} = 0, \qquad \text{then} \qquad A_x = A_y = A_z = 0.$$

Finally, by the Pythagorean theorem, the magnitude of vector **A** is

$$A = (A_x^2 + A_y^2 + A_z^2)^{1/2}. \tag{1.7a}$$

This resolution of a vector into its components can be carried out in a variety of coordinate systems, as shown in Chapter 2. Here we shall restrict ourselves to cartesian coordinates.

Equation 1.6 is actually an assertion that the three unit vectors **i**, **j**, and **k** *span* our real three-dimensional space: any constant vector may be written as a linear combination of **i**, **j**, and **k**. Since **i**, **j**, and **k** are linearly independent (no one is a linear combination of the other two), they form a *basis* for real three-dimensional space.

As a replacement of the graphical technique, addition and subtraction of vectors may now be carried out in terms of their components. For $\mathbf{A} = \mathbf{i}A_x + \mathbf{j}A_y + \mathbf{k}A_z$ and $\mathbf{B} = \mathbf{i}B_x + \mathbf{j}B_y + \mathbf{k}B_z$,

$$\mathbf{A} \pm \mathbf{B} = \mathbf{i}(A_x \pm B_x) + \mathbf{j}(A_y \pm B_y) + \mathbf{k}(A_z \pm B_z). \tag{1.7b}$$

Example 1.1.1

Let

$$\mathbf{A} = 6\mathbf{i} + 4\mathbf{j} + 3\mathbf{k}$$

$$\mathbf{B} = 2\mathbf{i} - 3\mathbf{j} - 3\mathbf{k}.$$

Then by Eq. 1.7*b*

$$\mathbf{A} + \mathbf{B} = 8\mathbf{i} + \mathbf{j}$$

and

$$\mathbf{A} - \mathbf{B} = 4\mathbf{i} + 7\mathbf{j} + 6\mathbf{k}.$$

EXERCISES

1.1.1 Show how to find **A** and **B**, given **A** + **B** and **A** − **B**.

1.1.2 The vector **A** whose magnitude is 10 units makes equal angles with the coordinate axes. Find A_x, A_y, and A_z.

1.1.3 Calculate the components of a unit vector which lies in the xy-plane and makes equal angles with the positive directions of the x- and y-axes.

1.1.4 A vector equation can be reduced to the form $\mathbf{A} = \mathbf{B}$. From this show that the one vector equation is equivalent to *three* scalar equations.
Assuming the validity of Newton's Second Law $\mathbf{F} = m\mathbf{a}$ as a *vector* equation this means that a_x depends only on F_x and is independent of F_y and F_z.

1.1.5 The vertices of a triangle A, B, and C are given by the points $(-1, 0, 2)$, $(0, 1, 0)$, and $(1, -1, 0)$, respectively. Find point D so that the figure **ABDC** forms a plane parallelogram.
Ans. $(2, 0, -2)$.

1.1.6 A triangle is defined by the vertices of three vectors, **A**, **B**, and **C** that extend from the origin. In terms of **A**, **B**, and **C** show that the *vector* sum of the successive sides of the triangle $(AB + BC + CA)$ is zero.

1.1.7 A sphere of radius a is centered at a point $\mathbf{r}_1$.
(a) Write out the algebraic equation for the sphere.
(b) Write out a *vector* equation for the sphere.

Ans. (a) $(x - x_1)^2 + (y - y_1)^2 + (z - z_1)^2 = a^2$.
(b) $\mathbf{r} = \mathbf{r}_1 + \mathbf{a}$.
(**a** takes on all directions but has a fixed magnitude, a.)

1.1.8 A corner reflector is formed by three mutually perpendicular reflecting surfaces.

Show that a ray of light incident upon the corner reflector (striking all three surfaces) is reflected back along a line parallel to the line of incidence.

Hint: Consider the effect of a reflection on the components of a vector describing the direction of the light ray.

1.2 Rotation of Coordinates

In the last section, vectors were defined or represented in two equivalent ways: (1) by specifying magnitude and direction, as with an arrow, and (2) by specifying the components. In this section, a third definition, in terms of behavior under rotation of the coordinate system, is given.

The definition of vector as a quantity with magnitude and direction breaks down in advanced work. On the one hand, we encounter quantities, such as elastic constants and index of refraction in anisotropic crystals, which have magnitude and direction *but* which are not vectors. On the other hand, our naïve approach is awkward to generalize, to extend to more complex quantities. We seek a new definition of vector, using our displacement vector **r** as a prototype.

There is an important physical basis for our development of a new definition. We describe our physical world by mathematics, but it and any physical predictions we may make must be *independent* of our mathematical analysis. Some writers like to compare the physical system to a building and the mathematical analysis to the scaffolding used to construct the building. In the end the scaffolding is stripped off and the building stands.

In our specific case we shall assume that space is isotropic; that is, there is no preferred direction or all directions are equivalent. Then the physical system being analyzed or the physical law being enunciated cannot and must not depend on our choice or *orientation* of our coordinate system.

Now we return to the concept of vector **r** as a geometric object independent of the coordinate system. Let us look at **r** in two different systems, one rotated in relation to the other.

For simplicity we consider first the two-dimensional case. If the coordinates (x, y) are rotated counterclockwise through an angle φ, *keeping* **r** *fixed*, we get the following relations between the components resolved in the original system

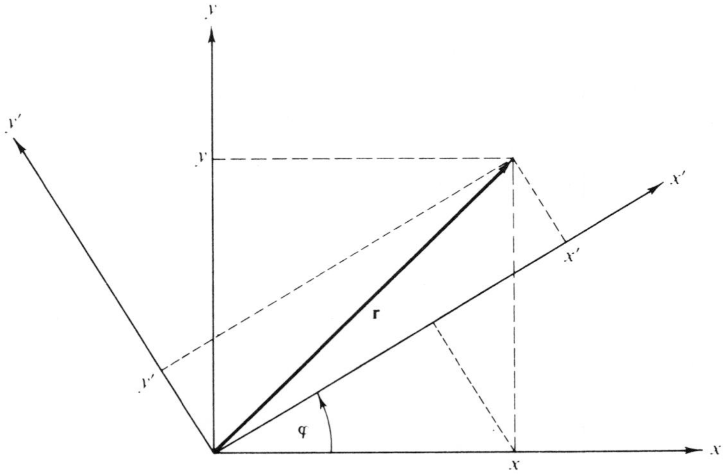

FIG. 1.6 Rotation of cartesian coordinates

(unprimed) and those resolved in the new rotated system (primed):

$$\begin{aligned} x' &= x \cos \varphi + y \sin \varphi, \\ y' &= -x \sin \varphi + y \cos \varphi \end{aligned} \tag{1.8}$$

We saw in Section 1.1 that a vector could be represented by the coordinates of a point; that is, the coordinates were proportional to the vector components. Hence the components of a vector must transform under rotation as coordinates of a point (such as **r**). Therefore, whenever any pair of quantities (A_x, A_y) in the xy-coordinate system is transformed into (A'_x, A'_y) by this rotation of the coordinate system with

$$\begin{aligned} A'_x &= A_x \cos \varphi + A_y \sin \varphi \\ A'_y &= -A_x \sin \varphi + A_y \cos \varphi, \end{aligned} \tag{1.9}$$

we *define*[1] A_x and A_y as the components of a vector **A**. Our vector now is defined in terms of the transformation of its components under rotation of the coordinate system. If A_x and A_y transform in the same way as x and y, the components of the two-dimensional displacement vector, they are the components of a vector **A**. If A_x and A_y do not show this behavior when the coordinates are rotated, they do not form a vector.

To complete the definition we must assign meaning to A'_x and A'_y in Eq. 1.9. Let us suppose that the components of a vector **A** are functions of the coordinates and perhaps of some other constant vector **c**.

$$\begin{aligned} A_x &= A_x(x, y, c_x, c_y), \\ A_y &= A_y(x, y, c_x, c_y). \end{aligned} \tag{1.9a}$$

In the primed coordinate system **A** has components A'_x and A'_y, which are also functions of position (in the primed system) and **c**.

$$\begin{aligned} A'_x &= A'_x(x', y', c'_x, c'_y) \\ A'_y &= A'_y(x', y', c'_x, c'_y). \end{aligned} \tag{1.9b}$$

If we use Eq. 1.8, the primed quantities x', y', c'_x, c'_y in Eq. 1.9*b* can be replaced by unprimed quantities *and* the angle of rotation φ. In order that no preferred system be created, the functional dependence of A'_x and A'_y upon the primed variables must be the same for all primed coordinate systems.

Now for the special case $\varphi = 0$, we have $x' = x$, $y' = y$, etc. Clearly

$$\begin{aligned} A_x &= A'_x, \\ A_y &= A'_y. \end{aligned}$$

Hence A'_x is the same function of x', y', c'_x, and c'_y as A_x is of the corresponding unprimed quantities and similarly for A'_y and A_y.

The vector components A_x and A_y satisfying the defining equations, Eq. 1.9,

[1] The corresponding definition of a scalar quantity is $S' = S$, i.e., invariant under rotation of the coordinates.

associate a magnitude A and a direction with each point in space. The magnitude is a scalar quantity, invariant to the rotation of the coordinate system. The direction (relative to the unprimed system) is likewise invariant to the rotation of the coordinate system (see Exercise 1.2.6). The result of all this is that the components of a vector may vary according to the rotation of the primed coordinate system. This is what Eq. 1.9 says. But the variation with angle is just such that the components in the rotated coordinate system A'_x and A'_y define a vector with the same magnitude and the same direction relative to the xy-axes as do A_x and A_y.

EXAMPLE 1.2.1

Given a pair of quantities $(-y, x)$, show that they are the components of a two-dimensional vector.

Since the question of whether or not these are the components of a vector depends, by definition, on their transformation properties, we investigate how they do transform. Let

$$\begin{aligned} V'_x &= -y \cos \varphi + x \sin \varphi, \\ V'_y &= y \sin \varphi + x \cos \varphi, \end{aligned} \tag{1.10}$$

using $V_x = -y$, $V_y = x$. By the arguments above (V'_x is the same function of x' and y' as V_x is of x and y), we have

$$\begin{aligned} V'_x &= -y', \\ V'_y &= x', \end{aligned} \tag{1.11}$$

and using Eq. 1.8 for the coordinate transformation to eliminate the primed variables, we see that the equalities hold. Equation 1.9, the defining equation of a two-dimensional vector, is satisfied. Therefore $(-y, x)$ are the components of a vector.

EXAMPLE 1.2.2

As an example of a quantity that, by our definition, is *not* a vector, consider $\mathbf{V} = \mathbf{i}x - \mathbf{j}y = (x, -y)$. Testing the transformation properties, we have, by Eq. 1.9,

$$\begin{aligned} V'_x &= x' = x \cos \varphi - y \sin \varphi, \\ V'_y &= -y' = -x \sin \varphi - y \cos \varphi. \end{aligned}$$

Clearly these equations are inconsistent with Eq. 1.8. Equation 1.9 is *not* satisfied. Hence $(x, -y)$ are *not* the components of a vector.

Many writers prefer to label the functions A_x and A_y satisfying Eq. 1.9 the components of a (two-dimensional) *vector field*. Distinguished from vector fields are *constant vectors* such as **i** and **j**, which do not transform so that **i**′ is the same function of x' and y' that it is of x and y. Indeed **i** is *not* a function of x and y at all; it is a constant. For a further note on constant vectors see Problem 1.2.1.

To go on to three and, later, four dimensions, it becomes convenient to use a more compact notation. Let

$$x \to x_1$$

$$y \to x_2$$

$$\begin{aligned} a_{11} &= \cos\varphi, & a_{12} &= \sin\varphi, \\ a_{21} &= -\sin\varphi, & a_{22} &= \cos\varphi. \end{aligned} \tag{1.12}$$

Then Eq. 1.8 becomes

$$\begin{aligned} x_1' &= a_{11}x_1 + a_{12}x_2, \\ x_2' &= a_{21}x_1 + a_{22}x_2. \end{aligned} \tag{1.13}$$

The coefficient a_{ij} may be interpreted as a direction cosine, the cosine of the angle between x_i' and x_j; that is,

$$a_{12} = \cos(x_1', x_2) = \sin\varphi,$$

$$a_{21} = \cos(x_2', x_1) = \cos\left(\varphi + \frac{\pi}{2}\right) = -\sin\varphi.$$

The advantage of the new notation[1] is that it permits us to use the summation symbol $\sum$ and to rewrite Eqs. 1.13 as

$$x_i' = \sum_{j=1}^{2} a_{ij}x_j, \qquad i = 1, 2. \tag{1.14}$$

Note that i remains as a parameter which gives rise to one equation when it is set equal to 1 and to a second equation when it is set equal to 2. The index j, of course, is a summation index, a dummy index, and just as with a variable of integration j may be replaced by any other convenient symbol.

The generalization to three, four, or N dimensions is very simple. The set of N quantities, V_j, is said to be the components of an N-dimensional vector, $\mathbf{V}$, if and only if the values in a rotated coordinate system are given by

$$V_i' = \sum_{j=1}^{N} a_{ij}V_j, \qquad i = 1, 2, \ldots, N. \tag{1.15}$$

As before, a_{ij} is the cosine of the angle between x_i' and x_j. Often the upper limit N and the corresponding range of i will not be indicated. It is taken for granted that the reader knows how many dimensions his space has.

From the definition of a_{ij} as the cosine of the angle between the positive x_i'

[1] The reader may wonder at the replacement of one parameter φ by four parameters a_{ij}. Clearly, the a_{ij} do not constitute a minimum set of parameters. For two dimensions the four a_{ij} are subject to the three constraints given in Eq. 1.18. The justification for the redundant set of direction cosines is the convenience it provides. Hopefully, this convenience will become more apparent in Chapters 3 and 4. For three dimensional rotations (9 a_{ij} but only three independent) alternate descriptions are provided by: (1) the Euler angles discussed in Section 4.3, (2) quaternions, and (3) the Cayley–Klein parameters. These alternatives have their respective advantages and disadvantages.

direction and the positive x_j direction we may write (cartesian coordinates)[1]

$$a_{ij} = \frac{\partial x'_i}{\partial x_j} = \frac{\partial x_j}{\partial x'_i}. \tag{1.16}$$

Note carefully that these are *partial derivatives*. By use of Eq. 1.16, Eq. 1.15 becomes

$$V'_i = \sum_{j=1}^{N} \frac{\partial x'_i}{\partial x_j} V_j = \sum_{j=1}^{N} \frac{\partial x_j}{\partial x'_i} V_j. \tag{1:17}$$

The direction cosines a_{ij} satisfy an *orthogonality condition*

$$\sum_i a_{ij} a_{ik} = \delta_{jk} \tag{1.18}$$

or, equivalently,

$$\sum_i a_{ji} a_{ki} = \delta_{jk}. \tag{1.19}$$

The symbol δ_{jk} is the Kronecker delta defined by

$$\begin{aligned} \delta_{jk} &= 1 \quad \text{for} \quad j = k, \\ \delta_{jk} &= 0 \quad \text{for} \quad j \neq k. \end{aligned} \tag{1.20}$$

The reader may easily verify that Eqs. 1.18 and 1.19 hold in the two-dimensional case by substituting in the specific a_{ij} from Eq. 1.12. The result is the well-known identity $\sin^2 \varphi + \cos^2 \varphi = 1$ for the nonvanishing case. To verify Eq. 1.18 in general form we may use the partial derivative forms of Eqs. 1.16 to obtain

$$\sum_i \frac{\partial x_j}{\partial x'_i} \frac{\partial x_k}{\partial x'_i} = \sum_i \frac{\partial x_j}{\partial x'_i} \frac{\partial x'_i}{\partial x_k} = \frac{\partial x_j}{\partial x_k}. \tag{1.21}$$

The last step follows by the standard rules for partial differentiation, assuming that x_j is a function of x'_1, x'_2, x'_3, etc. The final result, $\partial x_j / \partial x_k$, is equal to δ_{jk}, since x_j and x_k $(j \neq k)$ are assumed to be perpendicular (two or three dimensions) or orthogonal (for any number of dimensions). If $j = k$, the partial derivative is clearly equal to 1.

In redefining a vector in terms of how its components transform under a rotation of the coordinate system, two points deserve special emphasis:

1. This definition is developed because it is useful and appropriate in describing our physical world. Our vector equations will be independent of any particular coordinate system. (The coordinate system need not even be cartesian.) The vector equation can always be expressed in some particular coordinate system and, to obtain numerical results, the equation must ultimately be expressed in some specific coordinate system.

2. This definition is subject to a generalization that will open up the branch of mathematics known as tensor analysis (Chapter 3).

A qualification is also in order. The behavior of the vector components under

[1] Differentiate $x'_i = \sum a_{ik} x_k$ with respect to x_j. See the discussion following Eq. 1.21.

rotation of the coordinates is used in Section 1.3 to prove that a scalar product is a scalar, in Section 1.4 to prove that a vector product is a vector, and in Section 1.6 to show that the gradient of a scalar, $\nabla\psi$, is a vector. The remainder of this chapter proceeds on the basis of the less restrictive definitions of vector given in Section 1.1. This is consistent with the concept of vector or vector space (or linear space) of linear algebra. Here the name vector is applied to any objects, our three vector components, polynomials, or functions, etc., which may be added together and multiplied by numbers (scalars) as we have done in the two preceding sections. The collection of such objects is called a vector space or a linear space. It may have n dimensions, not just two or three, or it may be infinite-dimensional. The vectors may be real or complex. Much of modern quantum theory employs an infinite-dimensional vector space (Hilbert space) in which the unit vectors are replaced by orthonormal functions. Our Fourier series, Chapter 14, is an example of this.

EXERCISES

1.2.1 Given a *constant* vector **V** with components $V_x = 1$ and $V_y = 0$, show that the components of **V** in a rotated system are

$$V_{x'} = \cos\varphi,$$
$$V_{y'} = -\sin\varphi,$$

according to the vector transformation law. Clearly a φ dependence has entered and could be expected. We have singled out a preferred direction in space by introducing the constant vector. **V** is a vector according to the sense of Section 1.1, and we continue to accept it as a vector because of its usefulness.

1.2.2 Determine whether the following satisfy the vector transformation rule (Eq. 1.15).
(a) $(x - y,\ x + y,\ 0)$, for rotations about the z-axis.
(b) $(0,\ 2z + y,\ z - 2y)$, for rotations about the x-axis.
(c) $(y^2 + z^2,\ -xy,\ -xz)$, for rotations about each of the three coordinate axes.

1.2.3 (a) Show that $(xyc_x + y^2c_y,\ -x^2c_x - xyc_y)$ forms a vector. The quantities c_x and c_y are components of a constant vector **c**.
(b) Repeat for $(xyc_x - x^2c_y,\ y^2c_x - xyc_y)$.

1.2.4 A two-dimensional vector **V** has the form $(ax + by,\ cx + dy)$ with a, b, c, and d constants. Show that **V** is a linear combination of the radial vector $\mathbf{r} = \mathbf{i}x + \mathbf{j}y$ and the tangential vector $\mathbf{t} = \mathbf{i}y - \mathbf{j}x$.

$$\mathbf{V} = a\mathbf{r} + b\mathbf{t}.$$

Hint. The vector transformation law must hold (a) for all angles and (b) for all points, (x, y).

1.2.5 Prove the orthogonality condition $\sum_i a_{ji}a_{ki} = \delta_{jk}$. As a special case of this the direction cosines of Section 1.1 satisfy the relation

$$\cos^2\alpha + \cos^2\beta + \cos^2\gamma = 1,$$

a result that also follows from Eq. 1.7*a*.

1.2.6 (a) Show that the magnitude of a vector **A**, $A = (A_x^2 + A_y^2)^{1/2}$ is independent of the orientation of the rotated coordinate system,

$$(A_x^2 + A_y^2)^{1/2} = (A_x'^2 + A'^2)^{1/2}$$

independent of the rotation angle φ.
This independence of angle is expressed by saying that A is *invariant* under rotations.
(b) At a given point (x,y) **A** defines an angle α relative to the positive x-axis and α' relative to the positive x'-axis. The angle from x to x' is φ. Show that $\mathbf{A} = \mathbf{A}'$ defines the *same* direction in space when expressed in terms of its primed components, as in terms of its unprimed components, i.e., that

$$\alpha' = \alpha - \varphi.$$

1.3 Scalar or Dot Product

Having defined vectors, we now proceed to combine them. The laws for combining vectors must be mathematically consistent. From the possibilities that are consistent we select two that are both mathematically and physically interesting. A third possibility is introduced in Chapter 3, in which we form tensors.

The combination of $AB \cos \theta$, in which A and B are the magnitudes of two

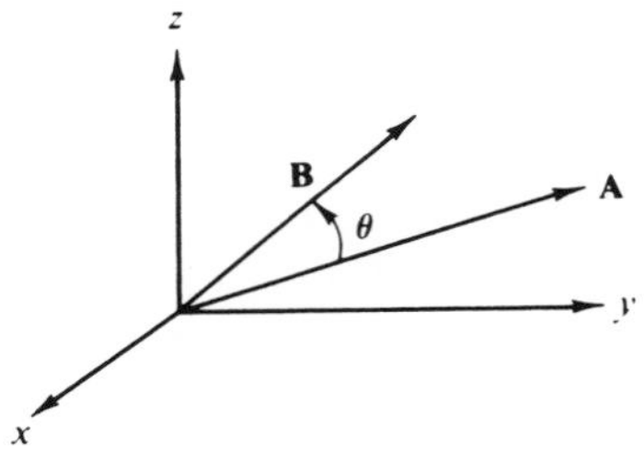

FIG. 1.7 Scalar product $\mathbf{A} \cdot \mathbf{B} = AB \cos \theta$

vectors and θ, the angle between them, occurs frequently in physics. For instance,

$$\text{work} = \text{force} \times \text{displacement} \times \cos \theta$$

is usually interpreted as displacement times the projection of the force along the displacement.

With such applications in mind, we define

$$\mathbf{A} \cdot \mathbf{B} = A_x B_x + A_y B_y + A_z B_z = \sum_i A_i B_i \tag{1.22}$$

as the scalar or dot product of **A** and **B**. We note that from this definition $\mathbf{A} \cdot \mathbf{B} = \mathbf{B} \cdot \mathbf{A}$. Our unit vectors **i**, **j**, and **k** satisfy the relations

$$\mathbf{i} \cdot \mathbf{i} = \mathbf{j} \cdot \mathbf{j} = \mathbf{k} \cdot \mathbf{k} = 1, \tag{1.22a}$$

whereas

$$\begin{aligned} \mathbf{i} \cdot \mathbf{j} = \mathbf{i} \cdot \mathbf{k} = \mathbf{j} \cdot \mathbf{k} = 0, \\ \mathbf{j} \cdot \mathbf{i} = \mathbf{k} \cdot \mathbf{i} = \mathbf{k} \cdot \mathbf{j} = 0. \end{aligned} \tag{1.22b}$$

If we reorient our axes and let **A** define a new x-axis,[1] then

$$A_x = A, \qquad A_y = 0, \qquad A_z = 0$$

and

$$B_x = B \cos \theta.$$

Then by Eq. 1.22

$$\mathbf{A} \cdot \mathbf{B} = AB \cos \theta, \tag{1.23}$$

which may be taken as a second definition of scalar product. This shows that work is the scalar product of force and displacement.

Example 1.3.1

For the two vectors **A** and **B** of Example 1.1.1, $\mathbf{A} = 6\mathbf{i} + 4\mathbf{j} + 3\mathbf{k}$, $\mathbf{B} = 2\mathbf{i} - 3\mathbf{j} - 3\mathbf{k}$,

$$\mathbf{A} \cdot \mathbf{B} = (12 - 12 - 9) = -9$$

by Eq. 1.22. In this case the projection of **A** on **B** (or **B** on **A**) is negative. Actually

$$|\mathbf{A}| = (36 + 16 + 9)^{1/2} = (61)^{1/2} = 7.81,$$

$$|\mathbf{B}| = (\ 4 + \ \ 9 + 9)^{1/2} = (22)^{1/2} = 4.69,$$

and $\cos \theta = -0.246$, $\theta = 104.2°$.

If $\mathbf{A} \cdot \mathbf{B} = 0$ and we know that $\mathbf{A} \neq 0$ and $\mathbf{B} \neq 0$, then from Eq. 1.23 $\cos \theta = 0$ or $\theta = 90°$, 270°, etc. The vectors **A** and **B** must be perpendicular. Alternately we may say **A** and **B** are orthogonal. The unit vectors **i**, **j**, and **k** are mutually orthogonal. To develop this notion of orthogonality one more step, suppose that **n** is a unit

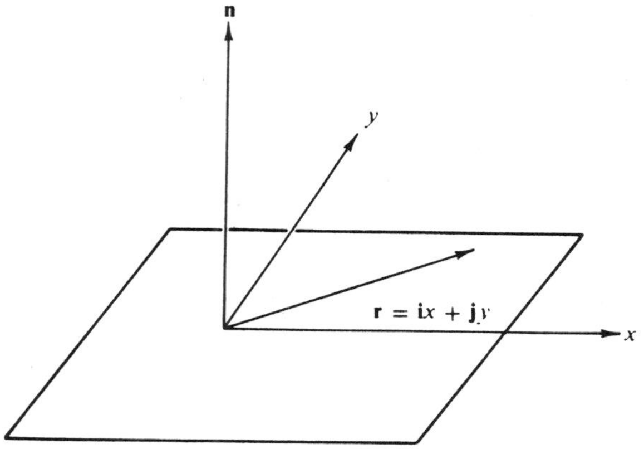

FIG. 1.8 A normal vector

[1] The invariance of **A** • **B** under rotation of the coordinates is proved later in this section.

vector and $\mathbf{r}$ is a nonzero vector in the xy-plane; that is, $\mathbf{r} = \mathbf{i}x + \mathbf{j}y$. If

$$\mathbf{n} \cdot \mathbf{r} = 0$$

for *all* choices of $\mathbf{r}$, then $\mathbf{n}$ must be perpendicular (orthogonal) to the xy-plane.

We have not yet shown that the word scalar is justified or that the scalar product is indeed a scalar quantity. To do this we investigate the behavior of $\mathbf{A} \cdot \mathbf{B}$ under a rotation of the coordinate system. By use of Eq. 1.15

$$\begin{aligned} A'_x B'_x + A'_y B'_y + A'_z B'_z &= \sum_i a_{xi} A_i \sum_j a_{xj} B_j \\ &\quad + \sum_i a_{yi} A_i \sum_j a_{yj} B_j \\ &\quad + \sum_i a_{zi} A_i \sum_j a_{zj} B_j . \end{aligned} \tag{1.24}$$

Using the indices k and l to sum over x, y, and z, we obtain

$$\sum_k A'_k B'_k = \sum_l \sum_i \sum_j a_{li} A_i a_{lj} B_j , \tag{1.25}$$

and, by rearranging the terms on the right-hand side,

$$\begin{aligned} \sum_k A'_k B'_k &= \sum_i \sum_j \sum_l (a_{li} a_{lj}) A_i B_j \\ &= \sum_i \sum_j \delta_{ij} A_i B_j \\ &= \sum_i A_i B_i . \end{aligned} \tag{1.26}$$

The last two steps follow by using Eq. 1.18, the orthogonality condition of the direction cosines, and Eq. 1.20, which defines the Kronecker delta. The effect of the Kronecker delta is to cancel all terms in a summation over either index except the term for which the indices are equal. In Eq. 1.26 its effect is to set $j = i$ and to eliminate the summation over j. Of course, we could equally well set $i = j$ and eliminate the summation over i. Equation 1.26 gives us

$$\sum_k A'_k B'_k = \sum_i A_i B_i , \tag{1.27}$$

which is just our definition of a scalar quantity, one that remains *invariant* under the rotation of the coordinate system.

In a similar approach which exploits this concept of invariance we take $\mathbf{C} = \mathbf{A} + \mathbf{B}$ and dot it into itself.

$$\begin{aligned} \mathbf{C} \cdot \mathbf{C} &= (\mathbf{A} + \mathbf{B}) \cdot (\mathbf{A} + \mathbf{B}) \\ &= \mathbf{A} \cdot \mathbf{A} + \mathbf{B} \cdot \mathbf{B} + 2\mathbf{A} \cdot \mathbf{B}. \end{aligned} \tag{1.28}$$

Since

$$\mathbf{C} \cdot \mathbf{C} = C^2, \tag{1.29}$$

the square of the magnitude of vector $\mathbf{C}$ and thus an invariant quantity, we see that

$$\mathbf{A} \cdot \mathbf{B} = \tfrac{1}{2}(C^2 - A^2 - B^2), \qquad \text{invariant.} \tag{1.30}$$

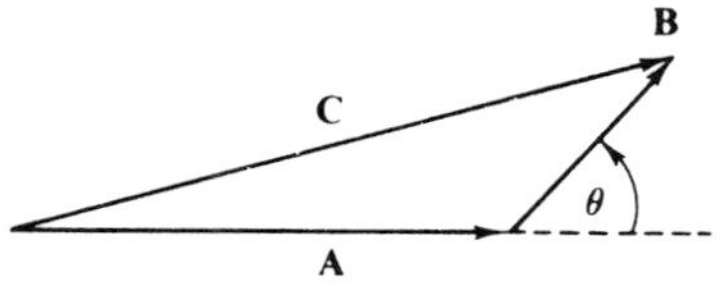

FIG. 1.9 The law of cosines

Since the right-hand side of Eq. 1.30 is invariant—that is, a scalar quantity—the left-hand side, $\mathbf{A} \cdot \mathbf{B}$, must also be invariant under rotation of the coordinate system. Hence $\mathbf{A} \cdot \mathbf{B}$ is a scalar.

Equation 1.28 is really another form of the law of cosines which is

$$C^2 = A^2 + B^2 + 2AB \cos \theta. \tag{1.31}$$

Comparing Eqs. 1.28 and 1.31, we have another verification of Eq. 1.23, or, if preferred, a vector derivation of the law of cosines.

An interesting illustration of the geometric interpretation of the scalar product is provided by an example from a branch of general relativity. Consider a four-dimensional sphere

$$x^2 + y^2 + z^2 + w^2 = 1$$

in x, y, z, w space. The surface of this four-dimensional sphere may be described by the vector $\mathbf{r} = (x, y, z, w)$ with the restriction that $|\mathbf{r}| = 1$. It is possible to construct a unit vector $\mathbf{t}$ that is tangential to this four-dimensional sphere over its entire surface. As one possible example,

$$\mathbf{t} = (y, -x, w, -z).$$

The reader may verify that

$$\mathbf{t} \cdot \mathbf{t} = 1,$$

therefore unit magnitude, and

$$\mathbf{t} \cdot \mathbf{r} = 0,$$

therefore tangential, over the entire sphere.

The two-dimensional analog exists (cf. Example 1.2.1), but there is no three-dimensional analog. Hair growing out of a sphere cannot be combed down all over. There will be a cowlick.

Our dot product, given by Eq. 1.22, may be generalized in two ways. Our space need not be restricted to three dimensions. In n-dimensional space, Eq. 1.22 applies with the sum running from 1 to n. n may be infinity, with the sum then a convergent infinite series (Section 5.2). The other generalization extends the concept of vector to embrace functions. The function analog of a dot or inner product appears in Section 9.4.

EXERCISES

1.3.1 What is the cosine of the angle between the vectors

$$\mathbf{A} = 3\mathbf{i} + 4\mathbf{j} + \mathbf{k}$$

and

$$\mathbf{B} = \mathbf{i} - \mathbf{j} + \mathbf{k}$$

Ans. $\cos\theta = 0, \qquad \theta = \frac{\pi}{2}.$

1.3.2 Two unit magnitude vectors $\mathbf{a}_i$ and $\mathbf{a}_j$ are required to be either parallel or perpendicular to each other. Show that $\mathbf{a}_i \cdot \mathbf{a}_j$ provides an interpretation of Eq. 1.18, the direction cosine orthogonality relation.

1.3.3 Given (1) that the dot product of a unit vector with itself is unity, and (2) that this relation is valid in all (rotated) coordinate systems, show that $\mathbf{i}' \cdot \mathbf{i}' = 1$ (with the primed system rotated 45° about the z-axis relative to the unprimed) implies that $\mathbf{i} \cdot \mathbf{j} = 0$.

1.3.4 The vector $\mathbf{r}$, starting at the origin, terminates at and specifies the point in space (x, y, z). Find the surface swept out by $\mathbf{r}$ if
(a) $(\mathbf{r} - \mathbf{a}) \cdot \mathbf{a} = 0$,
(b) $(\mathbf{r} - \mathbf{a}) \cdot \mathbf{r} = 0$.
The vector $\mathbf{a}$ is a constant (constant in magnitude and direction).

1.3.5

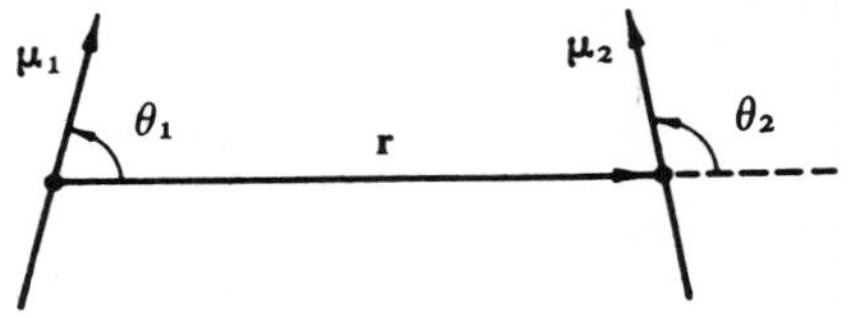

The interaction energy between two dipoles of moments $\boldsymbol{\mu}_1$ and $\boldsymbol{\mu}_2$ may be written in the vector form

$$V = -\frac{\boldsymbol{\mu}_1 \cdot \boldsymbol{\mu}_2}{r^3} + \frac{3(\boldsymbol{\mu}_1 \cdot \mathbf{r})(\boldsymbol{\mu}_2 \cdot \mathbf{r})}{r^5}$$

and in the scalar form

$$V = \frac{\mu_1 \mu_2}{r^3}(2\cos\theta_1\cos\theta_2 - \sin\theta_1\sin\theta_2\cos\varphi).$$

Here θ_1 and θ_2 are the angles of $\boldsymbol{\mu}_1$ and $\boldsymbol{\mu}_2$ relative to $\mathbf{r}$, while φ is the azimuth of $\boldsymbol{\mu}_2$ relative to the $\boldsymbol{\mu}_1 - \mathbf{r}$ plane. Show that these two forms are equivalent.
Hint. Eq. 12.198 will be helpful.

1.4 Vector or Cross Product

A second form of vector multiplication employs the sine of the included angle instead of the cosine. For instance, the angular momentum of a body is defined as

$$\text{angular momentum} = \text{radius arm} \times \text{linear momentum}$$
$$= \text{distance} \times \text{linear momentum} \times \sin\theta.$$

For convenience in treating problems relating to quantities such as angular momentum, torque, and angular velocity we define the vector or cross product as

$$\mathbf{C} = \mathbf{A} \times \mathbf{B},$$

with

$$C = AB\sin\theta. \tag{1.32}$$

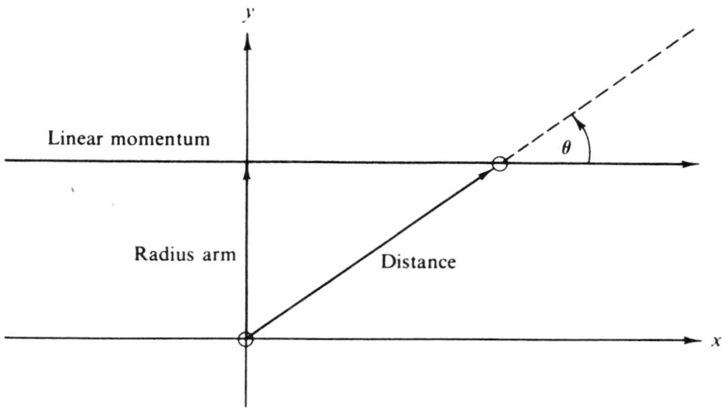

FIG. 1.10 Angular momentum

Unlike the preceding case of the scalar product, **C** is now a vector, and we assign it a direction perpendicular to the plane of **A** and **B** such that **A**, **B**, and **C** form a right-handed system. With this choice of direction we have

$$\mathbf{A} \times \mathbf{B} = -\mathbf{B} \times \mathbf{A}, \qquad \text{anticommutation.} \tag{1.32a}$$

From this definition of cross product we have

$$\mathbf{i} \times \mathbf{i} = \mathbf{j} \times \mathbf{j} = \mathbf{k} \times \mathbf{k} = 0, \tag{1.32b}$$

whereas

$$\mathbf{i} \times \mathbf{j} = \mathbf{k}, \qquad \mathbf{j} \times \mathbf{k} = \mathbf{i}, \qquad \mathbf{k} \times \mathbf{i} = \mathbf{j}$$

and

$$\mathbf{j} \times \mathbf{i} = -\mathbf{k}, \qquad \mathbf{k} \times \mathbf{j} = -\mathbf{i}, \qquad \mathbf{i} \times \mathbf{k} = -\mathbf{j}. \tag{1.32c}$$

The familiar magnetic induction **B** is usually defined by the vector product force equation[1]

$$\mathbf{F}_M = q\mathbf{v} \times \mathbf{B}.$$

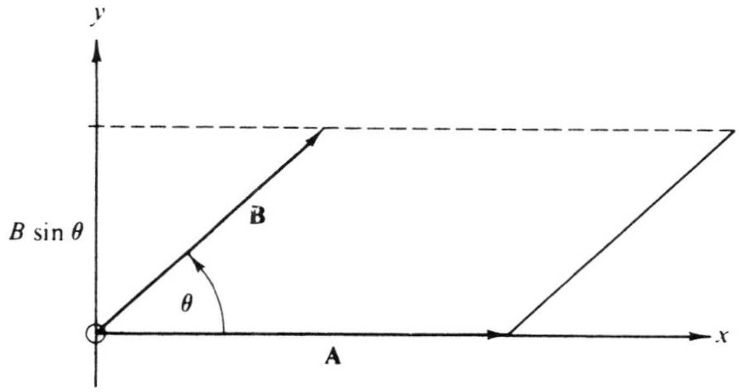

FIG. 1.11 Parallelogram representation of the vector product

[1] The electric field **E** is assumed here to be zero.

Here $\mathbf{v}$ is the velocity of the electric charge q and $\mathbf{F}_M$ is the resulting force on the moving charge.

The cross product has an important geometrical interpretation which we shall use in subsequent sections. In the parallelogram defined by $\mathbf{A}$ and $\mathbf{B}$ (Fig. 1.11) $B \sin \theta$ is the height if A is taken as the length of the base. Then $|\mathbf{A} \times \mathbf{B}| = AB \sin \theta$ is the *area* of the parallelogram. As a vector, $\mathbf{A} \times \mathbf{B}$ is the area of the parallelogram defined by $\mathbf{A}$ and $\mathbf{B}$, with the area vector normal to the plane of the parallelogram.

Parenthetically it might be noted that Eq. 1.32*c* and a modified Eq. 1.32*b* form the starting point for the development of *quaternions*. Equation 1.32*b* is replaced by $\mathbf{i} \times \mathbf{i} = \mathbf{j} \times \mathbf{j} = \mathbf{k} \times \mathbf{k} = -1$.

An alternate definition of the vector product $\mathbf{C} = \mathbf{A} \times \mathbf{B}$ consists in specifying the components of $\mathbf{C}$:

$$\begin{aligned} C_x &= A_y B_z - A_z B_y, \\ C_y &= A_z B_x - A_x B_z, \\ C_z &= A_x B_y - A_y B_x, \end{aligned} \tag{1.33}$$

or

$$C_i = A_j B_k - A_k B_j, \qquad i, j, k \text{ all different,} \tag{1.34}$$

and with cyclic permutation of the indices i, j, and k. The vector product $\mathbf{C}$ may be conveniently represented by a determinant[1]

$$\mathbf{C} = \begin{vmatrix} \mathbf{i} & \mathbf{j} & \mathbf{k} \\ A_x & A_y & A_z \\ B_x & B_y & B_z \end{vmatrix} \tag{1.35}$$

Expansion of the determinant across the top row reproduces the three components of $\mathbf{C}$ listed in Eq. 1.33.

Example 1.4.1

With $\mathbf{A}$ and $\mathbf{B}$ given in Example 1.1.1,

$$\mathbf{A} = 6\mathbf{i} + 4\mathbf{j} + 3\mathbf{k},$$

$$\mathbf{B} = 2\mathbf{i} - 3\mathbf{j} - 3\mathbf{k},$$

$$\begin{aligned} \mathbf{A} \times \mathbf{B} &= \begin{vmatrix} \mathbf{i} & \mathbf{j} & \mathbf{k} \\ 6 & 4 & 3 \\ 2 & -3 & -3 \end{vmatrix} \\ &= \mathbf{i}(-12 + 9) - \mathbf{j}(-18 - 6) + \mathbf{k}(-18 - 8) \\ &= -3\mathbf{i} + 24\mathbf{j} - 26\mathbf{k}. \end{aligned}$$

To show the equivalence of Eq. 1.32 and the component definition, Eq. 1.33,

[1] Cf. Section 4.1 for a summary of determinants.

let us form $\mathbf{A} \cdot \mathbf{C}$ and $\mathbf{B} \cdot \mathbf{C}$, using Eq. 1.33. We have

$$\begin{aligned}\mathbf{A} \cdot \mathbf{C} = \mathbf{A} \cdot (\mathbf{A} \times \mathbf{B}) &= A_x(A_y B_z - A_z B_y) \\ &\quad + A_y(A_z B_x - A_x B_z) \\ &\quad + A_z(A_x B_y - A_y B_x) = 0.\end{aligned} \tag{1.36}$$

Similarly,

$$\mathbf{B} \cdot \mathbf{C} = \mathbf{B} \cdot (\mathbf{A} \times \mathbf{B}) = 0. \tag{1.37}$$

Equations 1.36 and 1.37 show that $\mathbf{C}$ is perpendicular to both $\mathbf{A}$ and $\mathbf{B}$ ($\cos \theta = 0$, $\theta = \pm 90°$) and therefore perpendicular to the plane they determine. The positive direction is determined by considering special cases such as the unit vectors $\mathbf{i} \times \mathbf{j} = \mathbf{k}$ ($C_z = +A_x B_y$).

The magnitude is obtained from

$$\begin{aligned}(\mathbf{A} \times \mathbf{B}) \cdot (\mathbf{A} \times \mathbf{B}) &= A^2 B^2 - (\mathbf{A} \cdot \mathbf{B})^2 \\ &= A^2 B^2 - A^2 B^2 \cos^2 \theta \\ &= A^2 B^2 \sin^2 \theta.\end{aligned} \tag{1.38}$$

Hence

$$C = AB \sin \theta. \tag{1.39}$$

The big first step in Eq. 1.38 may be verified by expanding out in component form using Eq. 1.33 for $\mathbf{A} \times \mathbf{B}$ and Eq. 1.22 for the dot product. From Eqs. 1.36, 1.37, and 1.39 we see the equivalence of Eqs. 1.32 and 1.33, the two definitions of vector product.

There still remains the problem of verifying that $\mathbf{C} = \mathbf{A} \times \mathbf{B}$ is indeed a vector, that is, that it obeys Eq. 1.15, the vector transformation law. Starting in a rotated (primed system)

$$\begin{aligned}C_i' &= A_j' B_k' - A_k' B_j', \qquad i, j, \text{ and } k \text{ in cyclic order,} \\ &= \sum_l a_{jl} A_l \sum_m a_{km} B_m - \sum_l a_{kl} A_l \sum_m a_{jm} B_m \\ &= \sum_{l,m} (a_{jl} a_{km} - a_{kl} a_{jm}) A_l B_m.\end{aligned} \tag{1.40}$$

The combination of direction cosines in parentheses vanishes for $m = l$. We therefore have j and k taking on fixed values, dependent on the choice of i, and six combinations of l and m. If $i = 3$, then $j = 1$, $k = 2$ (cyclic order), and we have the following direction cosine combinations

$$\begin{aligned}a_{11} a_{22} - a_{21} a_{12} &= a_{33}, \\ a_{13} a_{21} - a_{23} a_{11} &= a_{32}, \\ a_{12} a_{23} - a_{22} a_{13} &= a_{31}\end{aligned} \tag{1.41}$$

and their negatives. Equations 1.41 are identities satisfied by the direction cosines. They may be verified with the use of determinants and matrices (cf. Exercise 4.3.3). Substituting back into Eq. 1.40,

$$\begin{aligned} C_3' &= a_{33}A_1B_2 + a_{32}A_3B_1 + a_{31}A_2B_3 \\ &\quad - a_{33}A_2B_1 - a_{32}A_1B_3 - a_{31}A_3B_2 \\ &= a_{31}C_1 + a_{32}C_2 + a_{33}C_3 \\ &= \sum_n a_{3n}C_n. \end{aligned} \tag{1.42}$$

By permuting indices to pick up C_1' and C_2', we see that Eq. 1.15 is satisfied and **C** is indeed a vector. It should be mentioned here that this vector nature of the cross product is an accident associated with the three-dimensional nature of ordinary space.[1] It will be seen in Chapter 3 that the cross product may also be treated as a second-rank antisymmetric tensor!

We now have two ways of multiplying vectors; a third form appears in Chapter 3. But what about division by a vector? It turns out that the ratio **B**/**A** is not uniquely specified (Exercise 4.2.19) unless **A** and **B** are also required to be parallel. Hence, division of one vector by another is not defined.

EXERCISES

1.4.1 Two vectors **A** and **B** are given by

$$\mathbf{A} = 2\mathbf{i} + 4\mathbf{j} + 6\mathbf{k},$$
$$\mathbf{B} = 3\mathbf{i} - 3\mathbf{j} - 5\mathbf{k}.$$

Compute the scalar and vector products $\mathbf{A} \cdot \mathbf{B}$ and $\mathbf{A} \times \mathbf{B}$.

1.4.2 Show that

(a) $(\mathbf{A} - \mathbf{B}) \cdot (\mathbf{A} + \mathbf{B}) = A^2 - B^2$,

(b) $(\mathbf{A} - \mathbf{B}) \times (\mathbf{A} + \mathbf{B}) = 2\mathbf{A} \times \mathbf{B}$.

The distributive laws needed here,

$$\mathbf{A} \cdot (\mathbf{B} + \mathbf{C}) = \mathbf{A} \cdot \mathbf{B} + \mathbf{A} \cdot \mathbf{C}$$

and

$$\mathbf{A} \times (\mathbf{B} + \mathbf{C}) = \mathbf{A} \times \mathbf{B} + \mathbf{A} \times \mathbf{C},$$

may easily be verified (if desired) by expansion in cartesian components.

1.4.3 Given the three vectors,

$$\mathbf{P} = 3\mathbf{i} + 2\mathbf{j} - \mathbf{k},$$
$$\mathbf{Q} = -6\mathbf{i} - 4\mathbf{j} + 2\mathbf{k},$$
$$\mathbf{R} = \mathbf{i} - 2\mathbf{j} - \mathbf{k},$$

find two that are perpendicular and two that are parallel or antiparallel.

1.4.4 Using the vectors

$$\mathbf{P} = \mathbf{i}\cos\theta + \mathbf{j}\sin\theta,$$
$$\mathbf{Q} = \mathbf{i}\cos\varphi - \mathbf{j}\sin\varphi,$$
$$\mathbf{R} = \mathbf{i}\cos\varphi + \mathbf{j}\sin\varphi,$$

[1] Specifically Eq. 1.41 holds only for three-dimensional space. Technically it is also possible to define a cross product in R^7, 7-dimensional space, but the cross product turns out to have unacceptable (pathological) properties.

prove the familiar trigonometric identities

$$\sin(\theta+\varphi) = \sin\theta\cos\varphi + \cos\theta\sin\varphi,$$
$$\cos(\theta+\varphi) = \cos\theta\cos\varphi - \sin\theta\sin\varphi.$$

1.4.5 (a) Find a vector **A** that is perpendicular to

$$\mathbf{U} = 2\mathbf{i} + \mathbf{j} - \mathbf{k}$$
$$\mathbf{V} = \mathbf{i} - \mathbf{j} + \mathbf{k}.$$

(b) What is **A** if, in addition to this requirement, we also demand that it have unit magnitude?

1.4.6 If four vectors **a**, **b**, **c**, and **d** all lie in the same plane, show that

$$(\mathbf{a}\times\mathbf{b})\times(\mathbf{c}\times\mathbf{d}) = 0.$$

Hint. Consider the directions of the cross-product vectors.

1.4.7 The coordinates of the three vertices of a triangle are (2, 1, 5), (5, 2, 8), and (4, 8, 2). Compute its area by vector methods.

1.4.8 The vertices of parallelogram *ABCD* are (1, 0, 0), (2, −1, 0), (0, −1, 1), and (−1, 0, 1) in order. Calculate the vector areas of triangle *ABD* and of triangle *BCD*. Are th^ two vector areas equal?

Ans. $\text{Area}_{ABD} = -\frac{1}{2}(\mathbf{i}+\mathbf{j}+2\mathbf{k})$.

1.4.9 The origin and the three vectors **A**, **B**, and **C** (all of which start at the origin) define a tetrahedron. Taking the *outward* direction as positive, calculate the total vector area of the four tetrahedral surfaces.

Note. In Section 1.11 this result is generalized to any closed surface.

1.4.10 Find the sides and angles of the spherical triangle *ABC* defined by the three vectors

$$\mathbf{A} = (1, 0, 0),$$
$$\mathbf{B} = \left(\frac{1}{\sqrt{2}}, 0, \frac{1}{\sqrt{2}}\right),$$

and

$$\mathbf{C} = \left(0, \frac{1}{\sqrt{2}}, \frac{1}{\sqrt{2}}\right).$$

Each vector starts from the origin.

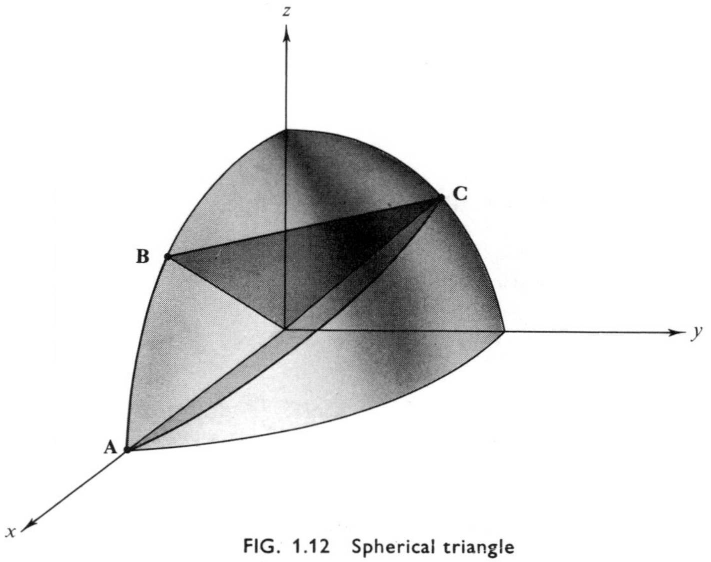

FIG. 1.12 Spherical triangle

1.4.11 Derive the law of sines:

$$\frac{\sin\alpha}{|\mathbf{A}|} = \frac{\sin\beta}{|\mathbf{B}|} = \frac{\sin\gamma}{|\mathbf{C}|}.$$

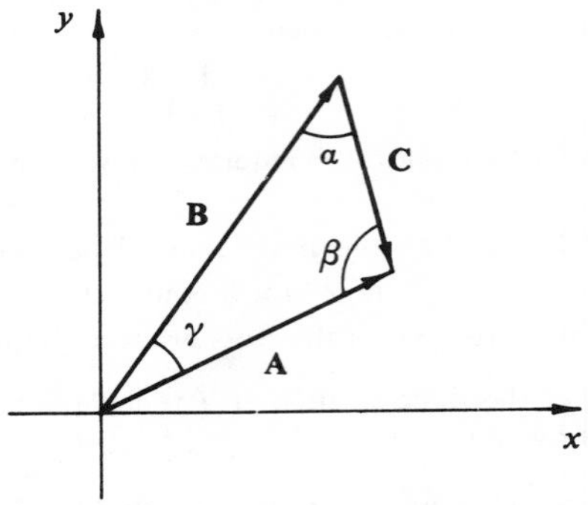

1.4.12 The magnetic induction **B** is *defined* by the Lorentz force equation

$$\mathbf{F} = q(\mathbf{v} \times \mathbf{B}).$$

Carrying out three experiments, we find that if

$$\mathbf{v} = \mathbf{i}, \qquad \frac{\mathbf{F}}{q} = 2\mathbf{k} - 4\mathbf{j},$$

$$\mathbf{v} = \mathbf{j}, \qquad \frac{\mathbf{F}}{q} = 4\mathbf{i} - \mathbf{k},$$

and

$$\mathbf{v} = \mathbf{k}, \qquad \frac{\mathbf{F}}{q} = \mathbf{j} - 2\mathbf{i}.$$

From the results of these three separate experiments calculate the magnetic induction **B**.

1.5 Triple Scalar Product, Triple Vector Product

Sections 1.3 and 1.4 cover the two types of multiplication of interest here. However, there are combinations of three vectors, $\mathbf{A} \cdot (\mathbf{B} \times \mathbf{C})$ and $\mathbf{A} \times (\mathbf{B} \times \mathbf{C})$, which occur with sufficient frequency to deserve further attention. The combination

$$\mathbf{A} \cdot (\mathbf{B} \times \mathbf{C})$$

is known as the triple scalar product. $\mathbf{B} \times \mathbf{C}$ yields a vector which, dotted into **A**, gives a scalar. We note that $(\mathbf{A} \cdot \mathbf{B}) \times \mathbf{C}$ represents a scalar crossed into a vector, an operation that is not defined. Hence, if we agree to exclude this undefined interpretation, the parentheses may be omitted and the triple scalar product written $\mathbf{A} \cdot \mathbf{B} \times \mathbf{C}$.

Using Eq. 1.33 for the cross product and Eq. 1.22 for the dot product,

$$\begin{aligned}\mathbf{A}\cdot\mathbf{B}\times\mathbf{C} &= A_x(B_yC_z - B_zC_y)\\ &\quad + A_y(B_zC_x - B_xC_z)\\ &\quad + A_z(B_xC_y - B_yC_x)\\ &= \mathbf{B}\cdot\mathbf{C}\times\mathbf{A} = \mathbf{C}\cdot\mathbf{A}\times\mathbf{B}\\ &= -\mathbf{A}\cdot\mathbf{C}\times\mathbf{B} = -\mathbf{C}\cdot\mathbf{B}\times\mathbf{A} = -\mathbf{B}\cdot\mathbf{A}\times\mathbf{C}, \qquad \text{etc.}\end{aligned} \tag{1.43}$$

The high degree of symmetry present in the component expansion should be noted. Every term contains the factors A_i, B_j, and C_k. If i, j, and k are in cyclic order (x, y, z), the sign is positive. If the order is anticyclic, the sign is negative. Further, the dot and the cross may be interchanged,

$$\mathbf{A}\cdot\mathbf{B}\times\mathbf{C} = \mathbf{A}\times\mathbf{B}\cdot\mathbf{C} \tag{1.44}$$

A convenient representation of the component expansion of Eq. 1.43 is provided by the determinant

$$\mathbf{A}\cdot\mathbf{B}\times\mathbf{C} = \begin{vmatrix} A_x & A_y & A_z \\ B_x & B_y & B_z \\ C_x & C_y & C_z \end{vmatrix} \tag{1.45}$$

The rules for interchanging rows and columns of a determinant[1] provide an immediate verification of the permutations listed in Eq. 1.43, whereas the symmetry of **A**, **B**, and **C** in the determinant form suggests the relation given in Eq. 1.44.

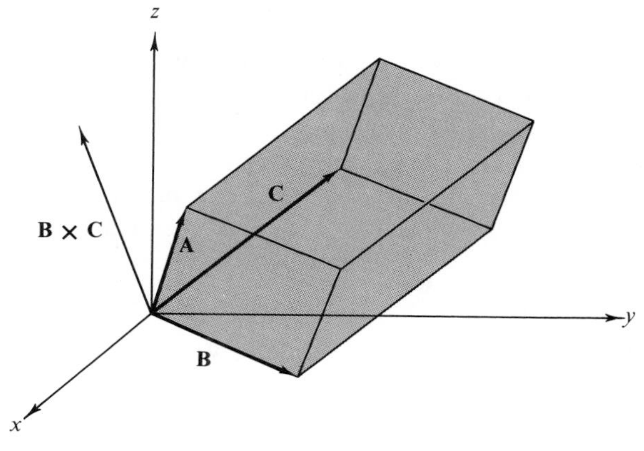

FIG. 1.13 Parallelepiped representation of triple scalar product

[1] Cf. Section 4.1 for a summary of the properties of determinants

The triple products encountered in Section 1.4, which showed that $\mathbf{A} \times \mathbf{B}$ was perpendicular to both $\mathbf{A}$ and $\mathbf{B}$, were special cases of the general result (Eq. 1.43).

The triple scalar product has a direct geometrical interpretation. The three vectors $\mathbf{A}$, $\mathbf{B}$, and $\mathbf{C}$ may be interpreted as defining a parallelepiped (Fig. 1.13).

$$\begin{aligned} |\mathbf{B} \times \mathbf{C}| &= BC \sin\theta \\ &= \text{area of parallelogram base.} \end{aligned} \tag{1.46}$$

The direction, of course, is normal to the base. Dotting $\mathbf{A}$ into this means multiplying the base area by the projection of $\mathbf{A}$ onto the normal, or base times height. Therefore

$$\mathbf{A} \cdot \mathbf{B} \times \mathbf{C} = \text{volume of parallelepiped defined by } \mathbf{A}, \mathbf{B}, \text{ and } \mathbf{C}.$$

EXAMPLE 1.5.1

For

$$\begin{aligned} \mathbf{A} &= \mathbf{i} + 2\mathbf{j} - \mathbf{k}, \\ \mathbf{B} &= \mathbf{j} + \mathbf{k}, \\ \mathbf{C} &= \mathbf{i} - \mathbf{j}. \end{aligned}$$

$$\mathbf{A} \cdot \mathbf{B} \times \mathbf{C} = \begin{vmatrix} 1 & 2 & -1 \\ 0 & 1 & 1 \\ 1 & -1 & 0 \end{vmatrix}. \tag{1.47}$$

By expansion by minors across the top row the determinant equals

$$1(0+1) - 2(0-1) - 1(0-1) = 4.$$

This is the volume of the parallelepiped defined by $\mathbf{A}$, $\mathbf{B}$, and $\mathbf{C}$. The reader should note that $\mathbf{A} \cdot \mathbf{B} \times \mathbf{C}$ may sometimes turn out to be negative! This problem and its interpretation are considered in Chapter 3.

The triple scalar product finds an interesting and important application in the construction of a reciprocal crystal lattice. Let $\mathbf{a}$, $\mathbf{b}$, and $\mathbf{c}$ (not necessarily mutually perpendicular) represent the vectors that define a crystal lattice. The distance from one lattice point to another may then be written

$$\mathbf{r} = n_a\mathbf{a} + n_b\mathbf{b} + n_c\mathbf{c}, \tag{1.48}$$

with n_a, n_b, and n_c taking on integral values. With these vectors we may form

$$\mathbf{a}' = \frac{\mathbf{b} \times \mathbf{c}}{\mathbf{a} \cdot \mathbf{b} \times \mathbf{c}}, \qquad \mathbf{b}' = \frac{\mathbf{c} \times \mathbf{a}}{\mathbf{a} \cdot \mathbf{b} \times \mathbf{c}}, \qquad \mathbf{c}' = \frac{\mathbf{a} \times \mathbf{b}}{\mathbf{a} \cdot \mathbf{b} \times \mathbf{c}}. \tag{1.48a}$$

We see that $\mathbf{a}'$ is perpendicular to the plane containing $\mathbf{b}$ and $\mathbf{c}$ and has a magnitude proportional to a^{-1}. In fact, we can readily show that

$$\mathbf{a}' \cdot \mathbf{a} = \mathbf{b}' \cdot \mathbf{b} = \mathbf{c}' \cdot \mathbf{c} = 1, \tag{1.48b}$$

whereas

$$\mathbf{a}' \cdot \mathbf{b} = \mathbf{a}' \cdot \mathbf{c} = \mathbf{b}' \cdot \mathbf{a} = \mathbf{b}' \cdot \mathbf{c} = \mathbf{c}' \cdot \mathbf{a} = \mathbf{c}' \cdot \mathbf{b} = 0. \tag{1.48c}$$

It is from Eqs. 1.48*b* and 1.48*c* that the name reciprocal lattice is derived. The mathematical space in which this reciprocal lattice exists is sometimes called a Fourier space, on the basis of relations to the Fourier analysis of Chapters 14 and 15. This reciprocal lattice is useful in problems involving the scattering of waves from the various planes in a crystal. Further details may be found in R. B. Leighton's *Principles of Modern Physics*, pp. 440–448 [New York: McGraw-Hill (1959)]. We encounter the reciprocal lattice again in an analysis of oblique coordinate systems, Section 4.4.

The second triple product of interest is $\mathbf{A} \times (\mathbf{B} \times \mathbf{C})$. Here the parentheses must be retained, as may be seen by considering the special case

$$\mathbf{i} \times (\mathbf{i} \times \mathbf{j}) = \mathbf{i} \times \mathbf{k} = -\mathbf{j} \tag{1.49}$$

but

$$(\mathbf{i} \times \mathbf{i}) \times \mathbf{j} = 0.$$

That the triple vector product is a vector follows from our discussion of vector product. Also, we see that the direction of the resulting vector is perpendicular to $\mathbf{A}$ and to $\mathbf{B} \times \mathbf{C}$. The plane defined by $\mathbf{B}$ and $\mathbf{C}$ is perpendicular to $\mathbf{B} \times \mathbf{C}$ and so $\mathbf{A} \times (\mathbf{B} \times \mathbf{C})$ lies in this plane. This means that $\mathbf{A} \times (\mathbf{B} \times \mathbf{C})$ will be a linear combination of $\mathbf{B}$ and $\mathbf{C}$. We find that

$$\mathbf{A} \times (\mathbf{B} \times \mathbf{C}) = \mathbf{B}(\mathbf{A} \cdot \mathbf{C}) - \mathbf{C}(\mathbf{A} \cdot \mathbf{B}), \tag{1.50}$$

a relation sometimes known as the *BAC-CAB* rule. This result may be verified by the direct though not very elegant method of expanding into cartesian components (cf. Exercise 1.5.1).

The *BAC-CAB* rule is probably the single most important vector identity. Because of its frequent use in problems and in future derivations, it probably should be memorized.

It might be noted here that as vectors are independent of the coordinates so a vector equation is independent of the particular coordinate system. The coordinate system only determines the components. If the vector equation can be established in cartesian coordinates, it is established and valid in any of the coordinate systems to be introduced in Chapter 2.

EXAMPLE 1.5.2

By using the three vectors given in Example 1.5.1 we obtain

$$\begin{aligned}\mathbf{A} \times (\mathbf{B} \times \mathbf{C}) &= (\mathbf{j} + \mathbf{k})(1 - 2) - (\mathbf{i} - \mathbf{j})(2 - 1)\\ &= -\mathbf{i} - \mathbf{k}\end{aligned}$$

by Eq. 1.50. In detail,

$$\mathbf{B} \times \mathbf{C} = \begin{vmatrix} \mathbf{i} & \mathbf{j} & \mathbf{k} \\ 0 & 1 & 1 \\ 1 & -1 & 0 \end{vmatrix} = \mathbf{i} + \mathbf{j} - \mathbf{k}$$

and

$$\mathbf{A} \times (\mathbf{B} \times \mathbf{C}) = \begin{vmatrix} \mathbf{i} & \mathbf{j} & \mathbf{k} \\ 1 & 2 & -1 \\ 1 & 1 & -1 \end{vmatrix} = -\mathbf{i} - \mathbf{k}.$$

Other, more complicated, products may be simplified by using these forms of the triple scalar and triple vector products.

EXERCISES

1.5.1 Verify the expansion of the triple vector product

$$\mathbf{A} \times (\mathbf{B} \times \mathbf{C}) = \mathbf{B}(\mathbf{A} \cdot \mathbf{C}) - \mathbf{C}(\mathbf{A} \cdot \mathbf{B})$$

by direct expansion in cartesian coordinates.

1.5.2 Show that the first step in Eq. 1.38, which is

$$(\mathbf{A} \times \mathbf{B}) \cdot (\mathbf{A} \times \mathbf{B}) = A^2B^2 - (\mathbf{A} \cdot \mathbf{B})^2,$$

is consistent with the *BAC-CAB* rule for a triple vector product.

1.5.3 Given the three vectors **A**, **B**, and **C**,

$$\mathbf{A} = \mathbf{i} + \mathbf{j},$$
$$\mathbf{B} = \mathbf{j} + \mathbf{k},$$
$$\mathbf{C} = \mathbf{i} - \mathbf{k}.$$

(a) Compute the triple scalar product, $\mathbf{A} \cdot \mathbf{B} \times \mathbf{C}$. Noting that $\mathbf{A} = \mathbf{B} + \mathbf{C}$, give a geometric interpretation of your result for the triple scalar product.

(b) Compute $\mathbf{A} \times (\mathbf{B} \times \mathbf{C})$.

1.5.4 Show that

$$\mathbf{a} \times (\mathbf{b} \times \mathbf{c}) + \mathbf{b} \times (\mathbf{c} \times \mathbf{a}) + \mathbf{c} \times (\mathbf{a} \times \mathbf{b}) = 0.$$

1.5.5 A vector **A** is decomposed into a radial vector $\mathbf{A}_r$ and a tangential vector $\mathbf{A}_t$. If $\mathbf{r}_0$ is a unit vector in the radial direction, show that

(a) $\mathbf{A}_r = \mathbf{r}_0(\mathbf{A} \cdot \mathbf{r}_0)$

and

(b) $\mathbf{A}_t = -\mathbf{r}_0 \times (\mathbf{r}_0 \times \mathbf{A})$.

1.5.6 Prove that a necessary and sufficient condition for the three (nonvanishing) vectors **A**, **B**, and **C** to be coplanar is the vanishing of the triple scalar product

$$\mathbf{A} \cdot \mathbf{B} \times \mathbf{C} = 0.$$

1.5.7 Three vectors **A**, **B**, and **C** are given by

$$\mathbf{A} = 3\mathbf{i} - 2\mathbf{j} + 2\mathbf{k},$$
$$\mathbf{B} = 6\mathbf{i} + 4\mathbf{j} - 2\mathbf{k},$$
$$\mathbf{C} = -3\mathbf{i} - 2\mathbf{j} - 4\mathbf{k}.$$

Compute the values of $\mathbf{A} \cdot \mathbf{B} \times \mathbf{C}$ and $\mathbf{A} \times (\mathbf{B} \times \mathbf{C})$, $\mathbf{C} \times (\mathbf{A} \times \mathbf{B})$ and $\mathbf{B} \times (\mathbf{C} \times \mathbf{A})$.

1.5.8 Vector $\mathbf{D}$ is a linear combination of three noncoplanar (and nonorthogonal) vectors:

$$\mathbf{D} = a\mathbf{A} + b\mathbf{B} + c\mathbf{C}.$$

Show that the coefficients are given by a ratio of triple scalar products,

$$a = \frac{\mathbf{D} \cdot \mathbf{B} \times \mathbf{C}}{\mathbf{A} \cdot \mathbf{B} \times \mathbf{C}}, \text{ etc.}$$

1.5.9 Show that

$$(\mathbf{A} \times \mathbf{B}) \cdot (\mathbf{C} \times \mathbf{D}) = (\mathbf{A} \cdot \mathbf{C})(\mathbf{B} \cdot \mathbf{D}) - (\mathbf{A} \cdot \mathbf{D})(\mathbf{B} \cdot \mathbf{C}).$$

1.5.10 Show that

$$(\mathbf{A} \times \mathbf{B}) \times (\mathbf{C} \times \mathbf{D}) = (\mathbf{A} \cdot \mathbf{B} \times \mathbf{D})\mathbf{C} - (\mathbf{A} \cdot \mathbf{B} \times \mathbf{C})\mathbf{D}.$$

1.5.11 For a *spherical* triangle such as pictured in Fig. 1.12, show that

$$\frac{\sin A}{\sin \overline{BC}} = \frac{\sin B}{\sin \overline{CA}} = \frac{\sin C}{\sin \overline{AB}}.$$

Here $\sin A$ is the sine of the included angle at A while $\overline{BC}$ is the side opposite (in radians). *Hint.* Exercise 1.5.10 will be useful.

1.5.12 Given

$$\mathbf{a}' = \frac{\mathbf{b} \times \mathbf{c}}{\mathbf{a} \cdot \mathbf{b} \times \mathbf{c}}, \quad \mathbf{b}' = \frac{\mathbf{c} \times \mathbf{a}}{\mathbf{a} \cdot \mathbf{b} \times \mathbf{c}}, \quad \mathbf{c}' = \frac{\mathbf{a} \times \mathbf{b}}{\mathbf{a} \cdot \mathbf{b} \times \mathbf{c}} \quad \text{and} \quad \mathbf{a} \cdot \mathbf{b} \times \mathbf{c} \neq 0,$$

show that

(a) $\mathbf{x}' \cdot \mathbf{y} = \delta_{xy}, \qquad (\mathbf{x}, \mathbf{y} = \mathbf{a}, \mathbf{b}, \mathbf{c}),$

(b) $\mathbf{a}' \cdot \mathbf{b}' \times \mathbf{c}' = (\mathbf{a} \cdot \mathbf{b} \times \mathbf{c})^{-1},$

(c) $\mathbf{a} = \dfrac{\mathbf{b}' \times \mathbf{c}'}{\mathbf{a}' \cdot \mathbf{b}' \times \mathbf{c}'}.$

1.5.13 If $\mathbf{x}' \cdot \mathbf{y} = \delta_{xy}$, $(\mathbf{x}, \mathbf{y} = \mathbf{a}, \mathbf{b}, \mathbf{c})$, prove that

$$\mathbf{a}' = \frac{\mathbf{b} \times \mathbf{c}}{\mathbf{a} \cdot \mathbf{b} \times \mathbf{c}}.$$

(This is the converse of Problem 1.5.12.)

1.5.14 Show that any vector $\mathbf{V}$ may be expressed in terms of the reciprocal vectors $\mathbf{a}', \mathbf{b}', \mathbf{c}'$ by

$$\mathbf{V} = (\mathbf{V} \cdot \mathbf{a})\mathbf{a}' + (\mathbf{V} \cdot \mathbf{b})\mathbf{b}' + (\mathbf{V} \cdot \mathbf{c})\mathbf{c}'.$$

1.5.15 An electric charge q_1 moving with velocity $\mathbf{v}_1$ produces a magnetic induction $\mathbf{B}$ given by

$$\mathbf{B} = \frac{\mu_0}{4\pi} q_1 \frac{\mathbf{v}_1 \times \mathbf{r}_0}{r^2} \quad \text{(mks units)},$$

where $\mathbf{r}_0$ points from q_1 to the point at which $\mathbf{B}$ is measured (Biot and Savart Law).

(a) Show that the magnetic force on a second charge q_2, velocity $\mathbf{v}_2$, is given by the triple vector product

$$\mathbf{F}_2 = \frac{\mu_0}{4\pi} \frac{q_1 q_2}{r^2} \mathbf{v}_2 \times (\mathbf{v}_1 \times \mathbf{r}_0).$$

(b) Write out the corresponding magnetic force $\mathbf{F}_1$ that q_2 exerts on q_1. Define your unit radial vector. How do $\mathbf{F}_1$ and $\mathbf{F}_2$ compare?

(c) Calculate $\mathbf{F}_1$ and $\mathbf{F}_2$ for the case of q_1 and q_2 moving along parallel trajectories side by side.

Ans. (b) $\mathbf{F}_1 = -\dfrac{\mu_0}{4\pi} \dfrac{q_1 q_2}{r^2} \mathbf{v}_1 \times (\mathbf{v}_2 \times \mathbf{r}_0)$. In general, there is no simple relation between $\mathbf{F}_1$ and $\mathbf{F}_2$. Specifically, Newton's third law, $\mathbf{F}_1 = -\mathbf{F}_2$, does not hold.

(c) $\mathbf{F}_1 = \frac{\mu_0}{4\pi}\frac{q_1 q_2}{r^2} v^2 \mathbf{r}_0 = -\mathbf{F}_2$. Mutual attraction.

1.6 Gradient, ∇

Suppose that $\varphi(x, y, z)$ is a scalar point function, that is, a function whose value depends on the values of the coordinates (x, y, z). As a scalar, it must have the same value at a given fixed point in space, independent of the rotation of our coordinate system, or

$$\varphi'(x_1', x_2', x_3') = \varphi(x_1, x_2, x_3). \tag{1.51}$$

By differentiating with respect to x_i' we obtain

$$\begin{aligned} \frac{\partial \varphi'(x_1', x_2', x_3')}{\partial x_i'} &= \frac{\partial \varphi(x_1, x_2, x_3)}{\partial x_i'} \\ &= \sum_j \frac{\partial \varphi}{\partial x_j}\frac{\partial x_j}{\partial x_i'} = \sum_j a_{ij}\frac{\partial \varphi}{\partial x_j} \end{aligned} \tag{1.52}$$

by the rules of partial differentiation and Eq. 1.16. But comparison with Eq. 1.17, the vector transformation law, now shows that we have *constructed* a vector with components $\partial\varphi/\partial x_j$. This vector we label the gradient of φ.

A convenient symbolism is

$$\nabla\varphi = \mathbf{i}\frac{\partial \varphi}{\partial x} + \mathbf{j}\frac{\partial \varphi}{\partial y} + \mathbf{k}\frac{\partial \varphi}{\partial z} \tag{1.53}$$

or

$$\nabla = \mathbf{i}\frac{\partial}{\partial x} + \mathbf{j}\frac{\partial}{\partial y} + \mathbf{k}\frac{\partial}{\partial z}. \tag{1.54}$$

$\nabla\varphi$ (or del φ) is our gradient of the scalar φ, whereas ∇ (del) itself is a vector differential operator (available to operate on or to differentiate a scalar φ). It should be emphasized that this operator is a hybrid creature that must satisfy both the laws for handling vectors and the laws of partial differentiation.

EXAMPLE 1.6.1

Calculate the gradient of $f(r) = f(\sqrt{x^2 + y^2 + z^2})$.

$$\nabla f(r) = \mathbf{i}\frac{\partial f(r)}{\partial x} + \mathbf{j}\frac{\partial f(r)}{\partial y} + \mathbf{k}\frac{\partial f(r)}{\partial z}.$$

Now $f(r)$ depends on x through the dependence of r on x. Therefore[1]

$$\frac{\partial f(r)}{\partial x} = \frac{df(r)}{dr} \cdot \frac{\partial r}{\partial x} = \frac{df}{dr} \cdot \frac{x}{r}.$$

Substituting into the equation for $\nabla f(r)$,

$$\begin{aligned}\nabla f(r) &= (\mathbf{i}x + \mathbf{j}y + \mathbf{k}z)\frac{1}{r}\frac{df}{dr} \\ &= \frac{\mathbf{r}}{r}\,\frac{df}{dr} \\ &= \mathbf{r}_0\,\frac{df}{dr}.\end{aligned}$$

Here $\mathbf{r}_0$ is a unit vector in the *positive* radial direction.

One immediate application of $\nabla\varphi$ is to dot it into an increment of length

$$d\mathbf{r} = \mathbf{i}\,dx + \mathbf{j}\,dy + \mathbf{k}\,dz. \tag{1.55}$$

Thus we obtain

$$\begin{aligned}(\nabla\varphi)\cdot d\mathbf{r} &= \frac{\partial\varphi}{\partial x}dx + \frac{\partial\varphi}{\partial y}dy + \frac{\partial\varphi}{\partial z}dz \\ &= d\varphi,\end{aligned} \tag{1.56}$$

the change in the scalar function φ corresponding to a change in position $d\mathbf{r}$. Now consider P and Q to be two points on a surface $\varphi(x, y, z) = C$, a constant. These points are chosen so that Q is a distance $d\mathbf{r}$ from P. Then, moving from P to Q, the change in $\varphi(x, y, z) = C$ is given by

$$\begin{aligned}d\varphi &= (\nabla\varphi)\cdot d\mathbf{r} \\ &= 0,\end{aligned} \tag{1.57}$$

since we stay on the surface $\varphi(x, y, z) = C$. This shows that $\nabla\varphi$ is perpendicular to $d\mathbf{r}$. Since $d\mathbf{r}$ may have any direction from P *as long as it stays in the surface* φ, point Q being restricted to the surface, but having arbitrary direction, $\nabla\varphi$ is seen to be normal to the surface φ = constant.

If we now permit $d\mathbf{r}$ to take us from one surface $\varphi = C_1$ to an adjacent surface $\varphi = C_2$,

$$\begin{aligned}d\varphi &= C_2 - C_1 = \Delta C \\ &= (\nabla\varphi)\cdot d\mathbf{r}.\end{aligned} \tag{1.58}$$

For a given $d\varphi$, $|d\mathbf{r}|$ is a minimum when it is chosen parallel to $\nabla\varphi$ ($\cos\theta = 1$); or, for a given $|d\mathbf{r}|$, the change in the scalar function φ is maximized by choosing $d\mathbf{r}$

[1] This is a special case of the chain rule of partial differentiation:

$$\frac{\partial f(r, \theta, \varphi)}{\partial x} = \frac{\partial f}{\partial r}\frac{\partial r}{\partial x} + \frac{\partial f}{\partial \theta}\frac{\partial \theta}{\partial x} + \frac{\partial f}{\partial \varphi}\frac{\partial \varphi}{\partial x}.$$

Here $\partial f/\partial\theta = \partial f/\partial\varphi = 0$, $\partial f/\partial r \to df/dr$.

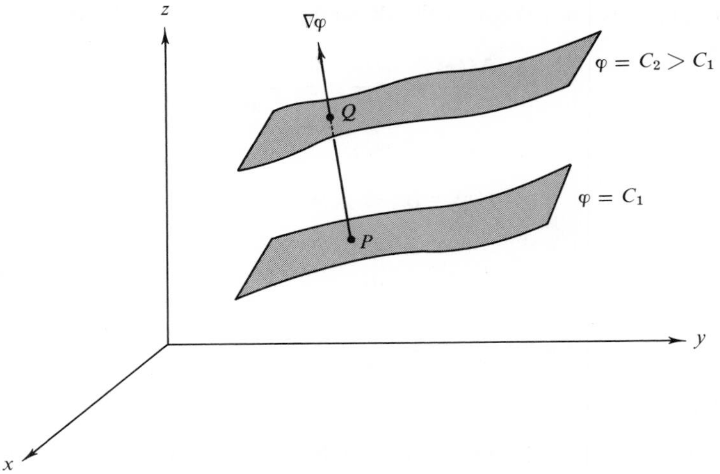

FIG. 1.14a Gradient

parallel to $\nabla\varphi$. *This identifies* $\nabla\varphi$ *as a vector having the direction of the maximum space rate of change of* φ, an identification that will be useful in Chapter 2 when we consider noncartesian coordinate systems.

EXAMPLE 1.6.2

As a specific example of the foregoing, and as an extension of Example 1.6.1, we consider the surfaces consisting of concentric spherical shells, Fig. 1.14b. We have

$$\varphi(x, y, z) = (x^2 + y^2 + z^2)^{1/2} = r_i = C_i$$

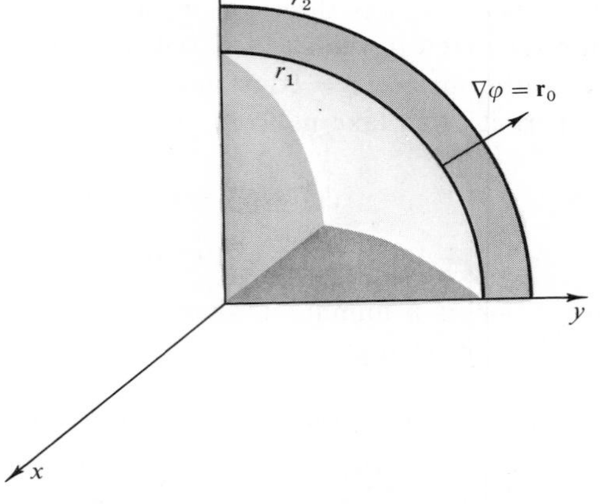

FIG. 1.14b Gradient for $\varphi(x, y, z) = (x^2 + y^2 + z^2)^{1/2}$, spherical shells: $(x^2 + y^2 + z^2)^{1/2} = r_2 = C_2$, $(x^2 + y^2 + z^2)^{1/2} = r_1 = C_1$

where r_i is the radius equal to C_i, our constant. $\Delta C = \Delta\varphi = \Delta r_i$, the distance between two shells. From Example 1.6.1

$$\nabla\varphi = \mathbf{r}_0.$$

The gradient of a scalar is of extreme importance in physics in expressing the relation between a force field and a potential field.

$$\text{Force} = -\nabla\,(\text{potential}). \tag{1.59}$$

This is illustrated by both gravitational and electrostatic fields, among others. The reader should satisfy himself that the minus sign in Eq. 1.59 results in water flowing downhill rather than uphill!

EXERCISES

1.6.1 If $S(x, y, z) = (x^2 + y^2 + z^2)^{-3/2}$, find
(a) ∇S at the point (1, 2, 3);
(b) the magnitude of the gradient of S, $|\nabla S|$ at (1, 2, 3); and
(c) the direction cosines of ∇S at (1, 2, 3).

1.6.2 (a) Find a unit vector perpendicular to the surface

$$x^2 + y^2 + z^2 = 3$$

at the point (1, 1, 1).
(b) Derive the equation of the plane tangent to the surface at (1, 1, 1).

Ans. (a) $(\mathbf{i} + \mathbf{j} + \mathbf{k})/\sqrt{3}$.
(b) $x + y + z = 3$.

1.6.3 Given a vector $\mathbf{r}_{12} = \mathbf{i}(x_1 - x_2) + \mathbf{j}(y_1 - y_2) + \mathbf{k}(z_1 - z_2)$, show that $\nabla_1 r_{12}$ (gradient with respect to x_1, y_1, and z_1 of the magnitude of $\mathbf{r}_{12}$) is a unit vector in the direction of $\mathbf{r}_{12}$.

1.6.4 If a vector function $\mathbf{F}$ depends on both space coordinates (x, y, z) and time t, show that

$$d\mathbf{F} = (d\mathbf{r} \cdot \nabla)\,\mathbf{F} + \frac{\partial \mathbf{F}}{\partial t}\,dt.$$

1.6.5 Show that $\nabla(uv) = v\nabla u + u\nabla v$, where u and v are differentiable scalar functions of x, y, and z.

1.6.6 (a) Show that a necessary and sufficient condition that $u(x, y, z)$ and $v(x, y, z)$ are related by some function $f(u, v) = 0$ is that $(\nabla u) \times (\nabla v) = 0$.
(b) If $u = u(x, y)$ and $v = v(x, y)$, show that the condition $(\nabla u) \times (\nabla v) = 0$ leads to the two-dimensional Jacobian

$$J\left(\frac{u, v}{x, y}\right) = \begin{vmatrix} \dfrac{\partial u}{\partial x} & \dfrac{\partial u}{\partial y} \\[2ex] \dfrac{\partial v}{\partial x} & \dfrac{\partial v}{\partial y} \end{vmatrix} = 0.$$

The functions u and v are assumed differentiable.

1.7 Divergence, $\nabla\cdot$

Differentiating a vector function is a simple extension of differentiating scalar quantities. Suppose $\mathbf{r}(t)$ describes the position of a satellite at some time t. Then, for differentiation with respect to time,

$$\frac{d\mathbf{r}(t)}{dt} = \lim_{\Delta t\to 0} \frac{\mathbf{r}(t+\Delta t) - \mathbf{r}(t)}{\Delta t}$$

$$= \mathbf{v}, \qquad \text{linear velocity.}$$

Graphically, we again have the slope of a curve, orbit, or trajectory, as shown in Fig. 1.15.

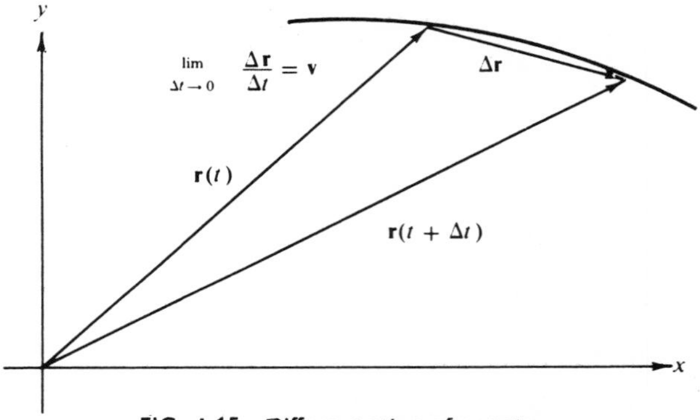

FIG. 1.15 Differentiation of a vector

If we resolve $\mathbf{r}(t)$ into its cartesian components, $d\mathbf{r}/dt$ always reduces directly to a vector sum of not more than three (for three-dimensional space) scalar derivatives. In other coordinate systems (Chapter 2) the situation is a little more complicated, for the unit vectors are no longer constant in direction. Differentiation with respect to the space coordinates is handled in the same way as differentiation with respect to time, as will be seen in the following paragraphs.

In Section 1.6 ∇ was defined as a vector operator. Now, paying careful attention to both its vector and its differential properties, we let it operate on a vector. First, as a vector we dot it into a second vector to obtain

$$\nabla\cdot\mathbf{V} = \frac{\partial V_x}{\partial x} + \frac{\partial V_y}{\partial y} + \frac{\partial V_z}{\partial z}, \tag{1.60}$$

known as the divergence of $\mathbf{V}$. This is a scalar, as discussed in Section 1.3.

EXAMPLE 1.7.1

Calculate $\nabla\cdot\mathbf{r}$.

$$\nabla\cdot\mathbf{r} = \left(\mathbf{i}\frac{\partial}{\partial x} + \mathbf{j}\frac{\partial}{\partial y} + \mathbf{k}\frac{\partial}{\partial z}\right)\cdot(\mathbf{i}x + \mathbf{j}y + \mathbf{k}z)$$

$$= \frac{\partial x}{\partial x} + \frac{\partial y}{\partial y} + \frac{\partial z}{\partial z},$$

or

$$\nabla \cdot \mathbf{r} = 3.$$

EXAMPLE 1.7.2

Generalizing Example 1.7.1,

$$\begin{aligned}
\nabla \cdot \mathbf{r} f(r) &= \frac{\partial}{\partial x}[x f(r)] + \frac{\partial}{\partial y}[y f(r)] + \frac{\partial}{\partial z}[z f(r)] \\
&= 3 f(r) + \frac{x^2}{r}\frac{df}{dr} + \frac{y^2}{r}\frac{df}{dr} + \frac{z^2}{r}\frac{df}{dr} \\
&= 3f(r) + r\frac{df}{dr}.
\end{aligned}$$

In particular, if $f(r) = r^{n-1}$,

$$\begin{aligned}
\nabla \cdot \mathbf{r} r^{n-1} &= \nabla \cdot \mathbf{r}_0 r^n \\
&= 3r^{n-1} + (n-1)r^{n-1} \\
&= (n+2)r^{n-1}.
\end{aligned}$$

This divergence vanishes for $n = -2$, an important fact in Section 1.14.

To develop a feeling for the physical significance of the divergence, consider $\nabla \cdot (\rho \mathbf{v})$ with $\mathbf{v}(x, y, z)$, the velocity of a compressible fluid and $\rho(x, y, z)$, its density at point (x, y, z). If we consider a small volume $dx\,dy\,dz$ (Fig. 1.16), the fluid flowing into this volume per unit time (positive x-direction) through the face $EFGH$ is (rate of flow in)$_{EFGH} = \rho v_x \, dy\, dz$. The rate of flow out (still positive x-direction) through face $ABCD$ is

$$(\text{rate of flow out})_{ABCD} = \left[\rho v_x + \frac{\partial}{\partial x}(\rho v_x)\, dx\right] dy\, dz,$$

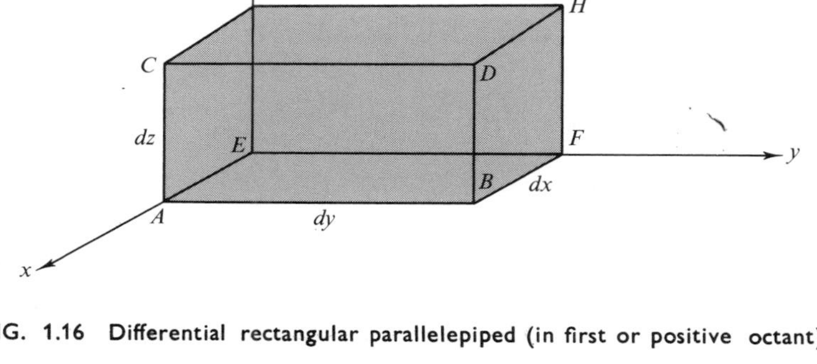

FIG. 1.16 Differential rectangular parallelepiped (in first or positive octant)

the derivative[1] allowing for the possibility of nonuniform density or velocity or both.[2] The net rate of flow *out*, for these two faces, is simply the difference between these two flows, or

$$\text{net rate of flow out}\Big|_x = \frac{\partial}{\partial x}(\rho v_x)\,dx\,dy\,dz.$$

Adding the net rate of flow out for the other four faces of our volume element,

$$\begin{aligned}\text{net flow out}\\ \text{(per unit time)} &= \left[\frac{\partial}{\partial x}(\rho v_x) + \frac{\partial}{\partial y}(\rho v_y) + \frac{\partial}{\partial z}(\rho v_z)\right] dx\,dy\,dz \\ &= \nabla \cdot (\rho \mathbf{v})\,dx\,dy\,dz. \qquad (1.61)\end{aligned}$$

Therefore the net flow of our compressible fluid out of the volume element $dx\,dy\,dz$ per unit volume per unit time is $\nabla \cdot (\rho\mathbf{v})$. Hence the name *divergence*. A direct application is in the continuity equation

$$\frac{\partial \rho}{\partial t} + \nabla \cdot (\rho \mathbf{v}) = 0, \qquad (1.62)$$

which simply states that a net flow out of the volume results in a decreased density inside the volume.

The combination $\nabla \cdot (f\mathbf{V})$, in which f is a scalar function and $\mathbf{V}$ a vector, may be written

$$\begin{aligned}\nabla \cdot (f\mathbf{V}) &= \frac{\partial}{\partial x}(fV_x) + \frac{\partial}{\partial y}(fV_y) + \frac{\partial}{\partial z}(fV_z) \\ &= \frac{\partial f}{\partial x}V_x + f\frac{\partial V_x}{\partial x} + \frac{\partial f}{\partial y}V_y + f\frac{\partial V_y}{\partial y} + \frac{\partial f}{\partial z}V_z + f\frac{\partial V_z}{\partial z} \\ &= (\nabla f)\cdot \mathbf{V} + f\nabla \cdot \mathbf{V}, \qquad (1.62a)\end{aligned}$$

which is just what we would expect for the derivative of a product.

If we have the special case of the divergence of a vector vanishing,

$$\nabla \cdot \mathbf{B} = 0, \qquad (1.63)$$

the vector $\mathbf{B}$ is said to be solenoidal, the term coming from the example in which $\mathbf{B}$ is the magnetic induction and Eq. 1.63 appears as one of Maxwell's equations.

[1] This expression gives the start of a Maclaurin expansion,

$$f(h) = f(0) + \frac{df}{dx}h + \frac{d^2f}{dx^2}\frac{h^2}{2!} + \cdots$$

(Section 5.6), in which $h = dx$. Each derivative is evaluated at the initial point $x = 0$.

[2] Strictly speaking, ρv_x is averaged over face *EFGH* and the expression $\rho v_x + (\partial/\partial x)(\rho v_x)\,dx$ is similarly averaged over face *ABCD*. Using an arbitrarily small differential volume, the averages reduce to the values used here.

EXERCISES

1.7.1 For a particle moving in a circular orbit $\mathbf{r} = \mathbf{i}r \cos \omega t + \mathbf{j}r \sin \omega t$,
(a) evaluate $\mathbf{r} \times \dot{\mathbf{r}}$.
(b) Show that $\ddot{\mathbf{r}} + \omega^2 \mathbf{r} = 0$.
The radius r and the angular velocity ω are constant. *Ans.* (*a*)$\mathbf{k}\omega r^2$.

1.7.2 Vector **A** satisfies the vector transformation law, Eq. 1.15. Show directly that its time derivative $d\mathbf{A}/dt$ also satisfies Eq. 1.15 and is therefore a vector.

1.7.3 Show, by differentiating components, that

(a) $\dfrac{d}{dt}(\mathbf{A} \cdot \mathbf{B}) = \dfrac{d\mathbf{A}}{dt} \cdot \mathbf{B} + \mathbf{A} \cdot \dfrac{d\mathbf{B}}{dt}$,

(b) $\dfrac{d}{dt}(\mathbf{A} \times \mathbf{B}) = \dfrac{d\mathbf{A}}{dt} \times \mathbf{B} + \mathbf{A} \times \dfrac{d\mathbf{B}}{dt}$,

just like the derivative of the product of two algebraic functions.

1.7.4 Prove $\nabla \cdot (\mathbf{a} \times \mathbf{b}) = \mathbf{b} \cdot \nabla \times \mathbf{a} - \mathbf{a} \cdot \nabla \times \mathbf{b}$.
Hint. Treat as a triple scalar product.

1.7.5 A rigid body is rotating with constant angular velocity $\boldsymbol{\omega}$. Show that the linear velocity $\mathbf{v}$ is solenoidal.

1.7.6 The electrostatic field of a point q is

$$\mathbf{E} = \frac{q}{4\pi\varepsilon_0} \cdot \frac{\mathbf{r}_0}{r^2}.$$

Calculate the divergence of **E**.

1.7.7 In Chapter 2 it will be seen that the *unit* vectors in noncartesian coordinate systems are usually functions of the coordinate variables, $\mathbf{a}_i = \mathbf{a}_i(q_1, q_2, q_3)$ but $|\mathbf{a}_i| = 1$. Show that either $\partial \mathbf{a}_i/\partial q_j = 0$ or that $\partial \mathbf{a}_i/\partial q_j$ is orthogonal to $\mathbf{a}_i$.

1.8 Curl, ∇×

Another possible operation with the vector operator ∇ is to cross it into a vector. We obtain

$$\nabla \times \mathbf{V} = \mathbf{i}\left(\frac{\partial}{\partial y} V_z - \frac{\partial}{\partial z} V_y\right) + \mathbf{j}\left(\frac{\partial}{\partial z} V_x - \frac{\partial}{\partial x} V_z\right) + \mathbf{k}\left(\frac{\partial}{\partial x} V_y - \frac{\partial}{\partial y} V_x\right)$$

$$= \begin{vmatrix} \mathbf{i} & \mathbf{j} & \mathbf{k} \\ \dfrac{\partial}{\partial x} & \dfrac{\partial}{\partial y} & \dfrac{\partial}{\partial z} \\ V_x & V_y & V_z \end{vmatrix}, \tag{1.64}$$

which is called the curl of **V**. In expanding this determinant form or in any operation with ∇, the derivative nature of ∇ must be considered. Specifically, $\mathbf{V} \times \nabla$ is defined only as an operator, another vector differential operator. It is certainly not equal, in general, to $-\nabla\times\mathbf{V}$.[1] In the case of Eq. 1.64, our determinant must be expanded *from the top down* so that we get the derivatives as shown in the middle portion of Eq. 1.64. If ∇ is crossed into the product of a scalar and a vector, we can show

$$\nabla\times(f\mathbf{V})\Big|_x=\left[\frac{\partial}{\partial y}(fV_z)-\frac{\partial}{\partial z}(fV_y)\right]$$
$$=\left(f\frac{\partial V_z}{\partial y}+\frac{\partial f}{\partial y}V_z-f\frac{\partial V_y}{\partial z}-\frac{\partial f}{\partial z}V_y\right)$$
$$=f\nabla\times\mathbf{V}|_x+(\nabla f)\times\mathbf{V}|_x. \qquad (1.65)$$

If we permute the coordinates $x\to y$, $y\to z$, $z\to x$ to pick up the y-component and then permute them a second time to pick up the z-component,

$$\nabla\times(f\mathbf{V})=f\nabla\times\mathbf{V}+(\nabla f)\times\mathbf{V}, \qquad (1.66)$$

which is the vector product analog of Eq. 1.62*a* .

EXAMPLE 1.8.1

Calculate $\nabla\times\mathbf{r}\,f(r)$
By Eq. 1.66

$$\nabla\times\mathbf{r}\,f(r)=f(r)\nabla\times\mathbf{r}+[\nabla f(r)]\times\mathbf{r}.$$

First,

$$\nabla\times\mathbf{r}=\begin{vmatrix}\mathbf{i} & \mathbf{j} & \mathbf{k}\\ \dfrac{\partial}{\partial x} & \dfrac{\partial}{\partial y} & \dfrac{\partial}{\partial z}\\ x & y & z\end{vmatrix}=0.$$

Second, using $\nabla f(r)=\mathbf{r}_0(df/dr)$ (Example 1.6.1),

$$\nabla\times\mathbf{r}f(r)=\frac{df}{dr}\mathbf{r}_0\times\mathbf{r}=0.$$

The vector product vanishes, since $\mathbf{r}=\mathbf{r}_0 r$ and $\mathbf{r}_0\times\mathbf{r}_0=0$.

The name curl has developed because in a sense $\nabla\times\mathbf{V}$ describes the rotation of the vector field $\mathbf{V}$ at the point at which the curl of $\mathbf{V}$ is evaluated. Suppose we have a rigid body in the xy-plane rotating about the z-axis with angular velocity $\boldsymbol{\omega}$. The linear velocity $\mathbf{v}$ at a given point described by $\mathbf{r}$ is given by

$$\mathbf{v}=\boldsymbol{\omega}\times\mathbf{r}. \qquad (1.67)$$

Let us calculate $\nabla\times\mathbf{v}$. We find

$$\nabla\times\mathbf{v}=\nabla\times(\boldsymbol{\omega}\times\mathbf{r}). \qquad (1.68)$$

[1] In this same spirit, if $\mathbf{A}$ is a differential operator, it is not necessarily true that $\mathbf{A}\times\mathbf{A}=0$. Specifically, for the quantum mechanical angular momentum *operator*, $\mathbf{L}=-i(\mathbf{r}\times\nabla)$, we find that $\mathbf{L}\times\mathbf{L}=i\mathbf{L}$.

By rearranging terms in Eq. 1.50 to respect the operator nature of ∇,

$$\nabla \times (\boldsymbol{\omega} \times \mathbf{r}) = \nabla \cdot \mathbf{r}\boldsymbol{\omega} - \nabla \cdot \boldsymbol{\omega}\mathbf{r}. \tag{1.69}$$

Here ∇ is dotted into the first vector, but as a differential operator it operates on *both* vectors. Expanding,

$$\begin{aligned} \nabla \times (\boldsymbol{\omega} \times \mathbf{r}) = \boldsymbol{\omega}\nabla \cdot \mathbf{r} + \mathbf{r} \cdot \nabla\boldsymbol{\omega} \\ -\mathbf{r}\nabla \cdot \boldsymbol{\omega} - \boldsymbol{\omega} \cdot \nabla\mathbf{r}. \end{aligned} \tag{1.70}$$

If $\boldsymbol{\omega}$ is constant, the second and third terms on the right side of Eq. 1.70 vanish. Further, we have

$$\nabla \cdot \mathbf{r} = 3 \tag{1.71}$$

from Example 1.7.1 and

$$\begin{aligned} \boldsymbol{\omega} \cdot \nabla\mathbf{r} &= \omega_x \frac{\partial}{\partial x}(\mathbf{i}x + \mathbf{j}y + \mathbf{k}z) \\ &\quad + \omega_y \frac{\partial}{\partial y}(\mathbf{i}x + \mathbf{j}y + \mathbf{k}z) \\ &\quad + \omega_z \frac{\partial}{\partial z}(\mathbf{i}x + \mathbf{j}y + \mathbf{k}z) \\ &= \mathbf{i}\omega_x + \mathbf{j}\omega_y + \mathbf{k}\omega_z \\ &= \boldsymbol{\omega}. \end{aligned} \tag{1.72}$$

By substituting Eqs. 1.71 and 1.72 into 1.70,

$$\nabla \times \mathbf{v} = \nabla \times (\boldsymbol{\omega} \times \mathbf{r}) = 2\boldsymbol{\omega}, \tag{1.73}$$

or the curl of the linear velocity of our rigid body is twice the angular velocity. In Section 1 12 an interpretation of $\nabla \times \mathbf{V}$ as a circulation measurable in terms of tiny paddlewheels will be developed. From the result $\boldsymbol{\omega} \cdot \nabla\mathbf{r} = \boldsymbol{\omega}$, Eq. 1.72, it appears the $\nabla\mathbf{r}$ is a unit operator in the sense that $\mathbf{A} \cdot (\nabla\mathbf{r}) = \mathbf{A}$ for any vector $\mathbf{A}$. This will become a formal result in Section 3.5 when we extend vector analysis to dyadics.

Whenever the curl of a vector $\mathbf{V}$ vanishes,

$$\nabla \times \mathbf{V} = 0. \tag{1.74}$$

$\mathbf{V}$ is labeled irrotational. The most important physical examples of irrotational vectors are the gravitational and electrostatic forces. In each case

$$\mathbf{V} = C\frac{\mathbf{r}_0}{r^2} = C\frac{\mathbf{r}}{r^3}, \tag{1.75}$$

where C is a constant and $\mathbf{r}_0$ is the unit vector in the outward radial direction. For the gravitational case we have $C = -Gm_1m_2$, given by Newton's law of universal gravitation. If $C = q_1q_2/4\pi\varepsilon_0$, we have Coulomb's law of electrostatics (mks units). The force $\mathbf{V}$ given in Eq. 1.75 may be shown to be irrotational by direct expansion into cartesian components as we did in Example 1.8.1. Another approach is developed in Chapter 2, in which we express $\nabla \times$, the curl, in terms of spherical polar coordinates.

For waves in an elastic medium, if the displacement $\mathbf{u}$ is irrotational, $\nabla \times \mathbf{u} = 0$, planes waves (or spherical waves at large distances) become longitudinal. If $\mathbf{u}$ is solenoidal, $\nabla \cdot \mathbf{u} = 0$, then the waves become transverse. A seismic disturbance will produce a displacement that may be resolved into a solenoidal part and an irrotational part (cf. Section 1.15). The irrotational part yields the longitudinal P (primary) earthquake waves. The solenoidal part gives rise to the slower transverse S (secondary) waves, Exercise 3.6.7.

Using the gradient, divergence, and curl, and of course the *BAC-CAB* rule, we may construct or verify a large number of useful vector identities. For verification, complete expansion into cartesian components is always a possibility. Sometimes if we use insight instead of routine shuffling of cartesian components, the verification process can be shortened drastically.

Remember that ∇ is a vector operator, a hybrid creature satisfying two sets of rules:

(a) vector rules, and

(b) partial differentiation rules—including differentiation of a product.

EXAMPLE 1.8.2. GRADIENT OF A DOT PRODUCT.

Verify that

$$\nabla(\mathbf{A} \cdot \mathbf{B}) = (\mathbf{B} \cdot \nabla)\mathbf{A} + (\mathbf{A} \cdot \nabla)\mathbf{B} + \mathbf{B} \times (\nabla \times \mathbf{A}) + \mathbf{A} \times (\nabla \times \mathbf{B}). \tag{1}$$

This particular example hinges on the recognition that $\nabla(\mathbf{A} \cdot \mathbf{B})$ is the type of term that appears in the *BAC-CAB* expansion of a triple vector product, Eq. 1.50. For instance

$$\mathbf{A} \times (\nabla \times \mathbf{B}) = \nabla(\mathbf{A} \cdot \mathbf{B}) - (\mathbf{A} \cdot \nabla)\mathbf{B} \tag{2}$$

with the ∇ differentiating only $\mathbf{B}$, not $\mathbf{A}$. From the commutativity of factors in a scalar product we may interchange $\mathbf{A}$ and $\mathbf{B}$ and write

$$\mathbf{B} \times (\nabla \times \mathbf{A}) = \nabla(\mathbf{A} \cdot \mathbf{B}) - (\mathbf{B} \cdot \nabla)\mathbf{A}, \tag{3}$$

now with ∇ differentiating only $\mathbf{A}$, not $\mathbf{B}$. Adding these two equations, we obtain ∇ differentiating the product $\mathbf{A} \cdot \mathbf{B}$ and the identity, Eq. (1).

This identity is used frequently in advanced electromagnetic theory. Exercise 1.8.12 is one simple illustration.

EXERCISES

1.8.1 Show, by rotating the coordinates, that the components of the curl of a vector transform as a vector.

Hint. The direction cosine identities of Eq. 1.41 are available as needed.

1.8.2 Show that $\mathbf{u} \times \mathbf{v}$ is solenoidal if $\mathbf{u}$ and $\mathbf{v}$ are each irrotational.

1.8.3 If A is irrotational, show that $\mathbf{A} \times \mathbf{r}$ is solenoidal.

1.8.4 A vector function $\mathbf{f}(x, y, z)$ is not irrotational but the product of $\mathbf{f}$ and a scalar function $g(x, y, z)$ is irrotational. Show that

$$\mathbf{f} \cdot \nabla \times \mathbf{f} = 0.$$

1.8.5 If (a) $\mathbf{V} = \mathbf{i}V_x(x, y) + \mathbf{j}V_y(x, y)$ and (b) $\nabla \times \mathbf{V} \neq 0$, prove that $\nabla \times \mathbf{V}$ is perpendicular to $\mathbf{V}$.

1.8.6 The quantum mechanical angular momentum operators are

$$L_x = -i\left(y\frac{\partial}{\partial z} - z\frac{\partial}{\partial y}\right),$$

$$L_y = -i\left(z\frac{\partial}{\partial x} - x\frac{\partial}{\partial z}\right),$$

$$L_z = -i\left(x\frac{\partial}{\partial y} - y\frac{\partial}{\partial x}\right).$$

Show that

$$L_x L_y - L_y L_x = iL_z,$$

hence

$$\mathbf{L} \times \mathbf{L} = i\mathbf{L}.$$

1.8.7 With the commutator bracket notation $[L_x, L_y] = L_x L_y - L_y L_x$, the angular momentum vector $\mathbf{L}$ satisfies $[L_x, L_y] = iL_z$, etc., or $\mathbf{L} \times \mathbf{L} = i\mathbf{L}$.
Two other vectors **a** and **b** commute with each other and with **L**, that is, $[\mathbf{a}, \mathbf{b}] = [\mathbf{a}, \mathbf{L}] = [\mathbf{b}, \mathbf{L}] = 0$. Show that

$$[\mathbf{a} \cdot \mathbf{L}, \mathbf{b} \cdot \mathbf{L}] = i(\mathbf{a} \times \mathbf{b}) \cdot \mathbf{L}.$$

1.8.8 Verify the vector identity

$$\nabla \times (\mathbf{A} \times \mathbf{B}) = (\mathbf{B} \cdot \nabla)\mathbf{A} - (\mathbf{A} \cdot \nabla)\mathbf{B} - \mathbf{B}(\nabla \cdot \mathbf{A}) + \mathbf{A}(\nabla \cdot \mathbf{B}).$$

1.8.9 As an alternative to the vector identity of Example 1.8.2 show that

$$\nabla(\mathbf{A} \cdot \mathbf{B}) = (\mathbf{A} \times \nabla) \times \mathbf{B} + (\mathbf{B} \times \nabla) \times \mathbf{A} + \mathbf{A}(\nabla \cdot \mathbf{B}) + \mathbf{B}(\nabla \cdot \mathbf{A}).$$

1.8.10 Verify the identity

$$\mathbf{A} \times (\nabla \times \mathbf{A}) = \tfrac{1}{2}\nabla(A^2) - (\mathbf{A} \cdot \nabla)\mathbf{A}.$$

1.8.11 If **A** and **B** are constant vectors, show that

$$\nabla(\mathbf{A} \cdot \mathbf{B} \times \mathbf{r}) = \mathbf{A} \times \mathbf{B}.$$

1.8.12 A distribution of electric currents creates a constant magnetic moment **m**. The force on **m** in an external magnetic induction **B** is given by

$$\mathbf{F} = \nabla \times (\mathbf{B} \times \mathbf{m}).$$

Show that

$$\mathbf{F} = \nabla(\mathbf{m} \cdot \mathbf{B}).$$

Note. Assuming no time dependence of the fields, Maxwell's equations yield $\nabla \times \mathbf{B} = 0$. Also $\nabla \cdot \mathbf{B} = 0$.

1.8.13 An electric dipole of moment $\mathbf{p}$ is located at the origin. The dipole creates an electric potential at $\mathbf{r}$ given by

$$\psi(\mathbf{r}) = \frac{\mathbf{p} \cdot \mathbf{r}}{4\pi\varepsilon_0 r^3}.$$

Find the electric field, $\mathbf{E} = -\nabla\psi$ at $\mathbf{r}$.

1.8.14 The vector potential $\mathbf{A}$ of a magnetic dipole, dipole moment $\mathbf{m}$, is given by $\mathbf{A}(\mathbf{r}) = (\mu_0/4\pi)(\mathbf{m} \times \mathbf{r}/r^3)$. Show that the magnetic induction $\mathbf{B} = \nabla \times \mathbf{A}$ is given by

$$\mathbf{B} = \frac{\mu_0}{4\pi} \frac{3\mathbf{r}_0(\mathbf{r}_0 \cdot \mathbf{m}) - \mathbf{m}}{r^3}.$$

Note. The limiting process leading to point dipoles is discussed in Section 12.1 for electric dipoles, Section 12.5 for magnetic dipoles.

1.8.15 The velocity of a two-dimensional flow of liquid is given by

$$\mathbf{V} = \mathbf{i}u(x,y) - \mathbf{j}v(x,y).$$

If the liquid is incompressible and the flow is irrotational show that

$$\frac{\partial u}{\partial x} = \frac{\partial v}{\partial y} \quad \text{and} \quad \frac{\partial u}{\partial y} = -\frac{\partial v}{\partial x}.$$

These are the Cauchy-Riemann conditions of Section 6.2

1.9 Successive Applications of ∇

We have now defined gradient, divergence, and curl to obtain vector, scalar, and vector quantities, respectively. Letting ∇ operate on each of these quantities, we obtain

(a) $\nabla \cdot \nabla\varphi$ (b) $\nabla \times \nabla\varphi$ (c) $\nabla\nabla \cdot \mathbf{V}$

(d) $\nabla \cdot \nabla \times \mathbf{V}$ (e) $\nabla \times (\nabla \times \mathbf{V})$,

all five expressions involving second derivatives and all five appearing in the second-order differential equations of mathematical physics, particularly in electromagnetic theory.

The first expression, $\nabla \cdot \nabla\varphi$, the divergence of the gradient, is named the Laplacian of φ. We have

$$\nabla \cdot \nabla\varphi = \left(\mathbf{i}\frac{\partial}{\partial x} + \mathbf{j}\frac{\partial}{\partial y} + \mathbf{k}\frac{\partial}{\partial z}\right) \cdot \left(\mathbf{i}\frac{\partial\varphi}{\partial x} + \mathbf{j}\frac{\partial\varphi}{\partial y} + \mathbf{k}\frac{\partial\varphi}{\partial z}\right)$$

$$= \frac{\partial^2\varphi}{\partial x^2} + \frac{\partial^2\varphi}{\partial y^2} + \frac{\partial^2\varphi}{\partial z^2}. \tag{1.76a}$$

When φ is the electrostatic potential, we have

$$\nabla \cdot \nabla\varphi = 0. \tag{1.76b}$$

which is Laplace's equation of electrostatics. Often the combination $\nabla \cdot \nabla$ is written ∇^2.

EXAMPLE 1.9.1

Calculate $\nabla \cdot \nabla\, g(r)$.

Referring to Examples 1.6.1 and 1.7.2,

$$\nabla \cdot \nabla\, g(r) = \nabla \cdot \mathbf{r}_0 \frac{dg}{dr}$$

$$= \frac{2}{r}\frac{dg}{dr} + \frac{d^2 g}{dr^2},$$

replacing $f(r)$ in Example 1.7.2 by $1/r \cdot dg/dr$. If $g(r) = r^n$, this reduces to

$$\nabla \cdot \nabla r^n = n(n+1)r^{n-2}.$$

This will vanish for $n = 0$ [$g(r) = $ constant] and for $n = -1$, that is, $g(r) = 1/r$ is a solution of Laplace's equation, $\nabla^2\, g(r) = 0$.

Expression (b) may be written

$$\nabla \times \nabla\varphi = \begin{vmatrix} \mathbf{i} & \mathbf{j} & \mathbf{k} \\ \dfrac{\partial}{\partial x} & \dfrac{\partial}{\partial y} & \dfrac{\partial}{\partial z} \\ \dfrac{\partial\varphi}{\partial x} & \dfrac{\partial\varphi}{\partial y} & \dfrac{\partial\varphi}{\partial z} \end{vmatrix}.$$

By expanding the determinant

$$\nabla \times \nabla\varphi = \mathbf{i}\left(\frac{\partial^2\varphi}{\partial y\,\partial z} - \frac{\partial^2\varphi}{\partial z\,\partial y}\right) + \mathbf{j}\left(\frac{\partial^2\varphi}{\partial z\,\partial x} - \frac{\partial^2\varphi}{\partial x\,\partial z}\right) + \mathbf{k}\left(\frac{\partial^2\varphi}{\partial x\,\partial y} - \frac{\partial^2\varphi}{\partial y\,\partial x}\right) \tag{1.77}$$

$$= 0,$$

assuming that the order of partial differentiation may be interchanged. This will be true as long as these second partial derivatives of φ are continuous functions. Then, from Eq. 1.77, the curl of a gradient is identically zero. All gradients, then, are irrotational.

Expression (d) is a triple scalar product which may be written

$$\nabla \cdot \nabla \times \mathbf{V} = \begin{vmatrix} \dfrac{\partial}{\partial x} & \dfrac{\partial}{\partial y} & \dfrac{\partial}{\partial z} \\ \dfrac{\partial}{\partial x} & \dfrac{\partial}{\partial y} & \dfrac{\partial}{\partial z} \\ V_x & V_y & V_z \end{vmatrix}. \tag{1.78}$$

Again, assuming continuity so that the order of differentiation is immaterial, we obtain

$$\nabla \cdot \nabla \times \mathbf{V} = 0. \tag{1.79}$$

The divergence of a curl vanishes or all curls are solenoidal. In Section 1.15 we shall see that vectors may be resolved into solenoidal and irrotational parts by Helmholtz' theorem.

The two remaining expressions satisfy a relation

$$\nabla \times (\nabla \times \mathbf{V}) = \nabla\nabla \cdot \mathbf{V} - \nabla \cdot \nabla\mathbf{V}. \tag{1.80}$$

This follows immediately from Eq. 1.50, the *BAC-CAB* rule, which we rewrite so that **C** appears at the extreme right of each term. The term $\nabla \cdot \nabla\mathbf{V}$ was not included in our list, but it may be *defined* by Eq. 1.80. If **V** is expanded in cartesian coordinates, so that the unit vectors are constant in direction as well as in magnitude, $\nabla \cdot \nabla\mathbf{V}$, a vector Laplacian, reduces to

$$\nabla \cdot \nabla\mathbf{V} = \mathbf{i}\nabla \cdot \nabla V_x + \mathbf{j}\nabla \cdot \nabla V_y + \mathbf{k}\nabla \cdot \nabla V_z,$$

a vector sum of ordinary scalar Laplacians. By expanding in cartesian coordinates, we may verify Eq. 1.80 as a vector identity.

EXAMPLE 1.9.2. ELECTROMAGNETIC WAVE EQUATION

One important application of this vector relation (Eq. 1.80), is in the derivation of the electromagnetic wave equation. In vacuum Maxwell's equations become

$$\nabla \cdot \mathbf{B} = 0, \tag{1.81a}$$

$$\nabla \cdot \mathbf{E} = 0, \tag{1.81b}$$

$$\nabla \times \mathbf{B} = \varepsilon_0\mu_0 \frac{\partial \mathbf{E}}{\partial t}, \tag{1.81c}$$

$$\nabla \times \mathbf{E} = -\frac{\partial \mathbf{B}}{\partial t}. \tag{1.81d}$$

Here **E** is the electric field, **B**, the magnetic induction, and ε_0 and μ_0, the electric and magnetic permittivities (mks units). Suppose we eliminate **B** from Eqs. 1.81*c* and 1.81*d*. We may do this by taking the curl of both sides of Eq. 1.81*d* and the time derivative of both sides of Eq. 1.81*c*. Since the space and time derivatives commute,

$$\frac{\partial}{\partial t}\nabla \times \mathbf{B} = \nabla \times \frac{\partial \mathbf{B}}{\partial t}, \tag{1.82}$$

and we obtain

$$\nabla \times (\nabla \times \mathbf{E}) = -\varepsilon_0\mu_0 \frac{\partial^2 \mathbf{E}}{\partial t^2}. \tag{1.83}$$

Application of Eqs. 1.80 and of 1.81*b* yields

$$\nabla \cdot \nabla\mathbf{E} = \varepsilon_0\mu_0 \frac{\partial^2 \mathbf{E}}{\partial t^2}, \tag{1.84}$$

the electromagnetic vector wave equation. Again, if **E** is expressed in cartesian coordinates, Eq. 1.84 separates into three scalar wave equations, each involving a scalar Laplacian.

EXERCISES

1.9.1 Verify Eq. 1.80

$$\nabla \times (\nabla \times \mathbf{V}) = \nabla\nabla \cdot \mathbf{V} - \nabla \cdot \nabla\mathbf{V}$$

by direct expansion in cartesian coordinates.

1.9.2 Show that the identity

$$\nabla \times (\nabla \times \mathbf{V}) = \nabla\nabla \cdot \mathbf{V} - \nabla \cdot \nabla\mathbf{V}$$

follows from the BAC-CAB rule for a triple vector product. Justify any alteration of the order of factors in the BAC and CAB terms.

1.9.3 Prove that $\nabla \times (\varphi \nabla \varphi) = 0$.

1.9.4 Prove that $(\nabla u) \times (\nabla v)$ is solenoidal where u and v are differentiable scalar functions.

1.9.5 φ is a scalar satisfying Laplace's equation, $\nabla^2\varphi = 0$. Show that $\nabla\varphi$ is *both* solenoidal and irrotational.

1.9.6 With ψ a scalar function show that

$$(\mathbf{r} \times \nabla) \cdot (\mathbf{r} \times \nabla)\psi = r^2 \nabla^2\psi - r^2 \frac{\partial^2\psi}{\partial r^2} - 2r \frac{\partial\psi}{\partial r}.$$

1.9.7 In a (nonrotating) isolated mass such as a star, the condition for equilibrium is

$$\nabla P + \rho\nabla\varphi = 0.$$

Here P is the total pressure, ρ the density, and φ the gravitational potential. Show that at any given point the normals to the surfaces of constant pressure and constant gravitational potential are parallel.

1.9.8 In the Pauli theory of the electron one encounters the expression

$$(\mathbf{p} - e\mathbf{A}) \times (\mathbf{p} - e\mathbf{A})\psi$$

where ψ is a scalar function. **A** is the magnetic vector potential related to the magnetic induction **B** by $\mathbf{B} = \nabla \times \mathbf{A}$. Given that $\mathbf{p} = -i\nabla$, show that this expression reduces to $ie\mathbf{B}\psi$.

1.9.9 Show that any solution of the equation

$$\nabla \times \nabla \times \mathbf{A} - k^2\mathbf{A} = 0$$

automatically satisfies the vector Helmholtz equation

$$\nabla^2\mathbf{A} + k^2\mathbf{A} = 0$$

and the solenoidal condition

$$\nabla \cdot \mathbf{A} = 0.$$

1.10 Vector Integration

The next step after differentiating vectors is to integrate them. Let us start with line integrals and then proceed to surface and volume integrals. In each case the method of attack will be to reduce the vector integral to scalar integrals with which the reader is assumed familiar. Using an increment of length $d\mathbf{r}$, we may encounter the line integrals

$$\int_c \varphi \, d\mathbf{r}, \tag{1.85a}$$

$$\int_c \mathbf{V} \cdot d\mathbf{r}, \tag{1.85b}$$

$$\int_c \mathbf{V} \times d\mathbf{r}, \tag{1.85c}$$

in each of which the integral is over some contour C that may be open or closed. With φ, a scalar, the first integral reduces immediately to

$$\int_c \varphi \, d\mathbf{r} = \mathbf{i} \int_c \varphi(x, y, z) \, dx + \mathbf{j} \int_c \varphi(x, y, z) \, dy + \mathbf{k} \int_c \varphi(x, y, z) \, dz. \tag{1.86}$$

This separation has employed the relation

$$\int \mathbf{i}\varphi \, dx = \mathbf{i} \int \varphi \, dx, \tag{1.87}$$

which is permissible because the cartesian unit vectors **i**, **j**, and **k** are constant in both magnitude and direction. Perhaps this is obvious here, but it will not be true in the noncartesian systems encountered in Chapter 2.

The three integrals on the right side of Eq. 1.86 are ordinary scalar integrals and, to avoid complications, we shall assume that they are Riemann integrals. Note, however, that the integral with respect to x cannot be evaluated unless y and z are known in terms of x and similarly for the integrals with respect to y and z. This simply means that the path of integration C must be specified. Unless the integrand has special properties that lead the integral to depend only on the value of the end points, the value will depend on the particular choice of contour C. For instance, if we choose the very special case $\varphi = 1$, Eq. 1.85*a* is just the vector distance from the start of contour C to the end point, in this case independent of the choice of path connecting fixed end points. With $d\mathbf{r} = \mathbf{i}\,dx + \mathbf{j}\,dy + \mathbf{k}\,dz$, the second and third forms also reduce to scalar integrals and, like Eq. 1.85*a*, are dependent, in general, on the choice of path. The form (Eq. 1.85*b*) is exactly the same as that encountered when we calculate the work done with a force that varies along the path,

$$W = \int \mathbf{F} \cdot d\mathbf{r}. \tag{1.88}$$

In this expression **F** is the force exerted on a particle or on an object to overcome the pre-existing force field, electrostatic, gravitational, and so on.

EXAMPLE 1.10.1

Integrate the scalar function $r^2 = x^2 + y^2$ from the origin to the point (1, 1) using a directed length increment $d\mathbf{r}$; that is, evaluate

$$\int_{0,0}^{1,1} (x^2 + y^2)\, d\mathbf{r}$$

If we expand $d\mathbf{r}$, the integral becomes

$$\int_{0,0}^{1,1} (x^2 + y^2)(\mathbf{i}\, dx + \mathbf{j}\, dy) = \mathbf{i} \int_{0,0}^{1,1} (x^2 + y^2)\, dx + \mathbf{j} \int_{0,0}^{1,1} (x^2 + y^2)\, dy.$$

Let us choose the path shown in Fig. 1.17. This means that $y = 0$ in the first integral

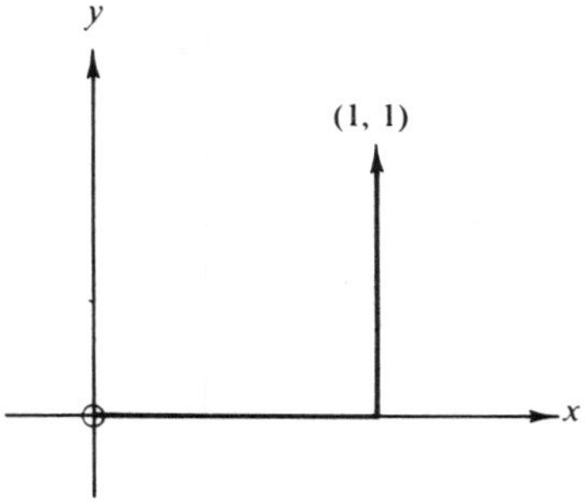

FIG. 1.17

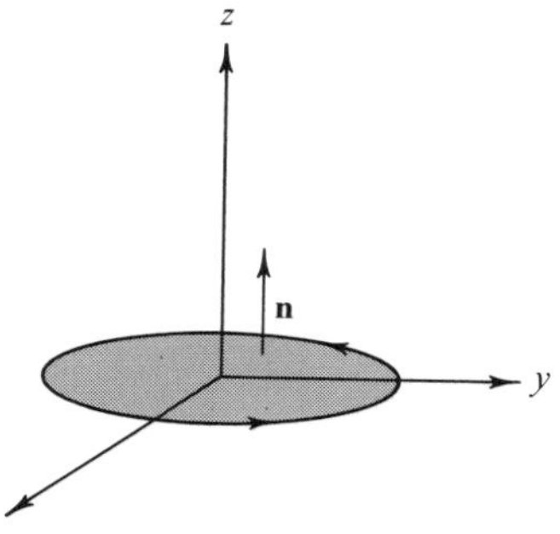

FIG. 1.18 Right-hand rule for the positive normal

and $x = 1$ in the second. By substituting into the integrals we get

$$\int_{0,0}^{1,1} (x^2 + y^2)\, d\mathbf{r} = \mathbf{i} \int_{0,y=0}^{1} (x^2 + y^2)\, dx + \mathbf{j} \int_{0,x=1}^{1} (x^2 + y^2)\, dy = \mathbf{i}\tfrac{1}{3} + \mathbf{j}\tfrac{4}{3}.$$

The reader may show that the path $(0, 0) \to (0, 1) \to (1, 1)$ yields $\mathbf{i}\frac{4}{3} + \mathbf{j}\frac{1}{3}$, whereas the path $x = y$ gives $\mathbf{i}\frac{2}{3} + \mathbf{j}\frac{2}{3}$. The value of the integral depends on the path chosen.

Surface integrals appear in the same forms as line integrals, the element of area also being a vector, $d\boldsymbol{\sigma}$.[1] Often this area element is written $\mathbf{n}\, dA$ in which $\mathbf{n}$ is a unit (normal) vector to indicate the positive direction.[2] There are two conventions for choosing the positive direction. First, if the surface is a closed surface, we agree to take the outward normal as positive. Second, if the surface is an open surface, the positive normal depends on the direction in which the perimeter of the open surface is traversed. If the right-hand fingers are placed in the direction of travel around the perimeter, the positive normal is indicated by the thumb of the right hand. As an illustration, a circle in the xy-plane (Fig. 1.18) mapped out from x to y to $-x$ to $-y$ and back to x will have its positive normal parallel to the positive z-axis (for our right-handed coordinate system). If the reader ever encounters one-sided surfaces,

[1] Recall that in Section 1.4 the area (of a parallelogram) represented a cross-product *vector*.

[2] While $\mathbf{n}$ always has unit length, its direction may well be a function of position.

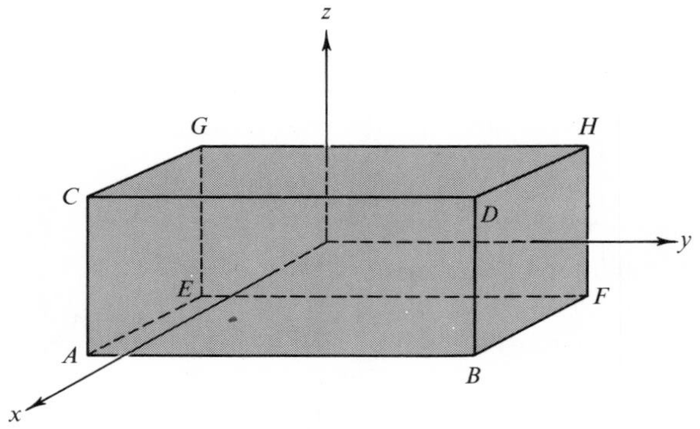

FIG. 1.19 Differential rectangular parallelepiped (origin at center)

such as Moebius strips, it is suggested that he either cut them and form reasonable, well-behaved surfaces or label them pathological and send them to the nearest mathematics department.

The surface integral $\int \mathbf{v} \cdot d\boldsymbol{\sigma}$ may be interpreted as a flow or flux through the given surface. This is really what we did in Section 1.7 to obtain the significance of the term divergence. This identification reappears in Section 1.11 as Gauss's theorem.

Volume integrals are somewhat simpler, for the volume element $d\tau$ is a scalar quantity.[1] We have

$$\int_V \mathbf{V}\, d\tau = \mathbf{i} \int_V V_x\, d\tau + \mathbf{j} \int_V V_y\, d\tau + \mathbf{k} \int_V V_z\, d\tau, \tag{1.89}$$

again reducing the vector integral to a vector sum of scalar integrals.

One interesting and significant application of our surface and volume integrals is their use in developing alternate definitions of our differential relations. We find

$$\nabla\varphi = \lim_{\int d\tau \to 0} \frac{\int \varphi\, d\boldsymbol{\sigma}}{\int d\tau}, \tag{1.90}$$

$$\nabla \cdot \mathbf{V} = \lim_{\int d\tau \to 0} \frac{\int \mathbf{V} \cdot d\boldsymbol{\sigma}}{\int d\tau}, \tag{1.91}$$

$$\nabla \times \mathbf{V} = \lim_{\int d\tau \to 0} \frac{\int d\boldsymbol{\sigma} \times \mathbf{V}}{\int d\tau}. \tag{1.92}$$

In these three equations $\int d\tau$ is the volume of a small region of space and $d\boldsymbol{\sigma}$ is the vector area element of this volume. The identification of Eq. 1.91 as the divergence of $\mathbf{V}$ was carried out in Section 1.7. Here we show that Eq. 1.90 is consistent with our earlier definition of $\nabla\varphi$ (Eq. 1.53). For simplicity we choose $\int d\tau$ to be the differential volume $dx\, dy\, dz$ (Fig. 1.19). This time we place the origin at the geometric center of our volume element. The area integral leads to six integrals, one

[1] Frequently the symbols d^3r and d^3x are used to denote a volume element in x (xyz or $x_1x_2x_3$) space.

for each of the six faces. Remembering that $d\boldsymbol{\sigma}$ is outward, $d\boldsymbol{\sigma}\cdot\mathbf{i} = -|d\boldsymbol{\sigma}|$ for surface $EFGH$, and $+|d\boldsymbol{\sigma}|$ for surface $ABCD$, we have

$$\int \varphi\, d\boldsymbol{\sigma} = -\mathbf{i}\int_{EFHG}\left(\varphi - \frac{\partial\varphi}{\partial x}\frac{dx}{2}\right)dy\,dz + \mathbf{i}\int_{ABDC}\left(\varphi + \frac{\partial\varphi}{\partial x}\frac{dx}{2}\right)dy\,dz$$

$$-\mathbf{j}\int_{AEGC}\left(\varphi - \frac{\partial\varphi}{\partial y}\frac{dy}{2}\right)dx\,dz + \mathbf{j}\int_{BFHD}\left(\varphi + \frac{\partial\varphi}{\partial y}\frac{dy}{2}\right)dx\,dz$$

$$-\mathbf{k}\int_{ABFE}\left(\varphi - \frac{\partial\varphi}{\partial z}\frac{dz}{2}\right)dx\,dy + \mathbf{k}\int_{CDHG}\left(\varphi + \frac{\partial\varphi}{\partial z}\frac{dz}{2}\right)dx\,dy.$$

Each integrand is evaluated at the origin with a correction included to correct for the displacement of the center of the face from the origin. Having chosen the total volume to be of differential size ($\int d\tau = dx\,dy\,dz$), we drop the integral signs on the right and obtain

$$\int \varphi\, d\boldsymbol{\sigma} = \left(\mathbf{i}\frac{\partial\varphi}{\partial x} + \mathbf{j}\frac{\partial\varphi}{\partial y} + \mathbf{k}\frac{\partial\varphi}{\partial z}\right)dx\,dy\,dz. \tag{1.93}$$

Dividing by

$$\int d\tau = dx\,dy\,dz,$$

Eq. 1.90 is verified.

This verification has been oversimplified in ignoring other correction terms beyond the first derivatives. These additional terms, which are introduced in Section 5.6 when the Taylor expansion is developed, vanish in the limit

$$\int d\tau \to 0\ (dx \to 0,\ dy \to 0,\ dz \to 0).$$

This, of course, is the reason for specifying in Eqs. 1.90, 1.91, and 1.92 that this limit be taken.

Verification of Eq. 1.92 follows these same lines exactly, using a differential volume $dx\,dy\,dz$.

EXERCISES

1.10.1 The force field acting on a two-dimensional linear oscillator may be described by

$$\mathbf{F} = -\mathbf{i}kx - \mathbf{j}ky.$$

Compare the work done moving against this force field when going from (1, 1) to (4, 4) by the following straight-line paths:

(a) $(1, 1) \to (4, 1) \to (4, 4)$

(b) $(1, 1) \to (1, 4) \to (4, 4)$

(c) $(1, 1) \to (4, 4)$ along $x = y$.

This means evaluating

$$-\int_{(1,1)}^{(4,4)} \mathbf{F}\cdot d\mathbf{r}$$

along each path.

1.10.2 Find the work done going around a unit circle in the xy-plane:
(a) counterclockwise from 0 to π,
(b) clockwise from 0 to $-\pi$,
doing work *against* a force field given by

$$\mathbf{F} = \frac{-\mathbf{i}y}{x^2+y^2} + \frac{\mathbf{j}x}{x^2+y^2}.$$

Note that the work done depends on the path.

1.10.3 Calculate the work you do in going from point (1, 1) to point (3, 3). The force *you exert* is given by

$$\mathbf{F} = \mathbf{i}(x-y) + \mathbf{j}(x+y).$$

Specify clearly the path you choose. Note that this force field is nonconservative.

1.10.4 Evaluate $\oint \mathbf{r} \cdot d\mathbf{r}$.

Note: the symbol $\oint$ means that the path of integration is a closed loop.

1.10.5 Evaluate

$$\frac{1}{3}\int_s \mathbf{r} \cdot d\boldsymbol{\sigma}$$

over the unit cube defined by the point (0, 0, 0) and the unit intercepts on the positive x-, y-, and z-axes. Note that (a) $\mathbf{r} \cdot d\boldsymbol{\sigma}$ is zero for three of the surfaces and (b) each of the three remaining surfaces contributes the same amount to the integral.

1.10.6 Show, by expansion of the surface integral, that

$$\lim_{\int d\tau \to 0} \frac{\int_s d\boldsymbol{\sigma} \times \mathbf{V}}{\int d\tau} = \nabla \times \mathbf{V}.$$

Hint. Choose the volume to be a differential volume, $dx\,dy\,dz$.

1.11 Gauss's Theorem

Here we verify a useful relation between a surface integral of a vector and the volume integral of a derivative of the vector. Let us assume that the vector $\mathbf{V}$ and $\nabla \cdot \mathbf{V}$ are continuous over the region of interest. Then Gauss's theorem states that

$$\int_S \mathbf{V} \cdot d\boldsymbol{\sigma} = \int_V \nabla \cdot \mathbf{V}\, d\tau. \tag{1.94}$$

We have seen (Eq. 1.61) that $\nabla \cdot \mathbf{V}$ may be interpreted as a net outflow of fluid per unit volume. Hence the right-hand side of Eq. 1.94 is the total rate of flow of fluid out of the volume over which we are integrating. Recognizing that the left side is also just a description of the flow out through the surface enclosing our volume, Gauss's theorem is proved.

For a detailed, rigorous proof along mathematical rather than physical lines the reader should consult Kellogg's *Foundations of Potential Theory* or some of the other references listed.

A frequently useful corollary of Gauss's theorem is a relation known as Green's theorem. If u and v are two scalar functions, we have the identities

$$\nabla \cdot (u \nabla v) = u \nabla \cdot \nabla v + (\nabla u) \cdot (\nabla v), \tag{1.95}$$

$$\nabla \cdot (v \nabla u) = v \nabla \cdot \nabla u + (\nabla v) \cdot (\nabla u). \tag{1.96}$$

Subtracting Eq. 1.96 from Eq. 1.95, integrating over a volume (u, v, and their derivatives, assumed continuous), and applying Eq. 1.94 (Gauss's theorem), we obtain

$$\int_V (u \nabla \cdot \nabla v - v \nabla \cdot \nabla u)\, d\tau = \int_S (u \nabla v - v \nabla u) \cdot d\boldsymbol{\sigma}. \tag{1.97}$$

This is Green's theorem. We use it for developing Green's functions, Chapters 8 and 16. An alternate form derived from Eq. 1.95 alone is

$$\int_S u \nabla v \cdot d\boldsymbol{\sigma} = \int_V u \nabla \cdot \nabla v\, d\tau + \int_V \nabla u \cdot \nabla v\, d\tau. \tag{1.98}$$

This is the form of Green's theorem used in Section 1.15.

Although Eq. 1.94 involving the divergence is by far the most important form of Gauss's theorem, volume integrals involving the gradient and the curl may also appear. Suppose

$$\mathbf{V}(x, y, z) = V(x, y, z)\mathbf{a}, \tag{1.99}$$

in which $\mathbf{a}$ is a vector with constant magnitude and constant but arbitrary direction. (You pick the direction, but once you have chosen it, hold it fixed.)

Equation 1.94 becomes

$$\begin{aligned} \mathbf{a} \cdot \int_S V\, d\boldsymbol{\sigma} &= \int_V \nabla \cdot \mathbf{a} V\, d\tau \\ &= \mathbf{a} \cdot \int_V \nabla V\, d\tau \end{aligned} \tag{1.100}$$

by Eq. 1.62a. This may be rewritten

$$\mathbf{a} \cdot \left[\int_S V\, d\boldsymbol{\sigma} - \int_V \nabla V\, d\tau\right] = 0. \tag{1.101}$$

Since $|\mathbf{a}| \neq 0$ and its direction is arbitrary, meaning that the cosine of the included angle cannot *always* vanish, the term in brackets must be zero. The result is

$$\int_S V\, d\boldsymbol{\sigma} = \int_V \nabla V\, d\tau. \tag{1.102}$$

In a similar manner, using $\mathbf{V} = \mathbf{a} \times \mathbf{P}$ in which $\mathbf{a}$ is a constant vector, we may show

$$\int_S d\boldsymbol{\sigma} \times \mathbf{P} = \int_V \nabla \times \mathbf{P}\, d\tau. \tag{1.103}$$

These last two forms of Gauss's theorem are used in the vector form of Kirchhoff diffraction theory. They may also be used to verify Eqs. 1.90 and 1.92.

Gauss's theorem may also be extended to dyadics or tensors (cf. Section 3.5).

EXERCISES

1.11.1 Prove that

$$\int_S d\boldsymbol{\sigma} = 0,$$

if S is a closed surface.

1.11.2 Show that

$$\frac{1}{3}\int_S \mathbf{r} \cdot d\boldsymbol{\sigma} = V,$$

where V is the volume enclosed by the closed surface S.

1.11.3 If $\mathbf{B} = \nabla \times \mathbf{A}$, show that

$$\int_S \mathbf{B} \cdot d\boldsymbol{\sigma} = 0$$

for any closed surface S.

1.11.4 Over some volume V let ψ be a solution of Laplace's equation (with the derivatives appearing there continuous). Prove that the integral over any closed surface in V of the normal derivative of ψ, ($\partial\psi/\partial n$, or $\nabla\psi \cdot \mathbf{n}$) will be zero.

1.11.5 In analogy to the integral definitions of gradient, divergence, and curl of Section 1.10, show that

$$\nabla^2\varphi = \lim_{\int d\tau \to 0} \frac{\int \nabla\varphi \cdot d\boldsymbol{\sigma}}{\int d\tau}.$$

1.11.6 The electric displacement vector $\mathbf{D}$ satisfies the Maxwell equation $\nabla \cdot \mathbf{D} = \rho$ where ρ is the charge density (per unit volume). At the boundary between two media there is a surface charge density σ (per unit area). Show that a boundary condition for $\mathbf{D}$ is

$$(\mathbf{D}_2 - \mathbf{D}_1) \cdot \mathbf{n} = \sigma.$$

$\mathbf{n}$ is a unit vector normal to the surface and out of medium 1.

Hint: Consider a *thin* pillbox as shown in the figure.

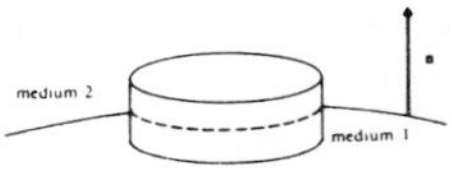

1.11.7 From Eq. 1.62(a) with $\mathbf{V}$ the electric field $\mathbf{E}$, and f the electrostatic potential φ, show that

$$\int \rho\varphi \, d\tau = \varepsilon_0 \int E^2 \, d\tau.$$

Hint: You may assume that φ vanishes at large r at least as fast as r^{-1}.

1.11.8 A particular steady state electric current distribution is localized in space. Choosing a bounding surface far enough out so that the current density $\mathbf{J}$ is zero everywhere on the surface, show that

$$\int \mathbf{J} \, d\tau = 0.$$

Hint: Take one component of $\mathbf{J}$ at a time. With $\nabla \cdot \mathbf{J} = 0$, show that $J_i = \nabla \cdot x_i \mathbf{J}$ and apply Gauss's theorem.

1.11.9 The creation of a *localized* system of electric currents (current density **J**) and magnetic fields may be shown to require an amount of work

$$W = \frac{1}{2}\int \mathbf{H} \cdot \mathbf{B}\, d\tau.$$

Transform this into

$$W = \frac{1}{2}\int \mathbf{J} \cdot \mathbf{A}\, d\tau.$$

A being the magnetic vector potential.

Hint: If the fields and currents are localized, a bounding surface may be taken far enough out so the integrals of the fields and currents over the surface yield zero.

1.12 Stokes's Theorem

Gauss's theorem relates the volume integral of a derivative of a function to an integral of the function over the closed surface bounding the volume. Here we consider an analogous relation between the surface integral of a derivative of a function and the line integral of the function, the path of integration being the perimeter bounding the surface. Let us start with a surface integral of a curl and transform it.

$$\int_S \nabla \times \mathbf{V} \cdot d\boldsymbol{\sigma} = \int_S \left(\frac{\partial V_x}{\partial z}\, d\sigma_y - \frac{\partial V_x}{\partial y}\, d\sigma_z + \frac{\partial V_y}{\partial x}\, d\sigma_z - \frac{\partial V_y}{\partial z}\, d\sigma_x + \frac{\partial V_z}{\partial y}\, d\sigma_x - \frac{\partial V_z}{\partial x}\, d\sigma_y \right), \tag{1.104}$$

expanding the integrand as a triple scalar product. The integration is over some given surface. Let us orient our cartesian coordinate axes so that this surface intersects the plane $x = c$, a constant, along the line AB as shown in Fig. 1.20. With the perimeter of the surface described by a line going through the $x = c$-plane in the negative direction at A and in the positive direction at B, $d\boldsymbol{\sigma}$ will be down as shown.

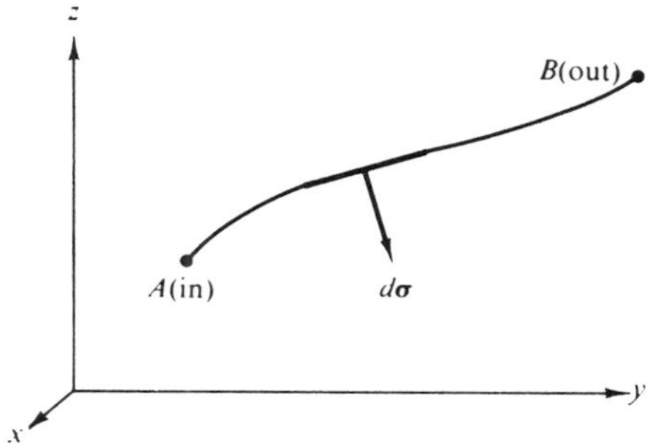

FIG. 1.20 Intersection of surface S with plane $x = c$

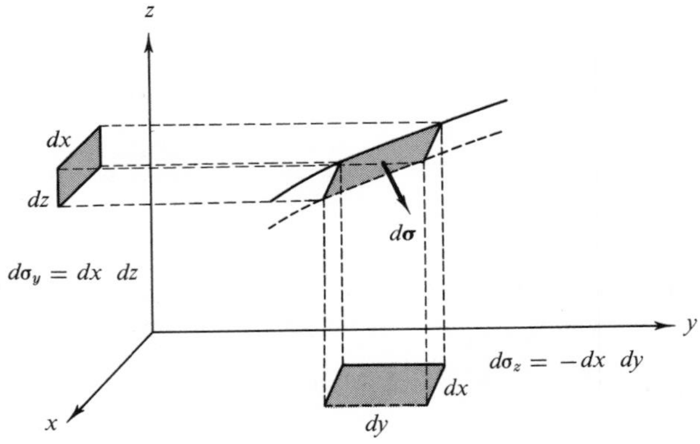

FIG. 1.21 Projection of $d\sigma$ on xy and on xz planes

In particular,

$$d\sigma_y = dx\,dz,$$
$$d\sigma_z = -dx\,dy. \tag{1.105}$$

The increment dx enters when we consider the strip of surface intercepted between the planes $x = c$ and $x = c + dx$. Integrating the derivatives of V_x over this strip, we obtain

$$\int_S \left(\frac{\partial V_x}{\partial z}\,d\sigma_y - \frac{\partial V_x}{\partial y}\,d\sigma_z\right) = \int_S \left(\frac{\partial V_x}{\partial y}\,dy + \frac{\partial V_x}{\partial z}\,dz\right) dx. \tag{1.106}$$

Since x is held constant in the integration from A to B,

$$\frac{\partial V_x}{\partial y}\,dy + \frac{\partial V_x}{\partial z}\,dz = dV_x \tag{1.107}$$

and our surface integral becomes

$$\int dx \int_A^B dV_x = \int V_x(x, y_B, z_B)\,dx - \int V_x(x, y_A, z_A)\,dx. \tag{1.107a}$$

Our choice of direction in describing the perimeter leads to

$$dx = d\lambda_x \quad \text{at } B, \qquad \text{right side of perimeter,}$$
$$dx = -d\lambda_x \quad \text{at } A, \qquad \text{left side of perimeter,}$$

where $d\boldsymbol{\lambda}$ is a vector differential length along the perimeter. Finally, letting x range far enough to cover the entire surface,

$$\int_S \left(\frac{\partial V_x}{\partial z}\,d\sigma_y - \frac{\partial V_x}{\partial y}\,d\sigma_z\right) = \oint V_x\,d\lambda_x. \tag{1.108}$$

The symbol $\oint$ indicates an integral around a closed path, in this case, the periphery of the surface. Either by cyclic permutation of the coordinates or by repeating the integration (derivatives of V_y and a plane $y = c$, etc.) we obtain similar results for the derivatives of V_y and of V_z. Adding

$$\int \nabla \times \mathbf{V} \cdot d\boldsymbol{\sigma} = \oint (V_x\, d\lambda_x + V_y\, d\lambda_y + V_z\, d\lambda_z)$$
$$= \oint \mathbf{V} \cdot d\boldsymbol{\lambda}. \tag{1.109}$$

This is Stokes's theorem. Again, the reader who desires greater mathematical rigor is referred to Kellogg's *Foundations of Potential Theory.*

As with Gauss's theorem, other relations between surface and line integrals are possible. We find

$$\int_S d\boldsymbol{\sigma} \times \nabla\varphi = \oint \varphi\, d\boldsymbol{\lambda} \tag{1.110}$$

and

$$\int_S (d\boldsymbol{\sigma} \times \nabla) \times \mathbf{P} = \oint d\boldsymbol{\lambda} \times \mathbf{P}. \tag{1.111}$$

Equation 1.110 may readily be verified by the substitution $\mathbf{V} = \mathbf{a}\varphi$ in which $\mathbf{a}$ is a vector of constant magnitude and of constant direction, as in Section 1.11. Substituting into Stokes's theorem, Eq. 1.109,

$$\int_S (\nabla \times \mathbf{a}\varphi) \cdot d\boldsymbol{\sigma} = -\int_S \mathbf{a} \times \nabla\varphi \cdot d\boldsymbol{\sigma}$$
$$= -\mathbf{a} \cdot \int_S \nabla\varphi \times d\boldsymbol{\sigma}. \tag{1.112}$$

For the line integral

$$\oint \mathbf{a}\varphi \cdot d\boldsymbol{\lambda} = \mathbf{a} \cdot \oint \varphi\, d\boldsymbol{\lambda}, \tag{1.113}$$

and we obtain

$$\mathbf{a} \cdot \left(\oint \varphi\, d\boldsymbol{\lambda} + \int_S \nabla\varphi \times d\boldsymbol{\sigma} \right) = 0. \tag{1.114}$$

Since the choice of direction of $\mathbf{a}$ is arbitrary, the expression in parentheses must vanish, thus verifying Eq. 1.110. Equation 1.111 may be derived in a similar manner by using $\mathbf{V} = \mathbf{a} \times \mathbf{P}$, in which $\mathbf{a}$ is again a constant vector

Returning to Eq. 1.109, $\oint \mathbf{V} \cdot d\boldsymbol{\lambda}$ may be interpreted as a flow of fluid, a circulation around a loop. If we take our surface to be a circle of area $\mathbf{k}\, d\sigma$,

$$|\nabla \times \mathbf{V}|\, d\sigma = \text{circulation around } d\sigma \text{ in } xy\text{-plane}.$$

This is the basis for measuring the curl of $\mathbf{V}$ by the rotation of a tiny paddlewheel. If the little paddlewheel does not rotate, the circulation is zero and $\mathbf{V}$, by Stokes's theorem, is irrotational. This circulation concept is the basis for the usual proof of Stokes's theorem. The original surface is divided into differential surface elements. Circulation currents along interior lines cancel and we are left with a circulation around the perimeter of the original surface.

The main application of Stokes's theorem is in potential theory, the subject of the next section.

EXERCISES

1.12.1 A vector $\mathbf{t} = -\mathbf{i}y + \mathbf{j}x$. With the help of Stokes's theorem, show that the integral around a continuous closed curve in the xy-plane

$$\tfrac{1}{2}\oint \mathbf{t} \cdot d\boldsymbol{\lambda} = \tfrac{1}{2}\oint (x\,dy - y\,dx) = A,$$

the area enclosed by the curve.

1.12.2 (a) From the calculation of the magnetic moment of a current loop, integrate around the perimeter of the current loop (in the xy-plane) and show that the scalar magnitude of

$$\oint \mathbf{r} \times d\mathbf{r}$$

is twice the area of the enclosed surface.
(b) The perimeter of an ellipse is described by $\mathbf{r} = \mathbf{i}a\cos\theta + \mathbf{j}b\sin\theta$. From part (a) show that the area of the ellipse is πab.

1.12.3 In the steady state the magnetic field $\mathbf{H}$ satisfies the Maxwell equation $\nabla \times \mathbf{H} = \mathbf{J}$, where $\mathbf{J}$ is the current density (per square meter). At the boundary between two media there is a surface current density $\mathbf{K}$ (per meter). Show that a boundary condition on $\mathbf{H}$ is

$$\mathbf{n} \times (\mathbf{H}_2 - \mathbf{H}_1) = \mathbf{K}.$$

$\mathbf{n}$ is a unit vector normal to the surface and out of medium 1.
Hint. Consider a narrow loop perpendicular to the interface as shown in the figure.

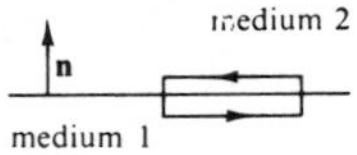

1.12.4 From Maxwell's equations, $\nabla \times \mathbf{H} = \mathbf{J}$ with $\mathbf{J}$ here the current density and $\mathbf{E} = 0$. Show from this that

$$\oint \mathbf{H} \cdot d\mathbf{r} = I$$

where I is the net electric current enclosed by the loop integral. These are the differential and integral forms of Ampere's law of magnetism.

1.12.5 A magnetic induction $\mathbf{B}$ is generated by electric current in a ring of radius R. Show that the *magnitude* of the vector potential $\mathbf{A}$ ($\mathbf{B} = \nabla \times \mathbf{A}$) at the ring is

$$|\mathbf{A}| = \frac{\varphi}{2\pi R}$$

where φ is the total magnetic flux passing through the ring.
Note: $\mathbf{A}$ is tangential to the ring.

1.12.6 Prove that

$$\int_S \nabla \times \mathbf{V} \cdot d\boldsymbol{\sigma} = 0,$$

if S is a closed surface.

1.12.7 Evaluate $\oint \mathbf{r} \cdot d\mathbf{r}$ (Exercise 1.10.4) by Stokes's theorem.

1.12.8 Prove that

$$\oint u \nabla v \cdot d\boldsymbol{\lambda} = -\oint v \nabla u \cdot d\boldsymbol{\lambda}.$$

1.12.9 Prove that

$$\oint u \nabla v \cdot d\boldsymbol{\lambda} = \int_S (\nabla u) \times (\nabla v) \cdot d\boldsymbol{\sigma}.$$

1.12.10 Use Stokes's theorem to construct an integral definition of $\nabla \times \mathbf{V}$.
Note. This is the approach used in Section 2.2 to develop the curl in curvilinear coordinates.

1.13 Potential Theory

Scalar potential. If a force over a given region of space S can be expressed as the negative gradient of a scalar function φ,

$$\mathbf{F} = -\nabla\varphi, \tag{1.115}$$

we will call φ a scalar potential. The force $\mathbf{F}$ appearing as the negative gradient of a single-valued scalar potential is labeled a *conservative* force. We want to know when a scalar potential function exists. To answer this question, we establish two other relations as equivalent to Eq. 1.115. These are

$$\nabla \times \mathbf{F} = 0 \tag{1.116}$$

and

$$\oint \mathbf{F} \cdot d\mathbf{r} = 0, \tag{1.117}$$

for every closed path in our region S. We proceed to show that each of these three equations implies the other two.

Let us start with

$$\mathbf{F} = -\nabla\varphi. \tag{1.118}$$

Then

$$\nabla \times \mathbf{F} = -\nabla \times \nabla\varphi = 0$$

by Eq. 1.77 or Eq. 1.115 implies Eq. 1.116. Turning to the line integral,

$$\oint \mathbf{F} \cdot d\mathbf{r} = -\oint \nabla\varphi \cdot d\mathbf{r}$$

$$= -\oint d\varphi, \tag{1.119}$$

using Eq. 1.56. Now $d\varphi$ integrates to give φ. Since we have specified a closed loop, the end points coincide and we get zero for every closed path in our region S for which Eq. 1.115 holds. It is important to note the restriction here that the potential be single-valued and that Eq. 1.115 hold for *all* points in S. This problem may arise in using a scalar magnetic potential, a perfectly valid procedure as long as no net current is encircled. As soon as we choose a path in space that encircles a net current, the scalar magnetic potential ceases to be single-valued and our analysis no longer applies.

Continuing this demonstration of equivalence, let us assume that Eq. 1.117 holds. If $\oint \mathbf{F} \cdot d\mathbf{r} = 0$ for all paths in S, we see that the value of the integral joining two distinct points A and B is independent of the path. Our premise is that

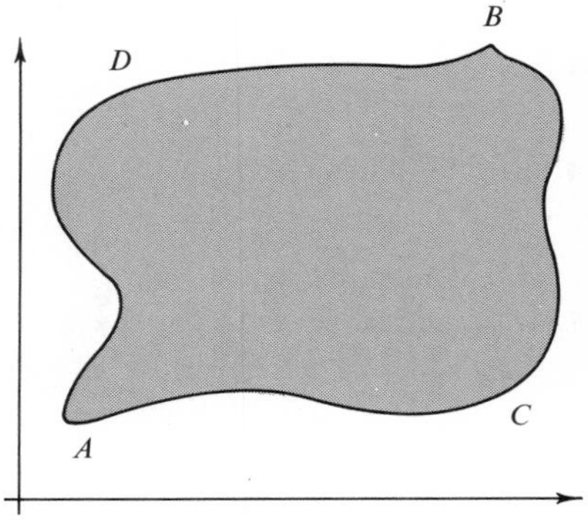

FIG. 1.22 Possible paths for doing work

$$\oint_{ACBDA} \mathbf{F} \cdot d\mathbf{r} = 0.$$

Therefore

$$\int_{ACB} \mathbf{F} \cdot d\mathbf{r} = -\int_{BDA} \mathbf{F} \cdot d\mathbf{r} = \int_{ADB} \mathbf{F} \cdot d\mathbf{r}, \tag{1.120}$$

reversing the sign by reversing the direction of integration. Physically, this means that the work done in going from A to B is independent of the path and that the work done in going around a closed path is zero. This is the reason for labeling such a force conservative: energy is conserved.

With the result shown in Eq.1.120, we have the work done dependent only on the end points, A and B. That is

$$\int_A^B \mathbf{F} \cdot d\mathbf{r} = \varphi(A) - \varphi(B). \tag{1.121}$$

The choice of sign on the right-hand side is arbitrary. The choice here is made to achieve agreement with Eq. 1.115 and to ensure that water will run downhill rather than uphill. For points A and B separated by a length $d\mathbf{r}$, Eq. 1.121 becomes

$$\begin{aligned} \mathbf{F} \cdot d\mathbf{r} &= -d\varphi \\ &= -\nabla\varphi \cdot d\mathbf{r} \end{aligned} \tag{1.122}$$

This may be rewritten

$$(\mathbf{F} + \nabla\varphi) \cdot d\mathbf{r} = 0, \tag{1.123}$$

and since $d\mathbf{r}$ is arbitrary, Eq. 1.115 must follow. If

$$\oint \mathbf{F} \cdot d\mathbf{r} = 0,$$

we may obtain Eq. 1.116 by using Stokes's theorem (Eq. 1.109).

$$\oint \mathbf{F} \cdot d\mathbf{r} = \int \nabla \times \mathbf{F} \cdot d\boldsymbol{\sigma}. \tag{1.124}$$

If we take the path of integration to be the perimeter of a differential area $d\boldsymbol{\sigma}$, the integrand in the surface integral must vanish. Hence Eq. 1.117 implies Eq. 1.116.

Finally, if $\nabla \times \mathbf{F} = 0$, we need only reverse our statement of Stokes's theorem

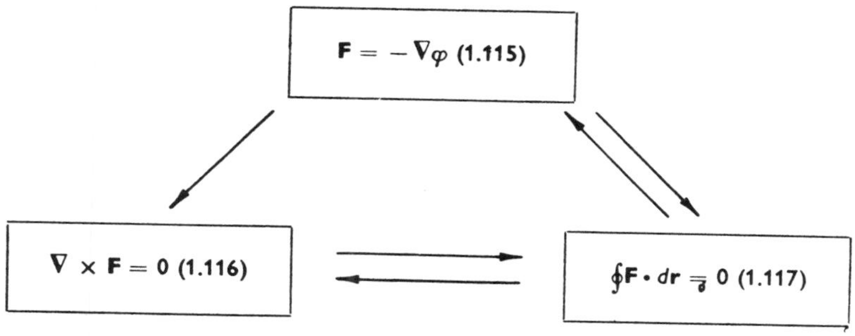

FIG. 1.23 Equivalent formulations

(Eq. 1.124) to derive Eq. 1.117. Then, by Eqs. 1.121 to 1.123 our initial statement $\mathbf{F} = -\nabla\varphi$ is derived. The equivalence is demonstrated.

To summarize, a single-valued scalar potential function φ exists if and only if $\mathbf{F}$ is irrotational or the work done around every closed loop is zero. The gravitational and electrostatic force fields given by Eq. 1.75 are irrotational and therefore conservative. Gravitational and electrostatic scalar potentials exist.

EXAMPLE 1.13.1

Find the scalar potential for the gravitational force on a unit mass m_1,

$$\mathbf{F}_G = -\frac{Gm_1m_2\mathbf{r}_0}{r^2} = -\frac{k\mathbf{r}_0}{r^2}, \qquad \text{radially } \textit{inward}.$$

By integrating Eq. 1.115 from infinity into position $\mathbf{r}$ we obtain

$$\varphi_G(r) - \varphi_G(\infty) = -\int_\infty^r \mathbf{F}_G \cdot d\mathbf{r} = +\int_r^\infty \mathbf{F}_G \cdot d\mathbf{r}. \tag{1.125}$$

By use of $\mathbf{F}_G = -\mathbf{F}_{\text{applied}}$, a comparison with Eq. 1.88 shows that the potential is the work done in bringing our unit mass in from infinity. (We can define only potential difference. Here we arbitrarily assign infinity to be a zero of potential.) The integral on the right-hand side of Eq. 1.125 is negative, meaning that $\varphi_G(r)$ is negative. Since $\mathbf{F}_G$ is radial, we obtain a contribution to φ only when $d\mathbf{r}$ is radial or

$$\varphi_G(r) = \int_r^\infty -\frac{k\,dr}{r^2} = -\frac{k}{r}$$

$$= -\frac{Gm_1m_2}{r}.$$

The final negative sign is a consequence of the attractive force of gravity.

EXAMPLE 1.13.2

Calculate the scalar potential for the *centrifugal* force per unit mass, $\mathbf{F}_C = \omega^2 r\mathbf{r}_0$, radially *outward*. Physically, this might be you on a large horizontal spinning disk

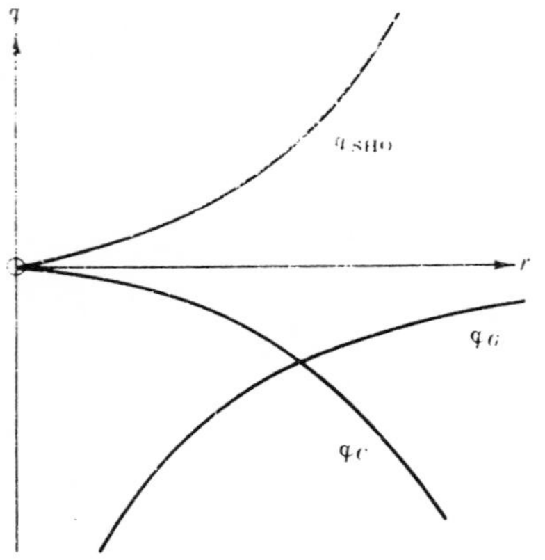

FIG. 1.24 Potential energy versus distance (gravitational, centrifugal, and simple harmonic oscillator)

at an amusement park. Proceeding as in Example 1.13.1, but integrating from the origin outward and taking $\varphi_C(0) = 0$,

$$\varphi_C(r) = -\int_0^r \mathbf{F}_C \cdot d\mathbf{r} = -\frac{\omega^2 r^2}{2}.$$

If we reverse signs, taking $\mathbf{F}_{SHO} = -k\mathbf{r}$, we obtain $\varphi_{SHO} = \frac{1}{2}kr^2$, the simple harmonic oscillatòr potential.

The gravitational, centrifugal, and simple harmonic oscillator potentials are shown in Fig. 1.24. Clearly the simple harmonic oscillator yields stability and describes a restoring force. The centrifugal potential describes an unstable situation.

In thermodynamics, which is sometimes called a search for exact differentials, we encounter equations of the form

$$df = P(x, y)\,dx + Q(x, y)\,dy. \tag{1.126}$$

The usual problem is to determine whether $\int (P(x, y)\,dx + Q(x, y)\,dy)$ depends only on the end points, that is, whether df is indeed an exact differential. The necessary and sufficient condition is that

$$df = \frac{\partial f}{\partial x}\,dx + \frac{\partial f}{\partial y}\,dy \tag{1.126a}$$

or that

$$P(x, y) = \frac{\partial f}{\partial x}, \qquad Q(x, y) = \frac{\partial f}{\partial y}. \tag{1.126b}$$

Equations 1.126*b* depend on the relation

$$\frac{\partial P(x, y)}{\partial y} = \frac{\partial Q(x, y)}{\partial x} \tag{1.126c}$$

being satisfied. This, however, is exactly analogous to Eq. 1.116, the requirement that **F** be irrotational. Indeed the z-component of Eq. 1.116 yields

$$\frac{\partial F_x}{\partial y} = \frac{\partial F_y}{\partial x}, \tag{1.126d}$$

Vector potential. In some branches of physics, especially electromagnetic theory, it is convenient to introduce a vector potential **A**, such that a (force) field **B** is given by

$$\mathbf{B} = \nabla \times \mathbf{A}. \tag{1.127}$$

Clearly, if Eq. 1.127 holds, $\nabla \cdot \mathbf{B} = 0$ by Eq. 1.79 and **B** is solenoidal. Here we want to develop a converse, to show that when **B** is solenoidal a vector potential **A** exists. We demonstrate the existence of **A** by actually calculating it. Suppose $\mathbf{B} = \mathbf{i}b_1 + \mathbf{j}b_2 + \mathbf{k}b_3$ and our unknown $\mathbf{A} = \mathbf{i}a_1 + \mathbf{j}a_2 + \mathbf{k}a_3$. By Eq. 1.127

$$\frac{\partial a_3}{\partial y} - \frac{\partial a_2}{\partial z} = b_1, \tag{1.128a}$$

$$\frac{\partial a_1}{\partial z} - \frac{\partial a_3}{\partial x} = b_2, \tag{1.128b}$$

$$\frac{\partial a_2}{\partial x} - \frac{\partial a_1}{\partial y} = b_3. \tag{1.128c}$$

Let us assume that the coordinates have been chosen so that **A** is parallel to the yz-plane, that is, $a_1 = 0$. Then

$$\begin{aligned} b_2 &= -\frac{\partial a_3}{\partial x} \\ b_3 &= \frac{\partial a_2}{\partial x}. \end{aligned} \tag{1.129}$$

Integrating,

$$\begin{aligned} a_2 &= \int_{x_0}^{x} b_3\, dx + f_2(y, z), \\ a_3 &= -\int_{x_0}^{x} b_2\, dx + f_3(y, z), \end{aligned} \tag{1.130}$$

where f_2 and f_3 are arbitrary functions of y and z but are *not* functions of x. Equation 1.128a becomes

$$\begin{aligned} \frac{\partial a_3}{\partial y} - \frac{\partial a_2}{\partial z} &= -\int_{x_0}^{x} \left(\frac{\partial b_2}{\partial y} + \frac{\partial b_3}{\partial z}\right) dx + \frac{\partial f_3}{\partial y} - \frac{\partial f_2}{\partial z} \\ &= \int_{x_0}^{x} \frac{\partial b_1}{\partial x}\, dx + \frac{\partial f_3}{\partial y} - \frac{\partial f_2}{\partial z}, \end{aligned} \tag{1.131}$$

using $\nabla \cdot \mathbf{B} = 0$. Integrating with respect to x, we obtain

$$\frac{\partial a_3}{\partial y} - \frac{\partial a_2}{\partial z} = b_1(x, y, z) - b_1(x_0, y, z) + \frac{\partial f_3}{\partial y} - \frac{\partial f_2}{\partial z}. \tag{1.132}$$

Remembering that f_3 and f_2 are arbitrary functions of y and z, we choose

$$f_2 = 0, \qquad f_3 = \int_{y_0}^{y} b_1(x_0, y, z)\, dy, \tag{1.133}$$

so that the right-hand side of Eq. 1.132 reduces to $b_1(x, y, z)$ in agreement with Equation 1.128*a*. With f_2 and f_3 given by Eq. 1.133, we can construct **A**.

$$\mathbf{A} = \mathbf{j}\int_{x_0}^{x} b_3(x, y, z)\, dx + \mathbf{k}\left[\int_{y_0}^{y} b_1(x_0, y, z)\, dy - \int_{x_0}^{x} b_2(x, y, z)\, dx\right]. \tag{1.134}$$

This is not quite complete. We may add any constant, since **B** is a derivative of **A**. What is much more important, we may add any gradient of a scalar function $\nabla\varphi$ without affecting **B** at all. Finally, the functions f_2 and f_3 are not unique. Other choices could have been made. It will be seen in Section 1.15 that we may still specify $\nabla \cdot \mathbf{A}$.

EXAMPLE 1.13.3

To illustrate the construction of a magnetic vector potential, we take the special but still important case of a constant magnetic induction

$$\mathbf{B} = \mathbf{k}B_z, \tag{1.135}$$

in which B_z is a constant. Equation 1.128 becomes

$$\frac{\partial a_3}{\partial y} - \frac{\partial a_2}{\partial z} = 0, \qquad \frac{\partial a_1}{\partial z} - \frac{\partial a_3}{\partial x} = 0, \qquad \frac{\partial a_2}{\partial x} - \frac{\partial a_1}{\partial y} = B_z. \tag{1.136}$$

If we assume that $a_1 = 0$, as before, then by Eq. 1.134

$$\mathbf{A} = \mathbf{j}\int^{x} B_z\, dx = \mathbf{j}xB_z, \tag{1.137}$$

setting a constant of integration equal to zero. It can readily be seen that this **A** satisfies Eq. 1.127.

To show that the choice $a_1 = 0$ was not sacred or at least not required, let us try setting $a_3 = 0$. From Eq. 1.136

$$\frac{\partial a_2}{\partial z} = 0, \tag{1.138a}$$

$$\frac{\partial a_1}{\partial z} = 0, \tag{1.138b}$$

$$\frac{\partial a_2}{\partial x} - \frac{\partial a_1}{\partial y} = B_z. \tag{1.138c}$$

We see a_1 and a_2 are independent of z or

$$a_1 = a_1(x, y), \qquad a_2 = a_2(x, y). \tag{1.139}$$

Equation 1.138c is satisfied if we take

$$a_2 = p \int^x B_z \, dx = pxB_z \tag{1.140}$$

and

$$a_1 = (p-1) \int^y B_z \, dy = (p-1)yB_z, \tag{1.141}$$

with p any constant. Then

$$\mathbf{A} = \mathbf{i}(p-1)yB_z + \mathbf{j}pxB_z. \tag{1.142}$$

Again, Eqs. 1.127, 1.135, and 1.142 are seen to be consistent. Comparison of Eqs. 1.137 and 1.142 shows immediately that $\mathbf{A}$ is not unique. The difference between Eqs. 1.137 and 1.142 and the appearance of the parameter p in Eq. 1.142 may be accounted for by rewriting Eq. 1.142 as

$$\begin{aligned} \mathbf{A} &= -\tfrac{1}{2}(\mathbf{i}y - \mathbf{j}x)B_z \\ &\quad + (p - \tfrac{1}{2})(\mathbf{i}y + \mathbf{j}x)B_z \\ &= -\tfrac{1}{2}(\mathbf{i}y - \mathbf{j}x)B_z + (p - \tfrac{1}{2})B_z \nabla\varphi \end{aligned} \tag{1.143}$$

with[1]

$$\varphi = xy. \tag{1.144}$$

The first term in $\mathbf{A}$ corresponds to the usual form

$$\mathbf{A} = \tfrac{1}{2}(\mathbf{B} \times \mathbf{r}) \tag{1.145}$$

for $\mathbf{B}$, a constant.

In many problems the magnetic vector potential $\mathbf{A}$ will be obtained from the current distribution that produces the magnetic induction $\mathbf{B}$. This means solving Poisson's (vector) equation (cf. Exercise 1.14.5).

EXERCISES

1.13.1 If a force $\mathbf{F}$ is given by

$$\mathbf{F} = (x^2 + y^2 + z^2)^n (\mathbf{i}x + \mathbf{j}y + \mathbf{k}z),$$

[1] $\varphi(x, y) = xy$ is clearly not a scalar function in the sense that the term was used in Section 1.3; that is, xy is not invariant to rotations about the z-axis. If such invariance is demanded, the parameter p must be set equal to $\tfrac{1}{2}$.

find
(a) $\nabla \cdot \mathbf{F}$,
(b) $\nabla \times \mathbf{F}$,
(c) a scalar potential $\varphi(x, y, z)$ so that $\mathbf{F} = -\nabla\varphi$.
(d) For what value of the exponent n does the scalar potential diverge at both the origin and infinity?

Ans. (a) $(2n+3)r^{2n}$ (c) $-\dfrac{1}{2n+2} r^{2n+2}, \quad n \neq -1$

(b) 0 (d) $n = -1, \quad \varphi = -\ln r$.

1.13.2 A sphere of radius a is uniformly charged (throughout its vôlume). Construct the electrostatic potential $\varphi(r)$ for $0 \leqslant r < \infty$. *Hint.* In Section 1.14 it is shown that the Coulomb force on a test charge at $r = r_0$ depends only on the charge at distances less than r_0 and is independent of the charge at distances greater than r_0. Note that this applies to a *spherically symmetric* charge distribution.

1.13.3 The usual problem in classical mechanics is to calculate the motion of a particle given the potential. For a uniform density (ρ_0), nonrotating massive sphere, Gauss's law of Section 1.14 leads to a gravitational force on a unit mass m_0 at a point r_0 produced by the attraction of the mass at $r \leqslant r_0$. The mass at $r > r_0$ contributes nothing to the force.
(a) Show that $\mathbf{F}/m_0 = -(4\pi G\rho_0/3)\mathbf{r}$, $0 \leqslant r \leqslant a$ where a is the radius of the sphere.
(b) Find the corresponding gravitational potential, $0 \leqslant r \leqslant a$.
(c) Imagine a vertical hole running completely through the center of the earth and out to the far side. Neglecting the rotation of the earth and assuming a uniform density $\rho_0 = 5.5$ gm/cm^3 calculate the nature of the motion of a particle dropped into the hole. What is its period?

1.13.4 A long straight wire carrying a current I produces a magnetic induction $\mathbf{B}$ with components

$$\mathbf{B} = \frac{\mu_0 I}{2\pi}\left(\frac{-y}{x^2+y^2}, \frac{x}{x^2+y^2}, 0\right).$$

Find a magnetic vector potential, $\mathbf{A}$.

Ans. $\mathbf{A} = -\mathbf{k}(\mu_0 I/4\pi)\ln(x^2+y^2)$.

(This solution is not unique.)

1.13.5 If

$$\mathbf{B} = \frac{\mathbf{r}_0}{r^2} = \left(\frac{x}{r^3}, \frac{y}{r^3}, \frac{z}{r^3}\right),$$

find a vector $\mathbf{A}$ such that $\nabla \times \mathbf{A} = \mathbf{B}$. One possible solution is

$$\mathbf{A} = \frac{\mathbf{i}yz}{r(x^2+y^2)} - \frac{\mathbf{j}xz}{r(x^2+y^2)}.$$

1.13.6 Show that the pair of equations

$$\mathbf{A} = \tfrac{1}{2}(\mathbf{B} \times \mathbf{r}),$$
$$\mathbf{B} = \nabla \times \mathbf{A},$$

is satisfied by any constant vector $\mathbf{B}$ (any orientation).

1.13.7 The magnetic induction $\mathbf{B}$ is related to the magnetic vector potential $\mathbf{A}$ by $\mathbf{B} = \nabla \times \mathbf{A}$. By Stokes' theorem

$$\int \mathbf{B} \cdot d\boldsymbol{\sigma} = \oint \mathbf{A} \cdot d\mathbf{r}.$$

Show that each side of this equation is invariant under the *gauge transformation*, $\mathbf{A} \to \mathbf{A} + \nabla\psi$.
Note. The complete gauge transformation is considered in Ex. 3.7.3.

1.13.8 With $\mathbf{E}$ the electric field and $\mathbf{A}$ the magnetic vector potential show that $[\mathbf{E} + \partial\mathbf{A}/\partial t]$ is irrotational and that therefore we may write

$$\mathbf{E} = -\nabla\varphi - \frac{\partial \mathbf{A}}{\partial t}.$$

1.13.9 The total force on a charge q moving with velocity $\mathbf{v}$ is

$$\mathbf{F} = q(\mathbf{E} + \mathbf{v} \times \mathbf{B}).$$

Using the scalar and vector potentials show that

$$\mathbf{F} = q\left[-\nabla\varphi - \frac{d\mathbf{A}}{dt} + \nabla(\mathbf{A} \cdot \mathbf{v})\right].$$

Note that we now have a total time derivative of $\mathbf{A}$ in place of the partial derivative of Ex. 1.13.8.

1.14 Gauss's Law, Poisson's Equation

Gauss's Law. Consider a point electric charge q at the origin of our coordinate system. This produces an electric field $\mathbf{E}$ given by

$$\mathbf{E} = \frac{q\mathbf{r}_0}{4\pi\varepsilon_0 r^2}. \tag{1.146}$$

We now derive Gauss's law which states that the surface integral

$$\int_S \mathbf{E} \cdot d\boldsymbol{\sigma} = \begin{cases} \dfrac{q}{\varepsilon_0}, \\ 0 \end{cases} \tag{1.147}$$

q/ε_0 if the closed surface S includes the origin (where q is located) and zero if the surface does not include the origin.

Using Gauss's theorem, Eq. 1.94 (and neglecting the $q/4\pi\varepsilon_0$),

$$\int_S \frac{\mathbf{r}_0 \cdot d\boldsymbol{\sigma}}{r^2} = \int_V \nabla \cdot \left(\frac{\mathbf{r}_0}{r^2}\right) d\tau = 0 \tag{1.148}$$

by Example 1.7.2, provided the surface S does not include the origin, where the integrands are not defined. This proves the second part of Gauss's law.

The first part, in which the surface S must include the origin, may be handled by surrounding the origin with a small sphere S' of radius δ. So that there will be no question what is inside and what is outside, imagine the volume outside the outer surface S and the volume inside surface $S'(r < \delta)$ connected by a small hole. This joins surfaces S and S', combining them into one single simply connected closed surface. Because the radius of the imaginary hole may be made vanishingly small, there is no additional contribution to the surface integral. Gauss's theorem now applies to the volume between S and S' without any difficulty. We have

$$\int_S \frac{\mathbf{r}_0 \cdot d\boldsymbol{\sigma}}{r^2} + \int_{S'} \frac{\mathbf{r}_0 \cdot d\boldsymbol{\sigma}'}{\delta^2} = 0. \tag{1.149}$$

We may evaluate the second integral, for $d\boldsymbol{\sigma}' = -\mathbf{r}_0\delta^2\, d\Omega$, in which $d\Omega$ is an element of solid angle. The minus sign appears because we agreed in Section 1.10 to have the positive normal $\mathbf{r}_0'$ *outward* from the volume. In this case our outward

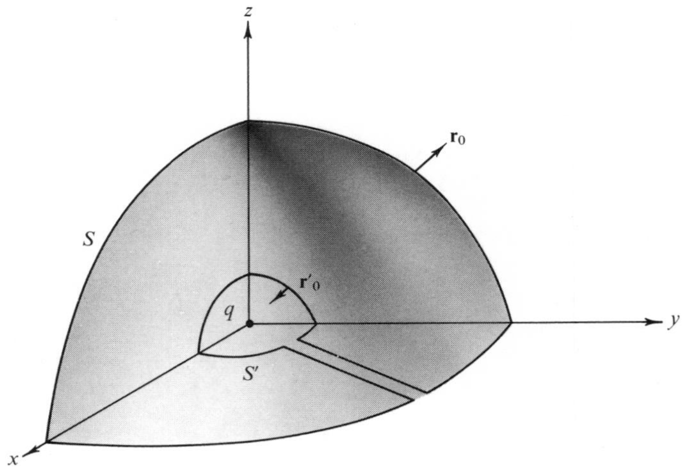

FIG. 1.25 Exclusion of the origin

$\mathbf{r}'_0$ is in the negative radial direction, $\mathbf{r}'_0 = -\mathbf{r}_0$. By integrating over all angles

$$\int_{S'} \frac{\mathbf{r}_0 \cdot d\boldsymbol{\sigma}'}{\delta^2} = -\int_{S'} \frac{\mathbf{r}_0 \cdot \mathbf{r}_0 \delta^2 \, d\Omega}{\delta^2} = -4\pi. \tag{1.150}$$

independent of the radius δ. With the constants from Eq. 1.146, this results in

$$\int_S \mathbf{E} \cdot d\boldsymbol{\sigma} = \frac{q}{4\pi\varepsilon_0} 4\pi = \frac{q}{\varepsilon_0}, \tag{1.151}$$

completing the proof of Gauss's law. Notice carefully that although the surface S may be spherical it *need not* be spherical.

Going just a bit further, we consider a distributed charge so that

$$q = \int_V \rho \, d\tau. \tag{1.152}$$

Equation 1.151 still applies, with q now interpreted as the total distributed charge enclosed by surface S.

$$\int_S \mathbf{E} \cdot d\boldsymbol{\sigma} = \int_V \frac{\rho}{\varepsilon_0} \, d\tau. \tag{1.153}$$

Using Gauss's theorem, we have

$$\int_V \nabla \cdot \mathbf{E} \, d\tau = \int_V \frac{\rho}{\varepsilon_0} \, d\tau. \tag{1.154}$$

Since our volume is completely arbitrary, the integrands must be equal or

$$\nabla \cdot \mathbf{E} = \frac{\rho}{\varepsilon_0}, \tag{1.155}$$

one of Maxwell's equations. If we reverse the argument, Gauss's law follows immediately from Maxwell's equation.

Poisson's equation. Replacing **E** by $-\nabla\varphi$, Eq. 1.155 becomes

$$\nabla \cdot \nabla \varphi = -\frac{\rho}{\varepsilon_0}, \tag{1.156}$$

which is Poisson's equation. For the condition $\rho = 0$ this reduces to an even more famous equation,

$$\nabla \cdot \nabla \varphi = 0, \tag{1.157}$$

Laplace's equation. We shall encounter Laplace's equation frequently in discussing various coordinate systems (Chapter 2) and in discussing the special functions of mathematical physics which appear as its solutions. Poisson's equation will be invaluable in developing the theory of Green's functions (Sections 8.6 and 16.5).

EXERCISES

1.14.1 Develop Gauss's Law for the two-dimensional case in which

$$\varphi = -q\frac{\ln \rho}{2\pi\varepsilon_0}, \quad \mathbf{E} = -\nabla\varphi = q\frac{\boldsymbol{\rho}_0}{2\pi\varepsilon_0 \rho}.$$

Here q is the charge at the origin or the line charge per unit length if our two-dimensional system is a unit thickness slice of a three-dimensional (circular cylindrical) system. The variable ρ is measured radially outward from the line charge. $\boldsymbol{\rho}_0$ is the corresponding unit vector (cf. Section 2.6).

1.14.2 (a) Show that Gauss's law follows from Maxwell's equation

$$\nabla \cdot \mathbf{E} = \frac{\rho}{\varepsilon_0}.$$

Here ρ is the usual charge density.

(b) Assuming that the electric field of a point charge q is spherically symmetric, show that Gauss's law implies the Coulomb inverse square expression

$$\mathbf{E} = \frac{q\mathbf{r}_0}{4\pi\varepsilon_0 r^2}.$$

1.14.3 Show that the value of the electrostatic potential φ at any point P is equal to the *average* of the potential over any spherical surface centered on P. There are no electric charges on or within the sphere.
Hint. Use Green's theorem, Eq. 1.97, with $u^{-1} = r$, the distance from P, and $v = \varphi$.

1.14.4 The energy stored in an electrostatic field may be written

$$W = \frac{1}{2}\int \rho(\mathbf{r})\varphi(\mathbf{r})\, d\tau.$$

where $\rho(\mathbf{r})$ is the charge density and $\varphi(\mathbf{r})$ the electrostatic potential. Transform this into

$$W = \tfrac{1}{2}\varepsilon_0 \int \mathbf{E}^2 d\tau$$

with **E** the electric field. This implies an energy *density* of $\frac{1}{2}\varepsilon_0 \mathbf{E}^2$. Assume that the charges are localized.

1.14.5 Using Maxwell's equations, show that for a static system (steady current) the magnetic vector potential **A** satisfies a vector Poisson equation,

$$\nabla^2 \mathbf{A} = -\mu \mathbf{J},$$

provided we require $\nabla \cdot \mathbf{A} = 0$.

1.15 Helmholtz' Theorem

In Section 1.13 it was emphasized that the choice of a magnetic vector potential **A** was not unique. The divergence of **A** was still undetermined. In this section two theorems about the divergence and curl of a vector are developed. The first theorem is as follows.

A vector is uniquely specified by giving its divergence and its curl within a region and its normal component over the boundary.

Let us take

$$\begin{aligned} \nabla \cdot \mathbf{V}_1 &= s, \\ \nabla \times \mathbf{V}_1 &= \mathbf{c}, \end{aligned} \tag{1.158}$$

where s may be interpreted as a source (charge) density and **c**, as a circulation (current) density. Assuming also that the normal component V_{1n} on the boundary is given, we want to show that $\mathbf{V}_1$ is unique. We do this by assuming the existence of a second vector $\mathbf{V}_2$, which satisfies Eq. 1.158 and has the same normal component over the boundary, and then showing that $\mathbf{V}_1 - \mathbf{V}_2 = 0$. Let

$$\mathbf{W} = \mathbf{V}_1 - \mathbf{V}_2 .$$

Then

$$\nabla \cdot \mathbf{W} = 0 \tag{1.159}$$

and

$$\nabla \times \mathbf{W} = 0. \tag{1.160}$$

Since **W** is irrotational we may write (by Section 1.13)

$$\mathbf{W} = -\nabla\varphi. \tag{1.161}$$

Substituting this into Eq. 1.159, we obtain

$$\nabla \cdot \nabla\varphi = 0, \tag{1.162}$$

Laplace's equation.

Now we draw upon Green's theorem in the form given in Eq. 1.98, letting u and v each equal φ. Since

$$W_n = V_{1n} - V_{2n} = 0 \tag{1.163}$$

on the boundary, Green's theorem reduces to

$$\int_V (\nabla\varphi) \cdot (\nabla\varphi)\, d\tau = \int_V \mathbf{W} \cdot \mathbf{W}\, d\tau = 0, \tag{1.164}$$

The quantity $\mathbf{W} \cdot \mathbf{W} = W^2$ is non-negative and so we must have

$$\mathbf{W} = \mathbf{V}_1 - \mathbf{V}_2 = 0 \tag{1.165}$$

everywhere. Thus $\mathbf{V}_1$ is unique, proving the theorem.

For our magnetic vector potential **A** the relation $\mathbf{B} = \nabla \times \mathbf{A}$ specifies the curl of **A**. Often for convenience we set $\nabla \cdot \mathbf{A} = 0$ (cf. Ex. 1.14.5). Then (with boundary conditions) **A** is fixed.

This theorem may be written as a uniqueness theorem for solutions of Laplace's equation, Ex. 1.15.1. In this form, this uniqueness theorem is of great importance in solving electrostatic and other Laplace equation boundary value problems. If we can find a solution of Laplace's equation that satisfies the necessary boundary conditions then our solution is the complete solution. Such boundary value problems are taken up in Sections 12.3 and 12.5.

Helmholtz' Theorem. The second theorem we shall prove is Helmholtz' theorem.

A vector **V** *satisfying Eq.* 1.158 *with both source and circulation densities vanishing at infinity may be written as the sum of two parts, one of which is irrotational, the other solenoidal.*

Helmholtz' theorem will clearly be satisfied if we may write **V** as

$$\mathbf{V} = -\nabla\varphi + \nabla \times \mathbf{A}, \tag{1.166}$$

$-\nabla\varphi$ being irrotational and $\nabla \times \mathbf{A}$ being solenoidal. We proceed to justify Eq. 1.166.

V is a known vector. Taking the divergence and curl

$$\nabla \cdot \mathbf{V} = s(\mathbf{r}) \tag{1.166a}$$

$$\nabla \times \mathbf{V} = \mathbf{c}(\mathbf{r}) \tag{1.166b}$$

with $s(\mathbf{r})$ and $\mathbf{c}(\mathbf{r})$ now known functions of position. From these two functions we construct a scalar potential $\varphi(\mathbf{r}_1)$,

$$\varphi(\mathbf{r}_1) = \frac{1}{4\pi}\int \frac{s(\mathbf{r}_2)}{r_{12}}\, d\tau_2, \tag{1.167a}$$

and a vector potential $\mathbf{A}(\mathbf{r}_1)$,

$$\mathbf{A}(\mathbf{r}_1) = \frac{1}{4\pi}\int \frac{\mathbf{c}(\mathbf{r}_2)}{r_{12}}\, d\tau_2. \tag{1.167b}$$

Here the argument $\mathbf{r}_1$ indicates (x_1, y_1, z_1), the field point; $\mathbf{r}_2$, the coordinates of the source point (x_2, y_2, z_2), whereas

$$r_{12} = [(x_1 - x_2)^2 + (y_1 - y_2)^2 + (z_1 - z_2)^2]^{1/2}. \tag{1.168}$$

When a direction is associated with r_{12}, the positive direction is taken to be away from the source toward the field point. Vectorially, $\mathbf{r}_{12} = \mathbf{r}_1 - \mathbf{r}_2$, as shown in Fig. 1.26. Of course, s and $\mathbf{c}$ must vanish sufficiently rapidly at large distances so that the integrals exist.

From the uniqueness theorem at the beginning of this section, **V** is uniquely specified by its divergence, s and curl, $\mathbf{c}$ (and boundary conditions). Returning to

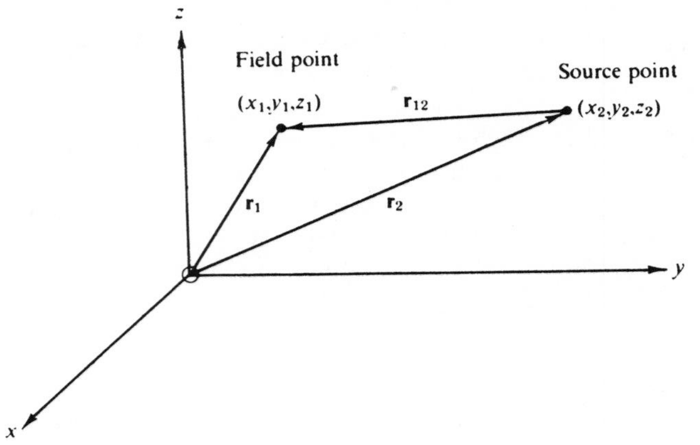

FIG. 1.26 Source and field points

Eq. 1.166

$$\nabla \cdot \mathbf{V} = - \nabla \cdot \nabla \varphi, \tag{1.169a}$$

the divergence of the curl vanishing and

$$\nabla \times \mathbf{V} = \nabla \times \nabla \times \mathbf{A}, \tag{1.169b}$$

the curl of the gradient vanishing. If we can show that

$$-\nabla \cdot \nabla \varphi(\mathbf{r}_1) = s(\mathbf{r}_1) \tag{1.169c}$$

and

$$\nabla \times \nabla \times \mathbf{A}(\mathbf{r}_1) = \mathbf{c}(\mathbf{r}_1), \tag{1.169d}$$

then $\mathbf{V}$ as given in Eq. 1.166 will have the proper divergence and curl. Our description will be internally consistent and Eq. 1.166 justified.

First we consider the divergence of $\mathbf{V}$:

$$\nabla \cdot \mathbf{V} = -\nabla \cdot \nabla \varphi = -\frac{1}{4\pi} \nabla \cdot \nabla \int \frac{s(\mathbf{r}_2)}{r_{12}} \, d\tau_2, \tag{1.170}$$

The Laplacian operator, $\nabla \cdot \nabla$ or ∇^2, operates on the field coordinates (x_1, y_1, z_1) and so commutes with the integration with respect to (x_2, y_2, z_2). We have

$$\nabla \cdot \mathbf{V} = -\frac{1}{4\pi} \int s(\mathbf{r}_2) \nabla_1^2 \left(\frac{1}{r_{12}}\right) d\tau_2. \tag{1.171}$$

In Section 1.14 we showed in the development of Gauss's law that

$$-\int \nabla \cdot \left(\frac{\mathbf{r}_0}{r^2}\right) d\tau = \int \nabla^2 \left(\frac{1}{r}\right) d\tau = \begin{cases} -4\pi \\ 0 \end{cases}. \tag{1.172}$$

depending on whether the integration included the origin $r = 0$. This result may be conveniently expressed by introducing the Dirac delta function, $\delta(r)$,[1]

$$\nabla^2\left(\frac{1}{r}\right) = -4\pi\,\delta(r). \tag{1.173}$$

This Dirac delta function is *defined* by its assigned properties

$$\delta(r) = 0, \qquad r \neq 0, \tag{1.174a}$$

$$\int f(r)\,\delta(r)\,d\tau = f(0), \tag{1.174b}$$

where $f(r)$ is any well-behaved function and the volume of integration includes the origin. As a special case of Eq. 1.174*b*,

$$\int \delta(r)\,d\tau = 1. \tag{1.175}$$

The quantity $\delta(r)$ is really not a function at all, since it is undefined (infinite) at $r=0$.

However, the crucial property, Eq. 1.174*b* can be developed rigorously as the limit of a *sequence* of functions, a distribution. This development appears in Section 8.6. Here we proceed to use the delta function in terms of its defining properties.

We must make two minor modifications in Eq. 1.173 before applying it. First, our source is at r_2, not at the origin. This means that the 4π in Gauss's law appears if and only if the surface includes the point $\mathbf{r} = \mathbf{r}_2$. To show this, Eq. 1.173 is rewritten:

$$\nabla^2\left(\frac{1}{r_{12}}\right) = -4\pi\,\delta(\mathbf{r}_1 - \mathbf{r}_2). \tag{1.176}$$

This shift of the source to $\mathbf{r}_2$ may be incorporated in the defining equations (1.174) as

$$\delta(\mathbf{r}_1 - \mathbf{r}_2) = 0, \qquad \mathbf{r}_1 \neq \mathbf{r}_2, \tag{1.177a}$$

$$\int f(\mathbf{r}_1)\,\delta(\mathbf{r}_1 - \mathbf{r}_2)\,d\mathbf{r}_1 = f(\mathbf{r}_2). \tag{1.177b}$$

Second, noting that differentiating r_{12}^{-1} twice with respect to x_2, y_2, z_2 is the same as differentiating *twice* with respect to x_1, y_1, z_1:

$$\begin{aligned}\nabla_1^2\left(\frac{1}{r_{12}}\right) = \nabla_2^2\left(\frac{1}{r_{12}}\right) &= -4\pi\,\delta(\mathbf{r}_1 - \mathbf{r}_2)\\ &= -4\pi\,\delta(\mathbf{r}_2 - \mathbf{r}_1).\end{aligned} \tag{1.178}$$

we could equally well have noted that from its defining properties

$$\delta(\mathbf{r}_1 - \mathbf{r}_2) = \delta(\mathbf{r}_2 - \mathbf{r}_1). \tag{1.179}$$

Rewriting Eq. 1.171 and using the Dirac delta function, Eq. 1.178, we may integrate to obtain

$$\begin{aligned}\nabla \cdot \mathbf{V} &= -\frac{1}{4\pi}\int s(\mathbf{r}_2)\nabla_2^2\left(\frac{1}{r_{12}}\right)d\tau_2\\ &= -\frac{1}{4\pi}\int s(\mathbf{r}_2)(-4\pi)\,\delta(\mathbf{r}_2 - \mathbf{r}_1)\,d\tau_2\\ &= s(\mathbf{r}_1).\end{aligned} \tag{1.180}$$

[1] Cf. Section 8.6 for a more extended treatment of the Dirac delta function.

The final step follows from Eq. 1.177*b* with the subscripts 1 and 2 exchanged. Our result, Eq. 1.180 shows that the assumed form of **V** and of the scalar potential φ are in agreement with the given divergence (Eq. 1.166*a*).

To complete the proof of Helmholtz' theorem we need to show that our assumptions are consistent with Eq. 1.166*a*, that is, that the curl of **V** is equal to $\mathbf{c}(\mathbf{r}_1)$. From Eq. 1.166

$$\begin{aligned}\nabla \times \mathbf{V} &= \nabla \times \nabla \times \mathbf{A} \\ &= \nabla\nabla \cdot \mathbf{A} - \nabla^2 \mathbf{A}. \qquad (1.181)\end{aligned}$$

The first term, $\nabla\nabla \cdot \mathbf{A}$ leads to

$$4\pi\nabla\nabla \cdot \mathbf{A} = \int \mathbf{c}(\mathbf{r}_2) \cdot \nabla_1 \nabla_1 \left(\frac{1}{r_{12}}\right) d\tau_2 \qquad (1.182)$$

by Eq. 1.167*b*. Again replacing the second derivatives with respect to x_1, y_1, z_1 by second derivatives with respect to x_2, y_2, z_2, we integrate each component[1] of Eq. 1.182 by parts:

$$\begin{aligned}4\pi\nabla\nabla \cdot \mathbf{A}\bigg|_x &= \int \mathbf{c}(\mathbf{r}_2) \cdot \nabla_2 \frac{\partial}{\partial x_2}\left(\frac{1}{r_{12}}\right) d\tau_2 \\ &= \int \nabla_2 \cdot \left[\mathbf{c}(\mathbf{r}_2) \frac{\partial}{\partial x_2}\left(\frac{1}{r_{12}}\right)\right] d\tau_2 - \int [\nabla_2 \cdot \mathbf{c}(\mathbf{r}_2)] \frac{\partial}{\partial x_2}\left(\frac{1}{r_{12}}\right) d\tau_2. \qquad (1.183)\end{aligned}$$

The second integral vanishes because the circulation density **c** is solenoidal.[2] The first integral may be transformed to a surface integral by Gauss's theorem. If **c** is bounded in space or vanishes faster than $1/r$ for large r, so that the integral in Eq. 1.167*b* exists, then by choosing a sufficiently large surface the first integral on the right-hand side of Eq. 1.183 also vanishes.
With $\nabla\nabla \cdot \mathbf{A} = 0$, Eq. 1.181 now reduces to

$$\nabla \times \mathbf{V} = -\nabla^2 \mathbf{A} = -\frac{1}{4\pi} \int \mathbf{c}(\mathbf{r}_2) \nabla_1^2 \left(\frac{1}{r_{12}}\right) d\tau_2. \qquad (1.184)$$

This is exactly like Eq. 1.171 except that the scalar $s(\mathbf{r}_2)$ is replaced by the vector circulation density $\mathbf{c}(\mathbf{r}_2)$. Introducing the Dirac delta function, as before, as a convenient way of carrying out the integration, it is seen that Eq. 1.184 reduces to Eq. 1.158. We see that our assumed form of **V**, given by Eq. 1.166, and of the vector potential **A**, given by Eq. 1.167*b*, are in agreement with Eq. 1.158 specifying the curl of **V**.

This completes the proof of Helmholtz' theorem, showing that a vector may be resolved into irrotational and solenoidal parts. Applied to the electromagnetic field, we have resolved our field vector **V** into an irrotational electric field **E**, derived from a scalar potential φ, and a solenoidal magnetic induction field **B**, derived from a vector potential **A**. The source density $s(\mathbf{r})$ may be interpreted as an electric charge density (divided by electric permittivity ε), whereas the circulation density $\mathbf{c}(\mathbf{r})$ becomes electric current density (times magnetic permeability μ).

[1] This avoids creating the *tensor* $\mathbf{c}(\mathbf{r}_2)\nabla_2$.

[2] Remember $\mathbf{c} = \nabla \times \mathbf{V}$ is known.

EXERCISES

1.15.1 Implicit in this section is a proof that a function $\psi(\mathbf{r})$ is uniquely specified by requiring it to (1) satisfy Laplace's equation and (2) satisfy a complete set of boundary conditions. Develop this proof explicitly.

1.15.2 (a) Assuming that $\mathbf{P}$ is a solution of the vector Poisson equation, $\nabla_1^2\mathbf{P}(\mathbf{r}_1) = -\mathbf{V}(\mathbf{r}_1)$,

$$\mathbf{P}(\mathbf{r}_1) = \frac{1}{4\pi}\int_V \frac{\mathbf{V}(\mathbf{r}_2)}{r_{12}}\,d\tau_2,$$

develop an alternate proof of Helmholtz' theorem, showing that $\mathbf{V}$ may be written as

$$\mathbf{V} = -\nabla\varphi + \nabla\times\mathbf{A},$$

where

$$\mathbf{A} = \nabla\times\mathbf{P},$$

and

$$\varphi = \nabla\cdot\mathbf{P}.$$

(b) Solving the vector Poisson equation, we find

$$\mathbf{P}(\mathbf{r}_1) = \frac{1}{4\pi}\int_V \frac{\mathbf{V}(\mathbf{r}_2)}{r_{12}}\,d\tau_2.$$

Show that this solution substituted into φ and $\mathbf{A}$ of part (a) leads to the expressions given for φ and $\mathbf{A}$ in Section 1.15.

REFERENCES

KELLOGG, O. D., *Foundations of Potential Theory*. New York: Dover (1953). Originally published 1929.

The classic text on potential theory.

LASS, H., *Vector and Tensor Analysis*. New York: McGraw-Hill (1950).

One of the clearest presentations of vector analysis. Written to be read and understood.

SCHWARTZ, M., S. GREEN, W. A. RUTLEDGE, *Vector Analysis with Applications to Geometry and Physics*. New York: Harper and Row (1960).

Excellent account of the elements of vector analysis with a wealth of applications in electromagnetism.

WREDE, R. C., *Introduction to Vector and Tensor Analysis*. New York: Wiley (1963).

Fine historical introduction. Excellent discussion of differentiation of vectors and applications to mechanics.

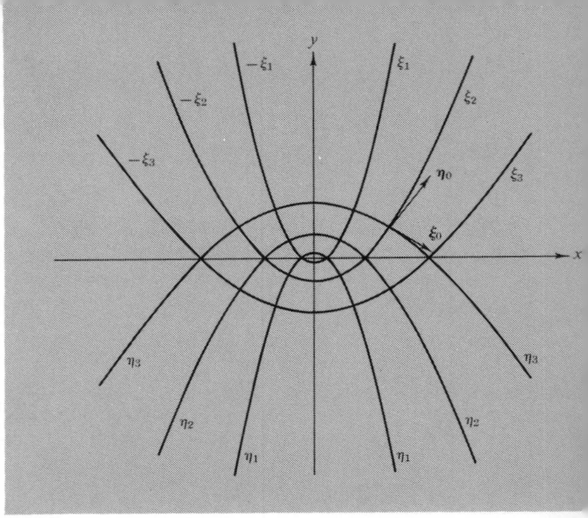

CHAPTER 2

COORDINATE SYSTEMS

In Chapter 1 we restricted ourselves almost completely to cartesian coordinate systems. A cartesian coordinate system offers the unique advantage that all three unit vectors, **i**, **j**, and **k**, are constant. We did introduce the radial distance **r** but even this was treated as a function of x, y, and z. Unfortunately, not all physical problems are well adapted to solution in cartesian coordinates. For instance, if we have a central force problem, $\mathbf{F} = \mathbf{r}_0 F(r)$, such as gravitational or electrostatic force, cartesian coordinates may be unusually inappropriate. Such a problem literally screams for the use of a coordinate system in which the radial distance is taken to be one of the coordinates, that is, spherical polar coordinates.

The point is that the coordinate system should be chosen to fit the problem, to exploit any constraint or symmetry present in it. Then, hopefully, it will be more readily soluble than if we had forced it into a cartesian framework. Quite often "more readily soluble" will mean that we have a partial differential equation that can be split into separate ordinary differential equations, often in "standard form" in the new coordinate system. This technique, the separation of variables, is discussed in Section 2.5.

We are primarily interested in coordinates in which the equation

$$\nabla^2 \psi + k^2 \psi = 0 \tag{2.1}$$

is separable. Equation 2.1 is much more general than it may appear. If

$k^2 = 0$	Eq. 2.1 →	Laplace's equation,
$k^2 = (+)$ constant		Helmholtz' equation,
$k^2 = (-)$ constant		Diffusion equation (space part),
$k^2 =$ constant $\times$ kinetic energy		Schrödinger wave equation.

It has been shown (L. P. Eisenhart, *Phys. Rev.* **45**, 427 (1934)) that there are eleven coordinate systems in which Eq. 2.1 is separable, all of which can be considered particular cases of the confocal ellipsoidal system. In addition, we shall touch briefly on three other systems that are useful in solving Laplace's equation.

Naturally there is a price that must be p^id for the use of a noncartesian coordinate system. We have not yet written expressions for gradient, divergence, or curl in any of the noncartesian coordinate systems. Such expressions are developed in very general form in Section 2.2. First we must develop a system of curvilinear coordinates, a general system that may be specialized to any of the fourteen particular systems of interest.

2.1 Curvilinear Coordinates

In cartesian coordinates we deal with three mutually perpendicular families of planes: $x =$ constant, $y =$ constant, and $z =$ constant. Imagine that we superimpose on this system three other families of surfaces. The surfaces of any one family need not be parallel to each other and they need not be planes. The three new families of surfaces need not be mutually perpendicular, but for simplicity we shall quickly impose this condition (Eq. 2.7). We may describe any point (x, y, z) as the intersection of three planes in cartesian coordinates or as the intersection of the three surfaces which form our new, curvilinear coordinates. Describing the curvilinear coordinate surfaces by $q_1 =$ constant, $q_2 =$ constant, $q_3 =$ constant, we may identify our point by (q_1, q_2, q_3) as well as by (x, y, z). This means that in principle we may write

$$\begin{aligned} x &= x(q_1, q_2, q_3), \\ y &= y(q_1, q_2, q_3), \\ z &= z(q_1, q_2, q_3), \end{aligned} \tag{2.2}$$

specifying x, y, z in terms of the q's and the inverse relations,

$$\begin{aligned} q_1 &= q_1(x, y, z), \\ q_2 &= q_2(x, y, z), \\ q_3 &= q_3(x, y, z). \end{aligned} \tag{2.3}$$

With each family of surface $q_i =$ constant, we can associate a unit vector $\mathbf{a}_i$ normal to the surface $q_i =$ constant and in the direction of increasing q_i.

The square of the distance between two neighboring points is given by

$$ds^2 = dx^2 + dy^2 + dz^2 = \sum_{ij} h_{ij}^2 \, dq_i \, dq_j. \tag{2.4}$$

The coefficients h_{ij}^2, which we now proceed to investigate, may be viewed as specifying the nature of the coordinate system (q_1, q_2, q_3). Collectively, these coefficients are referred to as the *metric*.

The first step in the determination of h_{ij}^2 is the partial differentiation of Eq. 2.2 which yields

$$dx = \frac{\partial x}{\partial q_1} dq_1 + \frac{\partial x}{\partial q_2} dq_2 + \frac{\partial x}{\partial q_3} dq_3, \tag{2.5}$$

and similarly for dy and dz. Squaring and substituting into Eq. 2.4, we have

$$h_{ij}^2 = \frac{\partial x}{\partial q_i}\frac{\partial x}{\partial q_j} + \frac{\partial y}{\partial q_i}\frac{\partial y}{\partial q_j} + \frac{\partial z}{\partial q_i}\frac{\partial z}{\partial q_j}. \tag{2.6}$$

At this point we limit ourselves to orthogonal (mutually perpendicular surfaces) coordinate systems which means (cf. Exercise 2.1.1)

$$h_{ij} = 0, \qquad i \neq j. \tag{2.7}$$

Now, to simplify the notation, we write $h_{ii} = h_i$ so that

$$ds^2 = (h_1\, dq_1)^2 + (h_2\, dq_2)^2 + (h_3\, dq_3)^2. \tag{2.8}$$

Our specific coordinate systems are described in subsequent sections by specifying these scale factors h_1, h_2, and h_3. Conversely, the scale factors may be conveniently identified by the relation

$$ds_i = h_i\, dq_i \tag{2.9}$$

for any given dq_i, holding the other q's constant. Note that the three curvilinear coordinates q_1, q_2, q_3 need not be lengths. The scale factors h_i may depend on the q's and they may have dimensions. The *product* $h_i\, dq_i$ must have dimensions of length.

From Eq. 2.9 we may immediately develop the area and volume elements

$$d\sigma_{ij} = ds_i\, ds_j = h_i h_j\, dq_i\, dq_j \tag{2.10}$$

and

$$d\tau = ds_1\, ds_2\, ds_3 = h_1 h_2 h_3\, dq_1\, dq_2\, dq_3. \tag{2.11}$$

The expressions in Eqs. 2.10 and 2.11 agree, of course, with the results of using the transformation equations, Eq. 2.2, and Jacobians.

EXERCISES

2.1.1 Show that limiting our attention to orthogonal coordinate systems implies that $h_{ij} = 0$ for $i \neq j$ (Eq. 2.7).

2.1.2 In the spherical polar coordinate system $q_1 = r, q_2 = \theta, q_3 = \varphi$. The transformation equations corresponding to Eq. 2.2 are

$$x = r \sin\theta \cos\varphi$$
$$y = r \sin\theta \sin\varphi$$
$$z = r \cos\theta.$$

(a) Calculate the spherical polar coordinate scale factors: h_r, h_θ, and h_φ.
(b) Check your calculated scale factors by the relation $ds_i = h_i\, dq_i$.

2.1.3 The u-, v-, z-coordinate system frequently used in electrostatics and in hydrodynamics is defined by

$$xy = u,$$
$$x^2 - y^2 = v,$$
$$z = z.$$

This u-, v-, z-system is orthogonal.

(a) In words, describe briefly the nature of each of the three families of coordinate surfaces.

(b) Sketch the system in the xy-plane showing the intersections of surfaces of constant u and surfaces of constant v with the xy-plane.

(c) Indicate the directions of the unit vector $\mathbf{u}_0$ and $\mathbf{v}_0$ in all four quadrants.

(d) Finally, is this u-, v-, z-system right-handed or left-handed?

2.1.4 A *two* dimensional system is described by the orthogonal coordinates q_1 and q_2. Show that the Jacobian

$$J\left(\frac{x, y}{q_1, q_2}\right) = h_1 h_2$$

in agreement with Eq. 2.10.

2.2 Differential Vector Operations

The starting point for developing the gradient, divergence, and curl operators in curvilinear coordinates is our interpretation of the gradient as the vector having the magnitude and direction of the maximum space rate of change (cf. Section 1.6). From this interpretation the component of $\nabla\psi(q_1, q_2, q_3)$ in the direction normal to the family of surfaces q_1 = constant is given by[1]

$$\nabla\psi\,|_1 = \frac{\partial\psi}{\partial s_1} = \frac{\partial\psi}{h_1\,\partial q_1}, \tag{2.12}$$

since this is the rate of change of ψ for varying q_1, holding q_2 and q_3 fixed. The quantity ds_1 is a differential length in the direction of increasing q_1 (cf. Eq. 2.9). In Section 2.1 we introduced a unit vector $\mathbf{a}_1$ to indicate this direction. By repeating Eq. 2.12 for q_2 and again for q_3 and adding vectorially the gradient becomes

$$\begin{aligned}\nabla\psi(q_1, q_2, q_3) &= \mathbf{a}_1 \frac{\partial\psi}{\partial s_1} + \mathbf{a}_2 \frac{\partial\psi}{\partial s_2} + \mathbf{a}_3 \frac{\partial\psi}{\partial s_3} \\ &= \mathbf{a}_1 \frac{\partial\psi}{h_1\,\partial q_1} + \mathbf{a}_2 \frac{\partial\psi}{h_2\,\partial q_2} + \mathbf{a}_3 \frac{\partial\psi}{h_3\,\partial q_3}.\end{aligned} \tag{2.13}$$

The divergence operator may be obtained from the second definition (Eq. 1.91) of Chapter 1 or equivalently from Gauss's theorem, Section 1.11. Let us use Eq. 1.91:

$$\nabla\cdot\mathbf{V}(q_1, q_2, q_3) = \lim_{\int d\tau\to 0} \frac{\int \mathbf{V}\cdot d\boldsymbol{\sigma}}{\int d\tau} \tag{2.14}$$

with a differential volume $h_1 h_2 h_3\, dq_1\, dq_2\, dq_3$. Note that the positive directions have been chosen so that (q_1, q_2, q_3) or $(\mathbf{a}_1, \mathbf{a}_2, \mathbf{a}_3)$ form a right-handed set.

[1] Here the use of φ to label a function is avoided because it is conventional to use this symbol to denote an azimuthal coordinate.

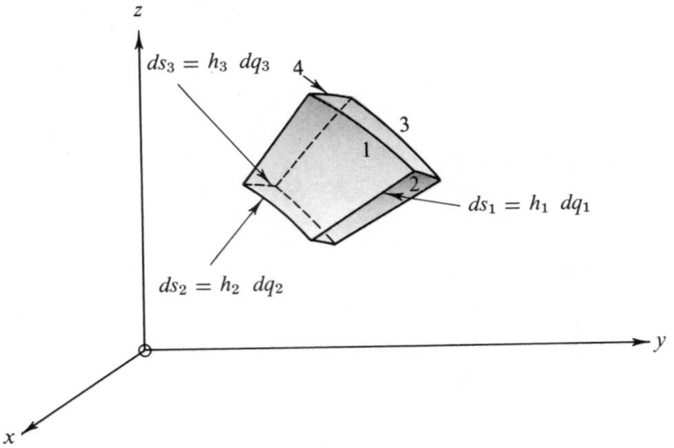

FIG. 2.1 Curvilinear volume element

The area integral for the two faces q_1 = constant is given by

$$\left[V_1 h_2 h_3 + \frac{\partial}{\partial q_1}(V_1 h_2 h_3)\, dq_1\right] dq_2\, dq_3 - V_1 h_2 h_3\, dq_2\, dq_3 = \frac{\partial}{\partial q_1}(V_1 h_2 h_3)\, dq_1\, dq_2\, dq_3 \tag{2.15}$$

exactly as in Sections 1.7 and 1.10.[1] Adding in the similar results for the other two pairs of surfaces, we obtain

$$\int \mathbf{V}(q_1, q_2, q_3) \cdot d\boldsymbol{\sigma} = \left[\frac{\partial}{\partial q_1}(V_1 h_2 h_3) + \frac{\partial}{\partial q_2}(V_2 h_3 h_1) + \frac{\partial}{\partial q_3}(V_3 h_1 h_2)\right] dq_1\, dq_2\, dq_3. \tag{2.16}$$

Division by our differential volume (Eq. 2.14) yields

$$\nabla \cdot \mathbf{V}(q_1, q_2, q_3) = \frac{1}{h_1 h_2 h_3}\left[\frac{\partial}{\partial q_1}(V_1 h_2 h_3) + \frac{\partial}{\partial q_2}(V_2 h_3 h_1) + \frac{\partial}{\partial q_3}(V_3 h_1 h_2)\right]. \tag{2.17}$$

In Eq. 2.17 V_i is the component of $\mathbf{V}$ in the $\mathbf{a}_i$-direction, increasing q_i, that is, $V_i = \mathbf{a}_i \cdot \mathbf{V}$.

We may obtain the Laplacian by combining Eqs. 2.13 and 2.17, using $\mathbf{V} = \nabla\psi(q_1, q_2, q_3)$. This leads to

$$\nabla \cdot \nabla\psi(q_1, q_2, q_3) = \frac{1}{h_1 h_2 h_3}\left[\frac{\partial}{\partial q_1}\left(\frac{h_2 h_3}{h_1}\frac{\partial\psi}{\partial q_1}\right) + \frac{\partial}{\partial q_2}\left(\frac{h_3 h_1}{h_2}\frac{\partial\psi}{\partial q_2}\right) + \frac{\partial}{\partial q_3}\left(\frac{h_1 h_2}{h_3}\frac{\partial\psi}{\partial q_3}\right)\right]. \tag{2.18a}$$

Finally, to develop $\nabla \times \mathbf{V}$, let us apply Stokes's theorem (Section 1.12) and, as

[1] Since we take the limit $dq_1, dq_2, dq_3 \to 0$, the second and higher order derivatives will drop out.

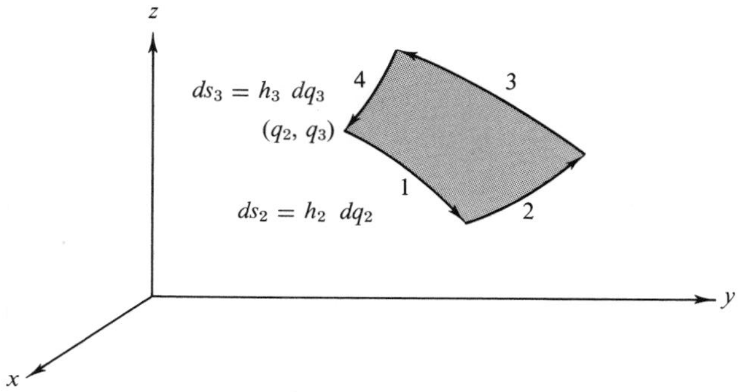

FIG. 2.2 Curvilinear surface element

with the divergence, take the limit as the surface area becomes vanishingly small. Working on one component at a time, we consider a differential surface element in the curvilinear surface q_1 = constant. From

$$\int_s \nabla \times \mathbf{V} \cdot d\boldsymbol{\sigma} = \nabla \times \mathbf{V}\,|_1\, h_2 h_3\, dq_2\, dq_3 \tag{2.18b}$$

Stokes's theorem yields

$$\nabla \times \mathbf{V}\,|_1\, h_2 h_3\, dq_2\, dq_3 = \oint \mathbf{V} \cdot d\boldsymbol{\lambda}, \tag{2.19}$$

with the line integral lying in the surface q_1 = constant. Following the loop (1, 2, 3, 4) of Fig. 2.2,

$$\begin{aligned}\oint \mathbf{V}(q_1, q_2, q_3) \cdot d\boldsymbol{\lambda} &= V_2 h_2\, dq_2 + \left[V_3 h_3 + \frac{\partial}{\partial q_2}(V_3 h_3)\, dq_2\right] dq_3 \\ &\quad - \left[V_2 h_2 + \frac{\partial}{\partial q_3}(V_2 h_2)\, dq_3\right] dq_2 - V_3 h_3\, dq_3 \\ &= \left[\frac{\partial}{\partial q_2}(h_3 V_3) - \frac{\partial}{\partial q_3}(h_2 V_2)\right] dq_2\, dq_3 .\end{aligned} \tag{2.20}$$

We pick up a positive sign when going in the positive direction on parts 1 and 2 and a negative sign on parts 3 and 4 because here we are going in the negative direction. From Eq. 2.19

$$\nabla \times \mathbf{V}\,|_1 = \frac{1}{h_2 h_3}\left[\frac{\partial}{\partial q_2}(h_3 V_3) - \frac{\partial}{\partial q_3}(h_2 V_2)\right]. \tag{2.21}$$

The remaining two components of $\nabla \times \mathbf{V}$ may be picked up by cyclic permutation of the indices. As in Chapter 1, it is often convenient to write the curl in determinant form:

$$\nabla \times \mathbf{V} = \frac{1}{h_1 h_2 h_3}\begin{vmatrix} \mathbf{a}_1 h_1 & \mathbf{a}_2 h_2 & \mathbf{a}_3 h_3 \\ \dfrac{\partial}{\partial q_1} & \dfrac{\partial}{\partial q_2} & \dfrac{\partial}{\partial q_3} \\ h_1 V_1 & h_2 V_2 & h_3 V_3 \end{vmatrix}. \tag{2.22}$$

This completes the determination of ∇, $\nabla\cdot$, $\nabla\times$, and the Laplacian ∇^2 in curvilinear coordinates. Armed with these general expressions, we proceed to study the eleven systems in which Eq. 2.1 is separable (cf. Section 2.5) for $k^2 \neq 0$ and three special coordinate systems (bipolar, toroidal, and bispherical coordinates).

EXERCISES

2.2.1 With $\mathbf{a}_1$ a unit vector in the direction of increasing q_1, show that

(a) $\nabla \cdot \mathbf{a}_1 = \dfrac{1}{h_1 h_2 h_3} \dfrac{\partial(h_2 h_3)}{\partial q_1}$

(b) $\nabla \times \mathbf{a}_1 = \dfrac{1}{h_1}\left[\mathbf{a}_2 \dfrac{\partial h_1}{h_3\,\partial q_3} - \mathbf{a}_3 \dfrac{\partial h_1}{h_2\,\partial q_2}\right]$

2.2.2 Show that the orthogonal unit vectors $\mathbf{a}_i$ may be defined by

$$\mathbf{a}_i = \frac{1}{h_i}\frac{\partial \mathbf{r}}{\partial q_i} \qquad (a)$$

In particular show that $\mathbf{a}_i \cdot \mathbf{a}_i = 1$ leads to an expression for h_i in agreement with Eq. 2.6.
Eq. (a) above may be taken as a starting point for deriving

$$\frac{\partial \mathbf{a}_i}{\partial q_j} = \mathbf{a}_j \frac{\partial h_j}{h_i\,\partial q_i}, \qquad i \neq j$$

and

$$\frac{\partial \mathbf{a}_i}{\partial q_i} = -\sum_{j\neq i} \mathbf{a}_j \frac{\partial h_i}{h_j\,\partial q_j}.$$

2.2.3 Develop arguments to show that ordinary dot and cross products (not involving ∇) in orthogonal curvilinear coordinates proceed as in cartesian coordinates *with no involvement of scale factors.*

2.2.4 Derive

$$\nabla\psi = \mathbf{a}_1 \frac{\partial\psi}{h_1\partial q_1} + \mathbf{a}_2 \frac{\partial\psi}{h_2\,\partial q_2} + \mathbf{a}_3 \frac{\partial\psi}{h_3\,\partial q_3}$$

by direct application of Eq. 1.90,

$$\nabla\psi = \lim_{\int d\tau\to 0} \frac{\int \psi\, d\sigma}{\int d\tau}.$$

Hint. Evaluation of the surface integral will lead to terms like $(h_1 h_2 h_3)^{-1}(\partial/\partial q_1)(\mathbf{a}_1 h_2 h_3)$. The results listed in Ex. 2.2.2 will be helpful.

2.3 Special Coordinate Systems—Rectangular Cartesian Coordinates

It has been emphasized that the choice of coordinate system may depend on constraints or symmetry conditions in the problem to be solved. It is perhaps convenient to list our fourteen systems, classifying them according to whether or not they have an axis of translation (perpendicular to a family of parallel plane surfaces) or an axis of rotational symmetry.

TABLE 2.1

Axis of Translation	Axis of Rotation	Neither
Cartesian (3 axes)		Confocal ellipsoidal
Circular cylindrical	Circular cylindrical	
	Spherical polar (3 axes)	
Elliptic cylindrical	Prolate spheroidal	
	Oblate spheroidal	
Parabolic cylindrical	Parabolic	
Bipolar	Toroidal	
	Bispherical	
		Conical
		Confocal paraboloidal

Table 2.1 contains fifteen entries—circular cylindrical coordinates with an axis of translation which is also an axis of rotational symmetry. The spacing in the table has been chosen to indicate relations between various coordinate systems. If we consider the two-dimensional version ($z = 0$) of a system with an axis of translation (left column) and rotate it about an axis of reflection symmetry, we generate the corresponding coordinate systems listed to the right in the center column. For instance, rotating the ($z = 0$)-plane of the elliptic cylindrical system about the major axis generates the prolate spheroidal system; rotating about the minor axis yields the oblate spheroidal system.

We do consider three systems with neither an axis of translation nor an axis of rotation. It might be noted that in this asymmetric group the confocal ellipsoidal system is sometimes taken as the most general system and almost all the others[1] are derived from it.

Rectangular cartesian coordinates. These are the cartesian coordinates on which Chapter 1 is based. In this simplest of all systems

$$\begin{aligned} h_1 &= h_x = 1, \\ h_2 &= h_y = 1, \\ h_3 &= h_z = 1. \end{aligned} \tag{2.23}$$

The families of coordinate surfaces are three sets of parallel planes: $x =$ constant, $y =$ constant, and $z =$ constant. The cartesian coordinate system is unique in that all its h_i's are constant. This will be a significant advantage in treating tensors in

[1] Excluding the bipolar system and its two rotational forms, toroidal and bispherical.

Chapter 3. Note also that the unit vectors, $\mathbf{a}_1$, $\mathbf{a}_2$, $\mathbf{a}_3$ or $\mathbf{i}$, $\mathbf{j}$, $\mathbf{k}$, have *fixed* directions.
From Eqs. 2.13, 2.17, 2.18, and 2.22 we reproduce the results of Chapter 1,

$$\nabla\psi = \mathbf{i}\frac{\partial\psi}{\partial x} + \mathbf{j}\frac{\partial\psi}{\partial y} + \mathbf{k}\frac{\partial\psi}{\partial z}, \tag{2.24}$$

$$\nabla\cdot\mathbf{V} = \frac{\partial V_x}{\partial x} + \frac{\partial V_y}{\partial y} + \frac{\partial V_z}{\partial z}, \tag{2.25}$$

$$\nabla\cdot\nabla\psi = \frac{\partial^2\psi}{\partial x^2} + \frac{\partial^2\psi}{\partial y^2} + \frac{\partial^2\psi}{\partial z^2}, \tag{2.26}$$

$$\nabla\times\mathbf{V} = \begin{vmatrix} \mathbf{i} & \mathbf{j} & \mathbf{k} \\ \dfrac{\partial}{\partial x} & \dfrac{\partial}{\partial y} & \dfrac{\partial}{\partial z} \\ V_x & V_y & V_z \end{vmatrix}. \tag{2.27}$$

2.4 Spherical Polar Coordinates (r, θ, φ)

Relabeling (q_1, q_2, q_3) as (r, θ, φ), the spherical polar coordinate system consists of the following:

1. Concentric spheres centered at the origin,

$$r = (x^2 + y^2 + z^2)^{1/2} = \text{constant}.$$

2. Right circular cones centered on the z-(polar) axis, vertices at the origin,

$$\theta = \arccos\frac{z}{(x^2 + y^2 + z^2)^{1/2}} = \text{constant}.$$

3. Half planes through the z-(polar) axis,

$$\varphi = \arctan\frac{y}{x} = \text{constant}.$$

By our arbitrary choice of definitions of θ, the polar angle, and φ, the azimuth angle, the z-axis is singled out for special treatment. The transformation equations corresponding to Eq. 2.2 are

$$\begin{aligned} x &= r\sin\theta\cos\varphi, \\ y &= r\sin\theta\sin\varphi, \\ z &= r\cos\theta, \end{aligned} \tag{2.28}$$

measuring θ from the positive z-axis and φ in the xy-plane from the positive x-axis. The ranges of values are $0 \le r < \infty$, $0 \le \theta \le \pi$, and $0 \le \varphi \le 2\pi$. From Eq. 2.6

$$\begin{aligned} h_1 &= h_r = 1, \\ h_2 &= h_\theta = r, \\ h_3 &= h_\varphi = r\sin\theta. \end{aligned} \tag{2.29}$$

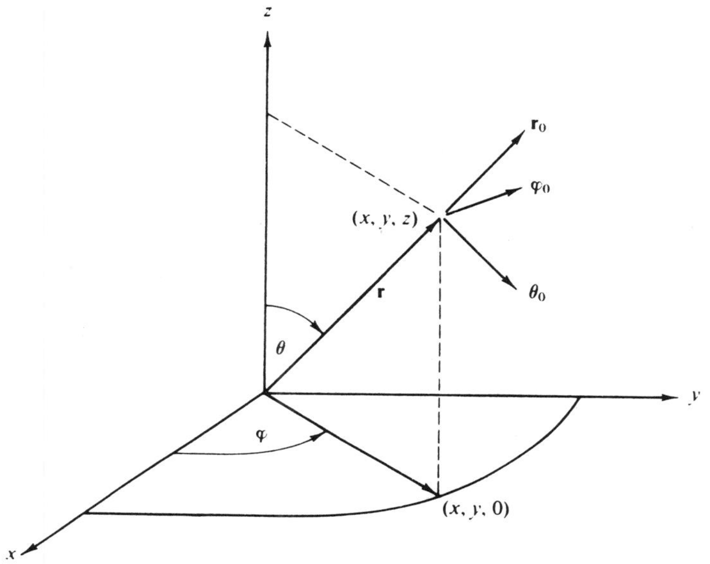

FIG. 2.3 Spherical polar coordinates

It must be emphasized that the unit vectors $\mathbf{r}_0$, $\boldsymbol{\theta}_0$, and $\boldsymbol{\varphi}_0$ vary in direction as the angles θ and φ vary. In terms of the fixed direction cartesian unit vectors **i**, **j**, and **k**,

$$\mathbf{r}_0 = \mathbf{i} \sin\theta \cos\varphi + \mathbf{j} \sin\theta \sin\varphi + \mathbf{k} \cos\theta,$$

$$\boldsymbol{\theta}_0 = \mathbf{i} \cos\theta \cos\varphi + \mathbf{j} \cos\theta \sin\varphi - \mathbf{k} \sin\theta,$$

$$\boldsymbol{\varphi}_0 = -\mathbf{i} \sin\varphi + \mathbf{j} \cos\varphi.$$

From Section 2.2, relabeling the curvilinear coordinate unit vectors $\mathbf{a}_1$, $\mathbf{a}_2$, $\mathbf{a}_3$ as $\mathbf{r}_0$, $\boldsymbol{\theta}_0$, and $\boldsymbol{\varphi}_0$,

$$\nabla\psi = \mathbf{r}_0 \frac{\partial\psi}{\partial r} + \boldsymbol{\theta}_0 \frac{1}{r}\frac{\partial\psi}{\partial\theta} + \boldsymbol{\varphi}_0 \frac{1}{r \sin\theta}\frac{\partial\psi}{\partial\varphi}, \tag{2.30}$$

$$\nabla \cdot \mathbf{V} = \frac{1}{r^2 \sin\theta}\left[\sin\theta \frac{\partial}{\partial r}\left(r^2 V_r\right) + r\frac{\partial}{\partial\theta}(\sin\theta\, V_\theta) + r\frac{\partial V_\varphi}{\partial\varphi}\right], \tag{2.31}$$

$$\nabla \cdot \nabla\psi = \frac{1}{r^2 \sin\theta}\left[\sin\theta \frac{\partial}{\partial r}\left(r^2 \frac{\partial\psi}{\partial r}\right) + \frac{\partial}{\partial\theta}\left(\sin\theta \frac{\partial\psi}{\partial\theta}\right) + \frac{1}{\sin\theta}\frac{\partial^2\psi}{\partial\varphi^2}\right], \tag{2.32}$$

$$\nabla \times \mathbf{V} = \frac{1}{r^2 \sin\theta}\begin{vmatrix} \mathbf{r}_0 & r\boldsymbol{\theta}_0 & r\sin\theta\boldsymbol{\varphi}_0 \\ \dfrac{\partial}{\partial r} & \dfrac{\partial}{\partial\theta} & \dfrac{\partial}{\partial\varphi} \\ V_r & rV_\theta & r\sin\theta V_\varphi \end{vmatrix}. \tag{2.33}$$

Occasionally the vector Laplacian $\nabla^2\mathbf{V}$ is needed in spherical polar coordinates. It is best obtained by using the vector identity (Eq. 1.80) of Chapter 1. For future reference

$$\nabla^2\mathbf{V}|_r = \left(-\frac{2}{r^2} + \frac{2}{r}\frac{\partial}{\partial r} + \frac{\partial^2}{\partial r^2} + \frac{\cos\theta}{r^2 \sin\theta}\frac{\partial}{\partial\theta} + \frac{1}{r^2}\frac{\partial^2}{\partial\theta^2} + \frac{1}{r^2\sin^2\theta}\frac{\partial^2}{\partial\varphi^2}\right)V_r$$

$$+\left(-\frac{2}{r^2}\frac{\partial}{\partial\theta} - \frac{2\cos\theta}{r^2\sin\theta}\right)V_\theta + \left(-\frac{2}{r^2\sin\theta}\frac{\partial}{\partial\varphi}\right)V_\varphi$$

$$= \nabla^2 V_r - \frac{2}{r^2}V_r - \frac{2}{r^2}\frac{\partial V_\theta}{\partial\theta} - \frac{2\cos\theta}{r^2\sin\theta}V_\theta - \frac{2}{r^2\sin\theta}\frac{\partial V_\varphi}{\partial\varphi}, \tag{2.34}$$

$$\nabla^2\mathbf{V}|_\theta = \nabla^2 V_\theta - \frac{1}{r^2\sin^2\theta}V_\theta + \frac{2}{r^2}\frac{\partial V_r}{\partial\theta} - \frac{2\cos\theta}{r^2\sin^2\theta}\frac{\partial V_\varphi}{\partial\varphi}, \tag{2.35}$$

$$\nabla^2\mathbf{V}|_\varphi = \nabla^2 V_\varphi - \frac{1}{r^2\sin^2\theta}V_\varphi + \frac{2}{r^2\sin\theta}\frac{\partial V_r}{\partial\varphi} + \frac{2\cos\theta}{r^2\sin^2\theta}\frac{\partial V_\theta}{\partial\varphi}. \tag{2.36}$$

These expressions for the components of $\nabla^2\mathbf{V}$ are undeniably messy, but sometimes they are needed. There is no guarantee that nature will always be simple.

Example 2.4.1

Using Eqs. 2.30–2.33, we can reproduce by inspection some of the results derived in Chapter 1 by laborious application of cartesian coordinates.

From Eq. 2.30

$$\nabla f(r) = \mathbf{r}_0 \frac{df}{dr}, \tag{2.37}$$

$$\nabla r^n = \mathbf{r}_0 n r^{n-1}.$$

From Eq. 2.31

$$\nabla \cdot \mathbf{r}_0 f(r) = \frac{2}{r}f(r) + \frac{df}{dr},$$

$$\nabla \cdot \mathbf{r}_0 r^n = (n+2)r^{n-1}. \tag{2.38}$$

From Eq. 2.32

$$\nabla^2 f(r) = \frac{2}{r}\frac{df}{dr} + \frac{d^2 f}{dr^2}, \tag{2.39}$$

$$\nabla^2 r^n = n(n+1)r^{n-2}. \tag{2.40}$$

Finally, from Eq. 2.33

$$\nabla \times \mathbf{r}_0 f(r) = 0. \tag{2.41}$$

Example 2.4.2

The computation of the magnetic vector potential of a single current loop in the xy plane involves the evaluation of

$$\mathbf{V} - \nabla \times [\nabla \times \boldsymbol{\varphi}_0 A_\varphi(r, \theta)]. \tag{2.41a}$$

In spherical polar coordinates this reduces as follows:

$$\mathbf{V} = \nabla \times \frac{1}{r^2 \sin\theta} \begin{vmatrix} \mathbf{r}_0 & r\boldsymbol{\theta}_0 & r\sin\theta\boldsymbol{\varphi}_0 \\ \dfrac{\partial}{\partial r} & \dfrac{\partial}{\partial\theta} & \dfrac{\partial}{\partial\varphi} \\ 0 & 0 & r\sin\theta A_\varphi(r,\theta) \end{vmatrix}$$

$$= \nabla \times \frac{1}{r^2 \sin\theta}\left[\mathbf{r}_0 \frac{\partial}{\partial\theta}(r\sin\theta A_\varphi) - r\boldsymbol{\theta}_0 \frac{\partial}{\partial r}(r\sin\theta A_\varphi)\right]. \qquad (2.41b)$$

Taking the curl a second time,

$$\mathbf{V} = \frac{1}{r^2\sin\theta}\begin{vmatrix} \mathbf{r}_0 & r\boldsymbol{\theta}_0 & r\sin\theta\boldsymbol{\varphi}_0 \\ \dfrac{\partial}{\partial r} & \dfrac{\partial}{\partial\theta} & \dfrac{\partial}{\partial\varphi} \\ \dfrac{1}{r^2\sin\theta}\dfrac{\partial}{\partial\theta}(r\sin\theta A_\varphi) & -\dfrac{1}{\sin\theta}\dfrac{\partial}{\partial r}(r\sin\theta A_\varphi) & 0 \end{vmatrix} \qquad (2.41c)$$

By expanding the determinant

$$\mathbf{V} = -\boldsymbol{\varphi}_0\left\{\frac{1}{r}\frac{\partial^2}{\partial r^2}(rA_\varphi) + \frac{1}{r^2}\frac{\partial}{\partial\theta}\left[\frac{1}{\sin\theta}\frac{\partial}{\partial\theta}(\sin\theta A_\varphi)\right]\right\}$$

$$= -\boldsymbol{\varphi}_0\left[\nabla^2 A_\varphi(r,\theta) - \frac{1}{r^2\sin^2\theta}A_\varphi(r,\theta)\right].$$

In Chapter 12 we shall see that $\mathbf{V}$ leads to the associated Legendre equation and that A_φ may be given by a series of associated Legendre polynomials.

EXERCISES

2.4.1 Resolve the spherical polar unit vectors into their cartesian components.

$$\mathbf{r}_0 = \mathbf{i}\sin\theta\cos\varphi + \mathbf{j}\sin\theta\sin\varphi + \mathbf{k}\cos\theta,$$
$$\boldsymbol{\theta}_0 = \mathbf{i}\cos\theta\cos\varphi + \mathbf{j}\cos\theta\sin\varphi - \mathbf{k}\sin\theta,$$
$$\boldsymbol{\varphi}_0 = -\mathbf{i}\sin\varphi + \mathbf{j}\cos\varphi.$$

2.4.2 (a) From the results of Exercise 2.4.1 calculate the partial derivatives of $\mathbf{r}_0$, $\boldsymbol{\theta}_0$, and $\boldsymbol{\varphi}_0$ with respect to r, θ, and φ.

(b) With ∇ given by

$$\mathbf{r}_0\frac{\partial}{\partial r} + \boldsymbol{\theta}_0\frac{1}{r}\frac{\partial}{\partial\theta} + \boldsymbol{\varphi}_0\frac{1}{r\sin\theta}\frac{\partial}{\partial\varphi}$$

(greatest space rate of change), use the results of part (a) to calculate $\nabla \cdot \nabla\psi$. Here is an alternate derivation of the Laplacian.

2.4.3 Resolve the cartesian unit vectors into their spherical polar components.

$$\mathbf{i} = \mathbf{r}_0 \sin\theta\cos\varphi + \boldsymbol{\theta}_0\cos\theta\cos\varphi - \boldsymbol{\varphi}_0\sin\varphi,$$
$$\mathbf{j} = \mathbf{r}_0 \sin\theta\sin\varphi + \boldsymbol{\theta}_0\cos\theta\sin\varphi + \boldsymbol{\varphi}_0\cos\varphi,$$
$$\mathbf{k} = \mathbf{r}_0\cos\theta - \boldsymbol{\theta}_0\sin\theta.$$

2.4.4 The direction of one vector is given by the angles θ_1 and φ_1. For a second vector the corresponding angles are θ_2 and φ_2. Show that the cosine of the included angle γ is given by

$$\cos\gamma = \cos\theta_1\cos\theta_2 + \sin\theta_1\sin\theta_2\cos(\varphi_1 - \varphi_2).$$

C.f. Fig. 12.15.

2.4.5 A vector $\mathbf{V}$ is tangential to the surface of a sphere. The curl of $\mathbf{V}$ is radial. What does this imply about the radial dependence of the spherical polar components of $\mathbf{V}$?

2.4.6 Modern physics lays great stress on the property of parity—whether a quantity remains invariant or changes sign under an inversion of the coordinate system.
(a) Show that the inversion (reflection through the origin) of a point (r, θ, φ) relative to *fixed* x-, y-, z-axes consists of the transformation

$$r \to r,$$
$$\theta \to \pi - \theta,$$
$$\varphi \to \pi + \varphi.$$

(b) Show that $\mathbf{r}_0$ and $\boldsymbol{\varphi}_0$ have odd parity (reversal of direction) and that $\boldsymbol{\theta}_0$ has even parity.

2.4.7 Eq. 1.72 was a demonstration that

$$\boldsymbol{\omega}\cdot\nabla\mathbf{r} = \boldsymbol{\omega},$$

using cartesian coordinates. Verify this result using *spherical polar coordinates*. In the language of dyadics (Section 3.5), $\nabla\mathbf{r}$ is the indemfactor, a unit dyadic.

2.4.8 A particle is moving through space. Find the spherical coordinate components of its velocity and acceleration:

$$v_r = \dot{r},$$
$$v_\theta = r\dot{\theta},$$
$$v_\varphi = r\sin\theta\dot{\varphi},$$
$$a_r = \ddot{r} - r\dot{\theta}^2 - r\sin^2\theta\dot{\varphi}^2,$$
$$a_\theta = r\ddot{\theta} + 2\dot{r}\dot{\theta} - r\sin\theta\cos\theta\dot{\varphi}^2,$$
$$a_\varphi = r\sin\theta\ddot{\varphi} + 2\dot{r}\sin\theta\dot{\varphi} + 2r\cos\theta\dot{\theta}\dot{\varphi}.$$

Hint.

$$\mathbf{r}(t) = \mathbf{r}_0(t)r(t)$$
$$= [\mathbf{i}\sin\theta(t)\cos\varphi(t) + \mathbf{j}\sin\theta(t)\sin\varphi(t) + \mathbf{k}\cos\theta(t)]r(t).$$

Note. Using the Lagrangian techniques of Section 17.3 these results may be obtained somewhat more elegantly. The dot in $\dot{r}$ means time derivative, $\dot{r} = dr/dt$. The notation was originated by Newton.

2.4.9 A particle m moves in response to a central force according to Newton's second law

$$m\ddot{\mathbf{r}} = \mathbf{r}_0 f(\mathbf{r}).$$

Show that $\mathbf{r}\times\dot{\mathbf{r}} = \mathbf{c}$, a constant and that the geometric interpretation of this leads to Kepler's second law.

2.4.10 Express $\partial/\partial x$, $\partial/\partial y$, $\partial/\partial z$ in spherical polar coordinates.

$$\frac{\partial}{\partial x} = \sin\theta\cos\varphi\frac{\partial}{\partial r} + \cos\theta\cos\varphi\frac{1}{r}\frac{\partial}{\partial\theta} - \frac{\sin\varphi}{r\sin\theta}\frac{\partial}{\partial\varphi},$$

$$\frac{\partial}{\partial y} = \sin\theta\sin\varphi\frac{\partial}{\partial r} + \cos\theta\sin\varphi\frac{1}{r}\frac{\partial}{\partial\theta} + \frac{\cos\varphi}{r\sin\theta}\frac{\partial}{\partial\varphi},$$

$$\frac{\partial}{\partial z} = \cos\theta\frac{\partial}{\partial r} - \sin\theta\frac{1}{r}\frac{\partial}{\partial\theta}.$$

Hint. Equate ∇_{xyz} and $\nabla_{r\theta\varphi}$.

2.4.11 From Exercise 2.4.10 show tnat

$$-i\left(x\frac{\partial}{\partial y} - y\frac{\partial}{\partial x}\right) = -i\frac{\partial}{\partial\varphi}.$$

This is the quantum mechanical operator corresponding to the z-component of angular momentum.

2.4.12 With the quantum mechanical angular momentum operator defined as $\mathbf{L} = -i(\mathbf{r}\times\nabla)$, show that

(a) $L_x + iL_y = e^{i\varphi}\left(\frac{\partial}{\partial\theta} + i\cot\theta\frac{\partial}{\partial\varphi}\right)$,

(b) $L_x - iL_y = -e^{-i\varphi}\left(\frac{\partial}{\partial\theta} - i\cot\theta\frac{\partial}{\partial\varphi}\right)$.

These are the raising and lowering operators of Sections 12.6 and 7.

2.4.13 Verify that $\mathbf{L}\times\mathbf{L} = i\mathbf{L}$ in spherical polar coordinates. $\mathbf{L} = -i(\mathbf{r}\times\nabla)$, the quantum mechanical angular momentum operator.

Hint: Use spherical polar coordinates for $\mathbf{L}$ but cartesian components for the cross product.

2.4.14 With $\mathbf{L} = -i\mathbf{r}\times\nabla$ verify the operator identities

(a) $\nabla = \mathbf{r}_0\frac{\partial}{\partial r} - i\frac{\mathbf{r}\times\mathbf{L}}{r^2}$,

(b) $\mathbf{r}\nabla^2 - \nabla\left(1 + r\frac{\partial}{\partial r}\right) = i\nabla\times\mathbf{L}$.

This latter identity is useful in relating angular momentum and Legendre's differential equation, Ex. 8.2.3.

2.4.15 Show that tne tollowing three forms (spherical coordinates) of $\nabla^2\psi(r)$ are equivalent.

(a) $\frac{1}{r^2}\frac{d}{dr}\left[r^2\frac{d\psi(r)}{dr}\right]$,

(b) $\frac{1}{r}\frac{d^2}{dr^2}\left[r\psi(r)\right]$,

(c) $\frac{d^2\psi(r)}{dr^2} + \frac{2}{r}\frac{d\psi(r)}{dr}$.

The second form is particularly convenient in establishing a correspondence between spherical polar and cartesian descriptions of a problem.

2.4.16 One model of the solar corona assumes that the steady-state equation of heat flow

$$\nabla\cdot(k\,\nabla T) = 0$$

is satisfied. Here, k, the thermal conductivity, is proportional to $T^{5/2}$. Assuming that the temperature T is proportional to r^n, show that the heat flow equation is satisfied by $T = T_0(r_0/r)^{2/7}$.

2.4.17 A certain force field is given by

$$\mathbf{F} = \mathbf{r}_0 \frac{2P\cos\theta}{r^3} + \boldsymbol{\theta}_0 \frac{P}{r^3}\sin\theta, \qquad r \geq P/2,$$

(in spherical polar coordinates).

(a) Examine $\nabla \times \mathbf{F}$ to see if a potential exists.

(b) Calculate $\oint \mathbf{F} \cdot d\boldsymbol{\lambda}$ for a unit circle in the plane $\theta = \pi/2$.
What does this indicate about the force being conservative or nonconservative?

(c) If you believe that $\mathbf{F}$ may be described by $\mathbf{F} = -\nabla\psi$, find ψ. Otherwise simply state that no acceptable potential exists.

2.4.18 (a) Show that $\mathbf{A} = -\boldsymbol{\varphi}_0(\cot\theta/r)$ is a solution of $\nabla \times \mathbf{A} = \mathbf{r}_0/r^2$.

(b) Show that this spherical polar coordinate solution agrees with the solution given for Exercise 1.13.5:

$$\mathbf{A} = \mathbf{i}\frac{yz}{r(x^2+y^2)} - \mathbf{j}\frac{xz}{r(x^2+y^2)}.$$

Note that the solution diverges for $\theta = 0, \pi$ corresponding to $x, y = 0$.

(c) Finally, show that $\mathbf{A} = -\boldsymbol{\theta}_0\varphi(\sin\theta/r)$ is a solution. Note that although this solution does not diverge ($r \neq 0$) it is no longer single-valued for all possible azimuth angles.

2.4.19 A magnetic vector potential is given by

$$\mathbf{A} = \frac{\mu_0}{4\pi}\frac{\mathbf{m} \times \mathbf{r}}{r^3}.$$

Show that this leads to the magnetic induction $\mathbf{B}$ of a point magnetic dipole, dipole moment $\mathbf{m}$.

Ans. For $\mathbf{m} = \mathbf{k}m$,

$$\nabla \times \mathbf{A} = \mathbf{r}_0 \frac{\mu_0}{4\pi}\frac{2m\cos\theta}{r^3} + \boldsymbol{\theta}_0 \frac{\mu_0}{4\pi}\frac{m\sin\theta}{r^3}.$$

Cf. Eqs. 12.146 and 12.147.

2.4.20 At large distances from its source, electric dipole radiation has fields

$$\mathbf{E} = a_E \sin\theta \frac{e^{i(kr-wt)}}{r}\boldsymbol{\theta}_0, \qquad \mathbf{B} = a_B \sin\theta \frac{e^{i(kr-wt)}}{r}\boldsymbol{\varphi}_0.$$

Show that Maxwell's equations

$$\nabla \times \mathbf{E} = -\frac{\partial \mathbf{B}}{\partial t} \quad \text{and} \quad \nabla \times \mathbf{B} = \varepsilon_0\mu_0\frac{\partial \mathbf{E}}{\partial t}$$

are satisfied, if we take

$$a_E/a_B = \omega/k = c = (\varepsilon_0\mu_0)^{-1/2}.$$

Hint. Since r is large, terms of order r^{-2} may be dropped.

2.5 Separation of Variables

In cartesian coordinates the Helmholtz equation (Eq. 2.1) becomes

$$\frac{\partial^2\psi}{\partial x^2} + \frac{\partial^2\psi}{\partial y^2} + \frac{\partial^2\psi}{\partial z^2} + k^2\psi = 0, \tag{2.42}$$

using Eq. 2.26 for the Laplacian. For the present let k^2 be a constant. Perhaps the simplest way of treating a partial differential equation such as 2.42 is to split it up into a set of ordinary differential equations. This may be done as follows: Let

$$\psi(x, y, z) = X(x)\,Y(y)\,Z(z) \tag{2.43}$$

and substitute back into Eq. 2.42. How do we know Eq. 2.43 is valid? The answer is very simple. We do not know it is valid! Rather we are proceeding in the spirit of let's try it and see if it works. If our attempt succeeds, then Eq. 2.43 will be justified. If it does not succeed, we shall find out soon enough and then we shall have to try another attack such as Green's functions, integral transforms, or brute force numerical analysis. With ψ assumed given by Eq. 2.43, Eq. 2.42 becomes

$$YZ\frac{d^2X}{dx^2} + XZ\frac{d^2Y}{dy^2} + XY\frac{d^2Z}{dz^2} + k^2XYZ = 0. \tag{2.44}$$

Dividing by $\psi = XYZ$ and rearranging terms, we obtain

$$\frac{1}{X}\frac{d^2X}{dx^2} = -k^2 - \frac{1}{Y}\frac{d^2Y}{dy^2} - \frac{1}{Z}\frac{d^2Z}{dz^2}. \tag{2.45}$$

Equation 2.45 exhibits one separation of variables. The left-hand side is a function of x alone, whereas the right-hand side depends only on y and z. So Eq. 2.45 is a sort of paradox. A function of x is equated to a function of y and z, but x, y, and z are all independent coordinates. This independence means that the behavior of x as an independent variable is not determined by y and z. The paradox is resolved by setting each side equal to a constant, a constant of separation. We choose[1]

$$\frac{1}{X}\frac{d^2X}{dx^2} = -l^2, \tag{2.46}$$

$$-k^2 - \frac{1}{Y}\frac{d^2Y}{dy^2} - \frac{1}{Z}\frac{d^2Z}{dz^2} = -l^2. \tag{2.47}$$

Now, turning our attention to Eq. 2.47,

$$\frac{1}{Y}\frac{d^2Y}{dy^2} = -k^2 + l^2 - \frac{1}{Z}\frac{d^2Z}{dz^2}, \tag{2.48}$$

and a second separation has been achieved. Here we have a function of y equated to a function of z and the same paradox appears. We resolve it as before by equating each side to another constant of separation, $-m^2$,

$$\frac{1}{Y}\frac{d^2Y}{dy^2} = -m^2, \tag{2.49}$$

$$\frac{1}{Z}\frac{d^2Z}{dz^2} = -k^2 + l^2 + m^2 = -n^2, \tag{2.50}$$

introducing a constant n^2 by $k^2 = l^2 + m^2 + n^2$ to produce a symmetric set of equations. Now we have three ordinary differential equations (2.46, 2.49, and

[1] The choice of sign, completely arbitrary here, will be fixed in specific problems by the need to satisfy specific boundary conditions.

2.50) to replace Eq. 2.42. Our assumption (Eq. 2.43) has succeeded and is thereby justified.

Our solution should be labeled according to the choice of our constants l, m, and n, that is,

$$\psi_{lmn}(x, y, z) = X_l(x)\, Y_m(y)\, Z_n(z). \tag{2.50a}$$

Subject to the conditions of the problem being solved and to the condition $k^2 = l^2 + m^2 + n^2$, we may choose l, m, and n as we like, and Eq. 2.50*a* will still be a solution of Eq. 2.1, provided $X_l(x)$ is a solution of Eq. 2.46, etc. We may develop the most general solution of Eq. 2.1 by taking a linear combination of solutions ψ_{lmn},

$$\Psi = \sum_{l,m,n} a_{lmn}\psi_{lmn}. \tag{2.50b}$$

The constant coefficients a_{lmn} are finally chosen to permit Ψ to satisfy the boundary conditions of the problem.

How is this possible? What is the justification for writing Eq. 2.50*b*? The justification is found in noting that $\nabla^2 + k^2$ is a linear (differential) operator. A linear operator $\mathscr{L}$ is defined as an operator with the following two properties:

$$\mathscr{L}(a\psi) = a\mathscr{L}\psi,$$

where a is a constant and

$$\mathscr{L}(\psi_1 + \psi_2) = \mathscr{L}\psi_1 + \mathscr{L}\psi_2.$$

As a consequence of these properties, any linear combination of solutions of a linear differential equation is also a solution. From its explicit form $\nabla^2 + k^2$ is seen to have these two properties (and is therefore a linear operator). Equation 2.50*b* then follows as a direct application of these two defining properties.[1]

A further generalization may be noted. The separation process just described would go through just as well for

$$k^2 = f(x) + g(y) + h(z) + k'^2, \tag{2.50c}$$

with k'^2 a new constant.

We would simply have

$$\frac{1}{X}\frac{d^2X}{dx^2} + f(x) = -l^2 \tag{2.50d}$$

replacing Eq. 2.46. The solutions X, Y, and Z would be different, but the technique of splitting the partial differential equation and of taking a linear combination of solutions would be the same.

In case the reader wonders what is going on here, this technique of separation of variables of a partial differential equation has been introduced to illustrate the usefulness of these coordinate systems. The solutions of the resultant ordinary differential equations are developed in Chapters 8 through 13.

Let us try to separate Eq. 2.1, again with k^2 constant, in spherical polar coordinates. Using Eq. 2.32, we obtain

[1] We are especially interested in linear operators because in quantum mechanics physical quantities are represented by linear operators operating in a complex, infinite dimensional Hilbert space.

$$\frac{1}{r^2 \sin\theta}\left[\sin\theta \frac{\partial}{\partial r}\left(r^2 \frac{\partial\psi}{\partial r}\right) + \frac{\partial}{\partial\theta}\left(\sin\theta \frac{\partial\psi}{\partial\theta}\right) + \frac{1}{\sin\theta}\frac{\partial^2\psi}{\partial\varphi^2}\right] = -k^2\psi. \tag{2.51}$$

Now, in analogy with Eq. 2.43 we try

$$\psi(r, \theta, \varphi) = R(r)\,\Theta(\theta)\,\Phi(\varphi). \tag{2.52}$$

By substituting back into Eq. 2.51 and dividing by $R\Theta\Phi$, we have

$$\frac{1}{Rr^2}\frac{d}{dr}\left(r^2\frac{dR}{dr}\right) + \frac{1}{\Theta r^2 \sin\theta}\frac{d}{d\theta}\left(\sin\theta\frac{d\Theta}{d\theta}\right) + \frac{1}{\Phi r^2 \sin^2\theta}\frac{d^2\Phi}{d\varphi^2} = -k^2. \tag{2.53}$$

Note that all derivatives are now ordinary derivatives rather than partials. By multiplying by $r^2 \sin^2\theta$ we can isolate $(1/\Phi)(d^2\Phi/d\varphi^2)$ to obtain

$$\frac{1}{\Phi}\frac{d^2\Phi}{d\varphi^2} = r^2\sin^2\theta\left[-k^2 - \frac{1}{r^2 R}\frac{d}{dr}\left(r^2\frac{dR}{dr}\right) - \frac{1}{r^2 \sin\theta\Theta}\frac{d}{d\theta}\left(\sin\theta\frac{d\Theta}{d\theta}\right)\right]. \tag{2.54}$$

Equation 2.54 relates a function of φ alone to a function of r and θ alone. Since r, θ, and φ are independent variables, we equate each side of Eq. 2.54 to a constant. Here a little consideration can simplify the later analysis. In almost all physical problems φ will appear as an azimuth angle. This suggests a periodic solution rather than an exponential. With this in mind, let us use $-m^2$ as the separation constant. Any constant will do, but this one will make life a little easier. Then

$$\frac{1}{\Phi}\frac{d^2\Phi(\varphi)}{d\varphi^2} = -m^2 \tag{2.55}$$

and

$$\frac{1}{r^2 R}\frac{d}{dr}\left(r^2\frac{dR}{dr}\right) + \frac{1}{r^2\sin\theta\Theta}\frac{d}{d\theta}\left(\sin\theta\frac{d\Theta}{d\theta}\right) - \frac{m^2}{r^2\sin^2\theta} = -k^2. \tag{2.56}$$

Multiplying Eq. 2.56 by r^2 and rearranging terms, we obtain

$$\frac{1}{R}\frac{d}{dr}\left(r^2\frac{dR}{dr}\right) + r^2k^2 = -\frac{1}{\sin\theta\Theta}\frac{d}{d\theta}\left(\sin\theta\frac{d\Theta}{d\theta}\right) + \frac{m^2}{\sin^2\theta}. \tag{2.57}$$

Again the variables are separated. We equate each side to a constant Q and finally obtain

$$\frac{1}{\sin\theta}\frac{d}{d\theta}\left(\sin\theta\frac{d\Theta}{d\theta}\right) - \frac{m^2}{\sin^2\theta}\Theta + Q\Theta = 0, \tag{2.58}$$

$$\frac{1}{r^2}\frac{d}{dr}\left(r^2\frac{dR}{dr}\right) + k^2R - \frac{QR}{r^2} = 0. \tag{2.59}$$

Once more we have replaced a partial differential equation of three variables by three ordinary differential equations. The solutions of these ordinary differential equations are discussed in Chapters 11 and 12. In Chapter 12, for example, Eq. 2.58 is identified as the associated Legendre equation in which the constant Q becomes $l(l+1)$; l is an integer.

Again, our most general solution may be written

$$\psi_{Qm}(r, \theta, \varphi) = \sum_{Q,m} R_Q(r)\,\Theta_{Qm}(\theta)\,\Phi_m(\varphi). \tag{2.60a}$$

The restriction that k^2 be a constant is unnecessarily severe. The separation process will still be possible for k^2 as general as

$$k^2 = f(r) + \frac{1}{r^2} g(\theta) + \frac{1}{r^2 \sin^2\theta} h(\varphi) + k'^2 \tag{2.60b}$$

In the hydrogen atom problem, one of the most important examples of the Schrödinger wave equation with a closed form solution, we have $k^2 = f(r)$. Equation 2.59 for the hydrogen atom becomes the associated Laguerre equation. Separation of variables and an investigation of the resulting ordinary differential equations are taken up again in Section 8.2. Now we return to an investigation of the remaining special coordinate systems.

EXERCISES

2.5.1 By letting the operator $\nabla^2 + k^2$ act on the general form $a_1 \psi_1(x, y, z) + a_2 \psi_2(x, y, z)$, show that it is linear, that is, that $(\nabla^2 + k^2)(a_1\psi_1 + a_2\psi_2) = a_1(\nabla^2 + k^2)\psi_1 + a_2(\nabla^2 + k^2)\psi_2$.

2.5.2 Verify that

$$\nabla^2\psi(r, \theta, \varphi) + \left[k^2 + f(r) + \frac{1}{r^2} g(\theta) + \frac{1}{r^2 \sin^2\theta} h(\varphi)\right] \psi(r, \theta, \varphi) = 0$$

is separable (in spherical polar coordinates). The functions f, g, and h are functions only of the variables indicated; k^2 is a constant.

2.5.3 An atomic (quantum mechanical) particle is confined inside a rectangular box of sides a, b, and c. The particle is described by a wave function ψ which satisfies the Schrödinger wave equation

$$-\frac{\hbar^2}{2m} \nabla^2\psi = E\psi.$$

The wave function is required to vanish at each surface of the box (but not to be identically zero). This condition imposes constraints on the separation constants and therefore on the energy E. What is the smallest value of E for which such a solution can be obtained?

Ans. $E = \dfrac{\pi^2\hbar^2}{2m}\left(\dfrac{1}{a^2} + \dfrac{1}{b^2} + \dfrac{1}{c^2}\right).$

2.6 Circular Cylindrical Coordinates (ρ, φ, z)

From Fig. 2.4 we obtain the transformation relations

$$\begin{aligned} x &= \rho \cos \varphi, \\ y &= \rho \sin \varphi, \\ z &= z. \end{aligned} \tag{2.61}$$

using ρ for the perpendicular distance from the z-axis and saving r for the distance from the origin. According to these equations or from the length elements the scale factors are

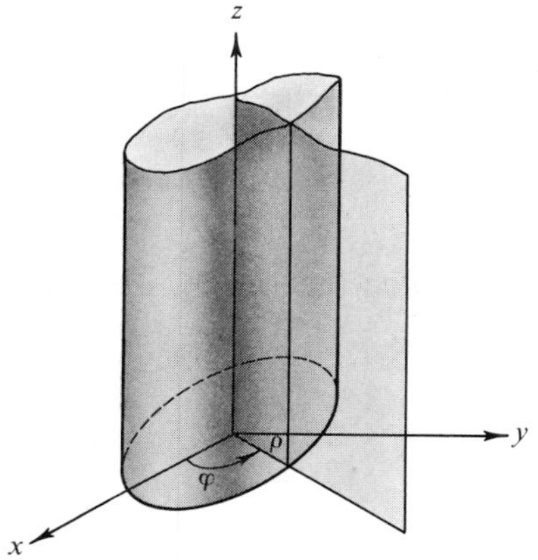

FIG. 2.4 Circular cylinder coordinates

$$h_1 = h_\rho = 1,$$
$$h_2 = h_\varphi = \rho, \tag{2.62}$$
$$h_3 = h_z = 1.$$

The families of coordinate surfaces shown in Fig. 2.4 are

1. Right circular cylinders having the z-axis as a common axis,

$$\rho = (x^2 + y^2)^{1/2} = \text{constant}.$$

2. Half planes through the z-axis,

$$\varphi = \tan^{-1}\left(\frac{y}{x}\right) = \text{constant}.$$

3. Planes parallel to the xy-plane, as in the cartesian system,

$$z = \text{constant}.$$

The limits on ρ, φ and z are

$$0 \leqslant \rho < \infty, \qquad 0 \leqslant \varphi \leqslant 2\pi, \quad \text{and} \quad -\infty < z < \infty.$$

From Eqs. 2.13, 2.17, 2.18, and 2.22,

$$\nabla\psi(\rho, \varphi, z) = \boldsymbol{\rho}_0 \frac{\partial\psi}{\partial\rho} + \boldsymbol{\varphi}_0 \frac{1}{\rho}\frac{\partial\psi}{\partial\varphi} + \mathbf{k}\frac{\partial\psi}{\partial z}, \tag{2.63}$$

$$\nabla\cdot\mathbf{V} = \frac{1}{\rho}\frac{\partial}{\partial\rho}(\rho V_\rho) + \frac{1}{\rho}\frac{\partial V_\varphi}{\partial\varphi} + \frac{\partial V_z}{\partial z}, \tag{2.64}$$

$$\nabla^2\psi = \frac{1}{\rho}\frac{\partial}{\partial\rho}\left(\rho\frac{\partial\psi}{\partial\rho}\right) + \frac{1}{\rho^2}\frac{\partial^2\psi}{\partial\varphi^2} + \frac{\partial^2\psi}{\partial z^2}, \tag{2.65}$$

$$\nabla\times\mathbf{V} = \frac{1}{\rho}\begin{vmatrix} \boldsymbol{\rho}_0 & \rho\boldsymbol{\varphi}_0 & \mathbf{k} \\ \dfrac{\partial}{\partial\rho} & \dfrac{\partial}{\partial\varphi} & \dfrac{\partial}{\partial z} \\ V_\rho & \rho V_\varphi & V_z \end{vmatrix}. \tag{2.66}$$

Finally, for problems such as circular wave guides or cylindrical cavity resonators the vector Laplacian $\nabla^2\mathbf{V}$ resolved in circular cylindrical coordinates is

$$\begin{aligned}
\nabla^2\mathbf{V}|_\rho &= \nabla^2 V_\rho - \frac{1}{\rho^2}V_\rho - \frac{2}{\rho^2}\frac{\partial V_\varphi}{\partial\varphi},\\
\nabla^2\mathbf{V}|_\varphi &= \nabla^2 V_\varphi - \frac{1}{\rho^2}V_\varphi + \frac{2}{\rho^2}\frac{\partial V_\rho}{\partial\varphi},\\
\nabla^2\mathbf{V}|_z &= \nabla^2 V_z.
\end{aligned} \tag{2.67}$$

The basic reason for the form of the z-component is that the z-axis is a cartesian axis, that is,

$$\begin{aligned}
\nabla^2(\boldsymbol{\rho}_0 V_\rho + \boldsymbol{\varphi}_0 V_\varphi + \mathbf{k}V_z) &= \nabla^2(\boldsymbol{\rho}_0 V_\rho + \boldsymbol{\varphi}_0 V_\varphi) + \mathbf{k}\,\nabla^2 V_z\\
&= \boldsymbol{\rho}_0 f(V_\rho, V_\varphi) + \boldsymbol{\varphi}_0 g(V_\rho, V_\varphi) + \mathbf{k}\,\nabla^2 V_z.
\end{aligned}$$

The operator ∇^2 operating on the $\boldsymbol{\rho}_0$, $\boldsymbol{\varphi}_0$ unit vectors stays in the $\boldsymbol{\rho}_0\,\boldsymbol{\varphi}_0$-plane. This behavior holds in all such cylindrical systems.

EXAMPLE 2.6.1 CYLINDRICAL RESONANT CAVITY

Consider a circular cylindrical cavity (radius a) with perfectly conducting walls. Electromagnetic waves will oscillate in such a cavity. If we assume our electric and magnetic fields have a time dependence $e^{-i\omega t}$, then Maxwell's equations lead to

$$\nabla \times \nabla \times \mathbf{E} = \omega^2 \varepsilon_0 \mu_0 \mathbf{E}. \qquad \text{(Cf. Example 1.9.2)} \tag{2.68}$$

With $\nabla \cdot \mathbf{E} = 0$ (vacuum, no charges),

$$\nabla^2\mathbf{E} + \alpha^2\mathbf{E} = 0,$$

where ∇^2 is the *vector* Laplacian and $\alpha^2 = \omega^2\varepsilon_0\mu_0$. In cylindrical coordinates E_z splits off, and we have the *scalar* Helmholtz equation

$$\nabla^2 E_z + \alpha^2 E_z = 0, \tag{2.69}$$

and the boundary condition $E_z\,(\rho = a) = 0$.

Using Eq. 2.65, Eq. 2.69 becomes

$$\frac{1}{\rho}\frac{\partial}{\partial\rho}\left(\rho\frac{\partial E_z}{\partial\rho}\right) + \frac{1}{\rho^2}\frac{\partial^2 E_z}{\partial\varphi^2} + \frac{\partial^2 E_z}{\partial z^2} + \alpha^2 E_z = 0. \tag{2.70}$$

We try

$$E_z(\rho, \varphi, z) = P(\rho)\Phi(\varphi)Z(z)$$

to obtain

$$\frac{1}{P\rho}\frac{d}{d\rho}\left(\rho\frac{dP}{d\rho}\right) + \frac{1}{\Phi\rho^2}\frac{d^2\Phi}{d\varphi^2} + \frac{1}{Z}\frac{d^2 Z}{dz^2} + \alpha^2 = 0. \tag{2.71}$$

Splitting off the z-dependence with a separation constant $-k^2$,

$$\frac{1}{Z}\frac{d^2Z}{dz^2} = -k^2.$$

For our cavity problem, sin kz and cos kz are the appropriate solutions (in that we can choose them to match the boundary conditions at the ends of the cavity). The exponentials $e^{\pm ikz}$ would be appropriate for a wave guide (traveling waves), cf. Section 11.3.

Using $\gamma^2 = \alpha^2 - k^2$, we isolate the φ dependence by multiplying by ρ^2. We set

$$\frac{1}{\Phi}\frac{d^2\Phi}{d\varphi^2} = -m^2,$$

with $\Phi(\varphi) = e^{\pm im\varphi}$, sin $m\varphi$, cos $m\varphi$. Then the remaining ρ dependence is

$$\rho\frac{d}{d\rho}\left(\rho\frac{dP}{d\rho}\right) + (\gamma^2\rho^2 - m^2)\,P = 0. \tag{2.72}$$

This is Bessel's equation. The solutions are developed in Chapter 11. This particular example, with Bessel functions, is continued as Example 11.1.2.

EXERCISES

2.6.1 Resolve the circular cylindrical unit vectors into their cartesian components.

$$\boldsymbol{\rho}_0 = \mathbf{i}\cos\varphi + \mathbf{j}\sin\varphi,$$
$$\boldsymbol{\varphi}_0 = -\mathbf{i}\sin\varphi + \mathbf{j}\cos\varphi,$$
$$\mathbf{k}_0 = \mathbf{k}.$$

2.6.2 Resolve the cartesian unit vectors into their circular cylindrical components.

$$\mathbf{i} = \boldsymbol{\rho}_0\cos\varphi - \boldsymbol{\varphi}_0\sin\varphi,$$
$$\mathbf{j} = \boldsymbol{\rho}_0\sin\varphi + \boldsymbol{\varphi}_0\cos\varphi,$$
$$\mathbf{k} = \mathbf{k}_0.$$

2.6.3 A particle is moving through space. Find the circular cylindrical components of its velocity and acceleration.

$$v_\rho = \dot{\rho}, \qquad a_\rho = \ddot{\rho} - \rho\dot{\varphi}^2,$$
$$v_\varphi = \rho\dot{\varphi}, \qquad a_\varphi = \rho\ddot{\varphi} + 2\dot{\rho}\dot{\varphi},$$
$$v_z = \dot{z}, \qquad a_z = \ddot{z}.$$

Hint.

$$\mathbf{r}(t) = \boldsymbol{\rho}_0(t)\rho(t) + \mathbf{k}z(t)$$
$$= [\mathbf{i}\cos\varphi(t) + \mathbf{j}\sin\varphi(t)]\rho(t) + \mathbf{k}z(t).$$

Note. $\dot{\rho} = d\rho/dt$, $\ddot{\rho} = d^2\rho/dt^2$, etc.

2.6.4 Show that the Helmholtz equation

$$\nabla^2\psi + k^2\psi = 0$$

is still separable in circular cylindrical coordinates if k^2 is generalized to $k^2 + f(\rho) + (1/\rho^2)g(\varphi) + h(z)$.

2.6.5 Solve Laplace's equation $\nabla^2\psi = 0$, in cylindrical coordinates for $\psi = \psi(\rho)$.

Ans. $\psi = k \ln \dfrac{\rho}{\rho_0}$

2.6.6 In right circular cylindrical coordinates a particular vector function is given by

$$\mathbf{V}(\rho, \varphi) = \boldsymbol{\rho}_0 V_\rho(\rho, \varphi) + \boldsymbol{\varphi}_0 V_\varphi(\rho, \varphi).$$

Show that $\nabla \times \mathbf{V}$ has only a z-component. Note that this result will hold for any vector confined to a surface $q_3 =$ constant as long as the products $h_1 V_1$ and $h_2 V_2$ are each independent of q_3.

2.6.7 A conducting wire along the z-axis carries a current I. The resulting magnetic vector potential is given by

$$\mathbf{A} = \mathbf{k}\frac{\mu I}{2\pi}\ln\left(\frac{1}{\rho}\right).$$

Show that the magnetic induction $\mathbf{B}$ is given by

$$\mathbf{B} = \boldsymbol{\varphi}_0 \frac{\mu I}{2\pi\rho}.$$

2.6.8 A force is described by

$$\mathbf{F} = -\mathbf{i}\frac{y}{x^2 + y^2} + \mathbf{j}\frac{x}{x^2 + y^2}.$$

(a) Express $\mathbf{F}$ in circular cylindrical coordinates.
Operating entirely in circular cylindrical coordinates for (b) and (c),
(b) calculate the curl of $\mathbf{F}$ and
(c) calculate the work done by $\mathbf{F}$ in encircling the unit circle once counterclockwise.
(d) How do you reconcile the results of (b) and (c)?

2.6.9 A transverse electromagnetic wave (TEM) in a coaxial wave guide has an electric field $\mathbf{E} = \mathbf{E}(\rho, \varphi)e^{i(kz-\omega t)}$ and a magnetic induction field of $\mathbf{B} = \mathbf{B}(\rho, \varphi)e^{i(kz-\omega t)}$. Since the wave is transverse neither $\mathbf{E}$ nor $\mathbf{B}$ has a z component. The two fields satisfy the *vector* Laplacian equation

$$\nabla^2\mathbf{E}(\rho, \varphi) = 0$$
$$\nabla^2\mathbf{B}(\rho, \varphi) = 0.$$

(a) Show that $\mathbf{E} = \boldsymbol{\rho}_0 E_0(a/\rho)e^{i(kz-\omega t)}$ and $\mathbf{B} = \boldsymbol{\varphi}_0 B_0(a/\rho)e^{i(kz-\omega t)}$ are solutions. Here a is the radius of the inner conductor.
(b) Assuming a vacuum inside the wave guide, verify that Maxwell's equations are

satisfied with

$$B_0/E_0 = k/\omega = \mu_0 \varepsilon_0(\omega/k) = 1/c.$$

2.6.10 A calculation of the magnetohydrodynamic pinch effect involves the evaluation of $(\mathbf{B} \cdot \nabla)\mathbf{B}$. If the magnetic induction $\mathbf{B}$ is taken to be $\mathbf{B} = \boldsymbol{\varphi}_0 B_\varphi(\rho)$, show that

$$(\mathbf{B} \cdot \nabla)\mathbf{B} = -\boldsymbol{\rho}_0 B_\varphi{}^2/\rho.$$

2.6.11 (a) Explain why ∇^2 in plane polar coordinates follows from ∇^2 in circular cylindrical coordinates with $z =$ constant.

(b) Explain why taking ∇^2 in spherical polar coordinates and restricting θ to $\pi/2$ does NOT lead to the plane polar form of ∇^2.

Note. $$\nabla^2(\rho, \varphi) = \frac{\partial^2}{\partial \rho^2} + \frac{1}{\rho}\frac{\partial}{\partial \rho} + \frac{1}{\rho^2}\frac{\partial^2}{\partial \varphi^2}.$$

2.7 Elliptic Cylindrical Coordinates (u, v, z)

One reasonable way of classifying the separable coordinate systems is to start with the confocal ellipsoidal system (Section 2.15) and derive the other systems as degenerate cases. Details of this procedure will be found in Morse and Feshbach's *Methods of Mathematical Physics*, Chapter 5. Here, to emphasize the application

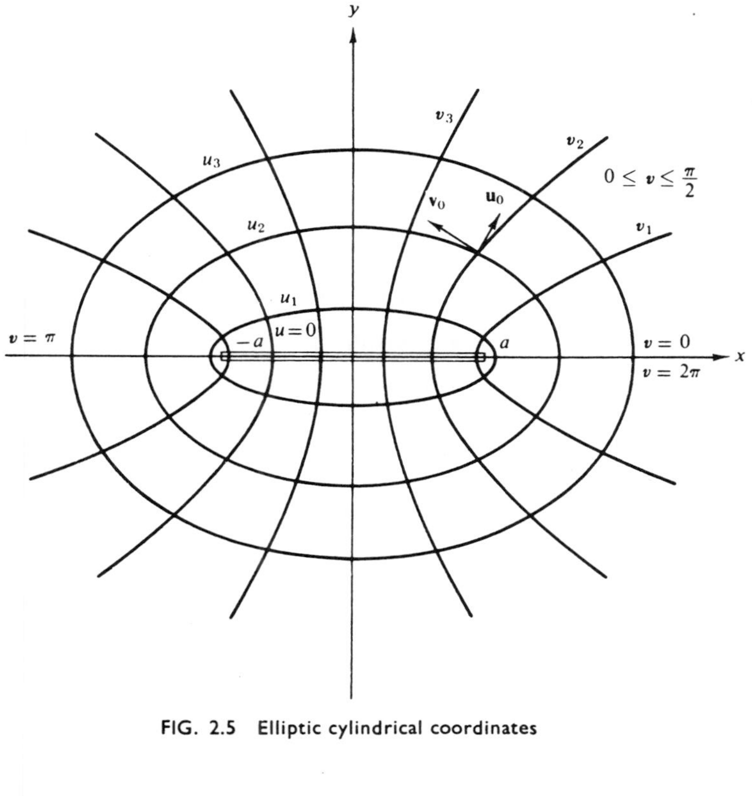

FIG. 2.5 Elliptic cylindrical coordinates

rather than a derivation, we take the coordinate systems in order of symmetry properties, proceeding with those that have an axis of translation. All of those with an axis of translation are essentially two-dimensional systems with a third dimension (the z-axis) tacked on.

For the elliptic cylindrical system we have

$$\begin{aligned} x &= a \cosh u \cos v, \\ y &= a \sinh u \sin v, \\ z &= z. \end{aligned} \tag{2.73}$$

The families of coordinate surfaces are the following:

1. Elliptic cylinders, $u = \text{constant}$, $0 \leqslant u < \infty$.
2. Hyperbolic cylinders, $v = \text{constant}$, $0 \leqslant v \leqslant 2\pi$.
3. Planes parallel to the xy-plane, $z = \text{constant}$, $-\infty < z < \infty$.

This may be seen by inverting Eq. 2.73. Squaring each side,

$$x^2 = a^2 \cosh^2 u \cos^2 v, \tag{2.74}$$

$$y^2 = a^2 \sinh^2 u \sin^2 v, \tag{2.75}$$

from which

$$\frac{x^2}{a^2 \cosh^2 u} + \frac{y^2}{a^2 \sinh^2 u} = 1, \tag{2.76}$$

$$\frac{x^2}{a^2 \cos^2 v} - \frac{y^2}{a^2 \sin^2 v} = 1. \tag{2.77}$$

Holding u constant, Eq. 2.76 yields a family of ellipses with the x-axis the major one. For $v = \text{constant}$, Eq. 2.77 gives hyperbolas with focal points on the x-axis.

The scale factors are

$$\begin{aligned} h_1 &= h_u = a(\sinh^2 u + \sin^2 v)^{1/2}, \\ h_2 &= h_v = a(\sinh^2 u + \sin^2 v)^{1/2}, \\ h_3 &= h_z = 1. \end{aligned} \tag{2.78}$$

We shall meet this system again as a two-dimensional system in Chapter 6 when we take up conformal mapping.

EXERCISES

2.7.1 Let $\cosh u = q_1$, $\cos v = q_2$, $z = q_3$. Find the new scale factors h_{q_1} and h_{q_2}.

$$h_{q_1} = a\left(\frac{q_1^2 - q_2^2}{q_1^2 - 1}\right)^{1/2},$$

$$h_{q_2} = a\left(\frac{q_1^2 - q_2^2}{1 - q_2^2}\right)^{1/2}.$$

2.7.2 Show that the Helmholtz equation in elliptic cylindrical coordinates separates into
(a) the linear oscillator equation for the z dependence,
(b) Mathieu's equation

$$\frac{d^2g}{dv^2} + (b - 2q\cos 2v)\, g = 0,$$

and
(c) Mathieu's modified equation

$$\frac{d^2f}{du^2} - (b - 2q\cosh 2u)\, f = 0.$$

2.8 Parabolic Cylindrical Coordinates (ξ, η, z)

The transformation equations,

$$\begin{aligned} x &= \xi\eta, \\ y &= \tfrac{1}{2}(\eta^2 - \xi^2), \\ z &= z, \end{aligned} \tag{2.79}$$

generate two sets of orthogonal parabolic cylinders (Fig. 2.6). By solving Eq. 2.79 for ξ and η we obtain the following:

1. Parabolic cylinders, ξ = constant,[1] $\quad -\infty < \xi < \infty$.
2. Parabolic cylinders, η = constant, $\quad 0 \leqslant \eta < \infty$.
3. Planes parallel to the xy-plane, z = constant, $\quad -\infty < z < \infty$.

From Eq. 2.6 the scale factors are

$$\begin{aligned} h_1 &= h_\xi = (\xi^2 + \eta^2)^{1/2}, \\ h_2 &= h_\eta = (\xi^2 + \eta^2)^{1/2}, \\ h_3 &= h_z = 1. \end{aligned} \tag{2.80}$$

2.9 Bipolar Coordinates (ξ, η, z)

This is an oddball coordinate system. It is not a degenerate case of the confocal ellipsoidal coordinates. Equation 2.1 is not completely separable in this system even for $k^2 = 0$ (cf. Exercise 2.9.2). It is included here as an example of how an unusual coordinate system may be chosen to fit a problem.

[1] The parabolic cylinder ξ = constant is invariant to the sign of ξ. We must let ξ (or η) go negative to cover negative values of x.

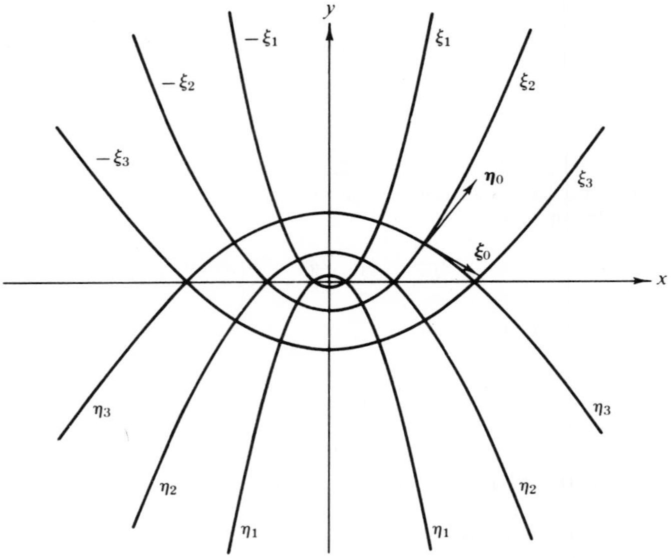

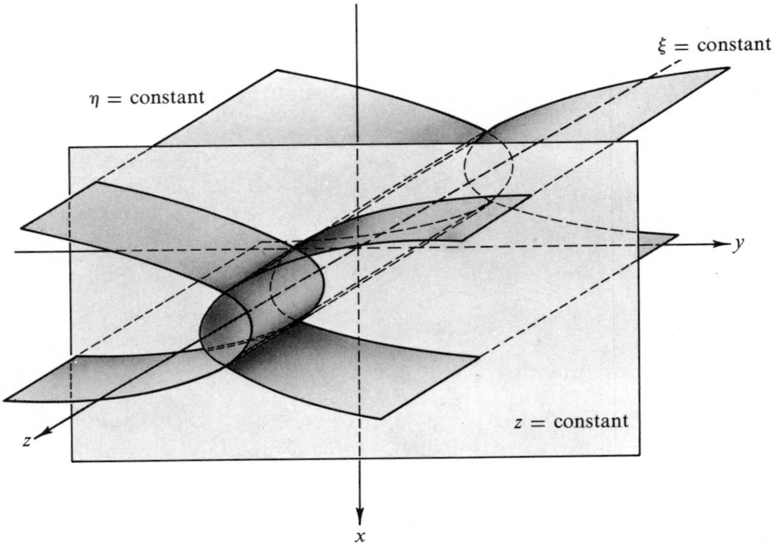

FIG. 2.6 Parabolic cylindrical coordinates. (*Top*) Cross section

The transformation equations are

$$x = \frac{a \sinh \eta}{\cosh \eta - \cos \xi}, \tag{2.81a}$$

$$y = \frac{a \sin \xi}{\cosh \eta - \cos \xi}, \tag{2.81b}$$

$$z = z. \tag{2.81c}$$

Dividing Eq. 2.81*a* by 2.81*b*, we obtain

$$\frac{x}{y} = \frac{\sinh \eta}{\sin \xi}. \tag{2.82}$$

Using Eq. 2.82 to eliminate ξ from Eq. 2.81*a*, we have

$$(x - a \coth \eta)^2 + y^2 = a^2 \operatorname{csch}^2 \eta. \tag{2.83}$$

Using Eq. 2.82 to eliminate η from Eq. 2.81*b*, we have

$$x^2 + (y - a \cot \xi)^2 = a^2 \csc^2 \xi. \tag{2.84}$$

From Eqs. 2.83 and 2.84 we may identify the coordinate surfaces as follows:

1. Circular cylinders, center at $y = a \cot \xi$,

$$\xi = \text{constant}, \qquad 0 \leqslant \xi \leqslant 2\pi.$$

2. Circular cylinders, center at $x = a \coth \eta$,

$$\eta = \text{constant}, \qquad -\infty < \eta < \infty.$$

3. Planes parallel to xy-plane,

$$z = \text{constant}, \qquad -\infty < z < \infty.$$

When $\eta \to \infty$, $\coth \eta \to 1$ and $\operatorname{csch} \eta \to 0$. Equation 2.83 has a solution $x = a$, $y = 0$. Similarly, when $\eta \to -\infty$, a solution is $x = -a$, $y = 0$, the circle degenerating to a point, the cylinder to a line. The family of circles (in the xy-plane) described by Eq. 2.84 passes through both of these points. This follows from noting that $x = \pm a$, $y = 0$ are solutions of Eq. 2.84 for any value of ξ.

The scale factors for the bipolar system are

$$\begin{aligned} h_1 = h_\xi &= \frac{a}{\cosh \eta - \cos \xi}, \\ h_2 = h_\eta &= \frac{a}{\cosh \eta - \cos \xi}, \\ h_3 = h_z &= 1. \end{aligned} \tag{2.85}$$

To see how the bipolar system may be useful let us start with the three points $(a, 0)$, $(-a, 0)$, and (x, y) and the two distance vectors ρ_1 and ρ_2 at angles of θ_1

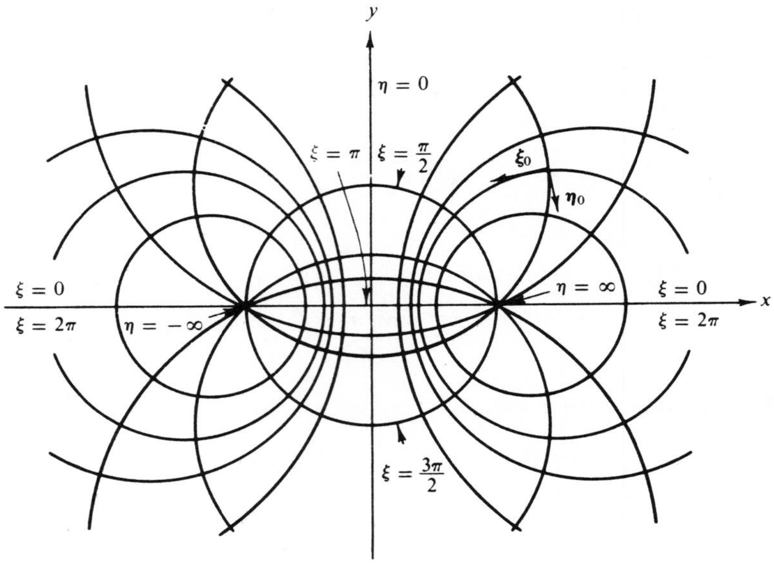

FIG. 2.7 Bipolar coordinates

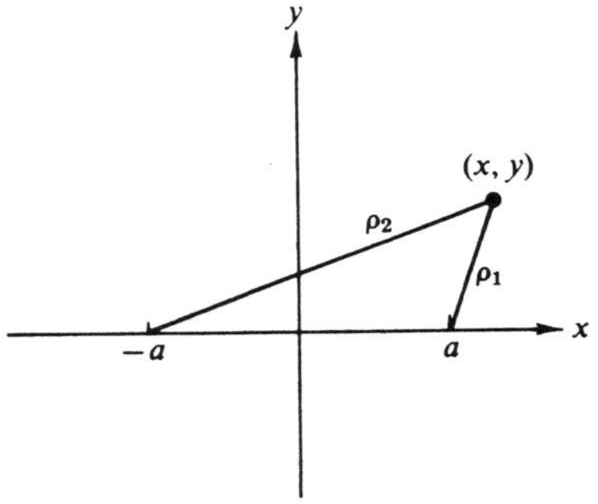

FIG. 2.8

and θ_2 from the positive x-axis. From Fig. 2.8

$$\rho_1^2 = (x - a)^2 + y^2, \qquad (2.86)$$
$$\rho_2^2 = (x + a)^2 + y^2,$$

and

$$\tan\theta_1 = \frac{y}{x - a}, \qquad (2.87)$$
$$\tan\theta_2 = \frac{y}{x + a}.$$

We define[1]

$$\eta_{12} = \ln\frac{\rho_2}{\rho_1}, \qquad (2.88a)$$

$$\xi_{12} = \theta_1 - \theta_2. \qquad (2.88b)$$

By taking $\tan\xi_{12}$ and Eq. 2.87

$$\tan\xi_{12} = \frac{\tan\theta_1 - \tan\theta_2}{1 + \tan\theta_1\tan\theta_2} \qquad (2.89)$$
$$= \frac{y/(x-a) - y/(x+a)}{1 + y^2/(x-a)(x+a)}.$$

[1] The notation ln is used to indicate $\log_e$.

From Eq. 2.89, Eq. 2.84 follows directly. This identifies ξ as $\xi_{12} = \theta_1 - \theta_2$. Solving Eq. 2.88*a* for ρ_2/ρ_1 and combining this with Eq. 2.86, we get

$$e^{2\eta_{12}} = \frac{\rho_2^2}{\rho_1^2} = \frac{(x+a)^2 + y^2}{(x-a)^2 + y^2} \tag{2.90}$$

Multiplication by $e^{-\eta_{12}}$ and use of the definitions of hyperbolic sine and cosine produces Eq. 2.83, which identifies η as $\eta_{12} = \ln(\rho_2/\rho_1)$. The following example exploits this identification.

EXAMPLE 2.9.1

An infinitely long straight wire carries a current I in the negative z-direction. A second wire, parallel to the first, carries a current I in the positive z-direction. Using

$$d\mathbf{A} = \frac{\mu_0}{4\pi} I \frac{d\boldsymbol{\lambda}}{r}, \tag{2.91}$$

find **A**, the magnetic vector potential, and **B**, the magnetic inductance.

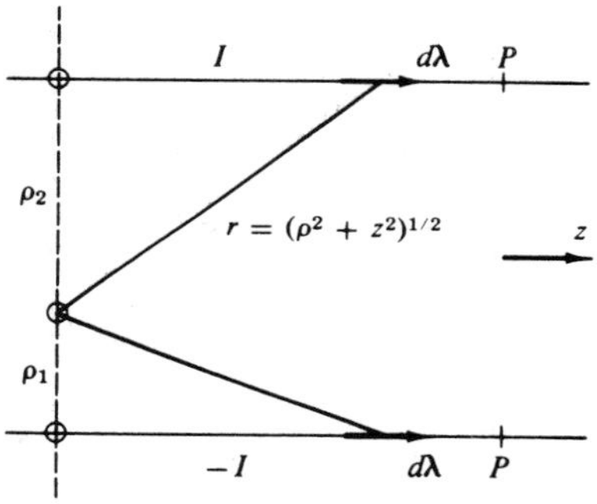

FIG. 2.9 Antiparallel electric currents

From Eq. 2.91 **A** has only a z-component. Integrating over each wire from 0 to P and taking the limit as $P \to \infty$, we obtain

$$A_z = \frac{\mu_0 I}{4\pi} \lim_{P\to\infty} \left(2\int_0^P \frac{dz}{\sqrt{\rho_2^2 + z^2}} - 2\int_0^P \frac{dz}{\sqrt{\rho_1^2 + z^2}} \right), \tag{2.92}$$

$$A_z = \frac{\mu_0 I}{4\pi} \lim_{P\to\infty} 2[\ln(z + \sqrt{\rho_2^2 + z^2})|_0^P - \ln(z + \sqrt{\rho_1^2 + z^2})|_0^P],$$

$$= \frac{\mu_0 I}{4\pi} \left(\lim_{P\to\infty} 2 \ln \frac{P + \sqrt{\rho_2^2 + P^2}}{P + \sqrt{\rho_1^2 + P^2}} - 2 \ln \frac{\rho_2}{\rho_1} \right). \tag{2.93}$$

This reduces to

$$A_z = -\frac{\mu_0 I}{2\pi} \ln \frac{\rho_2}{\rho_1} = -\frac{\mu_0 I}{2\pi} \eta. \tag{2.94}$$

So far there has been no need for bipolar coordinates. Now, however, let us calculate the magnetic inductance **B** from $\mathbf{B} = \nabla \times \mathbf{A}$. From Eqs. 2.22 and 2.85

$$\mathbf{B} = \frac{(\cosh\eta - \cos\xi)^2}{a^2} \begin{vmatrix} h_\xi \boldsymbol{\xi}_0 & h_\eta \boldsymbol{\eta}_0 & \mathbf{k} \\ \dfrac{\partial}{\partial \xi} & \dfrac{\partial}{\partial \eta} & \dfrac{\partial}{\partial z} \\ 0 & 0 & \dfrac{-\mu_0 I}{2\pi}\eta \end{vmatrix},$$

$$= -\boldsymbol{\xi}_0 \frac{(\cosh\eta - \cos\xi)}{a} \cdot \frac{\mu_0 I}{2\pi}. \tag{2.95}$$

The magnetic field has only a ξ_0-component. The reader is urged to try to compute **B** in some other coordinate system.

We shall return to bipolar coordinates in Sections 2.13 and 2.14 to derive the toroidal and bispherical coordinate systems.

EXERCISES

2.9.1 Show that specifying the radius of each of two parallel cylinders and the center to center distance fixes a particular bipolar coordinate system in the sense that η_1 (first circle), η_2 (second circle) and a are uniquely determined.

2.9.2 (a) Show that Laplace's equation, $\nabla^2 \psi(\xi, \eta, z,) = 0$ is *not* completely separable in bipolar coordinates.

(b) Show that a complete separation is possible if we require that $\psi = \psi(\xi, \eta)$, that is, if we restrict ourselves to a two-dimensional system.

2.9.3 Find the capacitance per unit length of two conducting cylinders of radii b and c and of infinite length, with axes parallel and a distance d apart.

$$C = \frac{2\pi\varepsilon_0}{\eta_1 - \eta_2}.$$

2.9.4 As a limiting case of Exercise 2.9.3, find the capacitance per unit length between a conducting cylinder and a conducting infinite plane parallel to the axis of the cylinder.

$$C = \frac{2\pi\varepsilon_0}{\eta_1}.$$

2.9.5 A parallel wire wave guide (transmission line) consists of two infinitely long conducting cylinders defined by $\eta = \pm\,\eta_1$.

(a) Show that

$$\eta_1 = \cosh^{-1}\left\{\frac{\text{center–center distance}}{\text{cylinder diameter}}\right\}.$$

(b) From Example 2.9.1 and Exercise 2.9.3 we expect a TEM mode with electric and magnetic fields of the form

$$\mathbf{E} = \boldsymbol{\eta}_0 \frac{1}{h_1} E_0 e^{i(kz-\omega t)}$$

$$\mathbf{H} = -\boldsymbol{\xi}_0 \frac{1}{h_1} H_0 e^{i(kz-\omega t)}.$$

Show that $E_0 = V_0/\eta_1$ where $2V_0$ is the maximum voltage difference between the cylinders.

(c) With $H_0 = (\varepsilon_0/\mu_0)^{1/2} E_0$, show that Maxwell's equations are satisfied.

(d) By integrating the time averaged Poynting vector

$$\mathbf{P} = \tfrac{1}{2}(\mathbf{E} \times \mathbf{H}^*)$$

calculate the rate at which energy is propagated along this transmission line.

Ans. Power $= 2\pi(\varepsilon_0/\mu_0)^{1/2}(V_0^2/\eta_1)$.

2.10 Prolate Spheroidal Coordinates (u, v, φ)

Let us start with the elliptic coordinates of Section 2.7 as a two-dimensional system. We can generate a three-dimensional system by rotating about the major or minor elliptic axes and introducing φ as an azimuth angle (Fig. 2.10). Rotating first about the major axis gives us prolate spheroidal coordinates with the following coordinate surfaces:

1. Prolate spheroids,

$$u = \text{constant}, \qquad 0 \leqslant u < \infty.$$

2. Hyperboloids of two sheets,

$$v = \text{constant}, \qquad 0 \leqslant v \leqslant \pi.$$

3. Half planes through the z-axis,

$$\varphi = \text{constant}, \qquad 0 \leqslant \varphi \leqslant 2\pi.$$

The transformation equàtions are

$$\begin{aligned} x &= a \sinh u \sin v \cos \varphi, \\ y &= a \sinh u \sin v \sin \varphi, \\ z &= a \cosh u \cos v. \end{aligned} \tag{2.96}$$

Note that we have permuted our cartesian axes to make the axis of rotational symmetry the z-axis. The scale factors for this system are

$$\begin{aligned} h_1 = h_u &= a(\sinh^2 u + \sin^2 v)^{1/2}, \\ &= a(\cosh^2 u - \cos^2 v)^{1/2}, \\ h_2 = h_v &= a(\sinh^2 u + \sin^2 v)^{1/2}, \\ h_3 = h_\varphi &= a \sinh u \sin v. \end{aligned} \tag{2.97}$$

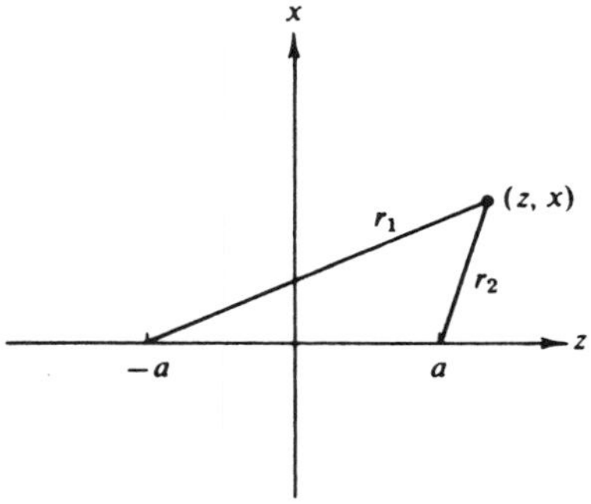

FIG. 2.11

The prolate spheroidal coordinates are rather important in physics, primarily because of their usefulness in treating "two-center" problems. The two centers will correspond to the two focal points, $(0, 0, a)$ and $(0, 0, -a)$, of the ellipsoids and hyperboloids of revolution. As shown in Fig. 2.11, we label the distance from the left focal point to the point (z, x), r_1, and the corresponding distance from the right focal point r_2.

$$r_1 + r_2 = \text{constant}, \qquad \text{for fixed } u.$$

The point (z, x) is described in terms of u and v by Eqs. 2.96. The azimuth is irrelevant here. From the properties of the

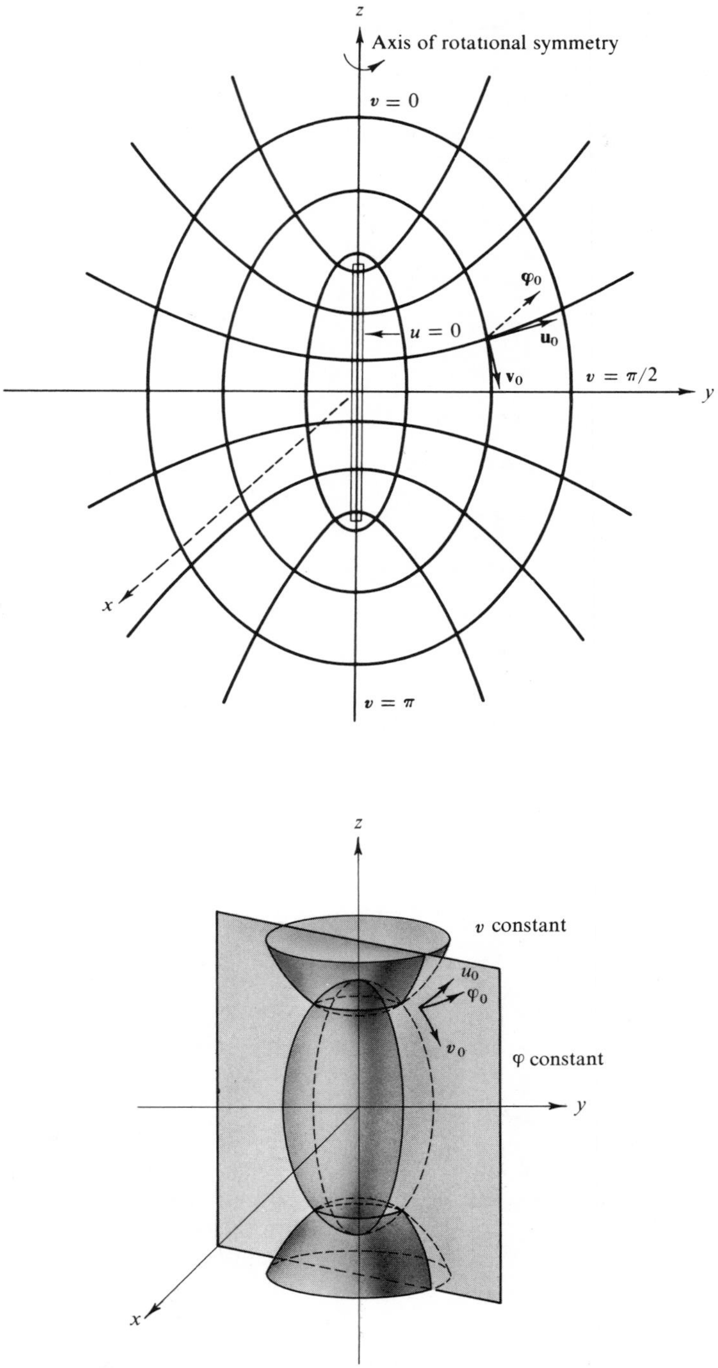

FIG. 2.10 Prolate spheroidal coordinates. (*Top*) Cross section

ellipse and hyperbola we know

$$\begin{aligned} r_1 + r_2 &= \text{constant}, \qquad \text{for fixed } u, \\ r_1 - r_2 &= \text{constant}, \qquad \text{for fixed } v. \end{aligned} \tag{2.98}$$

Using

$$\begin{aligned} r_1 &= [(a+z)^2 + x^2]^{1/2}, \\ r_2 &= [(a-z)^2 + x^2]^{1/2}, \end{aligned} \tag{2.99}$$

and Eq. 2.96, we find

$$\begin{aligned} r_1 &= a(\cosh u + \cos v), \\ r_2 &= a(\cosh u - \cos v), \end{aligned} \tag{2.100}$$

or

$$\begin{aligned} \frac{r_1 + r_2}{2a} &= \cosh u \\ \frac{r_1 - r_2}{2a} &= \cos v \end{aligned} \tag{2.101}$$

This means u is a function of the sum of the distances from the two centers, whereas v is a function of the difference of the distances from the two centers.

To facilitate this application of the coordinate system we change the variables by introducing

$$\begin{aligned} \xi_1 &= \cosh u, \qquad 1 \leqslant \xi_1 < \infty, \\ \xi_2 &= \cos v, \qquad -1 \leqslant \xi_2 \leqslant 1, \\ \xi_3 &= \varphi, \qquad 0 \leqslant \xi_3 \leqslant 2\pi. \end{aligned} \tag{2.102}$$

Note carefully that

$$h_{\xi_1} = h_{\cosh u} \neq h_u. \tag{2.103}$$

New variables involve new scale factors.

EXAMPLE 2.10.1

The hydrogen molecule ion is a system composed of two protons which we take to be fixed at the focal points and one electron. The Schrödinger wave equation for this system is

$$\frac{\hbar^2}{2M} \nabla^2 \psi - \frac{e^2}{r_1} \psi - \frac{e^2}{r_2} \psi + \frac{e^2}{r_{12}} \psi = E\psi. \tag{2.104}$$

The variables r_1 and r_2 are defined in Fig. 2.11, and r_{12}, the proton-proton distance, is just $2a$. The problem is to separate the variables in Eq. 2.104.

In choosing the prolate spheroidal coordinates, ξ_1, ξ_2, ξ_3, our first step is to calculate the scale factors. From Eqs. 2.96 and 2.102

$$h_{\xi_1} = a\left(\frac{\xi_1^2 - \xi_2^2}{\xi_1^2 - 1}\right)^{1/2}, \qquad h_{\xi_2} = a\left(\frac{\xi_1^2 - \xi_2^2}{1 - \xi_2^2}\right)^{1/2}, \tag{2.105}$$

$$h_{\xi_3} = a(\xi_1^2 - 1)^{1/2}(1 - \xi_2^2)^{1/2}.$$

Using these scale factors and Eq. 2.18*a*, we find

$$\nabla^2\psi = \frac{1}{a^2}\left\{\frac{1}{(\xi_1^2-\xi_2^2)}\frac{\partial}{\partial\xi_1}\left[(\xi_1^2-1)\frac{\partial\psi}{\partial\xi_1}\right] + \frac{1}{(\xi_1^2-\xi_2^2)}\frac{\partial}{\partial\xi_2}\left[(1-\xi_2^2)\frac{\partial\psi}{\partial\xi_2}\right] + \frac{1}{(\xi_1^2-1)(1-\xi_2^2)}\frac{\partial^2\psi}{\partial\xi_3^2}\right\}. \tag{2.106}$$

From Eq. 2.100

$$\frac{e^2}{r_1} + \frac{e^2}{r_2} = \frac{e^2 2a\xi_1}{a^2(\xi_1^2-\xi_2^2)}. \tag{2.107}$$

By substituting Eqs. 2.106 and 2.107 into Eq. 2.104 and using the now standard procedure,

$$\psi(\xi_1, \xi_2, \xi_3) = f_1(\xi_1)\, f_2(\xi_2)\, f_3(\xi_3), \tag{2.108}$$

we can quickly isolate the azimuthal (ξ_3) dependence to obtain

$$-\frac{\hbar^2}{2Ma^2}\left\{\frac{1}{(\xi_1^2-\xi_2^2)}\frac{1}{f_1}\frac{d}{d\xi_1}\left[(\xi_1^2-1)\frac{df_1}{d\xi_1}\right] + \frac{1}{(\xi_1^2-\xi_2^2)}\frac{1}{f_2}\frac{d}{d\xi_2}\left[(1-\xi_2^2)\frac{df_2}{d\xi_2}\right]\right\} - \frac{2e^2}{a}\frac{\xi_1}{(\xi_1^2-\xi_2^2)} - E'$$
$$= \frac{\hbar^2}{2Ma^2}\frac{1}{(\xi_1^2-1)(1-\xi_2^2)}\frac{1}{f_3}\frac{d^2 f_3}{d\xi_3^2}. \tag{2.109}$$

Here we have used $E' = E - e^2/r_{12}$, a constant. As in Sections 2.5 and 2.6, we set

$$\frac{1}{f_3}\frac{d^2 f_3}{d\xi_3^2} = -m^2. \tag{2.110}$$

Equation 2.109 may be simplified to yield

$$\frac{1}{f_1}\frac{d}{d\xi_1}\left[(\xi_1^2-1)\frac{df_1}{d\xi_1}\right] + \frac{1}{f_2}\frac{d}{d\xi_2}\left[(1-\xi_2^2)\frac{df_2}{d\xi_2}\right] + \frac{4Mae^2\xi_1}{\hbar^2} + \frac{2Ma^2E'}{\hbar^2}(\xi_1^2-\xi_2^2)$$
$$= m^2\left[\frac{1}{\xi_1^2-1} + \frac{1}{1-\xi_2^2}\right]. \tag{2.111}$$

The variables ξ_1 and ξ_2 separate by inspection, and we have one second-order differential equation for $f_1(\xi_1)$ and another for $f_2(\xi_2)$.

An example of the use of prolate spheroidal coordinates in electrostatics appears in Section 12.11.

EXERCISES

2.10.1 Using $\xi = \cosh u$, $\eta = \cos v$, show that the volume element in prolate spheroidal coordinates obtained by direct transformation of

$$d\tau = a^3 (\sinh^2 u + \sin^2 v) \sinh u \sin v \, du \, dv \, d\varphi$$

is

$$d\tau = -a^3(\xi^2 - \eta^2) \, d\xi \, d\eta \, d\varphi.$$

(The minus sign will be taken out by reversing the limits of integration over η.)

2.10.2 Using prolate spheroidal coordinates, set up the volume integral representing the volume of a given prolate ellipsoid using (a) u, v, φ and (b) ξ, η, φ. Evaluate the integrals and show that your results are equivalent to the usual result given in terms of the semi axes,

$$V = \frac{4}{3}\pi a_0^2 b_0,$$

where a_0 is the semiminor axis and b_0, the semimajor axis.

2.10.3 In the quantum mechanical analysis of the hydrogen molecule by the Heitler-London method we encounter the integral

$$I_{\text{HL}} = \frac{1}{\pi a_0^3} \int e^{-(r_1+r_2)/a_0} \, d\tau,$$

in which the volume integral is over all space. Introduce prolate spheroidal coordinates and evaluate the integral.

Ans. $I_{\text{HL}} = \left(1 + \frac{2a}{a_0} + \frac{4a^2}{3a_0^2}\right) e^{-2a/a_0}.$

2.11 Oblate Spheroidal Coordinates (u, v, φ)

When the elliptic coordinates of Section 2.7 (taken as a two-dimensional set) are rotated about the minor elliptic axis, we generate another three-dimensional spheroidal system, the oblate spheroidal coordinate system. Again φ is the azimuthal angle. The coordinate surfaces are the following:

1. Oblate spheroids,

$$u = \text{constant}, \qquad 0 \leqslant u < \infty.$$

2. Hyperboloids of one sheet,

$$v = \text{constant},[1] \qquad -\frac{\pi}{2} \leqslant v \leqslant \frac{\pi}{2}.$$

3. Half planes through the z-axis,

$$\varphi = \text{constant}, \qquad 0 \leqslant \varphi \leqslant 2\pi.$$

The transformation equations relating to cartesian coordinates may be written

$$x = a \cosh u \cos v \cos \varphi$$

$$y = a \cosh u \cos v \sin \varphi$$

$$z = a \sinh u \sin v.$$

[1] Note that v has a range of only π in contrast to the range of 2π for elliptic cylindrical coordinates (Section 2.7). The negative values of v generate negative values of z.

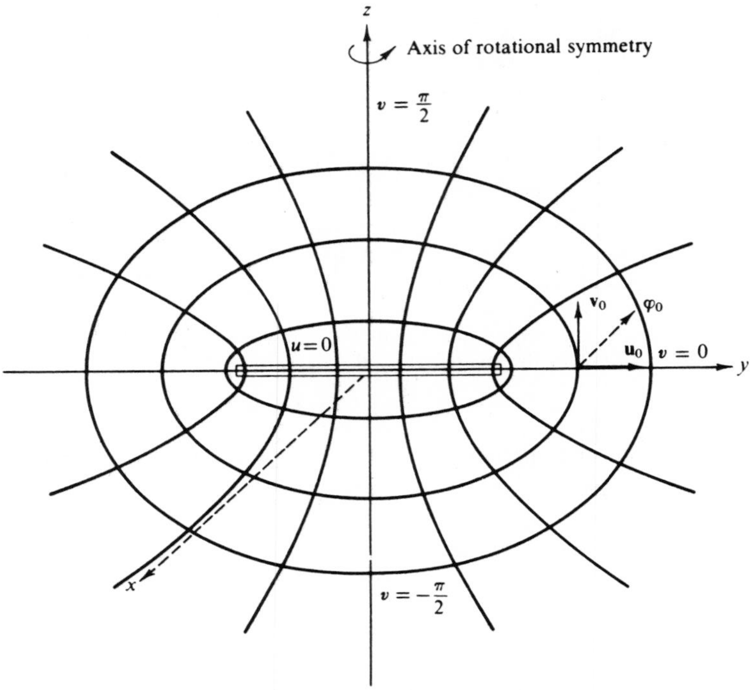

FIG. 2.12 Oblate spheroidal coordinates. Cross section

The scale factors become

$$\begin{aligned} h_1 = h_u &= a\,(\sinh^2 u + \sin^2 v)^{1/2} \\ &= a(\cosh^2 u - \cos^2 v)^{1/2}, \\ h_2 = h_v &= a(\sinh^2 u + \sin^2 v)^{1/2}, \\ h_3 = h_\varphi &= a \cosh u \cos v. \end{aligned} \tag{2.112}$$

Since holding u constant results in an oblate spheroid which is a good approximation to a planetary surface, this coordinate system has been useful in describing the earth's gravitational field. (J. P. Vinti, *Phys. Rev. Letters* **3**, 8 (1959)). Both prolate and oblate spheroidal coordinates are used in Section 12.11 to illustrate Legendre functions of the second kind.

Note carefully that if we require φ to advance from x to y as usual and if we insist on the order (u, v, φ), this system is left-handed! This will introduce an over-all (-1) in the expression for the curl. To get back to a right-handed system it is necessary to use only (v, u, φ),

$$\mathbf{v}_0 \times \mathbf{u}_0 = +\boldsymbol{\varphi}_0$$

or let $v \to (\pi/2) - v$ in the transformation equations.

EXERCISES

2.11.1 Separate Laplace's equation in oblate spheroidal coordinates. Solve the φ-dependent differential equation.

2.11.2 A thin conducting metal disk of radius a carries a total electric charge Q. Find the capacitance of the disk and the distribution of charge over the surface of the disk.

$$C = 8a\varepsilon_0,$$

$$\sigma = \frac{Q}{4\pi a\sqrt{a^2 - r^2}} \quad \text{(on each side).}$$

2.12 Parabolic Coordinates (ξ, η, φ)

In Section 2.8 two sets of orthogonal confocal parabolas were described. Imagine that we have taken the system shown in the xy-plane (Fig. 2.6) and have rotated about the y-axis the axis of symmetry for each set of parabolas. This generates two sets of orthogonal confocal paraboloids. By permuting the coordinates (cyclically) so that the axis of rotation is the z-axis we have the following:

1. Paraboloids about the positive z-axis,

$$\xi = \text{constant}, \qquad 0 \leqslant \xi < \infty.$$

2. Paraboloids about the negative z-axis,

$$\eta = \text{constant}, \qquad 0 \leqslant \eta < \infty.$$

3. Half planes through the z-axis,

$$\varphi = \text{constant}, \qquad 0 \leqslant \varphi \leqslant 2\pi.$$

Measuring the azimuth from the x-axis in the xy-plane, as usual, we obtain

$$\begin{aligned} x &= \xi\eta \cos \varphi, \\ y &= \xi\eta \sin \varphi, \\ z &= \tfrac{1}{2}(\eta^2 - \xi^2). \end{aligned} \tag{2.113}$$

From Eq. 2.113 we find the scale factors

$$\begin{aligned} h_1 &= h_\xi = (\xi^2 + \eta^2)^{1/2}, \\ h_2 &= h_\eta = (\xi^2 + \eta^2)^{1/2}, \\ h_3 &= h_\varphi = \xi\eta. \end{aligned} \tag{2.114}$$

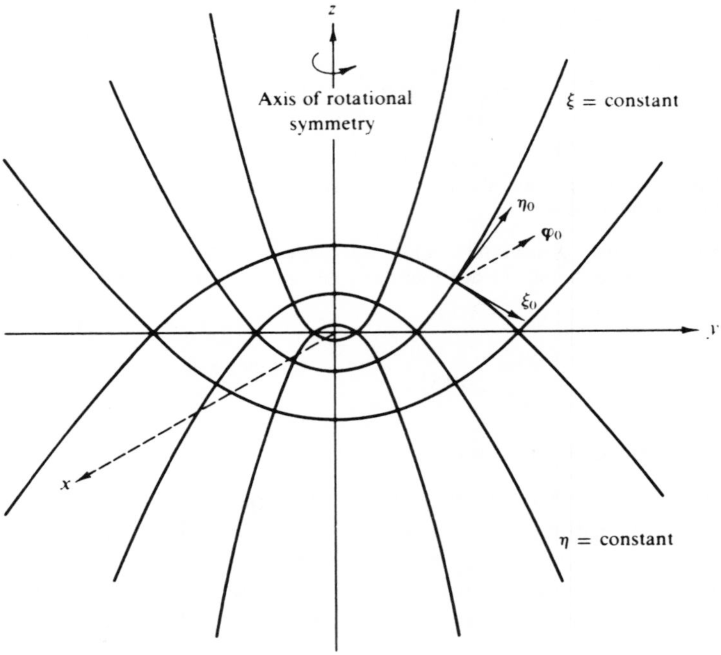

FIG. 2.13 Parabolic coordinates

From Fig. 2.13 it is seen that

$$\boldsymbol{\xi}_0 \times \boldsymbol{\eta}_0 = -\boldsymbol{\varphi}_0,$$

that is, the parabolic system (ξ, η, φ) given here is a left-handed system. Equations 2.113 imply that ξ and η each have dimensions of $(\text{length})^{1/2}$. For this reason some writers prefer to use $\xi^{1/2}$ in place of our ξ and $\eta^{1/2}$ in place of our η. Others have interchanged ξ and η.

The parabolic coordinates have found an application in the analysis of the Stark effect,[1] the shift of energy levels which results when an atom is placed in an *electric* field.

EXAMPLE 2.12.1 THE STARK EFFECT

The presence of the external electric field E_0, along the positive z-axis, adds a potential energy term $-eE_0z$ to the Schrödinger wave equation. For hydrogen we have

$$-\frac{\hbar^2}{2M}\nabla^2\psi - \frac{e^2}{r}\psi - eE_0z\psi = E\psi. \tag{2.115}$$

[1] H. A. Bethe and E. S. Salpeter. *Quantum Mechanics of One- and Two-Electron Atoms*. New York: Academic Press (1957).

Once more the problem is to separate the variables.

Using Eqs. 2.18*a* and 2.114, we obtain

$$\nabla^2\psi = \frac{1}{\xi\eta(\xi^2+\eta^2)}\left\{\frac{\partial}{\partial\xi}\left[\xi\eta\frac{\partial\psi}{\partial\xi}\right] + \frac{\partial}{\partial\eta}\left[\xi\eta\frac{\partial\psi}{\partial\eta}\right]\right\} + \frac{1}{\xi^2\eta^2}\frac{\partial^2\psi}{\partial\varphi^2}. \tag{2.116}$$

We also find

$$r = \frac{\xi^2+\eta^2}{2}. \tag{2.117}$$

Using Eqs. 2.116, 2.117, and $\psi = f(\xi)\,g(\eta)\,\Phi(\varphi)$, Eq. 2.115 becomes

$$\frac{\hbar^2}{2M}\frac{1}{(\xi^2+\eta^2)}\left[\frac{1}{\xi f}\frac{d}{d\xi}\left(\xi\frac{df}{d\xi}\right) + \frac{1}{\eta g}\frac{d}{d\eta}\left(\eta\frac{dg}{d\eta}\right)\right]$$
$$+\frac{\hbar^2}{2M}\frac{1}{\xi^2\eta^2}\frac{1}{\Phi}\frac{d^2\Phi}{d\varphi^2} + \frac{2e^2}{\xi^2+\eta^2} + \frac{eE_0}{2}(\eta^2-\xi^2) + E = 0. \tag{2.118}$$

Setting

$$\frac{1}{\Phi}\frac{d^2\Phi}{d\varphi^2} = -m^2, \tag{2.119}$$

Eq. 2.118 may readily be split into the two equations:

$$\frac{\hbar^2}{2M}\left[\frac{1}{\xi f}\frac{d}{d\xi}\left(\xi\frac{df}{d\xi}\right) - \frac{m^2}{\xi^2}\right] + E\xi^2 - \frac{eE_0\xi^4}{2} + A = 0 \tag{2.120}$$

and

$$\frac{\hbar^2}{2M}\left[\frac{1}{\eta g}\frac{d}{d\eta}\left(\eta\frac{dg}{d\eta}\right) - \frac{m^2}{\eta^2}\right] + E\eta^2 + \frac{eE_0\eta^4}{2} + B = 0. \tag{2.121}$$

The constants A and B are arbitrary except for the constraint $A + B = 2e^2$.

Other applications of parabolic coordinates are included in the problems.

EXERCISES

2.12.1 Find h_ξ^2, h_η^2, and h_φ^2 if the parabolic coordinates (ξ, η, φ) are related to the usual cartesian coordinates by

$$x = \sqrt{\xi\eta}\cos\varphi, \qquad y = \sqrt{\xi\eta}\sin\varphi,$$
$$z = \tfrac{1}{2}(\xi - \eta).$$

2.12.2 Using the ξ, η, φ defined in Ex. 2.12.1, derive the Stark effect equation corresponding to Eq. 2.120. The resulting equation appears in Ex. 8.4.11 (series solution) and 13.2.12 (Laguerre polynomials).

2.12.3 A concept of particular importance in atomic and nuclear physics is that of parity, the property of a wave function being either even or odd under inversion of the coordinates. In cartesian coordinates this inversion or parity operator P acting on (x, y, z) gives

$$P(x, y, z) = (-x, -y, -z).$$

Write out the corresponding operator equations in the following coordinate systems:
(a) Spherical polar (r, θ, φ)
(b) Circular cylindrical (ρ, φ, z)
(c) Prolate spheroidal (u, v, φ)
(d) Prolate spheroidal (ξ, η, φ)
(e) Oblate spheroidal (u, v, φ)
(f) Parabolic (ξ, η, φ)

2.12.4 (a) The wave equation for the hydrogenlike atom is

$$-\frac{\hbar^2}{2m}\nabla^2 u + Vu = Eu,$$

where V, the potential energy of the electron is

$$V = -\frac{Ze^2}{r}$$

and E is the total energy, a number. Show that variables can be separated by using parabolic coordinates.
(b) Show that the variables also separate in prolate spheroidal coordinates with the nucleus at one of the foci.

2.13 Toroidal Coordinates (ξ, η, φ)

This system is formed by rotating the xy-plane of the bipolar system (Section 2.9) about the y-axis of Fig. 2.7. The circles centered on the y-axis (ξ = constant) yield spheres, whereas the circles centered on the x-axis (η = constant) form toroids. By relabeling the coordinates so that the axis of rotation is again the z-axis, the transformation equations are

$$x = \frac{a \sinh \eta \cos \varphi}{\cosh \eta - \cos \xi},$$

$$y = \frac{a \sinh \eta \sin \varphi}{\cosh \eta - \cos \xi}, \tag{2.122}$$

$$z = \frac{a \sin \xi}{\cosh \eta - \cos \xi}.$$

From these equations the scale factors are

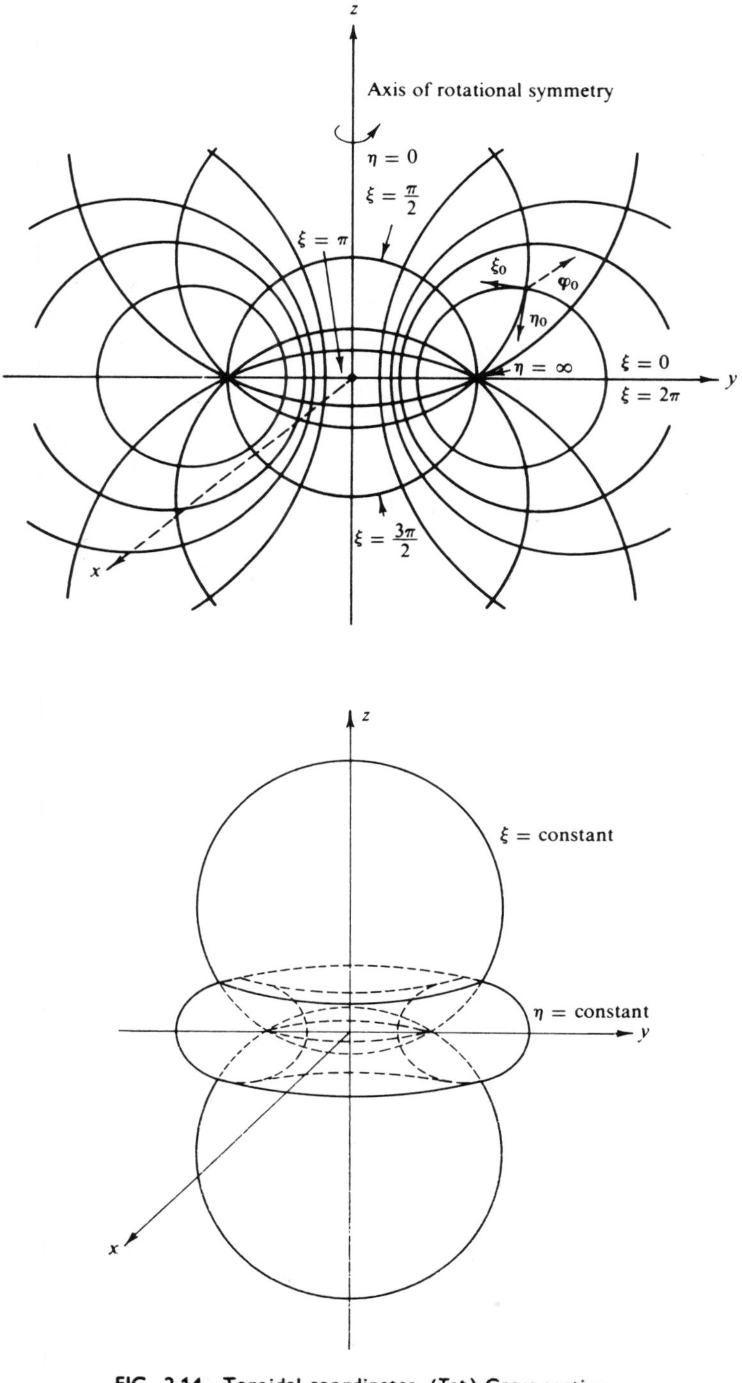

FIG. 2.14 Toroidal coordinates. (*Top*) Cross section

$$h_1 = h_\xi = \frac{a}{\cosh\eta - \cos\xi},$$

$$h_2 = h_\eta = \frac{a}{\cosh\eta - \cos\xi}, \tag{2.123}$$

$$h_3 = h_\varphi = \frac{a\sinh\eta}{\cosh\eta - \cos\xi}.$$

The coordinate surfaces formed by the rotation are the following:

1. Spheres centered at $(0, 0, a\cot\xi)$ with radii, $a|\csc\xi|$,

$$\xi = \text{constant}, \qquad 0 \leqslant \xi \leqslant 2\pi.$$

$$2az\cot\xi = x^2 + y^2 + z^2 - a^2. \tag{2.124}$$

2. Toroids,

$$\eta = \text{constant}, \qquad 0 \leqslant \eta < \infty.$$

The cross sections are circles displaced a distance $a\coth\eta$ from the z-axis and of radii $a\operatorname{csch}\eta$,

$$4a^2(x^2 + y^2)\coth^2\eta = (x^2 + y^2 + z^2 + a^2)^2. \tag{2.125}$$

3. Half planes through the z-axis,

$$\varphi = \text{constant}, \qquad 0 \leqslant \varphi \leqslant 2\pi.$$

Laplace's equation is not completely separable in toroidal coordinates. This coordinate system has some physical applications (such as describing vortex rings) but they are rare and the system is seldom used.

Again, as in the two preceding sections, note that (ξ, η, φ) yields a left-handed set. To transform to a right-handed system perhaps the simplest way is to take the coordinates in the order (η, ξ, φ).

EXERCISES

2.13.1 Show that the surface area of a toroid defined by Fig. 2.15 is $(2\pi a) \times (2\pi b) = 4\pi^2 ab$.

2.13.2 As a step in solving Laplace's equation in toroidal coordinates, assume the potential $\psi(\xi, \eta, \varphi)$ to have the form

$$\psi(\xi, \eta, \varphi) = \sqrt{\cosh\eta - \cos\xi}\; X(\xi)N(\eta)\Phi(\varphi).$$

Assume further that (a) $X(\xi) = \sin n\xi, \cos n\xi$, (b) $\Phi(\varphi) = \sin m\varphi, \cos m\varphi$, with n and m integers. What is the basis for these forms for $X(\xi)$ and $\Phi(\varphi)$? Show that Laplace's equation reduces to

$$\frac{1}{\sinh \eta} \frac{d}{d\eta} \left[\sinh \eta \frac{dN}{d\eta} \right] - \frac{m^2}{\sinh^2 \eta} N - (n^2 - \tfrac{1}{4})N = 0.$$

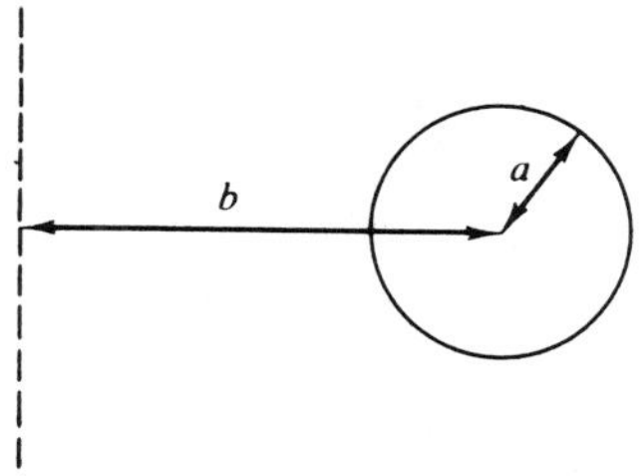

FIG. 2.15

2.14 Bispherical Coordinates (ξ, η, φ)

Returning to the bipolar coordinates of Section 2.9, a rotation of the xy-plane shown in Fig. 2.7, about the x-axis generates two families of orthogonal intersecting spheres. Adding planes of constant azimuth, this is our bispherical system with transformation equations:

$$\begin{aligned} x &= \frac{a \sin \xi \cos \varphi}{\cosh \eta - \cos \xi}, \\ y &= \frac{a \sin \xi \sin \varphi}{\cosh \eta - \cos \xi}, \\ z &= \frac{a \sinh \eta}{\cosh \eta - \cos \xi}. \end{aligned} \tag{2.126}$$

Once more the axis of rotation has been relabeled the z-axis. The scale factors become

$$\begin{aligned} h_1 = h_\xi &= \frac{a}{\cosh \eta - \cos \xi}, \\ h_2 = h_\eta &= \frac{a}{\cosh \eta - \cos \xi}, \\ h_3 = h_\varphi &= \frac{a \sin \xi}{\cosh \eta - \cos \xi}. \end{aligned} \tag{2.127}$$

The coordinate surfaces are the following:

1. A fourth-order surface of revolution about the z-axis,

$\xi = \text{constant}, \qquad 0 < \xi < \frac{\pi}{2}, \qquad$ "dimples" on z-axis,

$\xi = \frac{\pi}{2}, \qquad$ sphere,

$$\frac{\pi}{2} < \xi < \pi, \qquad \text{cusps on } z\text{-axis.}$$

2. Spheres of radius $a|\text{csch}\,\eta|$ centered at $(0, 0, a\coth\eta)$,

$$\eta = \text{constant}, \qquad -\infty < \eta < \infty.$$

3. Half planes through the z-axis,

$$\varphi = \text{constant}, \qquad 0 \leqslant \varphi \leqslant 2\pi.$$

Laplace's equation is partly separable in this system, though the general equation (2.1), $k^2 \neq 0$, is not separable. The bispherical coordinate system has been found to be useful in specialized electrostatic problems such as the capacitance between a conducting sphere and a nearby conducting plane (cf. Exercise 2.14.1).

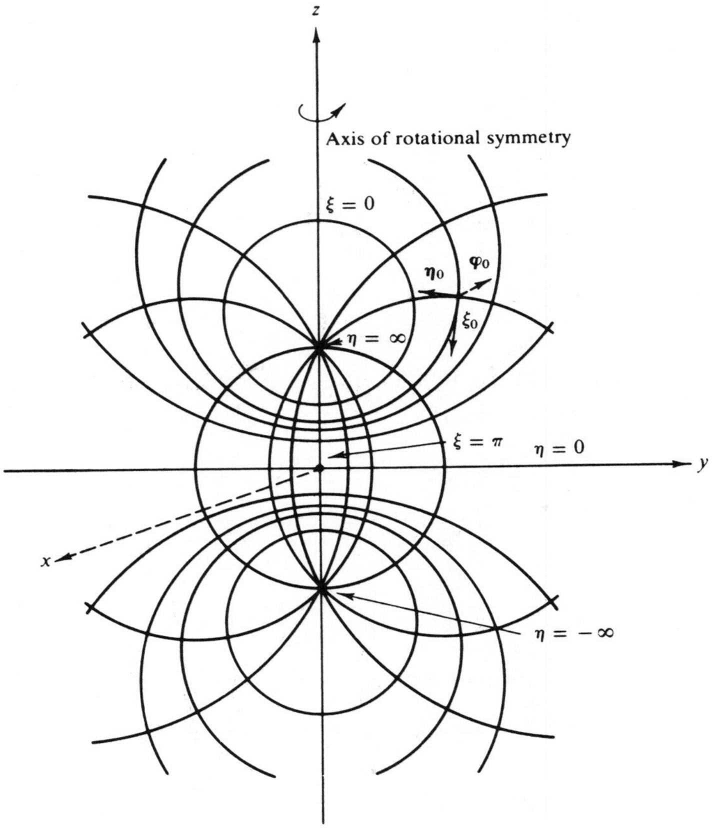

FIG. 2.16 Bispherical coordinates

EXERCISES

2.14.1 Show that Laplace's equation is separable in bispherical coordinates (to within a factor $\sqrt{\cosh \eta - \cos \xi}$).

Hint. Let $\psi(\xi, \eta, \varphi) = \sqrt{\cosh \eta - \cos \xi}\, X(\xi)N(\eta)\Phi(\varphi)$.

2.14.2 Using bispherical coordinates find the capacitance between a conducting sphere and (nonintersecting) conducting plane.

2.15 Confocal Ellipsoidal Coordinates (ξ_1, ξ_2, ξ_3)

This very general coordinate system has the following three families of coordinate surfaces:

1. Ellipsoids (no two axes are equal), ξ_1 = constant,

$$\frac{x^2}{a^2 - \xi_1} + \frac{y^2}{b^2 - \xi_1} + \frac{z^2}{c^2 - \xi_1} = 1. \tag{2.128}$$

2. Hyperboloids of one sheet, ξ_2 = constant,

$$\frac{x^2}{a^2 - \xi_2} + \frac{y^2}{b^2 - \xi_2} - \frac{z^2}{\xi_2 - c^2} = 1. \tag{2.129}$$

3. Hyperboloids of two sheets, ξ_3 = constant,

$$\frac{x^2}{a^2 - \xi_3} - \frac{y^2}{\xi_3 - b^2} - \frac{z^2}{\xi_3 - c^2} = 1. \tag{2.130}$$

The constants a, b, c are parameters which describe the ellipsoids and hyperboloids subject to the constraints

$$a^2 > \xi_3 > b^2 > \xi_2 > c^2 > \xi_1. \tag{2.131}$$

In Eqs. 2.128, 2.129, and 2.130, the minus signs resulting from these constraints were shown explicitly.

The transformation equations are

$$\begin{aligned} x^2 &= \frac{(a^2 - \xi_1)(a^2 - \xi_2)(a^2 - \xi_3)}{(a^2 - b^2)(a^2 - c^2)}, \\ y^2 &= \frac{(b^2 - \xi_1)(b^2 - \xi_2)(\xi_3 - b^2)}{(a^2 - b^2)(b^2 - c^2)}, \\ z^2 &= \frac{(c^2 - \xi_1)(\xi_2 - c^2)(\xi_3 - c^2)}{(a^2 - c^2)(b^2 - c^2)}. \end{aligned} \tag{2.132}$$

After an undue amount of algebra, the scale factors are found to be

$$h_1 = h_{\xi_1} = \frac{1}{2}\left[\frac{(\xi_2 - \xi_1)(\xi_3 - \xi_1)}{(a^2 - \xi_1)(b^2 - \xi_1)(c^2 - \xi_1)}\right]^{1/2}$$

$$h_2 = h_{\xi_2} = \frac{1}{2}\left[\frac{(\xi_3 - \xi_2)(\xi_2 - \xi_1)}{(a^2 - \xi_2)(b^2 - \xi_2)(c^2 - \xi_2)}\right]^{1/2} \tag{2.133}$$

$$h_3 = h_{\xi_3} = \frac{1}{2}\left[\frac{(\xi_3 - \xi_1)(\xi_3 - \xi_2)}{(a^2 - \xi_3)(\xi_3 - b^2)(\xi_3 - c^2)}\right]^{1/2}$$

As with the equations of the coordinate surfaces, the symmetry of this set has been sacrificed by requiring each factor to be positive.

From the transformation equations (2.132) it will be seen that a given point $P(\xi_1, \xi_2, \xi_3)$ corresponds to eight possible points $(\pm x, \pm y, \pm z)$, the cartesian coordinates appearing as squares. This eightfold multiplicity may be resolved by introducing an appropriate sign convention for ξ_1, ξ_2 and ξ_3 or by bringing in elliptic functions or related functions.

Although this coordinate system has been useful in problems of mathematical physics, its very generality makes it cumbersome and awkward to use. Since this text purports to be an introduction, we shall restrict ourselves to ellipsoids with axes of rotational symmetry.

2.16 Conical Coordinates (ξ_1, ξ_2, ξ_3)

This is one of the more unusual (and less useful) degenerate forms of the confocal ellipsoidal coordinate system of the preceding section. The coordinate surfaces are the following:

1. Spheres centered at the origin, radii ξ_1, $\xi_1 =$ constant,

$$x^2 + y^2 + z^2 = \xi_1^2. \tag{2.134}$$

2. Cones of elliptic cross section with apexes at the origin and axes along the z-axis, $\xi_2 =$ constant,

$$\frac{x^2}{\xi_2^2} + \frac{y^2}{\xi_2^2 - b^2} = \frac{z^2}{c^2 - \xi_2^2}. \tag{2.135}$$

3. Elliptic cones, apexes at the origin, axes along the x-axis, $\xi_3 =$ constant,

$$\frac{x^2}{\xi_3^2} = \frac{y^2}{b^2 - \xi_3^2} + \frac{z^2}{c^2 - \xi_3^2}. \tag{2.136}$$

As in Section 2.15, the parameters b and c satisfy constraints

$$c^2 > \xi_2^2 > b^2 > \xi_3^2. \tag{2.137}$$

Inverting the set of equations (2.134), (2.135), and (2.136), the transformation equations

$$x^2 = \left(\frac{\xi_1 \xi_2 \xi_3}{bc}\right)^2,$$

$$y^2 = \frac{\xi_1^2(\xi_2^2 - b^2)(b^2 - \xi_3^2)}{b^2(c^2 - b^2)}, \tag{2.138}$$

$$z^2 = \frac{\xi_1^2(c^2 - \xi_2^2)(c^2 - \xi_3^2)}{c^2(c^2 - b^2)}$$

are obtained. Then by Eq. 2.6 the scale factors are

$$h_1 = h_{\xi_1} = 1,$$

$$h_2 = h_{\xi_2} = \left[\frac{\xi_1^2(\xi_2^2 - \xi_3^2)}{(\xi_2^2 - b^2)(c^2 - \xi_2^2)}\right]^{1/2} \tag{2.139}$$

$$h_3 = h_{\xi_3} = \left[\frac{\xi_1^2(\xi_2^2 - \xi_3^2)}{(b^2 - \xi_3^2)(c^2 - \xi_3^2)}\right]^{1/2}$$

This really oddball coordinate system has been almost completely ignored. Recently, however, it was found useful to describe the angular momentum eigenfunctions of an asymmetric rotor.[1]

2.17 Confocal Parabolic Coordinates (ξ_1, ξ_2, ξ_3)

Except for the bipolar, toroidal, and bispherical coordinate systems, all the coordinate systems in this chapter are derivable from the confocal ellipsoidal coordinates (Section 2.15). The last of these degenerate or special systems is the confocal paraboloidal system. Here the coordinate surfaces are the following:

1. Confocal paraboloids of elliptic cross section extending along the negative z-axis, $\xi_1 =$ constant,

$$\frac{x^2}{a^2 - \xi_1} + \frac{y^2}{b^2 - \xi_1} + 2z + \xi_1 = 0. \tag{2.140}$$

2. Hyperbolic paraboloids, $\xi_2 =$ constant,

$$\frac{x^2}{a^2 - \xi_2} - \frac{y^2}{\xi_2 - b^2} + 2z + \xi_2 = 0. \tag{2.141}$$

3. Confocal paraboloids of elliptic cross section extending along the positive z-axis, $\xi_3 =$ constant,

$$\frac{x^2}{\xi_3 - a^2} + \frac{y^2}{\xi_3 - b^2} - 2z - \xi_3 = 0. \tag{2.142}$$

As in Sections 2.15 and 2.16, there are constraints on the parameters and variables

$$\xi_3 > a^2 > \xi_2 > b^2 > \xi_1. \tag{2.143}$$

[1] R. D. Spence, *Am. J. Phys.* **27**, 329 (1959).

The transformation equations are

$$x^2 = \frac{(a^2 - \xi_1)(a^2 - \xi_2)(\xi_3 - a^2)}{a^2 - b^2},$$

$$y^2 = \frac{(b^2 - \xi_1)(\xi_2 - b^2)(\xi_3 - b^2)}{a^2 - b^2}, \tag{2.144}$$

$$z = \tfrac{1}{2}(a^2 + b^2 - \xi_1 - \xi_2 - \xi_3)$$

with resulting scale factors

$$h_1 = h_{\xi_1} = \frac{1}{2}\left[\frac{(\xi_2 - \xi_1)(\xi_3 - \xi_1)}{(a^2 - \xi_1)(b^2 - \xi_1)}\right]^{1/2}$$

$$h_2 = h_{\xi_2} = \frac{1}{2}\left[\frac{(\xi_3 - \xi_2)(\xi_2 - \xi_1)}{(a^2 - \xi_2)(\xi_2 - b^2)}\right]^{1/2} \tag{2.145}$$

$$h_3 = h_{\xi_3} = \frac{1}{2}\left[\frac{(\xi_3 - \xi_1)(\xi_3 - \xi_2)}{(\xi_3 - a^2)(\xi_3 - b^2)}\right]^{1/2}$$

Applications of this system have been developed in electromagnetic theory[1] but within the scope of this book the system is of little interest.

REFERENCES

MORSE, P. M., and H. FESHBACH. *Methods of Theoretical Physics*. New York: McGraw-Hill (1953). Chapter 5 includes a description of most of the coordinate systems presented here. Note carefully that Morse and Feshbach are not above using left-handed coordinate systems even for cartesian coordinates. Elsewhere in this excellent (and difficult) book are many examples of the use of the various coordinate systems in solving physical problems.

[1] J. C. Maxwell. *A Treatise on Electricity and Magnetism*. Vol. I, 3rd Ed. Oxford: Oxford University Press (1904), Chapter X.

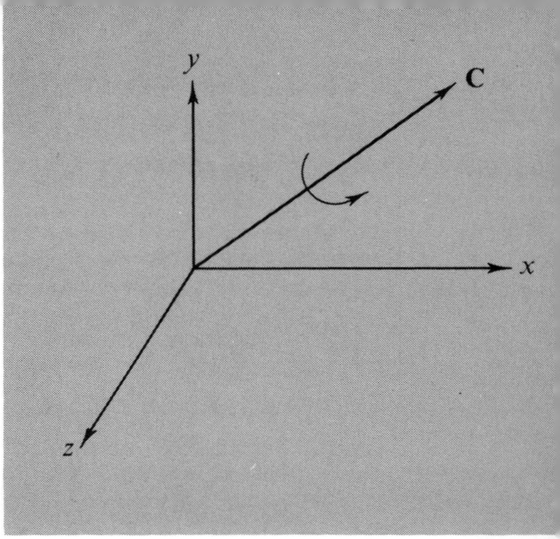

CHAPTER 3

TENSOR ANALYSIS

3.1 Introduction, Definitions

Tensors are important in many areas of physics, including general relativity and electromagnetic theory. One of the more prolific sources of tensor quantities is the anisotropic solid. Here the elastic, optical, electrical, and magnetic properties may well involve tensors. The elastic properties of the anisotropic solid are considered in some detail in Section 3.6. As an introductory illustration, let us consider the flow of electric current. We can write Ohm's law in the usual form

$$\mathbf{J} = \sigma \mathbf{E}, \tag{3.1}$$

with current density $\mathbf{J}$ and electric field $\mathbf{E}$, both vector quantities.[1] If we have an isotropic medium, σ, the conductivity, is a scalar, and for the x-component, for example,

$$J_1 = \sigma E_1. \tag{3.2}$$

However, if our medium is anisotropic, as in many crystals, or a plasma in the presence of a magnetic field, the current density in the x-direction may depend on the electric fields in the y- and z-directions as well as on the field in the x-direction. Assuming a linear relationship, we must replace Eq. 3.2 with

$$J_1 = \sigma_{11} E_1 + \sigma_{12} E_2 + \sigma_{13} E_3, \tag{3.3}$$

and, in general,

$$J_i = \sum_k \sigma_{ik} E_k. \tag{3.4}$$

For ordinary three-dimensional space the scalar conductivity σ has given way to a set of nine elements, σ_{ik}.

[1] Another example of this type of physical equation appears in Section 4.6.

$$\sigma \rightarrow \begin{pmatrix} \sigma_{11} & \sigma_{12} & \sigma_{13} \\ \sigma_{21} & \sigma_{22} & \sigma_{23} \\ \sigma_{31} & \sigma_{32} & \sigma_{33} \end{pmatrix} \tag{3.5}$$

This array of nine elements actually forms a tensor, as shown in Section 3.3.

In Chapter 1 a quantity that did not change under rotations of the coordinate system that is, an invariant quantity, was labeled a scalar. A quantity whose components transformed like those of the distance of a point from a chosen origin (Eq. 1.13, Section 1.2) was called a vector. This transformation property was adopted as the defining characteristic of a vector. There is a possible ambiguity in this definition (Eq. 3.6),

$$A_i' = \sum_j a_{ij} A_j, \tag{3.6}$$

in which a_{ij} is the cosine of the angle between the x_i'-axis and the x_j-axis.

If we start with our prototype vector $\mathbf{r}$, then

$$x_i' = \sum_j \frac{\partial x_i'}{\partial x_j} x_j \tag{3.7}$$

by partial differentiation. If we set

$$a_{ij} = \frac{\partial x_i'}{\partial x_j}, \tag{3.8}$$

Eqs. 3.6 and 3.7 are consistent. Any set of quantities A_j transforming according to

$$A_i' = \sum_j \frac{\partial x_i'}{\partial x_j} A_j \tag{3.9}$$

is defined as a contravariant vector.

However, we have already encountered a slightly different type of vector transformation. The gradient of a scalar, $\nabla\varphi$, defined by

$$\nabla\varphi = \mathbf{i}\frac{\partial\varphi}{\partial x_1} + \mathbf{j}\frac{\partial\varphi}{\partial x_2} + \mathbf{k}\frac{\partial\varphi}{\partial x_3} \tag{3.10}$$

(using x_1, x_2, x_3 for x, y, z), transforms as

$$\frac{\partial\varphi'}{\partial x_i'} = \sum_j \frac{\partial\varphi}{\partial x_j}\frac{\partial x_j}{\partial x_i'}, \tag{3.11}$$

using $\varphi = \varphi(x, y, z) = \varphi(x', y', z') = \varphi'$, φ being defined as a scalar quantity. Notice that this differs from Eq. 3.9 in that we have $\partial x_j/\partial x_i'$ instead of $\partial x_i'/\partial x_j$. Equation 3.11 is taken as the definition of a covariant vector with the gradient as the prototype.

In cartesian coordinates

$$\frac{\partial x_j}{\partial x_i'} = \frac{\partial x_i'}{\partial x_j} = a_{ij}, \tag{3.12}$$

and there is no difference between contravariant and covariant transformations. In other systems Eq. 3.12 in general does not apply, and the distinction between contravariant and covariant is real and must be observed. In the remainder of this section the components of a contravariant vector are denoted by a superscript, A^i, whereas a subscript is used for the components of a covariant vector A_i.

To remove some of the fear and mystery from the term tensor, let us rechristen a scalar as a tensor of rank zero and relabel a vector as a tensor of first rank. Then we proceed to define contravariant, mixed, and covariant tensors of second rank by the following equations

$$\begin{aligned} A'^{ij} &= \sum_{kl} \frac{\partial x'_i}{\partial x_k} \frac{\partial x'_j}{\partial x_l} A^{kl}, \\ B'^{i}{}_{j} &= \sum_{kl} \frac{\partial x'_i}{\partial x_k} \frac{\partial x_l}{\partial x'_j} B^{k}{}_{l}, \\ C'_{ij} &= \sum_{kl} \frac{\partial x_k}{\partial x'_i} \frac{\partial x_l}{\partial x'_j} C_{kl}. \end{aligned} \tag{3.13}$$

We see that A^{kl} is contravariant with respect to both indices, C_{kl} is covariant with respect to both indices, and $B^k{}_l$ transforms contravariantly with respect to the first index k but covariantly with respect to the second index l. Once again, if we are using cartesian coordinates, all three forms of the tensors of second rank, contravariant, mixed, and covariant are the same.

The second-rank tensor **A** (components A^{kl}) may be conveniently represented by writing out its components in a square array, (3 × 3 if we are in three-dimensional space)

$$\mathbf{A} = \begin{pmatrix} A^{11} & A^{12} & A^{13} \\ A^{21} & A^{22} & A^{23} \\ A^{31} & A^{32} & A^{33} \end{pmatrix}. \tag{3.14}$$

This does not mean that any square array of numbers or functions forms a tensor. The essential condition is that the components transform according to Eq.. 3.13.

This transformation requirement may be illustrated by examining in detail the two-dimensional tensor

$$\mathbf{T} = \begin{pmatrix} -xy & -y^2 \\ x^2 & xy \end{pmatrix}.$$

In a primed (rotated) coordinate system the $T^{11'}$-component must be $-x'y'$ as discussed in Section 1.2 for vectors. Proceeding exactly as in Section 1.2, we check to see if this is consistent with the defining equation. Eq. 3.13.

$$\begin{aligned} T^{11'} = -x'y' &\stackrel{?}{=} \sum_{kl} \frac{\partial x'_1}{\partial x_k} \frac{\partial x'_1}{\partial x_l} T^{kl} \\ &= \sum_{kl} a_{1k} a_{1l} T^{kl}, \end{aligned}$$

setting i and j equal to 1. We translate the left-hand side into unprimed coordinates by Eq. 3.7 or 1.8. Then with $a_{11} = \cos\theta$ and $a_{12} = \sin\theta$, and the given values of T^{kl} we obtain

$$-(x\cos\theta + y\sin\theta)(-x\sin\theta + y\cos\theta)$$
$$\stackrel{?}{=} \cos^2\theta T^{11} + \cos\theta\sin\theta T^{12} + \sin\theta\cos\theta T^{21} + \sin^2\theta T^{22}$$
$$= -xy\cos^2\theta - y^2\cos\theta\sin\theta + x^2\sin\theta\cos\theta + xy\sin^2\theta.$$

We have an identity showing that Eq. 3.13 is satisfied (for $T^{11'}$). Repetition of the other three components verifies that all transform in accordance with Eq. 3.13 and that **T** therefore is a second-rank tensor.

This transformation property is not something to be taken for granted. For instance, if one algebraic sign were changed, if T^{22} were $-xy$ instead of $+xy$, then

$$T^{11'} \neq \sum_{kl} a_{1k}a_{1l}T^{kl}$$

and the array

$$\begin{pmatrix} -xy & -y^2 \\ x^2 & -xy \end{pmatrix}$$

is not a tensor. It is not a tensor because it does not contain the required transformation properties.

The addition and subtraction of tensors is defined in terms of the individual elements just as for vectors. To add or subtract two tensors, the corresponding elements are added or subtracted. If

$$\mathbf{A} + \mathbf{B} = \mathbf{C}, \tag{3.15}$$

then

$$A^{ij} + B^{ij} = C^{ij}.$$

Of course, **A** and **B** must be tensors of the same rank and both expressed in a space of the same number of dimensions.

In tensor analysis it is customary to adopt a summation convention to put Eq. 3.13 and subsequent tensor equations in a more compact (and, for the beginning student, a more obscure) form. As long as we are distinguishing between contravariance and covariance, let us agree that when an index appears on one side of an equation, once as a superscript and once as a subscript, we automatically sum over that index. Then we may write the second expression in Eq. 3.13 as

$$B'^{i}{}_{j} = \frac{\partial x'_i}{\partial x_k}\frac{\partial x_l}{\partial x'_j} B^k_l; \tag{3.16}$$

with the summation of the right-hand side over k and l implied. This is the summation convention.

To illustrate the use of the summation convention and some of the techniques of tensor analysis, let us show that the now familiar Kronecker delta, δ_{kl}, is really a mixed tensor of second rank, δ^k_l. The question is, does δ^k_l transform according to Eq. 3.13? This is our criterion for calling it a tensor. We have, using the summation convention,

$$\delta^k_l \frac{\partial x'_i}{\partial x_k}\frac{\partial x_l}{\partial x'_j} = \frac{\partial x'_i}{\partial x_k}\frac{\partial x_k}{\partial x'_j} \tag{3.17}$$

by definition of the Kronecker delta. Now

$$\frac{\partial x'_i}{\partial x_k}\frac{\partial x_k}{\partial x'_j} = \frac{\partial x'_i}{\partial x'_j} \tag{3.18}$$

by direct partial differentiation of the right-hand side. However, x'_i and x'_j are independent coordinates, and therefore the variation of one with respect to the other must be zero if they are different, unity if they coincide; that is,

$$\frac{\partial x'_i}{\partial x'_j} = \delta'^i_j \tag{3.19}$$

Hence

$$\delta'^i_j = \frac{\partial x'_i}{\partial x_k}\frac{\partial x_l}{\partial x'_j}\delta^k_l,$$

showing that the δ^k_l are indeed the components of a mixed second-rank tensor. The Kronecker delta has one further interesting property. It has the same components in all of our rotated coordinate systems and is therefore called isotropic. In Section 3.4 we shall meet a third-rank isotropic tensor and three fourth-rank isotropic tensors. No isotropic first-rank tensor (vector) exists.

The order in which the indices appear in our description of a tensor is important. In general A^{mn} is independent of A^{nm}, but there are some cases of special interest. If

$$A^{mn} = A^{nm}, \tag{3.20}$$

we call the tensor symmetric. If, on the other hand,

$$A^{mn} = -A^{nm}, \tag{3.21}$$

the tensor is antisymmetric. Clearly every (second-rank) tensor can be resolved into symmetric and antisymmetric parts by the identity

$$A^{mn} = \tfrac{1}{2}(A^{mn} + A^{nm}) + \tfrac{1}{2}(A^{mn} - A^{nm}), \tag{3.22}$$

the first term on the right being a symmetric tensor, the second, an antisymmetric tensor. This resolution into symmetric and antisymmetric tensors will reappear in the theory of elasticity (Section 3.6). A similar resolution of functions into symmetric and antisymmetric parts is of extreme importance to quantum mechanics.

EXERCISES

3.1.1 (a) Prove that

$$\mathbf{A} = \begin{pmatrix} y^2 & -xy \\ -xy & x^2 \end{pmatrix} \quad \text{and} \quad \mathbf{B} = \begin{pmatrix} -xy & x^2 \\ -y^2 & xy \end{pmatrix}$$

are tensors.

(b) Prove that

$$\mathbf{C} = \begin{pmatrix} y^2 & xy \\ xy & x^2 \end{pmatrix} \quad \text{and} \quad \mathbf{D} = \begin{pmatrix} xy & y^2 \\ x^2 & -xy \end{pmatrix}$$

are *not* tensors.

3.1.2 The components of tensor **A** are equal to the corresponding components of tensor **B** in one particular coordinate system, i.e.

$$A_{ij}^{0} = B_{ij}^{0}$$

Show that tensor **A** is equal to tensor **B**, $A_{ij} = B_{ij}$, in *all* coordinate systems.

3.1.3 Show that if the components of any tensor of any rank vanish in some one particular coordinate system they vanish in all coordinate systems.

Note. This point takes on especial importance in the four-dimensional curved space of general relativity. If a quantity, expressed as a tensor, exists in one coordinate system, it exists in all coordinate systems and is not just a consequence of a *choice* of a coordinate system (as are centrifugal and Coriolis forces in Newtonian mechanics).

3.1.4 The first three components of a four-dimensional vector vanish in each of two reference frames. If the second reference frame is not merely a rotation of the first about the x_4 axis, that is, if at least one of the coefficients $a_{i4}(i = 1, 2, 3) \neq 0$, show that the fourth component vanishes in all reference frames.

Translated into relativistic mechanics this means that if momentum is conserved in two Lorentz frames then energy is conserved in all Lorentz frames.

3.1.5 From an analysis of the behavior of a general second-rank tensor under 90° and 180° rotations about the coordinate axes, show that an *isotropic* second-rank tensor must be a multiple of δ_{ij}.

3.1.6 The four-dimensional fourth-rank Riemann-Christoffel curvature tensor of general relativity, R_{iklm}, satisfies the symmetry relations

$$R_{iklm} = -R_{ikml} = -R_{kilm}$$

With the indices running from 1 to 4, show that the number of independent components is reduced from 256 to 36 and that the condition

$$R_{iklm} = R_{lmik}$$

further reduces the number of independent components to 21. Finally, if the components satisfy an identity $R_{iklm} + R_{ilmk} + R_{imkl} = 0$, show that the number of independent components is reduced to 20. *Note*. The final three-term identity furnishes new information only if all four indices are different. Then it reduces the number of independent components by one third.

3.1.7 T_{iklm} is antisymmetric with respect to all pairs of indices. How many independent components has it (three-dimensional space)?

3.2 Contraction, Direct Product

When dealing with vectors, we formed a scalar product (Section 1.3) by summing products of corresponding components:

$$\mathbf{A} \cdot \mathbf{B} = A_i B_i \qquad \text{(summation convention).} \tag{3.23}$$

The generalization of this expression in tensor analysis is a process known as contraction. Two indices, one covariant and the other contravariant, are set equal to each other and then (as implied by the summation convention), we sum over this repeated index. For example, let us contract the second-rank mixed tensor $B'^i{}_j$.

$$B'^i{}_j \rightarrow B'^i{}_i = \frac{\partial x'_i}{\partial x_k} \frac{\partial x_l}{\partial x'_i} B^k_l$$

$$= \frac{\partial x_l}{\partial x_k} B^k_l \tag{3.24}$$

by Eq. 3.18 and then by Eq. 3.19:

$$B'^i{}_i = \delta^l_k B^k_l = B^k_k. \tag{3.25}$$

Our contracted second-rank mixed tensor is invariant and therefore a scalar.[1] This is exactly what we obtained in Section 1.3 for the dot product of two vectors and Section 1.7 for the divergence of a vector. In general, the operation of contraction reduces the rank of a tensor by two.

The components of a covariant vector (first-rank tensor) a_i and those of a contravariant vector (first-rank tensor) b^j may be multiplied component by component to give the general term $a_i b^j$. This, by Eq. 3.13, is actually a second-rank tensor, for

$$a'_i b'^j = \frac{\partial x_k}{\partial x'_i} a_k \frac{\partial x'_j}{\partial x_l} b^l$$

$$= \frac{\partial x_k}{\partial x'_i} \frac{\partial x'_j}{\partial x_l} (a_k b^l). \tag{3.26}$$

Contracting,

$$a'_i b'^i = a_k b^k, \tag{3.27}$$

as in Eqs. 3.24 and 3.25 to give the regular scalar product.

The operation of adjointing two vectors a_i and b^j as in the last paragraph is known as forming the direct product. For the case of two vectors, the direct product is a tensor of second rank. In this sense we may attach meaning to $\nabla \mathbf{E}$, which was not defined within the framework of vector analysis. In general, the direct product of two tensors is a tensor of rank equal to the sum of the two initial ranks, that is,

$$A^i_j B^{kl} = C^{ikl}_j, \tag{3.28}$$

where C^{ikl}_j is a tensor of fourth rank.

So far the distinction between a covariant transformation and a contravariant transformation has been maintained because it does exist in noncartesian space and becausc it is of great importance in general relativity. At this point we refer the student interested in general relativity to the many excellent specialized texts in that field and restrict ourselves in the remainder of this chapter to cartesian coordinate systems. As noted in Section 3.1, the distinction between contravariance and covariance disappears and all indices are henceforth shown as subscripts. We restate the summation convention and the operation of contraction.

[1] In matrix analysis, this scalar is the *trace* of the matrix, Section 4.2.

Summation convention. When a subscript (letter, not number) appears twice on one side of an equation, summation with respect to that subscript is implied.

Contraction. Contraction consists of setting two unlike indices (subscripts) equal to each other and then summing as implied by the summation convention.

EXERCISES

3.2.1 If $T_{\cdots i}$ is a tensor of rank n, show that $\partial T_{\cdots i}/\partial x_j$ is a tensor of rank $n+1$ (cartesian coordinates). *Note*. In noncartesian coordinate systems the coefficients a_{ij} are, in general, functions of the coordinates, and the simple derivative of a tensor of rank n is not a tensor except in the special case of $n = 0$. In this case the derivative does yield a covariant vector (tensor of rank 1) by Eq. 3.11.

3.2.2 If $T_{ijk\cdots}$ is a tensor of rank n, show that $\sum \partial T_{ijk\cdots}/\partial x_j$ is a tensor of rank $n-1$ (cartesian coordinates).

3.2.3 The operator

$$\nabla^2 - \frac{1}{c^2}\frac{\partial^2}{\partial t^2}$$

may be written as

$$\sum_{i=1}^{4} \frac{\partial^2}{\partial x_i^2},$$

using $x_4 = ict$. This is the four-dimensional Laplacian, usually called the d'Alembertian and denoted by $\square^2$. Show that it is a *scalar* operator.

3.3 Quotient Rule

If A_i and B_j are vectors, we have seen in Section 3.2 how we may easily show that A_iB_j is a second-rank tensor. Here we are concerned with a variety of inverse relations. Consider such equations as

$$K_iA_i = B \tag{3.29a}$$

$$K_{ij}A_j = B_i \tag{3.29b}$$

$$K_{ij}A_{jk} = B_{ik} \tag{3.29c}$$

$$K_{ijkl}A_{ij} = B_{kl} \tag{3.29d}$$

$$K_{ij}A_k = B_{ijk} \tag{3.29e}$$

In each of these expressions **A** and **B** are known tensors of rank indicated by the number of indices and **A** is arbitrary. In each case K is an unknown quantity. We wish to establish the transformation properties of K. The quotient rule asserts that if the equation of interest holds in all (rotated) cartesian coordinate systems, K is a tensor of the indicated rank. Consider Eq. 3.29b as an illustration. In our primed coordinate system

$$K'_{ij}A'_j = B'_i = a_{ik}B_k, \tag{3.30}$$

using the vector transformation properties of $\mathbf{B}$. Since the equation holds in all rotated cartesian coordinate systems,

$$a_{ik}B_k = a_{ik}(K_{kl}A_l). \tag{3.31}$$

Now transforming $\mathbf{A}$ back into the primed coordinate system[1] (cf. Eq. 3.9),

$$K'_{ij}A'_j = a_{ik}K_{kl}a_{jl}A'_j \tag{3.32}$$

Rearranging,

$$(K'_{ij} - a_{ik}a_{jl}K_{kl})A'_j = 0. \tag{3.33}$$

This must hold for each value of the index i and for every primed coordinate system. Since the A'_j are arbitrary,[2] we conclude

$$K'_{ij} = a_{ik}a_{jl}K_{kl}, \tag{3.34}$$

which is our definition of a second-rank tensor.

The other equations may be treated the same way, giving rise to other forms of the quotient rule. One minor pitfall should be noted. The quotient rule does not necessarily apply if B is zero. The transformation properties of zero are indeterminate.

EXERCISES

3.3.1 The double summation $\sum_{ij} K_{ij}A_iB_j$ is invariant for any two vectors A_i and B_j. Prove that K_{ij} is a second-rank tensor. *Note.* In the form ds^2 (invariant) $= g_{ij}\,dx_i\,dx_j$, this result shows that g_{ij}, the "metric" is a tensor.

3.3.2 The equation $K_{ij}A_{jk} = B_{ik}$ holds for all orientations of the coordinate system. If $\mathbf{A}$ and $\mathbf{B}$ are second-rank tensors, show that $\mathbf{K}$ is a second-rank tensor also.

3.3.3 The exponential in a plane wave is $\exp[i(\mathbf{k}\cdot\mathbf{r} - \omega t)]$. We recognize $x_\mu = (x_1, x_2, x_3, ict)$ as a prototype vector in Minkowski space. If $\mathbf{k}\cdot\mathbf{r} - \omega t$ is a scalar under Lorentz transformations (Section 3.7), show that $k_\mu = (k_1, k_2, k_3, i\omega/c)$ is a vector in Minkowski space.

[1] Note carefully the order of the indices of the direction cosine a_{jl} in this *inverse* transformation. We have

$$A_l = \sum_j \frac{\partial x_l}{\partial x'_j} A'_j = \sum_j a_{jl}A'_j.$$

[2] We might, for instance, take $A'_1 = 1$ and $A'_m = 0$ for $m \neq 1$. Then the equation $K'_{i1} = a_{ik}a_{1l}K_{kl}$ follows immediately. The rest of Eq. 3.34 comes from other special choices of the arbitrary A'_j.

3.4 Pseudotensors, Dual Tensors

So far our coordinate transformations have been restricted to pure rotations. We now consider the effect of reflections or inversions. If we have transformation coefficients $a_{ij} = -\delta_{ij}$, then by Eq. 3.7

$$x_i = -x_i', \tag{3.35}$$

which is an inversion. Note carefully that this transformation changes our initial right-handed coordinate system into a left-handed coordinate system. Our prototype vector $\mathbf{r}$ with components (x_1, x_2, x_3) transforms to $\mathbf{r}' = (x_1', x_2', x_3') = (-x_1, -x_2, -x_3)$. This new vector $\mathbf{r}'$ has negative components, relative to the new transformed set of axes. As shown in Fig. 3.1, the result of reversing signs of

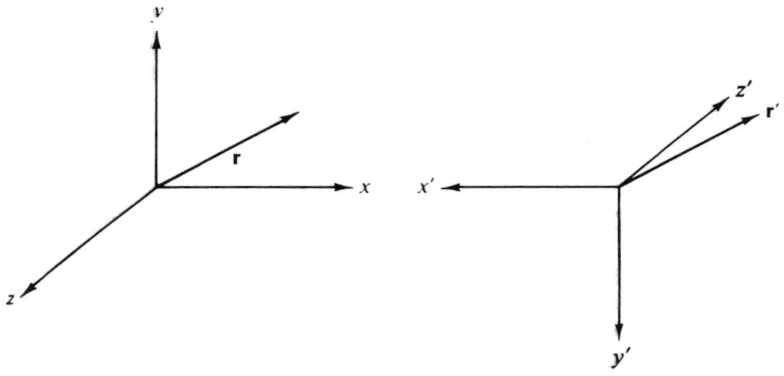

FIG. 3.1 Inversion of cartesian coordinates—polar vector

both axes and components is to leave the vector (an arrow in space) exactly as it was before the transformation was carried out. The distance vector $\mathbf{r}$ and all other vectors behaving this way under reflection or inversion of the coordinate system are called polar vectors.

A fundamental difference appears when we encounter a vector defined as the cross product of two polar vectors. Let $\mathbf{C} = \mathbf{A} \times \mathbf{B}$, where both $\mathbf{A}$ and $\mathbf{B}$ are polar

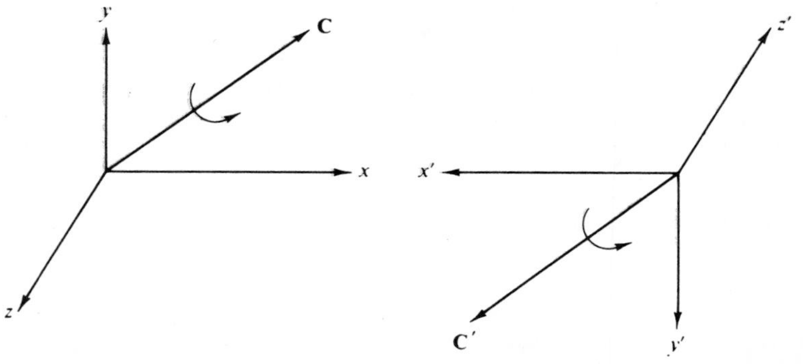

FIG. 3.2 Inversion of cartesian coordinates—axial vector

vectors. From Eq. 1.33 of Section 1.4 the components of **C** are given by

$$C_1 = A_2 B_3 - A_3 B_2. \tag{3.36}$$

and so on. Now, when the coordinate system is inverted, $A_i \to -A_i$, $B_j \to -B_j$ but from its definition $C_k \to +C_k$; that is, our cross-product vector, vector **C**, does *not* behave like a polar vector under inversion. To distinguish, we label it a pseudovector or axial vector (see Fig. 3.2). The term axial vector is frequently used because these cross products often arise from a description of rotation. Examples are

angular velocity, $\mathbf{v} = \boldsymbol{\omega} \times \mathbf{r}$,

angular momentum, $\mathbf{L} = \mathbf{r} \times \mathbf{p}$,

torque, $\mathbf{N} = \mathbf{r} \times \mathbf{f}$,

magnetic induction field **B**, $\dfrac{\partial \mathbf{B}}{\partial t} = -\nabla \times \mathbf{E}$.

Clearly, axial vectors occur frequently in elementary physics, although this fact is usually not pointed out. In a right-handed coordinate system an axial vector **C** has a sense of rotation associated with it given by a right-hand rule (cf. Section 1.4). In the inverted, left-handed system the sense of rotation is a left-handed rotation. This is indicated by the curved arrows in Fig. 3.2.

If we agree that the universe does not care whether we use a right- or left-handed coordinate system, then it does not make sense to add an axial vector to a polar vector. In the vector equation **A** = **B**, either both **A** and **B** are polar vectors or both must be axial vectors.[1] Similar restrictions apply to scalars and pseudoscalars and, in general, to the tensors and pseudotensors considered below.

In general, pseudovectors and pseudotensors will transform as

$$\begin{aligned} C_i' &= |a| a_{ij} C_j, \\ A_{ij}' &= |a| a_{ik} a_{jl} A_{kl}, \end{aligned} \tag{3.37}$$

where $|a|$ is the determinant[2] of the array of coefficients a_{mn}. In our inversion the determinant is

$$|a| = \begin{vmatrix} -1 & 0 & 0 \\ 0 & -1 & 0 \\ 0 & 0 & -1 \end{vmatrix} = -1. \tag{3.38}$$

For a reflection of one axis, the x-axis,

$$|a| = \begin{vmatrix} -1 & 0 & 0 \\ 0 & 1 & 0 \\ 0 & 0 & 1 \end{vmatrix} = -1, \tag{3.39}$$

[1] The big exception to this is in beta decay, weak interactions. Here, the universe does distinguish between right- and left-handed systems, and we do add polar and axial vector interactions.

[2] Determinants are described in Section 4.1.

and again the determinant $|a| = -1$. On the other hand, for all pure rotations the determinant $|a|$ is always $+1$. This is discussed further in Section 4.3. Often quantities that transform according to Eq. 3.37 are known as tensor densities. They are regular tensors as far as rotations are concerned, differing from tensors only in reflections or inversions of the coordinates, and then the only difference is the appearance of an additional minus sign from the determinant $|a|$.

In Chapter 1 the triple scalar product $S = \mathbf{A} \times \mathbf{B} \cdot \mathbf{C}$ was shown to be a scalar (under rotations). Now by considering the transformation given by Eq. 3.35 we see that $S \to -S$, proving that the triple scalar product is actually a pseudoscalar. This behavior was foreshadowed by the geometrical analogy of a volume. If all three parameters of the volume, length, depth, and height, change from positive distances to negative distances, the product of the three will be negative.

The pseudoscalar nature of the volume element has an interesting implication. In Maxwell's equation

$$\nabla \cdot \mathbf{E} = -\frac{\rho}{\varepsilon_0}$$

let us call ∇ and $\mathbf{E}$ polar vectors and ε_0, a scalar. Then ρ, the charge per unit volume, must be a scalar quantity; but volume has just been shown to be pseudoscalar. Hence electric charge is also a pseudoscalar quantity.

For future use it is convenient to introduce the three-dimensional Levi-Civita symbol ε_{ijk} defined by

$$\begin{aligned} &\varepsilon_{123} = \varepsilon_{231} = \varepsilon_{312} = 1, \\ &\varepsilon_{132} = \varepsilon_{213} = \varepsilon_{321} = -1, \\ &\text{all other } \varepsilon_{ijk} = 0. \end{aligned} \tag{3.40}$$

Suppose, now, that we have a third-rank pseudotensor δ_{ijk}, which in one particular coordinate system is equal to our ε_{ijk}. Then

$$\delta'_{ijk} = |a| a_{ip} a_{jq} a_{kr} \varepsilon_{pqr} \tag{3.41}$$

by definition of pseudotensor. Now

$$a_{1p} a_{2q} a_{3r} \varepsilon_{pqr} = |a| \tag{3.42}$$

by direct expansion of the determinant, showing that $\delta'_{123} = |a|^2 = 1 = \varepsilon_{123}$. Considering the other possibilities one by one, we find

$$\delta'_{ijk} = \varepsilon_{ijk} \tag{3.43}$$

for rotations and reflections. Hence our ε_{ijk} is a pseudotensor.[1] More than that, it is seen to be an isotropic pseudotensor with the same components in all rotated cartesian coordinate systems.

With any *antisymmetric* second-rank tensor C_{ij} (in three-dimensional space) we may associate a dual pseudovector C_i defined by

[1] The usefulness of ε_{ijk} extends far beyond this section. For instance, the matrices $\mathbf{M}_k$ of Ex. 4.2.15 were derived from $(\mathbf{M}_k)_{ij} = i\varepsilon_{ijk}$.

$$C_i = \tfrac{1}{2}\varepsilon_{ijk}C_{jk}. \tag{3.44}$$

Here the antisymmetric C_{jk} may be written

$$C_{jk} = \begin{pmatrix} 0 & C_{12} & -C_{31} \\ -C_{12} & 0 & C_{23} \\ C_{31} & -C_{23} & 0 \end{pmatrix}. \tag{3.45}$$

We know that C_i must transform as a vector under rotations from the double contraction of the fifth-rank (pseudo) tensor $\varepsilon_{ijk}C_{mn}$ but that it is really a pseudovector from the pseudo nature of ε_{ijk}. Specifically, the components of **C** are given by

$$(C_1, C_2, C_3) = (C_{23}, C_{31}, C_{12}). \tag{3.46}$$

Notice the cyclic order of the indices that comes from the cyclic order of the components of ε_{ijk}. This duality, given by Eq. 3.46, means that our three-dimensional vector product may literally be taken to be either a pseudovector or an antisymmetric second-rank tensor, depending on how we chose to write it out.

If we take three (polar) vectors **A**, **B**, and **C**, we may define

$$V_{ijk} = \begin{vmatrix} A_i & B_i & C_i \\ A_j & B_j & C_j \\ A_k & B_k & C_k \end{vmatrix} = A_iB_jC_k - A_iB_kC_j + \cdots \tag{3.47}$$

By an extension of the analysis of Section 3.1 each term $A_pB_qC_r$ is seen to be a third-rank tensor, making V_{ijk} a tensor of third rank. From its definition as a determinant it is totally antisymmetric, reversing sign under the interchange of any two indices, that is, the interchange of any two rows of the determinant. The dual quantity is

$$V = \frac{1}{3!}\varepsilon_{ijk}V_{ijk} \tag{3.48}$$

clearly a pseudoscalar. By expansion it is seen that

$$V = \begin{vmatrix} A_1 & B_1 & C_1 \\ A_2 & B_2 & C_2 \\ A_3 & B_3 & C_3 \end{vmatrix}, \tag{3.49}$$

our familiar triple scalar product.

For use in showing the covariance of Maxwell's equations, Section 3.7, we want to extend this dual vector analysis to four-dimensional space and, in particular, to indicate that the four-dimensional volume element, $dx_1\,dx_2\,dx_3\,dx_4$, is a pseudoscalar.

We introduce the Levi-Civita symbol ε_{ijkl}, the four-dimensional analog of ε_{ijk}. This quantity ε_{ijkl} is defined as totally antisymmetric in all four indices. If $(ijkl)$ is an even permutation of (1, 2, 3, 4), then ε_{ijkl} is defined as $+1$; if it is an odd

permutation, then ε_{ijkl} is -1. The Levi-Civita ε_{ijkl} may be proved a pseudotensor of rank 4 by analysis similar to that used for establishing the nature of ε_{ijk}. Introducing a fourth-rank tensor,

$$H_{ijkl} = \begin{vmatrix} A_i & B_i & C_i & D_i \\ A_j & B_j & C_j & D_j \\ A_k & B_k & C_k & D_k \\ A_l & B_l & C_l & D_l \end{vmatrix}, \tag{3.50}$$

built from the polar vectors **A**, **B**, **C**, and **D**, we may define the dual quantity

$$H = \frac{1}{4!}\, \varepsilon_{ijkl} H_{ijkl}. \tag{3.51}$$

We actually have a quadruple summation which reduces the rank to zero. From the pseudo nature of ε_{ijkl}, H is a pseudoscalar. Now we let **A**, **B**, **C**, and **D** be infinitesimal displacements along the four coordinate axes (Minkowski space),

$$\begin{aligned} \mathbf{A} &= (dx_1, 0, 0, 0) \\ \mathbf{B} &= (0, dx_2, 0, 0), \text{ etc.,} \end{aligned} \tag{3.52}$$

and

$$H = dx_1\, dx_2\, dx_3\, dx_4, \tag{3.53}$$

The four-dimensional volume element is now identified as a pseudoscalar. We use this result in Section 3.7. This result could have been expected from the results of the special theory of relativity. The Lorentz-Fitzgerald contraction of $dx_1\, dx_2\, dx_3$ just balances the time dilation of dx_4.

We slipped into this four dimensional space as a simple mathematical extension of the three-dimensional space and, indeed, we could just as easily have discussed 5-, 6-, or N-dimensional space. This is typical of the power of the component analysis. Physically, this four-dimensional space is usually taken as Minkowski space,

$$(x_1, x_2, x_3, x_4) = (x, y, z, ict), \tag{3.54}$$

where t is time. This is the merger of space and time achieved in special relativity. The transformations that describe the rotations in four-dimensional space are the Lorentz transformations of special relativity.

For some applications, particularly in the quantum theory of angular momentum, our cartesian tensors are not particularly convenient. In mathematical language our general second-rank tensor A_{ij} is reducible, which means that it can be decomposed into parts of lower tensor rank. Actually, we have already done this. From Eq. 3.25

$$A = A_{ii} \tag{3.55}$$

is a scalar quantity, the trace of A_{ij}.[1]

The antisymmetric portion

$$B_{ij} = \tfrac{1}{2}(A_{ij} - A_{ji}) \tag{3.56}$$

has just been shown to be equivalent to a (pseudo) vector, or

$$B_{ij} = C_k \qquad \text{cyclic permutation of } i,\ j,\ k. \tag{3.57}$$

By subtracting the scalar A and the vector C_k from our original tensor, we have an irreducible, symmetric, zero-trace second-rank tensor, S_{ij}, in which

$$S_{ij} = \tfrac{1}{2}(A_{ij} + A_{ji}) - \tfrac{1}{3}A\delta_{ij}, \tag{3.58}$$

with five independent components. Then, finally, our original cartesian tensor may be written

$$A_{ij} = \tfrac{1}{3}A\delta_{ij} + C_k + S_{ij}. \tag{3.59}$$

The three quantities A, C_k, and S_{ij} form spherical tensors of rank 0, 1, and 2, respectively, transforming like the spherical harmonics Y_L^M (Chapter 12) for $L = 0$, 1, and 2. Further details of such spherical tensors and their uses will be found in the book by Rose, cited in Chapter 12.

A specific example of the above reduction is furnished by the symmetric electric quadrupole tensor

$$Q_{ij} = \int (3x_i x_j - r^2\delta_{ij})\rho(x_1, x_2, x_3)\, d^3x.$$

The $-r^2\,\delta_{ij}$ term represents a subtraction of the scalar trace (the three $i = j$ terms). The resulting Q_{ij} has zero trace.

EXERCISES

3.4.1 An antisymmetric square array is given by

$$\begin{pmatrix} 0 & C_3 & -C_2 \\ -C_3 & 0 & C_1 \\ C_2 & -C_1 & 0 \end{pmatrix} = \begin{pmatrix} 0 & C_{12} & C_{13} \\ -C_{12} & 0 & C_{23} \\ -C_{13} & -C_{23} & 0 \end{pmatrix},$$

where (C_1, C_2, C_3) form a pseudovector. Assuming that the relation

$$C_i = \frac{1}{2!}\varepsilon_{ijk}C_{jk}$$

holds in all coordinate systems, prove that C_{jk} is a tensor. (This is another form of the quotient theorem).

[1] An alternate approach, using matrices, is given in Section 4.3. Cf. Exercise 4.3.8.

3.4.2 Show that the vector product is unique to three dimensional space, that is, that only in three dimensions can we establish a one-to-one correspondence between the components of an antisymmetric tensor (second-rank) and the components of a vector.

3.4.3 Show that
(a) $\delta_{ii} = 3$,
(b) $\delta_{ij}\varepsilon_{ijk} = 0$,
(c) $\varepsilon_{ipq}\varepsilon_{jpq} = 2\delta_{ij}$,
(d) $\varepsilon_{ijk}\varepsilon_{ijk} = 6$.

3.4.4 Show that

$$\varepsilon_{ijk}\varepsilon_{pqk} = \delta_{ip}\delta_{jq} - \delta_{iq}\delta_{jp} .$$

3.4.5 (a) Express the components of a cross product vector **C**, $\mathbf{C} = \mathbf{A} \times \mathbf{B}$, in terms of ε_{ijk} and the components of **A** and **B**.
(b) Use the antisymmetry of ε_{ijk} to show that $\mathbf{A} \cdot \mathbf{A} \times \mathbf{B} = 0$.

Ans. (a) $C_i = \varepsilon_{ijk} A_j B_k$.

3.4.6 (a) Show that the inertia tensor (matrix) of Section 4.6 may be written

$$I_{ij} = m(x_n x_n \delta_{ij} - x_i x_j)$$

for a particle of mass m at (x_1, x_2, x_3).
(b) Show that

$$I_{ij} = -M_{il} M_{lj} = -m\varepsilon_{ilk} x_k \varepsilon_{ljm} x_m$$

where $M_{il} = m^{1/2}\, \varepsilon_{ilk} x_k$. This is the contraction of two second-rank tensors and is identical with the matrix product of Section 4.2.

3.4.7 Write $\nabla \cdot \nabla \times \mathbf{A}$ and $\nabla \times \nabla\varphi$ in ε_{ijk} notation, so that it becomes obvious that each expression vanishes.

Ans.
$$\nabla \cdot \nabla \times \mathbf{A} = \varepsilon_{ijk} \frac{\partial}{\partial x_i} \frac{\partial}{\partial x_j} A_k$$

$$(\nabla \times \nabla\varphi)_i = \varepsilon_{ijk} \frac{\partial}{\partial x_j} \frac{\partial}{\partial x_k} \varphi.$$

3.4.8 Expressing cross products in terms of Levi-Civita symbols (ε_{ijk}), derive the *BAC-CAB* rule, Eq. 1.50.

3.4.9 Verify that each of the following fourth-rank tensors is isotropic, that is, that it has the same form independent of any rotation of the coordinate systems.
(a) $A_{ijkl} = \delta_{ij}\delta_{kl}$,
(b) $B_{ijkl} = \delta_{ik}\delta_{jl} + \delta_{il}\delta_{jk}$,
(c) $C_{ijkl} = \delta_{ik}\delta_{jl} - \delta_{il}\delta_{jk}$.

3.5 Dyadics

Occasionally, particularly in the older literature and older textbooks, the reader will see references to dyads or dyadics. The dyadic is a somewhat clumsy device for extending ordinary vector analysis to cover tensors of second rank.

If we adjoin two vectors **i** and **j** to form the combination **ij**, we have a dyad. Multiplication (scalar or vector) from the left involves the left-hand member of the pair and leaves the right-hand member strictly alone:

$$\begin{aligned}\mathbf{A}\cdot\mathbf{ij} &= [(\mathbf{i}A_x + \mathbf{j}A_y + \mathbf{k}A_z)\cdot\mathbf{i}]\mathbf{j}\\ &= A_x\mathbf{j}.\end{aligned} \tag{3.60}$$

Multiplication from the right is just the reverse; that is,

$$\begin{aligned}\mathbf{ij}\cdot\mathbf{A} &= \mathbf{i}[\mathbf{j}\cdot(\mathbf{i}A_x + \mathbf{j}A_y + \mathbf{k}A_z)]\\ &= \mathbf{i}A_y.\end{aligned} \tag{3.61}$$

From this we see that, in general, the operation of multiplication is noncommutative. It must be emphasized strongly that the **i** and **j** of the dyad **ij** are not operating on each other. If they had scalar coefficients, these would be multiplied together, but as far as the unit vectors are concerned there is no dot or cross product involved; they are just sitting there. As just shown, the order is significant $\mathbf{ij} \neq \mathbf{ji}$. We thus have a composite quantity which depends in part on the ordering. This dependence on ordering will reappear when we study matrices (Chapter 4) and complex quantities (Chapter 6), the complex number being literally an ordered pair of real numbers.

Extending this construction, we adjoin two vectors **A** and **B** to form

$$\begin{aligned}\mathbf{T} = \mathbf{AB} &= (\mathbf{i}A_x + \mathbf{j}A_y + \mathbf{k}A_z)(\mathbf{i}B_x + \mathbf{j}B_y + \mathbf{k}B_z)\\ &= \mathbf{ii}A_xB_x + \mathbf{ij}A_xB_y + \mathbf{ik}A_xB_z\\ &\quad + \mathbf{ji}A_yB_x + \mathbf{jj}A_yB_y + \mathbf{jk}A_yB_z\\ &\quad + \mathbf{ki}A_zB_x + \mathbf{kj}A_zB_y + \mathbf{kk}A_zB_z.\end{aligned} \tag{3.62}$$

The quantity $\mathbf{T} = \mathbf{AB}$ is a dyadic formed as shown from a combination of dyads. We have proved (Section 3.2) that this product of two vectors **AB** is a tensor of second rank. Hence, dyadics are tensors of second rank, written in a form that preserves the vector nature but obscures the tensor transformation properties.

It has already been noted that the multiplication of a vector and a dyadic is not commutative, but there is an important special case in which the operation is commutative. We take the dyadic **AB** and set

$$\mathbf{a}\cdot\mathbf{AB} = \mathbf{AB}\cdot\mathbf{a}, \tag{3.63}$$

where **a** is an arbitrary vector. If $\mathbf{a} = \mathbf{i}$, then

$$\begin{gathered}A_x\mathbf{B} = \mathbf{A}B_x\\ \mathbf{i}A_xB_x + \mathbf{j}A_xB_y + \mathbf{k}A_xB_z = \mathbf{i}A_xB_x + \mathbf{j}A_yB_x + \mathbf{k}A_zB_x.\end{gathered} \tag{3.64}$$

By equating components we obtain

$$\begin{aligned}A_xB_x &= A_xB_x,\\ A_xB_y &= A_yB_x,\\ A_xB_z &= A_zB_x,\end{aligned} \tag{3.65}$$

showing that $\mathbf{A} = c\mathbf{B}$, in which c is a constant. In other words, if multiplication with an arbitrary vector is commutative, the dyadic must be symmetric, and the coefficient of dyad **pq** equals the coefficient of dyad **qp**. Conversely, if the dyadic is symmetric, multiplication is commutative.

One of the most significant properties of a symmetric dyadic is that it can always be put in normal or diagonal form by proper choice of the coordinate axes:

$$\mathbf{T} \to \mathbf{ii}T_{xx} + \mathbf{jj}T_{yy} + \mathbf{kk}T_{zz}, \tag{3.66}$$

all the nondiagonal coefficients going to zero. The coordinate transformation that puts our dyadic in this diagonal form is known as the principal axis transformation. It is discussed at some length in Section 4.6.

There is an interesting and useful geometric interpretation of a symmetric dyadic. For simplicity let us assume that our symmetric dyadic **T** is already in its diagonal form. Then with **r**, the usual distance vector, we form the equation

$$\mathbf{r} \cdot \mathbf{T} \cdot \mathbf{r} = 1, \tag{3.67}$$

which limits the length of **r** according to its orientation. By expanding Eq. 3.67

$$(\mathbf{i}x + \mathbf{j}y + \mathbf{k}z) \cdot (\mathbf{ii}T_{xx} + \mathbf{jj}T_{yy} + \mathbf{kk}T_{zz}) \cdot (\mathbf{i}x + \mathbf{j}y + \mathbf{k}z) = 1$$

$$x^2T_{xx} + y^2T_{yy} + z^2T_{zz} = 1. \tag{3.68}$$

This defines an ellipsoid with semiaxes a, b, and c, given by

$$a = T_{xx}^{-1/2}, \qquad b = T_{yy}^{-1/2}, \qquad c = T_{zz}^{-1/2}. \tag{3.69}$$

Diagonalizing our dyadic corresponded to orienting the dyadic ellipsoid so that the ellipsoid axes were lined up with the coordinate axes.

If **U** is an antisymmetric dyadic,

$$U_{xx} = 0, \text{ etc.},$$

$$U_{xy} = -U_{yx}, \text{ etc.},$$

then for any vector **a**

$$\mathbf{a} \cdot \mathbf{U} = -\mathbf{U} \cdot \mathbf{a}. \tag{3.70}$$

Multiplication of a vector and an antisymmetric dyadic follows an anticommutation rule (cf. Exercise 3.5.4a).

Dyadics are rather awkward to handle in comparison with the usual tensor analysis (once the concept of transformation under coordinate rotation has been absorbed). They are quite unwieldy for representing third- or higher-rank tensors, so we shall return to tensor analysis and have nothing further to do with dyadic notation.

EXERCISES

3.5.1 If **A** and **B** transform as vectors, Eqs. 3.6 and 3.8, show that the dyadic **AB** satisfies the *tensor* transformation law, Eq. 3.13.

3.5.2 Show that $\mathbf{I} = \mathbf{ii} + \mathbf{jj} + \mathbf{kk}$ is a *unit dyadic* in the sense that for any vector **V**,

$$\mathbf{I} \cdot \mathbf{V} = \mathbf{V}.$$

The individual dyads **ii**, etc., are specific examples of the projection operators of quantum mechanics.

3.5.3 Show that $\nabla\mathbf{r}$ is equal to the unit dyadic, **I**.

3.5.4 If **U** is an antisymmetric dyadic and **V** a vector, show that
(a) $\mathbf{V} \cdot \mathbf{U} = -\mathbf{U} \cdot \mathbf{V}$
(b) $\mathbf{V} \cdot \mathbf{U} \cdot \mathbf{V} = 0.$

3.5.5 The two-dimensional vectors $\mathbf{r} = \mathbf{i}x + \mathbf{j}y$ and $\mathbf{t} = -y\mathbf{i} + \mathbf{j}x$ may be related by the tensor equation $\mathbf{r} \cdot \mathbf{U} = \mathbf{t}$.
(a) Find the tensor **U**, using our earlier component description of tensors.
(b) Find **U** and treat it as a dyadic.

3.5.6 In an investigation of the interaction of molecules a dyadic is formed from the unit relative distance vectors $\mathbf{e}_{12}$ given by

$$\mathbf{e}_{12} = \frac{\mathbf{r}_2 - \mathbf{r}_1}{|\mathbf{r}_2 - \mathbf{r}_1|}$$

For

$$\mathbf{U} = \mathbf{I} - 3\mathbf{e}_{12}\mathbf{e}_{12}$$

show that trace $\mathbf{U} \cdot \mathbf{U} = 6$.

I is the unit dyadic, that is $\mathbf{I} = \mathbf{ii} + \mathbf{jj} + \mathbf{kk}$.

3.5.7 Show that Gauss's theorem holds for dyadics, that

$$\int_s d\boldsymbol{\sigma} \cdot \mathbf{D} = \int_v \nabla \cdot \mathbf{D} \, d\tau.$$

3.5.8 Show that

$$\int_s d\boldsymbol{\sigma} \cdot \nabla\mathbf{E} + \int_s d\boldsymbol{\sigma} \times (\nabla \times \mathbf{E}) - \int_s d\boldsymbol{\sigma}\, \nabla \cdot \mathbf{E} = 0.$$

The function **E** is a vector function of position. The integration is over a simple closed surface. This improbable combination of surface integrals actually appears in the vector Kirchhoff diffraction theory.

3.5.9 Show that the following zero trace, symmetric, unit tensors

$$\mathbf{t}^0 = (2\mathbf{kk} - \mathbf{ii} - \mathbf{jj})/\sqrt{6}$$
$$\mathbf{t}^{\pm 1} = \mp\tfrac{1}{2}(\mathbf{ik} + \mathbf{ki}) - i\tfrac{1}{2}(\mathbf{jk} + \mathbf{kj})$$
$$\mathbf{t}^{\pm 2} = \tfrac{1}{2}(\mathbf{ii} - \mathbf{jj}) \pm i\,\tfrac{1}{2}(\mathbf{ij} + \mathbf{ji})$$

satisfy the double contraction relation

$$t_{ij}^{m*}\, t_{ij}^{n} = \delta_{mn}.$$

These unit tensors are used in defining *tensor* spherical harmonics, an extension of the vector spherical harmonics of Section 12.12. The tensor spherical harmonics, in turn, are helpful in describing gravitational waves.

3.6 Theory of Elasticity

When an elastic body is subjected to an external force or stress, it becomes deformed or strained. Our study of elasticity in terms of tensors falls naturally into three parts. First, a description of the strain or deformation of the elastic substance, second, a description of the force or stress that produces the deformation, and finally a generalized Hooke's law in tensor form, relating stress and strain.

Elastic strain: deformation. The deformation of our elastic body may be described by giving the change in relative position of the parts of the body when the body is subjected to some external stress. Consider a point P_0 at position $\mathbf{r}$ relative to some fixed origin and a second point Q_0 displaced from P_0 by a distance $\delta\mathbf{x}$. In the unstrained state the coordinates of Q_0 *relative to* P_0 are δx_i; in the strained state, when P_0 has been displaced a distance $\mathbf{u}$ to point P_1 and Q_0 a distance $\mathbf{v}$ to Q_1, the coordinates of Q_1 relative to P_1 are $\delta y_i = \delta x_i + \delta u_i$. The change in position

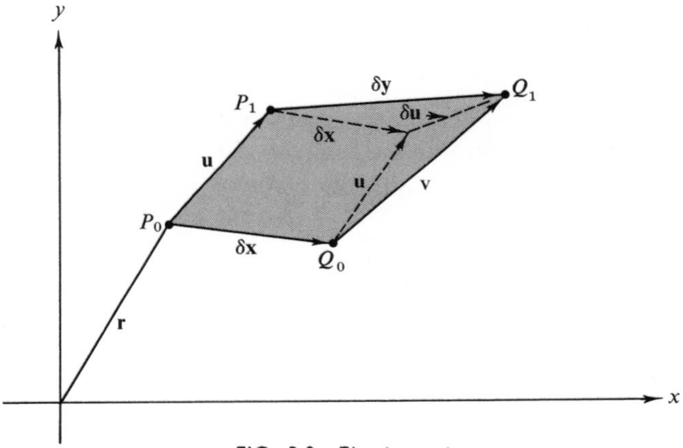

FIG. 3.3 Elastic strain

of Q *relative to* P is just δu_i. Neglecting second- and higher-order differentials,[1] a three-dimensional Taylor expansion yields

$$\delta\mathbf{u} = \mathbf{u}(\mathbf{r} + \delta\mathbf{x}) - \mathbf{u}(\mathbf{r}) = (\delta\mathbf{x} \cdot \nabla)\mathbf{u},$$

$$\delta u_i = \frac{\partial u_i}{\partial x_k}\,\delta x_k. \tag{3.71}$$

Since u_i is the component of a vector $\partial u_i/\partial x_k$ is an element of a second-rank tensor, $\nabla\mathbf{u}$. Resolving this tensor into symmetric and antisymmetric terms,

$$\delta u_i = \frac{1}{2}\left(\frac{\partial u_i}{\partial x_k} + \frac{\partial u_k}{\partial x_i}\right)\delta x_k - \frac{1}{2}\left(\frac{\partial u_k}{\partial x_i} - \frac{\partial u_i}{\partial x_k}\right)\delta x_k = \eta_{ik}\,\delta x_k - \xi_{ik}\,\delta x_k. \tag{3.72}$$

[1] The limitation to first-order terms is a rather severe limitation implying relative strains of no more than perhaps one percent in actual application.

The antisymmetric part, ξ_{ik}, may be identified as a pure rotation (and not a deformation at all).

From Section 3.4 we may associate an axial vector $\boldsymbol{\xi}$ with ξ_{ij}:

$$\boldsymbol{\xi} = \tfrac{1}{2}\nabla \times \mathbf{u}. \tag{3.73}$$

The displacement $\delta\mathbf{u}$ corresponding to the antisymmetric part $-\xi_{ik}\,\delta x_k$ becomes

$$\delta\mathbf{u} = \tfrac{1}{2}(\nabla \times \mathbf{u}) \times \delta\mathbf{x}. \tag{3.74}$$

This is a rotation about an instantaneous axis through P_0 in the direction $(\nabla \times \mathbf{u})$ through $|\nabla \times \mathbf{u}|$ radians.

The remaining symmetric part of our tensor η_{ij} is taken as a pure strain tensor. The diagonal elements $(\eta_{11}, \eta_{22}, \eta_{33})$ of η_{ik} represent stretches, whereas the nondiagonal elements represent shear strains.

This may be seen by considering Q_0 to be displaced from P_0 along the x-axis; $\delta\mathbf{x} = \mathbf{i}\,\delta x_1$. From Eq. 3.72

$$\begin{aligned} \delta u_1 &= \eta_{11}\,\delta x_1, \\ \delta u_2 &= \eta_{21}\,\delta x_1, \\ \delta u_3 &= \eta_{31}\,\delta x_1. \end{aligned} \tag{3.75}$$

Hence the displacement in the strained case is

$$\begin{aligned} \delta y_1 &= \delta x_1 + \delta u_1 = (1 + \eta_{11})\,\delta x_1, \\ \delta y_2 &= \delta u_2 = \eta_{21}\,\delta x_1, \\ \delta y_3 &= \delta u_3 = \eta_{31}\,\delta x_1. \end{aligned} \tag{3.76}$$

For our initial displacement, $\delta\mathbf{x} = \mathbf{i}\,\delta x_1$, the diagonal term η_{11} contributes to the 1 component of $\delta\mathbf{y}$ (stretching) and η_{21} and η_{31} contribute to δy_2 and δy_3, respectively, representing shears.

Stress-force. The stresses or forces must be defined carefully. Referring to Fig. 3.4, which shows a differential volume, the force in the x_i-direction acting on

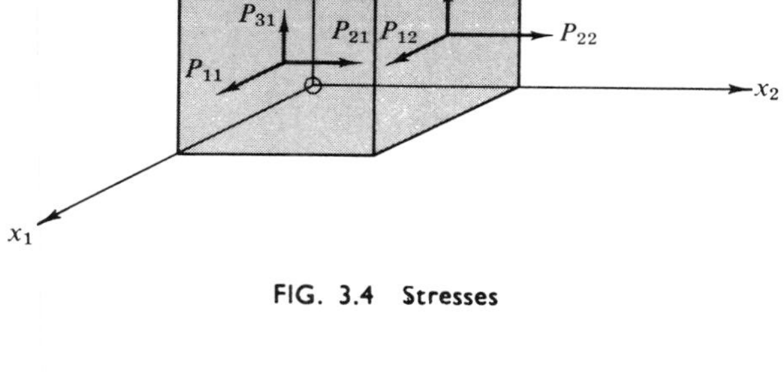

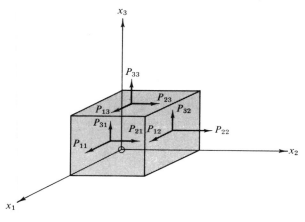

FIG. 3.4 Stresses

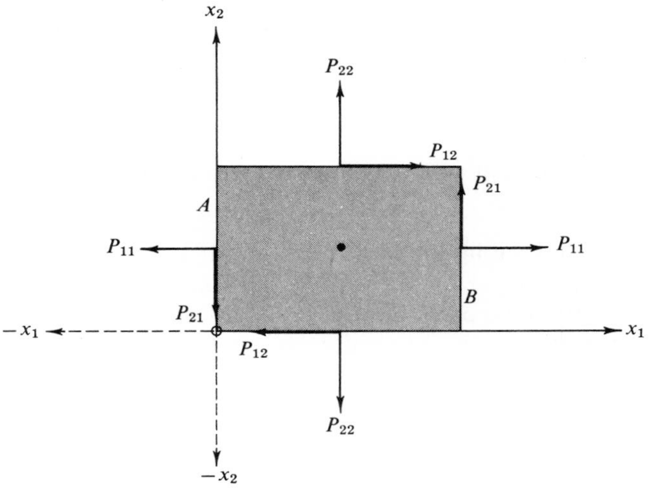

FIG. 3.5 Homogeneous stresses—sign reversal

the surface dA whose normal is in the x_j-direction is $P_{ij}\,dA$. The P_{ij}'s themselves are actually "pressures" in the sense of being force/area. Whenever the terms stress or force are used, it is understood that the P_{ij} are to be multiplied by the appropriate differential area. These are the forces acting *on* the small parallelepiped of Fig. 3.4. For clarity only the forces on the front three faces are shown. If we assume that the stresses are homogeneous, the forces on the opposite faces will be reversed in sign as shown in Fig. 3.5. Note that P_{21} is the shear (in the x_2-direction) applied to face A. For a homogeneous force face B must apply the same shearing stress P_{21} *to the outside medium.* The stress applied to face B by the outside medium is just the reverse, or P_{21} directed downward (in the $-x_2$-direction). Built into this argument are three assumptions that should be noted explicitly.

1. Homogeneous stress throughout the body.
2. Existence of static equilibrium.
3. Absence of body forces (such as gravity acting on the mass within the parallelepiped) and body torques (external magnetic field acting on magnetic domains).

These assumptions permit placement of a further restriction on the P_{ij}'s. Consider the net torque on the parallelepiped shown in Figs. 3.4 and 3.6 about the x_3-axis. The normal pressures P_{ii} exert no net torque. The shearing stresses P_{31} and P_{32} have zero moment arm. The shearing stresses P_{13} and P_{23} are balanced by equal and opposite stresses on the bottom face ($x_3 = 0$). Remaining are the torques

$$P_{21}(dx_2\,dx_3)\,dx_1$$

and

$$P_{12}(dx_1\,dx_3)\,dx_2, \tag{3.77}$$

which must balance,

$$P_{21}\,dx_1\,dx_2\,dx_3 = P_{12}\,dx_1\,dx_2\,dx_3$$

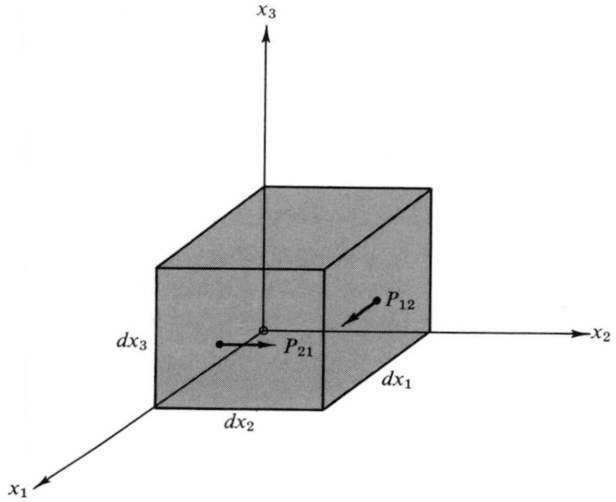

FIG. 3.6 Homogeneous stresses—balance of torques

in the absence of rotation about the x_3-axis. We have

$$P_{21} = P_{12} \tag{3.78}$$

and, in general, repeating the argument for the absence of rotation about x_2 and x_3,

$$P_{ij} = P_{ji}. \tag{3.79}$$

These are equalities of magnitude, not direction, which is given by the first index.

Thus the array of stresses (pressure) P_{ij} is symmetric. Now we show that this

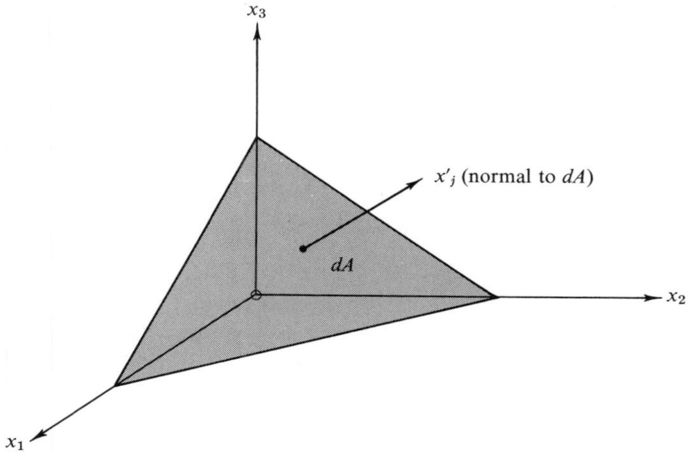

FIG. 3.7 Differential tetrahedron—balance of forces

array is a tensor. We form an infinitesimal tetrahedron with a slant face of area dA, normal in the x'_j-direction, as shown in Fig. 3.7. The forces on the slant face are $P'_{ij}\,dA$. The forces on the faces $x_1 = 0$, $x_2 = 0$, and $x_3 = 0$ are, respectively,

$$P_{m1}(a_{j1}\,dA),$$

$$P_{m2}(a_{j2}\,dA),$$

and

$$P_{m3}(a_{j3}\,dA),$$

where $a_{jk}\,dA$ is the area of the face $x_k = 0$, being given by the slant area dA projected onto the plane $x_k = 0$. Our a_{jk} is the usual direction cosine, the cosine of the angle between the x'_j- and the x_k-axes.

The force $P_{m1}a_{j1}\,dA$ is along x_m. Its component in the x'_i-direction is $a_{im}a_{j1}P_{m1}\,dA$ (no summation). If we now sum over m, the foregoing expression gives us the sum of the x'_i-components of the three forces on the back face, $x_1 = 0$, in the x'_i-direction. Finally, summing over all three faces, $x_k = 0$, the total force along x'_i is

$$a_{im}a_{jk}P_{mk}\,dA = P'_{ij}\,dA \tag{3.80}$$

for static equilibrium. Since the area dA is arbitrary, we have

$$P'_{ij} = a_{im}a_{jk}P_{mk}, \tag{3.81}$$

which by definition makes P_{mk} a tensor.

Stress-strain relations: Hooke's law. First, let us assume an isotropic elastic solid. Later we shall return to the general anisotropic case. Consider a uniform rod parallel to the x_1-axis.[1] By applying a small tensile stress P_{11},

$$E\eta_{11} = P_{11}, \tag{3.82a}$$

$$E\eta_{22} = -\sigma P_{11}, \tag{3.82b}$$

$$E\eta_{33} = -\sigma P_{11}, \tag{3.82c}$$

where E is Young's modulus and, σ is Poisson's ratio. If we add small tensile stresses P_{22} and P_{33}, Eq. 3.82a becomes

$$E\eta_{11} = P_{11} - \sigma P_{22} - \sigma P_{33}. \tag{3.83}$$

All through here we are limiting ourselves to *small* stresses and *small* strains so that the stress–strain relation will be linear. Eq. 3.83 may be rewritten

$$E\eta_{11} = (1 + \sigma)P_{11} - \sigma(P_{11} + P_{22} + P_{33}), \tag{3.84}$$

and similarly for $E\eta_{22}$ and $E\eta_{33}$.

Now η_{ij} and P_{ij} are tensors as proved earlier in this section. Because of the symmetry of our system their nondiagonal components are zero. To find the generalization of Eq. 3.84 in an arbitrarily oriented cartesian coordinate system we rotate the axes

[1] This special choice will start us off in a system identified in Chapter 4 as the principal axis system, the particular coordinate system in which the shearing strains vanish.

$$\eta'_{ij} = a_{ik}a_{jk}\eta_{kk},$$
$$P'_{ij} = a_{ik}a_{jk}P_{kk}. \tag{3.85}$$

If we multiply Eq. 3.84 by $a_{i1}a_{j1}$, the corresponding equation for $E\eta_{22}$ by $a_{i2}a_{j2}$, and the equation for $E\eta_{33}$ by $a_{i3}a_{j3}$ and add all three equations, we obtain

$$Ea_{ik}a_{jk}\eta_{kk} = (1+\sigma)a_{ik}a_{jk}P_{kk} - \sigma(P_{nn})a_{ik}a_{jk}. \tag{3.86}$$

Using Eqs. 3.85 and 3.18,

$$E\eta'_{ij} = (1+\sigma)P'_{ij} - \sigma(P'_{mm})\,\delta_{ij}, \tag{3.87}$$

where

$$(P'_{mm}) = (P_{nn}) = P_{11} + P_{22} + P_{33}, \tag{3.88}$$

the contracted (and, therefore invariant) tensor P_{ij}.

It is frequently more convenient to solve for the stresses P_{ik}. We may do this by setting $i = j$ and contracting

$$E\eta_{jj} = (1+\sigma)P_{jj} - 3\sigma P_{jj}$$
$$= (1-2\sigma)P_{jj}, \tag{3.89}$$

dropping the primes as superfluous. Substituting back into Eq. 3.87,

$$(1+\tau)P_{ij} = E\eta_{ij} + \frac{E\sigma}{1-2\sigma}\eta_{mm}\delta_{ij} \tag{3.90}$$

or

$$P_{ij} = 2\mu\eta_{ij} + \lambda\eta_{mm}\delta_{ij}, \tag{3.91}$$

where λ and μ, known as Lamé's constants, are given by

$$\lambda = \frac{\sigma E}{(1+\sigma)(1-2\sigma)}, \qquad \mu = \frac{E}{2(1+\sigma)}. \tag{3.92}$$

The constant μ may be identified as the rigidity or shear modulus. Consider a parallelepiped fixed to the $(x_3 = 0)$-plane with a tangential stress P_{12} applied. The displacement is $(\eta x_2, 0, 0)$. In terms of our strain tensor, $\eta_{ij} = 0$ except for $\eta_{12} = \eta_{21} = \frac{1}{2}\eta$. From Eq. 3.91

$$P_{12} = 2\mu \cdot \tfrac{1}{2}\eta = \mu\eta, \tag{3.93}$$

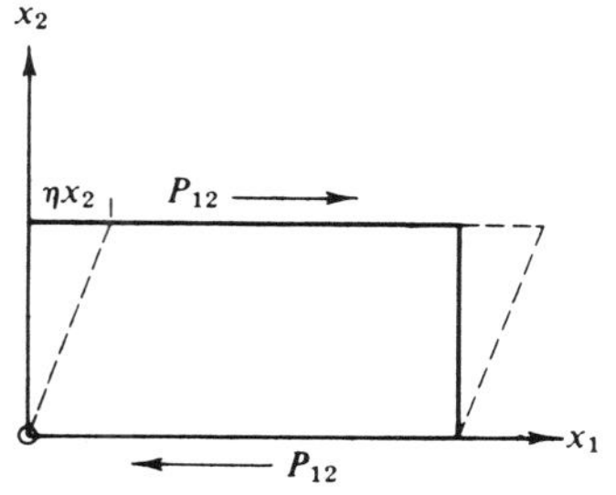

FIG. 3.8 Shear stress—shear strain

showing that μ is the ratio of shear stress to shear strain η.

If the strain is spherically symmetric, as in hydrostatic pressure,

$$\eta_{11} = \eta_{22} = \eta_{33}; \qquad \eta_{12} = \eta_{13} = \eta_{23} = 0. \tag{3.94}$$

Then

$$\begin{aligned} P_{11} &= 3\lambda\eta_{11} + 2\mu\eta_{11} \\ &= 3k\eta_{11}, \end{aligned} \tag{3.95}$$

where

$$k = \lambda + \tfrac{2}{3}\mu. \tag{3.96}$$

Since $3\eta_{11}$ is the relative change in volume to first-order, we identify k as the bulk modulus.

For the general case covering anisotropic as well as isotropic solids we express the linear stress-strain relation by a generalized Hooke's law.

$$P_{ij} = c_{ijkl}\eta_{kl}, \tag{3.97}$$

where c_{ijkl} is a fourth-rank tensor by the quotient theorem. Since the stress tensor P_{ij} is symmetric and the strain tensor η_{kl} is likewise symmetric,

$$c_{ijkl} = c_{jikl} = c_{ijlk} = c_{jilk}, \tag{3.98}$$

reducing the number of components from 81 (3^4) to 36. It may further be shown[1] that

$$c_{ijkl} = c_{klij}, \tag{3.99}$$

which further reduces the number of independent components to 21.

If we apply our general tensor relation (Eq. 3.97) to an isotropic body, the elastic constant tensor c_{ijkl} must be a linear combination of the most general isotropic tensors of fourth rank. Using the results of Exercise 3.4.9, we have

$$c_{ijkl} = a\delta_{ij}\delta_{kl} + b[\delta_{ik}\delta_{jl} + \delta_{il}\delta_{jk}] + c[\delta_{ik}\delta_{jl} - \delta_{il}\delta_{jk}]. \tag{3.100}$$

By substituting into Eq. 3.97,

$$P_{ij} = a\delta_{ij}\eta_{kk} + b(\eta_{ij} + \eta_{ji}) + c(\eta_{ij} - \eta_{ji}) \tag{3.101}$$

as before. Since η_{ij} is symmetric, this reduces to

$$P_{ij} = a\eta_{kk}\delta_{ij} + 2b\eta_{ij}$$

in complete agreement with Eq. 3.91.

EXERCISES

3.6.1 The three-dimensional fourth-rank stress-strain tensor c_{ijkl} satisfies

$$c_{ijkl} = c_{ijlk} = c_{jilk}.$$

[1] Cf. I. S. Sokolnikoff, *Mathematical Theory of Elasticity*. New York: McGraw-Hill (1956).

(a) Show that application of these symmetry conditions reduces the number of independent components or elements of c_{ijkl} from 81 to 36.
(b) If we further specify that

$$c_{ijkl} = c_{klij},$$

show that the number of independent components drops to 21.

3.6.2 (a) What arguments can you adduce to demonstrate that Poisson's ratio σ is nonnegative?
(b) Assuming that the shear modulus μ and the bulk modulus k are each nonnegative, set an upper limit to the value of Poisson's ratio.

Ans. (b) $\sigma < 1/2$.

3.6.3 Calculate the elastic potential energy (per unit volume) of an elastic isotropic body subjected to a small strain.

$$\textit{Ans.}\ \frac{W}{V} = \tfrac{1}{2}\lambda(\eta_{ii})^2 + \mu\eta_{ij}\eta_{ij}.$$

3.6.4 The potential energy of a strained elastic solid is given by

$$\text{P.E.} = \tfrac{1}{2}c_{ijkl}\eta_{ij}\eta_{kl}.$$

If the solid has *cubic* symmetry,
(a) Show that any c_{ijkl}, in which any subscript (1, 2, 3 or x, y, z) appears an *odd* number of times, vanishes, i.e.,

$$c_{1112} = 0$$

Hint. Reflect the coordinate appearing an odd number of times.
(b) Show that there remain three distinct nonzero elastic constants

$$c_{1111} = c_{2222} = c_{3333}$$

$$c_{1122} = c_{2211} = c_{1133}, \text{ etc.}$$

$$c_{1212} = c_{2121} = c_{1313}, \text{ etc.}$$

for a total of 15 elements.

3.6.5 If our elastic solid is *isotropic*, c_{ijkl} will have 15 nonvanishing components. Express these 15 components in terms of Young's modulus E and Poisson's ratio σ.

$$\textit{Ans.}\ c_{1111} = E\frac{(1-\sigma)}{(1+\sigma)(1-2\sigma)}.$$

$$c_{1122} = E\frac{\sigma}{(1+\sigma)(1-2\sigma)} = \lambda.$$

$$c_{1212} = E\frac{1}{2(1+\sigma)} = \mu.$$

3.6.6 The original strain tensor $\partial u_i/\partial x_k$ is reducible in the sense of Section 3.4. A partial reduction, the splitting off of the antisymmetric ξ_{ik}, is carried out in the first part of Section 3.6. Completing the reduction we may write

$$\begin{pmatrix} \eta_{11} & \eta_{12} & \eta_{13} \\ \eta_{21} & \eta_{22} & \eta_{23} \\ \eta_{31} & \eta_{32} & \eta_{33} \end{pmatrix} = \begin{pmatrix} \eta/3 & 0 & 0 \\ 0 & \eta/3 & 0 \\ 0 & 0 & \eta/3 \end{pmatrix} + \begin{pmatrix} \eta_{11}-\eta/3 & \eta_{12} & \eta_{13} \\ \eta_{21} & \eta_{22}-\eta/3 & \eta_{23} \\ \eta_{31} & \eta_{32} & \eta_{33}-\eta/3 \end{pmatrix}.$$

Here η is the contracted (scalar) η_{ii}.
(a) Show that the tensor $v_{ij} = \frac{1}{3}\eta\delta_{ij}$ describes a change in volume and no change of shape.
(b) Show that the second tensor $s_{ij} = \eta_{ij} - \frac{1}{3}\eta\delta_{ij}$ describes a change in shape (shear) with no change in volume to first order.

3.6.7 (a) Derive the equation for waves in an elastic medium

$$m\, \partial^2\mathbf{u}/\partial t^2 = (k + \tfrac{4}{3}\mu)\nabla\nabla \cdot \mathbf{u} - \mu\nabla \times \nabla \times \mathbf{u}.$$

Hint. Consider the net force on a unit cube (mass m).

(b) If the displacement **u** is irrotational, show that the elastic waves are propagated with a velocity $v = [(k + \frac{4}{3}\mu)/m]^{1/2}$ and are longitudinal (plane waves or spherical waves at large distances).

(c) If the displacement **u** is solenoidal show that the elastic waves are propagated with velocity $v = (\mu/m)^{1/2}$ and are transverse (plane waves or spherical waves at large distances).

3.7 Lorentz Covariance of Maxwell's Equations

The Lorentz transformations of special relativity involving space and time may be shown to be equivalent to rotations in a special four-dimensional space.[1] This space is Minkowski space with the fourth coordinate $x_4 = ict$. Basically, the reason for the inclusion of the imaginary i is to give a space in which the four-dimensional Pythagorean theorem has the same form as the usual three-dimensional cartesian case. This may be taken as the criterion of a cartesian system. Our immediate problem is to rewrite Maxwell's equations in tensor form in this Minkowski space. If we can do this, then, because of the transformation properties of our tensors, consistency with special relativity will be automatic. We have

$$\nabla \times \mathbf{E} = -\frac{\partial \mathbf{B}}{\partial t}, \tag{3.102a}$$

$$\nabla \times \mathbf{H} = \frac{\partial \mathbf{D}}{\partial t} + \rho\mathbf{v}, \tag{3.102b}$$

$$\nabla \cdot \mathbf{D} = \rho, \tag{3.102c}$$

$$\nabla \cdot \mathbf{B} = 0, \tag{3.102d}$$

and the relations

$$\mathbf{D} = \varepsilon_0\mathbf{E}, \qquad \mathbf{B} = \mu_0\mathbf{H}. \tag{3.103}$$

The symbols have their usual meanings as given in the introduction. For simplicity we assume vacuum ($\varepsilon = \varepsilon_0$, $\mu = \mu_0$).

In terms of scalar and magnetic vector potentials, we may write

$$\begin{aligned} \mathbf{B} &= \nabla \times \mathbf{A} \\ \mathbf{E} &= -\frac{\partial \mathbf{A}}{\partial t} - \nabla\varphi. \end{aligned} \tag{3.104}$$

Equation 3.104 specifies the curl of **A**; the divergence of **A** is still undefined (cf. Sections 1.13 and 1.15). We may, and for future convenience we do, impose the

[1] H. Goldstein, *Classical Mechanics*. Cambridge, Mass.: Addison-Wesley (1951). Chapter 6. The tensor equation for a photon $\Sigma x_\lambda^2 = 0$, independent of reference frame, leads to the Lorentz transformations.

further restriction on the vector potential **A**,

$$\nabla \cdot \mathbf{A} + \varepsilon_0 \mu_0 \frac{\partial \varphi}{\partial t} = 0. \tag{3.105}$$

This is the Lorentz relation. It will serve the purpose of uncoupling the differential equations for **A** and φ below. The potentials **A** and φ are not yet completely fixed. The freedom remaining is the topic of Ex. 3.7.3.

Now we rewrite the Maxwell equations in terms of the potentials **A** and φ. From Eqs. 3.102*c* and 3.104

$$\nabla^2 \varphi + \nabla \cdot \frac{\partial \mathbf{A}}{\partial t} = -\frac{\rho}{\varepsilon_0}, \tag{3.106}$$

whereas Eqs. 3.102*b* and 3.104 and Eq. 1.80 of Chapter 1 yield

$$\frac{\partial^2 \mathbf{A}}{\partial t^2} + \nabla \frac{\partial \varphi}{\partial t} + \frac{1}{\varepsilon_0 \mu_0} \{\nabla\nabla \cdot \mathbf{A} - \nabla^2 \mathbf{A}\} = \frac{\rho \mathbf{v}}{\varepsilon_0}. \tag{3.107}$$

Using the Lorentz relation, Eq. 3.105 and the relation $\varepsilon_0 \mu_0 = 1/c^2$,

$$\begin{aligned} \left[\nabla^2 - \frac{1}{c^2}\frac{\partial^2}{\partial t^2}\right]\mathbf{A} &= -\mu_0 \rho \mathbf{v}, \\ \left[\nabla^2 - \frac{1}{c^2}\frac{\partial^2}{\partial t^2}\right]\varphi &= -\frac{\rho}{\varepsilon_0}. \end{aligned} \tag{3.108}$$

Now the differential operator

$$\nabla^2 - \frac{1}{c^2}\frac{\partial^2}{\partial t^2}$$

becomes in Minkowski space

$$\sum_{\lambda=1}^{4} \frac{\partial^2}{\partial x_\lambda^2}.$$

Here we adopt Greek indices, as is customary in relativity theory, to indicate a summation from 1 to 4. This summation,

$$\sum_\lambda \frac{\partial^2}{\partial x_\lambda^2},$$

is a four-dimensional Laplacian, usually called the d'Alembertian and denoted by $\square^2$. It may readily be proved a scalar (cf. Exercise 3.2.3).

For convenience we define

$$\begin{aligned} A_1 &\equiv \frac{A_x}{\mu_0 c} = c\varepsilon_0 A_x, & A_3 &\equiv \frac{A_z}{\mu_0 c} = c\varepsilon_0 A_z, \\ A_2 &\equiv \frac{A_y}{\mu_0 c} = c\varepsilon_0 A_y, & A_4 &\equiv i\varepsilon_0 \varphi. \end{aligned} \tag{3.109}$$

If we further put

$$\frac{\rho v_x}{c} \equiv i_1, \qquad \frac{\rho v_y}{c} \equiv i_2, \qquad \frac{\rho v_z}{c} \equiv i_3, \qquad i\rho \equiv i_4, \tag{3.110}$$

then Eq. 3.108 may be written in the form

$$\sum_{\lambda} \frac{\partial^2}{\partial x_\lambda^2} A_\mu = -i_\mu . \tag{3.111}$$

Equation 3.111 looks like a tensor equation, but looks do not constitute proof. To prove that it is a tensor equation we start by investigating the transformation properties of the generalized current i_μ.

Since an electric charge element de is an invariant quantity, we have

$$de = \rho \, dx_1 \, dx_2 \, dx_3, \qquad \text{invariant}. \tag{3.112}$$

We saw in Section 3.4 that the four-dimensional volume element, $dx_1 \, dx_2 \, dx_3 \, dx_4$ was also invariant. Comparing this result, Eq. 3.53 with Eq. 3.112, we see that the charge density ρ must transform the same way as dx_4, the fourth component of a four-dimensional vector dx_λ. We put $i\rho = i_4$, with i_4 now established as the fourth component of a vector. The other parts of Eq. 3.110 may be expanded as

$$\begin{aligned} i_1 = \frac{\rho v_x}{c} = \frac{\rho}{c}\frac{dx_1}{dt} &= \frac{i\rho}{ic}\frac{dx_1}{dt} \\ &= i_4 \frac{dx_1}{dx_4}. \end{aligned} \tag{3.113}$$

Since we have just shown that i_4 transforms as dx_4, it means that i_1 transforms as dx_1. With similar results for i_2 and i_3, we have i_λ transforming as dx_λ, proving that i_λ is a vector, a four-dimensional vector in Minkowski space.

Equation 3.111, which follows directly from Maxwell's equations, Eq. 3.102, is assumed to hold in all cartesian systems (all Lorentz frames). Then, by the quotient rule, Section 3.3, A_μ is also a vector and Eq. 3.111 is a legitimate tensor equation.

Now, working backward, Eq. 3.104 may be written

$$\begin{aligned} i\varepsilon_0 E_j &= \frac{\partial A_j}{\partial x_4} - \frac{\partial A_4}{\partial x_j}, \qquad j = 1, 2, 3, \\ \frac{1}{\mu_0 c} B_i &= \frac{\partial A_k}{\partial x_j} - \frac{\partial A_j}{\partial x_k}, \qquad (i, j, k) = (1, 2, 3) \end{aligned} \tag{3.114}$$

and *cyclic* permutations.

We define a new tensor

$$\frac{\partial A_\lambda}{\partial x_\mu} - \frac{\partial A_\mu}{\partial x_\lambda} \equiv f_{\mu\lambda} = -f_{\lambda\mu},$$

an antisymmetric second-rank tensor, since A_λ is a vector. Written out explicitly,

$$f_{\mu\lambda} = \varepsilon_0 \begin{pmatrix} 0 & cB_z & -cB_y & -iE_x \\ -cB_z & 0 & cB_x & -iE_y \\ cB_y & -cB_x & 0 & -iE_z \\ iE_x & iE_y & iE_z & 0 \end{pmatrix}. \tag{3.115}$$

With this tensor we may write two of Maxwell's equations (3.102*b*) and (3.102*c*) combined as a tensor equation:

$$\frac{\partial f_{\lambda\mu}}{\partial x_\mu} = i_\lambda . \tag{3.116}$$

The left-hand side of Eq. 3.116 is a four-dimensional divergence of a tensor and therefore a vector. This, of course, is equivalent to contracting a third-rank tensor $\partial f_{\lambda\mu}/\partial x_\nu$ (cf. Exercises 3.2.1 and 3.2.2). The remaining two Maxwell equations (3.102*a*) and (3.102*d*) may be expressed in the tensor form

$$\frac{\partial f_{23}}{\partial x_1} + \frac{\partial f_{31}}{\partial x_2} + \frac{\partial f_{12}}{\partial x_3} = 0 \tag{3.117}$$

for Eq. 3.102*d* and three equations of the form

$$\frac{\partial f_{34}}{\partial x_2} + \frac{\partial f_{42}}{\partial x_3} + \frac{\partial f_{23}}{\partial x_4} = 0 \tag{3.118}$$

for Eq. 3.102*a*. (A second equation permutes 124, a third permutes 134.) Since

$$\frac{\partial f_{\lambda\mu}}{\partial x_\nu} \equiv t_{\lambda\mu\nu}$$

is a tensor (of third rank), Eqs. 3.102*a* and 3.102*d* are given by the tensor equation

$$t_{\lambda\mu\nu} + t_{\nu\lambda\mu} + t_{\mu\nu\lambda} = 0. \tag{3.119}$$

From Eqs. 3.117 and 3.118 the reader will understand that the indices λ, μ, and ν are supposed to be different. Actually, Eq. 3.119 automatically reduces to $0 = 0$ if any two indices coincide.

The establishment of the tensor equations (3.116) and (3.119) actually completes the demonstration of the Lorentz covariance of Maxwell's equations, but it is perhaps of interest to exploit the tensor properties of $f_{\lambda\mu}$ (Eq. 3.115). For the Lorentz transformation corresponding to motion along the $z(x_3)$-axis with velocity v, the "direction cosines" are[1]

$$a_{\lambda\mu} = \begin{pmatrix} 1 & 0 & 0 & 0 \\ 0 & 1 & 0 & 0 \\ 0 & 0 & \gamma & i\beta\gamma \\ 0 & 0 & -i\beta\gamma & \gamma \end{pmatrix}, \tag{3.120}$$

where

$$\beta = \frac{v}{c}$$

and

$$\gamma = (1 - \beta^2)^{-1/2} \tag{3.121}$$

[1] A group theoretic derivation of the Lorentz transformation appears in Section 4.12. See als Goldstein, Chapter 6.

Using the tensor transformation properties, we may calculate the electric and magnetic fields in the moving system in terms of the values in the original reference frame. From Eqs. 3.13, 3.115, and 3.120 we obtain

$$\begin{aligned} E'_x &= \frac{1}{\sqrt{1-\beta^2}}(E_x - vB_y), \\ E'_y &= \frac{1}{\sqrt{1-\beta^2}}(E_y + vB_x), \\ E'_z &= E_z, \end{aligned} \tag{3.122}$$

and

$$\begin{aligned} B'_x &= \frac{1}{\sqrt{1-\beta^2}}\left(B_x + \frac{v}{c^2}E_y\right), \\ B'_y &= \frac{1}{\sqrt{1-\beta^2}}\left(B_y - \frac{v}{c^2}E_x\right), \\ B'_z &= B_z. \end{aligned} \tag{3.123}$$

This coupling of **E** and **B** is to be expected. Consider, for instance, the case of zero electric field in the unprimed system

$$E_x = E_y = E_z = 0.$$

Clearly there will be no force on a stationary charged particle. When the particle is in motion with a *small* velocity **v** along the z-axis,[1] an observer on the particle sees fields (exerting a force on his charged particle) given by

$$E'_x = -vB_y,$$

$$E'_y = vB_x,$$

where **B** is a magnetic induction field in the unprimed system. These equations may be put in vector form

$$\mathbf{E}' = \mathbf{v} \times \mathbf{B}$$

or

$$\mathbf{F} = q\mathbf{v} \times \mathbf{B}, \tag{3.124}$$

which is usually taken as the operational definition of the magnetic induction **B**.

Finally, the tensor (or vector) properties allow us to construct a multitude of invariant quantities. One of the more important is the scalar product of the two four-dimensional vectors or four vectors A_λ and i_λ. We have

$$\begin{aligned} A_\lambda i_\lambda &= c\varepsilon_0 A_x \frac{\rho v_x}{c} + c\varepsilon_0 A_y \frac{\rho v_y}{c} \\ &\quad + c\varepsilon_0 A_z \frac{\rho v_z}{c} + i\varepsilon_0 \varphi i \rho \\ &= \varepsilon_0(\mathbf{A}\cdot\mathbf{J} - \rho\varphi), \qquad \text{invariant,} \end{aligned} \tag{3.125}$$

[1] If the velocity is not small (so that v^2/c^2 is negligible), a relativistic transformation of force is needed.

with **A** the usual magnetic vector potential and **J** the ordinary current density. The final term $\rho\varphi$ is the ordinary static electric coupling with dimensions of energy per unit volume. Hence our newly constructed scalar invariant is an energy density. The dynamic interaction of field and current is given by the product $\mathbf{A} \cdot \mathbf{J}$.

Other possible electromagnetic invariants appear in Exercise 3.7.8 and 3.7.10.

EXERCISES

3.7.1 Show that

$$\sum_\mu \frac{\partial i_\mu}{\partial x_\mu} = 0$$

is a statement of continuity of charge and current (cf. Section 1.7). If this equation is known to hold in all Lorentz reference frames, why does this not permit us to conclude that i_μ is a vector?

3.7.2 Write the Lorentz condition (Eq. 3.105), as a tensor equation in Minkowski space.

3.7.3 A gauge transformation consists of varying the scalar potential φ_1 and the vector potential $\mathbf{A}_1$ according to the relation

$$\varphi_2 = \varphi_1 + \frac{\partial\chi}{\partial t},$$

$$\mathbf{A}_2 = \mathbf{A}_1 - \nabla\chi.$$

The new function χ is required to satisfy the homogeneous wave equation.

$$\nabla^2\chi - \varepsilon_0\mu_0 \frac{\partial^2\chi}{\partial t^2} = 0.$$

Show the following:

(a) The Lorentz relation is unchanged.
(b) The new potentials satisfy the same inhomogeneous wave equations as did the original potentials.
(c) The fields **E** and **B** are unaltered.

The invariance of our electromagnetic theory under this transformation is called ***gauge invariance***.

3.7.4 A charged particle, charge q, mass m, obeys the Lorentz covariant equation

$$dp_\mu/d\tau = (q/\varepsilon_0 m_0 c) f_{\mu\nu} \mathbf{p}.$$

$\mathbf{p}$ is the four-dimensional momentum vector $(p_1, p_2, p_3, iE/c)$. τ is the proper time; $d\tau = dt\sqrt{1 - v^2/c^2}$, a Lorentz scalar. Show that the explicit space-time forms are

$$d\mathbf{p}/dt = q(\mathbf{E} + \mathbf{v} \times \mathbf{B})$$
$$dE/dt = q\mathbf{v} \cdot \mathbf{E}.$$

3.7.5 From the Lorentz transformation matrix elements (Eq. 3.120) derive the Einstein velocity addition law

$$u' = \frac{u - v}{1 - (uv/c^2)} \qquad \text{or} \qquad u = \frac{u' + v}{1 + (u'v/c^2)}$$

where $u = icdx_3/dx_4$ and $u' = icdx'_3/dx'_4$.

3.7.6 The dual of a four-dimensional second-rank tensor **B** may be defined by **B***, where the elements of the dual tensor are given by

$$B^*_{ij} = \frac{1}{2!}\varepsilon_{ijkl}B_{kl}.$$

Show that **B*** transforms as
(a) a second-rank tensor under rotations,
(b) a pseudotensor under inversions.
Note. The asterisk here does *not* mean complex conjugate.

3.7.7 Construct **f***, the dual of **f**, wnere **f** is the electromagnetic tensor given by Eq. 3.115.

$$\text{Ans. } \mathbf{f}^* = \varepsilon_0 \begin{pmatrix} 0 & -iE_z & iE_y & cB_x \\ iE_z & 0 & -iE_x & cB_y \\ -iE_y & iE_x & 0 & cB_z \\ -cB_x & -cB_y & -cB_z & 0 \end{pmatrix}$$

This corresponds to

$$c\mathbf{B} \to -i\mathbf{E},$$
$$-i\mathbf{E} \to c\mathbf{B}.$$

This transformation, sometimes called a "dual transformation," leaves Maxwell's equations in vacuum ($\rho = 0$) invariant.

3.7.8 As the quadruple contraction of a fourth-rank pseudotensor and two second-rank tensors $\varepsilon_{\mu\lambda\nu\sigma}f_{\mu\lambda}f_{\nu\sigma}$ is clearly a pseudoscalar. Evaluate it.

Ans. $-8i\varepsilon_0^2 c\mathbf{B}\cdot\mathbf{E}$.

3.7.9 (a) If an electromagnetic field is purely electric (or purely magnetic) in one particular Lorentz frame, show that **E** and **B** will be orthogonal in other Lorentz reference systems.

(b) Conversely, if **E** and **B** are orthogonal in one particular Lorentz frame, there exists a Lorentz reference system in which **E** (or **B**) vanishes. Find that reference system.

3.7.10 Show that $c^2B^2 - E^2$ is an invariant.

3.7.11 Since (dx_1, dx_2, dx_3, dx_4) is a vector, $dx_\mu\, dx_\mu$ is a scalar. Evaluate this scalar for a moving particle in two different coordinate systems: (a) a coordinate system fixed relative to you (lab system), and (b) acoordinate system moving with a moving particle (velocity v relative to you). With the time increment labelled $d\tau$ in the particle system and dt in the lab system, show that

$$d\tau = dt\sqrt{1 - v^2/c^2}.$$

$d\tau$ or τ is the proper time of the particle, a Lorentz invariant quantity.

REFERENCES

HEITLER, W. *The Quantum Theory of Radiation.* 2nd Ed., Oxford: Oxford University Press (1947).

JEFFREYS, Harold, *Cartesian Tensors.* Cambridge: Cambridge University Press (1952). This is an excellent discussion of cartesian tensors and their application to a wide variety of fields of classical physics.

LAWDEN, Derek F., *An Introduction to Tensor Calculus and Relativity.* New York: Wiley (Methuen monograph) (1962).

MOLLER, C., *The Theory of Relativity.* Oxford: Oxford University Press (1955). Most texts on general relativity include a discussion of tensor analysis. Chapter 4 develops tensor calculus, including the topic of dual tensors. The extension to noncartesian systems, as required by general relativity, is presented in Chapter 9.

PANOFSKY, W. K. H., and M. PHILLIPS. *Classical Electricity and Magnetism.* Reading, Mass.: Addison-Wesley (1955). The Lorentz covariance of Maxwell's equations is developed for both vacuum and material media. Panofsky and Phillips use contravariant and covariant tensors rather than Minkowski space. Discussions using Minkowski space are given by Heitler and Stratton.

SOKOLNIKOFF, I. S., *Tensor Analysis—Theory and Applications.* New York: Wiley (1951). Particularly useful for its extension of tensor analysis to non-Euclidean geometries.

SPAIN, B., *Tensor Calculus,* 3rd Ed. New York: Interscience Publishers. (1960). A useful concise presentation of tensors with applications to elasticity and relativity theory.

STRATTON, J. A., *Electromagnetic Theory.* New York: McGraw-Hill (1941).

TEMPLE, George, *Cartesian Tensors.* New York: Wiley (Methuen monograph) (1960).

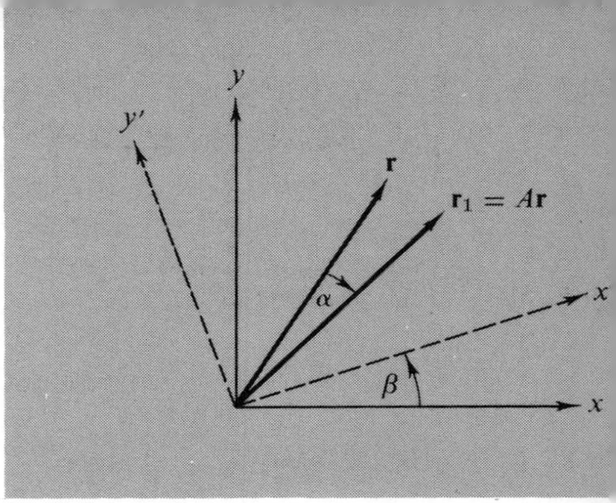

CHAPTER 4

DETERMINANTS, MATRICES, AND GROUP THEORY

4.1 Determinants

We begin our study of matrices by summarizing some properties of determinants, partly because determinants are useful in matrix analysis and partly to illustrate, by way of contrast, what matrices are not.

Properties. A determinant is a square array of numbers or functions which may be combined together according to the rule given below. We have

$$D = \begin{vmatrix} a_1 & b_1 & c_1 & \cdot & \cdot & \cdot \\ a_2 & b_2 & c_2 & \cdot & \cdot & \cdot \\ a_3 & b_3 & c_3 & \cdot & \cdot & \cdot \\ \cdot & \cdot & \cdot & \cdot & \cdot & \cdot \\ a_n & b_n & c_n & \cdot & \cdot & \cdot \end{vmatrix}. \tag{4.1}$$

The number of columns (and of rows) in the array is sometimes called the order of the determinant. In terms of its elements, a_i, b_j, etc., the value of the determinant D is

$$D = \sum_{i,j,k\cdots} \varepsilon_{ijk\ldots}\, a_i b_j c_k \cdots, \tag{4.2}$$

where $\varepsilon_{ijk\ldots}$, analogous to the Levi-Civita symbol of Section 3.4 is $+1$ for even permutations[1] of $(1, 2, 3, \ldots, n)$, -1 for odd permutations, and zero if any index is

[1] In a linear array *abcd* . . . , any single, simple transposition of adjacent elements yields an *odd* permutation of the original array: *abcd* → *bacd*. Two such transpositions yield an even permutation. In general, an odd number of such interchanges of adjacent elements results in an odd permutation; an even number of such transpositions yields an even permutation.

repeated.

Specifically, for the third order determinant, Eq. 4.2 leads to

$$D = +a_1b_2c_3 - a_1b_3c_2 + a_2b_3c_1 - a_2b_1c_3 + a_3b_1c_2 - a_3b_2c_1. \tag{4.4}$$

$$D = \begin{vmatrix} a_1 & b_1 & c_1 \\ a_2 & b_2 & c_2 \\ a_3 & b_3 & c_3 \end{vmatrix}, \tag{4.3}$$

The third-order determinant, then, is this particular linear combination of products. Each product contains one and only one element from each row and from each column. Each product is added in if the order represents an even permutation of rows (the columns being in a, b, c or 1, 2, 3 order) and subtracted if we have an odd permutation. Equation 4.3 may be considered shorthand notation for Eq. 4.4.

Several useful properties of the nth order determinants follow from Eq. 4.2. Again, to be specific, Eq. 4.4 for third-order determinants is used to illustrate these properties.

Laplacian development by minors. Equation 4.4 may be written

$$\begin{aligned} D &= a_1(b_2c_3 - b_3c_2) - a_2(b_1c_3 - b_3c_1) + a_3(b_1c_2 - b_2c_1) \\ &= a_1\begin{vmatrix} b_2 & c_2 \\ b_3 & c_3 \end{vmatrix} - a_2\begin{vmatrix} b_1 & c_1 \\ b_3 & c_3 \end{vmatrix} + a_3\begin{vmatrix} b_1 & c_1 \\ b_2 & c_2 \end{vmatrix}. \end{aligned} \tag{4.5}$$

In general, the nth order determinant may be expanded as a linear combination of the products of the elements of any row (or any column) and the $(n-1)$-order determinants formed by striking out the row and column of the original determinant in which the element appears. This reduced array (2×2 in this specific example) is called a "minor." If the element is in the ith row and the jth column, the sign associated with the product is $(-1)^{i+j}$. The minor with the sign $(-1)^{i+j}$ is called the "cofactor." If M_{ij} is used to designate the minor formed by omitting the ith row and the jth column and C_{ij} is the corresponding cofactor, Eq. 4.5 becomes

$$D = \sum_{i=1}^{3} (-1)^{i+1} a_i M_{i1} = \sum_{i=1}^{3} a_i C_{i1}.$$

In this case, expanding down the first column, we have $j = 1$ and the summation over i.

This Laplace expansion may be used to advantage in the evaluation of high-order determinants in which a lot of the elements are zero. For example, to find the value of the determinant

$$D = \begin{vmatrix} 0 & 1 & 0 & 0 \\ -1 & 0 & 0 & 0 \\ 0 & 0 & 0 & 1 \\ 0 & 0 & -1 & 0 \end{vmatrix}, \tag{4.6}$$

we expand across the top row to obtain

$$D = (-1)^{1+2} \cdot (1) \begin{vmatrix} -1 & 0 & 0 \\ 0 & 0 & 1 \\ 0 & -1 & 0 \end{vmatrix}. \tag{4.7}$$

Again, expanding across the top row, we get

$$\begin{aligned} D &= (-1) \cdot (-1)^{1+1} \cdot (-1) \begin{vmatrix} 0 & 1 \\ -1 & 0 \end{vmatrix} \\ &= \begin{vmatrix} 0 & 1 \\ -1 & 0 \end{vmatrix} = 1. \end{aligned} \tag{4.8}$$

This determinant D (Eq. 4.6) is formed from one of the Dirac matrices appearing in Dirac's relativistic electron theory.

Antisymmetry. The determinant changes sign if any two rows are interchanged or if any two columns are interchanged. This follows from the even-odd character of the Levi-Civita ε in Eq. 4.2 or explicitly from the form of Eqs. 4.3 and 4.4.[1]

This property was used in Section 3.4 to develop a totally antisymmetric linear combination. It is also frequently used in quantum mechanics in the construction of a many-particle wave function that, in accordance with the Pauli exclusion principle, will be antisymmetric under the interchange of any two identical spin $\frac{1}{2}$ particles (electrons, protons, neutrons, etc.).

As a special case of antisymmetry, any determinant with two rows equal or two columns equal equals zero.

If each element in a row or each element in a column is zero, the determinant is equal to zero.

If each element in a row or each element in a column is multiplied by a constant, the determinant is multiplied by that constant.

The value of a determinant is unchanged if a multiple of one row is added (column by column) to another row or if a multiple of one column is added (row by row) to another column.

We have

$$\begin{vmatrix} a_1 & b_1 & c_1 \\ a_2 & b_2 & c_2 \\ a_3 & b_3 & c_3 \end{vmatrix} = \begin{vmatrix} a_1 + kb_1 & b_1 & c_1 \\ a_2 + kb_2 & b_2 & c_2 \\ a_3 + kb_3 & b_3 & c_3 \end{vmatrix}. \tag{4.9}$$

Using the Laplace development on the right-hand side,

$$\begin{vmatrix} a_1 + kb_1 & b_1 & c_1 \\ a_2 + kb_2 & b_2 & c_2 \\ a_3 + kb_3 & b_3 & c_3 \end{vmatrix} = \begin{vmatrix} a_1 & b_1 & c_1 \\ a_2 & b_2 & c_2 \\ a_3 & b_3 & c_3 \end{vmatrix} + k \begin{vmatrix} b_1 & b_1 & c_1 \\ b_2 & b_2 & c_2 \\ b_3 & b_3 & c_3 \end{vmatrix}, \tag{4.10}$$

[1] The sign reversal is reasonably obvious for the interchange of two adjacent rows (or columns), this clearly being an odd permutation. The reader may wish to show that the interchange of *any* two rows is still an odd permutation.

then by the property of antisymmetry the second determinant on the right-hand side of Eq. 4.10 vanishes, verifying Eq. 4.9.

As a special case, a determinant is equal to zero if any two rows are proportional or any two columns are proportional.

Solution of a set of homogeneous equations. One of the major applications of determinants is in the establishment of a condition for the existence of a non-trivial solution for a set of linear homogeneous algebraic equations. Suppose we have three homogeneous equations with three unknowns (or n equations with n unknowns)

$$\begin{aligned} a_1x + b_1y + c_1z &= 0, \\ a_2x + b_2y + c_2z &= 0, \\ a_3x + b_3y + c_3z &= 0. \end{aligned} \tag{4.11}$$

The problem is to determine whether any solution, apart from the trivial one $x = 0$, $y = 0$, $z = 0$, exists.

By forming the determinant of the coefficients of Eq. 4.11 and then multiplying by x,

$$x\begin{vmatrix} a_1 & b_1 & c_1 \\ a_2 & b_2 & c_2 \\ a_3 & b_3 & c_3 \end{vmatrix} = \begin{vmatrix} a_1x & b_1 & c_1 \\ a_2x & b_2 & c_2 \\ a_3x & b_3 & c_3 \end{vmatrix} = \begin{vmatrix} a_1x + b_1y + c_1z & b_1 & c_1 \\ a_2x + b_2y + c_2z & b_2 & c_2 \\ a_3x + b_3y + c_3z & b_3 & c_3 \end{vmatrix}. \tag{4.12}$$

The final step follows from Eq. 4.9, but by Eq. 4.11 each element of the first column vanishes. Then

$$x\begin{vmatrix} a_1 & b_1 & c_1 \\ a_2 & b_2 & c_2 \\ a_3 & b_3 & c_3 \end{vmatrix} = \begin{vmatrix} 0 & b_1 & c_1 \\ 0 & b_2 & c_2 \\ 0 & b_3 & c_3 \end{vmatrix} = 0. \tag{4.13}$$

Therefore x (and y and z) must be zero *unless the determinant of the coefficients vanishes.* Conversely, we can show that if the determinant of the coefficients vanishes a nontrivial solution does indeed exist. This is used in Section 8.5 to establish the linear dependence or independence of a set of functions.

Solution of a set of nonhomogeneous equations. If our linear algebraic equations are nonhomogeneous, i.e., if the zeros on the right-hand side of Eq. 4.11 are replaced by d_1, d_2, and d_3 respectively, then from Eq. 4.12 we obtain[1] in place of Eq. 4.13.

[1] Exercise 1.5.10 gives the vector analog of Eq. 4.14.

$$x = \begin{vmatrix} d_1 & b_1 & c_1 \\ d_2 & b_2 & c_2 \\ d_3 & b_3 & c_3 \end{vmatrix} \Bigg/ \begin{vmatrix} a_1 & b_1 & c_1 \\ a_2 & b_2 & c_2 \\ a_3 & b_3 & c_3 \end{vmatrix} \tag{4.14}$$

For numerical work, this determinant solution, Eq. 4.14, is exceedingly unwieldy. The determinant may involve large numbers with alternate signs, and in the subtraction of two large numbers the relative error may soar to a point that makes the result worthless. Also, while the determinant method is illustrated here with three equations and three unknowns, we might easily have twenty equations with twenty unknowns. From the definition of determinant (Eq. 4.2), our nth order determinant will have $n!$ terms. If one were to ask a high-speed electronic computer to compute these $n!$ terms at the rate of one each microsecond, the computer would still take 20! microseconds or 77,000 years. There must be a better way.

In fact, there are better ways. One of the best is a straightforward elimination process often called Gauss elimination. To illustrate this technique, consider the following set of equations.

Example 4.1.4 Gauss Elimination

Solve

$$\begin{aligned} 3x + 2y + z &= 11 \\ 2x + 3y + z &= 13 \\ x + y + 4z &= 12. \end{aligned} \tag{4.15}$$

For convenience and for the optimum numerical accuracy, the equations are rearranged so that the largest coefficients run along the main diagonal (upper left to lower right). This has already been done in the above set.

The Gauss technique is to use the first equation to eliminate the first unknown, x, from the remaining equations. Then the (new) second equation is used to eliminate y from the last equation. In general, one works down through the set of equations, and then, with one unknown determined, one works back up to solve for each of the other unknowns in succession.

Dividing each row by its initial coefficient, Eqs. 4.15 become

$$\begin{aligned} x + 0.6667\,y + 0.3333\,z &= 3.6667 \\ x + 1.5000\,y + 0.5000\,z &= 6.5000 \\ x + 1.0000\,y + 4.0000\,z &= 12.0000. \end{aligned} \tag{4.16}$$

Now, using the first equation we eliminate x from the second and third:

$$\begin{aligned} x + 0.6667\,y + 0.3333\,z &= 3.6667 \\ 0.8333\,y + 0.1667\,z &= 2.8333 \\ 0.3333\,y + 3.6667\,z &= 8.3333, \end{aligned} \tag{4.17}$$

and

$$\begin{aligned} x + 0.6667\,y + \;\; 0.3333\,z &= \;\; 3.6667 \\ y + \;\; 0.2000\,z &= \;\; 3.4000 \\ y + 11.0000\,z &= 25.0000. \end{aligned} \tag{4.18}$$

Repeating the technique, the new second equation is used to eliminate y from the third equation:

$$\begin{aligned} x + 0.6667\,y + \;\; 0.3333\,z &= \;\; 3.6667 \\ y + \;\; 0.2000\,z &= \;\; 3.4000 \\ 10.8000\,z &= 21.6000, \end{aligned} \tag{4.19}$$

or

$$z = \;\; 2.0000.$$

Finally, working back up,

$$y + 0.2000 \times 2.0000 = 3.4000,$$

or

$$y = 3.0000.$$

Then with z and y determined,

$$x + 0.6667 \times 3.0000 + 0.3333 \times 2.0000 = 3.6667,$$

and

$$x = 1.0000.$$

The technique may not seem as elegant as Eq. 4.14, but it is well-adapted to modern computing machines, and it is far faster than the time spent with determinants.

A variation of this progressive elimination is known as Gauss-Jordan elimination. One starts as with the Gauss elimination above, but each new equation considered is used to eliminate a variable from *all* of the other equations, not just those below it. If we had used this Gauss-Jordan elimination above, Eq. 4.19 would become

$$\begin{aligned} x \qquad + 0.2000\,z &= 1.4000 \\ y + 0.2000\,z &= 3.4000 \\ z &= 2.0000, \end{aligned} \tag{4.20}$$

using the second equation of Eq. 4.18 to eliminate y from both the first and third equations. Then the third equation of Eq. 4.20 is used to eliminate z from the first and second, giving

$$\begin{aligned} x \qquad &= 1.0000 \\ y \qquad &= 3.0000 \\ z &= 2.0000. \end{aligned} \tag{4.21}$$

We return to this Gauss-Jordan technique in Section 4.2 for inverting matrices.

Another technique suitable for computers is the Gauss-Seidel iterative method. This method and the Gauss and Gauss-Jordan elimination techniques are discussed in considerable detail in *Mathematical Methods for Digital Computers* (A. Ralston, and H. S. Wilf, eds.). New York: Wiley (1967).

EXERCISES

4.1.1 Evaluate the following determinants

(a) $\begin{vmatrix} 1 & 0 & 1 \\ 0 & 1 & 0 \\ 1 & 0 & 0 \end{vmatrix}$

(b) $\dfrac{1}{\sqrt{2}}\begin{vmatrix} 0 & \sqrt{3} & 0 & 0 \\ \sqrt{3} & 0 & 2 & 0 \\ 0 & 2 & 0 & \sqrt{3} \\ 0 & 0 & \sqrt{3} & 0 \end{vmatrix}$

(c) $\begin{vmatrix} 1 & 2 & 0 \\ 3 & 1 & 2 \\ 0 & 3 & 1 \end{vmatrix}$

4.1.2 Test the set of linear homogeneous equations

$$x + 3y + 3z = 0,$$
$$x - y + z = 0,$$
$$2x + y + 3z = 0,$$

to see if it possesses a nontrivial solution.

4.1.3 Express the *components* of $\mathbf{A} \times \mathbf{B}$ as 2×2 determinants. Show then that the dot product $\mathbf{A} \cdot (\mathbf{A} \times \mathbf{B})$ yields a Laplacian expansion of a 3×3 determinant. Finally note that two rows of the 3×3 determinant are identical and hence $\mathbf{A} \cdot (\mathbf{A} \times \mathbf{B}) = 0$.

4.1.4 If C_{ij} is the cofactor of element a_{ij} (formed by striking out the ith row and jth column and including a sign $(-1)^{i+j}$), show that

(a) $\sum_i a_{ij}C_{ij} = \sum_i a_{ji}C_{ji} = |A|$, where $|A|$ is the determinant with the elements a_{ij},

(b) $\sum_i a_{ij}C_{ik} = \sum_i a_{ji}C_{ki} = 0, \qquad j \neq k.$

4.2 Matrices

Basic definitions. A matrix may be defined as a square or rectangular array of numbers or functions that obeys certain laws. This is a perfectly logical extension of familiar mathematical concepts. In arithmetic we deal with single numbers. In the theory of complex variables (Chapter 6) we deal with ordered pairs of numbers, $(1, 2) = 1 + 2i$, in which the ordering is important. We now consider numbers (or functions) ordered in a square or rectangular array. For convenience in later work the numbers are distinguished by two subscripts, the first indicating the row (horizontal) and the second indicating the column (vertical) in which the number appears. For instance, a_{13} is the matrix element in the first row, third column. Hence, if A is a matrix with m rows and n columns,

$$\mathsf{A} = \begin{pmatrix} a_{11} & a_{12} & \cdot & \cdot & \cdot & a_{1n} \\ a_{21} & a_{22} & \cdot & \cdot & \cdot & a_{2n} \\ \cdot & \cdot & \cdot & \cdot & \cdot & \cdot \\ a_{m1} & a_{m2} & \cdot & \cdot & \cdot & a_{mn} \end{pmatrix}. \tag{4.22}$$

Perhaps the most important thing to note is that the elements a_{ij} are not combined with one another. The matrix is not a determinant. It is not a single number; it is an ordered array of numbers. It makes no more sense to add or multiply all the a_{ij}'s together than it does to write $1 + 2i = 3$!

The matrix A, so far just an array of numbers, has the properties we assign to it. Literally, this means constructing a new form of mathematics. We postulate that matrices A, B, and C, with elements a_{ij}, b_{ij}, and c_{ij}, respectively, combine according to the following rules:

Equality. Matrix A = Matrix B if and only if $a_{ij} = b_{ij}$ for all values of i and j. This, of course, requires that A and B each be m by n arrays (m rows, n columns).

Addition. $\mathsf{A} + \mathsf{B} = \mathsf{C}$ if and only if $a_{ij} + b_{ij} = c_{ij}$ for all values of i and j, the elements combining according to the laws of ordinary algebra (or arithmetic if they are simple numbers). This means that $\mathsf{A} + \mathsf{B} = \mathsf{B} + \mathsf{A}$, commutation. Also, an associative law is satisfied $(\mathsf{A} + \mathsf{B}) + \mathsf{C} = \mathsf{A} + (\mathsf{B} + \mathsf{C})$.

Multiplication (by a scalar). The multiplication of matrix A by the scalar quantity α is defined as

$$\alpha\mathsf{A} = (\alpha\mathsf{A}),$$

in which the elements of $\alpha\mathsf{A}$ are αa_{ij}; that is, each element of matrix A is multiplied by the scalar factor. This is in striking contrast to the behavior of determinants in which the factor α multiplies only one column or one row and not every element of the entire determinant. A consequence of this scalar multiplication is that

$$\alpha\mathsf{A} = \mathsf{A}\alpha, \text{ commutation.}$$

Multiplication (matrix multiplication).

$$\mathsf{AB} = \mathsf{C} \quad \text{if and only if}^1 \quad c_{ij} = \sum_k a_{ik} b_{kj}. \tag{4.23}$$

The ij element of C is formed as a scalar product of the ith row of A with the jth column of B (which demands that A have the same number of columns (n) as B has rows). The dummy index k takes on all the values $1, 2, \cdots, n$ in succession, that is,

$$c_{ij} = a_{i1}b_{1j} + a_{i2}b_{2j} + a_{i3}b_{3j} \tag{4.24}$$

[1] Some authors follow the summation convention here (cf. Section 3.1).

for $n = 3$. Obviously the dummy index k may be replaced by any other symbol that is not already in use without altering Eq. 4.23. Perhaps the situation may be clarified by stating that Eq. 4.23 defines the method of combining certain matrices. This method of combination, to give it a label, is called matrix multiplication. To illustrate, consider two matrices

$$\sigma_1 = \begin{pmatrix} 0 & 1 \\ 1 & 0 \end{pmatrix} \quad \text{and} \quad \sigma_3 = \begin{pmatrix} 1 & 0 \\ 0 & -1 \end{pmatrix}. \tag{4.25}$$

The 11 element of the product, $(\sigma_1\sigma_3)_{11}$ is given by the sum of the products of elements of the first *row* of σ_1 with the corresponding elements of the first *column* of σ_3:

$$\begin{pmatrix} 0 & 1 \\ 1 & 0 \end{pmatrix} \begin{pmatrix} 1 & 0 \\ 0 & -1 \end{pmatrix} \rightarrow 0 \cdot 1 + 1 \cdot 0 = 0.$$

Continuing, we have

$$\sigma_1\sigma_3 = \begin{pmatrix} 0{\cdot}1 + 1{\cdot}0 & 0{\cdot}0 + 1{\cdot}(-1) \\ 1{\cdot}1 + 0{\cdot}0 & 1{\cdot}0 + 0{\cdot}(-1) \end{pmatrix} = \begin{pmatrix} 0 & -1 \\ 1 & 0 \end{pmatrix} \tag{4.26}$$

Here

$$(\sigma_1\sigma_3)_{ij} = \sigma_{1_{i1}}\sigma_{3_{1j}} + \sigma_{1_{i2}}\sigma_{3_{2j}}$$

Direct application of the definition of matrix multiplication shows that

$$\sigma_3\sigma_1 = \begin{pmatrix} 0 & 1 \\ -1 & 0 \end{pmatrix} \tag{4.27}$$

and by Eq. 4.23

$$\sigma_3\sigma_1 = -\sigma_1\sigma_3. \tag{4.28}$$

Except in special cases, matrix multiplication is not commutative.[1]

$$\mathsf{AB} \neq \mathsf{BA}. \tag{4.29}$$

However, from the definition of matrix multiplication we can show that an associative law does hold, $(\mathsf{AB})\mathsf{C} = \mathsf{A}(\mathsf{BC})$. There is also a distributive law, $\mathsf{A}(\mathsf{B} + \mathsf{C}) = \mathsf{AB} + \mathsf{AC}$.

Direct Product. A second way of multiplying matrices, known as the *direct* or *tensor product*, goes as follows. If A is an $(m \times m)$ matrix and B an $(n \times n)$ matrix, then the direct product is

$$\mathsf{A} \otimes \mathsf{B} = \mathsf{C}. \tag{4.30}$$

C is an $(mn \times mn)$ matrix with elements

$$C_{ik,jl} = A_{ij}B_{kl}. \tag{4.31}$$

[1] Commutation or the lack of it is conveniently described by the commutator bracket symbol, $[\mathsf{A}, \mathsf{B}] = \mathsf{AB} - \mathsf{BA}$. Equation 4.29 becomes $[\mathsf{A}, \mathsf{B}] \neq 0$.

For instance, if A and B are both 2×2 matrices,

$$\mathsf{A} \otimes \mathsf{B} = \begin{pmatrix} a_{11}\mathsf{B} & a_{12}\mathsf{B} \\ a_{21}\mathsf{B} & a_{22}\mathsf{B} \end{pmatrix} \tag{4.32}$$

$$= \begin{pmatrix} a_{11}b_{11} & a_{11}b_{12} & a_{12}b_{11} & a_{12}b_{12} \\ a_{11}b_{21} & a_{11}b_{22} & a_{12}b_{21} & a_{12}b_{22} \\ a_{21}b_{11} & a_{21}b_{12} & a_{22}b_{11} & a_{22}b_{12} \\ a_{21}b_{21} & a_{21}b_{22} & a_{22}b_{21} & a_{22}b_{22} \end{pmatrix}.$$

The direct product is associative but not commutative. As an example of the direct product, the Dirac matrices of Section 4.5 may be developed as direct products of the Pauli matrices and the unit matrix. Other examples appear in the construction of groups in group theory and in vector or Hilbert space in quantum theory.

Special Cases. A number of matrices are of special interest. If the matrix has one column and n rows, it is called a column vector, $\{x\}$ with components x_i, $i = 1, 2, \cdots, n$.

Similarly, if the matrix has one row and n columns, it is called a row vector, $[x]$ with components x_i, $i = 1, 2, \cdots, n$. Clearly, if A is an $n \times n$ matrix, $\{x\}$ an n-component column vector, and $[x]$ an n-component row vector,

$$\mathsf{A}\{x\} \quad \text{and} \quad [x]\mathsf{A}$$

are defined by Eq. 4.23, whereas

$$\mathsf{A}[x] \quad \text{and} \quad \{x\}\mathsf{A}$$

are not defined.

In the remainder of this chapter we shall confine our attention to column vectors, row vectors, and square matrices.

The unit matrix **1** has elements δ_{ij}, Kronecker delta, and the property that $\mathbf{1}\mathsf{A} = \mathsf{A}\mathbf{1} = \mathsf{A}$ for all A.

$$\mathbf{1} = \begin{pmatrix} 1 & 0 & 0 & 0 & \cdot & \cdot & \cdot \\ 0 & 1 & 0 & 0 & \cdot & \cdot & \cdot \\ 0 & 0 & 1 & 0 & \cdot & \cdot & \cdot \\ 0 & 0 & 0 & 1 & \cdot & \cdot & \cdot \\ \cdot & \cdot & \cdot & \cdot & \cdot & \cdot & \cdot \end{pmatrix}. \tag{4.33}$$

If all elements are zero, the matrix is called the null matrix and is denoted by O. For all A

$$\mathsf{OA} = \mathsf{AO} = \mathsf{O}.$$

$$\mathsf{O} = \begin{pmatrix} 0 & 0 & 0 & \cdot & \cdot & \cdot \\ 0 & 0 & 0 & \cdot & \cdot & \cdot \\ 0 & 0 & 0 & \cdot & \cdot & \cdot \\ \cdot & \cdot & \cdot & \cdot & \cdot & \cdot \end{pmatrix}. \tag{4.34}$$

It should be noted that it is possible for the product of two matrices to be the null matrix without either one being the null matrix. For example, if

$$\mathsf{A} = \begin{pmatrix} 1 & 1 \\ 0 & 0 \end{pmatrix} \quad \text{and} \quad \mathsf{B} = \begin{pmatrix} 1 & 0 \\ -1 & 0 \end{pmatrix},$$

$\mathsf{AB} = \mathsf{O}$. Once more the results of ordinary algebra do not apply directly.

Diagonal matrices. An important special type of matrix is the square matrix in which all the nondiagonal elements are zero. Specifically, if a 3×3 matrix A is diagonal,

$$\mathsf{A} = \begin{pmatrix} a_{11} & 0 & 0 \\ 0 & a_{22} & 0 \\ 0 & 0 & a_{33} \end{pmatrix}.$$

The physical interpretation of such diagonal matrices and the method of reducing matrices to this diagonal form are considered in Section 4.6. Here, we simply note a significant property of diagonal matrices—that multiplication of diagonal matrices is commutative,

$$\mathsf{AB} = \mathsf{BA}, \quad \text{if} \quad \mathsf{A} \quad \text{and} \quad \mathsf{B} \quad \text{are each diagonal.}$$

Trace. In any square matrix the sum of the diagonal elements is called the *trace*. One of its interesting and useful properties is that the trace of a product of two matrices A and B is independent of the order of multiplication:

$$\begin{aligned} \operatorname{trace}(\mathsf{AB}) &= \sum_i (\mathsf{AB})_{ii} = \sum_i \sum_j a_{ij} b_{ji} \\ &= \sum_j \sum_i b_{ji} a_{ij} = \sum_j (\mathsf{BA})_{jj} \\ &= \operatorname{trace}(\mathsf{BA}). \end{aligned} \tag{4.35}$$

This holds even though $\mathsf{AB} \neq \mathsf{BA}$.

Matrices are used extensively to represent the elements of groups (cf. Ex. 4.2.7 and Sections 4.7–4.12). The trace of the matrix representing the group element is known in group theory as the *character*. The reason for the special name and special attention is that while the matrices mav vary the trace or character remains invariant (cf. Ex. 4.3.8).

Matrix inversion. Consider matrix A to be known. The problem is to find the inverse matrix A^{-1} such that

$$\mathsf{AA}^{-1} = \mathsf{A}^{-1}\mathsf{A} = 1. \tag{4.36}$$

From Exercise 4.2.27

$$a_{ij}^{-1} = \frac{C_{ji}}{|\mathsf{A}|}, \tag{4.37}$$

with the assumption that the determinant of A (|A|) $\neq 0$. If it is zero, we label A singular. No inverse exists. This conclusion that we must require $|\mathsf{A}| \neq 0$ is about the only use of Eq. 4.37. As explained at the end of Section 4.1, this determinant form is *totally unsuited for numerical work* with large matrices.

There is a wide variety of alternative techniques. One of the best and most commonly used is the Gauss-Jordan matrix inversion technique. The theory is based on the results of Exercises 4.2.28 and 4.2.29, which show that there exist matrices M_L such that the product $\mathsf{M}_L\mathsf{A}$ will be A but with

(a) one row multiplied by a constant, or
(b) one row replaced by the original row minus a multiple of another row, or
(c) rows interchanged.

Other matrices M_R operating on the right ($\mathsf{A}\mathsf{M}_R$) can carry out the same operations on the *columns* of A.

This means that the matrix rows and columns may be altered (by matrix multiplication) as though we were dealing with determinants, so we can apply the Gauss-Jordan elimination techniques of Section 4.1 to the matrix elements. Hence, there exists a matrix M_L (or M_R) such that[1]

$$\mathsf{M}_L\mathsf{A} = 1. \tag{4.38}$$

Then $\mathsf{M}_L = \mathsf{A}^{-1}$. We determine M_L by carrying out the identical elimination operations on the unit matrix. Then

$$\mathsf{M}_L 1 = \mathsf{M}_L. \tag{4.39}$$

To clarify this, we consider a specific example.

EXAMPLE 4.2.1 GAUSS-JORDAN MATRIX INVERSION

We want to invert the matrix

$$\mathsf{A} = \begin{pmatrix} 3 & 2 & 1 \\ 2 & 3 & 1 \\ 1 & 1 & 4 \end{pmatrix}. \tag{4.40}$$

For convenience, we write A and 1 side by side and carry out the identical operations on each:

$$\begin{pmatrix} 3 & 2 & 1 \\ 2 & 3 & 1 \\ 1 & 1 & 4 \end{pmatrix} \quad \text{and} \quad \begin{pmatrix} 1 & 0 & 0 \\ 0 & 1 & 0 \\ 0 & 0 & 1 \end{pmatrix}. \tag{4.41}$$

[1] Remember that det (A) $\neq 0$.

To be systematic, we multiply each row to get $a_{k1} = 1$,

$$\begin{pmatrix} 1 & 0.6667 & 0.3333 \\ 1 & 1.5000 & 0.5000 \\ 1 & 1.0000 & 4.0000 \end{pmatrix} \quad \text{and} \quad \begin{pmatrix} 0.3333 & 0 & 0 \\ 0 & 0.5000 & 0 \\ 0 & 0 & 1 \end{pmatrix}. \tag{4.42}$$

Subtracting the first row from the second and third, we obtain

$$\begin{pmatrix} 1 & 0.6667 & 0.3333 \\ 0 & 0.8333 & 0.1667 \\ 0 & 0.3333 & 3.6667 \end{pmatrix} \quad \text{and} \quad \begin{pmatrix} 0.3333 & 0 & 0 \\ -0.3333 & 0.5000 & 0 \\ -0.3333 & 0 & 1 \end{pmatrix}. \tag{4.43}$$

Then we divide the second row (of *both* matrices) by 0.8333 and subtract 0.6667 times it from the first row, and 0.3333 times it from the third row. The results for both matrices are

$$\begin{pmatrix} 1 & 0 & 0.2000 \\ 0 & 1 & 0.2000 \\ 0 & 0 & 3.6000 \end{pmatrix} \quad \text{and} \quad \begin{pmatrix} 0.6000 & -0.4000 & 0 \\ -0.4000 & 0.6000 & 0 \\ -0.2000 & -0.2000 & 1 \end{pmatrix}. \tag{4.44}$$

We divide the third row (of both matrices) by 3.6. Then as the last step 0.2 times the third row is subtracted from each of the first two rows (of both matrices). Our final pair is

$$\begin{pmatrix} 1 & 0 & 0 \\ 0 & 1 & 0 \\ 0 & 0 & 1 \end{pmatrix} \quad \text{and} \quad \begin{pmatrix} 0.6111 & -0.3889 & -0.0556 \\ -0.3889 & 0.6111 & -0.0556 \\ -0.0556 & -0.0556 & 0.2778 \end{pmatrix}. \tag{4.45}$$

The check is to multiply the original A by the calculated A^{-1} to see if we really do get the unit matrix $\mathbf{1}$. The result to four decimal places is

$$\mathsf{AA}^{-1} = \begin{pmatrix} 0.9999 & -0.0001 & -0.0002 \\ -0.0001 & 0.9999 & -0.0002 \\ -0.0002 & -0.0002 & 1.0000 \end{pmatrix} \tag{4.46}$$

or $\mathbf{1}$, the unit matrix to within the round-off error (mostly from rounding off $-0.05555\cdots$ to -0.0556).

As with the Gauss-Jordan solution of simultaneous linear algebraic equations, this technique is well adapted to large computing machines. Indeed this Gauss-Jordan matrix inversion technique will probably be available in the program library as a subroutine.

EXERCISES

4.2.1 Show that matrix multiplication is associative, $(\mathsf{AB})\mathsf{C} = \mathsf{A}(\mathsf{BC})$.

4.2.2 Show that

$$(\mathsf{A} + \mathsf{B})(\mathsf{A} - \mathsf{B}) = \mathsf{A}^2 - \mathsf{B}^2$$

if and only if A and B commute,

$$[\mathsf{A}, \mathsf{B}] = 0.$$

4.2.3 Show that matrix A is a *linear operator* by showing that

$$\mathsf{A}(c_1\mathbf{r}_1 + c_2\mathbf{r}_2) = c_1\mathsf{A}\mathbf{r}_1 + c_2\mathsf{A}\mathbf{r}_2 .$$

It can be shown that an $n \times n$ matrix is the *most general* linear operator in n dimensional vector space. This means that every linear operator in this n dimensional vector space is equivalent to a matrix.

4.2.4 (a) Complex numbers, $a + ib$, with a and b real, may be represented by (or, are isomorphic with) 2×2 matrices:

$$a + ib \leftrightarrow \begin{pmatrix} a & b \\ -b & a \end{pmatrix}.$$

Show that this matrix representation is valid for (i) addition and (ii) multiplication.

(b) Find the matrix corresponding to $(a + ib)^{-1}$.

4.2.5 If A is an $n \times n$ matrix, show that

$$\det(-\mathsf{A}) = (-1)^n \det \mathsf{A}.$$

4.2.6 Matrix C is the matrix product of A and B. Show that the determinant of C is the product of the determinants of A and B.

$$\det \mathsf{C} = \det \mathsf{A} \times \det \mathsf{B}.$$

4.2.7 Given the three matrices

$$\mathsf{A} = \begin{pmatrix} -1 & 0 \\ 0 & -1 \end{pmatrix}, \quad \mathsf{B} = \begin{pmatrix} 0 & 1 \\ 1 & 0 \end{pmatrix}, \quad \text{and} \quad \mathsf{C} = \begin{pmatrix} 0 & -1 \\ -1 & 0 \end{pmatrix}.$$

Find all possible products of A, B, and C, two at a time, including squares. Express your answers in terms of A, B, and C, and **1**, the unit matrix.

These three matrices together with the unit matrix form a representation of a mathematical group, the *vierergruppe*.
Sections 4.7 and 4.8 (Group Theory) contain repeated references to this group.

4.2.8 Given

$$\mathsf{K} = \begin{pmatrix} 0 & 0 & i \\ -i & 0 & 0 \\ 0 & -1 & 0 \end{pmatrix},$$

show that

$$\mathsf{K}^n = \mathsf{KKK}\cdots(n \text{ factors}) = \mathbf{1}$$

(with the proper choice of n, $n \neq 0$).

4.2.9 Verify the Jacobi identity

$$[\mathsf{A}, [\mathsf{B}, \mathsf{C}]] = [\mathsf{B}, [\mathsf{A}, \mathsf{C}]] - [\mathsf{C}, [\mathsf{A}, \mathsf{B}]].$$

This is useful in matrix descriptions of elementary particles. As a mnemonic aid, the reader might note that the Jacobi identity has the same form as the *BAC-CAB* rule of Section 1.5.

4.2.10 Show that the matrices

$$\mathsf{A} = \begin{pmatrix} 0 & 1 & 0 \\ 0 & 0 & 0 \\ 0 & 0 & 0 \end{pmatrix}, \quad \mathsf{B} = \begin{pmatrix} 0 & 0 & 0 \\ 0 & 0 & 1 \\ 0 & 0 & 0 \end{pmatrix}, \quad \mathsf{C} = \begin{pmatrix} 0 & 0 & 1 \\ 0 & 0 & 0 \\ 0 & 0 & 0 \end{pmatrix}$$

satisfy the commutation relations

$$[\mathsf{A}, \mathsf{B}] = \mathsf{C}, \quad [\mathsf{A}, \mathsf{C}] = 0, \quad \text{and} \quad [\mathsf{B}, \mathsf{C}] = 0.$$

These matrices have been used in a Lie algebra approach to Hermite polynomials.[1]

4.2.11 Let

$$\mathsf{i} = \begin{pmatrix} 0 & 1 & 0 & 0 \\ -1 & 0 & 0 & 0 \\ 0 & 0 & 0 & 1 \\ 0 & 0 & -1 & 0 \end{pmatrix}, \quad \mathsf{j} = \begin{pmatrix} 0 & 0 & 0 & -1 \\ 0 & 0 & -1 & 0 \\ 0 & 1 & 0 & 0 \\ 1 & 0 & 0 & 0 \end{pmatrix}, \quad \text{and} \quad \mathsf{k} = \begin{pmatrix} 0 & 0 & -1 & 0 \\ 0 & 0 & 0 & 1 \\ 1 & 0 & 0 & 0 \\ 0 & -1 & 0 & 0 \end{pmatrix}$$

Show that

(a) $\mathsf{i}^2 = \mathsf{j}^2 = \mathsf{k}^2 = -\mathbf{1}$, where $\mathbf{1}$ is the unit matrix.

(b) $\mathsf{ij} = -\mathsf{ji} = \mathsf{k}$,

$\mathsf{jk} = -\mathsf{kj} = \mathsf{i}$,

$\mathsf{ki} = -\mathsf{ik} = \mathsf{j}$.

These three matrices (i, j, and k) plus the unit matrix $\mathbf{1}$ form a basis for *quaternions*. An alternate basis is provided by the four 2×2 matrices, $i\sigma_1$, $i\sigma_2$, $-i\sigma_3$, and $\mathbf{1}$, where the σ's are the Pauli spin matrices of Ex. 4.2.12.

4.2.12 The three Pauli spin matrices are

$$\sigma_1 = \begin{pmatrix} 0 & 1 \\ 1 & 0 \end{pmatrix}, \quad \sigma_2 = \begin{pmatrix} 0 & -i \\ i & 0 \end{pmatrix}, \quad \text{and} \quad \sigma_3 = \begin{pmatrix} 1 & 0 \\ 0 & -1 \end{pmatrix}.$$

Show that

(a) $\sigma_i^2 = \mathbf{1}$,

(b) $\sigma_i\sigma_j = i\sigma_k$, $\quad (i, j, k) = (1, 2, 3), (2, 3, 1), (3, 1, 2)$ (cyclic permutation),

(c) $\sigma_i\sigma_j + \sigma_j\sigma_i = 2\delta_{ij}\,\mathbf{1}$.

These matrices were used by Pauli in the nonrelativistic theory of electron spin.

4.2.13 Using the Pauli σ's of Exercise 4.2.12, show that

$$(\boldsymbol{\sigma} \cdot \mathbf{a})(\boldsymbol{\sigma} \cdot \mathbf{b}) = \mathbf{a} \cdot \mathbf{b}\,\mathbf{1} + i\boldsymbol{\sigma} \cdot (\mathbf{a} \times \mathbf{b}).$$

Here

$$\boldsymbol{\sigma} = \mathbf{i}\sigma_x + \mathbf{j}\sigma_y + \mathbf{k}\sigma_z$$

and $\mathbf{a}$ and $\mathbf{b}$ are ordinary vectors.

[1] Bruria Kaufman. "Special Functions of Mathematical Physics from the Viewpoint of Lie Algebra," *J. Mathematical Phys.* **7**, 447 (1966).

4.2.14 One description of spin 1 particles uses the matrices

$$M_x = \frac{1}{\sqrt{2}}\begin{pmatrix} 0 & 1 & 0 \\ 1 & 0 & 1 \\ 0 & 1 & 0 \end{pmatrix}, \quad M_y = \frac{1}{\sqrt{2}}\begin{pmatrix} 0 & -i & 0 \\ i & 0 & -i \\ 0 & i & 0 \end{pmatrix},$$

and

$$M_z = \begin{pmatrix} 1 & 0 & 0 \\ 0 & 0 & 0 \\ 0 & 0 & -1 \end{pmatrix}.$$

Show that

(a) $[M_x, M_y]$[1] $= iM_z$, etc. (cyclic permutation of indices).

Using the Levi-Civita symbol of Section 3.4, this may be written

$$[M_i, M_j] = i\varepsilon_{ijk} M_k.$$

(b) $M^2 \equiv M_x^2 + M_y^2 + M_z^2 = 2\mathbf{1}$,

where **1** is the unit matrix.

(c) $[M^2, M_i] = 0$,

$[M_z, L^+] = L^+$,

$[L^+, L^-] = 2M_z$

where

$L^+ \equiv M_x + iM_y$,

$L^- \equiv M_x - iM_y$.

4.2.15 Repeat Exercise 4.2.14 using an alternate representation,

$$M_x = \begin{pmatrix} 0 & 0 & 0 \\ 0 & 0 & -i \\ 0 & i & 0 \end{pmatrix}, \quad M_y = \begin{pmatrix} 0 & 0 & i \\ 0 & 0 & 0 \\ -i & 0 & 0 \end{pmatrix},$$

and

$$M_z = \begin{pmatrix} 0 & -i & 0 \\ i & 0 & 0 \\ 0 & 0 & 0 \end{pmatrix}.$$

In Section 4.10 these matrices appear as the *generators* of the rotation matrices.

4.2.16 Repeat Exercise 4.2.14, using the matrices for a spin of $\frac{3}{2}$,

$$M_x = \frac{1}{2}\begin{pmatrix} 0 & \sqrt{3} & 0 & 0 \\ \sqrt{3} & 0 & 2 & 0 \\ 0 & 2 & 0 & \sqrt{3} \\ 0 & 0 & \sqrt{3} & 0 \end{pmatrix}, \quad M_y = \frac{i}{2}\begin{pmatrix} 0 & -\sqrt{3} & 0 & 0 \\ \sqrt{3} & 0 & -2 & 0 \\ 0 & 2 & 0 & -\sqrt{3} \\ 0 & 0 & \sqrt{3} & 0 \end{pmatrix},$$

and

$$M_z = \frac{1}{2}\begin{pmatrix} 3 & 0 & 0 & 0 \\ 0 & 1 & 0 & 0 \\ 0 & 0 & -1 & 0 \\ 0 & 0 & 0 & -3 \end{pmatrix}.$$

4.2.17 An operator P commutes with J_x and J_y, the x and y components of an angular momentum operator. Show that P commutes with the third component of angular momentum,

[1] $[A, B] \equiv AB - BA$.

that is,

$$[\mathsf{P}, \mathsf{J}_z] = 0.$$

Hint: The angular momentum components must satisfy the commutation relation of Exercise 4.2.14(a).

4.2.18 The L^+ and L^- matrices of Ex. 4.2.14 are "ladder operators." L^+ operating on a system of spin projection m will raise the spin projection to $m+1$ if m is below its maximum. L^+ operating on $m_{\max}$ yields zero. L^- reduces the spin projection in unit steps in a similar fashion. Dividing by $\sqrt{2}$,

$$\mathsf{L}^+ = \begin{pmatrix} 0 & 1 & 0 \\ 0 & 0 & 1 \\ 0 & 0 & 0 \end{pmatrix}, \qquad \mathsf{L}^- = \begin{pmatrix} 0 & 0 & 0 \\ 1 & 0 & 0 \\ 0 & 1 & 0 \end{pmatrix}.$$

Show that

$$\begin{aligned} &\mathsf{L}^+\{-1\} = \{0\}, \quad \mathsf{L}^-\{-1\} = \text{null column vector}, \\ &\mathsf{L}^+\{0\} \ \ = \{1\}, \quad \mathsf{L}^-\{0\} = \{-1\}, \\ &\mathsf{L}^+\{1\} \ \ = \text{null column vector}, \ \mathsf{L}^-\{1\} = \{0\}. \end{aligned}$$

where

$$\{-1\} = \begin{pmatrix} 0 \\ 0 \\ 1 \end{pmatrix}, \quad \{0\} = \begin{pmatrix} 0 \\ 1 \\ 0 \end{pmatrix} \quad \text{and}\,\{1\} = \begin{pmatrix} 1 \\ 0 \\ 0 \end{pmatrix}$$

representing states of spin projection -1, 0, and 1, respectively.
Note: Differential operator analogs of these ladder operators appear in Ex. 12.6.7.

4.2.19 Vectors **A** and **B** are related by the tensor T

$$\mathbf{B} = \mathsf{T}\mathbf{A}.$$

Given **A** and **B** show that there is *no unique solution* for the components of T. This is why vector division **B**/**A** is undefined (apart from the special case of **A** and **B** parallel and T then a scalar).

4.2.20 We might ask for a vector $\mathbf{A}^{-1}$, an inverse of a given vector **A** in the sense that

$$\mathbf{A} \cdot \mathbf{A}^{-1} = \mathbf{A}^{-1} \cdot \mathbf{A} = 1.$$

Show that this relation does not suffice to define $\mathbf{A}^{-1}$ uniquely. **A** has literally an infinite number of inverses.

4.2.21 If A is diagonal, with all diagonal elements different, and A and B commute, show that B is diagonal.

4.2.22 If A and B are diagonal, show that A and B commute.

4.2.23 Show that trace (ABC) = trace (CBA) if any two of the three matrices commute.

4.2.24 With $\{x\}$ an N dimensional column vector and $[y]$ an N dimensional row vector, show that

$$\text{trace}\,(\{x\}[y]) = [y]\{x\}.$$

4.2.25 (a) If two nonsingular matrices anticommute, show that the trace of each one is zero. (Nonsingular means that the determinant of the matrix elements $\neq 0$.)
(b) For the conditions of part (a) to hold A and B must be $n \times n$ matrices with n *even*. Show that if n is *odd* a contradiction results.

4.2.26 If a matrix has an inverse, show that the inverse is unique.

4.2.27 If A^{-1} has elements

$$a_{ij}^{-1} = \frac{C_{ji}}{|\mathsf{A}|}$$

where C_{ji} is the jith cofactor of $|\mathsf{A}|$. Show that

$$\mathsf{A}^{-1}\mathsf{A} = 1.$$

Hence A^{-1} is the inverse of A (if $|\mathsf{A}| \neq 0$).

Note: In numerical work it sometimes happens that $|\mathsf{A}|$ is almost equal to 0. Then there is trouble.

4.2.28 Find the matrices M_L such that the product $\mathsf{M}_L\mathsf{A}$ will be A but with:
(a) the ith row multiplied by a constant k, ($a_{ij} \to ka_{ij}, j = 1, 2, 3, \ldots$).
(b) the ith row replaced by the original ith row minus a multiple of the mth row, ($a_{ij} \to a_{ij} - ka_{mj}, j = 1, 2, 3, \ldots$).
(c) the ith and mth rows interchanged, ($a_{ij} \to a_{mj}, a_{mj} \to a_{ij}, j = 1, 2, 3, \ldots$).

4.2.29 Find the matrices M_R such that the product $\mathsf{A}\mathsf{M}_R$ will be A but with:
(a) the ith column multiplied by a constant k, ($a_{ji} \to ka_{ji}, j = 1, 2, 3, \ldots$).
(b) the ith column replaced by the original ith column minus a multiple of the mth column, ($a_{ji} \to a_{ji} - ka_{jm}, j = 1, 2, 3, \ldots$).
(c) the ith and mth columns interchanged, ($a_{ji} \to a_{jm}, a_{jm} \to a_{ji}, j = 1, 2, 3, \ldots$).

4.2.30 Find the inverse of

$$\mathsf{A} = \begin{pmatrix} 3 & 2 & 1 \\ 2 & 2 & 1 \\ 1 & 1 & 4 \end{pmatrix}.$$

4.3 Orthogonal Matrices

Ordinary three-dimensional space may be described with the familiar cartesian coordinates (x, y, z). We consider a second set of cartesian coordinates (x', y', z') whose origin coincides with that of the first set but whose orientation is different. This section repeats portions of Chapters 1 and 3 in a slightly different context and with a different emphasis. In Chapters 1 and 3 attention was focused on the vector or tensor. Here emphasis is laid on the description of the coordinate rotation itself.

Direction cosines. A unit vector along the x'-axis ($\mathbf{i}'$) may be resolved into components along the x-, y-, and z-axes by the usual projection technique.

$$\mathbf{i}' = \mathbf{i}\cos(x', x) + \mathbf{j}\cos(x', y) + \mathbf{k}\cos(x', z). \tag{4.47}$$

For convenience these cosines, which are the direction cosines, are labeled

$$\begin{aligned} \cos(x', x) &= a_{11}, \\ \cos(x', y) &= a_{12}, \\ \cos(x', z) &= a_{13}. \end{aligned} \tag{4.48}$$

Continuing,

$$\cos(y', x) = a_{21}, \qquad (a_{21} \neq a_{12}),$$
$$\cos(y', y) = a_{22}, \text{ etc.} \tag{4.49}$$

Now, Eq. 4.47 may be rewritten

$$\mathbf{i}' = \mathbf{i}a_{11} + \mathbf{j}a_{12} + \mathbf{k}a_{13}$$

and also

$$\mathbf{j}' = \mathbf{i}a_{21} + \mathbf{j}a_{22} + \mathbf{k}a_{23},$$
$$\mathbf{k}' = \mathbf{i}a_{31} + \mathbf{j}a_{32} + \mathbf{k}a_{33}. \tag{4.50}$$

We may also go the other way by resolving **i**, **j**, and **k** into components in the primed system. Then

$$\mathbf{i} = \mathbf{i}'a_{11} + \mathbf{j}'a_{21} + \mathbf{k}'a_{31},$$
$$\mathbf{j} = \mathbf{i}'a_{12} + \mathbf{j}'a_{22} + \mathbf{k}'a_{32},$$
$$\mathbf{k} = \mathbf{i}'a_{13} + \mathbf{j}'a_{23} + \mathbf{k}'a_{33}. \tag{4.51}$$

Associating **i** and **i**′ with the subscript 1, **j** and **j**′ with the subscript 2, **k** and **k**′ with the subscript 3, we see that in each case the first subscript of a_{ij} refers to the primed unit vector (**i**′, **j**′, **k**′), whereas the second subscript refers to the unprimed unit vector (**i**, **j**, **k**).

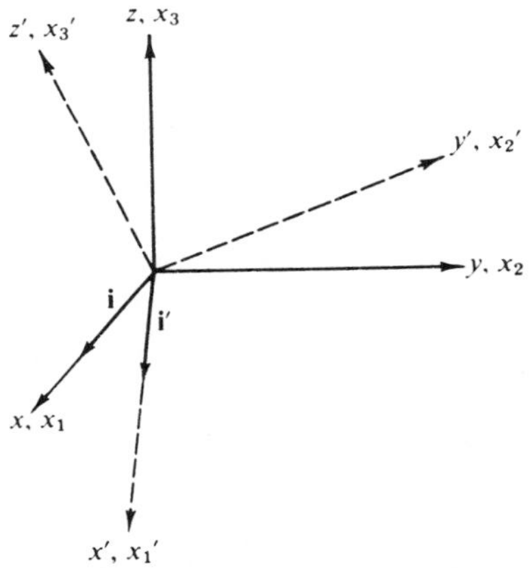

FIG. 4.1 Cartesian coordinate systems

Applications to vectors. If we consider a vector whose components are functions of the position in space, then

$$\mathbf{V}(x, y, z) = \mathbf{i}V_x + \mathbf{j}V_y + \mathbf{k}V_z$$
$$= \mathbf{V}'(x', y', z') = \mathbf{i}'V'_{x'} + \mathbf{j}'V'_{y'} + \mathbf{k}'V'_{z'}, \tag{4.52}$$

since the point may be given both by the coordinates (x, y, z) and the coordinates

(x', y', z'). Using Eq. 4.51 to eliminate **i**, **j**, and **k**, Eq. 4.52 may be separated into three scalar equations,

$$V'_{x'} = a_{11}V_x + a_{12}V_y + a_{13}V_z$$

$$V'_{y'} = a_{21}V_x + a_{22}V_y + a_{23}V_z \tag{4.53}$$

and

$$V'_{z'} = a_{31}V_x + a_{32}V_y + a_{33}V_z.$$

In particular, these relations will hold for the coordinates of a point (x, y, z) and (x', y', z'), giving

$$x' = a_{11}x + a_{12}y + a_{13}z,$$

$$y' = a_{21}x + a_{22}y + a_{23}z, \tag{4.54}$$

and

$$z' = a_{31}x + a_{32}y + a_{33}z.$$

It is convenient to change the notation slightly at this point.
Let

$$x \to x_1,$$

$$y \to x_2, \tag{4.55}$$

$$z \to x_3,$$

and similarly for the primed coordinates. In this notation the set of three equations (4.54) may be written

$$x'_i = \sum_{j=1}^{3} a_{ij}x_j, \tag{4.56}$$

where i takes on the values 1, 2, and 3 and the result is three *separate* equations.

Now let us set aside these results and try a different approach to the same problem. We consider two coordinate systems (x_1, x_2, x_3) and (x'_1, x'_2, x'_3) with a common origin and one point (x_1, x_2, x_3) in the unprimed system, (x'_1, x'_2, x'_3) in the primed system. Note the usual ambiguity. The same symbol x denotes both the coordinate axis and a particular distance along that axis. Since our system is linear, x'_i must be a linear combination of the x_i's. Let

$$x'_i = \sum_{j=1}^{3} a_{ij}x_j. \tag{4.57}$$

The a_{ij} may be identified as our old friends, the direction cosines. This identification is carried out for the two-dimensional case later.

If we have two sets of quantities (V_1, V_2, V_3) in the unprimed system and (V'_1, V'_2, V'_3) in the primed system, related in the same way as the coordinates of a point in the two different systems (Eq. 4.57),

$$V'_i = \sum_{j=1}^{3} a_{ij}V_j, \tag{4.58}$$

then, as in Section 1.2, the quantities (V_1, V_2, V_3) are defined as the components of a vector; that is, a vector is defined in terms of transformation properties of its components under a rotation of the coordinate system. In a sense the coordinates of a point have been taken as a prototype vector. The power and usefulness of this definition becomes apparent in Chapter 3, in which it is extended to define pseudovectors and tensors.

From Eq. 4.56 we can derive interesting information about the a_{ij}'s which describe the orientation of coordinate system (x'_1, x'_2, x'_3) relative to system (x_1, x_2, x_3). The length from the origin to the point is the same in both systems. Squaring, for convenience,

$$\begin{aligned}\sum_i x_i^2 &= \sum_i x_i'^2 \\ &= \sum_i \Big(\sum_j a_{ij}x_j\Big)\Big(\sum_k a_{ik}x_k\Big)\,^1 \\ &= \sum_{j,k} x_j x_k \sum_i a_{ij}a_{ik}.\end{aligned} \tag{4.59}$$

This can be true for all points if and only if

$$\sum_i a_{ij}a_{ik} = \delta_{jk}, \qquad j, k = 1, 2, 3. \tag{4.60}$$

Verification of Eq. 4.60, if needed, may be obtained by returning to Eq. 4.59 and setting $\mathbf{r} = (1, 0, 0)$, $(0, 1, 0)$, $(0, 0, 1)$, $(1, 1, 0)$, etc., to evaluate the nine relations given by equation 4.60. This process is valid, since Eq. 4.59 must hold for all $\mathbf{r}$ for a given set of a_{ij}. Equation 4.60, a consequence of requiring that the length remain constant (invariant) under rotation of the coordinate system, is called the *orthogonality condition*. The a_{ij}'s written as a matrix A, form an orthogonal matrix. Note carefully that Eq. 4.60 is *not* matrix multiplication. Rather it is interpreted later as a scalar product of two columns of A.

In matrix notation Eq. 4.56 becomes

$$\{x'\} = \mathsf{A}\{x\}. \tag{4.61}$$

Orthogonality conditions—two-dimensional case. A better feeling for the a_{ij}'s and the orthogonality condition may be gained by considering rotation in two dimensions in detail. (This can be thought of as a three-dimensional system with the x_1- x_2-axes rotated about x_3.) From Fig. 1.6, Chapter 1,

$$\begin{aligned} x'_1 &= x_1 \cos\varphi + x_2 \sin\varphi, \\ x'_2 &= -x_1 \sin\varphi + x_2 \cos\varphi. \end{aligned} \tag{4.62}$$

Therefore by Eq. 4.61

$$\mathsf{A} = \begin{pmatrix} \cos\varphi & \sin\varphi \\ -\sin\varphi & \cos\varphi \end{pmatrix}. \tag{4.63}$$

It is clear from Fig. 1.6 that

[1] Note that *two* independent indices j and k are used.

$$a_{11} = \cos\varphi = \cos(x_1', x_1),$$
$$a_{12} = \sin\varphi = \cos\left(\frac{\pi}{2} - \varphi\right) = \cos(x_1', x_2), \text{ etc.} \tag{4.64}$$

thus identifying the matrix elements a_{ij} as the direction cosines. Equation 4.60, the orthogonality condition, becomes

$$\sin^2\varphi + \cos^2\varphi = 1,$$
$$\sin\varphi\cos\varphi - \sin\varphi\cos\varphi = 0. \tag{4.65}$$

The extension to three dimensions (rotation of the coordinates through an angle φ counterclockwise about x_3) is simply

$$\mathsf{A} = \begin{pmatrix} \cos\varphi & \sin\varphi & 0 \\ -\sin\varphi & \cos\varphi & 0 \\ 0 & 0 & 1 \end{pmatrix}. \tag{4.66}$$

The $a_{33} = 1$ expresses the fact that $x_3' = x_3$, since the rotation has been about the x_3-axis. The zeros guarantee that x_1' and x_2' do not depend on x_3 and that x_3' does not depend on x_1 and x_2. In more sophisticated language, x_1 and x_2 span an invariant *subspace*, while x_3 forms an invariant *subspace* alone. The general form of A is reducible. Eq. 4.66 gives one possible decomposition.

Inverse matrix, A^{-1}. Returning to the general transformation matrix A, the inverse matrix A^{-1} is defined such that

$$\{x\} = \mathsf{A}^{-1}\{x'\}; \tag{4.67}$$

that is, A^{-1} describes the reverse of the rotation given by A and returns the coordinate system to its original position. Symbolically, Eqs. 4.61 and 4.67 combine to give

$$\{x\} = \mathsf{A}^{-1}\mathsf{A}\{x\}, \tag{4.68}$$

and since $\{x\}$ is arbitrary

$$\mathsf{A}^{-1}\mathsf{A} = 1 \tag{4.69}$$

the unit matrix. Similarly

$$\mathsf{A}\mathsf{A}^{-1} = 1, \tag{4.70}$$

using Eqs. 4.61 and 4.67 and eliminating $\{x\}$ instead of $\{x'\}$.

Transpose matrix, $\tilde{\mathsf{A}}$. We can determine the elements of our postulated inverse matrix A^{-1} by employing the orthogonality condition. Equation 4.60, the orthogonality condition, does *not* conform to our definition of matrix multiplication, but it can be put in the required form by *defining* a new matrix $\tilde{\mathsf{A}}$ such that

$$\tilde{a}_{ij} = a_{ji}; \tag{4.71}$$

that is, $\tilde{\mathsf{A}}$, called "A transpose," is formed from A by interchanging rows and columns. Equation 4.60 becomes

$$\tilde{\mathsf{A}}\mathsf{A} = 1. \tag{4.72}$$

This is a restatement of the orthogonality condition and may be taken as a definition of orthogonality. Multiplying Eq. 4.72 by A^{-1} from the right and using Eq. 4.70,

$$\tilde{\mathsf{A}} = \mathsf{A}^{-1}. \tag{4.73}$$

This important result that the inverse equals the transpose holds only for orthogonal matrices and indeed may be taken as a further restatement of the orthogonality condition.

Multiplying Eq. 4.73 by A from the left,

$$\mathsf{A}\tilde{\mathsf{A}} = 1 \tag{4.74}$$

or

$$\sum_i a_{ji} a_{ki} = \delta_{jk}, \tag{4.75}$$

which is still another form of the orthogonality condition.

It is now possible to see and understand why the term *orthogonal* is appropriate for these matrices. We have the general form

$$\mathsf{A} = \begin{pmatrix} a_{11} & a_{12} & a_{13} \\ a_{21} & a_{22} & a_{23} \\ a_{31} & a_{32} & a_{33} \end{pmatrix},$$

a matrix of direction cosines in which a_{ij} is the cosine of the angle between x_i' and x_j. Therefore a_{11}, a_{12}, a_{13} are the direction cosines of x_1' relative to x_1, x_2, x_3. These three elements of A *define* a unit length along x_1', that is, a unit vector $\mathbf{i}'$,

$$\mathbf{i}' = \mathbf{i}a_{11} + \mathbf{j}a_{12} + \mathbf{k}a_{13}.$$

The orthogonality relation (Eq. 4.75) is simply a statement that the unit vectors $\mathbf{i}'$, $\mathbf{j}'$, and $\mathbf{k}'$ are mutually perpendicular or orthogonal. Our orthogonal transformation matrix A rotates one orthogonal coordinate system into a second orthogonal coordinate system.

Euler Angles. Our transformation matrix A contains nine direction cosines. Clearly only three of these are independent, Eq. 4.60 providing six constraints. Equivalently, we may say that two parameters (θ and φ in spherical polar coordinates) are required to fix the axis of rotation. Then one additional parameter describes the amount of rotation about the specified axis. In the Lagrangian formulation of mechanics (Section 17.3) it is necessary to describe A using some set of three independent parameters rather than the redundant direction cosines. The usual choice of parameters is the Euler angles.[1]

The goal is to describe the orientation of a final rotated system (x_1''', x_2''', x_3''') relative to some initial coordinate system (x_1, x_2, x_3). The final system is developed in three steps—each step involving one rotation described by one Euler angle:

1. The x_1', x_2', x_3'-axes are rotated about the x_3-axis through an angle α counter-

[1] There are almost as many definitions of the Euler angles as there are authors. Here we follow the choice generally made by workers in the area of group theory and the quantum theory of angular momentum (cf. Section 4.9).

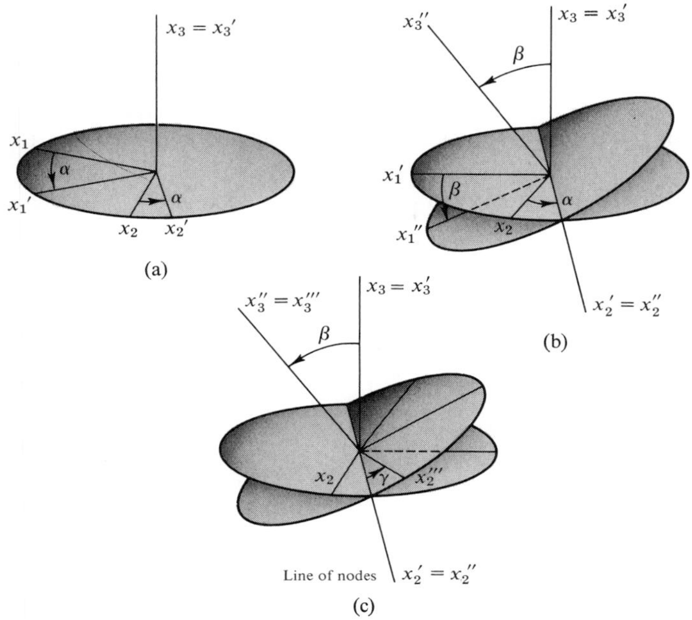

FIG. 4.2(a) Rotation about x_3 through angle α; (b) Rotation about x_2' through angle β; (c) Rotation about x_3'' through angle γ.

clockwise relative to x_1, x_2, x_3. (The x_3 and x_3'-axes coincide.)

2. The x_1'', x_2'', x_3''-axes are rotated about the x_2'-axis[2] through an angle β counterclockwise relative to x_1', x_2', x_3'. (The x_2' and the x_2''-axes coincide).

3. The third and final rotation is through an angle γ counterclockwise about the x_3''-axis, yielding the x_1''', x_2''', x_3''' system. (The x_3'' and x_3'''-axes coincide.)

The three matrices describing these rotations are:

$$\mathsf{R}_z(\alpha) = \begin{pmatrix} \cos\alpha & \sin\alpha & 0 \\ -\sin\alpha & \cos\alpha & 0 \\ 0 & 0 & 1 \end{pmatrix} \tag{4.76}$$

exactly like Eq. 4.66,

$$\mathsf{R}_y(\beta) = \begin{pmatrix} \cos\beta & 0 & -\sin\beta \\ 0 & 1 & 0 \\ \sin\beta & 0 & \cos\beta \end{pmatrix} \tag{4.77}$$

and

$$\mathsf{R}_z(\gamma) = \begin{pmatrix} \cos\gamma & \sin\gamma & 0 \\ -\sin\gamma & \cos\gamma & 0 \\ 0 & 0 & 1 \end{pmatrix} \tag{4.78}$$

[2] Many authors choose this second rotation to be about the x_1'-axis.

The total rotation is described by the triple matrix product.

$$\mathsf{A}(\alpha, \beta, \gamma) = \mathsf{R}_z(\gamma)\mathsf{R}_y(\beta)\mathsf{R}_z(\alpha). \tag{4.79}$$

(The component form of successive transformations is considered in Eqs. 4.89–4.92.)

Note the order: $\mathsf{R}_z(\alpha)$ operates first, then $\mathsf{R}_y(\beta)$, and finally $\mathsf{R}_z(\gamma)$. Direct multiplication gives

$$\mathsf{A}(\alpha, \beta, \gamma) = \begin{pmatrix} \cos\gamma\cos\beta\cos\alpha - \sin\gamma\sin\alpha & \cos\gamma\cos\beta\sin\alpha + \sin\gamma\cos\alpha & -\cos\gamma\sin\beta \\ -\sin\gamma\cos\beta\cos\alpha - \cos\gamma\sin\alpha & -\sin\gamma\cos\beta\sin\alpha + \cos\gamma\cos\alpha & \sin\gamma\sin\beta \\ \sin\beta\cos\alpha & \sin\beta\sin\alpha & \cos\beta \end{pmatrix}. \tag{4.80}$$

Equating $\mathsf{A}(a_{ij})$ with $\mathsf{A}(\alpha, \beta, \gamma)$, element by element, yields the direction cosines in terms of the three Euler angles. We could use this Euler angle identification to verify the direction cosine identities, Eq. 1.43, of Section 1.4, but the approach of Exercise 4.3.3 is much more elegant.

The Euler angle description of rotations forms a basis for developing the rotation group of Section 4.9.

It will be noticed that the matrices have been handled in two ways in the foregoing discussion: by their components and as single entities. Each technique has its own advantages. Both are useful.

Consider the evaluation of $(\mathsf{ST})^{-1}$ where ST is a (product) matrix which has an inverse. Then, clearly,

$$(\mathsf{ST})(\mathsf{ST})^{-1} = 1. \tag{4.81}$$

Multiplying first by S^{-1} and then by T^{-1} successively from the left,

$$(\mathsf{ST})^{-1} = \mathsf{T}^{-1}\mathsf{S}^{-1}. \tag{4.82}$$

The inverse of a product equals the product of the inverses *in reverse order*. This may be readily generalized to any number of factors.

On the other hand, the evaluation of $\widetilde{(\mathsf{ST})}$ may perhaps best be carried out by considering the components. Let $\mathsf{U} = \mathsf{ST}$, so that

$$\begin{aligned} u_{ik} &= \sum_j s_{ij} t_{jk} \\ &= \sum_j \tilde{t}_{kj} \tilde{s}_{ji}, \end{aligned} \tag{4.83}$$

using the definition of transpose. But

$$u_{ik} = \tilde{u}_{ki}, \tag{4.84}$$

and Eq. 4.83 may be written

$$\widetilde{(\mathsf{ST})} = \tilde{\mathsf{T}}\tilde{\mathsf{S}}. \tag{4.85}$$

The transpose of a product equals the product of the transposes *in reverse order*.

Note that in neither of the two illustrations have we required S or T to be orthogonal.

Symmetry properties. The transpose matrix is useful in a discussion of symmetry properties. If

$$\mathsf{A} = \tilde{\mathsf{A}}, \qquad a_{ij} = a_{ji}, \tag{4.86}$$

the matrix is called *symmetric*, whereas if

$$\mathsf{A} = -\tilde{\mathsf{A}}, \qquad a_{ij} = -a_{ji}, \tag{4.87}$$

it is called antisymmetric or skewsymmetric. The diagonal elements vanish. It is easy to show that any (square) matrix may be written as the sum of a symmetric matrix and an antisymmetric matrix. Consider the identity

$$\mathsf{A} = \tfrac{1}{2}[\mathsf{A} + \tilde{\mathsf{A}}] + \tfrac{1}{2}[\mathsf{A} - \tilde{\mathsf{A}}]. \tag{4.88}$$

$[\mathsf{A}+\tilde{\mathsf{A}}]$ is clearly symmetric, whereas $[\mathsf{A}-\tilde{\mathsf{A}}]$ is clearly antisymmetric. This is the matrix analog of Eq. 3.22, Chapter 3, for tensors.

Successive rotations, matrix multiplication. Returning to orthogonal matrices, let the coordinate rotation

$$\{x'\} = \mathsf{A}\{x\} \tag{4.89}$$

be followed by a second rotation given by matrix B such that

$$\{x''\} = \mathsf{B}\{x'\}. \tag{4.90}$$

In component form

$$\begin{aligned} x_i'' &= \sum_j b_{ij} x_j' \\ &= \sum_j b_{ij} \sum_k a_{jk} x_k \\ &= \sum_k \Big(\sum_j b_{ij} a_{jk}\Big) x_k . \end{aligned} \tag{4.91}$$

The summation over j is matrix multiplication defining a matrix $\mathsf{C} = \mathsf{BA}$ such that

$$x_i'' = \sum_k c_{ik} x_k . \tag{4.92}$$

Again the definition of matrix multiplication is found useful and indeed *this is the justification for its existence*. The physical interpretation is that the matrix product

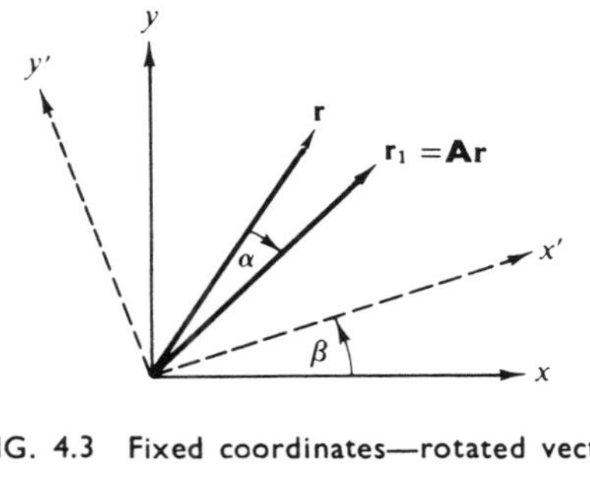

FIG. 4.3 Fixed coordinates—rotated vector

of the two matrices, BA, is the rotation that carries the unprimed system directly into the double-primed coordinate system.

So far we have interpreted the orthogonal matrix as rotating the coordinate system. This changes the components of a fixed vector (not rotating with the coordinates) (Fig. 1.6, Chapter 1). However, Eq. 4.89 may be interpreted equally well as a rotation of the *vector* in the *opposite* direction (Fig. 4.3).

Suppose we interpret matrix A as rotating a *vector* $\mathbf{r}$ into the position shown by $\mathbf{r}_1$.

$$\mathbf{r}_1 = \mathsf{A}\mathbf{r}. \tag{4.93}$$

Now let us rotate the *coordinates* by applying matrix B, which rotates (x, y, z) into (x', y', z')

$$\begin{aligned} \mathsf{B}\mathbf{r}_1 &= \mathsf{BA}\mathbf{r} \\ &= \mathsf{BA}(\mathsf{B}^{-1}\mathsf{B})\mathbf{r} \\ &= (\mathsf{BAB}^{-1})\mathsf{B}\mathbf{r}. \end{aligned} \tag{4.94}$$

$\mathsf{B}\mathbf{r}_1$ is just $\mathbf{r}_1$ in the new coordinate system with a similar interpretation holding for $\mathsf{B}\mathbf{r}$. Hence *in this new system* ($\mathsf{B}\mathbf{r}$) is rotated into position ($\mathsf{B}\mathbf{r}_1$) by the matrix BAB^{-1}; or, in other words, in the new system, the coordinates having been rotated by matrix B, A has the form A′, in which

$$\mathsf{A}' = \mathsf{BAB}^{-1}. \tag{4.95}$$

Similarity transformation. The transformation defined by Eq. 4.95 with B any matrix, not necessarily orthogonal, is known as a similarity transformation.

In component form Eq. 4.95 becomes

$$a'_{ij} = \sum_{k,l} b_{ik} a_{kl} b^{-1}_{lj}. \tag{4.96}$$

Now if B is orthogonal,

$$b^{-1}_{lj} = \tilde{b}_{lj} = b_{jl}, \tag{4.97}$$

and we have

$$a'_{ij} = \sum_{k,l} b_{ik} b_{jl} a_{kl}. \tag{4.98}$$

Relation to tensors. Comparing (4.98) with the equations of Section 3.1, we see that it is the definition of a tensor of second rank. Hence a matrix that transforms by an *orthogonal* similarity transformation is, by definition, a tensor. Clearly then, any *orthogonal* matrix A, interpreted as rotating a vector (Eq. 4.93), may be called a tensor. If, however, we consider the orthogonal matrix as a collection of fixed direction cosines, giving the new orientation of a coordinate system, there is no tensor transformation involved.

The symmetry and antisymmetry properties defined earlier are preserved under *orthogonal* similarity transformations. Let A be a symmetric matrix, $\mathsf{A} = \tilde{\mathsf{A}}$, and

$$\mathsf{A}' = \mathsf{BAB}^{-1}. \tag{4.99}$$

Now

$$\tilde{\mathsf{A}}' = \widetilde{\mathsf{BAB}^{-1}} = \widetilde{\mathsf{B}^{-1}}\tilde{\mathsf{A}}\tilde{\mathsf{B}} = \mathsf{B}\tilde{\mathsf{A}}\mathsf{B}^{-1}, \tag{4.100}$$

since B is orthogonal. But $\mathsf{A} = \tilde{\mathsf{A}}$. Therefore

$$\tilde{\mathsf{A}}' = \mathsf{BAB}^{-1} = \mathsf{A}', \tag{4.101}$$

showing that the property of symmetry is invariant under an orthogonal similarity transformation. In general, symmetry is *not* preserved under a nonorthogonal similarity transformation.

EXERCISES

Note. Assume all matrix elements are real.

4.3.1 Show that the product of two orthogonal matrices is orthogonal.
Note. This is a key step in showing that all $n \times n$ orthogonal matrices form a group (Section 4.9).

4.3.2 If A is orthogonal, show that its determinant has unit magnitude.

4.3.3 If A is orthogonal show that $a_{ij} = C_{ij}$, where C_{ij} is the *cofactor* of a_{ij}. This yields the identities of Eq. 1.41 used in Section 1.4 to show that a cross product of vectors is itself a vector.
Hint. Note Exercise 4.2.27.

4.3.4 Another set of Euler rotations in common use is
1. a rotation about the x_3-axis through an angle φ, counterclockwise,
2. a rotation about the x_1'-axis through an angle θ, counterclockwise, and
3. a rotation about the x_3''-axis through an angle ψ, counterclockwise.

If

$$\begin{aligned} \alpha &= \varphi - \pi/2 \qquad & \varphi &= \alpha + \pi/2 \\ \beta &= \theta & \theta &= \beta \\ \gamma &= \psi + \pi/2 & \psi &= \gamma - \pi/2, \end{aligned}$$

show that the final systems are identical.

4.3.5 Suppose the earth is moved (rotated) so that the north pole goes to 30° north, 20° West (original latitude and longitude system) and the 10° West meridian points due south.
(a) What are the Euler angles describing this rotation?
(b) Find the corresponding direction cosines.

$$\text{(b)}\ A = \begin{pmatrix} 0.9551 & -0.2552 & -0.1504 \\ 0.0052 & 0.5221 & -0.8529 \\ 0.2962 & 0.8138 & 0.5000 \end{pmatrix}$$

4.3.6 Verify that the Euler angle rotation matrix, Eq. 4.80, is invariant under the transformation

$$\alpha \to \alpha + \pi, \qquad \beta \to -\beta, \qquad \gamma \to \gamma - \pi.$$

4.3.7 Show that the trace of the product of a symmetric and an antisymmetric matrix is zero.

4.3.8 Show that the trace of a matrix remains invariant under similarity transformations.

4.3.9 Show that the determinant of a matrix remains invariant under similarity transformations.

4.3.10 Show that the property of antisymmetry is invariant under orthogonal similarity transformations.

4.3.11 Show that the sum of the squares of the elements of a matrix remains invariant under orthogonal similarity transformations.

4.3.12 As a generalization of Exercise 4.3.11, show that

$$\sum_{jk} S_{jk} T_{jk} = \sum_{l,m} S'_{lm} T'_{lm},$$

where the primed and unprimed elements are related by an orthogonal similarity transformation. This result is useful in deriving invariants in electromagnetic theory (cf. Section 3.7).

4.3.13 A rotation $\varphi_1 + \varphi_2$ about the z-axis is carried out as two successive rotations φ_1 and φ_2, each about the z-axis. Use the matrix representation of the rotations to derive the trigonometric identities:

$$\cos(\varphi_1 + \varphi_2) = \cos\varphi_1 \cos\varphi_2 - \sin\varphi_1 \sin\varphi_2$$

$$\sin(\varphi_1 + \varphi_2) = \sin\varphi_1 \cos\varphi_2 + \cos\varphi_1 \sin\varphi_2 .$$

4.4 Oblique Coordinates

Throughout this entire book so far—vector analysis, coordinate systems, tensor analysis, and now matrices—we have always taken our coordinates to be orthogonal. But sometimes the demands of a physical system force the use of a nonorthogonal or oblique system of coordinates. In describing the physical properties of a crystal, for example, it might be most convenient to use the coordinate system defined by the axes of this crystal—and these axes are often oblique.

Consider a coordinate system in which the noncoplanar unit vectors **a**, **b**, and **c** are *not orthogonal.* (For describing a crystal **a**, **b**, and **c** might not have unit magnitude either. The interatomic spacings would be more appropriate lengths.) Then an arbitrary vector may be written

$$\mathbf{V} = \mathbf{i}V_x + \mathbf{j}V_y + \mathbf{k}V_z = \mathbf{a}v_a + \mathbf{b}v_b + \mathbf{c}v_c = \mathbf{v} \tag{4.102}$$

V will denote the vector expressed in the usual rectangular cartesian system, while

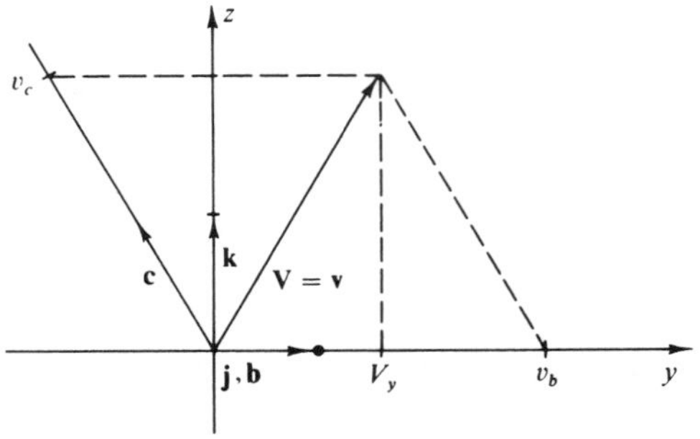

FIG. 4.4 $\mathbf{V} = \mathbf{j}V_y + \mathbf{k}V_z = \mathbf{b}v_b + \mathbf{c}v_c = \mathbf{v}$

$\mathbf{v}$ is the *same vector* expressed in the oblique coordinate system.

The special case (really two-dimensional) of $\mathbf{j}$, $\mathbf{k}$, $\mathbf{b}$, $\mathbf{c}$, and $\mathbf{V}$ all in the $x = 0$ plane is shown in Fig. 4.4. Note carefully that the components v_b and v_c are found by projecting the tip of $\mathbf{V}$ parallel to $\mathbf{c}$ for v_b and parallel to $\mathbf{b}$ for v_c. The general procedure for obtaining one component would be to pass a plane through the tip of $\mathbf{V}$ parallel to the plane defined by the other two unit vectors. With the components defined this way, the sum of the components is just $\mathbf{V}$ by the triangle or parallelogram laws of vector addition, Section 1.1.

We proceed from Eq. 4.102 exactly as in Section 4.3 with $\mathbf{a}$ instead of $\mathbf{i}'$, $\mathbf{b}$ instead of $\mathbf{j}'$ and $\mathbf{c}$ in place of $\mathbf{k}'$. From

$$\begin{aligned} \mathbf{a} &= \mathbf{i}a_x + \mathbf{j}a_y + \mathbf{k}a_z \\ \mathbf{b} &= \mathbf{i}b_x + \mathbf{j}b_y + \mathbf{k}b_z \\ \mathbf{c} &= \mathbf{i}c_x + \mathbf{j}c_y + \mathbf{k}c_z, \end{aligned} \tag{4.103}$$

equating cartesian components,

$$\begin{aligned} V_x &= a_x v_a + b_x v_b + c_x v_c \\ V_y &= a_y v_a + b_y v_b + c_y v_c \\ V_z &= a_z v_a + b_z v_b + c_z v_c. \end{aligned} \tag{4.104}$$

In matrix form

$$\mathbf{V} = \mathsf{P}\mathbf{v} \tag{4.105}$$

where

$$\mathsf{P} = \begin{pmatrix} a_x & b_x & c_x \\ a_y & b_y & c_y \\ a_z & b_z & c_z \end{pmatrix} \tag{4.106}$$

The transformation matrix P is *not* orthogonal, since the *column* vectors forming it, $\mathbf{a}$, $\mathbf{b}$, and $\mathbf{c}$ are not orthogonal.

Since

$$\mathbf{v} = \mathsf{P}^{-1}\mathbf{V} \tag{4.107}$$

we seek P^{-1}. The solution is actually developed in Section 1.5. The reciprocal lattice vectors

$$\mathbf{a}' = \frac{\mathbf{b} \times \mathbf{c}}{\mathbf{a} \times \mathbf{b} \cdot \mathbf{c}}, \qquad \mathbf{b}' = \frac{\mathbf{c} \times \mathbf{a}}{\mathbf{a} \times \mathbf{b} \cdot \mathbf{c}}, \qquad \mathbf{c}' = \frac{\mathbf{a} \times \mathbf{b}}{\mathbf{a} \times \mathbf{b} \cdot \mathbf{c}}, \tag{4.108}$$

taken as *row* vectors, form a matrix Q,

$$\mathsf{Q} = \begin{pmatrix} a'_x & a'_y & a'_z \\ b'_x & b'_y & b'_z \\ c'_x & c'_y & c'_z \end{pmatrix}. \tag{4.109}$$

It should be emphasized that **a**′, **b**′, and **c**′ are not orthogonal. Also they are not of unit length, and if **a**, **b**, and **c** have dimensions, then **a**′, **b**′, and **c**′ have *reciprocal* dimensions. If **a** is a length, **a**′ could be a wave number. From the properties developed in Section 1.5

$$\mathsf{PQ} = \mathsf{QP} = 1 \tag{4.110}$$

or

$$\mathsf{Q} = \mathsf{P}^{-1}, \qquad \mathsf{P} = \mathsf{Q}^{-1} \tag{4.111}$$

Exercise 4.4.1 outlines a slightly different, but equivalent, derivation of Q. From Eqs. 4.107 and 4.111

$$\mathbf{v} = \mathsf{Q}\mathbf{V} \tag{4.112}$$

Taking the transpose of Eqs. 4.105 and 4.112

$$[V] = [v]\tilde{\mathsf{P}}, \qquad [v] = [V]\tilde{\mathsf{Q}} \tag{4.113}$$

[ ] denoting a row vector—as in Section 4.2.

V may be resolved in the **a**′, **b**′, **c**′-space (the reciprocal lattice) exactly as in the **a**, **b**, **c**-space. From the primed analog of Eqs. 4.102–4.104

$$\mathbf{V} = \tilde{\mathsf{Q}}\mathbf{v}' \qquad \mathbf{v}' = \tilde{\mathsf{P}}\mathbf{V}, \tag{4.114}$$

and

$$[V] = [v']\mathsf{Q} \qquad [v'] = [V]\mathsf{P}. \tag{4.115}$$

The scalar product of two vectors **U** and **V** becomes

$$[U]\{V\} = [U]\mathsf{PQ}\{V\} = [u']\{v\} \tag{4.116}$$

from Eqs. 4.112 and 4.115. { } denotes a column vector. The square of a vector in oblique coordinates is not the sum of the squares of the components, but rather the sum of the products of an oblique component and the corresponding reciprocal lattice component.

If **U** and **V** in Eq. 4.116 above are the differential length $d\mathbf{R} = (dx, dy, dz)$, then

$$ds^2 = [dR]\{dR\} = [dr]\tilde{\mathsf{P}}\mathsf{P}\{dr\} \tag{4.117}$$

using Eqs. 4.105 and 4.113. ds^2 is the square of the distance element; $d\mathbf{r}$ is $d\mathbf{R}$ but resolved in the oblique coordinates. Reference to Eq. 2.4 identifies $\tilde{\mathsf{P}}\mathsf{P}$ as the metric of our oblique coordinates. The metric of the reciprocal lattice is $\mathsf{Q}\tilde{\mathsf{Q}}$

Further development of vector analysis, particularly of a vector calculus in oblique coordinates, is probably best considered a branch of noncartesian tensor analysis. $\mathbf{v} = (v_a, v_b, v_c)$ is a *contravariant* vector in the language of Section 3.1. The corresponding *covariant* components are (v_a', v_b', v_c') in the reciprocal lattice. From Eqs. 4.105, 4.112, and 4.114

$$\{v_i'\} = \tilde{\mathsf{P}}\mathsf{P}\{v_i\} \qquad \text{and} \qquad \{v_i\} = \mathsf{Q}\tilde{\mathsf{Q}}\{v_i'\}. \tag{4.118}$$

The metric $\tilde{\mathsf{P}}\mathsf{P}$ transforms the contravariant vector into covariant form. Its inverse, $\mathsf{Q}\tilde{\mathsf{Q}}$, transforms the covariant vector into contravariant form.

EXERCISES

4.4.1 From the result of Exercise 4.2.27 $q_{ij} = \dfrac{P_{ji}}{|\mathsf{P}|}$, derive the relation

$$\mathbf{a}' = \frac{\mathbf{b} \times \mathbf{c}}{\mathbf{a} \times \mathbf{b} \cdot \mathbf{c}}.$$

4.4.2 The vectors defining a particular system of oblique coordinates are

$$\mathbf{a} = \mathbf{i}, \qquad \mathbf{b} = \mathbf{j}, \qquad \text{and} \qquad \mathbf{c} = (\mathbf{j} + \mathbf{k})/\sqrt{2}.$$

(a) Find P, Q, and metric $\tilde{\mathsf{P}}\mathsf{P}$.
(b) If $\mathbf{V} = \mathbf{i} + 3\mathbf{j} + 2\mathbf{k}$, find $\mathbf{v}$ and $\mathbf{v}'$. Verify that

$$[v']\{v\} = V^2.$$

4.4.3 Show that
(a) $v_a' = \mathbf{a} \cdot \mathbf{V}$
(b) $v_a = \mathbf{a}' \cdot \mathbf{V}$.
Note that the lattice defining vectors $\mathbf{a}$, $\mathbf{a}'$, etc., need not have unit magnitude.

4.5 Hermitian Matrices, Unitary Matrices

Definitions. Up to this point it has been generally assumed that the matrix elements were real. For many calculations in classical physics real matrix elements will suffice. However, in quantum mechanics complex variables are inescapable because of the form of the basic commutation relations (or the form of the time-dependent Schrödinger equation). With this in mind, we generalize to the case of complex matrix elements. To handle these elements, it is convenient to define, or to label, some new properties.

1. Complex conjugate, A^*, formed by taking the complex conjugate ($i \to -i$) of each element, where $i = \sqrt{-1}$.

2. Adjoint, $A^\dagger$, formed by transposing A^*,

$$A^\dagger = \widetilde{A^*} = \tilde{A}^*. \tag{4.119}$$

3. Hermitian matrix. The matrix A is labeled Hermitian (or self-adjoint) if

$$A = A^\dagger. \tag{4.120}$$

In quantum mechanics (or matrix mechanics), matrices are usually constructed to be Hermitian.

4. Unitary matrix. Matrix U is labeled unitary if

$$U^\dagger = U^{-1}, \tag{4.121}$$

which represents a generalization of the concept of orthogonal matrix (cf. Eq. 4.73). If the matrix elements are complex, the physicist is almost always concerned with adjoint matrices, Hermitian matrices, and unitary matrices. Unitary matrices are especially important in quantum mechanics because they leave the length of a (complex) vector unchanged—analogous to the operation of an orthogonal matrix on a real vector. One important exception to this interest in unitary matrices is the group of Lorentz matrices, Sections 3.7 and 4.12. Using Minkowski space, these matrices are *orthogonal*, not *unitary*.

If the transforming matrix in a similarity transformation is unitary, the transformation is referred to as a unitary transformation,

$$A' = UAU^\dagger \tag{4.122}$$

Just as the product of two orthogonal matrices is found to be orthogonal (Exercise 4.3.1), so we can show that the product of two unitary matrices is unitary. Let U_1 and U_2 be unitary. Then

$$\begin{aligned} 1 &= (U_1U_2)(U_1U_2)^{-1} \\ &= U_1U_2U_2^{-1}U_1^{-1} \\ &= U_1U_2U_2^\dagger U_1^\dagger, \end{aligned} \tag{4.123}$$

using the unitary property. Since the operation of adjoint is the same as transpose (except for the complex conjugate),

$$(U_1U_2)^\dagger = U_2^\dagger U_1^\dagger, \tag{4.124}$$

(Exercise 4.5.2). Substituting into Eq. 4.123 we have

$$1 = (U_1U_2)(U_1U_2)^\dagger \tag{4.125}$$

Multiplying from the left by $(U_1U_2)^{-1}$, we obtain

$$(U_1U_2)^{-1} = (U_1U_2)^\dagger, \tag{4.126}$$

which shows that the product of two unitary matrices is itself unitary. This is one of the steps in demonstrating that the $n \times n$ unitary matrices form a group (Section

4.9). Other properties and applications of these concepts are included in the exercises at the end of this section.

Pauli matrices. Four by four complex matrices have been used extensively in relativistic theories of the electron. A convenient starting point for developing the 4×4 matrices is the set of three 2×2 Pauli matrices

$$\sigma_1 = \begin{pmatrix} 0 & 1 \\ 1 & 0 \end{pmatrix}, \quad \sigma_2 = \begin{pmatrix} 0 & -i \\ i & 0 \end{pmatrix}, \quad \sigma_3 = \begin{pmatrix} 1 & 0 \\ 0 & -1 \end{pmatrix} \tag{4.127}$$

These were introduced by W. Pauli to describe a particle of spin one half (nonrelativistic theory). It can readily be shown that (cf. Exercise 4.2.12) the Pauli σ's satisfy

$$\sigma_i \sigma_j + \sigma_j \sigma_i = 2\delta_{ij}\mathsf{1}, \qquad \text{anticommutation} \tag{4.128}$$

$$\sigma_i \sigma_j = i\sigma_k, \qquad \text{cyclic permutation of indices} \tag{4.129}$$

$$(\sigma_i)^2 = \mathsf{1}. \tag{4.130}$$

Dirac matrices. In 1927 P. A. M. Dirac extended this formalism. Dirac required a set of four anticommuting matrices. The three Pauli matrices plus the unit matrix form a complete set; that is, any constant 2×2 matrix M may be written

$$\mathsf{M} = c_0\mathsf{1} + c_1\sigma_1 + c_2\sigma_2 + c_3\sigma_3, \tag{4.131}$$

where c_0, c_1, c_2, and c_3 are constants. Hence the Pauli 2×2 matrices were inadequate; no fourth anticommuting matrix exists. We can show that 3×3 matrices likewise cannot furnish an anticommuting set of four matrices.

Turning to 4×4 matrices, we can build up a complete set as direct products of the Pauli matrices and the unit matrix. Let

$$\sigma_{i,\,\text{Dirac}} = \mathsf{1} \otimes \sigma_{i,\,\text{Pauli}} \tag{4.132}$$

$$\rho_{j,\,\text{Dirac}} = \sigma_{j,\,\text{Pauli}} \otimes \mathsf{1}. \tag{4.133}$$

For example,

$$\sigma_1 = \begin{pmatrix} \mathsf{1}(\sigma_1) & 0 \\ 0 & \mathsf{1}(\sigma_1) \end{pmatrix} = \begin{pmatrix} 0 & 1 & 0 & 0 \\ 1 & 0 & 0 & 0 \\ 0 & 0 & 0 & 1 \\ 0 & 0 & 1 & 0 \end{pmatrix}.$$

We can show that these 4×4 matrices satisfy the relations

$$\left.\begin{aligned} \sigma_i\sigma_j + \sigma_j\sigma_i &= 2\delta_{ij}\mathsf{1}, \\ \rho_i\rho_j + \rho_j\rho_i &= 2\delta_{ij}\mathsf{1}, \end{aligned}\right\} \qquad \text{anticommutation,} \tag{4.134}$$

$$\sigma_i\rho_j - \rho_j\sigma_i \equiv [\sigma_i, \rho_j] = 0, \qquad \text{commutation,} \tag{4.135}$$

and

$$\left.\begin{aligned} \sigma_i\sigma_j &= i\sigma_k, \\ \rho_i\rho_j &= i\rho_k, \end{aligned}\right\} \qquad \text{cyclic permutation of indices.} \tag{4.136}$$

TABLE 4.1

Dirac Matrices

	$\mathbf{1}$	σ_1	σ_2	σ_3
$\mathbf{1}$	$\begin{pmatrix} 1 & 0 & 0 & 0 \\ 0 & 1 & 0 & 0 \\ 0 & 0 & 1 & 0 \\ 0 & 0 & 0 & 1 \end{pmatrix}$ $\mathbf{1}$	$\begin{pmatrix} 0 & 1 & 0 & 0 \\ 1 & 0 & 0 & 0 \\ 0 & 0 & 0 & 1 \\ 0 & 0 & 1 & 0 \end{pmatrix}$ σ_1	$\begin{pmatrix} 0 & -i & 0 & 0 \\ i & 0 & 0 & 0 \\ 0 & 0 & 0 & -i \\ 0 & 0 & i & 0 \end{pmatrix}$ σ_2	$\begin{pmatrix} 1 & 0 & 0 & 0 \\ 0 & -1 & 0 & 0 \\ 0 & 0 & 1 & 0 \\ 0 & 0 & 0 & -1 \end{pmatrix}$ σ_3
ρ_1	$\begin{pmatrix} 0 & 0 & 1 & 0 \\ 0 & 0 & 0 & 1 \\ 1 & 0 & 0 & 0 \\ 0 & 1 & 0 & 0 \end{pmatrix}$ $\rho_1, -\gamma_5$	$\begin{pmatrix} 0 & 0 & 0 & 1 \\ 0 & 0 & 1 & 0 \\ 0 & 1 & 0 & 0 \\ 1 & 0 & 0 & 0 \end{pmatrix}$ α_1	$\begin{pmatrix} 0 & 0 & 0 & -i \\ 0 & 0 & i & 0 \\ 0 & -i & 0 & 0 \\ i & 0 & 0 & 0 \end{pmatrix}$ α_2	$\begin{pmatrix} 0 & 0 & 1 & 0 \\ 0 & 0 & 0 & -1 \\ 1 & 0 & 0 & 0 \\ 0 & -1 & 0 & 0 \end{pmatrix}$ α_3
ρ_2	$\begin{pmatrix} 0 & 0 & -i & 0 \\ 0 & 0 & 0 & -i \\ i & 0 & 0 & 0 \\ 0 & i & 0 & 0 \end{pmatrix}$ ρ_2, α_5	$\begin{pmatrix} 0 & 0 & 0 & -i \\ 0 & 0 & -i & 0 \\ 0 & i & 0 & 0 \\ i & 0 & 0 & 0 \end{pmatrix}$ γ_1	$\begin{pmatrix} 0 & 0 & 0 & -1 \\ 0 & 0 & 1 & 0 \\ 0 & 1 & 0 & 0 \\ -1 & 0 & 0 & 0 \end{pmatrix}$ γ_2	$\begin{pmatrix} 0 & 0 & -i & 0 \\ 0 & 0 & 0 & i \\ i & 0 & 0 & 0 \\ 0 & -i & 0 & 0 \end{pmatrix}$ γ_3
ρ_3	$\begin{pmatrix} 1 & 0 & 0 & 0 \\ 0 & 1 & 0 & 0 \\ 0 & 0 & -1 & 0 \\ 0 & 0 & 0 & -1 \end{pmatrix}$ $\rho_3, \alpha_4, \gamma_4, \beta$	$\begin{pmatrix} 0 & 1 & 0 & 0 \\ 1 & 0 & 0 & 0 \\ 0 & 0 & 0 & -1 \\ 0 & 0 & -1 & 0 \end{pmatrix}$ δ_1	$\begin{pmatrix} 0 & -i & 0 & 0 \\ i & 0 & 0 & 0 \\ 0 & 0 & 0 & i \\ 0 & 0 & -i & 0 \end{pmatrix}$ δ_2	$\begin{pmatrix} 1 & 0 & 0 & 0 \\ 0 & -1 & 0 & 0 \\ 0 & 0 & -1 & 0 \\ 0 & 0 & 0 & 1 \end{pmatrix}$ δ_3

It is now possible to set up a matrix multiplication table (Table 4.1).

Dirac originally chose to use the set of four matrices labeled $\alpha_1, \alpha_2, \alpha_3$, and α_4, where $\alpha_i = \rho_1\sigma_i$ and $\alpha_4 = \rho_3$. Today the set labelled γ_i, $i = 1, 2, 3, 4, 5$, is probably in more common use.

These 4×4 Dirac matrices may be referred to as E_{ij}, in which

$$\mathsf{E}_{ij} = \rho_i\sigma_j.$$

With the understanding that $\rho_0 = \sigma_0 = \mathbf{1}$, the unit matrix, we let the indices i and j range from 0 to 3. These 16 matrices E_{ij} have a number of interesting properties:

1. Det $\mathsf{E}_{ij} = +1$.
2. $\mathsf{E}_{ij}^2 = \mathbf{1}$.
3. $\mathsf{E}_{ij} = \mathsf{E}_{ij}^\dagger$; all are Hermitian and then, by property 2, unitary.
4. Trace $(\mathsf{E}_{ij}) = 0$ except for $\mathsf{E}_{00} = \mathbf{1}$ in which case trace $(\mathsf{E}_{00}) = 4$.

5. The 16 E_{ij} matrices *almost* form a mathematical group.[1] Any two of them multiplied together yield a member of the set within a factor of -1 or $\pm i$.

6. The 16 E_{ij} are linearly independent. No one can be written as a linear sum of the other 15.

7. The 16 E_{ij} form a complete set. Any 4×4 matrix (with constant elements) may be written as a linear combination of these 16,

$$A = \sum_{i,j=0}^{3} c_{ij} E_{ij},$$

where the coefficients c_{ij} are constants, real or complex.

Anticommuting sets. From these 16 Hermitian matrices we can form six anticommuting sets of five matrices each. Using the labels shown in Table 4.1 these sets are the following:

$$\begin{aligned}
&1.\ \alpha_1,\ \alpha_2,\ \alpha_3,\ \alpha_4,\ \alpha_5.\\
&2.\ \gamma_1,\ \gamma_2,\ \gamma_3,\ \gamma_4,\ \gamma_5.\\
&3.\ \delta_1,\ \delta_2,\ \delta_3,\ \rho_1,\ \rho_2.\\
&4.\ \alpha_1,\ \gamma_1,\ \delta_1,\ \sigma_2,\ \sigma_3.\\
&5.\ \alpha_2,\ \gamma_2,\ \delta_2,\ \sigma_1,\ \sigma_3.\\
&6.\ \alpha_3,\ \gamma_3,\ \delta_3,\ \sigma_1,\ \sigma_2.
\end{aligned} \tag{4.137}$$

Each E_{ij} (exclusive of the unit matrix) appears in *two* of the above sets. In addition to the set of α's, the set of γ's has been used extensively in relativistic quantum theory.

The largest completely commuting sets of Dirac matrices (including the unit matrix) have only four matrices.

The discussion of orthogonal matrices in Section 4.3 and unitary matrices in this section is only a bare beginning. The further extensions are of vital concern in modern "elementary" particle physics. With the Pauli and Dirac matrices, we can develop *spinors* for describing electrons, protons, and other spin one-half particles. The coordinate system rotations lead to $D^j(\alpha, \beta, \gamma)$, the rotation group, usually represented by matrices in which the elements are functions of the Euler angles describing the rotation. The special unitary group *SU*(3), (composed of 3×3 unitary matrices with determinant $+1$), has been used with considerable success to describe mesons and baryons. These extensions are considered further in Sections 4.9–4.11.

[1] The E_{ij} can be modified so that they satisfy the group property exactly, but then they are no longer Hermitian and unitary.

EXERCISES

4.5.1 Three angular momentum matrices satisfy the basic commutation relation

$$[\mathsf{J}_x, \mathsf{J}_y] = i\mathsf{J}_z$$

(and cyclic permutation of indices). If two of the matrices have real elements, show that the elements of the third must be pure imaginary.

4.5.2 Show that $(\mathsf{AB})^\dagger = \mathsf{B}^\dagger\mathsf{A}^\dagger$.

4.5.3 Matrix $\mathsf{C} = \mathsf{S}^\dagger\mathsf{S}$. Show that the trace is positive definite unless S is the null matrix in which case trace (C) = 0.

4.5.4 If A and B are Hermitian matrices, show that $(\mathsf{AB} + \mathsf{BA})$ and $i(\mathsf{AB} - \mathsf{BA})$ are also Hermitian.

4.5.5 Show that a Hermitian matrix remains Hermitian under unitary similarity transformations.

4.5.6 Two matrices A and B are each Hermitian. Find a necessary and sufficient condition for their product AB to be Hermitian.

Ans. $[\mathsf{A}, \mathsf{B}] = 0$.

4.5.7 Show that the inverse of a unitary matrix is unitary.

4.5.8 A particular similarity transformation yields

$$\mathsf{A}' = \mathsf{UAU}^{-1}$$
$$\mathsf{A}^{\dagger\prime} = \mathsf{UA}^\dagger\mathsf{U}^{-1}.$$

If the adjoint relationship is preserved ($\mathsf{A}^{\dagger\prime} = \mathsf{A}'^\dagger$) and det U = 1, show that U must be unitary.

4.5.9 Two matrices U and H are related by

$$\mathsf{U} = e^{ia\mathsf{H}}$$

with a real. (The exponential matrix function may be interpreted by a Maclaurin expansion. This will be done in Section 4.10.)
(a) If H is Hermitian, show that U is unitary.
(b) If U is unitary, show that H is Hermitian (to within terms of the form $2\pi n/a$).

4.5.10 Prove that the direct product of two unitary matrices is unitary.

4.5.11 Show that all 4 × 4 matrices (whose elements are complex numbers) may be formed as a linear combination of $\mathbf{1}$, γ_1, γ_2, γ_3, γ_4, and *products* of the γ's.

4.5.12 Denoting the 16 Dirac matrices by $\mathsf{E}_{ij} = \rho_i\sigma_j$ ($\rho_0 = \sigma_0 = 1$), show that
(a) $\mathsf{E}_{ij}^2 = \mathbf{1}$ for all i and j,
(b) $\mathsf{E}_{ij} = \mathsf{E}_{ij}^\dagger$ (Hermitian).
Hint. Use the known properties of ρ_i and σ_j.

4.5.13 Verify Eqs. 4.134–4.136 for the 4 × 4 σ and ρ matrices.

4.5.14 Using Eqs. 4.135 and 4.136, show that each of the six sets of Dirac matrices listed in Eq. 4.137 is actually an anticommuting set.

4.5.15 Using Eqs. 4.135 and 4.136 show that

(a) $\alpha_1\alpha_2\alpha_3\alpha_4\alpha_5 = +1$,

(b) $\gamma_1\gamma_2\gamma_3\gamma_4\gamma_5 = +1$.

4.5.16 If $M = \frac{1}{2}(1 + \gamma_5)$, show that

$$M^2 = M.$$

Note that γ_5 may be replaced by any other Dirac matrix (any E_{ij} of Table 4.1). If M is Hermitian then this result, $M^2 = M$, is the defining equation for a quantum mechanical projection operator.

4.5.17 Show that

$$\boldsymbol{\alpha} \times \boldsymbol{\alpha} = 2i\boldsymbol{\sigma}$$

where $\boldsymbol{\alpha}$ is a vector whose components are the α matrices,

$$\boldsymbol{\alpha} = (\alpha_1, \alpha_2, \alpha_3).$$

Note that if $\boldsymbol{\alpha}$ is a polar vector (Section 3.4) then $\boldsymbol{\sigma}$ is an axial vector.

4.5.18 Prove that the 16 Dirac matrices form a linearly independent set.
Hint: Assume the contrary. Let E_{mn} be a linear combination of the other E_{ij}'s. Multiply by E_{mn}. Take the trace and show that a contradiction results.

4.5.19 (a) If we assume that a given 4×4 matrix, A (with constant elements), can be written as a linear combination of the 16 Dirac matrices

$$A = \sum_{i,j=0}^{3} C_{ij} E_{ij},$$

show that

$$C_{mn} = \tfrac{1}{4} \operatorname{trace} (A E_{mn}).$$

(b) If A has one and only one nonvanishing element, show that there will be exactly four nonvanishing coefficients in its expansion.

(c) Expand

$$A = \begin{pmatrix} 1 & 0 & 0 & 0 \\ 0 & 0 & 0 & 0 \\ 0 & 0 & 0 & 0 \\ 0 & 0 & 0 & 0 \end{pmatrix}$$

in terms of the E_{ij}.

Ans. $A = \frac{1}{4}(E_{00} + E_{03} + E_{30} + E_{33})$
$= \frac{1}{4}(1 + \sigma_3 + \rho_3 + \delta_3)$.

4.5.20 If A is any one of the Dirac matrices (excluding the unit matrix), it will commute with 8 of the Dirac matrices and anticommute with the other 8. List the eight matrices that *anticommute* with γ_1.

Ans. $\sigma_2, \sigma_3, \rho_1, \alpha_1, \gamma_2, \gamma_3, \rho_3, \delta_1$.

4.5.21 For investigating questions of covariance under Lorentz transformations the Dirac electron theory is usually expressed in terms of γ_μ, $\mu = 1,2,3,4$. Show that these four matrices together with their products

(a) $\gamma_\mu\gamma_\nu$, $\mu \neq \nu$

(b) $\gamma_\mu\gamma_\nu\gamma_\lambda$, indices all different

(c) $\gamma_1\gamma_2\gamma_3\gamma_4$

and the unit matrix 1 reproduce all sixteen Dirac matrices (apart from constant factors).

4.5.22 Given $\mathbf{r}' = U\mathbf{r}$, with U a unitary matrix and $\mathbf{r}$ a (column) vector with complex elements, show that the norm (magnitude) of $\mathbf{r}$ is invariant under this operation.

4.6 Diagonalization of Matrices

Moment of inertia matrix. In many physical problems involving matrices it is desirable to carry out a (real) orthogonal similarity transformation or a unitary transformation to reduce the matrix to a diagonal form, nondiagonal elements all equal to zero. One particularly direct example of this is the moment of inertia matrix I, of a rigid body. From the definition of angular momentum **L** we have

$$\mathbf{L} = \mathsf{I}\boldsymbol{\omega} \tag{4.138}$$

$\boldsymbol{\omega}$ being the angular velocity. The inertia matrix I is found to have diagonal components

$$I_{xx} = \sum_i m_i(r_i^2 - x_i^2), \text{ etc.,} \tag{4.139}$$

the subscript i referring to the mass m_i. For the nondiagonal components we have the products of inertia

$$I_{xy} = -\sum_i m_i x_i y_i = I_{yx}. \tag{4.140}$$

By inspection, matrix I is symmetric. Also, since I appears in a physical equation of the form (4.138), which holds for all orientations of the coordinate system, it may be considered to be a tensor, (quotient rule, Section 3.3).

The problem now is to orient the coordinate axes in space so that the I_{xy} and the other nondiagonal elements will vanish. A consequence of this orientation and an indication of it is that if the angular velocity is along one such realigned axis the angular velocity and the angular momentum will be parallel.

Geometrical picture—ellipsoid. It is perhaps instructive to consider a geometrical picture of this problem. If the inertia matrix I is multiplied from each side

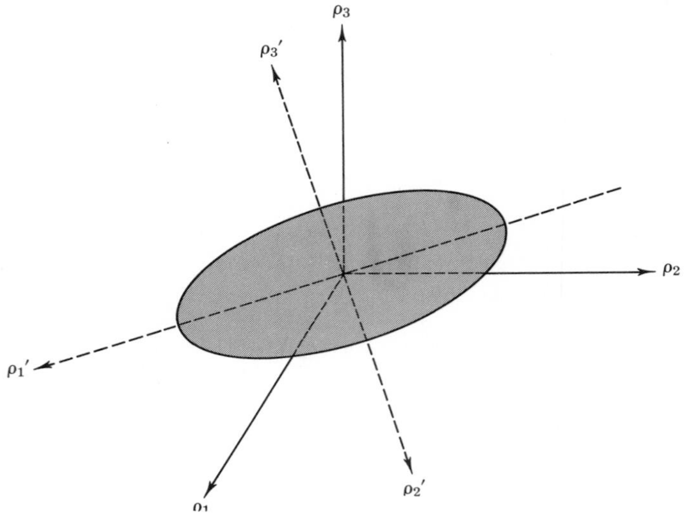

FIG. 4.5 Moment of inertia ellipsoid

by a unit vector of variable direction, $\mathbf{n} = (\alpha, \beta, \gamma)$,

$$[n]\mathsf{I}\{n\} = \mathrm{I}, \tag{4.141}$$

where I is a number (scalar) whose magnitude depends on the choice of direction of $\mathbf{n}$. Carrying out the multiplication,

$$\mathrm{I} = I_{xx}\alpha^2 + I_{yy}\beta^2 + I_{zz}\gamma^2 + 2I_{xy}\alpha\beta + 2I_{xz}\alpha\gamma + 2I_{yz}\beta\gamma. \tag{4.142}$$

To throw this into one of the standard forms for an ellipsoid we introduce

$$\boldsymbol{\rho} = \frac{\mathbf{n}}{\sqrt{\mathrm{I}}}, \tag{4.143}$$

in which $\boldsymbol{\rho}$ is variable in direction *and* magnitude. Equation 4.142 becomes

$$1 = I_{xx}\rho_1^2 + I_{yy}\rho_2^2 + I_{zz}\rho_3^2 + 2I_{xy}\rho_1\rho_2 + 2I_{xz}\rho_1\rho_3 + 2I_{yz}\rho_2\rho_3. \tag{4.144}$$

This is the general form of an ellipsoid relative to the coordinates ρ_1, ρ_2, ρ_3. However, from analytic geometry it is known that the coordinate axes can always be rotated to coincide with the axes of our ellipsoid. Then

$$1 = I_1\rho_1'^2 + I_2\rho_2'^2 + I_3\rho_3'^2, \tag{4.145}$$

in which $\rho_1', \rho_2', \rho_3'$ is the new set of coordinates.

Principal axes. In many elementary cases, especially when symmetry is present, these new axes, called the *principal axes*, can be found by inspection. We now proceed to develop a general method of finding the diagonal elements and the principal axes.

Hermitian matrices. First, an important theorem about the diagonal elements and the principal axes. In the equation

$$\mathsf{A}\mathbf{r} = \lambda\mathbf{r} \tag{4.146}$$

λ, a number (scalar), is known as the eigenvalue; $\mathbf{r}$, the corresponding vector, is the eigenvector.[1] The terms were introduced from the early German literature on quantum mechanics. We will now show that if A is a Hermitian matrix,[2] its eigenvalues are real and its eigenvectors orthogonal.

Let λ_i and λ_j be two eigenvalues and $\mathbf{r}_i$ and $\mathbf{r}_j$, the corresponding eigenvectors of A, a Hermitian matrix. Then

$$\mathsf{A}\mathbf{r}_i = \lambda_i\mathbf{r}_i \tag{4.147}$$

$$\mathsf{A}\mathbf{r}_j = \lambda_j\mathbf{r}_j. \tag{4.148}$$

Equation 4.147 is multiplied by $\mathbf{r}_j^\dagger$ to give

$$\mathbf{r}_j^\dagger\mathsf{A}\mathbf{r}_i = \lambda_i\mathbf{r}_j^\dagger\mathbf{r}_i. \tag{4.149}$$

[1] Equation 4.138 will take on this form when $\boldsymbol{\omega}$ is along one of the principal axes. Then $\mathbf{L} = \lambda\boldsymbol{\omega}$ and $\mathsf{I}\boldsymbol{\omega} = \lambda\boldsymbol{\omega}$. In the mathematics literature λ is usually called a *characteristic value*, $\boldsymbol{\omega}$ a *characteristic vector*.

[2] If A is real, the Hermitian requirement is replaced by a requirement of symmetry.

(Note that if $\mathbf{r}$ is a column vector $\mathbf{r}^\dagger$ is a row vector.) Equation 4.148 is multiplied by $\mathbf{r}_i^\dagger$ to give

$$\mathbf{r}_i^\dagger \mathsf{A}\mathbf{r}_j = \lambda_j \mathbf{r}_i^\dagger \mathbf{r}_j. \tag{4.150}$$

Taking the adjoint of this equation

$$\mathbf{r}_j^\dagger \mathsf{A}^\dagger \mathbf{r}_i = \lambda_j^* \mathbf{r}_j^\dagger \mathbf{r}_i \tag{4.151}$$

or

$$\mathbf{r}_j^\dagger \mathsf{A}\mathbf{r}_i = \lambda_j^* \mathbf{r}_j^\dagger \mathbf{r}_i, \tag{4.152}$$

since A is Hermitian. Subtracting Eq. 4.152 from Eq. 4.149, we obtain

$$(\lambda_i - \lambda_j^*)\mathbf{r}_j^\dagger \mathbf{r}_i = 0. \tag{4.153}$$

This is a general result for all possible combinations of i and j. First, let $i = j$. Then Eq. 4.153 becomes

$$(\lambda_i - \lambda_i^*)|\mathbf{r}_i|^2 = 0. \tag{4.154}$$

Since $|\mathbf{r}_i| = 0$ would be a trivial solution of Eq. 4.154, we conclude that

$$\lambda_i = \lambda_i^*, \tag{4.155}$$

or λ_i is real, for all i.

Second, for $i \neq j$, *and* $\lambda_i \neq \lambda_j$,

$$(\lambda_i - \lambda_j)\mathbf{r}_j^\dagger \mathbf{r}_i = 0 \tag{4.156}$$

or

$$\mathbf{r}_j^\dagger \mathbf{r}_i = 0, \tag{4.157}$$

which means that the eigenvectors of *distinct* eigenvalues are orthogonal, Eq. 4.157 being our generalization of orthogonality in this complex space.[3]

If $\lambda_i = \lambda_j$ (degenerate case), $\mathbf{r}_i$ is not automatically orthogonal to $\mathbf{r}_j$, but it may be *made* orthogonal. Consider the physical problem of the moment of inertia matrix again. If x_1 is an axis of rotational symmetry, then we will find that $\lambda_2 = \lambda_3$. Eigenvectors $\mathbf{r}_2$ and $\mathbf{r}_3$ are each perpendicular to the symmetry axis, $\mathbf{r}_1$, but they lie anywhere in the plane perpendicular to $\mathbf{r}_1$; that is, any linear combination of $\mathbf{r}_2$ and $\mathbf{r}_3$ is also an eigenvector. Consider $(a_2\mathbf{r}_2 + a_3\mathbf{r}_3)$ with a_2 and a_3 constants. Then

$$\begin{aligned}\mathsf{A}(a_2\mathbf{r}_2 + a_3\mathbf{r}_3) &= a_2\lambda_2\mathbf{r}_2 + a_3\lambda_3\mathbf{r}_3 \\ &= \lambda_2(a_2\mathbf{r}_2 + a_3\mathbf{r}_3),\end{aligned} \tag{4.158}$$

as is to be expected, for x_1 is an axis of rotational symmetry. Therefore, if $\mathbf{r}_1$ and $\mathbf{r}_2$ are fixed, $\mathbf{r}_3$ may simply be chosen to lie in the plane perpendicular to $\mathbf{r}_1$ and also perpendicular to $\mathbf{r}_2$. A general method of orthogonalizing solutions, the Schmidt process, is applied to functions in Section 9.3.

The foregoing is essentially an existence theorem. To determine the eigenvalues λ_i and the eigenvectors $\mathbf{r}_i$ actually we return to Eq. 4.146. Assuming $\mathbf{r}$ to be multiplied by the unit matrix, we may rewrite Eq. 4.146

$$(\mathsf{A} - \lambda 1)\mathbf{r} = 0, \tag{4.159}$$

[3] The corresponding theory for differential operators (Sturm-Liouville theory) appears in Section 9.2. The integral equation analog (Hilbert-Schmidt theory) is given in Section 16.4.

in which $\mathsf{1}$ is the unit matrix. This is a set of simultaneous, homogeneous, linear equations. By Section 4.1 it has nontrivial solutions only if the determinant of the coefficients vanishes,

$$|\mathsf{A} - \lambda\mathsf{1}| = 0. \tag{4.160}$$

Let us consider the case in which A is a 3×3 Hermitian matrix. Then

$$\begin{vmatrix} a_{11} - \lambda & a_{12} & a_{13} \\ a_{21} & a_{22} - \lambda & a_{23} \\ a_{31} & a_{32} & a_{33} - \lambda \end{vmatrix} = 0. \tag{4.161}$$

Because of its applications in astronomical theories Eq. 4.161 is usually called the *secular equation.*[1] Equation 4.161 yields a cubic equation in λ, which, of course, has three roots. By Eq. 4.155 we know that these roots are real. Substituting one root at a time back into Eq. 4.159, we can find the corresponding eigenvectors.

EXAMPLE 4.6.1

Let

$$\mathsf{A} = \begin{pmatrix} 0 & 1 & 0 \\ 1 & 0 & 0 \\ 0 & 0 & 0 \end{pmatrix}. \tag{4.162}$$

The secular equation is

$$\begin{vmatrix} -\lambda & 1 & 0 \\ 1 & -\lambda & 0 \\ 0 & 0 & -\lambda \end{vmatrix} = 0 \tag{4.163}$$

or

$$-\lambda(\lambda^2 - 1) = 0, \tag{4.164}$$

expanding by minors. The roots are $\lambda = -1, 0, 1$. To find the eigenvector corresponding to $\lambda = -1$, we substitute this value back into the eigenvalue equation Eq. 4.159

$$\begin{pmatrix} -\lambda & 1 & 0 \\ 1 & -\lambda & 0 \\ 0 & 0 & -\lambda \end{pmatrix} \begin{pmatrix} x \\ y \\ z \end{pmatrix} = \begin{pmatrix} 0 \\ 0 \\ 0 \end{pmatrix}. \tag{4.165}$$

With $\lambda = -1$, this yields

$$\begin{aligned} x + y &= 0, \\ z &= 0. \end{aligned} \tag{4.166}$$

Within an arbitrary scale factor, and an arbitrary sign (or phase factor), $\mathbf{r}_1 = (1, -1, 0)$. Note carefully that (for real $\mathbf{r}$ in ordinary space) the eigenvector singles out a line in space. The positive or negative sense is not determined. This in-

[1] This equation also appears in second order perturbation theory in quantum mechanics.

determinancy could be expected if we noted that Eq. 4.159 is homogeneous in $\mathbf{r}$. For convenience we will require that the eigenvectors be normalized to unity, $|\mathbf{r}_1| = 1$. With this choice of sign,

$$\mathbf{r}_1 = \left(\frac{1}{\sqrt{2}}, \frac{-1}{\sqrt{2}}, 0\right) \tag{4.167}$$

is fixed. For $\lambda = 0$, Eq. 4.159 yields

$$\begin{aligned} y &= 0, \\ x &= 0, \end{aligned} \tag{4.168}$$

$\mathbf{r}_2 = (0, 0, 1)$ is a suitable eigenvector. Finally for $\lambda = 1$, we get

$$\begin{aligned} -x + y &= 0, \\ z &= 0, \end{aligned} \tag{4.169}$$

or

$$\mathbf{r}_3 = \left(\frac{1}{\sqrt{2}}, \frac{1}{\sqrt{2}}, 0\right). \tag{4.170}$$

The orthogonality of $\mathbf{r}_1$, $\mathbf{r}_2$, and $\mathbf{r}_3$, corresponding to three distinct eigenvalues, may be easily verified.

Example 4.6.2

Consider

$$\mathsf{A} = \begin{pmatrix} 1 & 0 & 0 \\ 0 & 0 & 1 \\ 0 & 1 & 0 \end{pmatrix} \tag{4.171}$$

The secular equation is

$$\begin{vmatrix} 1-\lambda & 0 & 0 \\ 0 & -\lambda & 1 \\ 0 & 1 & -\lambda \end{vmatrix} = 0 \tag{4.172}$$

or

$$(1 - \lambda)(\lambda^2 - 1) = 0, \qquad \lambda = -1, 1, 1, \tag{4.173}$$

a degenerate case. If $\lambda = -1$, the eigenvalue equation (4.159) yields

$$\begin{aligned} 2x &= 0, \\ y + z &= 0. \end{aligned} \tag{4.174}$$

A suitable normalized eigenvector is

$$\mathbf{r}_1 = \left(0, \frac{1}{\sqrt{2}}, \frac{-1}{\sqrt{2}}\right). \tag{4.175}$$

For $\lambda = 1$, we get

$$-y + z = 0 \tag{4.176}$$

and no further information. We have an infinite number of choices. Suppose, as one possible choice, $\mathbf{r}_2$ is taken as

$$\mathbf{r}_2 = \left(0, \frac{1}{\sqrt{2}}, \frac{1}{\sqrt{2}}\right), \tag{4.177}$$

which clearly satisfies Eq. 4.176. Then $\mathbf{r}_3$ must be perpendicular to $\mathbf{r}_1$ and may be made perpendicular to $\mathbf{r}_2$ by

$$\mathbf{r}_3 = \mathbf{r}_1 \times \mathbf{r}_2 = (1, 0, 0). \tag{4.178}$$

Diagonalization. The equations, developed on page 196 for our existence theorem can be used to form a transformation matrix which will convert our Hermitian matrix A into diagonal form. Let R be a matrix formed from the three orthonormal *column* vectors $\mathbf{r}_1$, $\mathbf{r}_2$, and $\mathbf{r}_3$ in any desired order.

$$\mathsf{R} = \begin{pmatrix} x_1 & x_2 & x_3 \\ y_1 & y_2 & y_3 \\ z_1 & z_2 & z_3 \end{pmatrix}, \tag{4.179}$$

in which each column $\{x_i, y_i, z_i\}$ is an eigenvector $\mathbf{r}_i$. Since

$$\mathbf{r}_i^\dagger \mathbf{r}_j = \delta_{ij} \tag{4.180}$$

R is unitary (or simply orthogonal if A, and therefore $\mathbf{r}_i$ are real). Then, forming $\mathsf{R}^\dagger\mathsf{A}\mathsf{R}$, we have

$$\begin{aligned}
\mathsf{R}^\dagger\mathsf{A}\mathsf{R} &= \begin{pmatrix} [\ \mathbf{r}_1^*\ ] \\ [\ \mathbf{r}_2^*\ ] \\ [\ \mathbf{r}_3^*\ ] \end{pmatrix} \begin{pmatrix} & \mathsf{A} & \end{pmatrix} \left(\begin{pmatrix} \mathbf{r}_1 \end{pmatrix} \begin{pmatrix} \mathbf{r}_2 \end{pmatrix} \begin{pmatrix} \mathbf{r}_3 \end{pmatrix} \right) \\
&= \begin{pmatrix} [\ \mathbf{r}_1^*\ ] \\ [\ \mathbf{r}_2^*\ ] \\ [\ \mathbf{r}_3^*\ ] \end{pmatrix} \left(\begin{pmatrix} \lambda_1\mathbf{r}_1 \end{pmatrix} \begin{pmatrix} \lambda_2\mathbf{r}_2 \end{pmatrix} \begin{pmatrix} \lambda_3\mathbf{r}_3 \end{pmatrix} \right) \\
&= \begin{pmatrix} \lambda_1 & 0 & 0 \\ 0 & \lambda_2 & 0 \\ 0 & 0 & \lambda_3 \end{pmatrix}
\end{aligned} \tag{4.181}$$

Hence $\mathsf{R}^\dagger\mathsf{A}\mathsf{R}$ is a diagonal matrix with eigenvalues λ_i, the order of the eigenvalues corresponding to the order of the column vectors $\mathbf{r}_i$ in R. To develop the geometrical picture consider A, a real (symmetric) matrix with real eigenvalues and real eigenvectors. Matrix R corresponds to B^{-1} in Eq. 4.95 or better, $\tilde{\mathsf{R}}$ corresponds to B,

$\tilde{\mathsf{R}}$ being composed of the eigenvectors $\mathbf{r}_i$ written as row vectors.

$$\begin{pmatrix} [\mathbf{r}_1] \\ [\mathbf{r}_2] \\ [\mathbf{r}_3] \end{pmatrix} = \begin{pmatrix} x_1 & y_1 & z_1 \\ x_2 & y_2 & z_2 \\ x_3 & y_3 & z_3 \end{pmatrix} = \begin{pmatrix} b_{11} & b_{12} & b_{13} \\ b_{21} & b_{22} & b_{23} \\ b_{31} & b_{32} & b_{33} \end{pmatrix}. \tag{4.182}$$

Now, the row (b_{i1}, b_{i2}, b_{i3}), which defines a unit vector $\mathbf{r}_i$ in relation to the original coordinate system, specifies the three direction cosines of $\mathbf{r}_i$ with the original axes. Remembering that matrix B rotates the coordinate system into a new system in which (here) A is diagonal, we see that this new system is specified by the three eigenvectors $\mathbf{r}_i = (x_i, y_i, z_i)$. They are the unit vectors along the principal axes, the axes in relation to which A is diagonal.

The preceding analysis has the advantage of exhibiting and clarifying conceptual relationships in the diagonalization of matrices. However, for matrices larger than 3×3, or perhaps 4×4, the process rapidly becomes so cumbersome that we turn gratefully to high-speed computers and iterative techniques. One such technique is the Jacobi method for determining eigenvalues and eigenvectors of real symmetric matrices.

EXERCISES

4.6.1 Show that the eigenvalues of a matrix are unaltered if the matrix is transformed by a similarity transformation.
This property is not limited to symmetric or Hermitian matrices. It applies to unitary matrices and to any matrices that can be brought into diagonal form by a similarity transformation. As consequences of this invariance of the eigenvalues:
(a) the trace (sum of eigenvalues) is invariant under a similarity transformation, Ex. 4.3.8, and
(b) the determinant (product of eigenvalues) is invariant under a similarity transformation, Ex. 4.3.9.

4.6.2 As a converse of the theorem on page 195, show that if (a) the eigenvalues of a matrix are real and (b) the eigenvectors satisfy Eq. 4.180, $\mathbf{r}_i{}^\dagger\mathbf{r}_j = \delta_{ij}$, then the matrix is Hermitian.

4.6.3 Show that a real matrix that is not symmetric cannot be diagonalized by an orthogonal similarity transformation.
Hint. Assume that the nonsymmetric real matrix can be diagonalized and develop a contradiction.

4.6.4 The matrices representing the angular momentum components J_x, J_y, and J_z are all Hermitian. Show that the eigenvalues of J^2 where $J^2 = J_x{}^2 + J_y{}^2 + J_z{}^2$ are real and nonnegative.

4.6.5 The square of a particular Hermitian matrix is the unit matrix. Show that the eigenvalues of the original Hermitian matrix are all ± 1.
Note: The Pauli and Dirac matrices are specific examples.

4.6.6 A square matrix with zero determinant is labeled *singular*.

(a) If A is singular, show that there is at least one nonzero column vector $\mathbf{v}$ such that

$$\mathsf{A}\mathbf{v} = 0.$$

(b) If there is a nonzero vector $\mathbf{v}$ such that

$$\mathsf{A}\mathbf{v} = 0,$$

show that A is a singular matrix.

4.6.7 The same similarity transformation diagonalizes each of two matrices. Show that the original matrices must commute. (This is particularly important in the matrix (Heisenberg) formulation of quantum mechanics.)

4.6.8 Two Hermitian matrices A and B have the same eigenvalues. Show that A and B are related by a unitary similarity transformation.

4.6.9 Find the eigenvalues and an orthonormal (orthogonal and normalized) set of eigenvectors for the matrices of Exercise 4.2.14

4.6.10 Show that the inertia matrix for a single particle of mass m at (x, y, z) has zero determinant. Explain this result in terms of the invariance of the determinant of a matrix under similarity transformations (Exercise 4.3.9) and a possible rotation of the coordinate system.

4.6.11 A certain rigid body may be represented by three point masses:

$$\begin{aligned} m_1 &= 1 \quad \text{at} \quad (1, 1, -2)\\ m_2 &= 2 \quad \text{at} \quad (-1, -1, 0)\\ m_3 &= 1 \quad \text{at} \quad (1, 1, 2). \end{aligned}$$

(a) Find the inertia matrix.

(b) Diagonalize the inertia matrix obtaining the eigenvalues and the principal axes (as orthonormal eigenvectors).

4.6.12

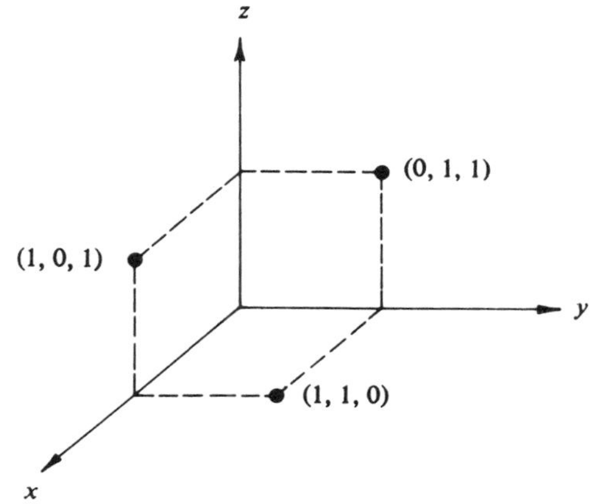

Unit masses are placed as shown in the figure.

(a) Find the moment of inertia matrix.

(b) Find the eigenvalues and a set of orthonormal eigenvectors.

(c) Explain the degeneracy in terms of the symmetry of the system.

Ans. $\mathsf{I} = \begin{pmatrix} 4 & -1 & -1\\ -1 & 4 & -1\\ -1 & -1 & 4 \end{pmatrix}$ $\quad \lambda_1 = 2$, $\quad \mathbf{r}_1 = (1/\sqrt{3}, 1/\sqrt{3}, 1/\sqrt{3})$, $\quad \lambda_2 = \lambda_3 = 5.$

4.6.13 Unit masses are at the eight corners of a cube $(\pm 1, \pm 1, \pm 1)$. Find the moment of inertia matrix and show that there is a triple degeneracy. This means that so far as moments of inertia are concerned, the cubic structure exhibits spherical symmetry.

4.6.14 Find the eigenvalues and corresponding orthonormal eigenvectors of the following matrices (as a numerical check, note that the sum of the eigenvalues equals the sum of the diagonal elements of the original matrix—Ex. 4.3.8):

$$\mathsf{A} = \begin{pmatrix} 1 & 0 & 1 \\ 0 & 1 & 0 \\ 1 & 0 & 1 \end{pmatrix}.$$

Ans. $\lambda = 0, 1, 2.$

4.6.15

$$\mathsf{A} = \begin{pmatrix} 1 & \sqrt{2} & 0 \\ \sqrt{2} & 0 & 0 \\ 0 & 0 & 0 \end{pmatrix}.$$

Ans. $\lambda = -1, 0, 2.$

4.6.16

$$\mathsf{A} = \begin{pmatrix} 1 & 1 & 0 \\ 1 & 0 & 1 \\ 0 & 1 & 1 \end{pmatrix}.$$

Ans. $\lambda = -1, 1, 2.$

4.6.17

$$\mathsf{A} = \begin{pmatrix} 1 & \sqrt{8} & 0 \\ \sqrt{8} & 1 & \sqrt{8} \\ 0 & \sqrt{8} & 1 \end{pmatrix}.$$

Ans. $\lambda = -3, 1, 5.$

4.6.18

$$\mathsf{A} = \begin{pmatrix} 1 & 0 & 0 \\ 0 & 1 & 1 \\ 0 & 1 & 1 \end{pmatrix}.$$

Ans. $\lambda = 0, 1, 2.$

4.6.19

$$\mathsf{A} = \begin{pmatrix} 1 & 0 & 0 \\ 0 & 1 & \sqrt{2} \\ 0 & \sqrt{2} & 0 \end{pmatrix}.$$

Ans. $\lambda = -1, 1, 2.$

4.6.20

$$\mathsf{A} = \begin{pmatrix} 0 & 1 & 0 \\ 1 & 0 & 1 \\ 0 & 1 & 0 \end{pmatrix}.$$

Ans. $\lambda = -\sqrt{2}, 0, \sqrt{2}.$

4.6.21

$$\mathsf{A} = \begin{pmatrix} 2 & 0 & 0 \\ 0 & 1 & 1 \\ 0 & 1 & 1 \end{pmatrix}.$$

Ans. $\lambda = 0, 2, 2.$

4.6.22

$$\mathsf{A} = \begin{pmatrix} 0 & 1 & 1 \\ 1 & 0 & 1 \\ 1 & 1 & 0 \end{pmatrix}.$$

Ans. $\lambda = -1, -1, 2.$

4.6.23

$$\mathsf{A} = \begin{pmatrix} 1 & -1 & -1 \\ -1 & 1 & -1 \\ -1 & -1 & 1 \end{pmatrix}.$$

Ans. $\lambda = -1, 2, 2.$

4.6.24

$$\mathsf{A} = \begin{pmatrix} 1 & 1 & 1 \\ 1 & 1 & 1 \\ 1 & 1 & 1 \end{pmatrix}.$$

Ans. $\lambda = 0, 0, 3.$

4.6.25
$$\mathsf{A} = \begin{pmatrix} 5 & 0 & 2 \\ 0 & 1 & 0 \\ 2 & 0 & 2 \end{pmatrix}.$$
Ans. $\lambda = 1, 1, 6$.

4.6.26
$$\mathsf{A} = \begin{pmatrix} 1 & 1 & 0 \\ 1 & 1 & 0 \\ 0 & 0 & 0 \end{pmatrix}.$$
Ans. $\lambda = 0, 0, 2$.

4.6.27
$$\mathsf{A} = \begin{pmatrix} 5 & 0 & \sqrt{3} \\ 0 & 3 & 0 \\ \sqrt{3} & 0 & 3 \end{pmatrix}.$$
Ans. $\lambda = 2, 3, 6$.

4.6.28 This is a continuation of Exercise 4 5.9 , where the unitary matrix U and the Hermitian matrix H are related by
$$\mathsf{U} = e^{ia\mathsf{H}}$$
(a) If trace H = 0, show that det U = +1.
(b) If det U = +1, show that trace H = 0.
Hint. H may be diagonalized by a similarity transformation. Then, interpreting the exponential by a Maclaurin expansion, U is also diagonal. The corresponding eigenvalues are given by $u_j = \exp(iah_j)$.
Note: These properties, and those of Ex. 4.5.9 , are vital in the development of the concept of *generators* in group theory—Section 4.10.

4.6.29 (a) Assuming a unitary matrix U to satisfy an eigenvalue equation $\mathsf{U}\mathbf{r} = \lambda\mathbf{r}$, show that the eigenvalues of the unitary matrix have unit magnitude.
This same result holds for orthogonal matrices.
(b) If $\mathbf{r}_1$ and $\mathbf{r}_2$ are the eigenvectors of a unitary matrix corresponding to two distinct eigenvalues, show that the eigenvectors are orthogonal in the sense that
$$\mathbf{r}_1{}^\dagger\mathbf{r}_2 = 0.$$

4.6.30 A non-Hermitian matrix A has eigenvalues λ_i and corresponding eigenvectors $\mathbf{u}_i$. The adjoint matrix $\mathsf{A}^\dagger$ has the same set of eigenvalues but *different* corresponding eigenvectors, $\mathbf{v}_i$. Show that the eigenfunctions form a *biorthogonal* set in the sense that
$$[v_i^*]\{u_j\} = 0 \qquad \text{for } \lambda_i^* \neq \lambda_j.$$

4.6.31 Show that every 2×2 matrix has two eigenvectors and corresponding eigenvalues. The eigenvectors are not necessarily orthogonal. The eigenvalues are not necessarily real.

4.6.32 As an illustration of Ex. 4.6.31 find the eigenvalues and corresponding eigenvectors for
$$\begin{pmatrix} 2 & 4 \\ 1 & 2 \end{pmatrix}.$$
Note that the eigenvectors are *not* orthogonal.
Ans. $\lambda_1 = 0$, $\mathbf{r}_1 = (2, -1)$;
$\lambda_2 = 4$, $\mathbf{r}_2 = (2, \ 1)$.

4.6.33 A matrix P is a projection operator satisfying the condition
$$\mathsf{P}^2 = \mathsf{P}.$$
Show that the corresponding eigenvalues $(\rho^2)_\lambda$ and ρ_λ satisfy the relation
$$(\rho^2)_\lambda = (\rho_\lambda)^2 = \rho_\lambda.$$
This means that the eigenvalues of P are 0 and 1.

4.7 Introduction to Group Theory

The theory of finite groups, developed originally as a branch of pure mathematics, can be a beautiful, fascinating toy. For the physicist, group theory, without any loss of its beauty, is also an extraordinarily useful tool for formalizing semi-intuitive concepts and for exploiting symmetries. Group theory becomes a useful tool for the development of crystallography and solid state physics when we introduce specific representations (matrices) and start calculating group characters (traces). A brief introduction to this area appears in Section 4.8. Perhaps even more important in physics is the extension of group theory to continuous groups[1] and the applications of these continuous groups to quantum theory and the particles of high energy physics. This is the topic of Sections 4.9–4.11.

As knowledge of our physical world expanded almost explosively in the first third of this century, Wigner and others realized that *invariance* was a key concept in understanding the new phenomena and in developing appropriate theories. The mathematical tool for treating invariants and symmetries is group theory. It represents a unification and formalization of principles such as parity and angular momentum widely used by physicists. Parity is related to invariance under inversion (Exercise 2.12.3). Conservation of angular momentum is a direct consequence of rotational symmetry, which means invariance under spacial rotations. While the formal techniques of group theory may not be necessary, these powerful mathematical techniques can save much labor. Group theory can produce a unification which (once grasped) leads to greater simplicity.

Definition of group. A *group G* may be defined as a set of objects or operations (called the *elements*) which may be combined or "multiplied" to form a well-defined product, and which satisfy the four conditions listed below. We label the set of elements $a, b, c, \ldots$:

1. If a and b are any two elements, then the product ab is also a member of the set.
2. The defined multiplication is associative, $(ab)c = a(bc)$.
3. There is a unit element I such that $Ia = aI = a$ for every element in the set.[2]
4. There must be an inverse or reciprocal of each element. The set must contain an element $b = a^{-1}$ such that $aa^{-1} = a^{-1}a = I$ for each element of the set.

In physics, these abstract conditions often take on direct physical meaning in terms of transformations of vectors, spinors, and tensors.

As a very simple, but not trivial, example of a group, consider the set $1, a, b, c$ which combine according to the group multiplication table[3]

	1	a	b	c
1	1	a	b	c
a	a	b	c	1
b	b	c	1	a
c	c	1	a	b

[1] These are groups with an infinite number of elements. Each element depends on one or more parameters, which vary continuously.

[2] Following Wigner, the unit element of a group is often labelled E, from the German *Einheit*, the unit.

[3] The order of the factors is row-column: $ab = c$ in the indicated example above.

To represent these group elements let

$$1 \to 1, \qquad a \to i, \qquad b \to -1, \qquad c \to -i, \tag{4.183}$$

combining by ordinary multiplication. Clearly the four group conditions are satisfied, and these four elements form a group. Since the multiplication of the group elements is commutative, the group is labelled *commutative* or *abelian* Our group is also a *cyclic group* in that the elements may be written as successive powers of one element, in this case i^n, $n = 0, 1, 2, 3$. Note that in writing out Eq. 4.183, we have selected a specific *representation* for this group of four objects.

We may recognize that the group elements $1, i, -1, -i$ may be interpreted as successive 90° rotations in the complex plane. Then from Eq. 4.63 we create the set of four 2 × 2 matrices

$$\begin{aligned} \mathsf{1} &= \begin{pmatrix} 1 & 0 \\ 0 & 1 \end{pmatrix} & \mathsf{A} &= \begin{pmatrix} 0 & -1 \\ 1 & 0 \end{pmatrix} \\ \mathsf{B} &= \begin{pmatrix} -1 & 0 \\ 0 & -1 \end{pmatrix} & \mathsf{C} &= \begin{pmatrix} 0 & 1 \\ -1 & 0 \end{pmatrix}. \end{aligned} \tag{4.184}$$

This set of four matrices forms a group with the law of combination being matrix multiplication. Here is a second representation, now in terms of matrices. A little matrix multiplication verifies that this representation too is abelian and cyclic. Clearly there is a correspondence of the two representations

$$1 \leftrightarrow 1 \leftrightarrow \mathsf{1} \qquad a \leftrightarrow i \leftrightarrow \mathsf{A} \qquad b \leftrightarrow -1 \leftrightarrow \mathsf{B} \qquad c \leftrightarrow -i \leftrightarrow \mathsf{C}. \tag{4.185}$$

Isomorphism, homomorphism. If the correspondence between the elements of two groups (or between two representations) is one-to-one, with each set of elements satisfying the same group multiplication table, then we say that the groups are *isomorphic*.[1] If the correspondence is two-to-one (or many-to-one) but still preserves the multiplicative relations, then the groups are *homomorphic*. In the case just considered, the two representations $(1, i, -1, -i)$ and (1, A, B, C) are isomorphic. The always possible, but trivial, representation (1, 1, 1, 1) would be homomorphic. A most important homomorphic correspondence between the groups O_3^+ and $SU(2)$ is developed in Section 4.9.

In contrast to this, there is no such correspondence between either of these representations and another group of four objects, the *vierergruppe*, Exercise 4.2.7. Confirming this, note that while the *vierergruppe* is abelian, it is noncyclic.

Matrix representations—reducible and irreducible. The representation of group elements by matrices is a very powerful technique and has been almost universally adopted among physicists. The use of matrices imposes no significant restriction. It can be shown that the elements of any finite group and of the continuous groups of Section 4.9 may be represented by matrices and, in particular,

[1] Suppose the elements of one group are labeled g_i, the elements of a second group h_i. Then $g_i \leftrightarrow h_i$, a one-to-one correspondence for all values of i. Also, if $g_i g_j = g_k$, and $h_i h_j = h_k$, then g_k and h_k must be corresponding elements.

by unitary matrices. In quantum mechanics, these unitary representations assume a special importance since unitary matrices can be diagonalized, and the eigenvalues can serve for the classification of quantum states.

If there exists a unitary transformation that will transform our original representation-matrices into a diagonal or block-diagonal form, for example,

$$\begin{pmatrix} r_{11} & r_{12} & r_{13} & r_{14} \\ r_{21} & r_{22} & r_{23} & r_{24} \\ r_{31} & r_{32} & r_{33} & r_{34} \\ r_{41} & r_{42} & r_{43} & r_{44} \end{pmatrix} \rightarrow \begin{pmatrix} p_{11} & p_{12} & 0 & 0 \\ p_{21} & p_{22} & 0 & 0 \\ 0 & 0 & q_{11} & q_{12} \\ 0 & 0 & q_{21} & q_{22} \end{pmatrix} \tag{4.186}$$

such that the smaller portions or submatrices are no longer coupled together, then the original representation is *reducible*. Equivalently we have

$$\mathsf{SRS}^{-1} = \begin{pmatrix} \mathsf{P} & \mathsf{O} \\ \mathsf{O} & \mathsf{Q} \end{pmatrix}. \tag{4.187}$$

If R is an $n \times n$ matrix, we might have P an $m \times m$ matrix, Q an $(n-m) \times (n-m)$ matrix. The O's are then rectangular matrices $m \times (n-m)$ and $(n-m) \times m$ with all elements zero. We may write this result as

$$\mathsf{R} = \mathsf{P} \oplus \mathsf{Q}, \tag{4.188}$$

and say that R has been decomposed into the representations P and Q. For instance, all representations of dimension greater than one of Abelian groups are reducible. If no such unitary transformation exists, the representation is *irreducible*. Among the Dirac matrices of Table 4.1, $\mathsf{1}, \sigma_1, \sigma_2, \sigma_3, \rho_3, \delta_1, \delta_2$, and δ_3 are in this reduced form. The topic of Exercise 4.7.1 is to show that the matrices $\mathsf{1}$, A, B, and C form a reducible representation—and to reduce them to the irreducible representations.

The irreducible representations play a role in group theory roughly analogous to the unit vectors of vector analysis. They are the simplest representations—all other representations may be built up from them.

Character. In Section 4.3, we see that a real matrix transforms under rotation of the coordinates by an orthogonal similarity transformation. Depending on the choice of reference frame, essentially the same matrix may take on an infinity of different forms. Likewise, our group representations may be put in an infinity of different forms using unitary transformations. But each such transformed representation is isomorphic with the original. From Exercise 4.3.8, the trace of each element (each matrix of our representation) is invariant under unitary transformations. Just because it is invariant, the trace (relabeled the *character*) assumes a role of some importance in group theory, particularly in applications to solid state physics.

Subgroup. It frequently happens that a subset of the group elements (including the unit element I) will by itself satisfy the four group requirements and is therefore a group. Such a subset is called a *subgroup*. The elements 1 and b of the four element group considered earlier form a subgroup. In Section 4.9, we consider O_3^+, the (continuous) group of all rotations in ordinary space. The rotations about any single axis form a subgroup of O_3^+. Numerous other examples of subgroups appear in the following sections.

In some instances, the similarity transform of each member x of the subgroup by all members g of the entire group is a member y of the subgroup,

$$y = gxg^{-1}. \tag{4.189}$$

Such a subgroup is called an *invariant subgroup* and is related to the multiplets of atomic and nuclear spectra and the particles discussed in Section 4.11. All subgroups of an abelian group are automatically invariant.

EXERCISES

4.7.1 Show that the matrices 1, A, B, and C of Eq. 4.184 are reducible. Reduce them.
Note: This means transforming A and C to diagonal form (by the *same* unitary transformation).

4.7.2 Possible operations on a crystal lattice include A_π (rotation by π), m (reflection), and i (inversion). These three operations combine as

$$A_\pi^2 = m^2 = i^2 = 1,$$

$$A_\pi \cdot m = i, \qquad m \cdot i = A_\pi, \qquad \text{and} \qquad i \cdot A_\pi = m.$$

Show that the group $(1, A_\pi, m, i)$ is isomorphic with the *vierergruppe*.

4.7.3 Four possible operations in the xy-plane are:

1. no change $\begin{cases} x \to x \\ y \to y \end{cases}$

2. inversion $\begin{cases} x \to -x \\ y \to -y \end{cases}$

3. reflection $\begin{cases} x \to -x \\ y \to y \end{cases}$

4. reflection $\begin{cases} x \to x \\ y \to -y. \end{cases}$

(a) Show that these four operations form a group.

(b) Show that this group is isomorphic with the *vierergruppe*.

(c) Set up a 2×2 matrix representation.

4.7.4 Rearrangement Theorem.
Given a group of n distinct elements $(I, a, b, c, \ldots n)$, show that the set of products $(aI, a^2, ab, ac, \ldots an)$ reproduces the n distinct elements in a new order.

4.8 Discrete Groups

In physics, groups usually appear as a set of operations that leave a system unchanged, invariant. This is an expression of symmetry. Indeed, a symmetry may be defined as the invariance of the Hamiltonian of a system under a group of transformations. Symmetry in this sense is important in classical mechanics, but it becomes even more important and more profound in quantum mechanics. In this section, we investigate the symmetry properties of sets of objects (atoms in a molecule or crystal). This provides additional illustrations of the group concepts of Section 4.7 and leads directly to dihedral groups. The dihedral groups in turn open up the study of the 32 point groups and 230 space groups which are of such importance in crystallography and solid state physics. It might be noted that it was through the study of crystal symmetries that the concepts of symmetry and group theory entered physics.

Two objects—two-fold symmetry axis. Consider first the two-dimensional system of two identical atoms in the xy-plane at (1, 0) and (−1, 0), Fig. 4.6. What

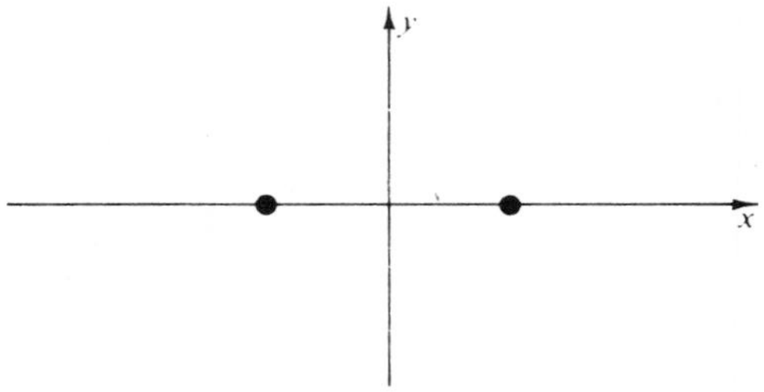

FIG. 4.6 Diatomic molecule H_2, N_2, O_2, Cl_2, etc

rotations[1] can be carried out that will leave this system invariant? The first candidate is, of course, the unit operator **1**. A rotation of π radians about the z-axis completes the list. So we have a rather uninteresting group of two members (1, −1). The z-axis is labeled a two-fold symmetry axis—corresponding to the two rotation

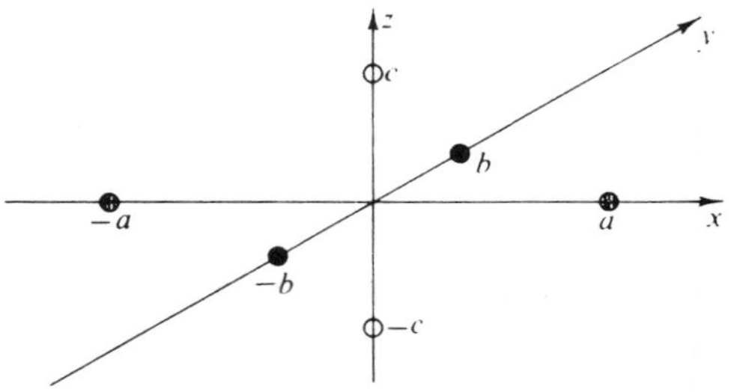

FIG 4.7 D_2 symmetry

[1] Here we deliberately exclude reflections and inversions. They must be brought in to develop the full set of 32 point groups.

angles 0 and π that leave the system invariant.

Our system becomes more interesting in three dimensions. Imagine now a molecule (or part of a crystal) with atoms of element X at $\pm a$ on the x-axis, atoms of element Y at $\pm b$ on the y-axis, and atoms of element Z at $\pm c$ on the z-axis as shown in Fig. 4.7. Clearly each axis is now a two-fold symmetry axis. Using $\mathsf{R}_x(\pi)$ to designate a rotation of π radians about the x-axis, we may set up a matrix representation of the rotations as in Section 4.3:

$$\mathsf{R}_x(\pi) = \begin{pmatrix} 1 & 0 & 0 \\ 0 & -1 & 0 \\ 0 & 0 & -1 \end{pmatrix} \qquad \mathsf{R}_y(\pi) = \begin{pmatrix} -1 & 0 & 0 \\ 0 & 1 & 0 \\ 0 & 0 & -1 \end{pmatrix}$$

$$\mathsf{R}_z(\pi) = \begin{pmatrix} -1 & 0 & 0 \\ 0 & -1 & 0 \\ 0 & 0 & 1 \end{pmatrix} \qquad \mathsf{1} = \begin{pmatrix} 1 & 0 & 0 \\ 0 & 1 & 0 \\ 0 & 0 & 1 \end{pmatrix}. \tag{4.190}$$

These four elements [$\mathsf{1}$, $\mathsf{R}_x(\pi)$, $\mathsf{R}_y(\pi)$, $\mathsf{R}_z(\pi)$] form an abelian group with a group multiplication table:

	1	$\mathsf{R}_x(\pi)$	$\mathsf{R}_y(\pi)$	$\mathsf{R}_z(\pi)$
1	1	R_x	R_y	R_z
$\mathsf{R}_x(\pi)$	R_x	1	R_z	R_y
$\mathsf{R}_y(\pi)$	R_y	R_z	1	R_x
$\mathsf{R}_z(\pi)$	R_z	R_y	R_x	1

Comparison with Exercises 4.2.7, 4.7.2, or 4.7.3 shows immediately that this group is the *vierergruppe*. The matrices of Eq. 4.190 are isomorphic with those of Ex. 4.2.7. Also, they are obviously reducible—being diagonal. The subgroups are (1, R_x), (1, R_y) and (1, R_z). They are invariant. It should be noted that a rotation of π about the y-axis and a rotation of π about the z-axis is equivalent to a rotation of π about the x-axis. $\mathsf{R}_z(\pi)\mathsf{R}_y(\pi) = \mathsf{R}_x(\pi)$. In symmetry terms, if y and z are two-fold symmetry axes, x is automatically a two-fold symmetry axis.

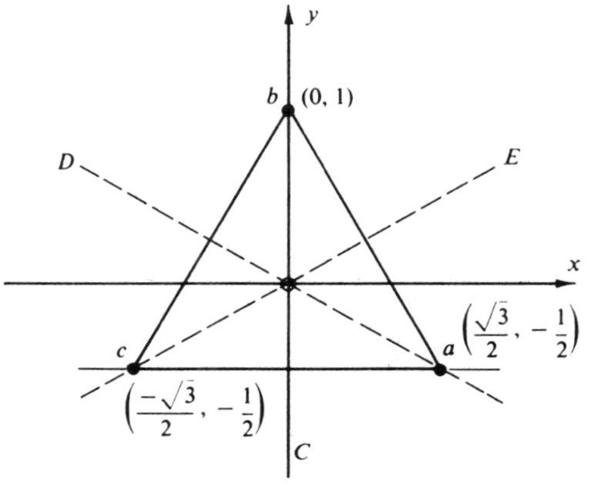

FIG. 4.8 Symmetry operations on an equilateral triangle

This symmetry group, the *vierergruppe* is often labeled D_2, the D signifying a dihedral group, and the subscript 2 signifying a two-fold symmetry axis (and no higher symmetry axis).

Three objects—three-fold symmetry axis. Consider now three identical atoms at the vertices of an equilateral triangle, Fig. 4.8. Rotations *of the triangle* of 0, $2\pi/3$, and $4\pi/3$ leave the triangle invariant. In matrix form, we have[1]

$$1 = \mathsf{R}_z(0) = \begin{pmatrix} 1 & 0 \\ 0 & 1 \end{pmatrix}$$

$$\mathsf{A} = \mathsf{R}_z(2\pi/3) = \begin{pmatrix} \cos 2\pi/3 & -\sin 2\pi/3 \\ \sin 2\pi/3 & \cos 2\pi/3 \end{pmatrix} = \begin{pmatrix} -1/2 & -\sqrt{3}/2 \\ \sqrt{3}/2 & -1/2 \end{pmatrix}$$

$$\mathsf{B} = \mathsf{R}_z(4\pi/3) = \begin{pmatrix} -1/2 & \sqrt{3}/2 \\ -\sqrt{3}/2 & -1/2 \end{pmatrix}. \tag{4.191}$$

The z-axis is a three-fold symmetry axis. $(1, \mathsf{A}, \mathsf{B})$ form a cyclic group, a subgroup of the complete six element group that follows.

In the xy-plane, there are three additional axes of symmetry—each atom (vertex) and the geometric center defining an axis. Each of these is a two-fold symmetry axis. These rotations may most easily be described within our two-dimensional framework by introducing reflections. The rotation of π about the C or y-axis, which means the interchanging of atoms a and c, is just a reflection of the x-axis:

$$\mathsf{C} = \mathsf{R}_C(\pi) = \begin{pmatrix} -1 & 0 \\ 0 & 1 \end{pmatrix}. \tag{4.192}$$

We may replace the rotation about the D-axis by a rotation of $4\pi/3$ (about our z-axis) followed by a reflection of the x-axis $(x \to -x)$:

$$\begin{aligned} \mathsf{D} = \mathsf{R}_D(\pi) &= \mathsf{CB} \\ &= \begin{pmatrix} -1 & 0 \\ 0 & 1 \end{pmatrix} \begin{pmatrix} -1/2 & \sqrt{3}/2 \\ -\sqrt{3}/2 & -1/2 \end{pmatrix} \\ &= \begin{pmatrix} 1/2 & -\sqrt{3}/2 \\ -\sqrt{3}/2 & -1/2 \end{pmatrix}. \end{aligned} \tag{4.193}$$

In a similar manner, the rotation of π about the E-axis interchanging a and b is replaced by a rotation of $2\pi/3$ (A) and then a reflection[2] of the x-axis $(x \to -x)$:

$$\begin{aligned} \mathsf{E} = \mathsf{R}_E(\pi) &= \mathsf{CA} \\ &= \begin{pmatrix} -1 & 0 \\ 0 & 1 \end{pmatrix} \begin{pmatrix} -1/2 & -\sqrt{3}/2 \\ \sqrt{3}/2 & -1/2 \end{pmatrix} \\ &= \begin{pmatrix} 1/2 & \sqrt{3}/2 \\ \sqrt{3}/2 & -1/2 \end{pmatrix}. \end{aligned} \tag{4.194}$$

[1] Note that here we are rotating the *triangle* counterclockwise relative to fixed coordinates.

[2] Note that, as a consequence of these reflections, $\det(\mathsf{C}) = \det(\mathsf{D}) = \det(\mathsf{E}) = -1$. The rotations A and B, of course, have a determinant of $+1$.

The complete group multiplication table is

	1	A	B	C	D	E
1	1	A	B	C	D	E
A	A	B	1	D	E	C
B	B	1	A	E	C	D
C	C	E	D	1	B	A
D	D	C	E	A	1	B
E	E	D	C	B	A	1

We have constructed a six element group and a 2×2 irreducible matrix representation of it. The only other distinct six element group is the cyclic group [1, R, R^2, R^3, R^4, R^5] with

$$R = \begin{pmatrix} \cos \pi/3 & -\sin \pi/3 \\ \sin \pi/3 & \cos \pi/3 \end{pmatrix} = \begin{pmatrix} 1/2 & -\sqrt{3}/2 \\ \sqrt{3}/2 & 1/2 \end{pmatrix}. \tag{4.195}$$

Our group [1, A, B, C, D, E] is labelled D_3 in crystallography, the dihedral group with a three-fold axis of symmetry. The three axes (C, D, and E) in the xy-plane automatically become two-fold symmetry axes. A consequence of this is that (1, C), (1, D), and (1, E) all form two element subgroups. None of these two element subgroups of D_3 is invariant.

There are two other irreducible representations of the symmetry group of the equilateral triangle: (i) the trivial (1, 1, 1, 1, 1, 1), and (ii) the almost as trivial (1, 1, 1, -1, -1, -1), the positive signs corresponding to proper rotations and the negative signs to improper rotations (involving a reflection). Both of these representations are homomorphic with D_3.

A general result for *finite* groups of h elements is that

$$\sum_i n_i^2 = h, \tag{4.196}$$

where n_i is the dimension of the matrices of the ith irreducible representation. Here we have $1^2 + 1^2 + 2^2 = 6$ for our three representations. No other irreducible representations of the symmetry group of three objects exist.

Dihedral groups, D_n. A dihedral group D_n with an n-fold symmetry axis implies n axes with angular separation of $2\pi/n$ radians. n is a positive integer, but otherwise unrestricted. If we apply the symmetry arguments to *crystal lattices*, then n is limited to 1, 2, 3, 4, and 6. The requirement of invariance of the crystal lattice under translations in the plane perpendicular to the n-fold axis excludes $n = 5$, 7, and higher values. Try to cover a plane completely with identical regular pentagons and with no overlapping. For individual molecules, this constraint does not exist, although the examples with $n > 6$ are rare. $n = 5$ is a real possibility. As an example, the symmetry group for ruthenocene, $(C_5H_5)_2Ru$, illustrated in Fig. 4.9 is D_5.[1]

[1] Actually the full technical label is D_{5h}, the h indicating invariance under a *reflection* of the five-fold axis.

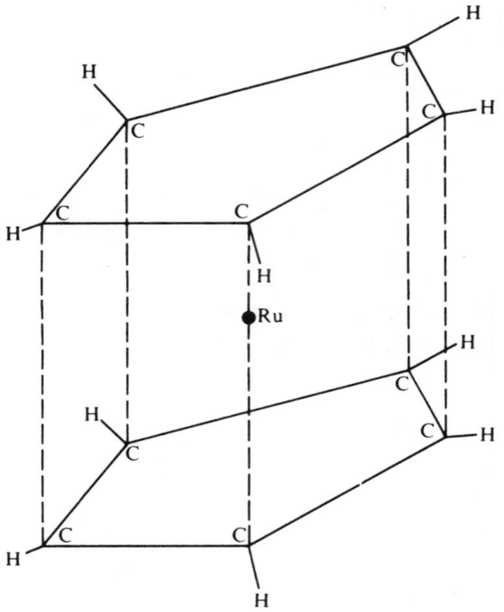

FIG. 4.9 Ruthenocene

Crystallographic point and space groups. The dihedral groups just considered are examples of the crystallographic point groups. A point group is composed of combinations of rotations and reflections (including inversions) that will leave some crystal lattice unchanged. Limiting the operations to rotations and reflections (including inversions) means that one point—the origin—remains *fixed*, hence the term point group. Including the cyclic groups, two cubic groups (tetrahedron and octahedron symmetries), and the improper forms (involving reflections), we come to a total of 32 point groups.

If, to the rotation and reflection operations that produced the point groups, we add the possibility of translations and still demand that some crystal lattice remain invariant, we come to the space groups. There are 230 distinct space groups, a number that is appalling except, possibly, to specialists in the field. For details (which can cover hundreds of pages), see the references.

EXERCISES

4.8.1 Show that the subgroup (1, A, B) of D_3 is an *invariant* subgroup.

4.8.2 Our group D_3 may be discussed as a *permutation* group of 3 objects. Matrix B, for instance, rotates a into b, b into c, and c into a. In three dimensions, this is

$$\begin{pmatrix} 0 & 1 & 0 \\ 0 & 0 & 1 \\ 1 & 0 & 0 \end{pmatrix} \begin{pmatrix} a \\ b \\ c \end{pmatrix} = \begin{pmatrix} b \\ c \\ a \end{pmatrix}.$$

(a) Develop analogous 3×3 representations for the other elements of D_3.
(b) Reduce your 3×3 representation to the 2×2 representation of this Section.
Note: The actual reduction of a reducible representation may be awkward. It is often easier to develop directly a new representation of the required dimension.

4.8.3 The elements of the dihedral group D_n may be written in the form

$$\mathsf{S}^\lambda \mathsf{R}_z^\mu(2\pi/n), \qquad \lambda = 0, 1$$
$$\mu = 0, 1, \ldots n-1,$$

where $\mathsf{R}_z(2\pi/n)$ represents a rotation of $2\pi/n$ about the n-fold symmetry axis, while S represents a rotation of π about an axis through the center of the regular polygon and one of its vertices.
For $\mathsf{S} = \mathsf{E}$ show that this form may describe the matrices A, B, C, and D of D_3.

4.8.4 Explain how the relation

$$\sum_i n_i^2 = h$$

applies to the *vierergruppe* ($h = 4$).

4.8.5 Develop the irreducible 2×2 matrix representation of the group of operations (rotations and reflections) that transform a square into itself. Give the group multiplication table.
Note: This is the symmetry group of a square and also the dihedral group, D_4.

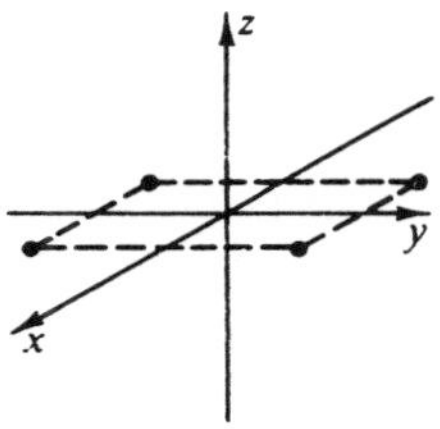

4.8.6 The permutation group of 4 objects contains $4! = 24$ elements. From Ex. 4.8.5, D_4, the symmetry group for a square, has far less than 24 elements. Explain the relation between D_4 and the permutation group of 4 objects.

4.8.7 In a simple cubic crystal, we might have identical atoms at $r = (la, ma, na)$, l, m, and n taking on all integral values.
(a) Show that each cartesian axis is a fourfold symmetry axis.
(b) The cubic group will consist of all operations (rotations, reflections, inversion) that leave the simple cubic crystal invariant. From a consideration of the permutation of the positive and negative coordinate axes, predict how many elements this cubic group will contain.

4.9 Continuous Groups

Infinite groups, Lie groups. All of the groups in the two preceding sections have contained a finite number of elements; 4 for the *vierergruppe*, 6 for D_3, etc. Here, we introduce groups with an infinite number of elements. The group element will contain one or more parameters that varies continuously over some range. The continuously varying parameter gives rise to a continuum of group elements. In contrast to the four member cyclic group $(1, i, -1, -i)$, we might have $e^{i\varphi}$, with φ varying continuously over the range $[0, 2\pi]$. The O_3^+ and SU(2) groups described below are additional examples.

Among the various mathematical possibilities, the continuous groups known as Lie groups are of particular interest. The characteristic of a Lie group is that the parameters of a product element are analytic functions[1] of the parameters of the factors. In the case of transformations, a rotation, for instance, we might write

$$x_i' = f_i(x_1, x_2, x_3, \theta) \tag{4.197}$$

(cf. Eq. 1.9). For this transformation group to be a Lie group, the functions f_i must be analytic functions of the parameter θ. This will be true for the O_3^+ and SU(2) groups considered here and in Section 4.10, for SU(3) encountered in Section 4.11, and for the Lorentz group of Section 4.12. All are Lie groups. The analytic nature of the functions (differentiability) allows us to develop the concept of *generator* (Section 4.10) and to reduce the study of the whole group to a study of the group elements in the neighborhood of the identity element.

If these parameters vary over *closed* intervals such as $[0, \pi]$, or $[0, 2\pi]$ for angles, the group is *compact*. An important property of this is that every representation of a compact group is equivalent to a unitary representation. In contrast to this, the homogeneous Lorentz group of Section 4.12 is not compact, and the representation $L(\mathbf{v})$ is not unitary.

Orthogonal Group, O_3^+. The set of $n \times n$ real orthogonal matrices forms a group. (Check to see that the group properties of Section 4.7 are satisfied.) Our $n \times n$ matrix has $n(n-1)/2$ *independent* parameters. For $n = 2$, there is only one independent parameter: one angle in Eq. 4.63. For $n = 3$, there are three independent parameters: the three Euler angles of Section 4.3.

We shall consider in some detail the set of 3×3 real orthogonal matrices with a determinant $+1$; rotations only, no reflections. This group is frequently labeled O_3^+, the $+$ indicating that the determinant is $+1$. From Section 4.3, the rotations about the coordinate axes are

$$\mathsf{R}_x(\varphi) = \begin{pmatrix} 1 & 0 & 0 \\ 0 & \cos\varphi & \sin\varphi \\ 0 & -\sin\varphi & \cos\varphi \end{pmatrix}, \qquad \mathsf{R}_y(\theta) = \begin{pmatrix} \cos\theta & 0 & -\sin\theta \\ 0 & 1 & 0 \\ \sin\theta & 0 & \cos\theta \end{pmatrix},$$

$$\mathsf{R}_z(\psi) = \begin{pmatrix} \cos\psi & \sin\psi & 0 \\ -\sin\psi & \cos\psi & 0 \\ 0 & 0 & 1 \end{pmatrix}. \tag{4.198}$$

We are following the conventions of Section 4.3. The rotations are counterclockwise

[1] Analytic, defined in Section 6.2, means having derivatives of all orders.

rotations of the coordinate system to a new orientation. Also from Section 4.3, the general member of O_3^+ is the Euler angle rotation

$$\mathsf{R}(\alpha, \beta, \gamma) = \mathsf{R}_z(\gamma)\mathsf{R}_y(\beta)\mathsf{R}_z(\alpha). \tag{4.199}$$

The relation of the O_3^+ group and orbital angular momentum is developed in Section 4.10. O_3^+ also appears in Section 4.11 leading into SU(3) and particle physics.

Special unitary group, SU(2). The set of $n \times n$ unitary matrices also forms a group. (Again, check to see that the group properties are satisfied.) This group is often labeled U(n). We impose the additional restriction that the determinant of the matrices be $+1$ and obtain the *special unitary* or *unitary unimodular group*, SU(n). Our $n \times n$ unitary, unit determinant matrix has $n^2 - 1$ independent parameters. For $n = 2$, there are 3 parameters—the same as for O_3^+. For $n = 3$, there are 8 parameters. This will become the Eightfold Way of Section 4.11.

For $n = 2$, we have SU(2) with a general group element

$$\mathsf{U} = \begin{pmatrix} a & b \\ -b^* & a^* \end{pmatrix}, \tag{4.200}$$

with $a^*a + b^*b = 1$. As indicated, a and b are complex. These parameters are often called the Cayley-Klein parameters, having been introduced by Cayley and Klein in connection with problems in mechanics. While not quite so obvious, an alternate general form is

$$\mathsf{U}(\xi, \eta, \zeta) = \begin{pmatrix} e^{i\xi}\cos\eta & e^{i\zeta}\sin\eta \\ -e^{-i\zeta}\sin\eta & e^{-i\xi}\cos\eta \end{pmatrix} \tag{4.201}$$

with the three parameters ξ, η, and ζ real.

Now let us determine the irreducible representations of SU(2). Returning to Eq. 4.200 U describes a transformation of a two component complex column vector (called a spinor):

$$\begin{pmatrix} u' \\ v' \end{pmatrix} = \begin{pmatrix} a & b \\ -b^* & a^* \end{pmatrix}\begin{pmatrix} u \\ v \end{pmatrix}, \tag{4.202}$$

or

$$\begin{aligned} u' &= au + bv, \\ v' &= -b^*u + a^*v. \end{aligned} \tag{4.203}$$

From the form of this result, if we were to start with a *homogeneous* polynomial of the nth degree in u and v and carry out the unitary transformation, Eq. 4.203, we would still have a homogeneous nth degree polynomial. The significance of this is that the $n + 1$ terms u^n, $u^{n-1}v$, $u^{n-2}v^2$, etc., belong to an $(n + 1)$-dimensional representation of our special unitary group.

To save algebraic juggling, we follow the choice of Wigner and let $n = 2j$ and consider the (monomial) function

$$f_m(u, v) = \frac{u^{j+m} v^{j-m}}{\sqrt{(j+m)!(j-m)!}}. \tag{4.204}$$

The denominator is a sort of normalizing factor that will make our representation unitary. If we take the action of U on $f_m(u, v)$ to be[1]

$$\mathsf{U} f_m(u, v) = f_m(u', v'), \tag{4.205}$$

then

$$\begin{aligned} \mathsf{U} f_m(u, v) &= f_m(au + bv, \ -b^*u + a^*v) \\ &= \frac{(au + bv)^{j+m}(-b^*u + a^*v)^{j-m}}{\sqrt{(j+m)!(j-m)!}}. \end{aligned} \tag{4.206}$$

Now the job is to express the right-hand side of Eq. 4.206 as a linear combination of terms of the form of $f_m(u, v)$. The coefficients in the linear combination will give us the desired representation. We expand the two binomials by the binomial theorem (Section 5.6), obtaining

$$\begin{aligned} (au + bv)^{j+m} &= \sum_{k=0}^{j+m} \frac{(j+m)!}{k!(j+m-k)!} a^{j+m-k} u^{j+m-k} b^k v^k \\ (-b^*u + a^*v)^{j-m} &= \sum_{l=0}^{j-m} (-1)^{j-m-l} \frac{(j-m)!}{l!(j-m-l)!} b^{*(j-m-l)} u^{j-m-l} a^{*l} v^l. \end{aligned} \tag{4.207}$$

Then

$$\begin{aligned} \mathsf{U} f_m(u, v) = \sum_{k=0}^{j+m} \sum_{l=0}^{j-m} (-1)^{j-m-l} & \frac{\sqrt{(j+m)!(j-m)!}}{k!\,l!(j+m-k)!(j-m-l)!} \\ & \times a^{j+m-k} a^{*l} b^k b^{*(j-m-l)} u^{2j-k-l} v^{k+l}. \end{aligned} \tag{4.208}$$

If we let $j - k - l = m'$,

$$u^{2j-k-l} v^{k+l} \to u^{j+m'} v^{j-m'} \tag{4.209}$$

matching the form of Eq. 4.204. Replacing the summation over l by a summation over m',

$$\mathsf{U} f_m(u, v) = \sum_{m'=-j}^{j} U_{mm'} f_{m'}(u, v) \tag{4.210}$$

where the matrix element $U_{mm'}$ is given by

$$\begin{aligned} U_{mm'} = \sum_{k=0}^{j+m} (-1)^{m'-m+k} & \frac{\sqrt{(j+m)!(j-m)!(j+m')!(j-m')!}}{k!(j-m'-k)!(j+m-k)!(m'-m+k)!} \\ & \times a^{j+m-k} a^{*(j-m'-k)} b^k b^{*(m'-m+k)}. \end{aligned} \tag{4.211}$$

[1] In Section 4.10, the transformation (rotation) of a *function* is defined in terms of the inverse rotation of the coordinates. Here, we use Eq. 4.205 since we are setting up a comparison with O_3^+ which is described in terms of rotations of the coordinates—Eq. 4.198.

The index k starts with zero and runs up to $j + m$, but the factorials[1] in the denominator guarantee that the coefficient will vanish if any exponent goes negative.

Equation 4.210 shows that the operation of $U_{mm'}$ on $f_{m'}$ (summed over m') is identical to U operating on f_m. On the basis of this correspondence, this isomorphism, we identify these matrices $U_{mm'}$ as representations of SU(2). Since m and m' each range from $-j$ to $+j$ in integral steps, our matrices $U_{mm'}$ have dimensions $(2j + 1) \times (2j + 1)$.

To be a little more specific about this—if $j = \frac{1}{2}$,

$$\mathsf{U} = \begin{array}{c} \\ m = \frac{1}{2} \\ m = -\frac{1}{2} \end{array} \begin{array}{c} \begin{array}{cc} m' = \frac{1}{2} & m' = -\frac{1}{2} \end{array} \\ \begin{pmatrix} a & b \\ -b^* & a^* \end{pmatrix} \end{array} \tag{4.212}$$

identical with Eq. 4.200. The cases for $j = 1$ and up are most conveniently handled with trigonometric functions, as shown below.

SU(2)—O_3^+ homomorphism. Our real orthogonal group O_3^+ clearly describes rotations in ordinary three-dimensional space. Just as clearly, it has the characteristic of leaving $x^2 + y^2 + z^2$ *invariant*.

The operation of SU(2) on a matrix is given by a unitary transformation, Eq. 4.122

$$\mathsf{M}' = \mathsf{U}\,\mathsf{M}\mathsf{U}^\dagger. \tag{4.213}$$

Taking M to be a 2×2 matrix, we note that any 2×2 matrix may be written as a linear combination of the unit matrix and the three Pauli matrices of Section 4.2. Let M be the zero trace matrix,

$$\mathsf{M} = x\sigma_1 + y\sigma_2 + z\sigma_3 = \begin{pmatrix} z & x - iy \\ x + iy & -z \end{pmatrix}, \tag{4.214}$$

the unit matrix not entering. Since the trace is invariant under a unitary transformation (Ex. 4.3.9), M' must have the same form,

$$\mathsf{M}' = x'\sigma_1 + y'\sigma_2 + z'\sigma_3 = \begin{pmatrix} z' & x' - iy' \\ x' + iy' & -z' \end{pmatrix}. \tag{4.215}$$

The determinant is also invariant under a unitary transformation (Ex. 4.3.13). Therefore,

$$-(x^2 + y^2 + z^2) = -(x'^2 + y'^2 + z'^2), \tag{4.216}$$

or $x^2 + y^2 + z^2$ is invariant under this operation of SU(2), just as with O_3^+. SU(2) must, therefore, describe a rotation. This suggests that SU(2) and O_3^+ may be isomorphic or homomorphic.

We approach the problem of what rotation SU(2) describes by considering special cases. Returning to Eq. 4.200 with one eye on Eq. 4.201, let $a = e^{i\xi}$ and $b = 0$,

[1] From Section 10.1, $(-n)! = \pm\infty$, for $n = 1, 2, 3, \ldots$.

or

$$\mathsf{U}_z = \begin{pmatrix} e^{i\xi} & 0 \\ 0 & e^{-i\xi} \end{pmatrix}. \tag{4.217}$$

Carrying out a unitary transformation on each of the three Pauli σ's,

$$\begin{aligned} \mathsf{U}_z\sigma_1\mathsf{U}_z^\dagger &= \begin{pmatrix} e^{i\xi} & 0 \\ 0 & e^{-i\xi} \end{pmatrix}\begin{pmatrix} 0 & 1 \\ 1 & 0 \end{pmatrix}\begin{pmatrix} e^{-i\xi} & 0 \\ 0 & e^{i\xi} \end{pmatrix} \\ &= \begin{pmatrix} 0 & e^{2i\xi} \\ e^{-2i\xi} & 0 \end{pmatrix}. \end{aligned} \tag{4.218}$$

We re-express this result in terms of the Pauli σ's to obtain

$$\mathsf{U}_z x\sigma_1\mathsf{U}_z^\dagger = x\cos 2\xi\sigma_1 - x\sin 2\xi\,\sigma_2. \tag{4.219}$$

Similarly,

$$\begin{aligned} \mathsf{U}_z y\sigma_2\,\mathsf{U}_z^\dagger &= y\sin 2\xi\,\sigma_1 + y\cos 2\xi\,\sigma_2 \\ \mathsf{U}_z z\sigma_3\,\mathsf{U}_z^\dagger &= z\sigma_3. \end{aligned} \tag{4.220}$$

From these double angle expressions, we see that we should start with a *half-angle*: $\xi = \alpha/2$. Then, from Eqs. 4.213–4.215, 4.219, and 4.220,

$$\begin{aligned} x' &= x\cos\alpha + y\sin\alpha \\ y' &= -x\sin\alpha + y\cos\alpha \\ z' &= z. \end{aligned} \tag{4.221}$$

The 2 × 2 unitary transformation using $\mathsf{U}_z(\alpha/2)$ is equivalent to the rotation operator $\mathsf{R}_z(\alpha)$ of Eq. 4.198.

The establishment of the correspondence of

$$\mathsf{U}_y(\beta/2) = \begin{pmatrix} \cos\beta/2 & \sin\beta/2 \\ -\sin\beta/2 & \cos\beta/2 \end{pmatrix} \tag{4.222}$$

and $\mathsf{R}_y(\beta)$, and of

$$\mathsf{U}_x(\varphi/2) = \begin{pmatrix} \cos\varphi/2 & i\sin\varphi/2 \\ i\sin\varphi/2 & \cos\varphi/2 \end{pmatrix} \tag{4.223}$$

and $\mathsf{R}_x(\varphi)$ are left as Ex. 4.9.7. The reader might note that $\mathsf{U}_k(\psi/2)$ has the general form

$$\mathsf{U}_k(\psi/2) = \mathbf{1}\cos\psi/2 + i\sigma_k\sin\psi/2 \tag{4.224}$$

where $k = x, y, z$. We return to this point in Section 4.10.

The correspondence

$$\mathsf{U}_z(\alpha/2) = \begin{pmatrix} e^{i\alpha/2} & 0 \\ 0 & e^{-i\alpha/2} \end{pmatrix} \leftrightarrow \begin{pmatrix} \cos\alpha & \sin\alpha & 0 \\ -\sin\alpha & \cos\alpha & 0 \\ 0 & 0 & 1 \end{pmatrix} = \mathsf{R}_z(\alpha) \tag{4.225}$$

is not a simple one-to-one correspondence. Specifically, as α in R_z ranges from 0 to 2π, the parameter in U_z, $\alpha/2$, goes from 0 to π. We find

$$\mathsf{R}_z(\alpha + 2\pi) = \mathsf{R}_z(\alpha)$$

$$\mathsf{U}_z(\alpha/2 + \pi) = \begin{pmatrix} -e^{i\alpha/2} & 0 \\ 0 & -e^{-i\alpha/2} \end{pmatrix} = -\mathsf{U}_z(\alpha/2). \tag{4.226}$$

Therefore, *both* $\mathsf{U}_z(\alpha/2)$ and $\mathsf{U}_z(\alpha/2 + \pi) = -\mathsf{U}_z(\alpha/2)$ correspond to $\mathsf{R}_z(\alpha)$. The correspondence is 2 to 1, or SU(2) and O_3^+ are *homomorphic*. This establishment of the correspondence between the representations of SU(2) and those of O_3^+ means that the known representations of SU(2) automatically provide us with the representations[1] of O_3^+

Combining the various rotations, a unitary transformation using

$$\mathsf{U}(\alpha, \beta, \gamma) = \mathsf{U}_z(\gamma/2)\mathsf{U}_y(\beta/2)\mathsf{U}_z(\alpha/2) \tag{4.227}$$

corresponds to the general Euler rotation $\mathsf{R}_z(\gamma)\mathsf{R}_y(\beta)\mathsf{R}_z(\alpha)$. By direct multiplication,

$$\mathsf{U}(\alpha, \beta, \gamma) = \begin{pmatrix} e^{i\gamma/2} & 0 \\ 0 & e^{-i\gamma/2} \end{pmatrix} \begin{pmatrix} \cos\beta/2 & \sin\beta/2 \\ -\sin\beta/2 & \cos\beta/2 \end{pmatrix} \begin{pmatrix} e^{i\alpha/2} & 0 \\ 0 & e^{-i\alpha/2} \end{pmatrix}$$

$$= \begin{pmatrix} e^{i(\gamma+\alpha)/2}\cos\beta/2 & e^{i(\gamma-\alpha)/2}\sin\beta/2 \\ -e^{-i(\gamma-\alpha)/2}\sin\beta/2 & e^{-i(\gamma+\alpha)/2}\cos\beta/2 \end{pmatrix}. \tag{4.228}$$

This is our alternate general form, Eq. 4.201, with

$$\xi = (\gamma + \alpha)/2, \qquad \eta = \beta/2, \qquad \zeta = (\gamma - \alpha)/2. \tag{4.229}$$

From Eq. 4.228, we may identify the parameters of Eq. 4.201 as

$$\begin{aligned} a &= e^{i(\gamma+\alpha)/2}\cos\beta/2 \\ b &= e^{i(\gamma-\alpha)/2}\sin\beta/2. \end{aligned} \tag{4.230}$$

With these, our SU(2) representation $U_{mm'}$ of Eq. 4.211 becomes

$$U_{mm'}(\alpha, \beta, \gamma) = \sum_{k=0}^{j+m} (-1)^k \frac{\sqrt{(j+m)!(j-m)!(j+m')!(j-m')!}}{k!(j-m'-k)!(j+m-k)!(m'-m+k)!} \times e^{im\gamma}\left(\cos\frac{\beta}{2}\right)^{2j+m-m'-2k}\left(-\sin\frac{\beta}{2}\right)^{m'-m+2k} e^{im'\alpha}. \tag{4.231}$$

Here are our irreducible representations in terms of the Euler angles. The importance of Eq. 4.231 is that it allows us to calculate the $(2j + 1) \times (2j + 1)$ irreducible representations of SU(2) for all j ($j = 0, \frac{1}{2}, 1, \frac{3}{2} \ldots$) and the irreducible representations of O_3^+ for integral orbital angular momentum j ($j = 0, 1, 2, \ldots$).

[1] While SU(2) has representations for integral and half odd integral values of j ($j = 0, \frac{1}{2}, 1, \frac{3}{2}, \ldots$), O_3^+ is limited to integral values of j ($j = 0, 1, 2, \ldots$). Further discussion of this point—the relation between O_3^+ and orbital angular momentum—appears in Sections 4.10 and 12.7.

Rotation matrix $\mathsf{D}^j(\alpha, \beta, \gamma)$. In the quantum mechanics literature, it is customary to take the adjoint of $U_{mm'}$ defining[1]

$$D^j_{m'm}(\alpha, \beta, \gamma) = U^*_{mm'}(\alpha, \beta, \gamma). \tag{4.232}$$

For $j = 0$,

$$\mathsf{D}^0(\alpha, \beta, \gamma) = 1. \tag{4.233}$$

For $j = \frac{1}{2}$,

$$\mathsf{D}^{1/2}(\alpha, \beta, \gamma) = \begin{matrix} & m = \frac{1}{2} & m = -\frac{1}{2} \\ m' = \frac{1}{2} & e^{-i\alpha/2} \cos \beta/2\, e^{-i\gamma/2} & -e^{-i\alpha/2} \sin \beta/2\, e^{i\gamma/2} \\ m' = -\frac{1}{2} & e^{i\alpha/2} \sin \beta/2\, e^{-i\gamma/2} & e^{i\alpha/2} \cos \beta/2\, e^{i\gamma/2} \end{matrix}. \tag{4.234}$$

For $j = 1$, Eqs. 4.231 and 4.232 lead to

$$\mathsf{D}^1(\alpha, \beta, \gamma) = \tag{4.235}$$

$$\begin{matrix} & m = 1 & m = 0 & m = -1 \\ m' = 1 & e^{-i\alpha} \dfrac{1 + \cos \beta}{2} e^{-i\gamma} & -e^{-i\alpha} \dfrac{\sin \beta}{\sqrt{2}} & e^{-i\alpha} \dfrac{1 - \cos \beta}{2} e^{i\gamma} \\ m' = 0 & \dfrac{\sin \beta}{\sqrt{2}} e^{-i\gamma} & \cos \beta & -\dfrac{\sin \beta}{\sqrt{2}} e^{i\gamma} \\ m' = -1 & e^{i\alpha} \dfrac{1 - \cos \beta}{2} e^{-i\gamma} & e^{i\alpha} \dfrac{\sin \beta}{\sqrt{2}} & e^{i\alpha} \dfrac{1 + \cos \beta}{2} e^{i\gamma} \end{matrix}$$

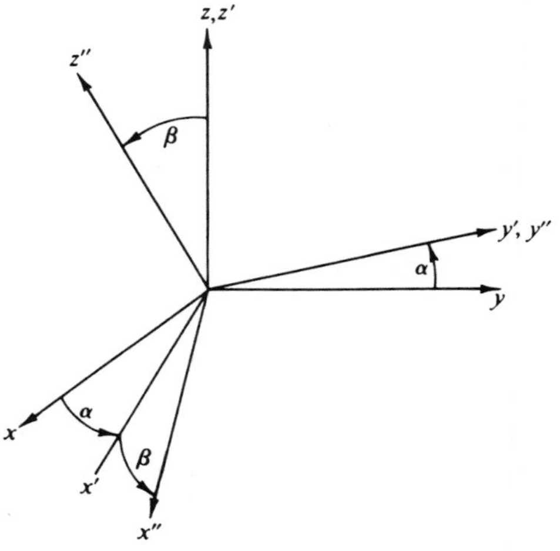

FIG. 4.10 Euler angle rotations ($\gamma = 0$)

[1] The reason for this is that $U_{mm'}$ is defined here in terms of rotations of coordinates. $D^j_{m'm}$ is used to rotate functions. Further discussion of this point appears in Section 4.10.

For $j = l$, integral, the operation of the rotation matrix D^l on the spherical harmonics (Section 12.6) is given by

$$Y_l^m(\theta', \varphi') = \sum_{m'} D^l_{m'm}(\alpha, \beta, 0) Y_l^{m'}(\theta, \varphi). \tag{4.236}$$

The first two Euler angles α and β define a new polar axis, z'' in Fig. 4.10 and a new zero of azimuth. (The third Euler angle γ corresponds to a rotation about the new polar axis and is irrelevant here.) The point $(\theta'\ \varphi')$ is the same point in space as (θ, φ), but measured relative to the rotated coordinate system rather than relative to the initial system.

Note the analogy with Eq. 4.205. The spherical harmonics are homogeneous functions of x, y, and z of order l.

For further details of the rotation matrix D^j, including a proof of the spherical harmonic addition theorem, the reader should consult the text by Rose cited at the end of this Chapter.

EXERCISES

4.9.1 Show that an $n \times n$ orthogonal matrix has $n(n-1)/2$ *independent* parameters.
Hint. The orthogonality condition, Eq. 4.60, provides constraints.

4.9.2 Show that an $n \times n$ special unitary matrix has $n^2 - 1$ *independent* parameters.
Hint. Each element may be complex—doubling the number of possible parameters. Some of the constraining equations are likewise complex—and count as two constraints.

4.9.3 The special linear group SL(2) consists of all 2×2 matrices (with complex elements) having a determinant of $+1$. Show that such matrices form a group.
Note. The SL(2) group can be related to the full Lorentz group, Section 4.12 much as the SU(2) group is related to O_3^+.

4.9.4 Show that R_z is (or is not) an invariant subgroup of O_3^+.

4.9.5 Prove that the general form of a 2×2 unitary, unimodular matrix is

$$\mathsf{U} = \begin{pmatrix} a & b \\ -b^* & a^* \end{pmatrix}$$

with $a^*a + b^*b = 1$.

4.9.6 Denoting the spinor (u, v) of Eq. 4.202 by s, show that $s^\dagger s = s'^\dagger s'$, the length of the spinor, is conserved under the transformation U.

4.9.7 (a) Show that $\mathsf{U}_y(\beta/2)$ corresponds to $\mathsf{R}_y(\beta)$.
(b) Show that $\mathsf{U}_x(\varphi/2)$ corresponds to $\mathsf{R}_x(\varphi)$.

4.9.8 (a) Show that the α and γ dependence of $\mathsf{D}^J(\alpha, \beta, \gamma)$ may be factored out such that

$$\mathsf{D}^J(\alpha, \beta, \gamma) = \mathsf{A}^J(\alpha)\mathsf{d}^J(\beta)\mathsf{C}^J(\gamma).$$

(b) Show that $\mathsf{A}^j(\alpha)$ and $\mathsf{C}^j(\gamma)$ are diagonal. Find the explicit forms.
(c) Show that $\mathsf{d}^j(\beta) = \mathsf{D}^j(0, \beta, 0)$.
Hint. Exs. 4.2.28 and 4.2.29 may be helpful.

4.9.9 By inspection of Eqs. 4.231 and 4.232, or the special cases, Eqs. 4.234 and 4.235,

$$\mathsf{D}^{j\dagger}(\alpha, \beta, \gamma) = \mathsf{D}^j(-\gamma, -\beta, -\alpha).$$

Explain why this should be so.

4.9.10 (a) Assuming that $\mathsf{D}^j(\alpha, \beta, \gamma)$ is unitary, show that

$$\sum_{m=-l}^{l} Y_l^{m*}(\theta_1, \varphi_1) Y_l^m(\theta_2, \varphi_2)$$

is a scalar quantity (invariant under rotations). This is a function analog of a scalar product of vectors.
(b) From part (a) derive the spherical harmonic addition theorem, Eq. 12.200:

$$P_l(\cos\gamma) = 4\pi(2l+1)^{-1} \sum_{m=-l}^{l} Y_l^{m*}(\theta_1, \varphi_1) Y_l^m(\theta_2, \varphi_2).$$

Hint. Set $\theta_1 = 0$ (which makes $\theta_2 = \gamma$).

4.10 Generators

Rotations and angular momentum. Two matrices U and H are related by the equation

$$\mathsf{U} = e^{ia\mathsf{H}} = 1 + ia\mathsf{H} + (ia\mathsf{H})^2/2! + \cdots. \tag{4.237}$$

Here a is a real parameter. The second portion of Eq. 4.237, a Maclaurin expansion, serves to interpret the exponential. From Section 4.5, if H is Hermitian, then U is unitary. Similarly if U is unitary, H is Hermitian.

Now, in the context of group theory H is labelled a *generator*, the generator of U. The relation of the generator to the rotation group O_3^+ is indicated schematically in Fig. 4.11.

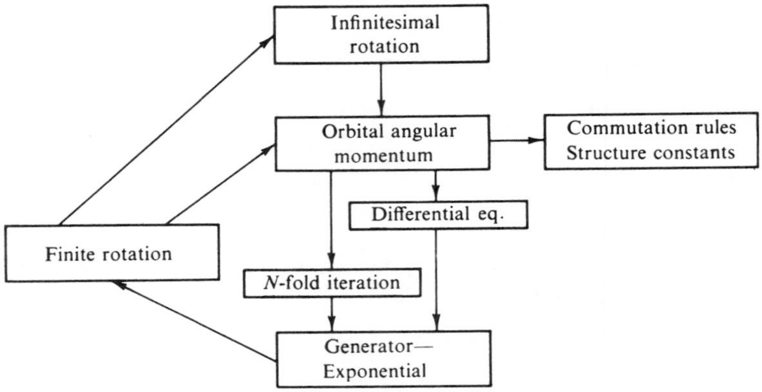

FIG. 4.11 Group-generator relationships

1. Starting with the left side of Fig. 4.11 the matrix describing a *finite rotation of the coordinates* through an angle φ counterclockwise about the z-axis is given by Eq. 4.76 as

$$\mathsf{R}_z(\varphi) = \begin{pmatrix} \cos\varphi & \sin\varphi & 0 \\ -\sin\varphi & \cos\varphi & 0 \\ 0 & 0 & 1 \end{pmatrix}. \tag{4.238}$$

2. Let the rotation described R_z be an infinitesimal rotation through an angle $\delta\varphi$. Then R_z may be written

$$\mathsf{R}_z(\delta\varphi) = 1 + i\delta\varphi\mathsf{M}_z \tag{4.239}$$

where

$$\mathsf{M}_z = \begin{pmatrix} 0 & -i & 0 \\ i & 0 & 0 \\ 0 & 0 & 0 \end{pmatrix}. \tag{4.240}$$

M_z and the corresponding matrices M_x and M_y appear in Exercise 4.2.15. M_z may also be obtained by differentiation. If we interpret the derivative of a matrix as the matrix of the derivatives, then

$$d\mathsf{R}_z/d\varphi\,|_{\varphi=0} = i\mathsf{M}_z. \tag{4.241}$$

From this point of view, Eq. 4.239 is a Maclaurin expansion of R_z with terms of order $(\delta\varphi)^2$ and beyond omitted. The validity of Eq. 4.241 is a consequence of the differentiability of Lie groups.

3. Our finite rotation φ may be compounded of successive infinitesimal rotations $\delta\varphi$.

$$\mathsf{R}_z(\delta\varphi_2 + \delta\varphi_1) = (1 + i\delta\varphi_2\mathsf{M}_z)(1 + i\delta\varphi_1\mathsf{M}_z). \tag{4.242}$$

Let $\delta\varphi = \varphi/N$ for N rotations, with $N \to \infty$. Then,

$$\mathsf{R}_z(\varphi) = \lim_{N\to\infty} [1 + (i\varphi/N)\mathsf{M}_z]^N = \exp i\varphi\,\mathsf{M}_z. \tag{4.243}$$

From this form, we identify M_z as the generator of the group $\mathsf{R}_z(\varphi)$, a subgroup of O_3^+. The actual reconstruction of $\mathsf{R}_z(\varphi)$ appears below. Two characteristics are worth noting:

a. M_z is Hermitian and $\mathsf{R}_z(\varphi)$ is unitary.
b. Trace $(\mathsf{M}_z) = 0$ and det $\mathsf{R}_z(\varphi) = +1$.

In direct analogy with M_z, M_x may be identified as the generator of R_x, the (sub) group of rotations about the x-axis. And then M_y generates R_y.

4. As indicated in Eq. 4.237, the exponential may be expanded to give

$$\exp i\varphi\mathsf{M}_z = 1 + i\varphi\mathsf{M}_z + (i\varphi\mathsf{M}_z)^2/2! + (i\varphi\mathsf{M}_z)^3/3! + \cdots$$

$$= \begin{pmatrix} 0 & 0 & 0 \\ 0 & 0 & 0 \\ 0 & 0 & 1 \end{pmatrix} + \begin{pmatrix} 1 & 0 & 0 \\ 0 & 1 & 0 \\ 0 & 0 & 0 \end{pmatrix} \{1 - \varphi^2/2! + \varphi^4/4! - \cdots\}$$
$$+ i\mathsf{M}_z\{\varphi - \varphi^3/3! + \varphi^5/5! - \cdots\}. \tag{4.244}$$

In the second equality above, the relations

$$\mathsf{M}_z^2 = \begin{pmatrix} 1 & 0 & 0 \\ 0 & 1 & 0 \\ 0 & 0 & 0 \end{pmatrix} \quad \text{and} \quad \mathsf{M}_z^3 = \mathsf{M}_z \tag{4.245}$$

have been used. Recognizing that the first series is $\cos\varphi$ and the second $\sin\varphi$, we have $\mathsf{R}_z(\varphi)$ as given in Eq. 4.238.

5. Returning to the infinitesimal level, our infinitesimal rotations commute:

$$[\mathsf{R}_x(\delta\varphi_x), \mathsf{R}_y(\delta\varphi_y)] = [\mathsf{R}_y(\delta\varphi_y), \mathsf{R}_z(\delta\varphi_z)]$$
$$= [\mathsf{R}_z(\delta\varphi_z), \mathsf{R}_x(\delta\varphi_x)] = 0 \tag{4.246}$$

and an infinitesimal rotation about an axis defined by a unit vector $\mathbf{n}$ becomes

$$\mathsf{R}_{\mathbf{n}}(\delta\varphi) = 1 + i(\delta\varphi_x\mathsf{M}_x + \delta\varphi_y\mathsf{M}_y + \delta\varphi_z\mathsf{M}_z)$$
$$= 1 + i\delta\varphi\, \mathbf{n}\cdot\mathsf{M}. \tag{4.247}$$

6. From Exercise 4.2.15 the generators satisfy the commutation relations

$$[\mathsf{M}_i, \mathsf{M}_j] = i\varepsilon_{ijk}\mathsf{M}_k. \tag{4.248}$$

Here ε_{ijk} is the totally antisymmetric Levi-Civita symbol of Section 3.4. A summation over k is implied, but there is only one nonvanishing term. The coefficient of M_k, $i\varepsilon_{ijk}$, is called the *structure constant*. The structure constants form the starting point for the development of a Lie algebra. As seen above, the group generators determine the structure constants. Conversely, it may be shown that the structure constants determine the group.

It might be noted that Eq. 4.248 has an infinite number of solutions. The three matrices M_x, M_y, M_z of Ex. 4.2.15 constitute one solution—corresponding to one unit of angular momentum. Other solutions, $(2l+1) \times (2l+1)$ matrices, with $l = 2, 3, 4, \ldots$ generate the other irreducible representations of the rotation group, O_3^+.

Rotation of functions. In all of the foregoing discussion, the matrices rotate the coordinates. Any physical system being described is held fixed. Now let us hold the coordinates fixed and rotate a function $\psi(x, y, z)$ relative to our fixed coordinates. With R to rotate the coordinates, we introduce an operator $\mathscr{R}$ to rotate functions. We *define* $\mathscr{R}$ by

$$\mathscr{R}\psi(x, y, z) = \psi'(x, y, z) \to \psi(\mathbf{x}') \tag{4.249}$$

with

$$\mathbf{x}' = \mathsf{R}\mathbf{x}. \tag{4.250}$$

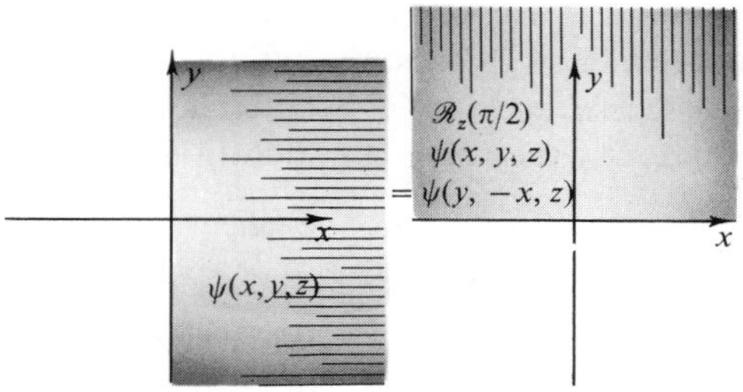

FIG. 4.12 Rotation of a function $\psi(x, y, z)$

In words, $\mathscr{R}$ operates on the function ψ, rotating ψ and creating a *new* function ψ'. This new function ψ' is numerically equal to $\psi(\mathbf{x}')$, where $\mathbf{x}'$ indicates that the *coordinates* have been rotated by R. For the special case of a rotation about the z-axis

$$\begin{aligned}\mathscr{R}_z(\varphi)\psi(x, y, z) = \ &\psi(x \cos \varphi + y \sin \varphi, \\ &-x \sin \varphi + y \cos \varphi, z).\end{aligned} \tag{4.251}$$

To get some understanding of the meaning of Eq. 4.251, consider the case $\varphi = \pi/2$. Then

$$\mathscr{R}_z\psi(x, y, z) = \psi(y, -x, z). \tag{4.252}$$

The function ψ may represent a wavefunction or some classical physical system.

Imagine that $\psi(x, y, z)$ is large when its *first* argument is large. Then $\mathscr{R}_z(\varphi = \pi/2)$ $\psi(x, y, z)$ will be large when the *first* argument of $\psi(y, -x, z)$ is large, i.e., when y is large. This is pictured in Fig. 4.12. The effect, then, of $\mathscr{R}_z$ is to rotate the pattern of the function ψ counterclockwise—the same as R would rotate the coordinate system.

Returning to Eq. 4.251, consider an infinitesimal rotation again, $\varphi \to \delta\varphi$. Then

$$\mathscr{R}_z(\delta\varphi)\psi(x, y, z) = \psi(x + y\,\delta\varphi, y - x\,\delta\varphi, z). \tag{4.253}$$

The right side may be expanded as a Taylor series (Section 5.6) to give

$$\begin{aligned}\mathscr{R}_z(\delta\varphi)\psi(x, y, z) &= \psi(x, y, z) - \delta\varphi\{x\,\partial\psi/\partial y - y\,\partial\psi/\partial x\} + O(\delta\varphi)^2 \\ &= (1 - i\,\delta\varphi L_z)\psi(x, y, z),\end{aligned} \tag{4.254}$$

the differential expression in curly brackets being iL_z. Since

$$\mathscr{R}_z(\varphi + \delta\varphi)\psi = \mathscr{R}_z(\delta\varphi)\mathscr{R}_z(\varphi)\psi = (1 - i\,\delta\varphi L_z)\mathscr{R}_z(\varphi)\psi, \tag{4.255}$$

we have (as an operator equation)

$$(\mathscr{R}_z(\varphi + \delta\varphi) - \mathscr{R}_z(\varphi))/\delta\varphi = -iL_z\mathscr{R}_z(\varphi). \tag{4.256}$$

The left side is just $d\mathscr{R}_z(\varphi)/d\varphi$, (for $\delta\varphi \to 0$). In this form, Eq. 4.256 integrates immediately to

$$\mathscr{R}_z(\varphi) = \exp -i\varphi L_z. \tag{4.257}$$

The constant of integration is fixed by the boundary condition $\mathscr{R}_z(0) = 1$.

Note the resemblance to Eq. 4.243 and the differences. R_z rotates the coordinates; $\mathscr{R}_z$ rotates functions. M_z is a matrix, L_z a differential operator. Note also that L_x, L_y, and L_z satisfy exactly the same commutation relation as M_x, M_y, and M_z,

$$[L_i, L_j] = i\varepsilon_{ijk}L_k \tag{4.258}$$

and yield the same structure constants.

Equations 4.257 and 4.243 might also be compared with two equations in Section 4.3: Eq. 4.89, in which **A** rotates coordinates counterclockwise, and Eq. 4.93, in which the *same* **A** rotates a vector *clockwise*. Here we have R rotating coordinates counterclockwise and $\mathscr{R}$ rotating functions *counterclockwise*. This is a consequence of the negative exponential in Eq. 4.257.

SU(2) and the Pauli matrices. The elements of the two dimensional special unitary group, SU(2), may be generated by

$$\exp \tfrac{1}{2}ia\sigma_1, \qquad \exp \tfrac{1}{2}ib\sigma_2 \qquad \text{and} \qquad \exp \tfrac{1}{2}ic\sigma_3 \tag{4.259}$$

where σ_1, σ_2, and σ_3 are the three Pauli spin matrices. The three parameters a, b, and c are real. Again, note that the σ's are Hermitian and have zero trace. The elements of SU(2), Eq. 4.259, are unitary and have a determinant of $+1$. It might be noted that the generators in diagonal form such as σ_3 lead to conserved quantum numbers.

The Pauli σ's satisfy commutation relations

$$[\sigma_i, \sigma_j] = 2i\varepsilon_{ijk}\sigma_k. \tag{4.260}$$

This differs from the **L** and M commutation relations, Eqs. 4.248 and 4.258 by a factor of 2. Let us therefore define $\mathsf{s}_i = \tfrac{1}{2}\sigma_i$, $i = 1, 2, 3$. Then

$$[\mathsf{s}_i, \mathsf{s}_j] = i\varepsilon_{ijk}\mathsf{s}_k \tag{4.261}$$

exactly like the angular momentum commutation relations[1], Eqs. 4.248 and 4.258, showing that the s_i, not σ_i, are the angular momentum operators. This is the reason for including the $\tfrac{1}{2}$'s in the generator-exponentials. Essentially this is the same as the adoption of the half angles in the investigation of the SU(2)-O_3^+ homomorphism in Section 4.9.

Expanding $\exp ia\mathsf{s}_1 = \exp \tfrac{1}{2}ia\sigma_1$ as a Maclaurin series

$$\begin{aligned}\exp \tfrac{1}{2}ia\sigma_1 &= \mathsf{1}\{1 - (a/2)^2/2! + (a/2)^4/4! - \cdots\} \\ &\quad + i\sigma_1\{(a/2) - (a/2)^3/3! + (a/2)^5/5! - \cdots\} \\ &= \begin{pmatrix} \cos a/2 & i \sin a/2 \\ i \sin a/2 & \cos a/2 \end{pmatrix} \\ &= \mathsf{1} \cos a/2 + i\sigma_1 \sin a/2,\end{aligned} \tag{4.262}$$

a special case of Eq. 4.224. The parameter a appears as an angle, the coefficient of

[1] These structure constants ($i\varepsilon_{ijk}$) lead to the SU(2) representations of dimension $2j+1$ for generators of dimension $2j+1$, $j = 0, \tfrac{1}{2}, 1, \tfrac{3}{2}, \ldots$. The *integral* j cases also lead to the representations of O_3^+, as discussed in Section 4.9.

an angular momentum matrix—like φ in Eq. 4.243. But in SU(2) form, the angle always appears as a half angle, $a/2$. Similarly (completing Eq. 4.224)

$$\begin{aligned} \exp \tfrac{1}{2} ib\sigma_2 &= \begin{pmatrix} \cos b/2 & \sin b/2 \\ -\sin b/2 & \cos b/2 \end{pmatrix} = 1 \cos b/2 + i\sigma_2 \sin b/2 \\ \exp \tfrac{1}{2} ic\sigma_3 &= \begin{pmatrix} \exp \tfrac{1}{2} ic & 0 \\ 0 & \exp -\tfrac{1}{2} ic \end{pmatrix} = 1 \cos c/2 + i\sigma_3 \sin c/2. \end{aligned} \tag{4.263}$$

With this identification of the exponentials, the general form of the SU(2) matrix may be written

$$\mathsf{U}(\alpha, \beta, \gamma) = (\exp \tfrac{1}{2} i\gamma\sigma_3)(\exp \tfrac{1}{2} i\beta\sigma_2)(\exp \tfrac{1}{2} i\alpha\sigma_3). \tag{4.264}$$

This reproduces Eq. 4.228 of Section 4.9. The selection of the Pauli matrices corresponds to the Euler angle rotations described in Sections 4.3 and 4.9.

Further examples of the infinitesimal rotation—exponentiation generator technique appear in Section 4.12.

EXERCISES

4.10.1 A *translation* operator $T(a)$ converts $\psi(x)$ to $\psi(x + a)$,

$$T(a)\psi(x) = \psi(x + a).$$

In terms of the (quantum mechanical) linear momentum operator $p_x = -i\, d/dx$, show that

$$T(a) = \exp iap_x.$$

Hint. Expand $\psi(x + a)$ as a Taylor series.

4.10.2 Consider the general SU(2) element Eq. 4.201 to be built up of three Euler rotations: (i) a rotation of $a/2$ about the z-axis, (ii) a rotation of $b/2$ about the new x-axis, and (iii) a rotation of $c/2$ about the new z-axis. (All rotations counterclockwise.) Using the Pauli σ generators show that these rotation angles are determined by

$$\begin{aligned} a &= \xi - \zeta + \pi/2 = \alpha + \pi/2 \\ b &= 2\eta \qquad\quad\;\; = \beta \\ c &= \xi + \zeta - \pi/2 = \gamma - \pi/2. \end{aligned}$$

Note: The angles a and b here are not the a and b of Eq. 4.200.

4.10.3 The angular momentum-exponential form of the Euler angle rotation operators is

$$\begin{aligned} \mathcal{R} &= \mathcal{R}_{z''}(\gamma)\mathcal{R}_{y'}(\beta)\mathcal{R}_z(\alpha) \\ &= \exp -i\gamma J_{z''} \exp -i\beta J_{y'} \exp -i\alpha J_z. \end{aligned}$$

Show that in terms of the *original axes*:

$$\mathcal{R} = \exp -i\alpha J_z \exp -i\beta J_y \exp -i\gamma J_z$$

Hint. The $\mathcal{R}$ operators transform as matrices. The rotation about the y'-axis (second Euler rotation) may be referred to the original y-axis by

$$\exp -i\beta J_{y'} = \exp -i\alpha J_z \exp -i\beta J_y \exp i\alpha J_z .$$

4.11 SU(2), SU(3), and Nuclear Particles

The application of group theory to "elementary" particles has been labeled by Wigner the third stage of group theory and physics. The first stage was the search for the 32 point groups and the 230 space groups giving crystal symmetries—Section 4.8. The second stage was a search for representations such as the representations of O_3^+ and SU(2)—Section 4.9. Now in this third stage, physicists are back to a search for groups.

In discussing the strongly interacting particles of high energy physics and the special unitary groups SU(2) and SU(3), it is helpful to look to angular momentum and the rotation group O_3^+ for an analogy. Suppose we have an electron in the spherically symmetric attractive potential of some atomic nucleus. The electron's Schrödinger wavefunction may be characterized by three quantum numbers n, l, and m. The energy, however, is $2l + 1$-fold degenerate, depending only on n and l[1]. The reason for this degeneracy may be stated in two equivalent ways:

1. The potential is spherically symmetric, independent of θ and φ, and
2. The Schrödinger Hamiltonian, $-(\hbar^2/2m_e)\nabla^2 + V(r)$ is *invariant* under ordinary spacial rotations (O_3^+).

As a consequence of the spherical symmetry of the potential, the angular momentum $\mathbf{L}$ is conserved. In Section 4.10, the cartesian components of $\mathbf{L}$ are identified as the generators of the rotation group O_3^+. Instead of representing L_x, L_y, and L_z by operators, let us use matrices. The exercises at the end of Section 4.2 provide examples for $l = \frac{1}{2}$, 1, and $\frac{3}{2}$. The L_i matrices are $(2l + 1) \times (2l + 1)$ matrices with the dimension the same as the number of the degenerate states.[2] These L_i matrices generate the $(2l + 1) \times (2l + 1)$ irreducible representations of O_3^+. The dimension $2l + 1$ is identified with the $2l + 1$ degenerate states.

The common way of eliminating this degeneracy is to introduce a constant magnetic induction $\mathbf{B}$. This leads to the Zeeman effect. This magnetic induction adds a term to the Schrödinger Hamiltonian that is *not* invariant under O_3^+. This is a symmetry-breaking term.

So much for the analogy. In the case of the strongly interacting particles (neutrons, protons, etc.) we cannot follow the analogy directly, because we do not yet fully understand the nuclear interaction. We do not know the Hamiltonian. So instead, let us run the analogy backwards.

[1] If the potential is a pure Coulomb potential, the energy depends only on n (cf. Section 13.2).

[2] With L_i a matrix, the Schrödinger wavefunction $\psi(r, \theta, \varphi)$ is replaced by a state vector—with $2l + 1$ components. Angular momentum and the $(2l + 1)$-fold degeneracy are discussed at some length in Section 12.7.

In the 1930's Heisenberg proposed that nuclear forces were charge independent, that the only two massive particles (baryons) known then, the neutron and proton were two different states of the *same* particle. Reference to Table 4.2 shows that they have almost the same mass. The fractional difference, $(m_n - m_p)/m_p \approx 0.0014$, is small, suggesting that the mass difference is produced by a small charge dependent perturbation. It was convenient to describe this near-degeneracy, by introducing a quantity **I** with z-projections $I_3 = \frac{1}{2}$ for the proton, $-\frac{1}{2}$ for the neutron. The name coined for **I** was isospin. Isospin had nothing to do with spin (the particle's intrinsic angular momentum) but the two component isospin state vector obeyed the same mathematical relations as the spin $J = \frac{1}{2}$ state vector, and in particular could be taken to be an eigenvector of the Pauli σ_3 matrix.

In the absence of charge dependent forces, isospin is conserved (the proton and neutron have the same mass) and we have a two-fold degeneracy. Equivalently, the unknown nuclear Hamiltonian must be invariant under the group generated by the isospin matrices. The isospin matrices are just the 3 Pauli matrices (2×2 matrices), and the group generated is the SU(2) group of Section 4.9, also 2×2 corresponding to our two-fold degeneracy.

By 1961 many more particles had been discovered (or created). The eight shown in Table 4.2 attracted particular attention.[1] It was convenient to describe them by

TABLE 4.2

Baryons with Spin 1/2 Even Parity

		Mass (MeV)	Y	I	I_3
	Ξ^-	1321.300			$-\frac{1}{2}$
Ξ			-1	$\frac{1}{2}$	
	Ξ^0	1314.900			$+\frac{1}{2}$
	Σ^-	1197.410			-1
Σ	Σ^0	1192.540	0	1	0
	Σ^+	1189.470			$+1$
Λ	Λ	1115.500	0	0	0
	n	939.550			$-\frac{1}{2}$
N			1	$\frac{1}{2}$	
	p	938.256			$+\frac{1}{2}$

[1] All masses are given in energy units, MeV.

characteristic quantum numbers, I for isospin, and Y for hypercharge. The particles may be grouped into charge or isospin multiplets. Then the hypercharge Y may be taken as twice the average charge of the multiplet. For the neutron-proton multiplet

$$Y = 2 \cdot \tfrac{1}{2}(0 + 1) = 1. \tag{4.265}$$

The hypercharge and isospin values are listed in Table 4.2.

From scattering and production experiments, it had become clear that both hypercharge Y and isospin I were conserved under strong (nuclear) interaction. Remember L (or l) is conserved under a spherically symmetric Hamiltonian. The eight particles thus appeared as an eight-fold degeneracy, but now with *two* quantities to be conserved. In 1961, Gell-Mann, and independently Ne'eman, suggested that the strong interaction should be invariant under the three-dimensional special unitary group, SU(3), i.e., should have SU(3) symmetry.

The choice of SU(3) was based first on the existence of two conserved quantities. This dictated a group of rank 2, a group, two of whose generators (and only two) commuted. Second, the group had to have an 8×8 representation to account for the 8 degenerate baryons. In a sense SU(3) was the simplest generalization of SU(2). Gell-Mann set up 8 generators: 3 for the components of isospin, 1 for hypercharge, and 4 additional ones. All are 3×3, zero trace matrices. As with O_3^+ and SU(2), there are an infinity of irreducible representations. An eight-dimensional one was associated with the eight particles of Table 4.2.[1]

We imagine the Hamiltonian for our 8 baryons to be composed of three parts

$$H = H_{\text{strong}} + H_{\text{medium}} + H_{\text{electromagnetic}}. \tag{4.266}$$

The first part, H_{strong}, possesses the SU(3) symmetry and leads to the eight-fold degeneracy. Introduction of a symmetry breaking interaction, H_{medium}, removes part of the degeneracy giving the four isospin multiplets Ξ, Σ, Λ and N. These are multiplets because H_{medium} still possesses SU(2) symmetry. Finally, the presence of charge-dependent forces splits the isospin multiplets and removes the last degeneracy. This imagined sequence is shown in Fig. 4.13.

Applying first order perturbation theory of quantum mechanics, simple relations among the baryon masses may be calculated. Also, intensity rules for decay and scattering processes may be obtained.

Perhaps the most spectacular success of this SU(3) model has been its prediction of new particles. In 1961, four K and three π mesons (all pseudoscalar; spin 0, odd parity) suggested another octet, similar to the baryon octet. The SU(3) theory predicted an eighth meson η^0, mass 563 MeV. The η^0 meson, experimentally determined mass 548 MeV, was found soon after. Groupings of nine of the heavier baryons (all with spin 3/2, even parity) suggested a 10-member group or decuplet. The missing tenth baryon was predicted to have a mass of about 1680 MeV and a negative charge. In 1964 the negatively charged Ω^-, mass 1675 ± 12 MeV, was discovered.

[1] This application of SU(3) has been called by Gell-Mann the "eight-fold way." Note the 8 independent parameters of SU(3) (from $n^2 - 1$), the 8 generators, the 8×8 representation associated with 8 particles. The name also refers to the Eight-fold Way of Buddha.

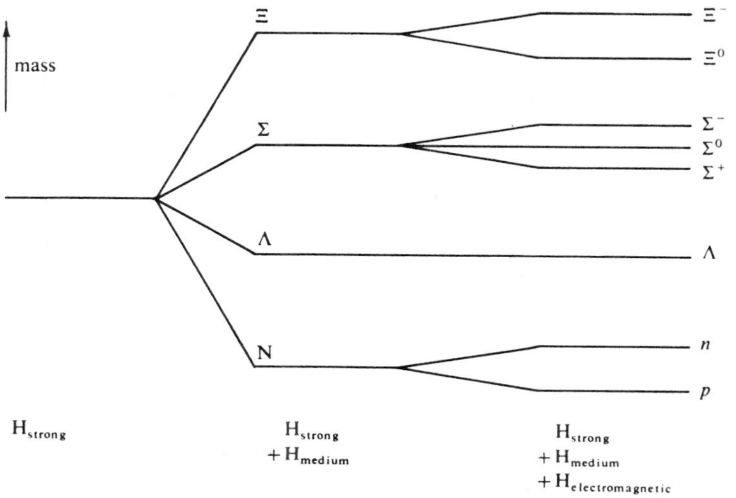

FIG. 4.13 Baryon mass splitting

Since the completion of this $3/2^+$ decuplet, a $5/2^-$ (odd parity) multiplet for baryons, and 1^- and 2^+ multiplets for mesons have been established.

The application of group theory to strongly interacting particles has been extended beyond SU(3). There has been an extensive investigation of SU(6) and of the more complex, higher dimensional groups. Great attention has been paid to the group generators and to the *structure constants* in the generator commutation relations (such as $i\varepsilon_{ijk}$ for orbital angular momentum). These structure constants define a *Lie algebra*. It is possible to associate space integrals of current densities with the group generators. This leads to a current algebra far beyond the scope of this discussion.

To keep group theory and its very real accomplishment in proper perspective, it should be emphasized that group theory identifies and formalizes symmetries. It classifies (and sometimes predicts) particles. But aside from saying that one part of the Hamiltonian has SU(2) symmetry and another part has SU(3) symmetry, group theory says *nothing* about the particle interactions. Remember that the statement that the atomic potential is spherically symmetric tells us nothing about the radial dependence of the potential or of the wavefunction.

4.12 Homogeneous Lorentz Group

Generalizing the approach to vectors of Section 1.2, scientists demand that our physical laws be covariant[1] under

(a) space and time translations,
(b) rotations in real, three-dimensional space, and
(c) Lorentz transformations.

[1] To be covariant means to have the same form in different coordinate systems so that there is *no preferred reference* system (cf. Sections 1.2 and 3.1).

The demand for covariance under translations is based on the homogeneity of space and time. Covariance under rotations is an assertion of the isotropy of space. The requirement of Lorentz covariance is based on acceptance of special relativity. All three of the above transformations together form the inhomogeneous Lorentz group or the Poincaré group. Here, we shall exclude translations. The space rotations and the Lorentz transformations together form a group—the homogeneous Lorentz group.

We first generate a subgroup, the Lorentz transformations in which the relative velocity $\mathbf{v}$ is along the $x = x_1$ axis. The generator may be determined by considering Lorentz space-time reference frames moving with a relative velocity δv, an infinitesimal.[1] The relations are similar to those for rotations in real space, Sections 1.2, 3.1, and 4.3, except that here the angle of rotation is pure imaginary (cf Section 3.7).

We shall work in Minkowski space with $x_4 = ict$. For an infinitesimal relative velocity δv, the space-time transformation is Galilean:

$$x_1' = x_1 - \delta v\, t = x_1 + i\,\delta\beta\, x_4. \tag{4.267}$$

Here, as usual, $\beta = v/c$. By symmetry, we also write

$$x_4' = x_4 + ia\;\delta\beta x_1 \tag{4.268}$$

with a a parameter that is fixed by the requirement that $x_1^2 + x_4^2$ be invariant,

$$x_1'^2 + x_4'^2 = x_1^2 + x_4^2. \tag{4.269}$$

Remember x_μ is the prototype four-dimensional vector in Minkowski space. Squaring and adding Eqs. 4.267 and 4.268 and discarding terms of order $(\delta\beta)^2$, we find $a = -1$. Eqs. 4.267 and 4.268 may be combined as a matrix equation

$$\begin{pmatrix} x_1' \\ x_4' \end{pmatrix}_{\delta\beta} = (1 + \delta\beta\sigma_2)\begin{pmatrix} x_1 \\ x_4 \end{pmatrix}. \tag{4.270}$$

σ_2 happens to be the negative of the Pauli matrix, σ_y.

The parameter $\delta\beta$ represents an infinitesimal change. Using the same techniques as in Section 4.10, we repeat the transformation N times to develop a *finite* transformation with the velocity parameter $\theta = N\,\delta\beta$. Then

$$\begin{pmatrix} x_1' \\ x_4' \end{pmatrix} = \left(1 + \frac{\theta\sigma_2}{N}\right)^N \begin{pmatrix} x_1 \\ x_4 \end{pmatrix}. \tag{4.271}$$

In the limit as $N \to \infty$,

$$\lim_{N\to\infty}\left(1 + \frac{\theta\sigma_2}{N}\right)^N = \exp\theta\sigma_2. \tag{4.272}$$

As in Section 4.10, the exponential is interpreted by a Maclaurin expansion

$$\exp\theta\sigma_2 = 1 + \theta\sigma_2 + (\theta\sigma_2)^2/2! + (\theta\sigma_2)^3/3! + \cdots. \tag{4.273}$$

[1] This derivation, with a slightly different metric, appears in an article by J. L. Strecker, *Am. J. Phys.* **35**, 12 (1967).

Noting that $\sigma_2^2 = 1$,

$$\exp \theta\sigma_2 = 1 \cosh \theta + \sigma_2 \sinh \theta. \tag{4.274}$$

Hence, our finite Lorentz transformation is

$$\begin{pmatrix} x_1' \\ x_4' \end{pmatrix} = \begin{pmatrix} \cosh \theta & i \sinh \theta \\ -i \sinh \theta & \cosh \theta \end{pmatrix} \begin{pmatrix} x_1 \\ x_4 \end{pmatrix}. \tag{4.275}$$

σ_2 has *generated* the representations of this special Lorentz transformation.

Cosh θ and sinh θ may be identified by considering the origin of the primed coordinate system, $x_1' = 0$, or $x_1 = vt$. Substituting into Eq. 4.275,

$$0 = x_1 \cosh \theta + x_4 \, \mathrm{i} \sinh \theta. \tag{4.276}$$

With $x_1 = vt$ and $x_4 = ict$.

$$\tanh \theta = \beta.$$

Using $1 - \tanh^2 \theta = (\cosh^2 \theta)^{-1}$,

$$\cosh \theta = (1 - \beta^2)^{-1/2} \equiv \gamma, \qquad \sinh \theta = \beta\gamma. \tag{4.277}$$

The matrix in Eq. 4.275 agrees with the $x_3 - x_4$ portion of the matrix in Eq. 3.120.

The preceding special case of the velocity parallel to one space axis is easy, but it does illustrate the infinitesimal velocity—exponentiation—generator technique. Now we apply this exact technique to derive the Lorentz transformation for the relative velocity $\mathbf{v}$ *not* parallel to any space axis.

Let $v_1 = \lambda|\mathbf{v}|$, $v_2 = \mu|\mathbf{v}|$, and $v_3 = \nu|\mathbf{v}|$ with λ, μ, and ν the direction cosines of $\mathbf{v}$. In analogy to Eq. 4.267 we write

$$\begin{aligned} x_1' &= x_1 + i\lambda \, \delta\beta \, x_4 \\ x_2' &= x_2 + i\mu \, \delta\beta \, x_4 \\ x_3' &= x_3 + i\nu \, \delta\beta \, x_4 . \end{aligned} \tag{4.278}$$

Again, by symmetry we try

$$x_4' = x_4 + ia_1\delta\beta x_1 + ia_2\delta\beta x_2 + ia_3\delta\beta x_3 . \tag{4.279}$$

From

$$\sum_{\xi=1}^{4} x_\xi'^2 = \sum_{\zeta=1}^{4} x_\zeta^2, \tag{4.280}$$

$$a_1 = -\lambda, \qquad a_2 = -\mu, \qquad \text{and} \qquad a_3 = -\nu.$$

Rewriting Eqs. 4.278 and 4.279 as a matrix equation:

$$\begin{pmatrix} x_1' \\ x_2' \\ x_3' \\ x_4' \end{pmatrix} = \begin{pmatrix} 1 & 0 & 0 & i\lambda\delta\beta \\ 0 & 1 & 0 & i\mu\delta\beta \\ 0 & 0 & 1 & i\nu\delta\beta \\ -i\lambda\delta\beta & -i\mu\delta\beta & -i\nu\delta\beta & 1 \end{pmatrix} \begin{pmatrix} x_1 \\ x_2 \\ x_3 \\ x_4 \end{pmatrix}. \tag{4.281}$$

Subtracting out 1 and removing $\delta\beta$ as a factor

$$\mathbf{x}' = (1 + \delta\beta\sigma)\mathbf{x}. \tag{4.282}$$

Here

$$\sigma = \begin{pmatrix} 0 & 0 & 0 & i\lambda \\ 0 & 0 & 0 & i\mu \\ 0 & 0 & 0 & i\nu \\ -i\lambda & -i\mu & -i\nu & 0 \end{pmatrix}. \tag{4.283}$$

By direct multiplication (with $\lambda^2 + \mu^2 + \nu^2 = 1$),

$$\sigma^2 = \begin{pmatrix} \lambda^2 & \lambda\mu & \lambda\nu & 0 \\ \lambda\mu & \mu^2 & \mu\nu & 0 \\ \lambda\nu & \mu\nu & \nu^2 & 0 \\ 0 & 0 & 0 & 1 \end{pmatrix} \tag{4.284}$$

and

$$\sigma^3 = \sigma. \tag{4.285}$$

As before, we iterate N times with $\theta = N\,\delta\beta$. Forming the exponential

$$\lim_{N\to\infty} (1 + \theta\sigma/N)^N = e^{\theta\sigma}$$
$$= 1 + \sigma \sinh\theta + \sigma^2(\cosh\theta - 1). \tag{4.286}$$

σ is our generator with the parameters λ, μ, and ν defininining the direction of the velocity built in. Writing out the second part of Eq. 4.286, the Lorentz transformation matrix in all its glory is

$$\mathsf{L}(\mathbf{v}) =$$

$$\begin{pmatrix} 1 + \lambda^2(\cosh\theta - 1) & \lambda\mu(\cosh\theta - 1) & \lambda\nu(\cosh\theta - 1) & i\lambda \sinh\theta \\ \lambda\mu(\cosh\theta - 1) & 1 + \mu^2(\cosh\theta - 1) & \mu\nu(\cosh\theta - 1) & i\mu \sinh\theta \\ \lambda\nu(\cosh\theta - 1) & \mu\nu(\cosh\theta - 1) & 1 + \nu^2(\cosh\theta - 1) & i\nu \sinh\theta \\ -i\lambda \sinh\theta & -i\mu \sinh\theta & -i\nu \sinh\theta & \cosh\theta \end{pmatrix}. \tag{4.287}$$

Again $\cosh\theta = (1 - \beta^2)^{-1/2} = \gamma$, $\sinh\theta = \beta\gamma$.

It is worth noting that the combination of Eqs. 4.286 and 4.287,

$$\mathsf{L}(\mathbf{v}) = e^{\theta\sigma} \tag{4.288}$$

is *not* in the exact form of Eq. 4.237. The exponent lacks the factor i, and $\mathsf{L}(\mathbf{v})$ is *not* unitary.

The matrices given by Eq. 4.275 for the case of $\mathbf{v} = \mathbf{i}v_x$ form a subgroup. The matrices of Eq. 4.287 do *not*. The product of two Lorentz transformation matrices, $\mathsf{L}(\mathbf{v}_1)$ and $\mathsf{L}(\mathbf{v}_2)$, yields a third Lorentz matrix $\mathsf{L}(\mathbf{v}_3)$—if the two velocities $\mathbf{v}_1$ and $\mathbf{v}_2$ are parallel. The resultant velocity $\mathbf{v}_3$ is related to $\mathbf{v}_1$ and $\mathbf{v}_2$ by the Einstein velocity addition law, Section 3.7. If $\mathbf{v}_1$ and $\mathbf{v}_2$ are not parallel, no such simple relation exists. Specifically, consider three reference frames S, S', and S'', with S and S' related by $\mathsf{L}(\mathbf{v}_1)$, and S' and S'' related by $\mathsf{L}(\mathbf{v}_2)$. If the velocity of S'' relative to the original system S is $\mathbf{v}_3$, S'' is *not* obtained from S by $\mathsf{L}(\mathbf{v}_3)$ alone. Rather we find that

$$\mathsf{L}(\mathbf{v}_3) = \mathsf{R}\mathsf{L}(\mathbf{v}_2)\mathsf{L}(\mathbf{v}_1) \tag{4.289}$$

where R is a 3×3 space rotation matrix embedded in our four-dimensional space-time. With $\mathbf{v}_1$ and $\mathbf{v}_2$ not parallel, the final system S'' is *rotated* relative to S. This rotation is the origin of the Thomas precession involved in spin-orbit coupling terms in atomic and nuclear physics. Because of its presence, the $\mathsf{L}(\mathbf{v})$ by themselves do not form a group.

EXERCISES

4.12.1 Obtain $\sigma(\lambda, \mu, \nu)$ by differentiating the final matrix, Eq. 4.287.

4.12.2 Two Lorentz transformations are carried out in succession: v_1 along the x-axis, then v_2 along the y-axis. Show that the resultant transformation (given by the product of these two successive transformations) *cannot* be put in the form of Eq. 4.287.
Note. The discrepancy corresponds to a rotation.

REFERENCES

AITKEN, A. C., *Determinants and Matrices*. New York: Interscience Publishers (1956). A readable introduction to determinants and matrices.

BICKLEY, W. G., and R. S. H. G. THOMPSON, *Matrices—Their Meaning and Manipulation*. Princeton, New Jersey: Van Nostrand (1964). A comprehensive account of the occurrence of matrices in physical problems, their analytic properties, and numerical techniques.

BUERGER, M. J., *Elementary Crystallography*. New York: Wiley (1956). A comprehensive discussion of crystal symmetries. Buerger develops all 32 point groups and all 230 space groups.

FALICOV, L. M., *Group Theory and Its Physical Applications*. Notes compiled by A. Luehrmann. Chicago: University of Chicago Press (1966). Group theory with an emphasis on applications to crystal symmetries and solid state physics.

GELL-MANN, M., and NE'EMAN, Y., *The Eightfold Way*. New York: Benjamin (1965). A collection of reprints of significant papers on SU(3) and the particles of high energy physics. The several introductory sections by Gell-Mann and Ne'eman are especially helpful.

HAMERMESH, M., *Group Theory and Its Application to Physical Problems*. Reading, Massachusetts: Addison-Wesley (1962). A detailed, rigorous account of both finite and continuous groups. The 32 point groups are developed. The continuous groups are treated with Lie algebra included. A wealth of applications to atomic and nuclear physics.

HIGMAN, B., *Applied Group-Theoretic and Matrix Methods*. New York: Dover (1964), Oxford: Oxford University Press (1955). A rather complete and unusually intelligible development of matrix analysis and group theory.

MICHAEL, A. D., *Matrix and Tensor Calculus with Applications to Mechanics, Elasticity and Aeronautics*. New York: Wiley (1947). An unusually clear exposition of matrices and tensors and the development of a calculus of matrices and tensors.

RAM, B., *Am. J. Phys.* **35**, 16 (1967). An excellent discussion of the application of SU(3) to the strongly interacting particles (baryons).

ROSE, M. E., *Elementary Theory of Angular Momentum*. New York: Wiley (1957). As part of the development of the quantum theory of angular momentum, Rose includes a detailed and readable account of the rotation group.

WIGNER, E. P., *Group Theory and Its Application to the Quantum Mechanics of Atomic Spectra*. Translated by J. J. Griffin. New York and London: Academic Press (1959). This is the classic reference on group theory for the physicist. The rotation group is treated in considerable detail. There are a wealth of applications to atomic physics.

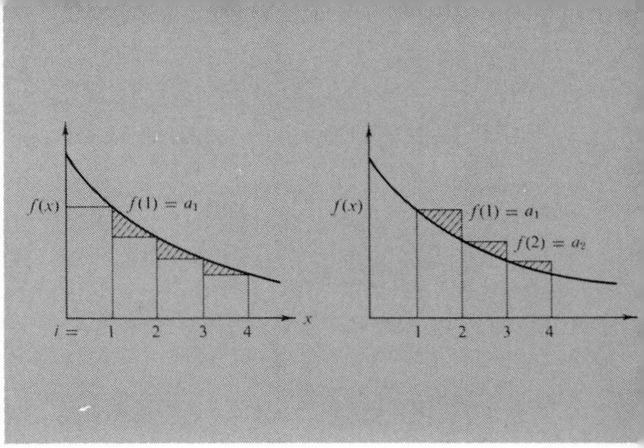

CHAPTER 5

INFINITE SERIES

5.1 Fundamental Concepts

Infinite series, literally summations of an infinite number of terms, occur frequently in both pure and applied mathematics. They may be used by the pure mathematician to define functions as a fundamental approach to the theory of functions. They may be used also for calculating accurate values of transcendental constants and transcendental functions. In the mathematics of science and engineering infinite series are ubiquitous, for they appear in the evaluation of integrals (Sections 5.6 and 5.7), in the solution of differential equations (Sections 8.4 and 8.5), and as Fourier series (Chapter 14) and compete with integral representations for the description of a host of special functions (Chapters 11, 12, and 13).

Right at the start we face the problem of attaching meaning to the sum of an infinite number of terms. The usual approach is by partial sums. If we have a sequence of finite terms $u_1, u_2, u_3, u_4, u_5, \ldots$, we define the ith partial sum as

$$s_i = \sum_{n=1}^{i} u_n. \tag{5.1}$$

This is a finite summation and offers no difficulties. If the partial sums s_i converge to a (finite) limit as $i \to \infty$,

$$\lim_{i \to \infty} s_i = S, \tag{5.2}$$

the infinite series $\sum_{n=1}^{\infty} u_n$ is said to be *convergent* and to have the value S. Note carefully that we reasonably, plausibly, but still arbitrarily *define* the infinite series as equal to S. The reader should also note that a necessary condition for this convergence to a limit is that $\lim_{n \to \infty} u_n = 0$. This condition, however, is not sufficient to guarantee convergence.

Our partial sums s_i may not converge to a single limit but may oscillate, as in the case

$$\sum_{n=1}^{\infty} u_n = 1 - 1 + 1 - 1 + 1 + \cdots - (-1)^n + \cdots. \tag{5.3}$$

Clearly, $s_i = 1$ for i odd but 0 for i even. There is no convergence to a limit, and series such as this one are labeled oscillatory.

For the series

$$\sum_{n=1}^{\infty} 1 + 2 + 3 + \cdots + n + \cdots \tag{5.4}$$

we have

$$s_n = \frac{n(n+1)}{2}. \tag{5.5}$$

As $n \to \infty$

$$\lim_{n\to\infty} s_n = \infty. \tag{5.6}$$

Whenever the sequence of partial sums diverges (approaches $\pm\infty$), the infinite series is said to *diverge*. Often the term divergent is extended to include oscillatory series as well.

Because we evaluate the partial sums by ordinary arithmetic, the convergent series, defined in terms of a limit of the partial sums, assume a position of supreme importance. Two examples may clarify the nature of convergence or divergence of a series and will also serve as a basis for a further detailed investigation in the next section.

The geometrical sequence, starting with a and with a ratio r ($r \geqslant 0$), is given by

$$a, ar, ar^2, ar^3, \ldots, ar^{n-1}, \ldots$$

The nth partial sum is given by

$$s_n = a\,\frac{1 - r^n}{1 - r}. \tag{5.7}$$

Taking the limit as $n \to \infty$,

$$\lim_{n\to\infty} s_n = \frac{a}{1 - r}, \qquad \text{for } r < 1. \tag{5.8}$$

Hence, by definition, the infinite geometric series converges for $r < 1$ and is given by

$$\sum_{n=1}^{\infty} ar^{n-1} = \frac{a}{1 - r}. \tag{5.9}$$

On the other hand, if $r \geqslant 1$, the necessary condition $u_n \to 0$ is not satisfied and the infinite series diverges.

As a second and more involved example, we consider the harmonic series

$$\sum_{n=1}^{\infty} n^{-1} = 1 + \frac{1}{2} + \frac{1}{3} + \frac{1}{4} + \cdots + \frac{1}{n} + \cdots. \tag{5.10}$$

We have the $\lim_{n\to\infty} u_n = \lim_{n\to\infty} 1/n = 0$, but this is not sufficient to guarantee convergence. If we group the terms (no change in order) as

$$1 + \tfrac{1}{2} + (\tfrac{1}{3} + \tfrac{1}{4}) + (\tfrac{1}{5} + \tfrac{1}{6} + \tfrac{1}{7} + \tfrac{1}{8}) + (\tfrac{1}{9} + \cdots + \tfrac{1}{16}) + \cdots, \tag{5.11}$$

it will be seen that each pair of parentheses encloses p terms of the form

$$\frac{1}{p+1} + \frac{1}{p+2} + \cdots + \frac{1}{p+p} > \frac{p}{2p} = \frac{1}{2}. \tag{5.12}$$

Forming partial sums by adding the parenthetical groups one by one, we obtain

$$\begin{aligned} s_1 &= 1, & s_4 &> \frac{5}{2}, \\ s_2 &= \frac{3}{2}, & s_5 &> \frac{6}{2}, \\ s_3 &> \frac{4}{2}, & s_n &> \frac{n+1}{2}. \end{aligned} \tag{5.13}$$

The harmonic series considered in this way is certainly divergent.[1] An alternate and independent demonstration of its divergence appears in Section 5.2.

Using the binomial theorem[2] (Section 5.6), we may expand the function $(1 + x)^{-1}$:

$$\frac{1}{1+x} = 1 - x + x^2 - x^3 + \cdots + (-x)^{n-1} + \cdots. \tag{5.14}$$

If we let $x \to 1$, this series becomes

$$1 - 1 + 1 - 1 + 1 - 1 + \cdots, \tag{5.15}$$

a series that we labeled oscillatory earlier in this section. Although it does not converge in the usual sense, it is possible to attach meaning to this series. Euler, for example, assigned a value of $\frac{1}{2}$ to this oscillatory sequence on the basis of the correspondence between this series and the well-defined function $(1 + x)^{-1}$. Unfortunately, such correspondence between series and function is not unique and this approach must be refined. Other methods of assigning a meaning to a divergent or oscillatory series, methods of defining a sum, have been developed. In general, however, this aspect of infinite series is of relatively little interest to the scientist or the engineer. An exception to this statement, the very important asymptotic or semiconvergent series, is considered in Section 5.11.

[1] The (finite) harmonic series appears in an interesting note on the maximum stable displacement of a stack of coins, "The Leaning Tower of Lire," P. R. Johnson, *Amer. J. Phys.* **23**, 240 (1955).

[2] Actually Eq. 5.14 may be taken as an identity and verified by multiplying both sides by $1 + x$.

EXERCISES

5.1.1 Show that

$$\sum_{n=1}^{\infty} \frac{1}{(2n-1)(2n+1)} = \frac{1}{2}.$$

Hint. Show (by mathematical induction) that $s_m = m/(2m+1)$.

5.1.2 Show that

$$\sum_{n=1}^{\infty} \frac{1}{n(n+1)} = 1.$$

5.2 Convergence Tests

Although nonconvergent series may be useful in certain special cases, (cf. Section 5.11), we usually insist, as a matter of convenience if not necessity, that our series be convergent. It therefore becomes a matter of extreme importance to be able to tell whether a given series is convergent. We shall develop a number of possible tests, starting with the simple and relatively insensitive tests and working up to the more complicated but quite sensitive tests.

For the present let us consider a series of positive terms, $a_n > 0$, postponing negative terms until the next section.

Comparison test. If, term by term, a series of terms $u_n \leqslant a_n$, in which the a_n form a convergent series, the series $\sum_n u_n$ is also convergent.

If term by term a series of terms $v_n \geqslant b_n$, in which the b_n form a divergent series, the series $\sum_n v_n$ is also divergent. Note that comparisons of u_n with b_n or v_n with a_n yield no information.

For the convergent series a_n we already have the geometric series, whereas the harmonic series will serve as the divergent series b_n. As other series are identified as either convergent or divergent, they may be used for the known series in this comparison test

EXAMPLE 5.2.1

Test $\sum_n n^{-p}$, $p = 0.999$, for convergence. Since $n^{-0.999} > n^{-1}$, and $b_n = n^{-1}$ forms the divergent harmonic series, the comparison test shows that $\sum_n n^{-0.999}$ is divergent. Generalizing, $\sum_n n^{-p}$ is seen to be divergent for all $p \leqslant 1$.

Cauchy root test. If $(a_n)^{1/n} \leqslant r < 1$ for all sufficiently large n, with r independent of n, then $\sum_n a_n$ is convergent. If $(a_n)^{1/n} \geq 1$ for all sufficiently large n, then $\sum_n a_n$ is divergent.

The first part of this test is verified easily by raising $(a_n)^{1/n} \leq r$ to the nth power. We get

$$a_n \leq r^n < 1.$$

Since r^n is just the nth term in a convergent geometric series, $\sum_n a_n$ is convergent by the comparison test. Conversely, if $(a_n)^{1/n} \geq 1$, then $a_n \geq 1$ and the series must diverge. This root test is particularly useful in establishing the properties of power series (Section 5.7).

D'Alembert or Cauchy ratio test. If $a_{n+1}/a_n \leq r < 1$ for all sufficiently large n, and r is independent of n, then $\sum_n a_n$ is convergent. If $a_{n+1}/a_n \geq 1$ for all sufficiently large n, then $\sum_n a_n$ is divergent.

Convergence is proved by direct comparison with the geometric series $a_n(1 + r + r^2 + \cdots)$. In the second part $a_{n+1} \geq a_n$ and divergence should be reasonably obvious. Although not quite so sensitive as the Cauchy root test, this D'Alembert ratio test is one of the easiest to apply and is widely used. An alternate statement of the ratio test is in the form of a limit:

If

$$\begin{aligned} \lim_{n\to\infty} \frac{a_{n+1}}{a_n} &< 1, && \text{convergence,} \\ &> 1, && \text{divergence,} \\ &= 1, && \text{indeterminant.} \end{aligned} \tag{5.16}$$

Because of this final indeterminant possibility, the ratio test is likely to fail at crucial points, and more delicate, more sensitive tests are necessary.

The alert reader may wonder how this indeterminacy arose. Actually it was concealed in the first statement $a_{n+1}/a_n \leq r < 1$. We might encounter $a_{n+1}/a_n < 1$ but be unable to choose an $r < 1$ *and independent of n* such that $a_{n+1}/a_n \leq r$ for all sufficiently large n. An example is provided by the harmonic series

$$\frac{a_{n+1}}{a_n} = \frac{n}{n+1} < 1. \tag{5.17}$$

Since

$$\lim_{n\to\infty} \frac{a_{n+1}}{a_n} = 1, \tag{5.18}$$

no fixed ratio $r < 1$ exists and the ratio test fails.

EXAMPLE 5.2.2

Test $\sum_n n/2^n$ for convergence.

$$\frac{a_{n+1}}{a_n} = \frac{(n+1)/2^{n+1}}{n/2^n} = \frac{1}{2} \cdot \frac{n+1}{n}. \tag{5.19}$$

Since

$$\frac{a_{n+1}}{a_n} \leqslant \frac{3}{4} \quad \text{for } n \geqslant 2, \tag{5.20}$$

we have convergence. Alternatively,

$$\lim_{n\to\infty} \frac{a_{n+1}}{a_n} = \frac{1}{2} \tag{5.21}$$

and again—convergence.

Cauchy or Maclaurin integral test. This is another sort of comparison test in which we compare a series with an integral. Geometrically we compare the area of a series of unit-width rectangles with the area under a curve.

Let $f(x)$ be a continuous, monotonic decreasing function in which $f(n) = a_n$. Then $\sum_n a_n$ converges if $\int_1^\infty f(x)\,dx$ is finite and diverges if the integral is infinite. For the ith partial sum

$$s_i = \sum_{n=1}^{i} a_n = \sum_{n=1}^{i} f(n). \tag{5.22}$$

But

$$s_i > \int_1^{i+1} f(x)\,dx \tag{5.23}$$

by Fig. 5.1a, $f(x)$ being monotonic decreasing. On the other hand, from Fig. 5.1b,

$$s_i - a_1 < \int_1^i f(x)\,dx, \tag{5.24}$$

in which the series is represented by the inscribed rectangles. Taking the limit as $i \to \infty$, we have

$$\int_1^\infty f(x)\,dx < \sum_{n=1}^{\infty} a_n < \int_1^\infty f(x)\,dx + a_1. \tag{5.25}$$

Hence the infinite series converges or diverges as the corresponding integral converges or diverges.

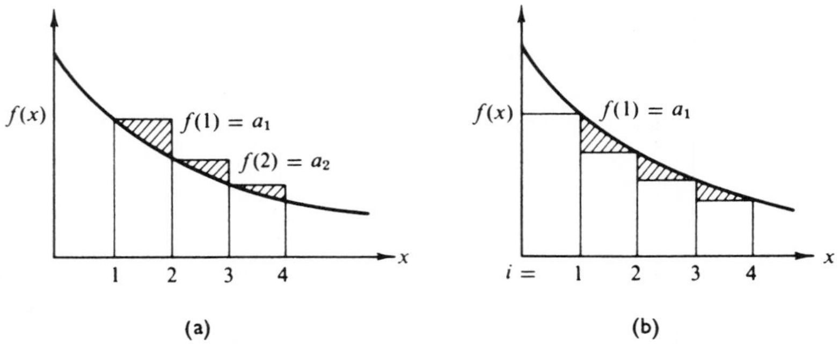

FIG. 5.1(a) Comparison of integral and sum-blocks leading. (b) Comparison of integral and sum-blocks lagging

EXAMPLE 5.2.3

The Riemann zeta function is defined by

$$\zeta(p) = \sum_{n=1}^{\infty} n^{-p}. \tag{5.26}$$

We may take $f(x) = x^{-p}$ and then

$$\int_1^{\infty} x^{-p}\,dx = \left.\frac{x^{-p+1}}{-p+1}\right|_1^{\infty}, \qquad p \neq 1$$

$$= \ln x\,\Big|_1^{\infty}, \qquad p = 1. \tag{5.27}$$

The integral and therefore the series are divergent for $p \leqslant 1$, convergent for $p > 1$. This, incidentally, is an independent proof that the harmonic series ($p = 1$) diverges and diverges logarithmically. The sum of the first million terms $\sum^{1{,}000{,}000} n^{-1}$, is only 14.392 726 ... .

This integral comparison may also be used to set an upper limit to the Euler-Mascheroni constant[1] defined by

$$\gamma = \lim_{n\to\infty}\left(\sum_{m=1}^{n} m^{-1} - \ln n\right). \tag{5.28}$$

Returning to partial sums,

$$s_n = \sum_{m=1}^{n} m^{-1} - \ln n < \int_1^n \frac{dx}{x} - \ln n + 1. \tag{5.29}$$

Evaluating the integral on the right, $s_n < 1$ for all n and therefore $\gamma < 1$. Actually the Euler-Mascheroni constant is 0.577 215 66 ... .

Kummer's test. This is the first of three tests which are somewhat more difficult to apply than the preceding tests. Their importance lies in their power and sensitivity. Frequently, at least one of the three will work when the simpler easier tests are indecisive. It must be remembered, however, that these tests, like those previously discussed, are ultimately based on comparisons. It can be shown that there is no most slowly converging convergent series and no most slowly diverging divergent series. This means that all of the convergence tests given here, including Kummer's, may fail sometime.

We consider a series of positive terms u_i and a sequence of finite positive constants a_i. If

$$a_n \frac{u_n}{u_{n+1}} - a_{n+1} \geqslant C > 0 \tag{5.30}$$

for all $n \geqslant N$, some fixed number,[2] then $\sum_{i=1}^{\infty} u_i$ converges. If

$$a_n \frac{u_n}{u_{n+1}} - a_{n+1} \leqslant 0 \tag{5.31}$$

[1] This is the notation of "Handbook of Mathematical Functions." National Bureau of Standards, Applied Mathematics Series-55 (*AMS*-55).

[2] With u_m finite, the partial sum s_N will always be finite for N finite. The convergence or divergence of a series depends on the behavior of the last infinity of terms, *not* on the first N terms.

and $\sum_{i=1}^{\infty} a_i^{-1}$ diverges, then $\sum_{i=1}^{\infty} u_i$ diverges.

The proof of this powerful test is remarkably simple. From Eq. 5.30, with C some positive constant,

$$\begin{aligned} Cu_{N+1} &\leqslant a_N u_N - a_{N+1} u_{N+1} \\ Cu_{N+2} &\leqslant a_{N+1} u_{N+1} - a_{N+2} u_{N+2} \\ &\cdots\cdots\cdots\cdots\cdots \\ Cu_n &\leqslant a_{n-1} u_{n-1} - a_n u_n \end{aligned} \tag{5.32}$$

Adding and dividing by C, $(C \neq 0)$,

$$\sum_{i=N+1}^{n} u_i \leqslant \frac{a_N u_N}{C} - \frac{a_n u_n}{C}. \tag{5.33}$$

Hence for the partial sum, s_n,

$$\begin{aligned} s_n &\leqslant \sum_{i=1}^{N} u_i + \frac{a_N u_N}{C} - \frac{a_n u_n}{C} \\ &< \sum_{i=1}^{N} u_i + \frac{a_N u_N}{C}, \qquad \text{a constant, independent of } n. \end{aligned} \tag{5.34}$$

The partial sums therefore have an upper bound. With zero as an obvious lower bound, the series $\sum_i u_i$ must converge.

Divergence is shown as follows. From Eq. 5.31

$$a_n u_n \geqslant a_{n-1} u_{n-1} \geqslant \cdots \geqslant a_N u_N. \tag{5.35}$$

Thus

$$u_n \geqslant \frac{a_N u_N}{a_n} \tag{5.36}$$

and

$$\sum_{i=N+1}^{\infty} u_i \geqslant a_N u_N \sum_{i=N+1}^{\infty} a_i^{-1}. \tag{5.37}$$

By the comparison test $\sum_i u_i$ diverges.

Equations 5.30 and 5.31 are often given in a limit form:

$$\lim_{n\to\infty} \left(a_n \frac{u_n}{u_{n+1}} - a_{n+1} \right) = C. \tag{5.38}$$

Thus for $C > 0$ we have convergence, whereas for $C < 0$ (and $\sum_i a_i^{-1}$ divergent) we have divergence. It is perhaps useful to show the equivalence of Eq. 5.38 and Eqs. 5.30 and 5.31 and to show why indeterminacy creeps in when the limit $C = 0$. From the definition of limit

$$\left| a_n \frac{u_n}{u_{n+1}} - a_{n+1} - C \right| < \varepsilon \tag{5.39}$$

for all $n \geqslant N$ and all $\varepsilon > 0$, no matter how small ε may be. When the absolute value signs are removed,

$$C - \varepsilon < a_n \frac{u_n}{u_{n+1}} - a_{n+1} < C + \varepsilon. \tag{5.40}$$

Now if $C > 0$, Eq. 5.30 follows from ε sufficiently small. On the other hand, if $C < 0$, Eq. 5.31 follows. However, if $C = 0$, the center term $a_n(u_n/u_{n+1}) - a_{n+1}$ may be either positive or negative and the proof fails. The primary use of Kummer's test is to prove other tests such as Raabe's (cf. also Exercise 5.2.3).

If the positive constants a_n of Kummer's test are chosen $a_n = n$, we have Raabe's test.

Raabe's test. If $u_n > 0$ and if

$$n\left(\frac{u_n}{u_{n+1}} - 1\right) \geqslant P > 1 \tag{5.41}$$

for all $n \geqslant N$, where N is a positive integer independent of n, then $\sum_i u_i$ converges. If

$$n\left(\frac{u_n}{u_{n+1}} - 1\right) \leqslant 1, \tag{5.42}$$

then $\sum_i u_i$ diverges.

The limit form of Raabe's test is

$$\lim_{n\to\infty} n\left(\frac{u_n}{u_{n+1}} - 1\right) = P. \tag{5.43}$$

We have convergence for $P > 1$, divergence for $P < 1$, and no test for $P = 1$ exactly as with the Kummer test. This indeterminacy is pointed up by Exercise 5.2.4, which presents a convergent series and a divergent series with both series yielding $P = 1$ in Eq. 5.43.

Raabe's test is more sensitive than the d'Alembert ratio test because $\sum_{n=1}^{\infty} n^{-1}$ diverges more slowly than $\sum_{n=1}^{\infty} 1$. We obtain a still more sensitive test (and one that is relatively easy to apply) by choosing $a_n = n \ln n$. This is Gauss's test.

Gauss's test. If $u_n > 0$ for all finite n and

$$\frac{u_n}{u_{n+1}} = 1 + \frac{h}{n} + \frac{B(n)}{n^2}, \tag{5.44}$$

in which $B(n)$ is a bounded function of n for $n \to \infty$, then $\sum_i u_i$ converges for $h > 1$ and diverges for $h \leqslant 1$.

This is an extremely sensitive test of series convergence. It will work for all series the physicist is likely to encounter. For $h > 1$ or $h < 1$ the proof follows directly from Raabe's test

$$\lim_{n\to\infty} n\left[1 + \frac{h}{n} + \frac{B(n)}{n^2} - 1\right] = \lim_{n\to\infty}\left[h + \frac{B(n)}{n}\right] = h. \tag{5.45}$$

If $h = 1$, Raabe's test fails. However, if we return to Kummer's test and use $a_n = n \ln n$, Eq. 5.38 leads to

$$\lim_{n\to\infty}\left\{n\ln n\left[1+\frac{1}{n}+\frac{B(n)}{n^2}\right]-(n+1)\ln(n+1)\right\}$$

$$=\lim_{n\to\infty}\left[n\ln n\cdot\frac{(n+1)}{n}-(n+1)\ln(n+1)\right]$$

$$=\lim_{n\to\infty}(n+1)\left[\ln n-\ln n-\ln\left(1+\frac{1}{n}\right)\right]. \tag{5.46}$$

Borrowing a result from Section 5.6 (which is not dependent on Gauss's test), we have

$$\lim_{n\to\infty} -(n+1)\ln\left(1+\frac{1}{n}\right)=\lim_{n\to\infty} -(n+1)\left(\frac{1}{n}-\frac{1}{2n^2}+\frac{1}{3n^3}\cdots\right)$$

$$=-1<0. \tag{5.47}$$

Hence we have divergence for $h = 1$. This is an example of a successful application of Kummer's test in which Raabe's test had failed.

EXAMPLE 5.2.4

The recurrence relation for the series solution of Legendre's equation (Section 8.4) may be put in the form

$$\frac{a_{2j+2}}{a_{2j}}=\frac{2j(2j+1)-l(l+1)}{(2j+1)(2j+2)}. \tag{5.48}$$

This is equivalent to u_{2j+2}/u_{2j} for $x = +1$. For $j \gg l$,[1]

$$\frac{u_{2j}}{u_{2j+2}}\to\frac{(2j+1)(2j+2)}{2j(2j+1)}=\frac{2j+2}{2j}$$

$$=1+\frac{1}{j}. \tag{5.49}$$

By Eq. 5.44 the series is divergent. Later we shall demand that the Legendre series be finite at $x = 1$. We shall eliminate the divergence by setting the parameter $n = 2j_0$, an even integer.

All of this section so far has been concerned with establishing convergence as an abstract mathematical property. In practice, the *rate* of convergence may be of extreme importance. We might, for instance, be attempting to estimate the convergence limit. A method of increasing the *rate* of convergence has been given by Kummer. If we have a (slowly converging) series S and a known series S', we may write

$$S = S' + (S - S').$$

The point of this simple-minded rearrangement is that by proper choice of S' the new series $(S - S')$ may converge *more rapidly* than the original S.

[1] The n dependence enters $B(n)$ but does not affect h.

EXAMPLE 5.2.5

Let $S = \sum_{n=1}^{\infty} n^{-3}$. In Section 5.9 this is identified as $\zeta(3)$. Choose

$$S' = \sum_{n=2}^{\infty} \frac{1}{(n-1)n(n+1)} = \frac{1}{4}.$$

The numerical value of S' follows from an extension of Ex. 5.1.2
Then

$$\sum_{n=1}^{\infty} n^{-3} = \frac{1}{4} + 1 + \sum_{n=2}^{\infty} \left(\frac{1}{n^3} - \frac{1}{n^3 - n}\right) \tag{5.50}$$

$$= \frac{5}{4} - \sum_{n=2}^{\infty} \frac{1}{n^3(n^2 - 1)},$$

which converges as n^{-5}, obviously faster than the original n^{-3}.

For problems in numerical approximation of a series limit Kummer's method can be very useful indeed.

EXERCISES

5.2.1 (a) Prove that if

$$\lim_{n \to \infty} n^p u_n \to A < \infty; \quad p > 1$$

the series $\sum_{n=1}^{\infty} u_n$ converges.

(b) Prove that if

$$\lim_{n \to \infty} n u_n = A > 0$$

the series diverges. (The test fails for $A = 0$.)

These two tests, known as limit tests, are often convenient for establishing the convergence or divergence of a series. They may be treated as comparison tests, comparing with

$$\sum n^{-q}, \qquad 1 \leqslant q < p.$$

5.2.2 If

$$\lim_{n \to \infty} \frac{b_n}{a_n} = K,$$

a constant with $0 < K < \infty$, show that $\sum b_n$ converges or diverges with $\sum a_n$.
Hint. If $\sum a_n$ converges, use

$$b_n' = \frac{1}{2K} b_n.$$

If $\sum a_n$ diverges, use

$$b_n'' = \frac{2}{K} b_n.$$

5.2.3 Show that the complete d'Alembert ratio test follows directly from Kummer's test with $a_i = 1$.

5.2.4 Show that Raabe's test is indecisive for $P = 1$ by establishing that $P = 1$ for the series

(a) $u_n = \dfrac{1}{n \ln n}$ and that this series diverges.

(b) $u_n = \dfrac{1}{n(\ln n)^2}$ and that this series converges.

5.2.5 Gauss's test is often given in the form of a test of the ratio

$$\frac{u_n}{u_{n+1}} = \frac{n^2 + a_1 n + a_0}{n^2 + b_1 n + b_0}.$$

For what values of the parameters a_1 and b_1 is there convergence? Divergence?

5.2.6 Test for convergence

(a) $\sum\limits_{n=2}^{\infty} (\ln n)^{-1}$

(b) $\sum\limits_{n=1}^{\infty} \dfrac{n!}{10^n}$.

(c) $\sum\limits_{n=1}^{\infty} \dfrac{1}{2n(2n-1)}$.

(d) $\sum\limits_{n=1}^{\infty} [n(n+1)]^{-1/2}$.

(e) $\sum\limits_{n=0}^{\infty} \dfrac{1}{2n+1}$.

5.2.7 Test for convergence

(a) $\sum\limits_{n=1}^{\infty} \dfrac{1}{n(n+1)}$

(b) $\sum\limits_{n=2}^{\infty} \dfrac{1}{n \ln n}$

(c) $\sum\limits_{n=1}^{\infty} \dfrac{1}{n 2^n}$

(d) $\sum\limits_{n=1}^{\infty} \ln\left(1 + \dfrac{1}{n}\right)$

(e) $\sum\limits_{n=1}^{\infty} \dfrac{1}{n \cdot n^{1/n}}$

5.2.8 For what values of p and q will be the following series converge?

$$\sum_{n=2}^{\infty} \frac{1}{n^p (\ln n)^q},$$

Ans. Convergent for $\begin{cases} p > 1, & \text{all } q, \\ p = 1, & q > 1, \end{cases}$

divergent for $\begin{cases} p < 1, & \text{all } q, \\ p = 1, & q \leqslant 1. \end{cases}$

5.2.9 Determine the range of convergence for Gauss's hypergeometric series

$$F(\alpha, \beta, \gamma; x) = 1 + \frac{\alpha\beta}{1!\,\gamma} x + \frac{\alpha(\alpha+1)\beta(\beta+1)}{2!\,\gamma(\gamma+1)} x^2 + \cdots.$$

Hint. Gauss developed Gauss's test for the specific purpose of establishing the convergence of this series. *Ans.* Convergent for $-1 < x < 1$ and for $x = \pm 1$ if $\gamma > \alpha + \beta$.

5.2.10 A simple machine calculation yields

$$\sum_{n=1}^{100} n^{-3} = 1.202\,007.$$

Show that

$$1.202\,056 \leqslant \sum_{n=1}^{\infty} n^{-3} \leqslant 1.202\,057.$$

Hint. Use integrals to set upper and lower bounds on $\sum_{n=101}^{\infty} n^{-3}$.

Comment. A more exact value for summation $\sum_{1}^{\infty} n^{-3}$ is 1.202 056 903 ⋯ .

5.2.11 Set upper and lower bounds on $\sum_{n=1}^{1,000,000} n^{-1}$, assuming that (a) the Euler-Mascheroni constant is known. *Ans*. $14.392\,726 < \sum_{n=1}^{1,000,000} n^{-1} < 14.392\,727$.
(b) The Euler-Mascheroni constant is unknown.

5.2.12 Given $\sum_{n=1}^{1,000} n^{-1} = 7.485\,470 \cdots$, set upper and lower bounds on the Euler-Mascheroni constant. *Ans*. $0.5767 < \gamma < 0.5778$.

5.2.13 (From Olbers's paradox.) Assume a static universe in which the stars are uniformly distributed. Divide all space into shells of constant thickness; the stars in any one shell by themselves subtend a solid angle of ω_0. *Allowing for the blocking out of distant stars by nearer stars*, show that the total net solid angle subtended by all stars, shells extending to infinity, is *exactly* 4π. (Therefore the night sky should be ablaze with light.)

5.2.14 Test for convergence

$$\sum_{n=1}^{\infty}\left[\frac{1\cdot 3\cdot 5\cdots(2n-1)}{2\cdot 4\cdot 6\cdots(2n)}\right]^2 = \frac{1}{4}+\frac{9}{64}+\frac{25}{256}+\cdots.$$

5.2.15 The Legendre series, $\sum_{j\text{ even}} u_j(x)$, satisfies the recurrence relations

$$u_{j+2}(x) = \frac{(j+1)(j+2)-l(l+1)}{(j+2)(j+3)}\, x^2\, u_j(x)$$

in which the index j is even and l is some constant (but, in this problem, *not* a non-negative odd integer). Find the range of values of x for which this Legendre series is convergent. Test the endpoints carefully. *Ans*. $-1 < x < 1$.

5.2.16 A series solution (Section 8.4) of the Chebyshev equation leads to successive terms having the ratio

$$\frac{u_{j+2}(x)}{u_j(x)} = \frac{(k+j)^2-n^2}{(k+j+1)(k+j+2)}\, x^2.$$

with $k = 0$ and $k = 1$. Test for convergence at $x = \pm 1$. *Ans*. Convergent.

5.2.17 A series solution for the ultraspherical (Gegenbauer) function $C_n^{\alpha}(x)$ leads to the recurrence relation

$$a_{j+2} = a_j\, \frac{(k+j)(k+j+2\alpha)-n(n+2\alpha)}{(k+j+1)(k+j+2)}.$$

Investigate the convergence of each of these series at $x = \pm 1$ as a function of the parameter α. *Ans*. Convergent for $\alpha < \frac{1}{2}$.
Divergent for $\alpha \geq \frac{1}{2}$.

5.2.18 A series expansion of the incomplete beta function (Section 10.4) yields

$$B_x(p,q) = x^p\left\{\frac{1}{p}+\frac{1-q}{p+1}\,x+\frac{(1-q)(2-q)}{2!(p+2)}\,x^2+\cdots\right.$$
$$\left.+\frac{(1-q)(2-q)\cdots(n-q)}{n!(p+n)}\,x^n+\cdots\right\}.$$

Given that $0 \leqslant x \leqslant 1$, $p > 0$, and $q > 0$, test this series for convergence. What happens at $x = 1$?

5.2.19 Show that the following series is convergent.

$$\sum_{s=0}^{\infty} \frac{(2s-1)!!}{(2s)!!(2s+1)}$$

[*Note:* $(2s-1)!! = (2s-1)(2s-3)\cdots 3\cdot 1$ with $(-1)!! = 1$. $(2s)!! = (2s)(2s-2)\cdots 4\cdot 2$ with $0!! = 1$. The series appears as a series expansion of $\sin^{-1}(1)$ and equals $\pi/2$.]

5.3 Alternating Series

In Section 5.2 we limited ourselves to series of positive terms. Now, in contrast, we consider infinite series in which the signs alternate. The partial cancellation due to alternating signs makes convergence more rapid and much easier to identify. We shall prove the Leibnitz criterion, a general condition for the convergence of an alternating series.

Leibnitz criterion. Consider the series $\sum_{n=1}^{\infty}(-1)^{n+1}a_n$ with $a_n > 0$. If a_n is monotonic decreasing (for sufficiently large n) and $\lim_{n\to\infty} a_n = 0$, then the series converges.

To prove this, we examine the even partial sums

$$\begin{aligned} s_{2n} &= a_1 - a_2 + a_3 - \cdots - a_{2n}, \\ s_{2n+2} &= s_{2n} + (a_{2n+1} - a_{2n+2}). \end{aligned} \tag{5.51}$$

Since $a_{2n+1} > a_{2n+2}$, we have

$$s_{2n+2} > s_{2n}. \tag{5.52}$$

On the other hand,

$$s_{2n+2} = a_1 - (a_2 - a_3) - (a_4 - a_5) - \cdots - a_{2n+2}. \tag{5.53}$$

Hence, with each pair of terms $a_{2p} - a_{2p+1} > 0$,

$$s_{2n+2} < a_1. \tag{5.54}$$

With the even partial sums bounded $s_{2n} < s_{2n+2} < a_1$ and the terms a_n decreasing monotonically and approaching zero, this alternating series converges.

One further important result can be extracted from the partial sums. From the difference between the series limit S and the partial sum s_n

$$\begin{aligned} S - s_n &= a_{n+1} - a_{n+2} + a_{n+3} - a_{n+4} + \cdots \\ &= a_{n+1} - (a_{n+2} - a_{n+3}) - (a_{n+4} - a_{n+5}) - \cdots \end{aligned} \tag{5.55}$$

or

$$S - s_n < a_{n+1}. \tag{5.56}$$

Equation 5.56 says that the error in cutting off an alternating series after n terms is less than a_{n+1}, the first term dropped. A knowledge of the error obtained this way may be of great practical importance.

Absolute convergence. Given a series of terms u_n in which u_n may vary in sign, if $\sum |u_n|$ converges, then $\sum u_n$ is said to be absolutely convergent. If $\sum u_n$ converges but $\sum |u_n|$ diverges, the convergence is called conditional.

The alternating harmonic series is a simple example of this conditional convergence. We have

$$\sum_{n=1}^{\infty}(-1)^{n-1}n^{-1} = 1 - \frac{1}{2} + \frac{1}{3} - \frac{1}{4} + \cdots + \frac{1}{n} - \cdots, \tag{5.57}$$

convergent by the Leibnitz criterion; but

$$\sum_{n=1}^{\infty} n^{-1} = 1 + \frac{1}{2} + \frac{1}{3} + \frac{1}{4} + \cdots + \frac{1}{n} + \cdots \tag{5.58}$$

has been shown to be divergent in Sections 5.1 and 5.2.

The reader will note that all the tests developed in Section 5.2 assume a series of positive terms. Therefore all the tests in that section guarantee absolute convergence.

EXERCISES

5.3.1 (a) From the electrostatic two hemisphere problem (Exercise 12.3.14) we obtain the series

$$\sum_{s=0}^{\infty} (-1)^s(4s+3)\frac{(2s-1)!!}{(2s+2)!!}.$$

Test for convergence.

(b) The corresponding series for the surface charge density is

$$\sum_{s=0}^{\infty} (-1)^s(4s+3)\frac{(2s-1)!!}{(2s)!!}.$$

Test for convergence. The !! notation is explained in Section 10.1.

5.3.2 Show by direct numerical computation that the sum of the first ten terms of

$$\lim_{x\to 1} \ln(1+x) = \ln 2 = \sum_{n=1}^{\infty}(-1)^{n-1}n^{-1}$$

differs from ln 2 by less than the eleventh term: $\ln 2 = 0.69314\,71806 \cdots$.

5.3.3 In Ex. 5.2.9 the hypergeometric series is shown convergent for $x = \pm 1$, if $\gamma > \alpha + \beta$. Show that there is conditional convergence for $x = -1$ for γ down to $\gamma > \alpha + \beta - 1$. *Hint*. The asymptotic behavior of the factorial function is given by Stirling's series, Section 10.3.

5.4 Algebra of Series

The establishment of absolute convergence is important because it can be proved that absolutely convergent series may be handled according to the ordinary familiar rules of algebra or arithmetic.

1. If an infinite series is absolutely convergent, the series sum is independent of the order in which the terms are added.
2. The series may be multiplied with another absolutely convergent series. The limit of the product will be the product of the individual series limits. The product series, a double series, will also converge absolutely.

No such guarantees can be given for conditionally convergent series. Again consider the alternating harmonic series. If we write

$$1 - \tfrac{1}{2} + \tfrac{1}{3} - \tfrac{1}{4} + \cdots = 1 - (\tfrac{1}{2} - \tfrac{1}{3}) - (\tfrac{1}{4} - \tfrac{1}{5}) - \cdots, \tag{5.59}$$

it is clear that the sum

$$\sum_{n=1}^{\infty} (-1)^{n-1} n^{-1} < 1. \tag{5.60}$$

However, if we rearrange the terms slightly, we may make the alternating harmonic series converge to $\frac{3}{2}$. We regroup the terms of Eq. 5.59, taking

$$\begin{aligned}(1 + \tfrac{1}{3} + \tfrac{1}{5}) - (\tfrac{1}{2}) + (\tfrac{1}{7} + \tfrac{1}{9} + \tfrac{1}{11} + \tfrac{1}{13} + \tfrac{1}{15}) - (\tfrac{1}{4})\\ + (\tfrac{1}{17} + \cdots + \tfrac{1}{25}) - (\tfrac{1}{6}) + (\tfrac{1}{27} + \cdots + \tfrac{1}{35}) - (\tfrac{1}{8}) + \cdots.\end{aligned} \tag{5.61}$$

Treating the terms grouped in parenthesis as single terms for convenience, we obtain the partial sums

$s_1 = 1.5333$	$s_2 = 1.0333$
$s_3 = 1.5218$	$s_4 = 1.2718$
$s_5 = 1.5143$	$s_6 = 1.3476$
$s_7 = 1.5103$	$s_8 = 1.3853$
$s_9 = 1.5078$	$s_{10} = 1.4078$

From this tabulation of s_n and the plot of s_n versus n in Fig. 5.2 the convergence to $\frac{3}{2}$ is fairly clear. We have rearranged the terms, taking positive terms until the partial sum was equal to or greater than $\frac{3}{2}$, then adding in negative terms until the partial sum just fell below $\frac{3}{2}$, and so on. As the series extends to infinity, all of the original terms will eventually appear, but the partial sums of this rearranged alternating harmonic series converge to $\frac{3}{2}$.

By a suitable rearrangement of terms a conditionally convergent series may be made to converge to any desired value or even to diverge. This statement is sometimes given as Riemann's theorem. Obviously, conditionally convergent series

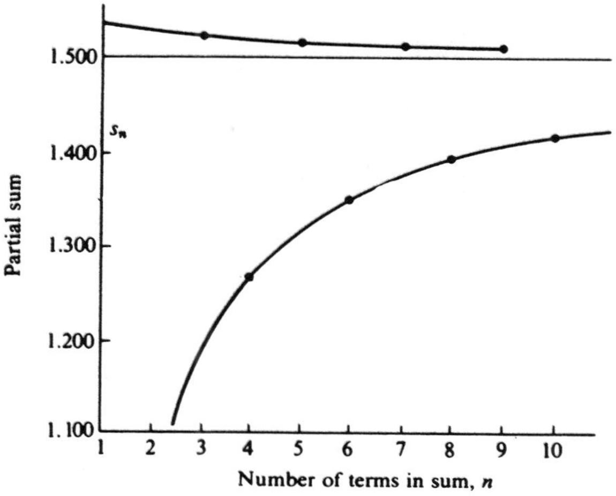

FIG. 5.2 Alternating harmonic series—terms rearranged to give convergence to 1.5

must be treated with caution.

Another aspect of the rearrangement of series appears in the treatment of double series

$$\sum_{m=0}^{\infty} \sum_{n=0}^{\infty} a_{n,m}.$$

Let us substitute

$$n = q \geqslant 0,$$
$$m = p - q \geqslant 0$$
$$(q \leqslant p).$$

This results in the identity

$$\sum_{m=0}^{\infty} \sum_{n=0}^{\infty} a_{n,m} = \sum_{p=0}^{\infty} \sum_{q=0}^{p} a_{q,p-q} \tag{5.62}$$

The substitution

$$n = s \geqslant 0,$$
$$m = r - 2s \geqslant 0$$
$$\left(s \leqslant \frac{r}{2}\right),$$

leads to

$$\sum_{m=0}^{\infty} \sum_{n=0}^{\infty} a_{n,m} = \sum_{r=0}^{\infty} \sum_{s=0}^{[r/2]} a_{s,r-2s} \tag{5.63}$$

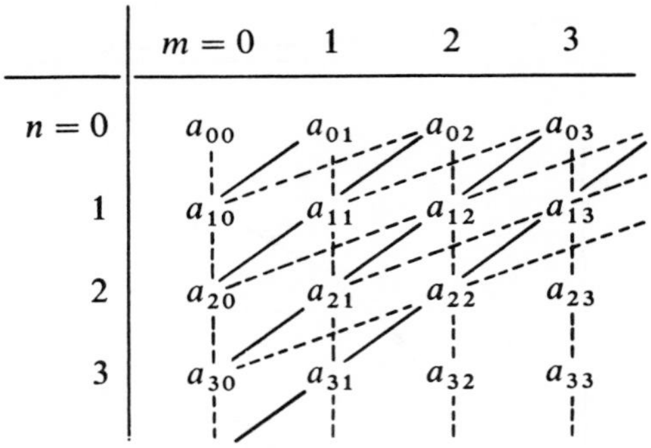

FIG. 5.3 Double series—summation over n indicated by vertical dashed lines

	$p = 0$	1	2	3
$q = 0$	a_{00}	a_{01}	a_{02}	a_{03}
1		a_{10}	a_{11}	a_{12}
2			a_{20}	a_{21}
3				a_{30}

FIG. 5.4 Double series—again the first summation is represented by vertical dashed lines but these vertical lines correspond to diagonals in Fig. 5.3

	$r = 0$	1	2	3	4
$s = 0$	a_{00}	a_{01}	a_{02}	a_{03}	a_{04}
1			a_{10}	a_{11}	a_{12}
2					a_{20}

FIG. 5.5 Double Series. The summation over s corresponds to a summation along the almost horizontal slanted lines in Fig. 5.3.

with $[r/2] = r/2$ for r even, $(r-1)/2$ for r odd. Equations 5.62 and 5.63 are represented in Fig. 5.3.

Equations 5.62 and 5.63 are clearly rearrangements of the array of coefficients a_{nm}, rearrangements that are valid as long as we have absolute convergence.

The combination of Eqs. 5.62 and 5.63,

$$\sum_{p=0}^{\infty}\sum_{q=0}^{p} a_{q,p-q} = \sum_{r=0}^{\infty}\sum_{s=0}^{[r/2]} a_{s,r-2s}, \tag{5.64}$$

is used in Section 12.1 in the determination of the series form of the Legendre polynomials.

EXERCISE

5.4.1 Given the series

$$\ln(1+x) = x - \frac{x^2}{2} + \frac{x^3}{3} - \frac{x^4}{4}\cdots, \qquad -1 < x \le 1,$$

show that

(a) $\ln(1-x) = -x - \dfrac{x^2}{2} - \dfrac{x^3}{3} - \dfrac{x^4}{4}\cdots, \qquad -1 \leqslant x < 1.$

(b) $\ln\left(\dfrac{1+x}{1-x}\right) = 2\left(x + \dfrac{x^3}{3} + \dfrac{x^5}{5} + \cdots\right), \qquad -1 < x < 1.$

The original series, $\ln(1+x)$, appears in an analysis of binding energy in crystals. It is $\frac{1}{2}$ the Madelung constant ($2\ln 2$) for a chain of atoms. The second series (b) is useful in normalizing the Legendre polynomials (Section 12.3) and in developing a second solution for Legendre's differential equation (Section 12.10).

5.5 Series of Functions

We extend our concept of infinite series to include the possibility that each term u_n may be a function of some variable, $u_n = u_n(x)$. The partial sums become functions of the variable x

$$s_n(x) = u_1(x) + u_2(x) + \cdots + u_n(x), \tag{5.65}$$

as does the series sum, defined as the limit of the partial sums

$$\sum_{n=1}^{\infty} u_n(x) = S(x) = \lim_{n\to\infty} s_n(x). \tag{5.66}$$

So far we have concerned ourselves with the behavior of the partial sums as a function of n. Now we consider how the foregoing quantities depend on x. The key concept here is that of uniform convergence.

Uniform convergence. If for any small $\varepsilon > 0$ there exists a number N, *independent of* x in the interval $[a, b]$ $(a \leqslant x \leqslant b)$ such that

$$|S(x) - s_n(x)| < \varepsilon, \qquad \text{for all } n \geqslant N, \tag{5.67}$$

the series is said to be uniformly convergent in the interval $[a, b]$. This says that for our series to be uniformly convergent, it must be possible to find a finite N so that the tail of the infinite series, $|\sum_{i=N+1}^{\infty} u_i(x)|$, will be less than an arbitrarily small ε for all x in the given interval.

This condition, Eq. 5.67, which defines uniform convergence, is illustrated in Fig. 5.6. The point is that no matter how small ε is taken to be we can always choose n large enough so that the absolute magnitude of the difference between $S(x)$ and $s_n(x)$ is less than ε for all x, $a \leqslant x \leqslant b$. If this cannot be done, then $\sum u_n(x)$ is *not* uniformly convergent in $[a, b]$.

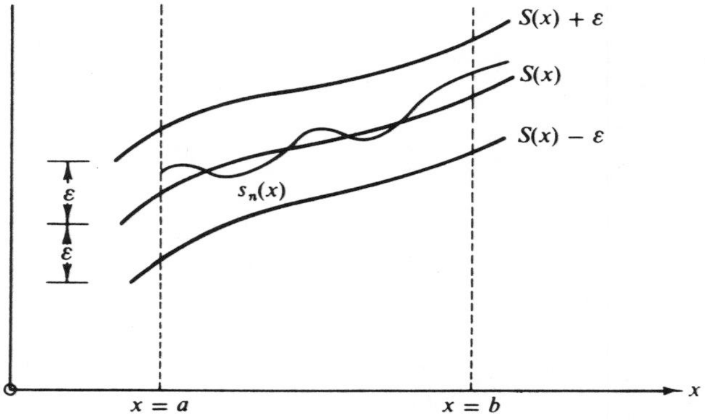

FIG. 5.6 Uniform convergence

EXAMPLE 5.5.1

$$\sum_{n=1}^{\infty} u_n(x) = \sum_{n=1}^{\infty} \frac{x}{[(n-1)x+1][nx+1]} \tag{5.68}$$

The partial sum $s_n(x) = nx(nx+1)^{-1}$ as may be verified by mathematical induction. By inspection this expression for $s_n(x)$ holds for $n = 1, 2$. We assume it holds for n terms and then prove it holds for $n + 1$ terms.

$$\begin{aligned} s_{n+1}(x) &= s_n(x) + \frac{x}{[nx+1][(n+1)x+1]} \\ &= \frac{nx}{[nx+1]} + \frac{x}{[nx+1][(n+1)x+1]} \\ &= \frac{(n+1)x}{(n+1)x+1}, \end{aligned}$$

completing the proof.

Letting n approach infinity, we obtain

$$S(0) = \lim_{n \to \infty} s_n(0) = 0,$$

$$S(x \neq 0) = \lim_{n \to \infty} s_n(x \neq 0) = 1.$$

We have a discontinuity in our series limit at $x = 0$. However, $s_n(x)$ is a continuous function of x, $0 \leqslant x \leqslant 1$, for all finite n. Equation 5.67 with ε sufficiently small, will be violated for all *finite* n. Our series does not converge uniformly.

Weierstrass M test. The most commonly encountered test for uniform convergence is the Weierstrass M test. If we can construct a series of numbers $\sum_1^\infty M_i$, in which $M_i \geqslant |u_i(x)|$ for all x in the interval $[a, b]$ and $\sum_1^\infty M_i$ is convergent, our series $\sum_1^\infty u_i(x)$ will be *uniformly* convergent in $[a, b]$.

The proof of this Weierstrass M test is direct and simple. Since $\sum_i M_i$ converges, some number N exists such that for $n + 1 \geqslant N$,

$$\sum_{i=n+1}^{\infty} M_i < \varepsilon. \tag{5.69}$$

This follows from our definition of convergence. Then, with $|u_i(x)| \leqslant M_i$ for all x in the interval $a \leqslant x \leqslant b$,

$$\sum_{i=n+1}^{\infty} |u_i(x)| < \varepsilon. \tag{5.70}$$

Hence

$$|S(x) - s_n(x)| = \left| \sum_{i=n+1}^{\infty} u_i(x) \right| < \varepsilon, \tag{5.71}$$

and by definition $\sum_{i=1}^{\infty} u_i(x)$ is uniformly convergent in $[a, b]$. Since we have specified absolute values in the statement of the Weierstrass M test, the series $\sum_{i=1}^{\infty} u_i(x)$ is also seen to be *absolutely* convergent.

The reader should note carefully that uniform convergence and absolute convergence are independent properties. Neither implies the other. For specific examples,

$$\sum_{n=1}^{\infty} \frac{(-1)^n}{n + x^2}, \qquad -\infty < x < \infty \tag{5.72}$$

and

$$\sum_{n=1}^{\infty} (-1)^{n-1} \frac{x^n}{n} = \ln(1 + x), \qquad 0 \leqslant x \leqslant 1 \tag{5.73}$$

converge uniformly in the indicated intervals but do not converge absolutely. On the other hand,

$$\begin{aligned} \sum_{n=0}^{\infty} (1 - x)x^n &= 1, \qquad 0 \leqslant x < 1 \\ &= 0, \qquad x = 1 \end{aligned} \tag{5.74}$$

converges absolutely but does not converge uniformly in [0, 1].

From the definition of uniform convergence we may show that any series

$$f(x) = \sum_{n=1}^{\infty} u_n(x) \tag{5.75}$$

cannot converge uniformly in any interval that includes a discontinuity of $f(x)$.

Since the Weierstrass M test establishes both uniform and absolute convergence, it will necessarily fail for series that are uniformly but conditionally convergent. A somewhat more delicate test for uniform convergence has been given by Abel. If

$$u_n(x) = a_n f_n(x),$$
$$\sum a_n = A, \quad \text{convergent},$$

and the functions $f_n(x)$ are monotonic $[f_{n+1}(x) \leqslant f_n(x)]$ and bounded, $0 \leqslant f_n(x) \leqslant M$, for all x in $[a, b]$, then $\sum u_n(x)$ converges uniformly in $[a, b]$.

This test is expecially useful in analyzing power series (cf. Section 5.7). Details of the proof of Abel's test and other tests for uniform convergence are given in the references listed at the end of this chapter.

Uniformly convergent series have three particularly useful properties.

1. If the individual terms $u_n(x)$ are continuous, the series sum

$$f(x) = \sum_{n=1}^{\infty} u_n(x) \tag{5.76}$$

is also continuous.

2. If the individual terms $u_n(x)$ are continuous, the series may be integrated term by term. The sum of the integrals is equal to the integral of the sum.

$$\int_a^b f(x)\,dx = \sum_{n=1}^{\infty} \int_a^b u_n(x)\,dx. \tag{5.77}$$

3. The derivative of the series sum $f(x)$ equals the sum of the individual term derivatives,

$$\frac{d}{dx} f(x) = \sum_{n=1}^{\infty} \frac{d}{dx} u_n(x), \tag{5.78}$$

provided the following conditions are satisfied.

$$u_n(x) \quad \text{and} \quad \frac{du_n(x)}{dx} \quad \text{are continuous in } [a, b].$$

$$\sum_{n=1}^{\infty} \frac{du_n(x)}{dx} \quad \text{is uniformly convergent in } [a, b].$$

Term-by-term integration of a uniformly convergent series[1] requires only continuity of the individual terms. This condition is almost always satisfied in physical

[1] Term-by-term integration *may* also be valid in the absence of uniform convergence.

applications. Term-by-term differentiation of a series is often not valid because more restrictive conditions must be satisfied. Indeed, we shall encounter cases in Chapter 14, Fourier Series, in which term-by-term differentiation of a uniformly convergent series leads to a divergent series.

EXERCISES

5.5.1 Find the range of *uniform* convergence of

(a) $\sum_{n=1}^{\infty} \frac{(-1)^{n-1}}{n^x}$

(b) $\sum_{n=1}^{\infty} \frac{1}{n^x}$

Ans. (a) $1 \leqslant x < \infty$.
(b) $1 < s \leqslant x < \infty$.

5.5.2 For what range of x is the geometric series $\sum_{n=0}^{\infty} x^n$ uniformly convergent?

Ans. $-1 < -s \leqslant x \leqslant s < 1$

5.5.3 For what range of positive values of x is $\sum_{n=0}^{\infty} 1/(1+x^n)$

(a) Convergent? (b) Uniformly convergent?

5.5.4 If the series of the coefficients $\sum a_n$ and $\sum b_n$ are absolutely convergent, show that the Fourier series

$$\sum(a_n \cos nx + b_n \sin nx)$$

is *uniformly* convergent for $-\infty < x < \infty$.

5.6 Taylor's Expansion

This is an expansion of a function into an infinite series or into a finite series plus a remainder term. The coefficients of the successive terms of the series involve the successive derivatives of the function. We have already used Taylor's expansion in the establishment of a physical interpretation of divergence (Section 1.7) and in other sections of Chapters 1 and 2. Now we derive the Taylor expansion.

We assume that our function $f(x)$ has a continuous nth derivative[1] in the interval $a \leqslant x \leqslant b$. Then, integrating this nth derivative n times,

$$\int_a^x f^{(n)}(x)\,dx = f^{(n-1)}(x)\Big|_a^x = f^{(n-1)}(x) - f^{(n-1)}(a)$$

$$\int_a^x \left(\int_a^x f^{(n)}(x)\,dx\right) dx = \int_a^x [f^{(n-1)}(x) - f^{(n-1)}(a)]\,dx \tag{5.79}$$

$$= f^{(n-2)}(x) - f^{(n-2)}(a) - (x-a)f^{(n-1)}(a).$$

[1] Taylor's expansion may be derived under slightly less restrictive conditions, cf. *Methods of Mathematical Physics*, Jeffreys and Jeffreys, Section 1.133.

Continuing, we obtain

$$\iiint_a^x f^{(n)}(x)\,(dx)^3 = f^{(n-3)}(x) - f^{(n-3)}(a) - (x-a)\,f^{(n-2)}(a) - \frac{(x-a)^2}{2} f^{(n-1)}(a). \quad (5.80)$$

Finally, on integrating for the nth time,

$$\int_a^x \cdots \int f^{(n)}(x)\,(dx)^n = f(x) - f(a) - (x-a)\,f'(a) - \frac{(x-a)^2}{2!} f''(a) - \cdots - \frac{(x-a)^{n-1}}{(n-1)!} f^{(n-1)}(a). \quad (5.81)$$

Note that this expression is exact. No terms have been dropped, no approximations made. Now, solving for $f(x)$, we have

$$f(x) = f(a) + (x-a)f'(a) + \frac{(x-a)^2}{2!} f''(a) + \cdots + \frac{(x-a)^{n-1}}{(n-1)!} f^{(n-1)}(a) + R_n. \quad (5.82)$$

The remainder, R_n, is given by the n-fold integral

$$R_n = \int_a^x \cdots \int f^{(n)}(x)\,(dx)^n. \quad (5.83)$$

This remainder, Eq. 5.83, may be put into perhaps more intelligible form by using the mean value theorem of differential calculus.

$$\int_a^x g(x)\,dx = (x-a)\,g(\xi), \quad (5.84)$$

with $a \leqslant \xi \leqslant x$. By integrating n times we get the Lagrangian form of the remainder:

$$R_n = \frac{(x-a)^n}{n!} f^{(n)}(\xi). \quad (5.85)$$

With Taylor's expansion in this form we are not concerned with any questions of infinite series convergence. This series is finite, and the only questions concern the magnitude of the remainder.

When the function $f(x)$ is such that

$$\lim_{n\to\infty} R_n = 0, \quad (5.86)$$

Eq. 5.82 becomes Taylor's series

$$f(x) = f(a) + (x-a)\,f'(a) + \frac{(x-a)^2}{2!} f''(a) + \cdots = \sum_{n=0}^{\infty} \frac{(x-a)^n}{n!} f^{(n)}(a).^{1} \quad (5.87)$$

[1] Note that $0! = 1$ (cf. Section 10.1).

An equivalent operator form of this Taylor expansion appears in Exercise 4.10.1.

Maclaurin theorem. If we expand about the origin ($a = 0$), Eq. 5.87 is known as Maclaurin's series

$$f(x) = f(0) + x f'(0) + \frac{x^2}{2!} f''(0) + \cdots$$
$$= \sum_{n=0}^{\infty} \frac{x^n}{n!} f^{(n)}(0). \tag{5.88}$$

An immediate application of the Maclaurin series (or the Taylor series) is in the expansion of various transcendental functions into infinite series.

EXAMPLE 5.6.1

Let $f(x) = e^x$. Differentiating,

$$f^{(n)}(0) = 1 \tag{5.89}$$

for all n, $n = 1, 2, 3, \ldots$. Then, by Eq. 5.88, we have

$$e^x = 1 + x + \frac{x^2}{2!} + \frac{x^3}{3!} + \cdots$$
$$= \sum_{n=0}^{\infty} \frac{x^n}{n!}. \tag{5.90}$$

This is the series expansion of the exponential function. Some authors use this series to define the exponential function.

Although this series is clearly convergent for all x, we should check the remainder term, R_n. By Eq. 5.85 we have

$$R_n = \frac{x^n}{n!} f^{(n)}(\xi)$$
$$= \frac{x^n}{n!} e^{\xi}, \qquad 0 \leqslant \xi \leqslant x. \tag{5.91}$$

Therefore

$$R_n \leqslant \frac{x^n e^x}{n!} \tag{5.92}$$

and

$$\lim_{n \to \infty} R_n = 0 \tag{5.93}$$

for all *finite* values of x, which indicates that this Maclaurin expansion of e^x is valid over the range $-\infty < x < \infty$.

EXAMPLE 5.6.2

Let $f(x) = \ln(1 + x)$. By differentiating we obtain

$$f'(x) = (1 + x)^{-1},$$
$$f^{(n)}(x) = (-1)^{n-1}(n - 1)!\,(1 + x)^{-n}. \tag{5.94}$$

The Maclaurin expansion (Eq. 5.88) yields

$$\ln(1+x) = x - \frac{x^2}{2} + \frac{x^3}{3} - \frac{x^4}{4} + \cdots + R_n$$

$$= \sum_{p=1}^{n} (-1)^{p-1} \frac{(x)^p}{p} + R_n. \tag{5.95}$$

In this case our remainder is given by

$$R_n = \frac{x^n}{n!} f^{(n)}(\xi), \qquad 0 \leqslant \xi \leqslant x$$

$$\leqslant \frac{x^n}{n}, \qquad 0 \leqslant \xi \leqslant x \leqslant 1. \tag{5.96}$$

Now the remainder approaches zero as n is increased indefinitely, provided $0 \leqslant x \leqslant 1$.[1] As an infinite series

$$\ln(1+x) = \sum_{n=1}^{\infty} (-1)^{n-1} \frac{x^n}{n}, \tag{5.97}$$

which converges for $-1 < x \leqslant 1$. The range $-1 < x < 1$ is easily established by the d'Alembert ratio test (Section 5.2). Convergence at $x = 1$ follows by the Leibnitz criterion (Section 5.3). In particular, at $x = 1$, we have

$$\ln 2 = 1 - \tfrac{1}{2} + \tfrac{1}{3} - \tfrac{1}{4} + \tfrac{1}{5} - \cdots$$

$$= \sum_{n=1}^{\infty} (-1)^{n-1} n^{-1}, \tag{5.98}$$

the conditionally convergent alternating harmonic series.

Binomial theorem. A second, extremely important application of the Taylor and Maclaurin expansions is the derivation of the binomial theorem for negative and/or nonintegral powers.

Let $f(x) = (1+x)^m$, in which m may be negative and is not limited to integral values. Direct application of Eq. 5.88 gives

$$(1+x)^m = 1 + mx + \frac{m(m-1)}{2!} x^2 + \cdots + R_n. \tag{5.99}$$

For this function the remainder is

$$R_n = \frac{x^n}{n!} (1+\xi)^{m-n} \times m(m-1)\cdots(m-n+1) \tag{5.100}$$

and ξ lies between 0 and x, $0 \leqslant \xi \leqslant x$. Now, for $n > m$, $(1+\xi)^{m-n}$ is a maximum for $\xi = 0$. Therefore

$$R_n \leqslant \frac{x^n}{n!} \times m(m-1)\cdots(m-n+1). \tag{5.101}$$

[1] This range can easily be extended to $-1 < x \leqslant 1$ but not to $x = -1$.

Note that the m dependent factors do not yield a zero unless m is a non-negative integer; R_n tends to zero as $n \to \infty$ if x is restricted to the range $0 \leqslant x < 1$.

The binomial expansion therefore is shown to be

$$(1+x)^m = 1 + mx + \frac{m(m-1)}{2!}x^2 + \frac{m(m-1)(m-2)}{3!}x^3 + \cdots. \quad (5.102)$$

In other, equivalent notation

$$(1+x)^m = \sum_{n=0}^{\infty} \frac{m!}{n!(m-n)!} x^m$$

$$= \sum_{n=0}^{\infty} \binom{m}{n} x^m. \quad (5.103)$$

The quantity $\binom{m}{n}$, which equals $m!/n!(m-n)!$ is called a *binomial coefficient*.

Although we have only shown that the remainder vanishes,

$$\lim_{n\to\infty} R_n = 0,$$

for $0 \leqslant x < 1$, the series in Eq. 5.102 actually may be shown to be convergent for the extended range $-1 < x < 1$.

EXAMPLE 5.6.3

The total relativistic energy of a particle is

$$E = mc^2\left(1 - \frac{v^2}{c^2}\right)^{-1/2} \quad (5.104)$$

Compare this equation with the classical kinetic energy, $\frac{1}{2}mv^2$.

By Eq. 5.102 with $x = -v^2/c^2$ and $m = -\frac{1}{2}$ we have

$$E = mc^2\left[1 - \frac{1}{2}\left(-\frac{v^2}{c^2}\right) + \frac{(-1/2)(-3/2)}{2!}\left(-\frac{v^2}{c^2}\right)^2 + \frac{(-1/2)(-3/2)(-5/2)}{3!}\left(-\frac{v^2}{c^2}\right)^3 + \cdots\right]$$

or

$$E = mc^2 + \frac{1}{2}mv^2 + \frac{3}{8}mv^2 \cdot \frac{v^2}{c^2} + \frac{5}{16}mv^2 \cdot \left(\frac{v^2}{c^2}\right)^2 + \cdots. \quad (5.105)$$

The first term, mc^2, is identified as the rest mass energy. Then

$$E_{\text{kinetic}} = \frac{1}{2}mv^2\left[1 + \frac{3}{4}\frac{v^2}{c^2} + \frac{5}{8}\left(\frac{v^2}{c^2}\right)^2 + \cdots\right]. \quad (5.106)$$

For particle velocity $v \ll c$, the velocity of light, the expression in the brackets reduces to unity and we see that the kinetic portion of the total relativistic energy agrees with the classical result.

For polynomials we can generalize the binomial expansion to

$$(a_1 + a_2 + \cdots + a_m)^n = \sum \frac{n!}{n_1!n_2!\ldots n_m!} a_1^{n_1} a_2^{n_2} \cdots a_m^{n_m},$$

where the summation includes all different combinations of $n_1, n_2, \ldots, n_m$ with $\sum_{i=1}^m n_i = n$. Here n_i and n are all integral. This generalization finds considerable use in statistical mechanics.

Taylor expansion—more than one variable. If the function f has more than one independent variable, say $f = f(x, y)$, the Taylor expansion becomes

$$\begin{aligned} f(x, y) = f(a, b) &+ (x-a)\frac{\partial f}{\partial x} + (y-b)\frac{\partial f}{\partial y} \\ &+ \frac{1}{2!}\left[(x-a)^2 \frac{\partial^2 f}{\partial x^2} + 2(x-a)(y-b)\frac{\partial^2 f}{\partial x\,\partial y} + (y-b)^2 \frac{\partial^2 f}{\partial y^2}\right] \\ &+ \frac{1}{3!}\left[(x-a)^3 \frac{\partial^3 f}{\partial x^3} + 3(x-a)^2(y-b)\frac{\partial^3 f}{\partial x^2\,\partial y}\right. \\ &\left. + 3(x-a)(y-b)^2 \frac{\partial^3 f}{\partial x\,\partial y^2} + (y-b)^3 \frac{\partial^3 f}{\partial y^3}\right] + \cdots, \end{aligned} \tag{5.107}$$

with all derivatives evaluated at the point (a, b). Using $\alpha_j t = x_j - x_{j0}$, we may write the Taylor expansion for m independent variables in the symbolic form

$$f(x_j) = \sum_{n=0}^{\infty} \frac{t^n}{n!}\left(\sum_{i=1}^{m} \alpha_i \frac{\partial}{\partial x_i}\right)^n f(x_k)\Bigg|_{x_k = x_{k0}} \tag{5.108}$$

A convenient vector form is

$$\psi(\mathbf{r} + \mathbf{a}) = \sum_{n=0}^{\infty} \frac{1}{n!}(\mathbf{a} \cdot \nabla)^n \psi(\mathbf{r}). \tag{5.109}$$

EXERCISES

5.6.1 Show that

(a) $\sin x = \sum_{n=0}^{\infty} (-1)^n \frac{x^{2n+1}}{(2n+1)!}$, (b) $\cos x = \sum_{n=0}^{\infty} (-1)^n \frac{x^{2n}}{(2n)!}$.

In Section 6.1 e^{ix} is *defined* by a series expansion such that

$$e^{ix} = \cos x + i \sin x.$$

This is the basis for the polar representation of complex quantities. As a special case we find, with $x = \pi$,

$$e^{i\pi} = -1.$$

5.6.2 Derive a series expansion of cot x in increasing powers of x by dividing cos x by sin x. *Note*. The resultant series which starts with $1/x$ is actually a Laurent series (Section 6.5). Although the two series for sin x and cos x were valid for all x, the convergence of the series for cot x is limited by the zeros of the denominator, sin x.

5.6.3 (a) Expand $(1 + x)\ln(1 + x)$ in a Maclaurin series. Find the limits on x for convergence.
(b) From the results for part (a) show that

$$\ln 2 = \tfrac{1}{2} + \tfrac{1}{2}\sum_{n=1}^{\infty} \frac{(-1)^{n+1}}{n(n+1)}.$$

Ans. (a) $(1 + x)\ln(1 + x) = x + \sum_{n=2}^{\infty}(-1)^n \dfrac{x^n}{n(n-1)}, \quad -1 < x \leq 1.$

5.6.4 The Raabe test for $\sum(n \ln n)^{-1}$ leads to

$$\lim_{n\to\infty} n\left[\frac{(n+1)\ln(n+1)}{n \ln n} - 1\right].$$

Show that this limit is unity (which means that the Raabe test here is indeterminant).

5.6.5 Show by series expansion that

$$\frac{1}{2}\ln\frac{\eta_0 + 1}{\eta_0 - 1} = \coth^{-1}\eta_0, \qquad |\eta_0| > 1.$$

This identity is used to obtain a second solution for Legendre's equation (Section 12.10).

5.6.6 Show that $f(x) = x^{1/2}$ (a) has no Maclaurin expansion but (b) has a Taylor expansion about any point $x_0 \neq 0$. Find the range of convergence of the Taylor expansion about $x = x_0$.

5.6.7 Let x be an approximation for a zero of $f(x)$ and Δx, the correction. Show that by neglecting terms of order $(\Delta x)^2$

$$\Delta x = -\frac{f(x)}{f'(x)}.$$

This is Newton's formula for finding a root.

5.6.8 Expand a function $\Phi(x, y, z)$ by Taylor's expansion. Evaluate $\bar{\Phi}$, the average value of Φ, averaged over a small cube of side a centered on the origin and show that the Laplacian of Φ is a measure of deviation of $\bar{\Phi}$ from $\Phi(0, 0, 0)$.

5.6.9 The ratio of two differentiable functions $f(x)$ and $g(x)$, takes on the indeterminate form 0/0 at $x = x_0$. Using Taylor expansions prove L'Hospital's rule

$$\lim_{x\to x_0}\frac{f(x)}{g(x)} = \lim_{x\to x_0}\frac{f'(x)}{g'(x)}.$$

5.6.10 With $n > 1$, show that

(a) $\dfrac{1}{n} - \ln\left(\dfrac{n}{n-1}\right) < 0, \qquad$ (b) $\dfrac{1}{n} - \ln\left(\dfrac{n+1}{n}\right) > 0.$

Use these inequalities to show that the limit defining the Euler-Mascheroni constant is finite.

5.6.11 Expand $(1 - 2tz + t^2)^{-1/2}$ in powers of t. Assume that t is small. Collect the coefficients of t^0, t^1, and t^2.

Ans. $a_0 = P_0(z) = 1,$

$a_1 = P_1(z) = z,$

$a_2 = P_2(z) = \tfrac{1}{2}(3z^2 - 1),$

where $a_n = P_n(z)$, the nth Legendre polynomial.

5.6.12 Using the double factorial notation of Section 10.1 show that

$$(1+x)^{-m/2}=\sum_{n=0}^{\infty}(-1)^n\frac{(m+2n-2)!!}{2^n n!\,(m-2)!!}x^n,$$

for $m=1, 2, 3, \ldots$.

5.6.13 Using binomial expansions compare the three Doppler shift formulas:

(a) $\nu'=\nu\left(1\mp\frac{v}{c}\right)^{-1}$ moving source; (c) $\nu'=\nu\left(1\pm\frac{v}{c}\right)\left(1-\frac{v^2}{c^2}\right)^{-1/2}$, relativistic;

(b) $\nu'=\nu\left(1\pm\frac{v}{c}\right)$, moving observer;

Note that the relativistic formula agrees with the classical formulas if terms of order v^2/c^2 can be neglected.

5.6.14 In the theory of general relativity there are various ways of relating (defining) a velocity of recession of a galaxy to its red shift, δ. Milne's model (kinematic relativity) gives

(a) $v_1=c\,\delta(1+\frac{1}{2}\delta)$,

(b) $v_2=c\,\delta(1+\frac{1}{2}\delta)(1+\delta)^{-2}$ (c) $1+\delta=\left[\frac{1+v_3/c}{1-v_3/c}\right]^{1/2}$

1. Show that for $\delta\ll 1$ (and $v_3/c\ll 1$) all three formulas reduce to $v=c\delta$.
2. Compare the three velocities through terms of order δ^2.

5.6.15 The relativistic sum w of two velocities u and v is given by

$$\frac{w}{c}=\frac{u/c+v/c}{1+uv/c^2}.$$

If

$$\frac{v}{c}=\frac{u}{c}=1-\alpha,$$

where $0\leqslant\alpha\leqslant 1$, find w/c in powers of α through terms in α^3.

5.6.16 The displacement x of a particle of rest mass m_0, resulting from a constant force $m_0 g$ along the x-axis, is

$$x=\frac{c^2}{g}\left\{\left[1+\left(g\frac{t}{c}\right)^2\right]^{1/2}-1\right\},$$

including relativistic effects. Find the displacement x as a power series in time t. Compare with the classical result

$$x=\tfrac{1}{2}gt^2.$$

5.6.17 By use of Dirac's relativistic theory the fine structure formula of atomic spectroscopy is given by

$$E=mc^2\left[1+\frac{\gamma^2}{(s+n-|k|)^2}\right]^{-1/2},$$

where

$$s=(|k|^2-\gamma^2)^{1/2},\qquad k=\pm 1, \pm 2, \pm 3, \ldots.$$

Expand in powers of γ^2 through order γ^4. ($\gamma^2=Ze^2/\hbar c$, with Z the atomic number.) This expansion is useful in comparing the predictions of the Dirac electron theory with those of a relativistic Schrödinger electron theory. Experimental results support the Dirac theory.

5.6.18 In a head-on proton–proton collision, the ratio of the kinetic energy in the center of mass system to the incident kinetic energy is

$$R=\frac{\sqrt{2mc^2(E_k+2mc^2)}-2mc^2}{E_k}.$$

Find the value of this ratio of kinetic energies for
(a) $E_k \ll mc^2$ (nonrelativistic)
(b) $E_k \gg mc^2$ (extreme-relativistic)

Ans. (a) $\frac{1}{2}$, (b) 0. The latter answer is a sort of law of diminishing returns for high energy particle accelerators (with stationary targets).

5.6.19 With binomial expansions

$$\frac{x}{1-x} = \sum_{n=1}^{\infty} x^n, \quad \frac{x}{x-1} = \frac{1}{1-x^{-1}} = \sum_{n=0}^{\infty} x^{-n}.$$

Adding these two series $\sum_{n=-\infty}^{\infty} x^n = 0$.

Hopefully we can agree that this is nonsense but what has gone wrong?

5.7 Power Series

The power series is a special and extremely useful type of infinite series of the form

$$f(x) = a_0 + a_1 x + a_2 x^2 + a_3 x^3 + \cdots$$
$$= \sum_{n=0}^{\infty} a_n x^n \tag{5.110}$$

where the coefficients a_i are constants, independent of x.[1]

Convergence. Equation 5.110 may readily be tested for convergence by either the Cauchy root test or the d'Alembert ratio test (Section 5.2). If

$$\lim_{n\to\infty} \frac{a_{n+1}}{a_n} = R^{-1}, \tag{5.111}$$

the series converges for $-R < x < R$. This is the interval or radius of convergence. Since the root and ratio tests fail when the limit is unity, the end points of the interval require special attention.

For instance, if $a_n = n^{-1}$, then $R = 1$ and, from Sections 5.1, 5.2, and 5.3, the series converges for $x = -1$ but diverges for $x = +1$. If $a_n = n!$, then $R = 0$ and the series diverges for all $x \neq 0$.

Uniform and absolute convergence. Suppose our power series (Eq. 5.110) has been found convergent for $-R < x < R$; then it will be uniformly and absolutely convergent in any *interior* interval, $-S \leqslant x \leqslant S$, where $0 < S < R$.

This may be proved directly by the Weierstrass M test (Section 5.5) by using $M_i = |a_i|\, S^i$.

[1] Equation 5.110 may be rewritten with $z = x + iy$, replacing x. The following sections will then yield uniform convergence, integrability, and differentiability in a region of a complex plane in place of an interval on the x-axis.

Continuity. Since each of the terms $u_n(x) = a_n x^n$ is a continuous function of x and $f(x) = \sum a_n x^n$ converges uniformly for $-S \leqslant x \leqslant S$, $f(x)$ must be a continuous function in the interval of uniform convergence.

This behavior is to be contrasted with the strikingly different behavior of Fourier series (Chapter 14), in which the Fourier series is used frequently to represent discontinuous functions such as sawtooth and square waves.

Differentiation and integration. With $u_n(x)$ continuous and $\sum a_n x^n$ uniformly convergent, we find that the differentiated series is a power series with continuous functions and the same radius of convergence as the original series. The new factors introduced by differentiation (or integration) do not affect either the root or the ratio test. Therefore our power series may be differentiated or integrated as often as desired within the interval of uniform convergence (Exercise 5.7.14).

In view of the rather severe restrictions placed on differentiation (Section 5.5), this is a remarkable and valuable result.

Uniqueness theorem. In the preceding section, using the Maclaurin series, we expanded e^x and $\ln(1 + x)$ into infinite series. In the succeeding chapters functions are frequently represented or perhaps defined by infinite series. We now establish that the power-series representation is unique.

If

$$\begin{aligned} f(x) &= \sum_{n=0}^{\infty} a_n x^n, \qquad -R_a < x < R_a \\ &= \sum_{n=0}^{\infty} b_n x^n, \qquad -R_b < x < \end{aligned} \tag{5.112}$$

with overlapping intervals of convergence, including the origin, then

$$a_n = b_n \tag{5.113}$$

for all n; that is, we assume two (different) power-series representations and then proceed to show that the two are actually identical.

From Eq. 5.112

$$\sum_{n=0}^{\infty} a_n x^n = \sum_{n=0}^{\infty} b_n x^n, \qquad -R < x < R, \tag{5.114}$$

where R is the smaller of R_a, R_b. By setting $x = 0$ to eliminate all but the constant terms, we obtain

$$a_0 = b_0. \tag{5.115}$$

Now, exploiting the differentiability of our power series, we differentiate Eq. 5.113, getting

$$\sum_{n=1}^{\infty} n a_n x^{n-1} = \sum_{n=1}^{\infty} n b_n x^{n-1}. \tag{5.116}$$

We again set $x = 0$ to isolate the new constant terms and find

$$a_1 = b_1. \tag{5.117}$$

By repeating this process n times, we get

$$a_n = b_n, \tag{5.118}$$

which shows that the two series coincide. Therefore our power-series representation is unique.

This will be a crucial point in Section 8.4, in which we use a power series to develop solutions of differential equations. This uniqueness of power series appears frequently in theoretical physics. The establishment of perturbation theory in quantum mechanics is one example. The power-series representation of functions is often useful in evaluating indeterminate forms, particularly when l'Hospital's rule may be awkward to apply (Exercise 5.7.10).

EXAMPLE 5.7.1

Evaluate

$$\lim_{x \to 0} \frac{1 - \cos x}{x^2}. \tag{5.119}$$

Replacing $\cos x$ by its Maclaurin series expansion,

$$\begin{aligned}\frac{1 - \cos x}{x^2} &= \frac{1 - (1 - x^2/2! + x^4/4! - \cdots)}{x^2} \\ &= \frac{x^2/2! - x^4/4! + \cdots}{x^2} \\ &= 1/2! - x^2/4! + \cdots.\end{aligned}$$

Letting $x \to 0$,

$$\lim_{x \to 0} \frac{1 - \cos x}{x^2} = \frac{1}{2}. \tag{5.120}$$

Reversion (inversion) of power series. Suppose we are given a series

$$\begin{aligned}y - y_0 &= a_1(x - x_0) + a_2(x - x_0)^2 + \cdots \\ &= \sum_{n=1}^{\infty} a_n(x - x_0)^n.\end{aligned} \tag{5.121}$$

This gives $(y - y_0)$ in terms of $(x - x_0)$. However, it may be desirable to have an explicit expression for $(x - x_0)$ in terms of $(y - y_0)$. We may solve Eq. 5.121 for $x - x_0$ by reversion (or inversion) of our series. Assume that

$$x - x_0 = \sum_{n=1}^{\infty} b_n(y - y_0)^n, \tag{5.122}$$

with the b_n to be determined in terms of the assumed known a_n. A brute-force approach, which is perfectly adequate for the first few coefficients, is simply to substitute Eq. 5.121 into Eq. 5.122. By equating coefficients of $(x - x_0)^n$ on both sides of Eq. 5.122, since the power series is unique, we obtain

$$\begin{aligned} b_1 &= \frac{1}{a_1}, \\ b_2 &= -\frac{a_2}{a_1^3}, \\ b_3 &= \frac{1}{a_1^5}(2a_2^2 - a_1 a_3), \\ b_4 &= \frac{1}{a_1^7}(5a_1 a_2 a_3 - a_1^2 a_4 - 5a_2^3), \qquad \text{etc.} \end{aligned} \tag{5.123}$$

Some of the higher coefficients are listed by Dwight.[1] In Section 7.3 a more general and much more elegant approach is developed by the use of complex variables.

EXERCISES

5.7.1 The classical Langevin theory of paramagnetism leads to an expression for the magnetic polarization

$$P(x) = C\left(\frac{\cosh x}{\sinh x} - \frac{1}{x}\right).$$

Expand $P(x)$ as a power series for small x (low fields, high temperature).

5.7.2 The depolarizing factor L for an oblate ellipsoid in a uniform electric field parallel to the axis of rotation (Section 12.11) is

$$L = \frac{1}{\varepsilon_0}(1 + \zeta_0^2)(1 - \zeta_0 \cot^{-1} \zeta_0),$$

where ζ_0 defines an oblate ellipsoid in oblate spheroidal coordinates (ξ, ζ, φ). Show that

$$\lim_{\zeta_0 \to \infty} L = \frac{1}{3\varepsilon_0} \quad \text{(sphere)},$$

$$\lim_{\zeta_0 \to 0} L = \frac{1}{\varepsilon_0} \quad \text{(thin sheet)}.$$

5.7.3 The corresponding depolarizing factor (Exercise 5.7.2) for a prolate ellipsoid is

$$L = \frac{1}{\varepsilon_0}(\eta_0^2 - 1)\left(\frac{1}{2}\eta_0 \ln \frac{\eta_0 + 1}{\eta_0 - 1} - 1\right).$$

Show that

$$\lim_{\eta_0 \to \infty} L = \frac{1}{3\varepsilon_0} \quad \text{(sphere)},$$

$$\lim_{\eta_0 \to 1} L = 0 \quad \text{(long needle)}.$$

[1] H. B. Dwight, *Tables of Integrals and Other Mathematical Data.* 4th Ed. New York: Macmillan (1961). (Cf. Formula No. 50.)

5.7.4 The analysis of the diffraction pattern of a circular opening involves

$$\int_0^{2\pi} \cos(c \cos \varphi)\, d\varphi.$$

Expand the integrand in a series and integrate by using

$$\int_0^{2\pi} \cos^{2n}\varphi\, d\varphi = \frac{(2n)!}{2^{2n}(n!)^2} \cdot 2\pi,$$

$$\int_0^{2\pi} \cos^{2n+1}\varphi\, d\varphi = 0.$$

The result is 2π times the Bessel function $J_0(c)$.

5.7.5 Neutrons are created (by a nuclear reaction) inside a hollow sphere of radius R. The newly created neutrons are uniformly distributed over the spherical volume. Assuming that all directions are equally probable (isotropy), what is the average distance a neutron will travel before striking the surface of the sphere? Assume straight line motion, no collisions.

(a) Show that

$$\bar{r} = \tfrac{3}{2}R \int_0^1 \int_0^\pi \sqrt{1 - k^2 \sin^2 \theta}\, k^2\, dk \sin\theta\, d\theta.$$

(b) Expand the integrand as a series and integrate to obtain

$$\bar{r} = R\left[1 - 3 \sum_{n=1}^{\infty} \frac{1}{(2n-1)(2n+1)(2n+3)}\right].$$

(c) Show that the sum of this infinite series is $\frac{1}{12}$, giving $\bar{r} = \frac{3}{4}R$. *Hint.* Show that $s_n = \frac{1}{12} - [4(2n+1)(2n+3)]^{-1}$ by mathematical induction. Then let $n \to \infty$.

5.7.6 Given that

$$\int_0^1 \frac{dx}{1+x^2} = \tan^{-1} x \Big|_0^1 = \frac{\pi}{4},$$

expand the integrand into a series and integrate term by term obtaining

$$\frac{\pi}{4} = 1 - \tfrac{1}{3} + \tfrac{1}{5} - \tfrac{1}{7} + \tfrac{1}{9} - \cdots + (-1)^n \frac{1}{2n+1} + \cdots,$$

which is Leibnitz' formula for π. Compare the convergence (or lack of it) of the integrand series and the integrated series at $x = 1$.

Leibnitz' formula converges so slowly that it is quite useless for numerical work; π has been computed to 100,000 decimals [D. Shanks and J. W. Wrench, Jr., Mathematics of Computation **16**, 76 (1962)] by using expressions such as

$$\pi = 24 \tan^{-1}\tfrac{1}{8} + 8 \tan^{-1}\tfrac{1}{57} + 4 \tan^{-1}\tfrac{1}{239},$$

$$\pi = 48 \tan^{-1}\tfrac{1}{18} + 32 \tan^{-1}\tfrac{1}{57} - 20 \tan^{-1}\tfrac{1}{239}.$$

5.7.7 Expand the incomplete factorial function

$$\int_0^x e^{-t} t^n\, dt$$

in a series of powers of x for small values of x. What is the range of convergence of the resulting series? Why was x specified to be small?

Ans. $$\int_0^x e^{-t} t^n\, dt = x^{n+1}\left[\frac{1}{(n+1)} - \frac{x}{(n+2)} + \frac{x^2}{2!(n+3)} - \cdots \frac{(-1)^p x^p}{p!(n+p+1)} + \cdots\right].$$

5.7.8 An analysis of the Gibbs phenomenon of Section 14.5 leads to the expression

$$\frac{2}{\pi} \int_0^\pi \frac{\sin \xi}{\xi}\, d\xi.$$

Find the numerical value of this expression to four significant figures.

Ans. 1.179.

5.7.9 Derive the series expansion of the incomplete beta function

$$B_x(p, q) = \int_0^x t^{p-1}(1-t)^{q-1}\,dt$$

$$= x^p\left\{\frac{1}{p} + \frac{1-q}{p+1}x + \cdots + \frac{(1-q)\cdots(n-q)}{n!(p+n)}x^n + \cdots\right\}$$

for $0 \leqslant x \leqslant 1$, $p > 0$ and $q > 0$ (if $x = 1$).

5.7.10 Evaluate

(a) $\lim_{x\to 0} \frac{\sin(\tan x) - \tan(\sin x)}{x^7}$,

(b) $\lim_{x\to 0} x^{-n} j_n(x)$ for $n = 3$,

where $j_n(x)$ is a spherical Bessel function (Section 11.7) defined by

$$j_n(x) = (-1)^n x^n \left(\frac{d}{x\,dx}\right)^n \left(\frac{\sin x}{x}\right).$$

Ans. (a) $-\frac{1}{30}$,

(b) $\frac{1}{1\cdot 3\cdot 5\cdots(2n+1)} \to \frac{1}{105}$ for $n = 3$.

5.7.11 Neutron transport theory gives the following expression for the inverse neutron diffusion length of k:

$$\frac{a-b}{k}\tanh^{-1}\left(\frac{k}{a}\right) = 1.$$

By series inversion or otherwise, determine k^2 as a series of powers of b/a. Give the first two terms of the series.

Ans. $k^2 = 3ab\left(1 - \frac{4}{5}\frac{b}{a}\right)$.

5.7.12 Develop a series expansion of $\sinh^{-1} x$ in powers of x by

(a) reversion of the series for $\sinh y$,

(b) a direct Maclaurin expansion.

5.7.13 A function $f(z)$ is represented by a *descending* power series

$$f(z) = \sum_{n=0}^{\infty} a_n z^{-n}, \qquad R \leq z < \infty.$$

Show that this series expansion is unique, i.e., if $f(z) = \sum_{n=0}^{\infty} b_n z^{-n}$, $R \leq z < \infty$, then $a_n = b_n$ for all n.

5.7.14 A power series given by

$$f(x) = \sum_{n=0}^{\infty} a_n x^n$$

converges for $-R < x < R$. Show that the differentiated series and the integrated series have the same interval of convergence. (Do not bother about the end points $x = \pm R$.)

5.7.15 Assuming that $f(x)$ may be expanded in a power series about the origin, $f(x)=\sum_{n=0}^{\infty} a_n x^n$, with some nonzero range of convergence. Use the techniques employed in proving uniqueness of series to show that your assumed series is a Maclaurin series:

$$a_n = \frac{1}{n!} f^{(n)}(0).$$

5.8 Elliptic Integrals

Elliptic integrals are included here partly as an illustration of the use of power series and partly for their own intrinsic interest. This interest includes the occurrence of elliptic integrals in physical problems (Example 5.8.1 and Exercise 5.8.4) and applications in mathematical problems.

EXAMPLE 5.8.1 PERIOD OF A SIMPLE PENDULUM

For small amplitude oscillations our pendulum has simple harmonic motion with a period $T = 2\pi(l/g)^{1/2}$. For a maximum amplitude θ_M large enough so that $\sin\theta_M \neq \theta_M$, Newton's second law of motion and Lagrange's equation (Section 17.7) lead to a nonlinear differential equation ($\sin\theta$ is a nonlinear function of θ), so we turn to a different approach.

The swinging mass m has a kinetic energy of $\frac{1}{2}ml^2(d\theta/dt)^2$ and a potential energy of $-mgl\cos\theta$ ($\theta = \frac{1}{2}\pi$ taken for the arbitrary zero of potential energy). Since $d\theta/dt = 0$ at $\theta = \theta_M$, the conservation of energy principle gives

$$\tfrac{1}{2}\, ml^2\left(\frac{d\theta}{dt}\right)^2 - mgl\cos\theta = -mgl\cos\theta_M. \tag{5.124}$$

Solving for $d\theta/dt$ we obtain

$$\frac{d\theta}{dt} = \pm\left(\frac{2g}{l}\right)^{1/2}(\cos\theta - \cos\theta_M)^{1/2} \tag{5.125}$$

with the mass m cancelling out. We take t to be zero when $\theta = 0$ and $d\theta/dt > 0$. An integration from $\theta = 0$ to $\theta = \theta_M$ yields

$$\int_0^{\theta_M}(\cos\theta - \cos\theta_M)^{-1/2}\,d\theta = \left(\frac{2g}{l}\right)^{1/2}\int_0^t dt = \left(\frac{2g}{l}\right)^{1/2} t. \tag{5.126}$$

This is $\frac{1}{4}$ of a cycle, and therefore the time t is $\frac{1}{4}$ of the period, T. We note that $\theta \leqslant \theta_M$, and with a bit of clairvoyance we try the half-angle substitution

$$\sin\left(\frac{\theta}{2}\right) = \sin\left(\frac{\theta_M}{2}\right)\sin\varphi. \tag{5.127}$$

With this, Eq. 5.126 becomes

$$T = 4\left(\frac{l}{g}\right)^{1/2}\int_0^{\pi/2}\left(1 - \sin^2\left(\frac{\theta_M}{2}\right)\sin^2\varphi\right)^{-1/2} d\varphi. \tag{5.128}$$

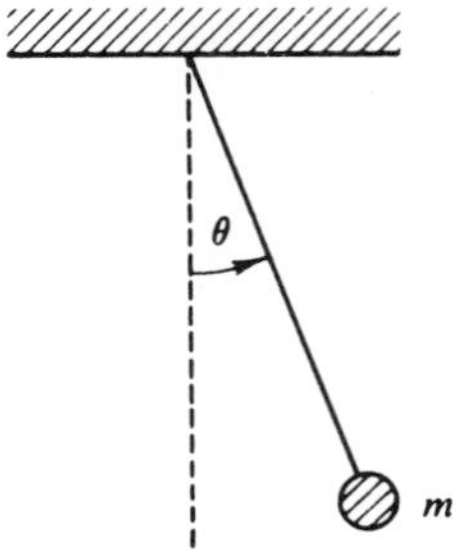

FIG. 5.7 Simple pendulum

While not an obvious improvement over Eq. 5.126, the integral now defines the complete elliptic integral of the first kind, $K(\sin\theta_M/2)$. From the series expansion, the period of our pendulum may be developed as a power series—powers of $\sin\theta_M/2$:

$$T = 2\pi\left(\frac{l}{g}\right)^{1/2}\left\{1 + \frac{1}{4}\sin^2\frac{\theta_M}{2} + \frac{9}{64}\sin^4\frac{\theta_M}{2} + \cdots\right\}. \tag{5.129}$$

Definitions. Generalizing Example 5.8.1 to include the upper limit as a variable, the *elliptic integral of the first kind* is defined as

$$F(\varphi\backslash\alpha) = \int_0^{\varphi}(1 - \sin^2\alpha\sin^2\theta)^{-1/2}\,d\theta \tag{5.130a}$$

or

$$F(x\,|\,m) = \int_0^x [(1 - t^2)(1 - mt^2)]^{-1/2}\,dt, \qquad 0 \leqslant m < 1. \tag{5.130b}$$

(This is the notation of *AMS*-55.) For $\varphi = \pi/2$, $x = 1$, we have the *complete elliptic integral of the first kind*

$$\begin{aligned} K(m) &= \int_0^{\pi/2}(1 - m\sin^2\theta)^{-1/2}\,d\theta \\ &= \int_0^1 [(1 - t^2)(1 - mt^2)]^{-1/2}\,dt \end{aligned} \tag{5.131}$$

with $m = \sin^2\alpha$, $0 \leqslant m < 1$.

The *elliptic integral of the second kind* is defined by

$$E(\varphi\backslash\alpha) = \int_0^{\varphi}(1 - \sin^2\alpha\sin^2\theta)^{1/2}\,d\theta \tag{5.132a}$$

or

$$E(x\,|\,m) = \int_0^x \left(\frac{1 - mt^2}{1 - t^2}\right)^{1/2} dt, \qquad 0 \leqslant m \leqslant 1. \tag{5.132b}$$

Again for the case $\varphi = \pi/2$, $x = 1$, we have the *complete elliptic integral of the second kind*

$$E(m) = \int_0^{\pi/2} (1 - m \sin^2 \theta)^{1/2}\, d\theta$$
$$= \int_0^1 \left(\frac{1 - mt^2}{1 - t^2}\right)^{1/2} dt, \qquad 0 \leqslant m \leqslant 1. \tag{5.133}$$

Exercise 5.8.1 is an example of its occurrence. Fig. 5.8 shows the behavior of $K(m)$ and $E(m)$. Extensive tables are available in *AMS*-55.

Series Expansion. For our range $0 \leqslant m < 1$, the denominator of $K(m)$ may be expanded by the binomial series,

$$(1 - m \sin^2 \theta)^{-1/2} = 1 + \frac{1}{2} m \sin^2 \theta + \frac{3}{8} m^2 \sin^4 \theta + \cdots$$
$$= \sum_{n=0}^{\infty} \frac{(2n-1)!!}{(2n)!!} m^n \sin^{2n} \theta. \tag{5.134}$$

For any closed interval $[0, m_{\max}]$, $m_{\max} < 1$, this series is uniformly convergent and may be integrated term-by-term. From Ex. 10.4.9

$$\int_0^{\pi/2} \sin^{2n} \theta\, d\theta = \frac{(2n-1)!!}{(2n)!!} \cdot \frac{\pi}{2}. \tag{5.135}$$

Hence,

$$K(m) = \frac{\pi}{2}\left\{1 + \left(\frac{1}{2}\right)^2 m + \left(\frac{1 \cdot 3}{2 \cdot 4}\right)^2 m^2 + \left(\frac{1 \cdot 3 \cdot 5}{2 \cdot 4 \cdot 6}\right)^2 m^3 + \cdots\right\}. \tag{5.136}$$

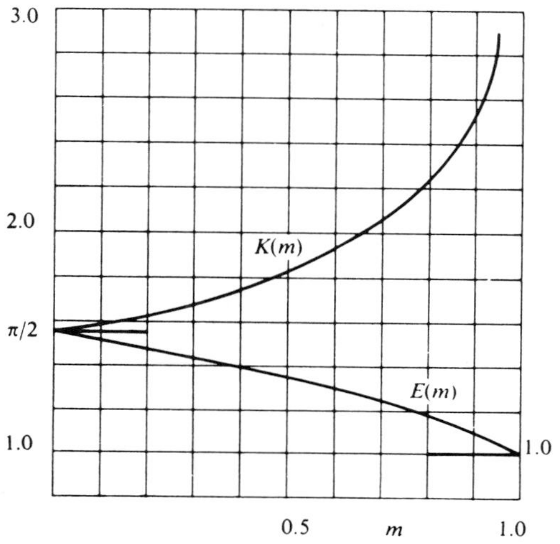

FIG. 5.8 Complete elliptic integrals, $K(m)$ and $E(m)$

Similarly,

$$E(m) = \frac{\pi}{2}\left\{1 - \left(\frac{1}{2}\right)^2 \frac{m}{1} - \left(\frac{1 \cdot 3}{2 \cdot 4}\right)^2 \frac{m^2}{3} - \left(\frac{1 \cdot 3 \cdot 5}{2 \cdot 4 \cdot 6}\right)^2 \frac{m^3}{5} - \cdots\right\}. \quad (5.137)$$

(Exercise 5.8.2). In Section 13.4 these series are identified as hypergeometric functions, and we have

$$K(m) = \frac{\pi}{2}\, {}_2F_1\left(\tfrac{1}{2}, \tfrac{1}{2}, 1; m\right) \quad (5.138)$$

$$E(m) = \frac{\pi}{2}\, {}_2F_1\left(-\tfrac{1}{2}, \tfrac{1}{2}, 1; m\right). \quad (5.139)$$

Limiting Values. From the series, Eqs. 5.136 and 5.137 or from the defining integrals,

$$\lim_{m \to 0} K(m) = \frac{\pi}{2}, \quad (5.140)$$

$$\lim_{m \to 0} E(m) = \frac{\pi}{2}. \quad (5.141)$$

For $m \to 1$, the series expansions are of little use. However, the integrals yield

$$\lim_{m \to 1} K(m) = \infty, \quad (5.142)$$

the integral diverging logarithmically, and

$$\lim_{m \to 1} E(m) = 1. \quad (5.143)$$

The elliptic integrals have been used extensively in the past for evaluating integrals. For instance, integrals of the form

$$I = \int_0^x R(t, \sqrt{a_4 t^4 + a_3 t^3 + a_2 t^2 + a_1 t + a_0})\, dt,$$

where R is a rational function of t and of the radical, may be expressed in terms of elliptic integrals. Jahnke and Emde, Chapter 5, give pages of such transformations. With high speed computers available for direct numerical evaluation, interest in these elliptic integral techniques has declined, However, elliptic integrals still remain of interest because of their appearance in physical problems—Exercises 5.8.4 and 5.8.5.

EXERCISES

5.8.1 The ellipse $x^2/a^2 + y^2/b^2 = 1$ may be represented parametrically by $x = a \sin \theta$, $y = b \cos \theta$. Show that the length of arc within the first quadrant is

$$a \int_0^{\pi/2} (1 - m \sin^2 \theta)^{1/2} \, d\theta = aE(m).$$

Here $$0 \leqslant m = (a^2 - b^2)/a^2 \leqslant 1.$$

5.8.2 Derive the series expansion

$$E(m) = \frac{\pi}{2} \left\{ 1 - \left(\frac{1}{2}\right)^2 \frac{m}{1} - \left(\frac{1 \cdot 3}{2 \cdot 4}\right)^2 \frac{m^2}{3} - \cdots \right\}$$

$$= \frac{\pi}{2} \left\{ 1 - \sum_{n=1}^{\infty} \left[\frac{(2n-1)!!}{(2n)!!} \right]^2 \frac{m^n}{(2n-1)} \right\}.$$

5.8.3 Show that

$$\lim_{m \to 0} \frac{(K - E)}{m} = \frac{\pi}{4}.$$

5.8.4 A circular loop of wire in the xy-plane, as shown, carries a current I. Given that the vector potential is

$$A_\varphi(\rho, \varphi, z) = \frac{a\mu_0 I}{2\pi} \int_0^{\pi} \frac{\cos \alpha \, d\alpha}{(a^2 + \rho^2 + z^2 - 2a\rho \cos \alpha)^{1/2}}$$

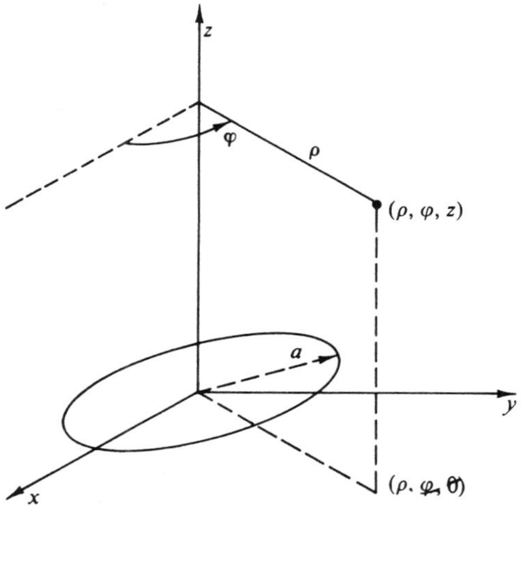

show that

$$A_\varphi(\rho, \varphi, z) = \frac{\mu_0 I}{\pi k}\left(\frac{a}{\rho}\right)^{1/2}\left[\left(1 - \frac{k^2}{2}\right)K - E\right]$$

where

$$k^2 = \frac{4a\rho}{(a+\rho)^2 + z^2}.$$

5.8.5 An analysis of the magnetic vector potential of a circular current loop leads to the expression

$$f(k^2) = k^{-2}[(2 - k^2)K(k^2) - 2E(k^2)]$$

where $K(k^2)$ and $E(k^2)$ are the complete elliptic integrals of the first and second kinds. Show that for $k^2 \ll 1$ ($r \gg$ radius of loop)

$$f(k^2) \approx \frac{\pi k^2}{16}.$$

5.9 Bernoulli Numbers

The Bernoulli numbers were introduced by Jacques (James, Jacob) Bernoulli. There are several equivalent definitions, but extreme care must be taken, for some authors introduce variations in numbering or in algebraic signs. One relatively simple approach is to define the Bernoulli numbers by the series

$$\frac{x}{e^x - 1} = \sum_{n=0}^{\infty} \frac{B_n x^n}{n!}. \tag{5.144}$$

By differentiating this power series repeatedly and then setting $x = 0$, we obtain

$$B_n = \left[\frac{d^n}{dx^n}\left(\frac{x}{e^x - 1}\right)\right]_{x=0}. \tag{5.145}$$

Specifically,

$$B_1 = \frac{d}{dx}\left(\frac{x}{e^x - 1}\right)\bigg|_{x=0} = \frac{1}{e^x - 1} - \frac{xe^x}{(e^x - 1)^2}\bigg|_{x=0}$$
$$= -\frac{1}{2}, \tag{5.146}$$

as may be seen by series expansion of the denominators.

Since these derivatives are awkward to evaluate, we may introduce instead a series expansion into the defining expression (Eq. 5.144), to obtain

$$\frac{1}{x}\left(x + \frac{x^2}{2!} + \frac{x^3}{3!} + \cdots\right)\left(B_0 + B_1 x + B_2 \frac{x^2}{2!} + \cdots\right) = 1. \tag{5.147}$$

Using the power-series uniqueness theorem (Section 5.7) with the coefficient of x^0 equal to unity and the coefficient of $x^n (n \neq 0)$ equal to zero, we obtain

$$B_0 = 1$$

$$\frac{1}{2!}B_0 + B_1 = 0, \qquad B_1 = -\frac{1}{2} \tag{5.148}$$

$$\frac{1}{3!}B_0 + \frac{1}{2!}B_1 + \frac{B_2}{2!} = 0, \qquad B_2 = \frac{1}{6}.$$

Continuing,

$$B_4 = -\frac{1}{30} \qquad B_{2n+1} = 0 \qquad n = 1, 2, 3 \cdots,$$

$$B_6 = \frac{1}{42} \tag{5.149}$$

$$B_8 = -\frac{1}{30}$$

$$B_{10} = \frac{5}{66}$$

$$B_{12} = -\frac{691}{2730}, \text{ etc.}$$

An alternate (and equivalent) definition of B_{2n} is given by the expression

$$x \cot x = \sum_{n=0}^{\infty}(-1)^n B_{2n}\frac{(2x)^{2n}}{(2n)!}, \qquad -\pi < x < \pi. \tag{5.150}$$

Using the method of residues (Section 7.3) or working from the infinite product representation of sin x (Section 5.10), we find that

$$B_{2n} = \frac{(-1)^{n-1}\,2(2n)!}{(2\pi)^{2n}}\sum_{p=1}^{\infty} p^{-2n}, \qquad n = 1, 2, 3 \cdots. \tag{5.151}$$

This representation of the Bernoulli numbers was discovered by Euler. It is readily seen from Eq. 5.151 that $|B_{2n}|$ increases without limit as $n \to \infty$. Numerical values have been calculated by Glaisher.[1] Illustrating the divergent behavior of the Bernoulli numbers, we have

$$\begin{aligned} B_{20} &= -5.291 \times 10^2 \\ B_{200} &= -3.647 \times 10^{215}. \end{aligned} \tag{5.152}$$

Some authors prefer to define the Bernoulli numbers with a modified version of Eq. 5.151 by using

$$B_n = \frac{2(2n)!}{(2\pi)^{2n}}\sum_{p=1}^{\infty} p^{-2n} \tag{5.153}$$

the subscript being just half of our subscript and all signs are positive. Again, when using other texts or references, the reader must check carefully to see exactly how the Bernoulli numbers are defined.

[1] *Trans. Cambridge Phil. Soc.* **XII**, 390 (1871–1879).

The Bernoulli numbers occur frequently in number theory. The von Standt-Clausen theorem states that

$$B_{2n} = A_n - \frac{1}{p_1} - \frac{1}{p_2} - \frac{1}{p_3} - \cdots - \frac{1}{p_k}, \tag{5.154}$$

in which A_n is an integer and $p_1, p_2, \cdots, p_k$ are prime numbers which exceed by 1 a divisor of $2n$. It may readily be verified that this holds for

$$\begin{aligned} &B_6(A_3 = 1,\ p = 2,\ 3,\ 7), \\ &B_8(A_4 = 1, p = 2, 3, 5), \\ &B_{10}(A_5 = 1, p = 2, 3, 11), \end{aligned} \tag{5.155}$$

and other special cases.

The Bernoulli numbers appear in the summation of integral powers of the integers,

$$\sum_{j=1}^{N} j^p, \qquad p \text{ integral},$$

and in numerous series expansion of the transcendental functions, including

$$\begin{gathered} \tan x, \\ \cot x, \\ \csc x, \\ \ln|\sin x|, \\ \ln|\cos x|, \\ \ln|\tan x|, \\ \tanh x, \\ \coth x, \end{gathered}$$

and

$$\operatorname{csch} x.$$

For example

$$\tan x = x + \frac{x^3}{3} + \frac{2}{15}x^5 + \cdots + \frac{(-1)^{n-1}2^{2n}(2^{2n}-1)B_{2n}}{(2n)!}x^{2n-1} + \cdots. \tag{5.156}$$

The Bernoulli numbers are likely to come in such series expansions because of the defining equations (5.144) and (5.150) and because of their relation to the Riemann zeta function

$$\zeta(2n) = \sum_{p=1}^{\infty} p^{-2n}. \tag{5.157}$$

Bernoulli functions. If Eq. 5.144 is generalized slightly, we have

$$\frac{xe^{xs}}{e^x - 1} = \sum_{n=0}^{\infty} B_n(s)\frac{x^n}{n!} \tag{5.158}$$

defining the *Bernoulli functions*, $B_n(s)$. Clearly,

$$B_n(0) = B_n, \qquad n = 0, 1, 2, \ldots, \tag{5.159}$$

the Bernoulli function evaluated at zero, equals the corresponding Bernoulli number. Two particularly important properties of the Bernoulli functions follow from the defining relation: a differentiation relation

$$B_n'(s) = nB_{n-1}(s), \qquad n = 1, 2, 3 \ldots, \tag{5.160}$$

and a symmetry relation

$$B_n(1) = (-1)^n B_n(0), \qquad n = 0, 1, 2, \ldots. \tag{5.161}$$

These relations are used in the development of the Euler-Maclaurin integration formula.

Euler-Maclaurin integration formula. One use of the Bernoulli functions is in the derivation of the Euler-Maclaurin integration formula. This formula is used in Section 10.3 for the development of an asymptotic expression for the factorial function—Stirling's series.

The technique is repeated integration by parts using Eq. 5.160 to create new derivatives. We start with

$$\int_0^1 f(x)\,dx = \int_0^1 f(x)B_0(x)\,dx. \tag{5.162}$$

From Eq. 5.160 and Exercise 5.9.2

$$B_1'(x) = B_0(x) = 1. \tag{5.163}$$

Substituting $B_1'(x)$ into Eq. 5.162 and integrating by parts,

$$\begin{aligned}\int_0^1 f(x)\,dx &= f(1)B_1(1) - f(0)B_1(0) - \int_0^1 f'(x)B_1(x)\,dx \\ &= \tfrac{1}{2}[f(1) + f(0)] - \int_0^1 f'(x)B_1(x)\,dx.\end{aligned} \tag{5.164}$$

Again using Eq. 5.160

$$B_1(x) = \tfrac{1}{2}B_2'(x) \tag{5.165}$$

and integrating by parts

$$\begin{aligned}\int_0^1 f(x)\,dx = \frac{1}{2}&[f(1) + f(0)] \\ &- \frac{1}{2!}[f'(1)B_2(1) - f'(0)B_2(0)] \\ &+ \frac{1}{2!}\int_0^1 f^{(2)}(x)B_2(x)\,dx.\end{aligned} \tag{5.166}$$

Using the relations

$$\begin{aligned} B_{2n}(1) &= B_{2n}(0) = B_{2n}, \qquad & n &= 0, 1, 2, \ldots \\ B_{2n+1}(1) &= B_{2n+1}(0) = 0, \qquad & n &= 1, 2, 3, \ldots \end{aligned} \tag{5.167}$$

and continuing this process

$$\int_0^1 f(x)\,dx = \frac{1}{2}[f(1) + f(0)] - \sum_{p=1}^{q} \frac{1}{(2p)!} B_{2p}[f^{(2p-1)}(1) - f^{(2p-1)}(0)] + \frac{1}{(2q)!}\int_0^1 f^{(2q)}(x)B_{2q}(x)\,dx. \tag{5.168a}$$

This is the Euler-Maclaurin integration formula. It assumes that the function $f(x)$ has the required derivatives.

The range of integration in Eq. 5.168*a* may be shifted from [0, 1] to [1, 2] by replacing $f(x)$ by $f(x + 1)$. Adding such results up to $[n - 1, n]$,

$$\int_0^n f(x)\,dx = \tfrac{1}{2}f(0) + f(1) + f(2) + \cdots + f(n-1) + \tfrac{1}{2}f(n) - \sum_{p=1}^{q} \frac{1}{(2p)!} B_{2p}[f^{(2p-1)}(n) - f^{(2p-1)}(0)] + \text{remainder term}. \tag{5.168b}$$

This is the form used in Ex. 5.9.5 for summing positive powers of integers and in Section 10.3 for the derivation of Stirling's formula.

Riemann zeta function. This series $\sum_{p=1}^{\infty} p^{-2n}$ was used as a comparison series for testing convergence (Section 5.2) and in Eq. 5.151 as one definition of the Bernoulli numbers, B_{2n}. It also serves to define the Riemann zeta function by

$$\zeta(s) \equiv \sum_{n=1}^{\infty} n^{-s}, \qquad s > 1. \tag{5.169}$$

Figure 5.9 is a plot of $\zeta(s) - 1$. An integral expression for this Riemann zeta function appears in Section 10.2 as part of the development of the gamma function.

Another interesting expression for the Riemann zeta function may be derived as follows:

$$\zeta(s)(1 - 2^{-s}) = 1 + \frac{1}{2^s} + \frac{1}{3^s} + \dots - \left(\frac{1}{2^s} + \frac{1}{4^s} + \frac{1}{6^s} + \cdots\right), \tag{5.170}$$

eliminating all the n^{-s} where n is a multiple of 2. Then

$$\zeta(s)(1 - 2^{-s})(1 - 3^{-s}) = 1 + \frac{1}{3^s} + \frac{1}{5^s} + \frac{1}{7^s} + \frac{1}{9^s} + \cdots - \left(\frac{1}{3^s} + \frac{1}{9^s} + \frac{1}{15^s} + \cdots\right), \tag{5.171}$$

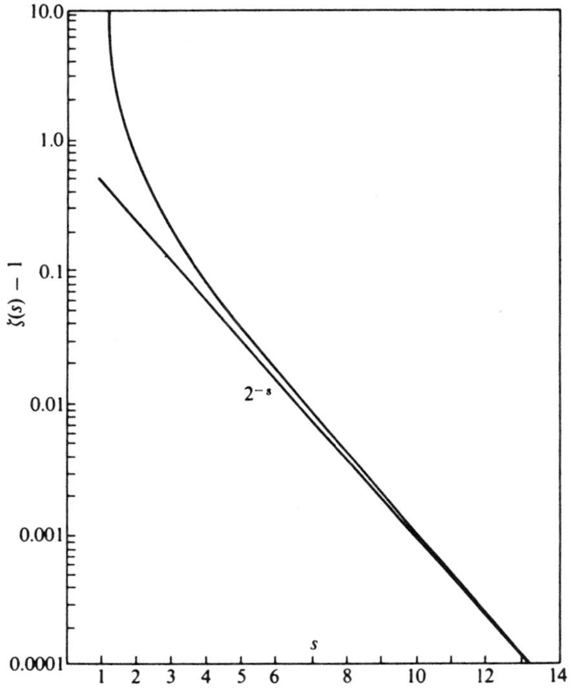

FIG. 5.9 Riemann zeta function, $\zeta(s) - 1$ versus s

eliminating all the remaining terms in which n is a multiple of 3. Continuing, $\zeta(s)(1 - 2^{-s})(1 - 3^{-s})(1 - 5^{-s})\cdots(1 - P^{-s})$, where P is a prime number, and all terms n^{-s}, in which n is a multiple of any integer up through P, are cancelled out. As $P \to \infty$,

$$\zeta(s)(1 - 2^{-s})(1 - 3^{-s})\cdots(1 - P^{-s}) = \zeta(s) \prod_{p(\text{prime})=2}^{\infty} (1 - P^{-s})$$
$$= 1. \tag{5.172}$$

Therefore

$$\zeta(s) = \left[\prod_{p(\text{prime})=2}^{\infty} (1 - P^{-s})\right]^{-1} \tag{5.173}$$

giving $\zeta(s)$ as an infinite product.

This cancellation procedure has a clear application in numerical computation. Eq. 5.170 will give $\zeta(s)(1 - 2^{-s})$ to the same accuracy as Eq. 5.169 gives $\zeta(s)$, but with only half as many terms. (In either case, a correction would be made for the neglected tail of the series by the Maclaurin integral test technique—replacing the series by an integral, Section 5.2.)

A table of values of the Riemann zeta function is included in Section 10.2.

EXERCISES

5.9.1 Show that

$$\tan x = \sum_{n=1}^{\infty} \frac{(-1)^{n-1}2^{2n}(2^{2n}-1)B_{2n}}{(2n)!} x^{2n-1}, \qquad -\frac{\pi}{2} < x < \frac{\pi}{2}.$$

Hint: $\tan x = \cot x - 2 \cot 2x$.

5.9.2 The Bernoulli numbers generated in Eq. 5.144 may be generalized to Bernoulli polynomials,

$$\frac{xe^{xs}}{e^x - 1} = \sum_{n=0}^{\infty} B_n(s) \frac{x^n}{n!}.$$

Show that

$$B_0(s) = 1$$
$$B_1(s) = s - 1/2$$
$$B_2(s) = s^2 - s + 1/6.$$

Note that $B_n(0) = B_n$, the Bernoulli number.

5.9.3 Show that $B_n'(s) = n\, B_{n-1}(s), \qquad n = 1, 2, 3, \ldots.$
Hint: Differentiate the equation in Exercise 5.9.2.

5.9.4 Show that

$$B_n(1) = (-1)^n B_n(0).$$

Hint. Go back to the generating function, Eq. 5.158 or Exercise 5.9.2.

5.9.5 The Euler-Maclaurin integration formula may be used for the evaluation of finite series:

$$\sum_{m=1}^{n} f(m) = \int_1^n f(x)\, dx + \frac{1}{2} f(1) + \frac{1}{2} f(n) + \frac{B_2}{2!} [f'(n) - f'(1)] + \cdots.$$

Show that

(a) $\sum_{m=1}^{n} m = \frac{1}{2} n(n+1).$

(b) $\sum_{m=1}^{n} m^2 = \frac{1}{6} n(n+1)(2n+1).$

(c) $\sum_{m=1}^{n} m^3 = \frac{1}{4} n^2(n+1)^2.$

(d) $\sum_{m=1}^{n} m^4 = \frac{1}{30} n(n+1)(2n+1)(3n^2+3n-1).$

5.9.6 From

$$B_{2n} = (-1)^{n-1} \frac{2(2n)!}{(2\pi)^{2n}} \zeta(2n).$$

Show that

(a) $\zeta(2) = \dfrac{\pi^2}{6}$ (d) $\zeta(8) = \dfrac{\pi^8}{9450}$

(b) $\zeta(4) = \dfrac{\pi^4}{90}$ (e) $\zeta(10) = \dfrac{\pi^{10}}{93{,}555}$

(c) $\zeta(6) = \dfrac{\pi^6}{945}$

5.9.7 Planck's black-body radiation law involves the integral

$$\int_0^\infty \frac{x^3\, dx}{e^x - 1}.$$

Show that this equals $6\,\zeta(4)$. From Exercise 5.9.6

$$\zeta(4) = \frac{\pi^4}{90}.$$

5.9.8 Prove that

$$\int_0^\infty \frac{x^n e^x\, dx}{(e^x - 1)^2} = n!\, \zeta(n).$$

Assuming n to be real, show that each side of the equation diverges if $n = 1$. Hence the above equation carries the condition $n > 1$.
Integrals such as this appear in the quantum theory of transport effects—thermal and electrical conductivity.

5.9.9 The Bloch-Gruneissen approximation for the resistance in a monovalent metal is

$$\rho = C \frac{T^5}{\Theta^6} \int_0^{\Theta/T} \frac{x^5\, dx}{(e^x - 1)(1 - e^{-x})},$$

where Θ is the Debye temperature characteristic of the metal.

(a) For $T \to \infty$ show that

$$\rho \approx \frac{C}{4} \cdot \frac{T}{\Theta^2}.$$

(b) For $T \to 0$, show that

$$\rho \approx 5!\, \zeta(5) C \frac{T^5}{\Theta^6}.$$

5.9.10 Show that

(a) $\displaystyle\int_0^1 \frac{\ln(1+x)}{x}\, dx = \frac{1}{2} \zeta(2)$ (b) $\displaystyle\lim_{a \to 1} \int_0^a \frac{\ln(1-x)}{x}\, dx = \zeta(2).$

From Exercise 5.9.6, $\zeta(2) = \pi^2/6$. Note that the integrand in part (b) diverges for $a = 1$ but that the *integrated* series is convergent.

5.9.11 The integral

$$\int_0^1 [\ln(1-x)]^2 \frac{dx}{x}$$

appears in the fourth-order correction to the magnetic moment of the electron. Show that it equals $2\,\zeta(3)$. *Hint*. Let $1 - x = e^{-t}$.

5.9.12 Show that

$$\int_0^\infty \frac{(\ln z)^2}{1 + z^2}\, dz = 4 \left(1 - \frac{1}{3^3} + \frac{1}{5^3} - \frac{1}{7^3} + \cdots\right).$$

By contour integration (Exercise 7.2.13), this may be shown equal to $\pi^3/8$.

5.9.13 For "small" values of x

$$\ln(x!) = -\gamma x + \sum_{n=2}^{\infty} (-1)^n \frac{\zeta(n)}{n} x^n,$$

where γ is the Euler-Mascheroni constant and $\zeta(n)$ the Riemann zeta function. For what values of x does this series converge? *Ans.* $-1 < x < 1$.
Note that if $x = 1$, we obtain

$$\gamma = \sum_{n=2}^{\infty} (-1)^n \frac{\zeta(n)}{n},$$

a series for the Euler-Mascheroni constant.

5.9.14 Show that the series expansion of $\ln(x!)$ (Exercise 5.9.13) may be written as

$$\text{(a) } \ln(x!) = \frac{1}{2}\ln\left(\frac{\pi x}{\sin \pi x}\right) - \gamma x - \sum_{n=1}^{\infty} \frac{\zeta(2n+1)}{2n+1} x^{2n+1}$$

$$\text{(b) } \ln(x!) = \frac{1}{2}\ln\left(\frac{\pi x}{\sin \pi x}\right) - \frac{1}{2}\ln\left(\frac{1+x}{1-x}\right) + (1-\gamma)x - \sum_{n=1}^{\infty} [\zeta(2n+1) - 1] \frac{x^{2n+1}}{2n+1}.$$

Determine the range of convergence of each of these expressions.

5.10 Infinite Products

Consider a succession of positive factors $f_1 \cdot f_2 \cdot f_3 \cdot f_4 \cdots f_n$, $(f_i > 0)$. Using capital pi to indicate product, as capital sigma indicates a sum, we have

$$f_1 \cdot f_2 \cdot f_3 \cdots f_n = \prod_{i=1}^{n} f_i. \tag{5.174}$$

We define p_n, a partial product, in analogy with s_n the partial sum,

$$p_n = \prod_{i=1}^{n} f_i \tag{5.175}$$

and then investigate the limit

$$\lim_{n\to\infty} p_n = P. \tag{5.176}$$

If P is finite (but not zero), we say the infinite product is convergent. If P is infinite *or zero*, the infinite product is labeled divergent.

Since the product will diverge to infinity if

$$\lim_{n=\infty} f_n > 1 \tag{5.177}$$

or to zero for

$$\lim_{n=\infty} f_n < 1, \qquad (\text{and } > 0), \tag{5.178}$$

it is convenient to write our infinite product as

$$\prod_{n=1}^{\infty} (1 + a_n).$$

The condition $a_n \to 0$ is then a necessary (but not sufficient) condition for convergence.

The infinite product may be related to an infinite series by the obvious method of taking the logarithm

$$\ln \prod_{n=1}^{\infty}(1 + a_n) = \sum_{n=1}^{\infty} \ln(1 + a_n). \tag{5.179}$$

A more useful relationship is stated by the following theorem.

Convergence of infinite product. If $0 \leqslant a_n < 1$, the infinite products $\prod_{n=1}^{\infty}(1 + a_n)$ and $\prod_{n=1}^{\infty}(1 - a_n)$ converge if $\sum_{n=1}^{\infty} a_n$ converges and diverge if $\sum_{n=1}^{\infty} a_n$ diverges.

Considering the term $1 + a_n$, we see from Eq. 5.90

$$1 + a_n \leqslant e^{a_n}. \tag{5.180}$$

Therefore for the partial product p_n

$$p_n \leqslant e^{s_n} \tag{5.181}$$

and, letting $n \to \infty$,

$$\prod_{n=1}^{\infty}(1 + a_n) \leqslant \exp \sum_{n=1}^{\infty} a_n, \tag{5.182}$$

thus establishing an upper bound for the infinite product.

To develop a lower bound we note that

$$p_n = 1 + \sum_{i=1}^{n} a_i + \sum_{i=1}^{n} \sum_{j=1}^{n} a_i a_j + \cdots, \qquad > s_n, \tag{5.183}$$

since $a_i \geqslant 0$. Hence

$$\prod_{n=1}^{\infty}(1 + a_n) \geqslant \sum_{n=1}^{\infty} a_n. \tag{5.184}$$

If the infinite sum remains finite, the infinite product will also. If the infinite sum diverges, so will the infinite product.

The case of $\prod(1 - a_n)$ is complicated by the negative signs, but a proof that depends on the foregoing proof may be developed by noting that for $a_n < \frac{1}{2}$ (remember $a_n \to 0$ for convergence)

$$(1 - a_n) \leqslant (1 + a_n)^{-1}$$

and

$$(1 - a_n) \geqslant (1 + 2a_n)^{-1}. \tag{5.185}$$

sin x, cos, and gamma function. The reader will recognize that an nth order polynomial $P_n(x)$ with n real roots may be written as a product of n factors

$$P_n(x) = (x - x_1)(x - x_2) \cdots (x - x_n) = \prod_{i=1}^{n}(x - x_i). \tag{5.186}$$

In much the same way we may expect that a function with an infinite number of roots may be written as an infinite product, one factor for each root. This is indeed the case for the trigonometric functions. We have the two very useful infinite product representations,

$$\sin x = x \prod_{n=1}^{\infty}\left(1 - \frac{x^2}{n^2\pi^2}\right) \tag{5.187}$$

$$\cos x = \prod_{n=1}^{\infty} \left[1 - \frac{4x^2}{(2n-1)^2\pi^2}\right]. \tag{5.188}$$

The most convenient and perhaps most elegant derivation of these two expressions is by the use of complex variables. The derivation is given in Section 7.3. By our theorem of convergence, Eqs. 5.187 and 5.188 are convergent for all finite values of x. Specifically, for the infinite product for $\sin x$, $a_n = x^2/n^2\pi^2$,

$$\sum_{n=1}^{\infty} a_n = \frac{x^2}{\pi^2} \sum_{n=1}^{\infty} n^{-2} = \frac{x^2}{\pi^2}\zeta(2)$$

$$= \frac{x^2}{6} \tag{5.189}$$

by Exercise 5.9.6. The series corresponding to Eq. 5.188 behaves in a similar manner.

Equation 5.187 leads to two interesting results. First, if we set $x = \pi/2$, we obtain

$$1 = \frac{\pi}{2} \prod_{n=1}^{\infty} \left[1 - \frac{1}{(2n)^2}\right] = \frac{\pi}{2} \prod_{n=1}^{\infty} \left[\frac{(2n)^2 - 1}{(2n)^2}\right]. \tag{5.190}$$

Solving for $\pi/2$,

$$\frac{\pi}{2} = \prod_{n=1}^{\infty} \left[\frac{(2n)^2}{(2n-1)(2n+1)}\right] = \frac{2\cdot 2}{1\cdot 3} \cdot \frac{4\cdot 4}{3\cdot 5} \cdot \frac{6\cdot 6}{5\cdot 7} \cdots, \tag{5.191}$$

which is Wallis' famous formula for $\pi/2$.

The second result involves the gamma or factorial function (Section 10.1). One definition of the gamma function is

$$\Gamma(x) = \left[x e^{\gamma x} \prod_{r=1}^{\infty} \left(1 + \frac{x}{r}\right) e^{-x/r}\right]^{-1}, \tag{5.192}$$

where γ is the usual Euler-Mascheroni constant (cf. Section 5.2). If we take the product of $\Gamma(x)$ and $\Gamma(-x)$, Eq. 5.192 leads to

$$\begin{aligned}\Gamma(x)\Gamma(-x) &= -\left[x e^{\gamma x} \prod \left(1 + \frac{x}{r}\right) e^{-x/r} x e^{-\gamma x} \prod \left(1 - \frac{x}{r}\right) e^{x/r}\right]^{-1} \\ &= -\left[x^2 \prod_{r=1}^{\infty} \left(1 - \frac{x^2}{r^2}\right)\right]^{-1}\end{aligned} \tag{5.193}$$

Using Eq. 5.187 with x replaced by πx, this becomes

$$\Gamma(x)\Gamma(-x) = -\frac{\pi}{x \sin \pi x}. \tag{5.194}$$

Anticipating a recurrence relation developed in Section 10.1, $-x\Gamma(-x) = \Gamma(1-x)$.

Eq. 5.194 may be written

$$\Gamma(x)\Gamma(1-x) = \frac{\pi}{\sin \pi x}. \tag{5.195}$$

This will be useful in treating the gamma function (Chapter 10).

Strictly speaking, we should check the range of x for which Eq. 5.192 is convergent. Clearly, individual factors will vanish for $x = 0, -1, -2, \ldots$. The proof that the infinite product converges for all other (finite) values of x is left as Ex. 5.10.7.

EXERCISES

5.10.1 Using

$$\ln \prod_{n=1}^{\infty} (1 \pm a_n) = \sum_{n=1}^{\infty} \ln (1 \pm a_n)$$

and the Maclaurin expansion of $\ln (1 \pm a_n)$, show that the infinite product $\prod_{n=1}^{\infty} (1 \pm a_n)$ converges or diverges with the infinite series $\sum_{n=1}^{\infty} a_n$.

5.10.2 Show that the infinite product representations of $\sin x$ and $\cos x$ are consistent with the identity $2 \sin x \cos x = \sin 2x$.

5.10.3 Determine the limit to which

$$\prod_{n=2}^{\infty} \left(1 + \frac{(-1)^n}{n}\right)$$

converges.

5.10.4 Show that

$$\prod_{n=2}^{\infty} \left[1 - \frac{2}{n(n+1)}\right] = \frac{1}{3}.$$

5.10.5 Using the infinite product representations of $\sin x$, show that

$$x \cot x = 1 - 2 \sum_{m,n=1}^{\infty} \left(\frac{x}{n\pi}\right)^{2m},$$

hence that the Bernoulli number

$$B_{2n} = (-1)^{n-1} \frac{2(2n)!}{(2\pi)^{2n}} \zeta(2n),$$

5.10.6 Verify the Euler identity

$$\prod_{p=1}^{\infty} (1 + z^p) = \prod_{q=1}^{\infty} (1 - z^{2q-1})^{-1}, \qquad |z| < 1.$$

5.10.7 Show that $\prod_{r=1}^{\infty} (1 + x/r)e^{-x/r}$ converges for all finite x (except for the zeros of $1 + x/r$).

Hint. Write the nth factor as $1 + a_n$.

5.11 Asymptotic or Semiconvergent Series

Incomplete gamma function. The nature of an asymptotic series is perhaps best illustrated by a specific example. Suppose that we have the logarithmic integral function[1]

$$Ei(x) = \int_{-\infty}^{x} \frac{e^u}{u} \, du, \tag{5.196}$$

or

$$-Ei(-x) = \int_{x}^{\infty} \frac{e^{-u}}{u} \, du = E_1(x), \tag{5.197}$$

to be evaluated for large values of x. Better still, let us take a generalization of the incomplete factorial function (incomplete gamma function),[2]

$$I(x, p) = \int_{x}^{\infty} e^{-u} u^{-p} \, du, \tag{5.198}$$

in which x and p are positive. Again we seek to evaluate it for large values of x.

Integrating by parts, we obtain

$$\begin{aligned} I(x, p) &= \frac{e^{-x}}{x^p} - p \int_{x}^{\infty} e^{-u} u^{-p-1} \, du \\ &= \frac{e^{-x}}{x^p} - \frac{pe^{-x}}{x^{p+1}} + p(p+1) \int_{x}^{\infty} e^{-u} u^{-p-2} \, du. \end{aligned} \tag{5.199}$$

Continuing to integrate by parts, we develop the series

$$I(x, p) = e^{-x}\left(\frac{1}{x^p} - \frac{p}{x^{p+1}} + \frac{p(p+1)}{x^{p+2}} - \cdots\right) + (-1)^n \frac{(p+n-1)!}{(p-1)!} \int_{x}^{\infty} e^{-u} u^{-p-n} \, du. \tag{5.200}$$

This is a remarkable series. Checking the convergence by the d'Alembert ratio test, we find

$$\begin{aligned} \lim_{n\to\infty} \frac{|u_{n+1}|}{|u_n|} &= \lim_{n\to\infty} \frac{(p+n)!}{(p+n-1)!} \cdot \frac{1}{x} \\ &= \lim_{n\to\infty} \frac{p+n}{x} \\ &= \infty, \end{aligned} \tag{5.201}$$

for all finite values of x. Therefore our series as an infinite series diverges everywhere! Before discarding Eq. 5.200 as worthless, let us see how well a given partial sum approximates the incomplete factorial function, $I(x, p)$.

$$I(x, p) - s_n(x, p) = (-1)^{n+1} \frac{(p+n)!}{(p-1)!} \int_{x}^{\infty} e^{-u} u^{-p-n-1} \, du. \tag{5.202}$$

[1] This function occurs frequently in astrophysical problems involving gas with a Maxwell-Boltzmann energy distribution.

[2] See also Section 10.5.

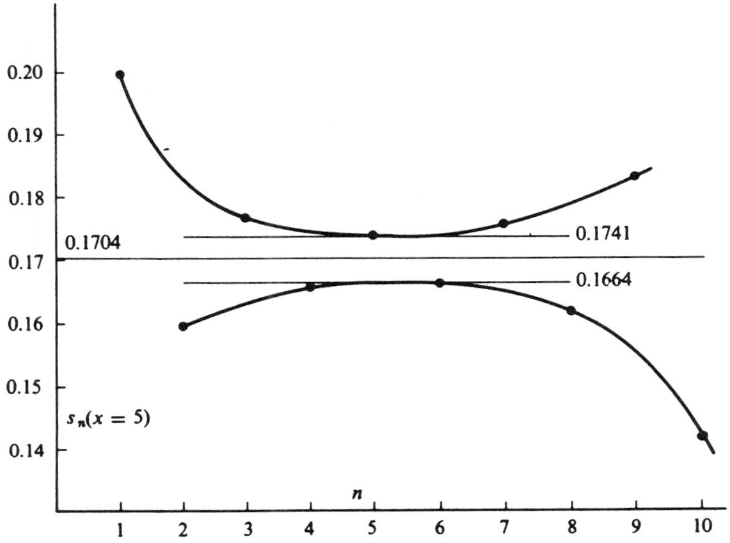

FIG. 5.10 Partial sums of $e^x E_1(x)|_{x=5}$

In absolute value

$$|I(x, p) - s_n(x, p)| \leqslant \frac{(p+n)!}{(p-1)!} \int_x^\infty u^{-p-n-1}\, du$$

$$\leqslant \frac{(p+n-1)!}{(p-1)!} \cdot \frac{1}{x^{p+n}}. \tag{5.203}$$

This means that if we take x large enough, our partial sum s_n is an arbitrarily good approximation to the desired function, $I(x, p)$. Our divergent series (Eq. 5.200), is therefore perfectly good for computations. For this reason it is sometimes called a semiconvergent series.

Since the remainder $R_n(x, p)$ alternates in sign, the successive partial sums give alternately upper and lower bounds for $I(x, p)$. The behavior of the series (with $p = 1$) as a function of the number of terms included is shown in Fig. 5.10. We have

$$e^x E_1(x) = e^x \int_x^\infty \frac{e^{-u}}{u}\, du$$

$$= \frac{1}{x} - \frac{1!}{x^2} + \frac{2!}{x^3} - \frac{3!}{x^4} + \cdots, \tag{5.204}$$

which is evaluated at $x = 5$. For a given value of x the successive upper and lower bounds given by the partial sums first converge and then diverge. The optimum determination of $e^x E_1(x)$ is then given by the closest approach of the upper and lower bounds, that is, between $s_4 = s_6 = 0.1664$ and $s_5 = 0.1741$ for $x = 5$. Therefore

$$0.1664 \leqslant e^x E_1(x)|_{x=5} \leqslant 0.1741. \tag{5.205}$$

Actually, from tables,

$$e^x E_1(x)|_{x=5} = 0.1704, \tag{5.206}$$

within the limits established by our asymptotic expansion. Inclusion of additional terms in the series expansion beyond the optimum point literally reduces the accuracy of the representation.

As x is increased, the spread between the lowest upper bound and the highest lower bound will diminish. By taking x large enough $e^x E_1(x)$ may be computed to any desired degree of accuracy. Other properties of $E_1(x)$ are derived and discussed in Section 10.5.

Cosine and sine integrals. Asymptotic series may also be developed from definite integrals—if the integrand has the required behavior. As an example, the cosine and sine integrals (Section 10.5) are defined by

$$Ci(x) = -\int_x^\infty \frac{\cos t}{t}\, dt, \tag{5.207}$$

$$si(x) = -\int_x^\infty \frac{\sin t}{t}\, dt. \tag{5.208}$$

Combining these with regular trigonometric functions, we may define

$$\begin{aligned} f(x) &= Ci(x) \sin x - si(x) \cos x = \int_0^\infty \frac{\sin y}{y + x}\, dy, \\ g(x) &= -Ci(x) \cos x - si(x) \sin x = \int_0^\infty \frac{\cos y}{y + x}\, dy, \end{aligned} \tag{5.209}$$

with the new variable $y = t - x$. Going to complex variables, Section 6.1,

$$\begin{aligned} g(x) + if(x) &= \int_0^\infty \frac{e^{iy}}{y + x}\, dy \\ &= \int_0^\infty \frac{ie^{-xu}}{1 + iu}\, du, \end{aligned} \tag{5.210}$$

in which $u = -iy/x$. The limits of integration, 0 to ∞, rather than 0 to $-i\infty$, may be justified by Cauchy's theorem, Section 6.3. Rationalizing the denominator and equating real part to real part and imaginary part to imaginary part,

$$\begin{aligned} g(x) &= \int_0^\infty \frac{ue^{-xu}}{1 + u^2}\, du, \\ f(x) &= \int_0^\infty \frac{e^{-xu}}{1 + u^2}\, du. \end{aligned} \tag{5.211}$$

For convergence of the integrals we must require that $\mathscr{R}(x) > 0$.[1]

Now, to develop the asymptotic expansions, let $v = xu$ and expand the factor $[1 + (v/x)^2]^{-1}$ by the binomial theorem.[2] We have

[1] $\mathscr{R}(x)$ = Real Part of (complex) x (cf. Section 6.1).

[2] This step is valid for $v < x$. The contributions from $v \geqslant x$ will be negligible (for large x) because of the negative exponential.

$$f(x) \approx \frac{1}{x}\int_0^\infty e^{-v}\sum_{n=0}^{\infty}(-1)^n \frac{v^{2n}}{x^{2n}}\,dv = \frac{1}{x}\sum_{n=0}^{\infty}(-1)^n\frac{(2n)!}{x^{2n}}$$

$$g(x) \approx \frac{1}{x^2}\int_0^\infty e^{-v}\sum_{n=0}^{\infty}(-1)^n \frac{v^{2n+1}}{x^{2n}}\,dv = \frac{1}{x^2}\sum_{n=0}^{\infty}(-1)^n\frac{(2n+1)!}{x^{2n}}. \qquad (5.212)$$

From Eqs. 5.209 and 5.212,

$$Ci(x) \approx \frac{\sin x}{x}\sum_{n=0}^{\infty}(-1)^n\frac{(2n)!}{x^{2n}} - \frac{\cos x}{x^2}\sum_{n=0}^{\infty}(-1)^n\frac{(2n+1)!}{x^{2n}}$$

$$si(x) \approx -\frac{\cos x}{x}\sum_{n=0}^{\infty}(-1)^n\frac{(2n)!}{x^{2n}} - \frac{\sin x}{x^2}\sum_{n=0}^{\infty}(-1)^n\frac{(2n+1)!}{x^{2n}} \qquad (5.213)$$

the desired asymptotic expansions.

This technique of expanding the integrand of a definite integral and integrating term by term is applied in Section 11.6 to develop an asymptotic expansion of the modified Bessel function K_ν and in Section 13.5 for expansions of the two confluent hypergeometric functions $M(a, c; x)$ and $U(a, c; x)$.

Definition of asymptotic series. The behavior of these series (Eqs. 5.200 and 5.213) is consistent with the defining properties of an asymptotic series. Following Poincaré, we take

$$x^n R_n(x) = x^n[f(x) - s_n(x)], \qquad (5.214)$$

where

$$s_n(x) = a_0 + \frac{a_1}{x} + \frac{a_2}{x^2} + \cdots + \frac{a_n}{x^n}. \qquad (5.215)$$

The asymptotic expansion of $f(x)$ has the properties that

$$\lim_{x\to\infty} x^n R_n(x) = 0, \qquad \text{for fixed } n, \qquad (5.216)$$

and

$$\lim_{n\to\infty} x^n R_n(x) = \infty, \qquad \text{for fixed } x.[1] \qquad (5.217)$$

With conditions (5.216) and (5.217) satisfied, we write

$$f(x) \approx \sum_{n=0}^{\infty} a_n x^{-n}. \qquad (5.218)$$

Note the use of $\approx$ in place of $=$. The function $f(x)$ is equal to the series only in the limit as $x \to \infty$.

Asymptotic expansions of two functions may be multiplied together and the result will be an asymptotic expansion of the product of the two functions.

The asymptotic expansion of a given function $f(t)$ may be integrated term by term (just as in a uniformly convergent series of continuous functions) from $x \leqslant t < \infty$ and the result will be an asymptotic expansion of $\int_x^\infty f(t)\,dt$. Term-by-

[1] This excludes convergent series of inverse powers of x. Some writers feel that this distinction, this exclusion, is artificial and unnecessary.

term differentiation, however, is valid only under very special conditions.

Some functions do not possess an asymptotic expansion; e^x is an example of such a function. However, if a function has an asymptotic expansion, it has only one. The correspondence is not one-to-one; many functions may have the same asymptotic expansion.

One of the most useful and powerful methods of generating asymptotic expansions, the method of steepest descents, will be developed in Section 7.4. Applications include the derivation of Stirling's formula for the (complete) factorial function (Section 10.3) and the asymptotic forms of the various Bessel functions (Section 11.6). Asymptotic series occur fairly often in mathematical physics. One of the earliest and still important approximation treatments of quantum mechanics, the *WKB* expansion, is an asymptotic series.

Applications to computing. Asymptotic series are frequently used in the computations of functions by modern highspeed electronic computers. This is the case for the Neumann functions $N_0(x)$ and $N_1(x)$, and the modified Bessel functions $I_n(x)$ and $K_n(x)$. The relevant asymptotic series are given as Eqs. 11.137, 11.145, and 11.147. A further discussion of these functions is included in Section 11.6. The asymptotic series for the exponential integral, Eq. 5.204; for the Fresnel integrals, Ex. 5.11.2, and for the Gauss error function, Ex. 5.11.4, are used for the evaluation of these integrals for large values of the argument. How large the argument should be depends upon what accuracy is required. In actual practice, a finite portion of the asymptotic series is telescoped using Chebyshev techniques to optimize the accuracy as discussed in Section 13.3.

EXERCISES

5.11.1 Stirling's formula for the logarithm of the factorial function is

$$\ln(x!) = \frac{1}{2}\ln 2\pi + \left(x + \frac{1}{2}\right)\ln x - x - \sum_{n=1} \frac{B_{2n}}{(2n)(2n-1)} x^{1-2n}.$$

The B_{2n} are the Bernoulli numbers (Section 5.8). Show that Stirling's formula is an *asymptotic* expansion.

5.11.2 Integrating by parts, develop asymptotic expansions of the Fresnel integrals.

$$\text{(a)}\ C(x) = \int_0^x \cos\frac{\pi u^2}{2}\,du, \qquad \text{(b)}\ S(x) = \int_0^x \sin\frac{\pi u^2}{2}\,du.$$

These integrals appear in the analysis of a knife-edge diffraction pattern.

5.11.3 Rederive the asymptotic expansions of $Ci(x)$ and $si(x)$ by repeated integration by parts.

Hint. $$Ci(x) + i\,si(x) = -\int_x^\infty \frac{e^{it}}{t}\,dt.$$

5.11.4 Derive the asymptotic expansion of the Gauss error function

$$\operatorname{erf}(x) = \frac{2}{\sqrt{\pi}} \int_0^x e^{-t^2}\, dt$$

$$= 1 - \frac{e^{-x^2}}{\sqrt{\pi x}}\left(1 - \frac{1}{2x^2} + \frac{1 \cdot 3}{2^2 x^4} - \frac{1 \cdot 3 \cdot 5}{2^3 x^6} + \cdots\right).$$

Hint. $\operatorname{erf}(x) = 1 - \operatorname{erfc}(x) = 1 - \dfrac{2}{\sqrt{\pi}} \displaystyle\int_x^\infty e^{-t^2}\, dt.$

Normalized so that $\operatorname{erf}(\infty) = 1$, this function plays an important role in probability theory. It may be expressed in terms of the Fresnel integrals (Ex. 5.11.2), the incomplete gamma functions (Section 10.5), and the confluent hypergeometric functions (Section 13.5).

5.11.5 The asymptotic expressions for the various Bessel functions, Section 11.6, contain the series

$$P_\nu(z) \sim 1 + \sum_{n=1}^{\infty} (-1)^n \frac{\prod_{s=1}^{2n} [4\nu^2 - (2s-1)^2]}{(2n)!(8z)^{2n}},$$

$$Q_\nu(z) \sim \sum_{n=1}^{\infty} (-1)^{n+1} \frac{\prod_{2n-1}^{s=1} [\nu^2 - (2s-1)^2]}{(2n-1)!(8z)^{2n-1}}.$$

Show that these two series are indeed asymptotic series.

REFERENCES

DAVIS, H. T., *Tables of Higher Mathematical Functions*. Bloomington, Indiana: Principia Press (1935). Volume II contains extensive information on Bernoulli numbers and polynomials.

HARDY, G. H., *Divergent Series*. Oxford: Clarendon Press (1956). A standard, comprehensive work on methods of treating divergent series. Hardy includes an instructive account of the gradual development of the concepts of convergence and divergence.

HYSLOP, J. M., *Infinite Series*. 5th Ed. New York: Interscience Publishers (1959). An excellent presentation of infinite series and infinite products.

KNOPP, Konrad, *Theory and Application of Infinite Series*. London: Blackie and Son (reprinted 1946). This is a thorough, comprehensive, and authoritative work, which covers infinite series and products. Proofs of almost all of the statements not proved in Chapter 5 will be found in this book.

MANGULIS, V., *Handbook of Series for Scientists and Engineers*. New York and London: Academic Press (1965). A most convenient and useful collection of series. Includes algebraic functions, Fourier series, and series of the special functions: Bessel, Legendre, etc.

SMAIL, L. L., *Elements of the Theory of Infinite Processes*. New York: McGraw-Hill (1923). An older reference but still one of the best for independent study of infinite series and infinite products.

SOKOLNIKOFF, I. S., and R. M. REDHEFFER, *Mathematics of Physics and Modern Engineering*. New York: McGraw-Hill (1958). A long Chapter 2 (101 pages) presents infinite series in a thorough but very readable form. Extensions to the solutions of differential equations, to complex series, and to Fourier series are included.

The topic of infinite series is treated in most texts on advanced calculus. Among the better references is

WIDDER, David V., *Advanced Calculus*, 2nd Ed. Englewood Cliffs, New Jersey: Prentice-Hall (1961).

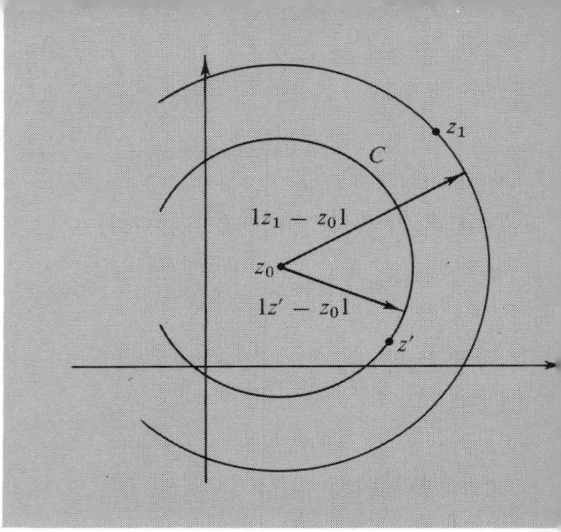

CHAPTER 6

FUNCTIONS OF A COMPLEX VARIABLE I

ANALYTIC PROPERTIES CONFORMAL MAPPING

The imaginary numbers are a wonderful flight of God's spirit; they are almost an amphibian between being and not being.
GOTTFRIED WILHELM VON LEIBNITZ, 1702

We turn now to a study of functions of a complex variable. In this area we develop some of the most powerful and widely useful tools in all of mathematical analysis. To indicate, at least partly, why complex variables are important, we mention briefly three areas of application.

1. For many pairs of functions u and v, both u and v satisfy Laplace's equation

$$\nabla^2\psi = \frac{\partial^2\psi(x, y)}{\partial x^2} + \frac{\partial^2\psi(x, y)}{\partial y^2} = 0.$$

Hence either u or v may be used to describe a two-dimensional electrostatic potential. The other function which gives a family of curves orthogonal to those of the first function, may then be used to describe the electric field $\mathbf{E}$. A similar situation holds for the hydrodynamics of an ideal fluid in irrotational motion. The function u might describe the velocity potential, whereas the function v would then be the stream function.

In many cases in which the functions u and v are unknown mapping or transforming in the complex plane permits us to create a coordinate system tailored to the particular problem.

2. It will be seen in Chapter 8 that the second-order differential equations of interest in physics may be solved by power series. The same power series may be used in the complex plane to replace x by the complex variable z. The dependence of the solution $f(z)$ at a given z_0 on the behavio of $f(z)$ elsewhere gives us greater insight into the behavior of our solution and a powerful tool (analytic continuation) for extending the region in which our solution is valid.

3. Integrals in the complex plane have a wide variety of useful applications.

a. Evaluating definite integrals.
b. Inverting power series.
c. Forming infinite products.
d. Obtaining solutions of differential equations for large values of the variable (asymptotic solutions).
e. Investigating the stability of potentially oscillatory systems.

6.1 Complex Algebra

A complex number is nothing more than an ordered pair of two ordinary numbers, (a, b) or $a + ib$, in which i is $(-1)^{1/2}$. Similarly, a complex variable is an ordered pair of two real variables,

$$z = (x, y) = x + iy. \tag{6.1}$$

The reader will see that the ordering is significant, that in general $a + ib$ is not equal to $b + ia$ and $x + iy$ is not equal to $y + ix$.[1]

It is frequently convenient to employ a graphical representation of the complex variable. By plotting x, the real part of z, as the abscissa and y, the imaginary part of z, as the ordinate, we have the complex plane or Argand plane shown in Fig. 6.1. If we assign specific values to x and y, then z corresponds to a point (x, y) in the plane. In terms of the ordering mentioned before, it is obvious that the point (x, y) does not coincide with the point (y, x) except for the special case of $x = y$.

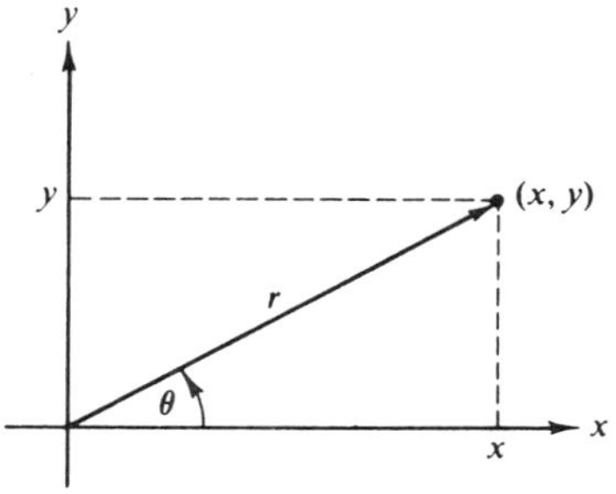

FIG. 6.1 Complex plane—Argand diagram

[1] The algebra of complex numbers, $a + ib$, is isomorphic with that of matrices of the form

$$\begin{pmatrix} a & b \\ -b & a \end{pmatrix} \qquad \text{(cf. Ex. 4.2.4).}$$

In Chapter 1, the points in the xy-plane are identified with the two-dimensional displacement vector $\mathbf{r} = \mathbf{i}x + \mathbf{j}y$. As a result, two-dimensional vector analogs can be developed for much of our complex analysis. Ex. 6.1.1 is one simple example; Cauchy's theorem, Section 6.3, is another.

Further, from Fig. 6.1, we may write

$$x = r\cos\theta$$
$$y = r\sin\theta \tag{6.2}$$

and

$$z = r(\cos\theta + i\sin\theta) \tag{6.3}$$

Using a result that is suggested (but not rigorously proved)[1] by Section 5.6, we have the very useful polar representation

$$z = re^{i\theta}. \tag{6.4}$$

In this representation r is called the modulus or magnitude of z ($r = |z|$) and the angle θ is labeled the argument or phase of z.

Analytically or graphically, using the vector analogy it may be shown that the modulus of the sum of two complex numbers is no greater than the sum of the moduli and no less than the difference, Ex. 6.1.2,

$$|z_1| - |z_2| \leqslant |z_1 + z_2| \leqslant |z_1| + |z_2|. \tag{6.5}$$

Because of the vector analogy, these are called the triangle inequalities.

Using the polar form, Eq. 6.4, the magnitude of a product is the product of the magnitudes,

$$|z_1 \cdot z_2| = |z_1| \cdot |z_2|. \tag{6.6}$$

Also,

$$\arg(z_1 \cdot z_2) = \arg z_1 + \arg z_2. \tag{6.7}$$

From our complex variable z complex functions $f(z)$ or $w(z)$ may be constructed. These complex functions may then be resolved into real and imaginary parts

$$w(z) = u(x, y) + i\,v(x, y), \tag{6.8}$$

in which the separate functions $u(x, y)$ and $v(x, y)$ are pure real. For example, if $f(z) = z^2$, we have

$$f(z) = (x + iy)^2$$
$$= (x^2 - y^2) + i\,2\,xy.$$

The real part of a function $f(z)$ will be labeled $\mathscr{R}f(z)$, while the imaginary part will be labeled $\mathscr{I}f(z)$. In Eq. 6.8 above,

$$\mathscr{R}w(z) = u(x, y),$$
$$\mathscr{I}w(z) = v(x, y).$$

[1] Strictly speaking, Chapter 5 was limited to real variables. However, we can define e^z as $\sum_{n=0}^{\infty} z^n/n!$ for complex z. The development of power series expansions for complex functions is taken up in Section 6.5 (Laurent expansion).

In all these steps, complex number, variable, and function, the operation of replacing i by $-i$ is called "taking the complex conjugate." The complex conjugate of z is denoted by z^* where[1]

$$z^* = x - iy. \tag{6.9}$$

The complex variable z and its complex conjugate z^* are mirror images of each other

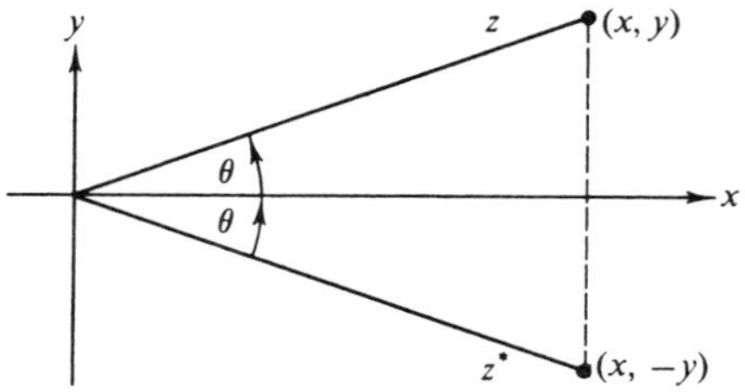

FIG. 6.2 Complex conjugate points

reflected in the x-axis, that is, inversion of the y-axis (cf. Fig. 6.2). The product zz^* leads to

$$\begin{aligned} zz^* = (x + iy)(x - iy) &= x^2 + y^2 \\ &= r^2. \end{aligned} \tag{6.10}$$

Hence

$$(zz^*)^{1/2} = |z|,$$

the magnitude of z.

Functions of a complex variable. All of the elementary functions of real variables may be extended into the complex plane—replacing the real variable x by the complex variable z. This is an example of the analytic continuation mentioned in Section 6.5. The extremely important relation, Eq. 6.4, is an illustration of this. Moving into the complex plane opens up new opportunities for analysis.

Example 6.1.1 De Moivre's Formula

If Eq. 6.3 is raised to the nth power, we have

$$e^{in\theta} = (\cos\theta + i\sin\theta)^n. \tag{6.11}$$

Expanding the exponential now with argument $n\theta$,

$$\cos n\theta + i\sin n\theta = (\cos\theta + i\sin\theta)^n. \tag{6.12}$$

This is De Moivre's formula.

Now if the right-hand side of Eq. 6.12 is expanded by the binomial theorem, we obtain $\cos n\theta$ as a series of powers of $\cos\theta$ and $\sin\theta$, Exercise 6.1.4.

Numerous other examples of relations among the exponential, hyperbolic, and trigonometric functions in the complex plane appear in the exercises.

[1] The complex conjugate is often denoted by $\bar{z}$.

Occasionally there are complications. The logarithm of a complex variable may be expanded using the polar representation

$$\begin{aligned} \ln z &= \ln re^{i\theta} \\ &= \ln r + i\theta. \end{aligned} \tag{6.13a}$$

This is not complete. To the phase angle θ, we may add any integral multiple of 2π without changing z. Hence Eq. 6.13a should read

$$\begin{aligned} \ln z &= \ln re^{i(\theta + 2n\pi)} \\ &= \ln r + i(\theta + 2n\pi). \end{aligned} \tag{6.13b}$$

The parameter n may be any integer. This means that $\ln z$ is a multivalued function having an infinite number of values for a single pair of real values r and θ. To avoid ambiguity, we usually agree to set $n = 0$ and limit the phase to an interval of length 2π such as $(-\pi, \pi)$. The line in the z-plane which is not crossed, the negative real axis in this case, is labeled a *cut line*.

Further discussion of these functions, including the logarithm, appears in Section 6.6.

EXERCISES

6.1.1 The complex quantities $a = u + iv$ and $b = x + iy$ may also be represented as two-dimensional vectors, $\mathbf{a} = \mathbf{i}u + \mathbf{j}v$, $\mathbf{b} = \mathbf{i}x + \mathbf{j}y$. Show that

$$a^*b = \mathbf{a} \cdot \mathbf{b} + i\mathbf{k} \cdot \mathbf{a} \times \mathbf{b}.$$

6.1.2 Prove algebraically that

$$|z_1| - |z_2| \leqslant |z_1 + z_2| \leqslant |z_1| + |z_2|.$$

Interpret this result in terms of vectors.

6.1.3 Show that complex numbers have square roots and that the square roots are contained in the complex plane. What are the square roots of i?

6.1.4 Show that

(a) $\cos n\theta = \cos^n \theta - \binom{n}{2}\cos^{n-2}\theta \sin^2\theta + \binom{n}{4}\cos^{n-4}\theta \sin^4\theta - \cdots.$

(b) $\sin n\theta = \binom{n}{1}\cos^{n-1}\theta \sin\theta - \binom{n}{3}\cos^{n-3}\theta \sin^3\theta + \binom{n}{5}\cos^{n-5}\theta \sin^5\theta - \cdots.$

Note. The quantities $\binom{n}{m}$ are binomial coefficients: $\binom{n}{m} = \dfrac{n!}{(n-m)!\, m!}$

6.1.5 Prove that

(a) $\sum_{n=0}^{N-1} \cos nx = \dfrac{\sin N(x/2)}{\sin x/2} \cos (N-1)\dfrac{x}{2}$, (b) $\sum_{n=0}^{N-1} \sin nx = \dfrac{\sin N(x/2)}{\sin x/2} \sin (N-1)\dfrac{x}{2}$.

These series occur in the analysis of the multiple-slit diffraction pattern. Another application is the analysis of the Gibbs phenomenon, Section 14.5.
Hint. Parts (a) and (b) may be combined to form a *geometric* series (cf. Section 5.1).

6.1.6 For $-1 < p < 1$ prove that

(a) $\sum_{n=0}^{\infty} p^n \cos nx = \dfrac{1 - p\cos x}{1 - 2p\cos x + p^2}$, (b) $\sum_{n=0}^{\infty} p^n \sin nx = \dfrac{p \sin x}{1 - 2p\cos x + p^2}$.

These series occur in the theory of the Fabry-Perot interferometer.

6.1.7 Assume that the trigonometric functions and the hyperbolic functions are defined for complex argument by the appropriate power series

$$\sin z = \sum_{n=1,\text{odd}}^{\infty} (-1)^{(n-1)/2} \frac{z^n}{n!} = \sum_{s=0}^{\infty} (-1)^s \frac{z^{2s+1}}{(2s+1)!},$$

$$\cos z = \sum_{n=0,\text{even}}^{\infty} (-1)^{n/2} \frac{z^n}{n!} = \sum_{s=0}^{\infty} (-1)^s \frac{z^{2s}}{(2s)!},$$

$$\sinh z = \sum_{n=1,\text{odd}}^{\infty} \frac{z^n}{n!} = \sum_{s=0}^{\infty} \frac{z^{2s+1}}{(2s+1)!},$$

$$\cosh z = \sum_{n=0,\text{even}}^{\infty} \frac{z^n}{n!} = \sum_{s=0}^{\infty} \frac{z^{2s}}{(2s)!},$$

(a) show that

$$i \sin z = \sinh iz, \qquad \sin iz = i \sinh z,$$

$$\cos z = \cosh iz, \qquad \cos iz = \cosh z.$$

(b) Verify that familiar functional relations, such as

$$\cosh z = \frac{e^z + e^{-z}}{2},$$

$$\sin(z_1 + z_2) = \sin z_1 \cos z_2 + \sin z_2 \cos z_1,$$

still hold in the complex plane.

6.1.8 Using the identities

$$\cos z = \frac{e^{iz} + e^{-iz}}{2},$$

$$\sin z = \frac{e^{iz} - e^{-iz}}{2i},$$

established from comparison of power series, show that
(a) $\sin(x + iy) = \sin x \cosh y + i \cos x \sinh y,$
$\cos(x + iy) = \cos x \cosh y - i \sin x \sinh y,$
(b) $|\sin z|^2 = \sin^2 x + \sinh^2 y,$
$|\cos z|^2 = \cos^2 x + \sinh^2 y.$
This demonstrates that we may have $\sin z$, $\cos z > 1$ in the complex plane.

6.1.9 From the identities in Ex. 6.1.7 and 6.1.8 show that
(a) $\sinh(x + iy) = \sinh x \cos y + i \cosh x \sin y,$
$\cosh(x + iy) = \cosh x \cos y + i \sinh x \sin y,$
(b) $|\sinh z|^2 = \sinh^2 x + \sin^2 y,$
$|\cosh z|^2 = \sinh^2 x + \cos^2 y.$

6.1.10 Show that

(a) $\tanh(z/2) = \dfrac{\sinh x + i \sin y}{\cosh x + \cos y}$

(b) $\coth(z/2) = \dfrac{\sinh x - i \sin y}{\cosh x - \cos y}.$

These relations will be useful in conformal mapping problems involving bipolar coordinates—Section 6.7 and 6.8.

6.1.11 Find all the zeros of

(a) $\sin z$, (c) $\sinh z$,
(b) $\cos z$, (d) $\cosh z$.

6.1.12 Show that

(a) $\sin^{-1} z = -i \ln(iz \pm \sqrt{1 - z^2})$, (d) $\sinh^{-1} z = \ln(z + \sqrt{z^2 + 1})$,
(b) $\cos^{-1} z = -i \ln(z \pm \sqrt{z^2 - 1})$, (e) $\cosh^{-1} z = \ln(z + \sqrt{z^2 - 1})$,
(c) $\tan^{-1} z = \dfrac{i}{2} \ln\left(\dfrac{i + z}{i - z}\right)$, (f) $\tanh^{-1} z = \dfrac{1}{2} \ln\left(\dfrac{1 + z}{1 - z}\right).$

Hint. 1. Express the trigonometric and hyperbolic functions in terms of exponentials. 2. Solve for the exponential and then for the exponent.

6.1.13 In the quantum theory of the photoionization we encounter the identity

$$\left(\frac{ia - 1}{ia + 1}\right)^{ib} = e^{-2b\cot^{-1} a},$$

in which a and b are real. Verify this identity.

6.1.14 A plane wave of light of angular frequency ω is represented by

$$e^{i\omega(t - nx/c)}$$

In a certain substance the simple real index of refraction n is replaced by the complex quantity $n - ik$. What is the effect of k on the wave? What does k correspond to physically? The generalization of a quantity from real to complex form occurs frequently in physics. Examples range from the complex Young's modulus of viscoelastic materials to the complex potential of the "cloudy crystal ball" model of the atomic nucleus.

6.1.15 We see that for the angular momentum components defined in Exercise 2.4.12

$$(L_x - iL_y) \neq (L_x + iL_y)^*$$

Explain why this occurs.

6.1.16 Show that the *phase* of $f(z) = u + iv$ is equal to the imaginary part of the logarithm of $f(z)$. Ex. 10.2.8 depends upon this result.

6.2 Cauchy-Riemann Conditions

Having established complex functions of a complex variable, we now proceed to differentiate them. The derivative of $f(z)$, like that of a real function, is defined by

$$\lim_{\delta z \to 0} \frac{f(z + \delta z) - f(z)}{z + \delta z - z} = \lim_{\delta z \to 0} \frac{\delta f(z)}{\delta z} = \frac{df}{dz} \quad \text{or} \quad f'(z), \tag{6.14}$$

provided that the limit is *independent* of the particular approach to the point z. For real variables we require that the right-hand limit ($x \to x_0$ from above) and the left-hand limit ($x \to x_0$ from below) be equal for the derivative $df(x)/dx$ to exist at $x = x_0$. Now, with z (or z_0) some point in a *plane*, our requirement that the limit be independent of the direction of approach is very restrictive.

Consider increments δx and δy of the variables x and y, respectively. Then

$$\delta z = \delta x + i\,\delta y. \tag{6.15}$$

Also

$$\delta f = \delta u + i\,\delta v, \tag{6.16}$$

so that

$$\frac{\delta f}{\delta z} = \frac{\delta u + i\,\delta v}{\delta x + i\,\delta y} \tag{6.17}$$

Let us take the limit indicated in Eq. 6.14 by two different approaches as shown in Fig. 6.3. First, with $\delta y = 0$, we let $\delta x \to 0$. Equation 6.14 yields

$$\begin{aligned} \lim_{\delta z \to 0} \frac{\delta f}{\delta z} &= \lim_{\delta x \to 0} \left(\frac{\delta u}{\delta x} + i\,\frac{\delta v}{\delta x} \right) \\ &= \frac{\partial u}{\partial x} + i\,\frac{\partial v}{\partial x}, \end{aligned} \tag{6.18}$$

assuming the partial derivatives exist. For a second approach, we set $\delta x = 0$ and then let $\delta y \to 0$. This leads to

$$\begin{aligned} \lim_{\delta z \to 0} \frac{\delta f}{\delta z} &= \lim_{\delta y \to 0} \left(-i\,\frac{\delta u}{\delta y} + \frac{\delta v}{\delta y} \right) \\ &= -i\,\frac{\partial u}{\partial y} + \frac{\partial v}{\partial y}. \end{aligned} \tag{6.19}$$

If we are to have a derivative df/dz, Eqs. 6.18 and 6.19 must be identical. Equating real parts to real parts and imaginary parts to imaginary parts (like components of vectors), we obtain

$$\frac{\partial u}{\partial x} = \frac{\partial v}{\partial y}, \qquad \frac{\partial u}{\partial y} = -\frac{\partial v}{\partial x}. \tag{6.20}$$

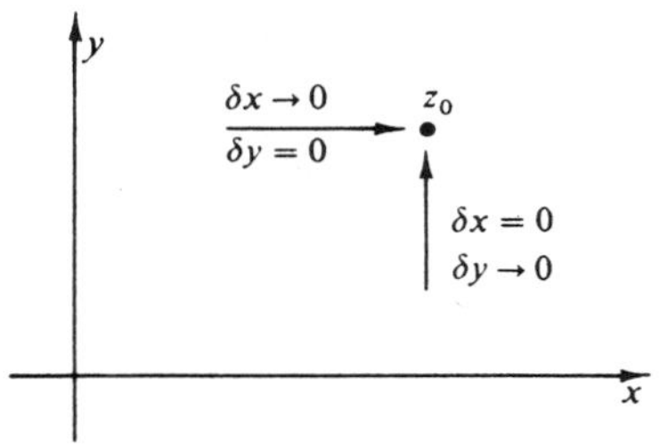

FIG. 6.3 Alternate approaches to z_0

These are the famous Cauchy-Riemann conditions. They were discovered by Cauchy and used extensively by Riemann in his theory of analytic functions. These Cauchy-Riemann conditions are necessary for the existence of a derivative of $f(z)$, that is, if df/dz exists, the Cauchy-Riemann conditions must hold.

Conversely, if the Cauchy-Riemann conditions are satisfied and the partial derivatives are continuous, the derivative exists. This may be shown by writing

$$\delta f = \left(\frac{\partial u}{\partial x} + i\frac{\partial v}{\partial x}\right)\delta x + \left(\frac{\partial u}{\partial y} + i\frac{\partial v}{\partial y}\right)\delta y \tag{6.21}$$

and then

$$\frac{\delta f}{\delta z} = \frac{\partial u/\partial x + i(\partial v/\partial x)}{1 + i(\delta y/\delta x)}\left[1 + i\frac{\delta y}{\delta x}\left(\frac{\partial v/\partial y - i(\partial u/\partial y)}{\partial u/\partial x + i(\partial v/\partial x)}\right)\right], \tag{6.22}$$

having divided numerator and denominator by the differential of x. Using the Cauchy-Riemann conditions (Eq. 6.20), we have

$$\frac{\partial v/\partial y - i(\partial u/\partial y)}{\partial u/\partial x + i(\partial v/\partial x)} = 1 \tag{6.23}$$

and

$$\frac{\delta f}{\delta z} = \frac{\partial u}{\partial x} + i\frac{\partial v}{\partial x}, \tag{6.24}$$

which shows that $\lim \delta f/\delta z$ is independent of the direction of approach in the complex plane as long as the partial derivatives are continuous.

It is well worth noting that the Cauchy-Riemann conditions guarantee that the curves $u = c_1$ will be orthogonal to the curves $v = c_2$ (cf. Section 2.1). This is fundamental in application to potential problems in a variety of areas of physics. If $u = c_1$ is a line of electric force, then $v = c_2$ is an equipotential line (surface), and vice versa. A further implication for potential theory is developed in Ex. 6.2.1.

Analytic functions. Finally, if $f(z)$ is differentiable at $z = z_0$ and in some small region around z_0, we say that $f(z)$ is *analytic* at $z = z_0$. If $f(z)$ is analytic everywhere in the (finite) complex plane we shall call it an *entire* function. Our theory of complex variables here is essentially one of analytic functions of complex variables, which points up the crucial importance of the Cauchy-Riemann conditions. The concept of analyticity carried on in advanced theories of modern physics plays a crucial role in dispersion theory (of elementary particles). If $f'(z)$ does not exist at $z = z_0$, then z_0 is labeled a singular point and consideration of it is postponed until Section 7.1.

To illustrate the Cauchy-Riemann conditions, consider two very simple examples.

EXAMPLE 6.2.1

Let $f(z) = z^2$. Then the real part $u(x, y) = x^2 - y^2$ and the imaginary part $v(x, y) = 2xy$. Following Eq. 6.20,

$$\frac{\partial u}{\partial x} = 2x = \frac{\partial v}{\partial y}, \qquad \frac{\partial u}{\partial y} = -2y = -\frac{\partial v}{\partial x}.$$

We see that $f(z) = z^2$ satisfies the Cauchy-Riemann conditions throughout the complex plane. Since the partial derivatives are clearly continuous, we conclude that $f(z) = z^2$ is analytic.

EXAMPLE 6.2.2

Let $f(z) = z^*$. Now $u = x$ and $v = -y$. Applying the Cauchy-Riemann conditions, we obtain

$$\frac{\partial u}{\partial x} = 1 \neq \frac{\partial v}{\partial y}.$$

The Cauchy-Riemann conditions are *not* satisfied and $f(z) = z^*$ is not an analytic function of z. It is interesting to note that $f(z) = z^*$ is continuous, thus providing an example of a function that is everywhere continuous but nowhere differentiable.

The derivative of a real function of a real variable is essentially a local characteristic, in that it provides information about the function only in a local neighborhood—for instance, as a truncated Taylor expansion. The existence of a derivative of a function of a complex variable has much more far-reaching implications. The real and imaginary parts of our analytic function must separately satisfy Laplace's equation. This is Exercise 6.2.1. Further, our analytic function is guaranteed derivatives of all orders, Section 6.4. In this sense, the derivative not only governs the local behavior of the complex function, but controls the distant behavior as well.

EXERCISES

6.2.1 The functions $u(x, y)$ and $v(x, y)$ are the real and imaginary parts, respectively, of an analytic function $w(z)$.

(a) Assuming that the required derivatives exist, show that

$$\nabla^2 u = \nabla^2 v = 0.$$

(b) Show that

$$\frac{\partial u}{\partial x}\frac{\partial u}{\partial y} + \frac{\partial v}{\partial x}\frac{\partial v}{\partial y} = 0.$$

and give a geometric interpretation.

6.2.2 Having shown that the real part $u(x, y)$ and the imaginary part $v(x, y)$ of an analytic function $w(z)$ each satisfy Laplace's equation, show that $u(x, y)$ and $v(x, y)$ cannot have either a maximum or a minimum in the interior of any region in which $w(z)$ is analytic. (They can have saddle points.)

6.2.3 Let $A = \partial^2 w/\partial x^2$, $B = \partial^2 w/\partial x\, \partial y$, $C = \partial^2 w/\partial y^2$. From the calculus of functions of *two* variables, $w(x, y)$, we have a saddle point if

$$B^2 - AC > 0.$$

With $f(z) = u(x, y) + iv(x, y)$, apply the Cauchy-Riemann conditions and show that neither $u(x, y)$ nor $v(x, y)$ has a maximum or a minimum in a finite region of the complex plane.

6.2.4 Find the analytic functions

$$w(z) = u(x, y) + i\, v(x, y)$$

if

(a) $u(x, y) = x^3 - 3xy^2$,

(b) $v(x, y) = e^{-y} \sin x$.

6.2.5 If there is some common region in which $w_1 = u(x, y) + iv(x, y)$ and $w_2 = w_1^* = u(x, y) - i\, v(x, y)$ are both analytic, prove that $u(x, y)$ and $v(x, y)$ are constants.

6.2.6 Using $f(re^{i\theta}) = R(r, \theta)e^{i\Theta(r,\theta)}$, in which $R(r, \theta)$ and $\Theta(r, \theta)$ are differentiable real functions of r and θ, show that the Cauchy-Riemann conditions in polar coordinates become

(a) $\dfrac{\partial R}{\partial r} = \dfrac{R}{r}\dfrac{\partial \Theta}{\partial \theta}$,

(b) $\dfrac{\partial R}{r\,\partial \theta} = -R\dfrac{\partial \Theta}{\partial r}$.

6.2.7 As an extension of Ex. 6.2.6 show that $\Theta(r, \theta)$ satisfies Laplace's equation in polar coordinates, Eq. 2.65 (without the final term).

6.2.8 Two dimensional irrotational fluid flow is conveniently described by a complex potential $f(z) = u(x, y) + iv(x, y)$. We shall label the real part $u(x, y)$, the velocity potential and the imaginary part $v(x, y)$, the stream function. The fluid velocity $\mathbf{V}$ is given by $\mathbf{V} = \nabla u$. If $f(z)$ is analytic

(a) show that $df/dz = V_x - iV_y$,

(b) show that $\nabla \cdot \mathbf{V} = 0$ (no sources or sinks),

(c) show that $\nabla \times \mathbf{V} = 0$ (irrotational, nonturbulent flow).

6.3 Cauchy's Integral Theorem

Contour integrals. With differentiation under control, we turn to integration. The integral of a complex variable over a contour in the complex plane may be defined in close analogy to the (Riemann) integral of a real function integrated along the real x-axis.

We divide the contour $z_0 z_0'$ into n intervals by picking $n - 1$ intermediate points $z_1, z_2, \ldots$ on the contour. Consider the sum

$$S_n = \sum_{j=1}^{n} f(\zeta_j)(z_j - z_{j-1}) \tag{6.25}$$

where ζ_j is a point on the curve between z_j and z_{j-1}. Now let $n \to \infty$ with

$$|z_j - z_{j-1}| \to 0$$

for all j. If the $\lim_{n\to\infty} S_n$ exists and is independent of the details of choosing the points z_j and ζ_j, then

$$\lim_{n\to\infty} \sum_{j=1}^{n} f(\zeta_j)(z_j - z_{j-1}) = \int_{z_0}^{z_0'} f(z)\, dz. \tag{6.26}$$

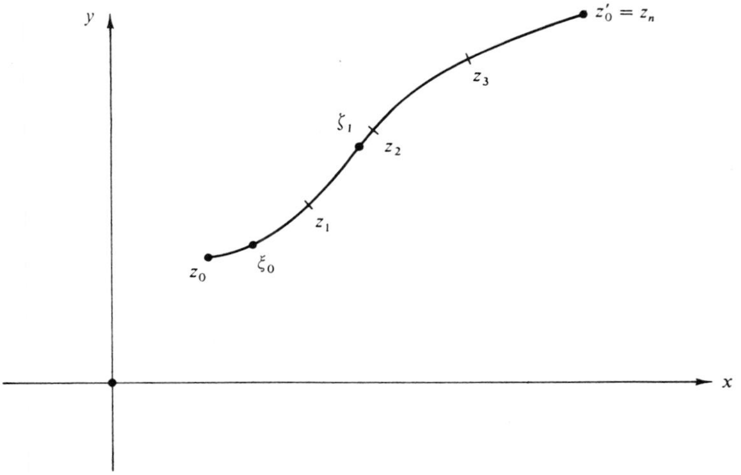

FIG. 6.4

The right-hand side of Eq. 6.26 is called the contour integral of $f(z)$ (along the specified contour C from $z = z_0$ to $z = z_0'$).

Stokes's theorem proof. Cauchy's integral theorem is the first of two basic theorems in the theory of the behavior of functions of a complex variable. First, a proof under relatively restrictive conditions—conditions that are intolerable to the mathematician developing a beautiful abstract theory but that are usually satisfied in physical problems.

If a function $f(z)$ is analytic (therefore single-valued) and its partial derivatives are continuous throughout some simply connected region R,[1] for every closed path

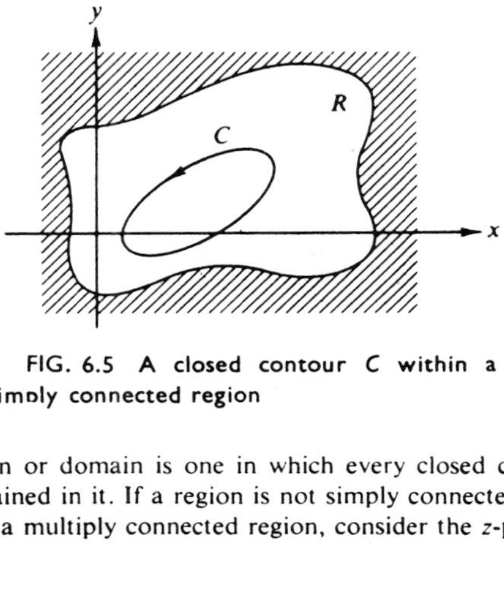

FIG. 6.5 A closed contour C within a simply connected region

[1]A simply connected region or domain is one in which every closed contour in that region encloses only the points contained in it. If a region is not simply connected, it is called multiply connected. As an example of a multiply connected region, consider the z-plane with the interior of the unit circle *excluded*.

C in R the line integral of $f(z)$ around C is zero or

$$\int_C f(z)\,dz = \oint_C f(z)\,dz = 0. \tag{6.27}$$

The symbol $\oint$ is used to emphasize that the path is closed. The reader will recall that in Section 1.13 such a function $f(z)$, identified as a force, was labeled conservative.

In this form the Cauchy integral theorem may be proved by direct application of Stokes's theorem (Section 1.12). With $f(z) = u(x, y) + i\,v(x, y)$ and $dz = dx + i\,dy$,

$$\begin{aligned}\oint_C f(z)\,dz &= \oint_C (u + iv)(dx + i\,dy)\\ &= \oint_C (u\,dx - v\,dy) + i\oint(v\,dx + u\,dy).\end{aligned} \tag{6.28}$$

These two line integrals may be converted to surface integrals by Stokes's theorem, a procedure that is justified if the partial derivatives are continuous within C. Using

$$\mathbf{V} = \mathbf{i}V_x + \mathbf{j}V_y.$$

we have

$$\oint_C (V_x\,dx + V_y\,dy) = \int\left(\frac{\partial V_y}{\partial x} - \frac{\partial V_x}{\partial y}\right)dx\,dy. \tag{6.29}$$

For the first integral in the last part of Eq. 6.28 let $u = V_x$ and $v = -V_y$.[1] Then

$$\begin{aligned}\oint_C (u\,dx - v\,dy) &= \oint_C (V_x\,dx + V_y\,dy)\\ &= \int\left(\frac{\partial V_y}{\partial x} - \frac{\partial V_x}{\partial y}\right)dx\,dy\\ &= -\int\left(\frac{\partial v}{\partial x} + \frac{\partial u}{\partial y}\right)dx\,dy.\end{aligned} \tag{6.30}$$

For the second integral on the right side of Eq. 6.28, we let $u = V_y$ and $v = V_x$. Using Stokes's theorem again, we obtain

$$\oint (v\,dx + u\,dy) = \int\left(\frac{\partial u}{\partial x} - \frac{\partial v}{\partial y}\right)dx\,dy. \tag{6.31}$$

On application of the Cauchy-Riemann conditions which must hold, since $f(z)$ is assumed analytic, each integrand vanishes and

$$\begin{aligned}\oint f(z)\,dz &= -\int\left(\frac{\partial v}{\partial x} + \frac{\partial u}{\partial y}\right)dx\,dy + i\int\left(\frac{\partial u}{\partial x} - \frac{\partial v}{\partial y}\right)dx\,dy\\ &= 0.\end{aligned} \tag{6.32}$$

[1] In the proof of Stokes's theorem, Section 1.12, V_x and V_y are any two functions (with continuous partial derivatives).

Cauchy-Goursat proof. This complete the proof of Cauchy's integral theorem. However, the proof is marred from a theoretical point of view by the need for continuity of the first partial derivatives. Actually, as shown by Goursat, this condition is not essential. An outline of the Goursat proof is as follows. We subdivide the region inside the contour C into a network of small squares as indicated in Fig. 6.6. Then

$$\oint_C f(z)\,dz = \sum_j \oint_{C_j} f(z)\,dz \tag{6.33}$$

all integrals along interior lines cancelling out. To attack the $\oint_{C_j} f(z)\,dz$ we construct the function

$$\delta_j(z, z_j) = \frac{f(z) - f(z_j)}{z - z_j} - \left.\frac{df(z)}{dz}\right|_{z=z_j} \tag{6.34}$$

with z_j an interior point of the jth subregion. Note that $[f(z) - f(z_j)]/(z - z_j)$ is an approximation to the derivative at $z = z_j$. Equivalently, we may note that if $f(z)$ had a Taylor expansion (which we have not yet proved), then $\delta_j(z, z_j)$ would be of order $z - z_j$, approaching zero as the network was made finer. We may make

$$|\delta_j(z, z_j)| < \varepsilon, \tag{6.35}$$

where ε is an arbitrarily chosen small positive quantity.

Solving Eq. 6.34 for $f(z)$ and integrating around C_j

$$\oint_{C_j} f(z)\,dz = \oint_{C_j} (z - z_j)\delta_j(z, z_j)\,dz \tag{6.36}$$

the integrals of the other terms vanishing. When Eqs. 6.35 and 6.36 are combined,

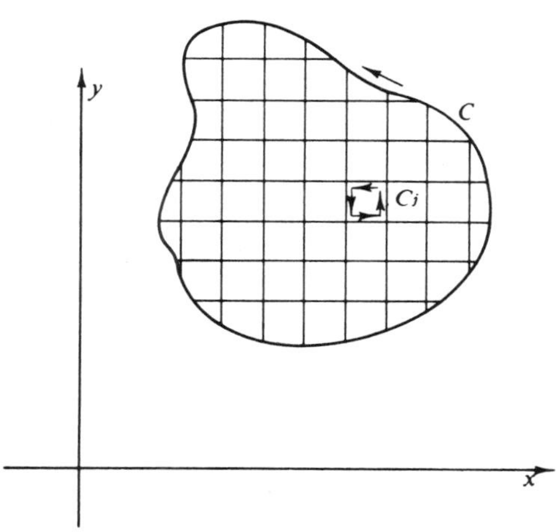

FIG. 6.6 Cauchy-Goursat contours

one may show that

$$\left| \sum_j \oint_{C_j} f(z)\,dz \right| < A\varepsilon \tag{6.37}$$

where A is a term of the order of the area of the enclosed region. Since ε is arbitrary, we let $\varepsilon \to 0$ and conclude that:

If a function $f(z)$ is analytic on and within a closed path C,

$$\oint_C f(z)\,dz = 0. \tag{6.38}$$

Details of the proof of this significantly more general and more powerful form can be found in Churchill and in the other references cited. Actually, we can still prove the theorem for $f(z)$ analytic within the interior of C and only continuous on C.

The consequence of the Cauchy integral theorem is that for analytic functions the line integral is a function only of its end points;

$$\int_{z_1}^{z_2} f(z)\,dz = F(z_2) - F(z_1) = -\int_{z_2}^{z_1} f(z)\,dz, \tag{6.39}$$

again exactly like the case of a conservative force, Section 1.13.

Multiply connected regions. The original statement of our theorem demanded a simply connected region. This restriction may easily be relaxed by the creation of a barrier, a cut line. Consider the multiply connected region of Fig. 6.7, in which $f(z)$ is not defined for the interior R'. Cauchy's integral theorem is not valid for the contour C, as shown, but we can construct a contour C' for which the theorem holds. We cut from the interior forbidden region R' to the forbidden region exterior to R and then run a new contour C', as shown in Fig. 6.8.

The new contour C' through $ABDEFGA$ never crosses the cut line which literally converts R into a simply connected region. The three-dimensional analog of this technique was used in Section 1.14 to prove Gauss's law. By Eq. 6.39

$$\int_G^A f(z)\,dz = -\int_E^D f(z)\,dz, \tag{6.40}$$

$f(z)$ having been continuous across the cut line and line segments DE and GA

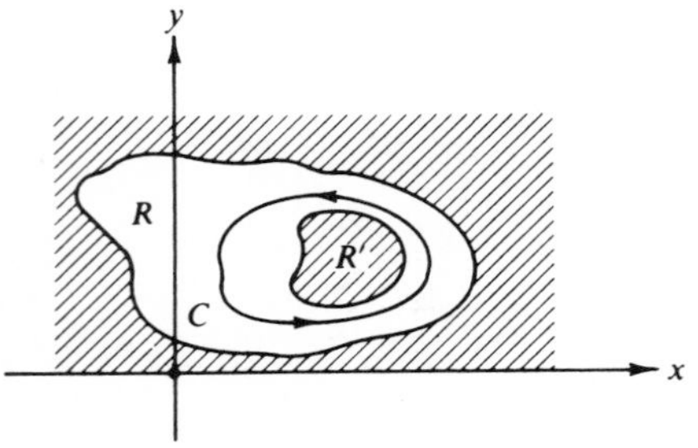

FIG. 6.7 A closed contour C in a multiply connected region

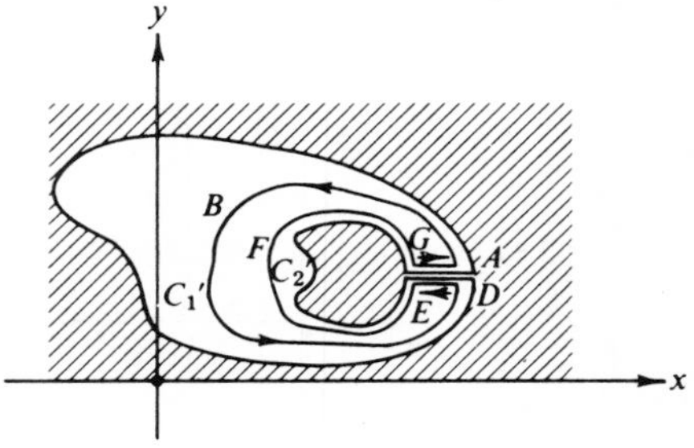

FIG. 6.8 Conversion of a multiply connected region into a simply connected region

arbitrarily close together. Then

$$\oint_{C'} f(z)\,dz = \int_{ABD} f(z)\,dz + \int_{EFG} f(z)\,dz$$
$$= 0 \qquad (6.41)$$

by Cauchy's integral theorem, with region R now simply connected. Applying Eq. 6.39 once again with $ABD \to C_1'$ and $EFG \to -C_2'$, we obtain

$$\oint_{C'_1} f(z)\,dz = \oint_{C'_2} f(z)\,dz, \qquad (6.42)$$

in which C_1' and C_2' are both traversed in the same (counterclockwise) direction.

EXERCISES

6.3.1 Prove that

$$\left|\int_C f(z)\,dz\right| \leqslant |f|_{\max} \cdot L$$

where $|f|_{\max}$ is the maximum value of $|f(z)|$ along the contour C and L is the length of the contour.

6.3.2 Verify that

$$\int_{0,0}^{1,1} z^*\,dz$$

does depend on the path by evaluating the integral for the two paths shown in Fig. 6.9. Recall that $f(z) = z^*$ is not an analytic function of z and that Cauchy's integral theorem therefore does not apply.

6.3.3 Show that

$$\oint \frac{dz}{z^2 + z} = 0,$$

in which the contour C is a circle defined by $|z| = R > 1$. In Section 7.2 it is shown that the integral yields $2\pi i$ for $R < 1$.

6.4 Cauchy's Integral Formula

As in the preceding section, we consider a function $f(z)$ that is analytic on a closed contour C and within the interior region bounded by C. We seek to prove that

$$\oint_C \frac{f(z)}{z - z_0}\,dz = 2\pi i\, f(z_0), \qquad (6.43)$$

in which z_0 is some point in the interior region bounded by C. This is the second of the two basic theorems mentioned in Section 6.2.

Although $f(z)$ is assumed analytic, the integrand is $f(z)/(z - z_0)$ and this is *not* analytic at $z = z_0$. If the contour is deformed as shown in Fig. 6.10 (or Fig. 6.8, Section 6.3), Cauchy's integral theorem applies. By Eq. 6.42

$$\oint_C \frac{f(z)}{z - z_0}\,dz = \oint_{C_2} \frac{f(z)}{z - z_0}\,dz, \tag{6.44}$$

where C is the original outer contour and C_2 is the circle surrounding the point z_0 traversed in a *counterclockwise* direction. Let $z = z_0 + re^{i\theta}$, using the polar representation because of the circular shape of the path around z_0. Here r is small and will eventually be made to approach zero. We have

$$\oint_{C_2} \frac{f(z)}{z - z_0}\,dz = \oint_{C_2} \frac{f(z_0 + re^{i\theta})}{re^{i\theta}}\,rie^{i\theta}\,d\theta.$$

Taking the limit as $r \to 0$,

$$\oint_{C_2} \frac{f(z)}{z - z_0}\,dz = i\,f(z_0)\oint_{C_2} d\theta$$

$$= 2\pi i\,f(z_0), \tag{6.45}$$

since $f(z)$ is analytic and therefore continuous at $z = z_0$. This proves the Cauchy integral formula.

Here is a remarkable result. The value of an analytic function $f(z)$ is given at an interior point $z = z_0$ once the values on the boundary C are specified. This is closely analogous to a two-dimensional form of Gauss's law (Section 1.14) in which the magnitude of an interior line charge would be given in terms of the cylindrical surface integral of the electric field **E**.

A further analogy is the determination of a function in real space by an integral of the function and the corresponding Green's function (and their derivatives) over the bounding surface. Kirchhoff diffraction theory is an example of this.

Cauchy's integral formula may be used to obtain an expression for the derivative of $f(z)$. From Eq. 6.43, with $f(z)$ analytic,

$$\frac{f(z_0 + \delta z) - f(z_0)}{\delta z} = \frac{1}{2\pi i\,\delta z}\left(\oint \frac{f(z)}{z - z_0 - \delta z}\,dz - \oint \frac{f(z)}{z - z_0}\,dz\right).$$

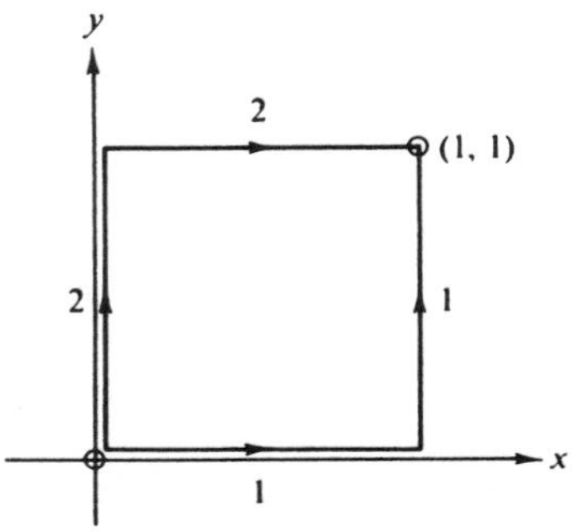

FIG. 6.9

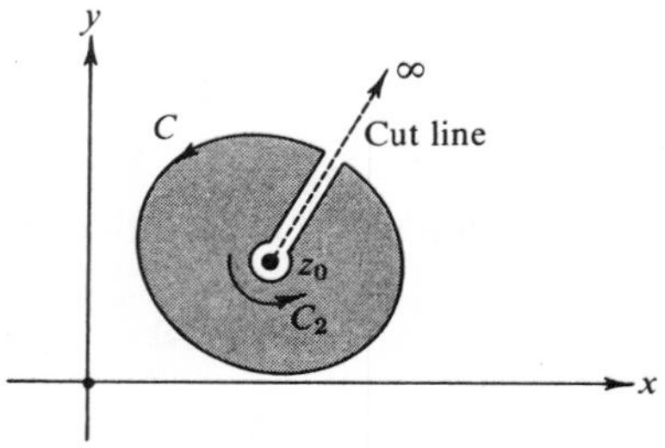

FIG. 6.10 Exclusion of a singular point

Then, by definition of derivative (Eq. 6.14),

$$f'(z_0) = \lim_{\delta z \to 0} \frac{1}{2\pi i\, \delta z} \oint \frac{\delta z\, f(z)}{(z - z_0 - \delta z)(z - z_0)}\, dz$$

$$= \frac{1}{2\pi i} \oint \frac{f(z)}{(z - z_0)^2}\, dz. \tag{6.46}$$

The alert reader will see that this result could have been obtained by differentiating Eq. 6.43 under the integral sign with respect to z_0. This formal or turning-the-crank approach is valid, but the justification for it is contained in the analysis just presented.

This process of differentiation may be repeated to obtain

$$f^{(n)}(z_0) = \frac{n!}{2\pi i} \oint \frac{f(z)\, dz}{(z - z_0)^{n+1}}; \tag{6.47}$$

that is, the requirement that $f(z)$ be analytic not only guarantees a first derivative but derivatives of *all* orders as well! The derivatives of $f(z)$ are automatically analytic. The reader should notice that this statement assumes the Goursat version of the Cauchy integral theorem. This is why Goursat's contribution is so significant in the development of the theory of complex variables.

Morera's theorem. A further application of Cauchy's integral formula is in the proof of Morera's theorem, which is the converse of Cauchy's integral theorem. The theorem states the following:

If a function $f(z)$ is continuous in a simply connected region R and $\oint_C f(z)\, dz = 0$ for every closed contour C within R, then $f(z)$ is analytic throughout R.

Let us integrate $f(z)$ from z_1 to z_2. Since every closed path integral of $f(z)$ vanishes, the integral is independent of path and depends only on its end points. We label the result of the integration $F(z)$, with

$$F(z_2) - F(z_1) = \int_{z_1}^{z_2} f(z)\, dz. \tag{6.48}$$

As an identity,

$$\frac{F(z_2) - F(z_1)}{z_2 - z_1} - f(z_1) = \frac{\int_{z_1}^{z_2} [f(t) - f(z_1)]\, dt}{z_2 - z_1}, \tag{6.49}$$

using t as another complex variable. Now we take the limit as $z_2 \to z_1$.

$$\lim_{z_2 \to z_1} \frac{\int_{z_1}^{z_2} [f(t) - f(z_1)]\, dt}{z_2 - z_1} = 0, \tag{6.50}$$

since $f(t)$ is continous.[1] Therefore

$$\lim_{z_2 \to z_1} \frac{F(z_2) - F(z_1)}{z_2 - z_1} = F'(z)\Big|_{z = z_1} = f(z_1) \tag{6.51}$$

[1] We can quote the mean value theorem of calculus here.

by definition of derivative (Eq. 6.14). We have proved that $F'(z)$ at $z = z_1$ exists and equals $f(z_1)$. Since z_1 is any point in R, we see that $F(z)$ is analytic. Then by Cauchy's integral formula (cf. Eq. 6.47) $F'(z) = f(z)$ is also analytic, proving Morera's theorem.

Drawing once more on our electrostatic analog, $f(z)$ might be used to represent the electrostatic field **E**. If the net charge within every closed region in R is zero (Gauss's law), the charge density is everywhere zero in R. Alternatively, in terms of the analysis of Section 1.13, $f(z)$ represents a conservative force (by definition of conservative), and then we find that it is always possible to express it as the derivative of a potential function $F(z)$.

Verification of Fourier inverse transform. Let $f(z)$ and $g(z)$ be analytic functions of z in the region of interest. The two functions are related by

$$f_a(z) = \frac{1}{\sqrt{2\pi}} \int_{-a}^{a} e^{izw} g(w)\, dw. \tag{6.52}$$

In Chapter 15 $f(z)$ is labeled the Fourier transform[1] of $g(w)$.

We now proceed to show, using the Cauchy integral formula, that

$$g(t) = \frac{1}{\sqrt{2\pi}} \int_{-\infty}^{\infty} e^{-izt} f_a(z)\, dz. \tag{6.53}$$

First, note that $g(w)$ is analytic. This permits a deformation of the contour of Eq. 6.52 off of the real axis. The two possibilities shown in Fig. 6.11 are used below to give convergence and to create a contour integral. Next, we evaluate $g(t)$ by substituting $f_a(z)$ from Eq. 6.52 into Eq. 6.53, obtaining

$$I = \frac{1}{2\pi} \int_{-\infty}^{\infty} e^{-izt}\, dz \int_{-a}^{a} e^{izw} g(w)\, dw.$$

The z integration is split into two parts and the order of integration is interchanged so that

$$I = \frac{1}{2\pi} \int_{-a\,C_1}^{a} g(w)\, dw \int_{-\infty}^{0} e^{iz(w-t)}\, dz + \frac{1}{2\pi} \int_{-a\,C_2}^{a} g(w)\, dw \int_{0}^{\infty} e^{iz(w-t)}\, dz, \tag{6.54}$$

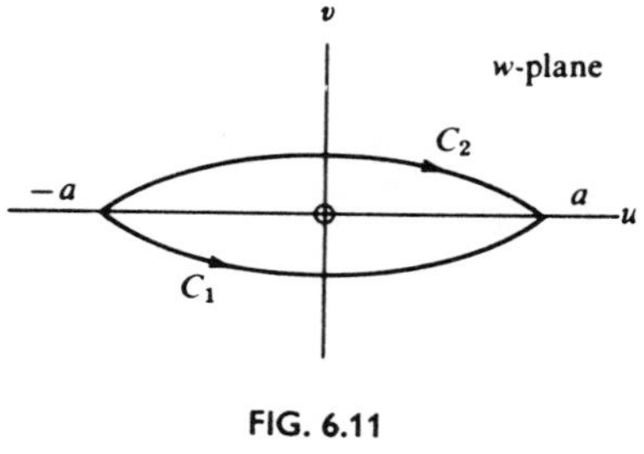

FIG. 6.11

[1] We shall take $\lim a \to \infty$.

In the first term on the right the w integration is carried over the contour C_1, whereas C_2 is chosen for the w integration in the second term. Integrating with respect to z first (because we can do it) by inspection, we obtain

$$I = \frac{1}{2\pi i}\int_{C_1} g(w)\,dw \left|\frac{e^{iz(w-t)}}{w-t}\right|_{-\infty}^{0} + \frac{1}{2\pi i}\int_{C_2} g(w)\,dw \left|\frac{e^{iz(w-t)}}{w-t}\right|_{0}^{\infty}. \tag{6.55}$$

The exponential may be written as $e^{-zv}\,e^{iz(u-t)}$, in which w is replaced by $u + iv$. Now, because of the choice of contour in the w-plane, we have $\mathscr{I}(w) = v$, negative for C_1, positive for C_2. Our choice of contour guarantees that $e^{iz(w-t)}$ vanishes as $e^{-\infty}$ for $z \to \pm\infty$. Equation 6.55 becomes

$$I = \frac{1}{2\pi i}\left[\int_{C_1}\frac{g(w)}{w-t}\,dw - \int_{C_2}\frac{g(w)}{w-t}\,dw\right]$$
$$= \frac{1}{2\pi i}\oint \frac{g(w)}{w-t}\,dw.$$

For $-a < t < a$, t is within the closed contour, and by Cauchy's integral formula

$$I = \frac{1}{\sqrt{2\pi}}\int_{-\infty}^{\infty} e^{-izt} f_a(z)\,dz = g(t). \tag{6.56}$$

This result (Eq. 6.56) (with $a \to \infty$) is the Fourier inverse transform.

Note that we have assumed $f(z)$ and $g(z)$ analytic. This restriction is actually more severe than necessary for the validity of the Fourier transform equations.

EXERCISES

6.4.1 Show that

$$\oint_C (z - z_0)^n\,dz = \begin{cases} 2\pi i, & n = -1, \\ 0, & n \neq -1, \end{cases}$$

where the contour C encircles the point $z = z_0$ in a positive (counterclockwise) sense. The exponent n is an integer.

6.4.2 Show that

$$\frac{1}{2\pi i}\oint z^{m-n-1}\,dz, \qquad m \text{ and } n \text{ integers}$$

(with the contour encircling the origin once counterclockwise), is a representation of the Kronecker delta $\delta_{m,n}$.

6.4.3 Solve Exercise 6.3.3 by separating the integrand into partial fractions and then applying Cauchy's integral theorem for multiply connected regions.

Note. Partial fractions are explained in Section 15.7 in connection with Laplace transforms.

6.4.4 Assuming that $f(z)$ is analytic on and within a closed contour C and that the point z_0 is within C, show that

$$\oint_C \frac{f'(z)}{(z - z_0)}\,dz = \oint_C \frac{f(z)}{(z - z_0)^2}\,dz.$$

6.4.5 Show that

$$|f^{(n)}(z_0)| \leqslant \frac{Mn!}{R^n},$$

where R is the radius of a circle centered at $z = z_0$ and M is the maximum value of $|f(z)|$ on that circle. Assume that $f(z)$ is analytic on and within the circle.

6.4.6 If $f(z)$ is analytic and bounded [$|f(z)| \leqslant M$, a constant] for all z, show that $f(z)$ must be a constant. This is Liouville's theorem.

6.4.7 Fundamental Theorem of Algebra. As a corollary of Liouville's theorem, Ex. 6.4.6, show that every polynomial equation,

$$P(z) = a_0 + a_1 z + \cdots + a_n z^n = 0$$

has at least one root. Here $n > 0$ and $a_n \neq 0$.
Hint: Consider $f(z) = 1/P(z)$.
Note: Once the above result is established we can divide out the root and repeat the process for the resulting $n - 1$ degree polynomial. This leads to the conclusion that $P(z)$ has exactly n roots.

6.4.8 (a) A function $f(z)$ is analytic within a closed contour C (and continuous on C). If $f(z) \neq 0$ within C and $|f(z)| \geqslant M$ on C, show that

$$|f(z)| \geqslant M$$

for all points within C. *Hint.* Consider $w(z) = 1/f(z)$.

(b) If $f(z) = 0$ within the contour C, show that the foregoing result does not hold, that it is possible to have $|f(z)| = 0$ at one or more points in the interior with $|f(z)| > 0$ over the entire bounding contour. Cite a specific example of an analytic function that behaves this way.

6.5 Laurent Expansion

Taylor expansion. The Cauchy integral formula of the preceding section opens up the way for another derivation of Taylor's series (Section 5.6) but this time for functions of a complex variable. Suppose we are trying to expand $f(z)$ about $z = z_0$ and we have $z = z_1$ as the nearest point on the Argand diagram for which $f(z)$ is not analytic. We construct a circle C centered at $z = z_0$ with radius $|z' - z_0| < |z_1 - z_0|$. Since z_1 was assumed to be the nearest point at which $f(z)$

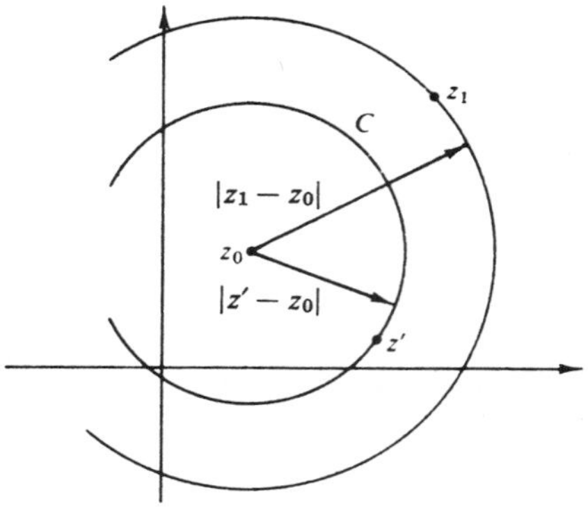

FIG. 6.12

was not analytic, $f(z)$ is necessarily analytic on and within C.

From Equation 6.43, the Cauchy integral formula,

$$\begin{aligned} f(z) &= \frac{1}{2\pi i} \oint_C \frac{f(z')\,dz'}{z'-z} \\ &= \frac{1}{2\pi i} \oint_C \frac{f(z')\,dz'}{(z'-z_0)-(z-z_0)} \\ &= \frac{1}{2\pi i} \oint_C \frac{f(z')\,dz'}{(z'-z_0)[1-(z-z_0)/(z'-z_0)]} \end{aligned} \tag{6.57}$$

Here z' is a point on the contour C and z is any point interior to C. It is not quite rigorously legal to expand the denominator of the integrand in Eq. 6.57 by the binomial theorem, for we have not yet proved the binomial theorem for complex variables. Instead, we note the identity

$$\frac{1}{1-t} = 1 + t + t^2 + t^3 + \cdots = \sum_{n=0}^{\infty} t^n, \tag{6.58}$$

which may easily be verified by multiplying both sides by $1 - t$. The infinite series, following the methods of Section 5.2, is convergent for $|t| < 1$.

Now for point z interior to C, $|z - z_0| < |z' - z_0|$, and, using Eq. 6.58, Eq. 6.57 becomes

$$f(z) = \frac{1}{2\pi i} \oint_C \sum_{n=0}^{\infty} \frac{(z-z_0)^n f(z')\,dz'}{(z'-z_0)^{n+1}} \tag{6.59}$$

Interchanging the order of integration and summation (valid since Eq. 6.58 is uniformly convergent for $|t| < 1$), we obtain

$$f(z) = \frac{1}{2\pi i} \sum_{n=0}^{\infty} (z-z_0)^n \oint_C \frac{f(z')\,dz'}{(z'-z_0)^{n+1}}. \tag{6.60}$$

Referring to Eq. 6.47, we get

$$f(z) = \sum_{n=0}^{\infty} (z-z_0)^n \frac{f^{(n)}(z_0)}{n!}, \tag{6.61}$$

which is our desired Taylor expansion. Note that it is based only on the assumption that $f(z)$ is analytic for $|z - z_0| < |z_1 - z_0|$. Just as for real variable power series (Section 5.7), this expansion is unique for a given z_0.

Schwarz reflection principle. From the binomial expansion of $g(z) = (z - x_0)^n$ for integral n, it is easy to see that the complex conjugate of the function is the function of the complex conjugate,

$$g^*(z) = (z - x_0)^{n*} = (z^* - x_0)^n = g(z^*). \tag{6.62}$$

This leads us to the Schwarz reflection principle:

If a function $f(z)$ is (1) *analytic over some region including the real axis, and* (2) *real when z is real, then*

$$f^*(z) = f(z^*). \tag{6.63}$$

Expanding $f(z)$ about some (nonsingular) point x_0 on the real axis,

$$f(z) = \sum_{n=0}^{\infty} (z - x_0)^n \frac{f^{(n)}(x_0)}{n!} \tag{6.64}$$

by Eq. 6.61. Since $f(z)$ is analytic at $z = x_0$, this Taylor expansion exists. Since $f(z)$ is real when z is real, $f^{(n)}(x_0)$ must be real for all n. Then using Eq. 6.62, Eq. 6.63, the Schwarz reflection principle, follows immediately. Exercise 6.5.6 is another form of this principle.

Analytic continuation. In the foregoing discussion we assumed that $f(z)$ has an isolated nonanalytic or singular point at $z = z_1$ (Fig. 6.12). For a specific example of this behavior consider

$$f(z) = \frac{1}{1 + z}, \tag{6.65}$$

which becomes infinite at $z = -1$. Therefore $f(z)$ is nonanalytic at $z_1 = -1$ or $z_1 = -1$ is our singular point. By Eq. 6.61 or the binomial theorem for complex functions that follows directly from it,

$$\frac{1}{1+z} = 1 - z + z^2 - z^3 + \cdots = \sum_{n=0}^{\infty} (-1)^n z^n \tag{6.66}$$

convergent for $|z| < 1$. If we label this circle of convergence C_1, Eq. 6.66 holds for $f(z)$ in the interior of C, which we label region S_1.

The situation is that $f(z)$ expanded about the origin holds only in S_1 (and on C_1 excluding $z_1 = -1$), but we know from the form of $f(z)$ that it is well defined and analytic elsewhere in the complex plane outside S_1. Analytic continuation is a process of extending the region in which a function such as the series in Eq. 6.66 is defined. For instance, suppose we expand $f(z)$ about the point $z = i$. We have

$$\begin{aligned} f(z) = \frac{1}{1 + z} &= \frac{1}{1 + i + (z - i)} \\ &= \frac{1}{1 + i}\left(1 + \frac{z - i}{1 + i}\right)^{-1} \end{aligned} \tag{6.67}$$

By Eq. 6.61 again or 6.66

$$f(z) = \frac{1}{1 + i}\left[1 - \frac{z - i}{1 + i} + \left(\frac{z - i}{1 + i}\right)^2 - \cdots\right], \tag{6.68}$$

convergent for $|z - i| < |1 + i| = \sqrt{2}$. Our circle of convergence is C_2 and the region bounded by C_2 is labeled S_2. Now $f(z)$ is defined by the expansion (Eq. 6.68) for S_2, which overlaps S_1 and extends out further in the complex plane.[1] This

[1] One of the most powerful and beautiful results of the more abstract theory of functions of a complex variable is that if two analytic functions coincide in any region, such as the overlap of S_1 and S_2, or coincide on any line segment they are the same function in the sense that they will coincide everywhere as long as they are both well defined. In this case the agreement of the expansions (Eqs. 6.66 and 6.68) over the region common to S_1 and S_2 would establish the identity of the functions these expansions represent. Then Eq. 6.68 would represent an analytic continuation or extension of $f(z)$ into regions not covered by Eq. 6.66. We could equally well say that $f(z) = 1/(1 + z)$ is itself an analytic continuation of either of the series given by Eqs. 6.66 and 6.68.

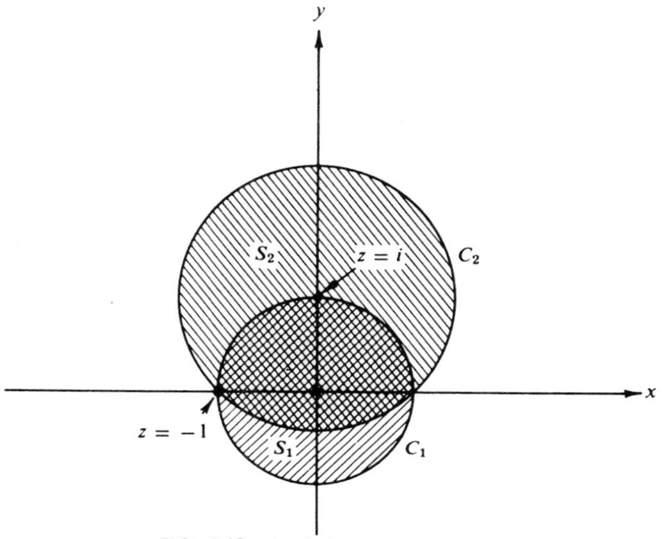

FIG. 6.13 Analytic continuation

extension is an analytic continuation, and when we have only isolated singular points to contend with the function can be extended indefinitely. As an example, we shall use a recurrence relation in Section 10.1 to extend the factorial function around the isolated singular points, $z = -n, n = 1, 2, 3 \cdots$. As another example, the hypergeometric equation is satisfied by the hypergeometric function defined by the series, Eq. 13.99, for $|z| < 1$. The integral representation given in Ex. 13.4.8 permits a continuation over the entire complex plane.

All of our elementary functions, e^z, $\sin z$, etc., can be extended into the complex plane (cf. Exercise 6.1.7). For instance, they can be *defined* by power-series expansions such as

$$e^z = 1 + \frac{z}{1!} + \frac{z^2}{2!} + \cdots = \sum_{n=0}^{\infty} \frac{z^n}{n!} \tag{6.69}$$

for the exponential. Such definitions agree with the real variable definitions along the real x-axis and literally constitute an analytic continuation of the corresponding real functions into the complex plane.

Laurent series. We frequently encounter functions that are analytic in an annular region, say of inner radius r and outer radius R, as shown in Fig. 6.14. Drawing an imaginary cut line to convert our region into a simply connected region, we apply Cauchy's integral formula, and for two circles, C_2 and C_1, centered at $z = z_0$ and with radii r_2 and r_1 respectively, where $r < r_2 < r_1 < R$, we have[1]

$$f(z) = \frac{1}{2\pi i} \oint_{C_1} \frac{f(z')\,dz'}{z' - z} - \frac{1}{2\pi i} \oint_{C_2} \frac{f(z')\,dz'}{z' - z}. \tag{6.70}$$

Note carefully that in Eq. 6.70 an explicit minus sign has been introduced so that

[1] We may take r_2 arbitrarily close to r and r_1 arbitrarily close to R, maximizing the area enclosed between C_1 and C_2.

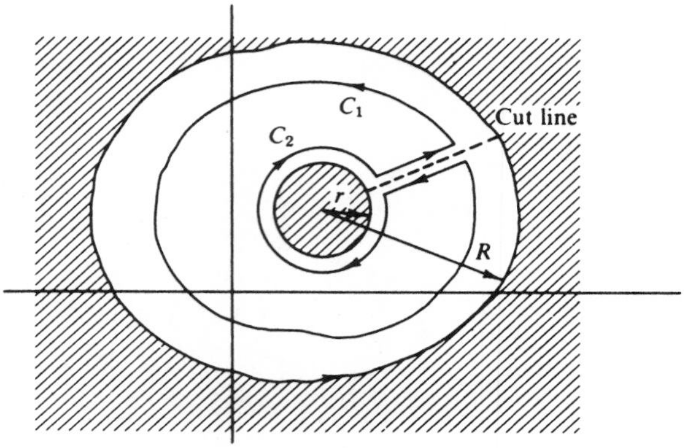

FIG. 6.14 Laurent expansion, $f(z)$ analytic for $r < |z - z_0| < R$

contour C_2 (like C_1) is to be traversed in the *positive* (counterclockwise) sense. The treatment of Eq. 6.70 now proceeds exactly like that of Eq. 6.57 in the development of the Taylor series. Each denominator is written as $(z' - z_0) - (z - z_0)$ and expanded by the binomial theorem which now follows from the Taylor series (Eq. 6.61).

Noting that for C_1, $|z' - z_0| > |z - z_0|$ while for C_2, $|z' - z_0| < |z - z_0|$, we find

$$f(z) = \frac{1}{2\pi i} \sum_{n=0}^{\infty} (z - z_0)^n \oint_{C_1} \frac{f(z')\,dz'}{(z' - z_0)^{n+1}} + \frac{1}{2\pi i} \sum_{n=1}^{\infty} (z - z_0)^{-n} \oint_{C_2} (z' - z_0)^{n-1} f(z')\,dz'. \tag{6.71}$$

The minus sign of Eq. 6.70 has been absorbed by the binomial expansion. Labeling the first series S_1 and the second S_2,

$$S_1 = \frac{1}{2\pi i} \sum_{n=0}^{\infty} (z - z_0)^n \oint_{C_1} \frac{f(z')\,dz'}{(z' - z_0)^{n+1}}, \tag{6.72}$$

which is the regular Taylor expansion, convergent for $|z - z_0| < |z' - z_0| = r_1$, that is, for all z *interior* to the larger circle, C_1. For the second series in Eq. 6.71 we have

$$S_2 = \frac{1}{2\pi i} \sum_{n=1}^{\infty} (z - z_0)^{-n} \oint_{C_2} (z' - z_0)^{n-1} f(z')\,dz' \tag{6.73}$$

convergent for $|z - z_0| > |z' - z_0| = r_2$, that is, for all z *exterior* to the smaller circle C_2. Remember, C_2 now goes counterclockwise.

These two series may be combined into one series[1] (a Laurent series) by

$$f(z) = \sum_{n=-\infty}^{\infty} a_n (z - z_0)^n, \tag{6.74}$$

where

$$a_n = \frac{1}{2\pi i} \oint_C \frac{f(z')\,dz'}{(z' - z_0)^{n+1}}. \tag{6.75}$$

[1] Replace n by $-n$ in S_2 and add.

Here C is any contour within the annular region $r < |z - z_0| < R$ encircling z_0 once in a counterclockwise sense. If we assume that such an annular region of convergence does exist, Eq. 6.74 is the Laurent series or Laurent expansion of $f(z)$.

The use of the cut line (Fig. 6.14) is convenient in converting the annular region into a simply connected region. Since our function is analytic in this annular region (and therefore single-valued), the cut line is not essential and, indeed, does not appear in the final result, Eq. 6.75. In contrast to this, functions with branch points must have cut lines—Section 7.1.

Numerous examples of Laurent series appear in Chapter 7. We limit ourselves here to one simple example to illustrate the application of Eq. 6.74.

EXAMPLE 6.5.1

Let $f(z) = [z(z-1)]^{-1}$. If we choose $z_0 = 0$, then $r = 0$ and $R = 1$, $f(z)$ diverging at $z = 1$. From Eqs. 6.75 and 6.74

$$\begin{aligned} a_n &= \frac{1}{2\pi i}\oint \frac{dz'}{(z')^{n+2}(z'-1)} \\ &= \frac{-1}{2\pi i}\oint \sum_{m=0}^{\infty} (z')^m \frac{dz'}{(z')^{n+2}}. \end{aligned} \tag{6.76}$$

Again interchanging the order of summation and integration (uniformly convergent series),

$$a_n = -\frac{1}{2\pi i}\sum_{m=0}^{\infty}\oint \frac{dz'}{(z')^{n+2-m}}. \tag{6.77}$$

If we employ the polar form (or cf. Exercise 6.4.2),

$$\begin{aligned} a_n &= -\frac{1}{2\pi i}\sum_{m=0}^{\infty}\oint \frac{rie^{i\theta}\,d\theta}{r^{n+2-m}e^{i(n+2-m)\theta}} \\ &= -\frac{1}{2\pi i}\cdot 2\pi i \sum_{m=0}^{\infty}\delta_{n+2-m,1}. \end{aligned} \tag{6.78}$$

In other words,

$$a_n = \begin{cases} -1 & \text{for} \quad n \geqslant -1, \\ 0 & \text{for} \quad n < -1. \end{cases} \tag{6.79}$$

The Laurent expansion (Eq. 6.74) becomes

$$\begin{aligned} \frac{1}{z(z-1)} &= -\frac{1}{z} - 1 - z - z^2 - z^3 - \cdots \\ &= -\sum_{n=-1}^{\infty} z^n. \end{aligned} \tag{6.80}$$

For this simple function the Laurent series can, of course, be obtained by a direct binomial expansion.

The Laurent series differs from the Taylor series by the obvious feature of negative powers of $(z - z_0)$. For this reason the Laurent series will always diverge at least at $z = z_0$ and perhaps as far out as some distance r (Fig. 6.14).

EXERCISES

6.5.1 Develop the Taylor expansion of $\ln(1 + z)$.

Ans. $\sum_{n=1}^{\infty} (-1)^{n-1} \frac{z^n}{n}$.

6.5.2 Derive the binomial expansion

$$(1 + z)^m = 1 + mz + \frac{m(m-1)}{1 \cdot 2} z^2 + \cdots$$

$$= \sum_{n=0}^{\infty} \binom{m}{n} z^n$$

for m any real number. The expansion is convergent for $|z| < 1$.

6.5.3 A function $f(z)$ is analytic on and within the unit circle. Also $|f(z)| \leqslant 1$ for $|z| \leqslant 1$ and $f(0) = 0$. Show that $|f(z)| \leqslant |z|$ for $|z| \leqslant 1$.
Hint. One approach is to show that $f(z)/z$ is analytic and then express $[f(z_0)/z_0]^n$ by the Cauchy integral formula. Finally, consider absolute magnitudes and take the nth root. This exercise is sometimes called Schwarz's theorem.

6.5.4 If $f(z)$ is a real function of the complex variable z and the Laurent expansion about the origin, $f(z) = \sum a_n z^n$, has $a_n = 0$ for $n < -N$, show that all of the coefficients, a_n, are real.

6.5.5 A function $f(z) = u(x, y) + i\,v(x, y)$ satisfies the conditions for the Schwarz reflection principle. Show that
(a) u is an even function of y;
(b) v is an odd function of y.

6.5.6 A function $f(z)$ can be expanded in a Laurent series about the origin with the coefficients a_n real. Show that the complex conjugate of this function of z is the same function of the complex conjugate of z; that is,

$$f^*(z) = f(z^*).$$

Verify this explicitly for
(a) $f(z) = z^n$, n an integer,
(b) $f(z) = \sin z$.
If $f(z) = iz$, $(a_1 = i)$, show that the foregoing statement does *not* hold.

6.5.7 Prove that the Laurent expansion of a given function about a given point is unique; that is, if

$$f(z) = \sum_{n=-N}^{\infty} a_n (z - z_0)^n = \sum_{n=-N}^{\infty} b_n (z - z_0)^n,$$

show that $a_n = b_n$ for all n. *Hint.* Use the Cauchy integral formula.

6.5.8 (a) Develop a Laurent expansion of $f(z) = [z(z-1)]^{-1}$ about the point $z = 1$ valid for small values of $|z-1|$. Specify the exact range over which your expansion holds. This is an analytic continuation of Eq. 6.80.
(b) Determine the Laurent expansion of $f(z)$ about $z = 1$ but for $|z-1|$ large.

6.6 Mapping

In the preceding sections we have defined analytic functions and developed some of their main features. From these developments the integral relations of Chapter 7 follow directly. Here we introduce some of the more geometric aspects of functions of complex variables, aspects that will be useful in visualizing the integral operations in Chapter 7 and that are valuable in their own right in solving Laplace's equation in two-dimensional systems.

In ordinary analytic geometry we may take $y = f(x)$ and then plot y versus x. Our problem here is more complicated, for z is a function of two variables x and y. We shall use the notation

$$w = f(z) = u(x, y) + i\,v(x, y). \tag{6.81}$$

Then for a point in the z-plane (specific values for x *and* y) there may correspond specific values for $u(x, y)$ and $v(x, y)$ which then yield a point in the w-plane. As points in the z-plane transform or are mapped into points in the w-plane, so lines or areas in the z-plane will be mapped into lines or areas in the w-plane. Our immediate purpose is to see how lines and areas map from the z-plane to the w-plane for a number of simple functions.

Translation.

$$w = z + z_0 \tag{6.82}$$

The function w is equal to the variable z plus a constant, $z_0 = x_0 + iy_0$. By Eqs. 6.1 and 6.81

$$\begin{aligned} u &= x + x_0, \\ v &= y + y_0, \end{aligned} \tag{6.83}$$

representing a pure translation of the coordinate axes as shown in Fig. 6.15.

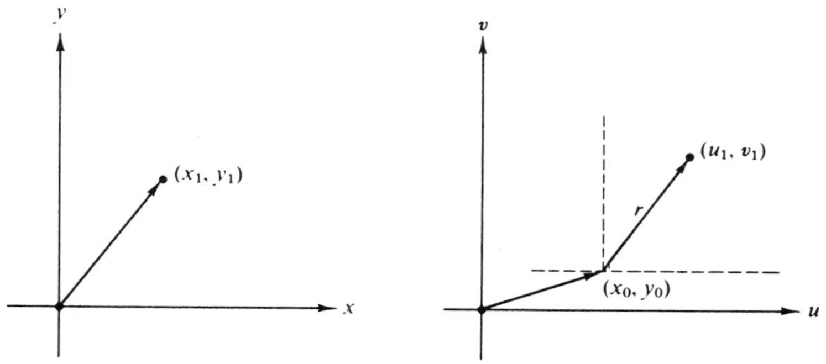

FIG. 6.15 Translation

Rotation.

$$w = zz_0. \tag{6.84}$$

Here it is convenient to return to the polar representation, using

$$w = \rho e^{i\varphi}, \qquad z = re^{i\theta}, \quad \text{and} \quad z_0 = r_0 e^{i\theta_0}, \tag{6.85}$$

then

$$\rho e^{i\varphi} = rr_0 e^{i(\theta+\theta_0)} \tag{6.86}$$

or

$$\rho = rr_0,$$
$$\varphi = \theta + \theta_0. \tag{6.87}$$

Two things have occurred. First, the modulus r has been modified, either expanded or contracted, by the factor r_0. Second, the argument θ has been increased by the additive constant θ_0. This represents a *rotation* of the complex variable through an angle θ_0. For the special case of $z_0 = i$, we have a pure rotation through $\pi/2$ radians.

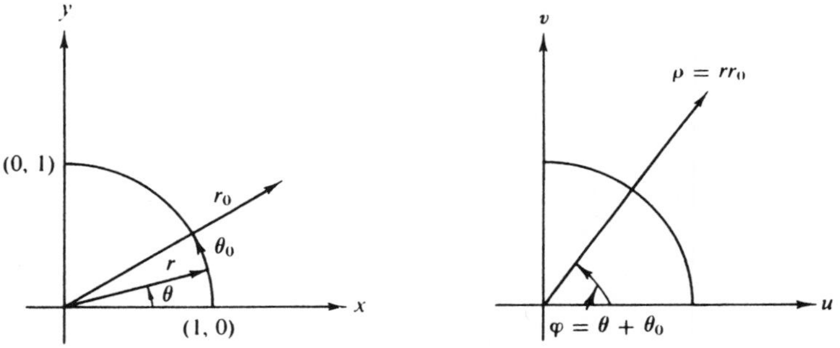

FIG. 6.16 Rotation

Inversion.

$$w = \frac{1}{z}. \tag{6.88}$$

Again, using the polar form,

$$\rho e^{i\varphi} = \frac{1}{re^{i\theta}} = \frac{1}{r} e^{-i\theta} \tag{6.89}$$

shows that

$$\rho = \frac{1}{r}, \qquad \varphi = -\theta. \tag{6.90}$$

The first part of Eq. 6.90 shows the inversion clearly. The interior of the unit circle is mapped onto the exterior and vice versa. In addition, the second part of Eq. 6.90 shows that the polar angle is reversed in sign. Equation 6.88 therefore also involves a reflection of the y-axis exactly like the complex conjugate equation.

To see how lines in the z-plane transform into the w-plane it is convenient to return to the cartesian form:

$$u + iv = \frac{1}{x + iy}. \tag{6.91}$$

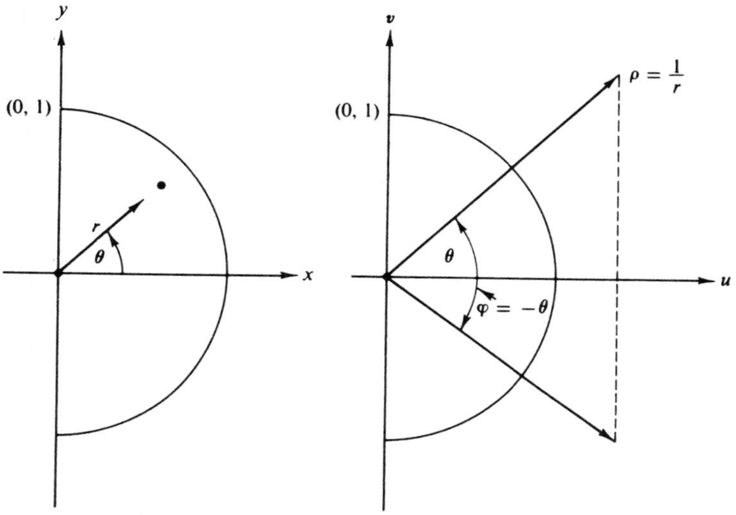

FIG. 6.17 Inversion

Rationalizing the right-hand side by multiplying numerator and denominator by z^* and then equating the real parts and the imaginary parts, we have

$$u = \frac{x}{x^2 + y^2}, \qquad x = \frac{u}{u^2 + v^2}$$
$$v = -\frac{y}{x^2 + y^2}, \qquad y = -\frac{v}{u^2 + v^2}. \tag{6.92}$$

A circle centered at the origin in the z-plane has the form

$$x^2 + y^2 = r^2 \tag{6.93}$$

and by Eq. 6.92 transforms into

$$\frac{u^2}{(u^2 + v^2)^2} + \frac{v^2}{(u^2 + v^2)^2} = r^2. \tag{6.94}$$

Simplifying Eq. 6.94, we obtain

$$u^2 + v^2 = \frac{1}{r^2} = \rho^2, \tag{6.95}$$

which describes a circle in the w-plane also centered at the origin.

The horizontal line $y = c_1$ transforms into

$$\frac{-v}{u^2 + v^2} = c_1 \tag{6.96}$$

or

$$u^2 + v^2 + \frac{v}{c_1} + \frac{1}{(2c_1)^2} = \frac{1}{(2c_1)^2}, \tag{6.97}$$

which describes a circle in the w-plane of radius $1/2c_1$ and centered at $u = 0$, $v = -1/2c_1$ (Fig. 6.18).

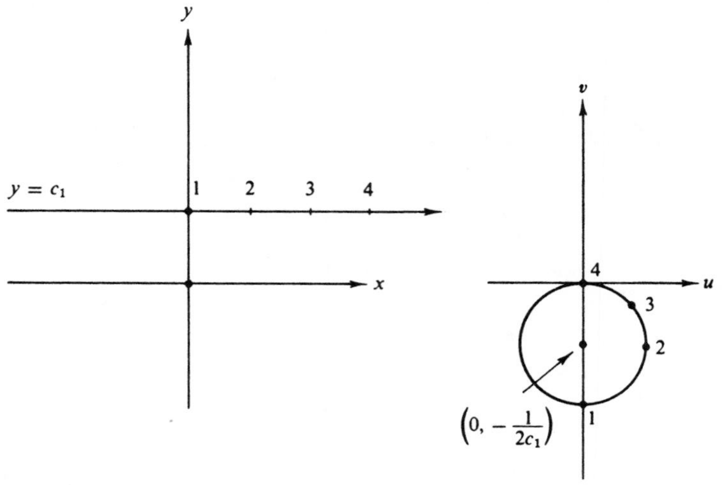

FIG. 6.18 Inversion, line ↔ circle

The reader may pick up the other three possibilities, $x = \pm c_1$, $y = -c_1$, by rotating the xy-axes. In general, any straight line or circle in the z-plane will transform into a straight line or a circle in the w-plane (cf. Exercise 6.6.1).

The three transformations just discussed have all involved one-to-one correspondence of points in the z-plane to points in the w-plane. Now to illustrate the variety of transformations possible and the problems that can arise we introduce first a two-to-one correspondence and then a many-to-one correspondence. Finally, we take up the inverses of these two transformations.

Consider first the transformation

$$w = z^2, \tag{6.98}$$

which leads to

$$\rho = r^2, \qquad \varphi = 2\theta. \tag{6.99}$$

Clearly our transformation is nonlinear, for the modulus is squared, but the significant feature of Eq. 6.99 is that the phase angle or argument is doubled. This means that the

$$\text{first quadrant of } z, \quad 0 \leqslant \theta < \frac{\pi}{2} \to \text{upper half plane of } w, \quad 0 \leqslant \varphi < \pi,$$

$$\text{upper half plane of } z, \quad 0 \leqslant \theta < \pi \to \text{whole plane of } w, \quad 0 \leqslant \varphi < 2\pi.$$

The lower half plane of z maps into the already covered entire plane of w, thus covering the w-plane a *second* time. This is our two-to-one correspondence, two distinct points in the z-plane, z_0 and $z_0 e^{i\pi} = -z_0$, corresponding to the single point $w = z_0^2$.

In cartesian representation

$$u + iv = (x + iy)^2$$
$$= x^2 - y^2 + i2xy, \tag{6.100}$$

leading to

$$u = x^2 - y^2,$$
$$v = 2xy. \tag{6.101}$$

Hence the lines $u = c_1$, $v = c_2$ in the w-plane correspond to $x^2 - y^2 = c_1$, $2xy = c_2$, rectangular (and orthogonal) hyperbolas in the z-plane (Fig. 6.19). To every point

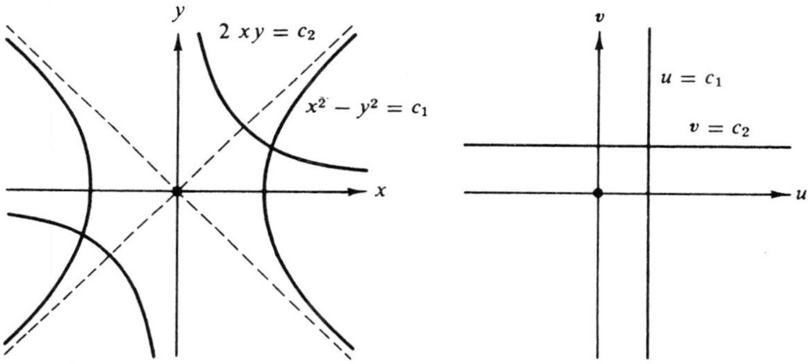

FIG. 6.19 Mapping—hyperbolic coordinates

on the hyperbola $x^2 - y^2 = c_1$ in the right half plane, $x > 0$, one point on the line $u = c_1$ corresponds and vice versa. However, every point on the line $u = c_1$ also corresponds to a point on the hyperbola $x^2 - y^2 = c_1$ in the left half plane, $x < 0$, as already explained.

It will be shown in Section 6.7 that if lines in the w-plane are orthogonal the corresponding lines in the z-plane are also orthogonal, as long as the transformation is analytic. Since $u = c_1$ and $v = c_2$ are constructed perpendicular to each other, the corresponding hyperbolas in the z-plane are orthogonal. We have literally constructed a new orthogonal system of hyperbolic lines (or surfaces if we add an axis perpendicular to x and y). Exercise 2.1.3 was an analysis of this system. It might be noted that if the hyperbolic lines are electric or magnetic lines of force, then we have a quadrupole lens useful in focusing beams of high-energy particles.

The transformation

$$w = e^z \tag{6.102}$$

leads to

$$\rho e^{i\varphi} = e^{x+iy} \tag{6.103}$$

or

$$\rho = e^x,$$
$$\varphi = y. \tag{6.104}$$

If y ranges from $0 \leqslant y < 2\pi$ (or $-\pi \leqslant y < \pi$), then φ covers the same range. But this is the whole w-plane. In other words, a horizontal strip in the z-plane of width 2π maps into the entire w-plane. Further, any point $x + i(y + 2n\pi)$, in which n is any integer, maps into the same point (by Eq. 6.104), in the w-plane. We have a many-(infinitely many)-to-one correspondence.

The inverse of the fourth transformation (Eq. 6.98) is

$$w = z^{1/2}. \tag{6.105}$$

From the relation

$$\rho e^{i\varphi} = r^{1/2}\, e^{i\theta/2}, \tag{6.106}$$

and

$$2\varphi = \theta, \tag{6.107}$$

we now have two points in the w-plane (arguments φ and $\varphi + \pi$) corresponding to one point in the z-plane(except for the point $z = 0$). Or, to put it another way, θ and $\theta + 2\pi$ correspond to φ and $\varphi + \pi$, two distinct points in the w-plane. This is the complex variable analog of the simple real variable equation $y^2 = x$, in which two values of y, plus and minus, correspond to each value of x.

The important point here is that we can make the function w of Eq. 6.105 a single-valued function instead of a double-valued function if we agree to restrict θ to a range such as $0 \leqslant \theta < 2\pi$. This may be done by agreeing never to cross the line $\theta = 0$ in the z-plane (Fig. 6.20). Such a line of demarcation is called a cut line. The point of termination ($z = 0$, here) in a multivalued function is known as a branch point. It is a form of a singular point (cf. Section 7.1), $f(z)$ not being analytic at $z = 0$.

Any line running from $z = 0$ out to infinity would serve equally well. The purpose of the cut line is to restrict the argument of z. The points z_0 and $z_0 e^{2\pi i}$ coincide in the z-plane but yield different points w and $we^{i\pi} = -w$ in the w-plane. Hence in the absence of a cut line the function $w = z^{1/2}$ is ambiguous.

We shall encounter branch points and cut lines frequently in Chapter 7.

Finally, as the inverse of the fifth transformation (Eq. 6.102) we have

$$w = \ln z. \tag{6.108}$$

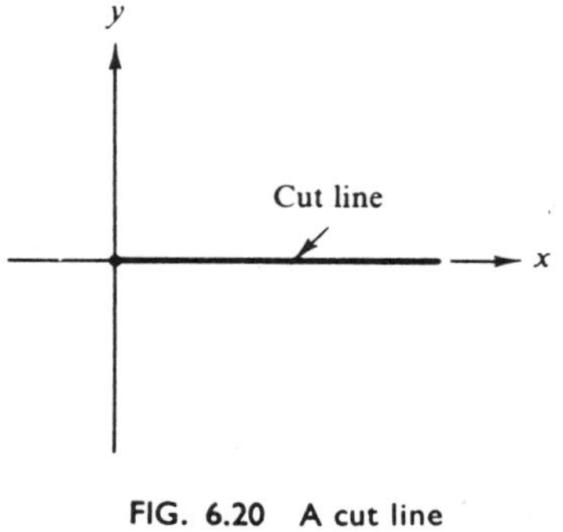

FIG. 6.20 A cut line

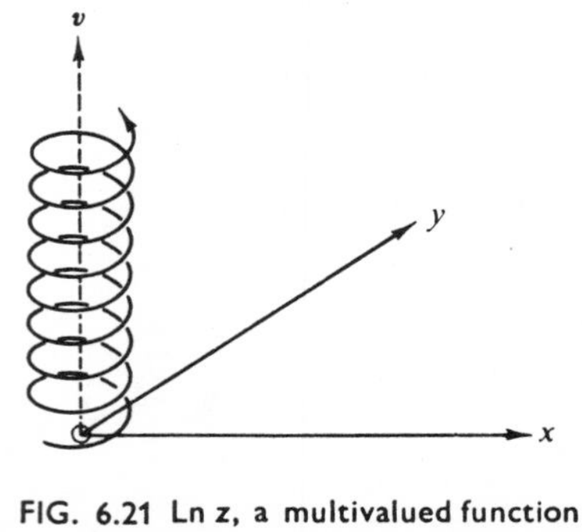

FIG. 6.21 Ln z, a multivalued function

By expanding it we obtain

$$u + iv = \ln re^{i\theta}$$
$$= \ln r + i\theta. \tag{6.109}$$

For a given point z_0 in the z-plane the argument θ is unspecified within an integral multiple of 2π. This means that

$$v = \theta + 2n\pi \tag{6.110}$$

and as in the exponential transformation we have an infinitely many-to-one correspondence.

Equation 6.108 has a nice physical representation. If we go around the unit circle in the z-plane, $r = 1$ and by Eq. 6.109 $u = \ln r = 0$; but $v = \theta$, and θ is steadily increasing and continues to increase as θ continues, past 2π. The behavior in the w-plane as we go around and around the unit circle in the z-plane is like the advance of a screw as it is rotated or the ascent of a person walking up a spiral staircase.

As in the preceding example, we make the correspondence unique (and Eq. 6.108 unambiguous) by restricting θ to a range such as $0 \leqslant \theta < 2\pi$ by taking the line $\theta = 0$ (positive real axis) as a cut line. This is equivalent to taking one and only one complete turn of the spiral staircase.

It is because of the multivalued nature of ln z that the contour integral

$$\oint \frac{dz}{z} = 2\pi i, \qquad \neq 0,$$

integrating about the origin. This property appears in Exercises 6.4.1 and 6.4.2 and is the basis for the entire calculus of residues (Chapter 7).

EXERCISES

6.6.1 How do circles centered on the origin in the z-plane transform for

(a) $w_1(z) = z + \dfrac{1}{z}$, (b) $w_2(z) = z - \dfrac{1}{z}$ $(z \neq 0)$?

What happens when $|z| \to 1$?

6.6.2 What part of the z-plane corresponds to the interior of the unit circle in the w-plane if

(a) $w = \dfrac{z-1}{z+1}$, (b) $w = \dfrac{z-i}{z+i}$?

6.6.3 Discuss the transformations

(a) $w(z) = \sin z$, (c) $w(z) = \sinh z$,

(b) $w(z) = \cos z$, (d) $w(z) = \cosh z$.

Show how the lines $x = c_1$, $y = c_2$ map into the w-plane. Note that the last three transformations can be obtained from the first one by appropriate translation and/or rotation.

6.6.4 Show that the function

$$w(z) = (z^2 - 1)^{1/2}$$

is analytic if we take
(a) $-1 \leqslant x \leqslant 1, \quad y = 0$
or
(b) $-\infty < x \leqslant -1$ and $1 \leqslant x < \infty, \quad y = 0$
as cut lines.

6.6.5 Show that negative numbers have logarithms in the complex plane. In particular, find $\ln(-1)$.
Ans. $\ln(-1) = i\pi$.

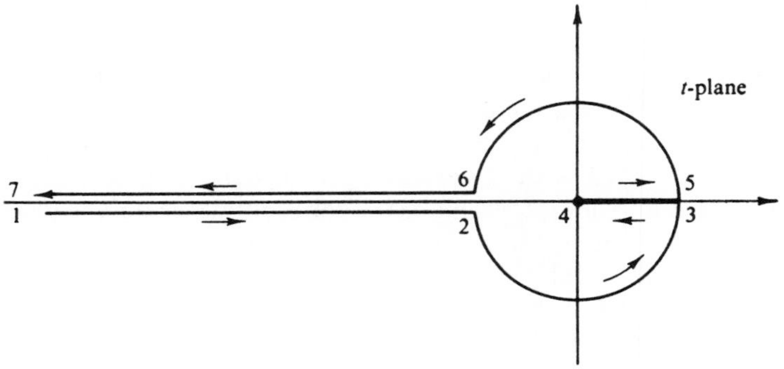

FIG. 6.22 Bessel function integration contour

6.6.6 An integral representation of the Bessel function follows the contour in the t-plane shown above. Map this contour into the θ-plane with $t = e^{\theta}$. Many additional examples of mapping are given in Chapters 11, 12, and 13.

6.7 Conformal Mapping

In Section 6.6 hyperbolas were mapped into straight lines and straight lines were mapped into circles. Yet in all these transformations one feature stayed constant. This constancy was a result of all of the transformations of Section 6.6 being analytic.

As long as $w = f(z)$ is an analytic function, we have

$$\frac{df}{dz} = \frac{dw}{dz} = \lim_{\Delta z \to 0} \frac{\Delta w}{\Delta z}. \tag{6.111}$$

Assuming that this equation is in polar form, we may equate modulus to modulus and argument to argument. For the latter (assuming that $df/dz \neq 0$)

$$\begin{aligned} \arg \lim_{\Delta z \to 0} \frac{\Delta w}{\Delta z} &= \lim_{\Delta z \to 0} \arg \frac{\Delta w}{\Delta z} \\ &= \lim_{\Delta z \to 0} \arg \Delta w - \lim_{\Delta z \to 0} \arg \Delta z \\ &= \arg \frac{df}{dz} = \alpha, \end{aligned} \tag{6.112}$$

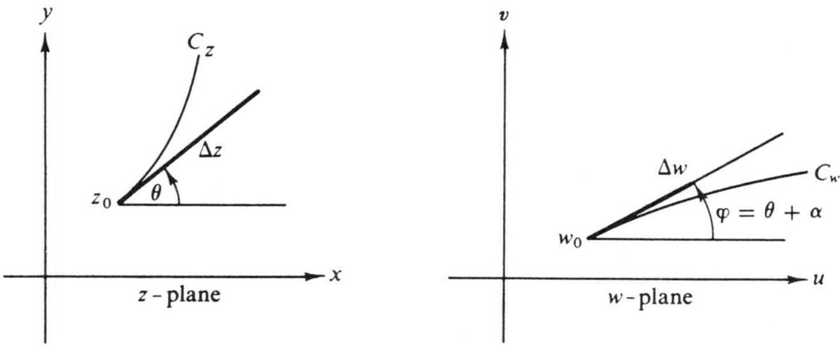

FIG. 6.23 Conformal mapping—preservation of angles

where α, the argument of the derivative, may depend on z but is a constant for a fixed z, independent of the direction of approach. To see the significance of this, consider two curves, C_z in the z-plane and the corresponding curve C_w in the w-plane (Fig. 6.23). The increment Δz is shown at an angle of θ relative to the real (x) axis whereas the corresponding increment Δw forms an angle of φ with the real (u) axis. From Eq. 6.112

$$\varphi = \theta + \alpha, \tag{6.113}$$

or any line in the z-plane is rotated through an angle α in the w-plane as long as w is an analytic transformation and the derivative is not zero.[1]

Since this result holds for any line through z_0, it will hold for a pair of lines. Then for the angle between these two lines

$$\varphi_2 - \varphi_1 = (\theta_2 + \alpha) - (\theta_1 + \alpha) = \theta_2 - \theta_1, \tag{6.114}$$

which shows that the included angle is preserved under an analytic transformation. Such angle-preserving transformations are called *conformal*. The rotation angle α will, in general, depend on z. In addition $|f'(z)|$ will, in general, be a function of z. Thus, although angles are preserved, the coordinate lines may be deformed. Indeed this is the basis for creating new coordinate systems as discussed later in this section.

The importance of conformal transformations to scientists and engineers is in their application to the solution of Laplace's equation for problems of electrostatics, hydrodynamics, heat flow, and so on.[2] Suppose we have the physically absurd but mathematically simple problem of determining the electric field lines and the equipotential lines (surfaces) between two hyperbolic surfaces with cross sections $x^2 - y^2 = c_1$ at potential V_1 and $x^2 - y^2 = c_2$ at potential V_2[3] (Fig. 6.24). We know that there will be equipotential lines and electric fields perpendicular to

[1] If $df/dz = 0$, its argument or phase is undefined and the (analytic) transformation will not necessarily preserve angles.

[2] Applications of conformal mapping also include the construction of geographical maps in which angles must be preserved for use in navigation.

[3] Our complex variable analysis is two-dimensional. The electrostatic force is three-dimensional. We reconcile them by considering a cross-sectional plane of a cylindrical system; that is, we assume a dimension t perpendicular to x and y and assume that *nothing* depends on t.

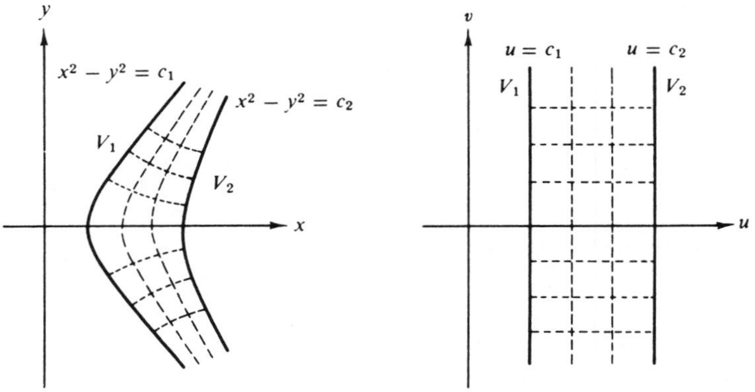

FIG. 6.24 Conformal mapping—hyperbolic coordinates

them, roughly as shown in Fig. 6.24, but with these unusual boundary conditions let us pretend that the analytical forms of the equipotentials and of the force lines are unknown. This is equivalent to saying that we do not know the solutions of Laplace's equation, $\nabla^2\psi = 0$, that satisfy these boundary conditions.

If we transform to the w-plane by $w = z^2$, (cf. Eq. 6.98 et seq.), the equipotential line V_1 becomes the vertical line $u = c_1$ and the equipotential line V_2 becomes the vertical line $u = c_2$. In the w-plane Laplace's equation may be solved by inspection to yield the rectangular net $u = c_i$ for the equipotentials and $v = c_j$ for the electric field lines. By transforming back to the z-plane we find that all the right angles are preserved, since $z = w^{1/2}$ is also analytic (except at $w = 0$). In analytic form we have

$$\begin{aligned} u &= x^2 - y^2 = c_i \quad \text{equipotentials,} \\ v &= 2xy = c_j \quad \text{electric field lines.} \end{aligned} \tag{6.115}$$

The net result is that a problem that we at least pretended to be unable to solve in the z-plane has been transformed into the w-plane. Here, with a custom-tailored coordinate system literally designed to simplify this particular problem, the problem has been solved by inspection. Finally, the solution is transformed back to the original z-plane as Eq. 6.115.

At the risk of belaboring the obvious, let us prove that a solution of Laplace's equation in the w-plane $\Psi(u, v)$ is still a solution when transformed back into the z-plane, provided only that $w = f(z)$ is analytic. As a fringe benefit, the proof points up the usefulness of the Cauchy-Riemann conditions. We have

$$\begin{aligned} &z\text{-plane, coordinates } x, y, \psi = \psi(x, y), \\ &w\text{-plane, coordinates } u, v, \Psi = \Psi(u, v), \end{aligned} \tag{6.116}$$

where the curve $\Psi(u, v) = a$ in the w-plane corresponds to the curve $\psi(x, y) = a$ in the z-plane; that is,

$$\Psi(u, v) = \Psi[u(x, y), v(x, y)] = \psi(x, y). \tag{6.117}$$

Differentiating $\Psi(u, v)$ with respect to x, we obtain

$$\frac{\partial \Psi(u, v)}{\partial x} = \frac{\partial u}{\partial x}\frac{\partial \Psi}{\partial u} + \frac{\partial v}{\partial x}\frac{\partial \Psi}{\partial v}$$

$$\frac{\partial^2 \Psi}{\partial x^2} = \frac{\partial^2 u}{\partial x^2}\frac{\partial \Psi}{\partial u} + \left(\frac{\partial u}{\partial x}\right)^2 \frac{\partial^2 \Psi}{\partial u^2} + 2\frac{\partial u}{\partial x}\frac{\partial v}{\partial x}\frac{\partial^2 \Psi}{\partial u\,\partial v} + \frac{\partial^2 v}{\partial x^2}\frac{\partial \Psi}{\partial v} + \left(\frac{\partial v}{\partial x}\right)^2 \frac{\partial^2 \Psi}{\partial v^2}. \tag{6.118}$$

A similar result holds for $\partial^2\Psi/\partial y^2$. Then

$$\frac{\partial^2 \psi(x, y)}{\partial x^2} + \frac{\partial^2 \psi(x, y)}{\partial y^2} = \frac{\partial^2 \Psi(u, v)}{\partial x^2} + \frac{\partial^2 \Psi(u, v)}{\partial y^2} = \left(\frac{\partial^2 u}{\partial x^2} + \frac{\partial^2 u}{\partial y^2}\right)\frac{\partial \Psi}{\partial u}$$
$$+ \left(\frac{\partial^2 v}{\partial x^2} + \frac{\partial^2 v}{\partial y^2}\right)\frac{\partial \Psi}{\partial v} + 2\left(\frac{\partial u}{\partial x}\frac{\partial v}{\partial x} + \frac{\partial u}{\partial y}\frac{\partial v}{\partial y}\right)\frac{\partial^2 \Psi}{\partial u\,\partial v} \tag{6.119}$$
$$+ \left[\left(\frac{\partial u}{\partial x}\right)^2 + \left(\frac{\partial u}{\partial y}\right)^2\right]\frac{\partial^2 \Psi}{\partial u^2} + \left[\left(\frac{\partial v}{\partial x}\right)^2 + \left(\frac{\partial v}{\partial y}\right)^2\right]\frac{\partial^2 \Psi}{\partial v^2}.$$

On the right-hand side of Eq. 6.119, the first two parentheses vanish, for u and v both satisfy Laplace's equation when w is analytic (cf. Exercise 6.2.1). The third parentheses yields zero from the Cauchy-Riemann conditions, and the same conditions show that the two square brackets are equal. The result is that

$$\nabla^2 \psi(x, y) = \left[\left(\frac{\partial u}{\partial x}\right)^2 + \left(\frac{\partial u}{\partial y}\right)^2\right]\nabla^2 \Psi(u, v)$$
$$= 0, \tag{6.120}$$

since $\Psi(u, v)$ is known to satisfy Laplace's equation in the w-plane. Therefore, when a solution of Laplace's equation is subjected to an analytic transformation, it remains a solution of Laplace's equation. Finally, by Ex. 1.15.2, a solution of Laplace's equation satisfying a complete set of boundary conditions is unique. Our transformed solution (satisfying our boundary conditions) is the solution.

EXAMPLE 6.7.1

A perhaps more realistic and more interesting example of this technique is provided by a conducting circular cylinder parallel to an infinite metal plane (Fig. 6.25).

The transformation relating the z- and w-planes is

$$z = ia \tan \frac{w}{2}$$
$$= a \tanh \frac{iw}{2}, \tag{6.121}$$

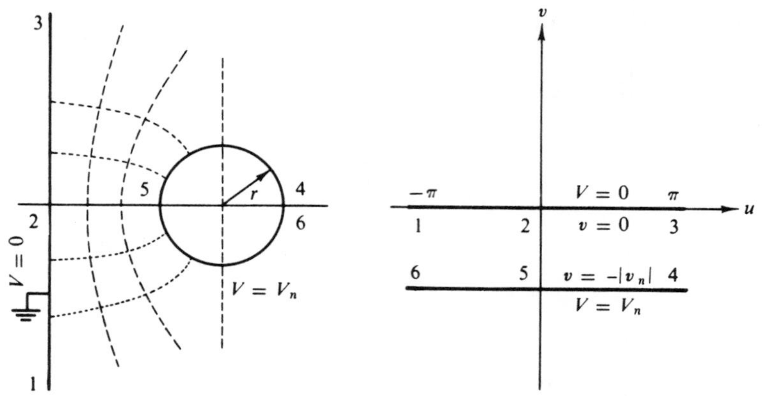

FIG. 6.25 Infinite cylinder parallel to plane

as the reader may verify.[1] With the help of Ex. 6.1.10, Eq. 6.121 leads to

$$x = -a \frac{\sinh v}{\cosh v + \cos u},$$

$$y = a \frac{\sin u}{\cosh v + \cos u}.$$

Eliminating u, we see that the circle is centered at $x = d = -a \coth v_n$, $y = 0$ and has a radius of $r = -a \operatorname{csch} v_n$. This means that $v_n = -\cosh^{-1}(d/r)$, choosing the negative sign since v_n is negative. In the w-plane, by inspection,[2] the electrostatic potential V is

$$V = +V_n \frac{v}{v_n}, \qquad -|v_n| \leqslant v \leqslant 0$$

$$= \frac{-vV_n}{\cosh^{-1}(d/r)}. \tag{6.122}$$

Now, if we invert Eq. 6.121, obtain v in terms of z, and substitute into Eq. 6.121, we will have the equipotentials in the z-plane. Solving Eq. 6.121 for w, we obtain

[1] Verification is not too difficult, but how about derivation? Some of the transformation needed may be built up from combinations of elementary functions. Equation 6.121 could be built up from exponentials and trigonometric functions with some difficulty, The usual "derivation" of a transform comes from a dictionary of transforms. Among the better, more complete dictionaries of transforms are Z. Nehari, *Conformal Mapping*. New York: McGraw-Hill (1952), and H. Kober, *Dictionary of Conformal Representations*. New York: Dover (1952).

[2] This deserves more than just a quick glance. The boundary lines (surfaces) in the w-plane are *finite*, $-\pi \leqslant u \leqslant \pi$. However, $u = \pi$, $v = v$ corresponds to the same point in the z-plane as $u = -\pi$, $v = v$, so that there are no end effects after all. The coordinate u is defined only for $-\pi \leqslant u \leqslant \pi$.

$$w = 2\tan^{-1}\left(-\frac{iz}{a}\right),$$

$$v = -\mathscr{I}\left[2\tan^{-1}\left(\frac{iz}{a}\right)\right] \tag{6.123}$$

$$= -\mathscr{R}\left[2\tanh^{-1}\left(\frac{z}{a}\right)\right].$$

Therefore the potential in the right-hand z-plane ($x \geqslant 0$) is

$$V = +\mathscr{R}\,\frac{[2\tanh^{-1}(z/a)]V_n}{\cosh^{-1}(d/a)} \tag{6.124}$$

all negative signs cancelling. This can be put in better form. Comparison with Section 2.9 shows that Eq. 6.121 *generates* a bipolar coordinate system. The equipotentials are the coordinate circles[1]

$$(x + a\coth v)^2 + y^2 = a^2\,\mathrm{csch}^2 v. \tag{6.125}$$

The electrostatic problem is not quite complete. There is still the question of the capacitance between the two equipotential surfaces (per unit length normal to the xy-plane). We have

$$w(z) = u(x, y) + i\,v(x, y), \tag{6.126}$$

with $u(x, y)$ describing the equipotentials and its conjugate function $v(x, y)$ describing the field lines or vice versa. The choice is arbitrary but to be consistent with Eq. 6.123, which is a specific form of Eq. 6.126, let us take $v(x, y)$ to describe the equipotentials and its conjugate function $u(x, y)$ to describe the field lines.

The capacitance per unit length is defined by

$$C = \frac{\text{charge per unit length}}{\text{potential difference}} = \frac{q}{v_2 - v_1}. \tag{6.127}$$

The electric charge per unit length (on one conductor) may be calculated, starting from Gauss's law (Section 1.14).

$$q = \varepsilon_0 \int \mathbf{E} \cdot d\boldsymbol{\sigma} = -\varepsilon_0 \int \nabla v \cdot d\boldsymbol{\sigma} \tag{6.128}$$

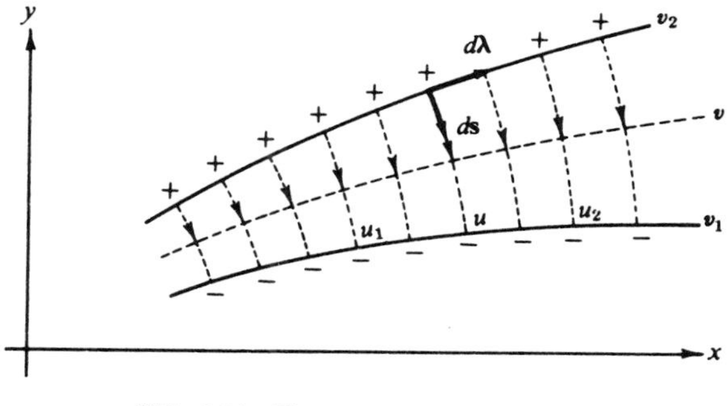

FIG. 6.26 Electrostatic capacitance

[1] This is the final step in identifying r and d in terms of v_n.

Now

$$\nabla v = \mathbf{i}\frac{\partial v}{\partial x} + \mathbf{j}\frac{\partial v}{\partial y}$$

$$= -\mathbf{i}\frac{\partial u}{\partial y} + \mathbf{j}\frac{\partial u}{\partial x} \tag{6.129}$$

from the Cauchy-Riemann conditions. If we replace the area element $d\boldsymbol{\sigma}$ by

$$d\boldsymbol{\sigma} = -\mathbf{k} \times d\boldsymbol{\lambda}, \tag{6.130}$$

where $d\boldsymbol{\lambda}$ is the direction of increasing u (Fig. 6.26) and $\mathbf{k}$ is given by $\mathbf{i} \times \mathbf{j}$,

$$q = \varepsilon_0 \int_{u_1}^{u_2} \left(-\mathbf{i}\frac{\partial u}{\partial y} + \mathbf{j}\frac{\partial u}{\partial x}\right) \cdot \mathbf{k} \times d\boldsymbol{\lambda}$$

$$= \varepsilon_0 \int_{u_1}^{u_2} \left(\mathbf{i}\frac{\partial u}{\partial x} + \mathbf{j}\frac{\partial u}{\partial y}\right) \cdot d\boldsymbol{\lambda}$$

$$= \varepsilon_0(u_2 - u_1). \tag{6.131}$$

When Eq. 6.131 for the charge per unit length (between u_1 and u_2) is substituted into Eq. 6.127, we obtain

$$C = \varepsilon_0 \frac{u_2 - u_1}{v_2 - v_1}. \tag{6.132}$$

If we apply Eq. 6.132 to the foregoing plane and cylinder problem, the capacitance per unit length is found to be

$$C = \varepsilon_0 \frac{2\pi}{(-)v_n} = \frac{2\pi\varepsilon_0}{\cosh^{-1}(d/r)}. \tag{6.133}$$

in agreement with Ex. 2.9.3.

One final example of the use of conformal transformations.

EXAMPLE 6.7.2

A potential difference $2V_0$ is established between two surfaces of semicircular cross sections shown in Fig. 6.27. We want the potential at any point between the curved surfaces.

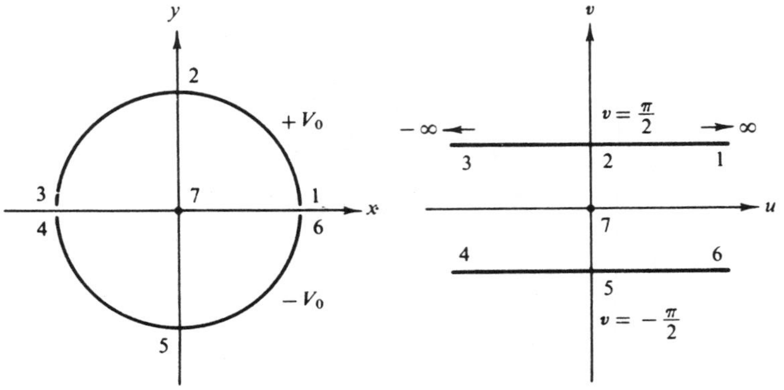

FIG. 6.27 Charged, conducting, semicircular cylinders

The semicircles in the z-plane ($x^2 + y^2 = 1$, unit radius) may be mapped into parallel straight lines in the w-plane by[1]

$$w = \ln\left(\frac{1+z}{1-z}\right). \tag{6.134}$$

We have

$$u + iv = \ln\frac{1 + x + iy}{1 - x - iy} = \ln\frac{1 - x^2 - y^2 + 2iy}{1 - 2x + x^2 + y^2}. \tag{6.135}$$

Replacing $(1 - x^2 - y^2 + 2iy)/(1 - 2x + x^2 + y^2)$ by $Re^{i\alpha}$,

$$u + iv = \ln R + i\alpha, \tag{6.136}$$

where

$$v = \alpha = \tan^{-1}\frac{2y}{1 - (x^2 + y^2)}. \tag{6.137}$$

To map the semicircular boundaries

$$\begin{aligned} \lim_{\substack{x^2+y^2\to 1\\ y>0}} \tan^{-1}\frac{2y}{1-(x^2+y^2)} &= \frac{\pi}{2},\\ \lim_{\substack{x^2+y^2\to 1\\ y<0}} \tan^{-1}\frac{2y}{1-(x^2+y^2)} &= -\frac{\pi}{2}, \end{aligned} \tag{6.138}$$

as shown in Fig. 6.27. The potential is obtained (by inspection) as

$$\begin{aligned} V &= \frac{2V_0}{\pi} v \\ &= \frac{2V_0}{\pi}\tan^{-1}\left[\frac{2y}{1-(x^2+y^2)}\right] \\ &= \frac{2V_0}{\pi}\tan^{-1}\left(\frac{2r\sin\theta}{1-r^2}\right), \qquad 0 \leqslant r \leqslant 1. \end{aligned} \tag{6.139}$$

This particular problem may be solved easily by separating variables in circular cylindrical coordinates (cf. Ex. 14.3.5). We obtain an infinite series solution for the potential. A possible advantage of the conformal mapping approach is that the solution is given in closed form.

The reader may show that the range of u is $-\infty < u < \infty$, which means that the capacitance is infinite!

[1] Solving for z, $z = \tanh w/2$, so that we are really using Eq. 6.121 again in a different form and in a different context.

It must be noted that, while the conformal transformation technique of solving Laplace's equation is elegant and sometimes very useful, still its usefulness is sharply limited. The technique may be applied to two-dimensional problems (and to three-dimensional problems when the third dimension is an axis of translational symmetry—no dependence upon position along this axis). Even so, the analytical solution is feasible only for cases with a moderate degree of symmetry or simplicity. Numerical integration techniques, essential for three-dimensional problems, may be applied even more easily to the two-dimensional problems.

EXERCISES

6.7.1 Expand $f(z)$ in a Taylor series about the point $z = z_0$ where $f'(z_0) = 0$. (Angles not preserved.) Show that if the first $n - 1$ derivatives vanish but $f^{(n)}(z_0) \neq 0$, then angles in the z-plane with vertices at $z = z_0$ appear in the w-plane multiplied by n.

6.7.2 Develop the transformations that create each of the four cylindrical coordinate systems:

(a) Circular cylindrical $\quad x = \rho \cos \varphi,$
$\quad y = \rho \sin \varphi.$

(b) Elliptic cylindrical $\quad x = a \cosh u \cos v,$
$\quad y = a \sinh u \sin v.$

(c) Parabolic cylindrical $\quad x = \xi\eta,$
$\quad y = \frac{1}{2}(\eta^2 - \xi^2).$

(d) Bipolar
$$x = \frac{a \sinh \eta}{\cosh \eta - \cos \xi},$$
$$y = \frac{a \sin \xi}{\cosh \eta - \cos \xi}.$$

Note. These transformations are not necessarily analytic.

6.7.3 In the transformation
$$e^z = \frac{a - w}{a + w}$$
how do the coordinate lines in the z-plane transform? What coordinate system of Chapter 2 have you constructed?

6.7.4 In an electrostatic problem the equipotentials are given by $u(x, y) = c_i$, in which u is the real part of $w(z)$ and $w(z)$ is analytic. Show that the magnitude of the electric field $\mathbf{E}$ is given by
$$|\mathbf{E}| = \left|\frac{dw(z)}{dz}\right|.$$

6.7.5 Show how the transformation Eq. 6.108
$$w = a \ln z$$
can represent the (two-dimensional) field of a line charge, the line being perpendicular to the xy-plane. Renormalize your transformation so that it describes unit charge (per unit length.)

6.7.6 A conducting cylinder of radius a carries a charge (per unit length) of Q. An outer concentric conducting cylinder of radius b carries a charge (per unit length) of $-Q$.

(a) Find a transformation $w = w(z)$ that will map the equipotential cylinders into parallel line segments.

(b) Find the potential V and the electric field $\mathbf{E}$ as a function of position in the w-plane.

(c) Transforming back to the z-plane, show that

$$V(\rho) = -\frac{Q}{2\pi\varepsilon_0}\ln(\rho/a).$$

$$\mathbf{E}(\rho) = \boldsymbol{\rho}_0 \frac{Q}{2\pi\varepsilon_0\rho}.$$

(d) Show that the capacitance (per unit length) is

$$C = \frac{2\pi\varepsilon_0}{\ln(b/a)}.$$

6.7.7 Two cylindrical conductors, one of radius r_1 and the other of radius r_2, are separated by a center-to-center distance D. What is their capacitance per unit length if

(a) $D > r_1 + r_2$.

(b) $0 < D < |r_1 - r_2|$.

(c) For $D \gg r_1 + r_2$, show that the capacitance per unit length becomes

$$C \approx \pi\varepsilon_0/\ln[D/(r_1 r_2)^{1/2}].$$

As a check, compare with the solution given for the same problem in bipolar coordinates, Ex. 2.9.3.

6.7.8 (a) Find the electrostatic potential *exterior* to the semicircular cylinders of Example 6.7.2.

(b) Calculate the charge density $\sigma = \varepsilon_0 E_r$ ($r = 1$, exterior)

Ans. (a) $V = \dfrac{2V_0}{\pi}\tan^{-1}\left(\dfrac{2r\sin\theta}{r^2 - 1}\right), \quad r \geq 1.$

(b) $\sigma = -\varepsilon_0 \left.\dfrac{\partial V}{\partial r}\right|_{r=1} = \dfrac{2\varepsilon_0 V_0}{\pi}\csc\theta.$

6.7.9 A thin, flat conducting strip of width $2a$ is raised to a potential V_0 above ground. Find the following:

(a) The lines of constant potential.

(b) The electric field lines. (These are sketched in Fig. 6.28.)

(c) The distribution of electric charge as a function of distance from the center line of the strip.

(d) The capacitance per unit length.

6.7.10 The two analytic transformations

$$z = a\sin w$$

$$z = -a\tanh\frac{iw}{2}$$

develop sets of equipotentials and field lines corresponding to $u = c_i$ and $v = c_j$, with the line segment $-a \leqslant x \leqslant a$, $y = 0$ as a possible equipotential when $v = 0$.

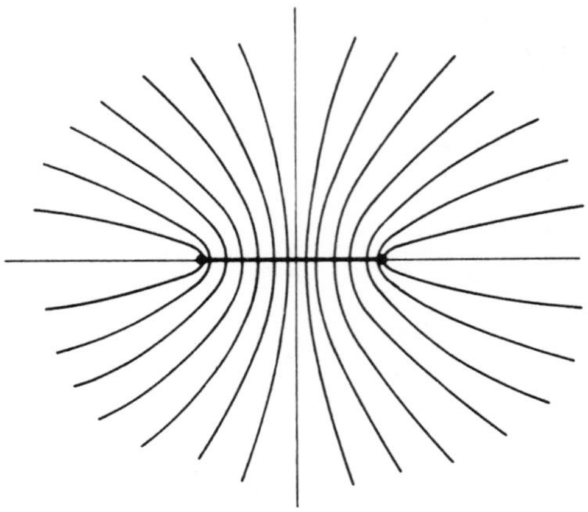

FIG. 6.28 Electric field of charged conducting strip

(a) Sketch the orthogonal system resulting from each of the two transformations.
(b) What is the difference between these two cases from a physical point of view?

6.7.11 Show that the transformation

$$w = e^z + z$$

maps the infinite lines $y = \pm\pi$ into semi-infinite lines $u \leqslant u_0$, $v = \pm\pi$. This is equivalent to transforming an infinite or edgeless parallel plate capacitor (z-plane) into a parallel plate capacitor (w-plane). Sketch the equipotentials in the w-plane near the edge of the capacitor plates.

Find the electric field at the plane midway between the two plates (at $v = 0$) as a function of u.

6.7.12 (a) A conducting cylinder of unit radius is placed in a uniform electric field, $\mathbf{E}_0$. The axis of the cylinder is perpendicular to the direction of the original field. Using the transformation

$$w = z + \frac{1}{z},$$

find the perturbed electrostatic potential.

6.7.13 Reinterpret Ex. 6.7.12 in terms of the irrotational flow of fluid past a circular obstacle.

6.7.14 In terms of fluid flow show how

$$w = a \ln z$$

may be interpreted as a source at the origin ($\nabla \cdot \mathbf{V} \neq 0$). Note that the flow is still irrotational, $\nabla \times \mathbf{V} = 0$.

Hint: Show that the net flow of fluid across any closed curve is $\oint dv$, where v is $\mathscr{I}(w)$.
The derivation of this is similar to the calculation of electrostatic capacitance in this section.

6.8 Schwarz-Christoffel Transformation

The Schwarz-Christoffel transformation may be set up to map the real axis of the w-plane into any desired polygon in the z-plane, the upper half ($v > 0$) of the

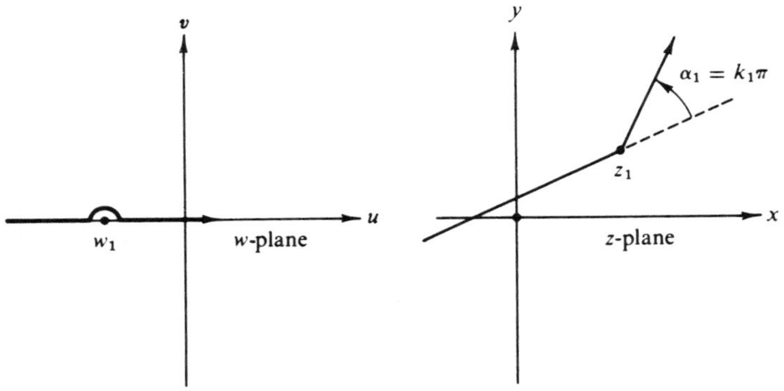

FIG. 6.29 Schwarz-Christoffel transformation

w-plane being mapped into the interior of the polygon. The inverse transformation then maps the z-plane polygon into the upper half of the w-plane. This provides us with a systematic way of deriving a useful conformal transformation for polygon-type figures, including various degenerate cases.

To develop the Schwarz-Christoffel transformation, consider first the function $z'(w)$:

$$\frac{dz}{dw} = A(w - w_1)^{-k_1}, \tag{6.140}$$

in which A is a complex constant, k_1 is a real constant, and w_1 is a point on the real (u) axis. The phase or argument is given by

$$\arg \frac{dz}{dw} = \begin{cases} \arg A - k_1\pi, & w < w_1, \\ \arg A, & w > w_1, \end{cases} \tag{6.141}$$

for w moving along the real axis. We have $\arg(w - w_1) = \pi$ for $w < w_1$, and $\arg(w - w_1) = 0$ for $w > w_1$; that is, as w moves along the u-axis in the positive sense the argument of dz/dw discontinuously jumps by an amount $k_1\pi$ when w passes point w_1. Referring to Section 6.7, especially Fig. 6.23, we note that $\arg \Delta w = 0$ constant. Therefore

$$\arg \Delta z = \begin{cases} \arg A - k_1\pi, & w < w_1, \\ \arg A, & w > w_1. \end{cases} \tag{6.142}$$

Since A is a constant (at our disposal), $\arg A$ is a constant and z therefore is a pair of straight-line segments which form an exterior angle $\alpha_1 = k_1\pi$.

From the factor $(w - w_1)^{-k_1}$, Eq. 6.140, we obtain one vertex of a polygon. By including n factors of this form we may construct an n vertex polygon:

$$\frac{dz}{dw} = A(w - w_1)^{-k_1}(w - w_2)^{-k_2} \cdots (w - w_n)^{-k_n}, \tag{6.143}$$

with the constraint

$$\sum_{i=1}^{n} k_i = 2, \tag{6.144}$$

so that the sum of the exterior angles will be 2π.[1]

By integrating Eq. 6.143 we have

$$z = A \int^{w} (w - w_1)^{-k_1}(w - w_2)^{-k_2} \cdots (w - w_n)^{-k_n}\, dw + B. \tag{6.145}$$

The complex constant A permits us to rotate and orient our z-plane polygon as desired and the complex constant of integration B is available for any needed translational displacement. In the z-plane the polygon may be specified by giving the locations of the n vertices which means fixing $2n$ constants. However, Eq. 6.145 has $2n + 4$ parameters,

n	w_i's	(The w_i's are real.)
n	k_i's	
4	A and B	
$2n + 4$		

from which we deduct one for the constraint of Eq. 6.144 to yield $2n + 3$ parameters. Hence three of the singularities w_i may be chosen arbitrarily, after which the remainder are uniquely determined. Usually the three arbitrary w_i's are chosen to facilitate the evaluation of the integral (Eq. 6.145).

Example 6.8.1

Map the real w-axis into the "triangle" shown in Fig. 6.30. From the diagram

$$\begin{aligned} \alpha_1 &\to \frac{\pi}{2}, & k_1 &\to \frac{1}{2} \\ \alpha_2 &= \frac{\pi}{2}, & k_2 &= \frac{1}{2} \\ \alpha_3 &\to \pi, & k_3 &\to 1. \end{aligned} \tag{6.146}$$

For convenience we let $w_1 = -1$, $w_2 = +1$, and $w_3 \to \infty$, with A available to compensate for w_3. Then, by Eq. 6.145,

$$\begin{aligned} z &= A \int^{w} (w + 1)^{-1/2}(w - 1)^{-1/2}\, dw + B \\ &= A \int^{w} (w^2 - 1)^{-1/2} + B. \end{aligned} \tag{6.147}$$

Integrating

$$\begin{aligned} z &= A \cosh^{-1} w + B \\ w &= \cosh\left(\frac{z - B}{A}\right). \end{aligned} \tag{6.148}$$

The constants A and B are evaluated by noting that when $w = -1$, $z = ia$, and

[1] If treated in terms of the closed contours of Chapter 7, this guarantees that the infinite semicircle in the upper half w-plane which closes the contour will contribute nothing to the integral.

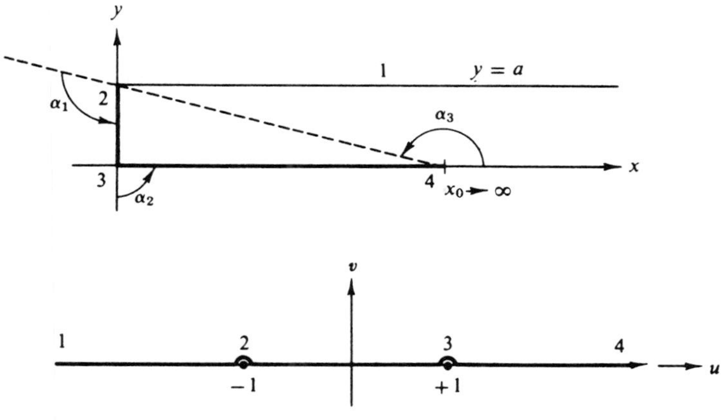

FIG. 6.30 Schwarz-Christoffel mapping of a degenerate triangle

when $w = +1$, $z = 0$. From the latter condition, $B = 0$. From the former

$$\cosh \frac{ia}{A} = \cos \frac{a}{A} = -1, \tag{6.149}$$

which shows that $A = a/\pi$. Hence

$$w = \cosh \frac{\pi z}{a}. \tag{6.150}$$

The upper half of the w-plane is mapped into the interior of the semi-infinite strip.

EXAMPLE 6.8.2

Map the real w-axis into the "triangle" shown in Fig. 6.31. From this diagram

$$\begin{aligned} \alpha_1 &\to \pi, & k_1 &\to 1, \\ \alpha_3 &\to \pi, & k_3 &\to 1, \end{aligned} \tag{6.151}$$

and, since $\alpha_2 \to 0$, $k_2 \to 0$, and this factor may be dropped. As in the preceding example, we arbitrarily place the singularities at -1 and $+1$. Then

$$\begin{aligned} z &= A \int^w (w-1)^{-1}(w+1)^{-1}\,dw + B \\ &= \frac{A}{2} \ln \frac{w-1}{w+1} + B. \end{aligned} \tag{6.152}$$

The constants A and B are evaluated by requiring that $w = 0$ correspond to $z = 0$, or

$$\frac{A}{2} \ln(-1) + B = \frac{i\pi A}{2} + B = 0. \tag{6.153}$$

Also, as $w \to \pm\infty$, $z = ia$ or

$$ia = \frac{A}{2} \ln 1 + B = B. \tag{6.154}$$

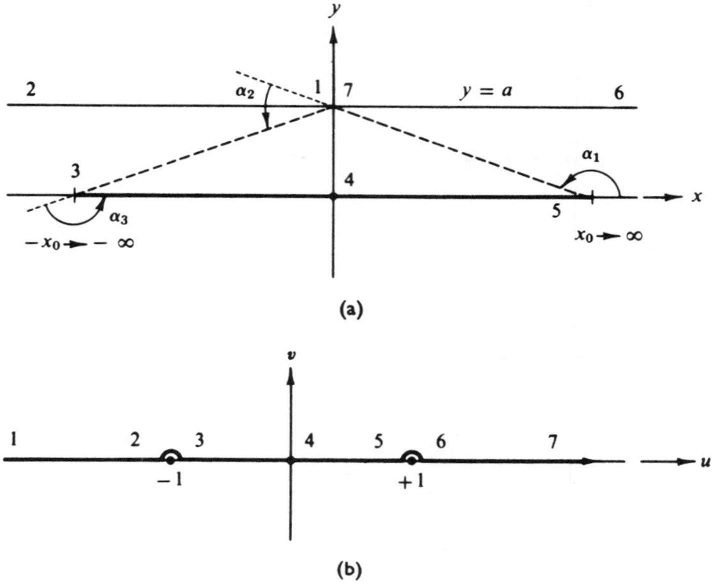

FIG. 6.31 Schwarz-Christoffel mapping of a second degenerate triangle

Therefore

$$z = -\frac{a}{\pi}\ln\left(\frac{w-1}{w+1}\right) + ia. \tag{6.155}$$

It should be noted that the choice of singularities ($w_3 = -1$, $w_1 = +1$) forces the correspondence of

$$\begin{aligned} -1 < u < 1, \quad & v = 0 \quad \text{and} \quad -\infty < x < \infty, \quad y = 0, \\ -\infty < u < -1, \quad & v = 0 \quad \text{and} \quad -\infty < x < 0, \quad y = a, \\ 1 < u < \infty, \quad & v = 0 \quad \text{and} \quad 0 < x < \infty, \quad y = a, \end{aligned}$$

shown by the position numbers on the z- and w-planes. A different choice of singularities would have led to a different transformation and a different correspondence. In some exercises Eq. 6.155 may be used to take the Schwarz-Christoffel transformation of a polygon and map it into two parallel infinite lines so that Laplace's equation may be solved by inspection.

EXERCISES

6.8.1 Using the Schwarz-Christoffel procedure, map the "polygon" bounded by the three lines $x = -a$, $y = 0$, $x = +a$ into the u-axis (w-plane) with singularities at $u = \pm 1$, $v = 0$.

6.8.2 The real axis in the z-plane between $x = -a$ and $x = +a$ is held at a potential of 100 volts. The remainder of the real axis is grounded. Find the potential for any point of $x = 0$. What is the capacitance (per unit length) of this system?

6.8.3 Derive the transformation $w = z^2$ (which maps the first quandrant of the z-plane into the upper half of the w-plane) by using the Schwarz-Christoffel transformation.

6.8.4 Find the electrostatic potential at any point $x > 0$, $y > 0$, if the real (x) axis is grounded, $V = 0$, and the imaginary (y) axis is held at $V = 100$.

6.8.5 Derive the parallel plate capacitor transformation

$$w = e^z + z$$

(Exercise 6.7.11) by using the Schwarz-Christoffel transformation.

REFERENCES

Churchill, R. V., *Complex Variables and Applications*. 2nd Ed. New York: McGraw-Hill (1960). For both the beginning and the advanced student this is an excellent text. It is readable and quite complete. A detailed proof of the Cauchy-Goursat theorem is given in Chapter 5.

Lass, H., *Elements of Pure and Applied Mathematics*, New York: McGraw-Hill (1957). Chapter 4 is a useful presentation of complex-variable theory.

Morse, P. M., and Feshbach, H., *Methods of Theoretical Physics*. New York: McGraw-Hill (1953). Chapter 4 is a presentation of portions of the theory of functions of a complex variable of interest to theoretical physicists.

Smith, L. P., *Mathematical Methods for Scientists and Engineers*. New York: Prentice-Hall (1953). Smith uses complex variables extensively in the development of power series, series of functions, theory of functions, conformal mapping, and calculus of residues. Chapter 9 includes some involved but realistic applications of the Schwarz-Christoffel transformation.

Sokolnikoff, I. S., and Redheffer, R. M., *Mathematics of Physics and Modern Engineering*. New York: McGraw-Hill (1958). Chapter 7 covers complex variables.

Spiegel, M.R., *Theory and Problems of Complex Variables*. New York: Schaum (1964). An excellent summary of the theory of complex variables for scientists.

Watson, G. N., *Complex Integration and Cauchy's Theorem*. New York: Hafner (orig. 1917). A short work containing a rigorous development of the Cauchy integral theorem and integral formula. Applications to the calculus of residues are included. *Cambridge Tracts in Mathematics, and Mathematical Physics*, No. 15.

Other references are given at the end of Chapter 15.

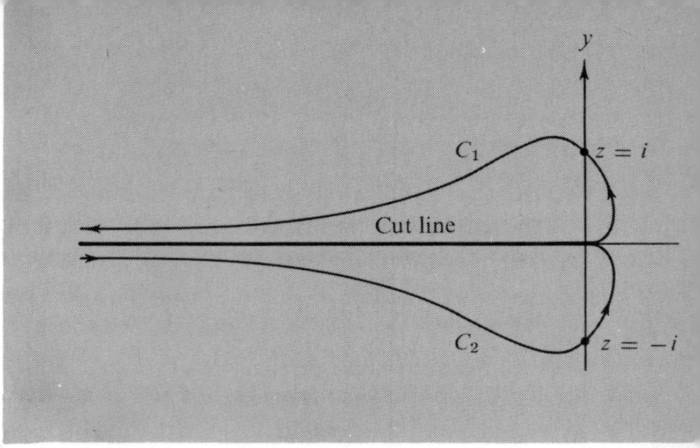

CHAPTER 7

FUNCTIONS OF A COMPLEX VARIABLE II

CALCULUS OF RESIDUES

7.1 Singularities

In this chapter we return to the line of analysis that started with the Cauchy-Riemann conditions in Chapter 6 and led on through the Laurent expansion (Section 6.5). The Laurent expansion represents a generalization of the Taylor series in the presence of singularities. We define the point z_0 as an isolated singular point of the function $f(z)$ if $f(z)$ is not analytic at $z = z_0$ but is analytic at neighboring points.

Poles. In the Laurent expansion of $f(z)$ about z_0

$$f(z) = \sum_{n=-\infty}^{\infty} a_n(z - z_0)^n. \tag{7.1}$$

If $a_n = 0$ for $n < -m < 0$ and $a_{-m} \neq 0$, we say that z_0 is a pole of order m. For instance, if $m = 1$, that is, if $a_{-1}/(z - z_0)$ is the first nonvanishing term in the Laurent series, we have a pole of order one, often called a simple pole.

If, on the other hand, the summation continues to $n = -\infty$, the z_0 is a pole of infinite order and is called an essential singularity. These essential singularities have many pathological features. For instance, we can show that in any small neighborhood of an essential singularity of $f(z)$ the function $f(z)$ comes arbitrarily close to any (and therefore every) preselected complex quantity w_0.[1] Literally, the entire

[1] This theorem is due to Picard. A proof is given by E. C. Titchmarsh, *The Theory of Functions*, 2nd Ed. New York: Oxford University Press (1939).

w-plane is mapped into the neighborhood of the point z_0. One point of fundamental difference between a pole of finite order and an essential singularity is that a pole of order m can be removed by multiplying $f(z)$ by $(z - z_0)^m$. This obviously cannot be done for an essential singularity.

The behavior of $f(z)$ as $z \to \infty$ is defined in terms of the behavior of $f(1/t)$ as $t \to 0$. Consider the function

$$\sin z = \sum_{n=0}^{\infty} \frac{(-1)^n z^{2n+1}}{(2n+1)!}. \tag{7.2}$$

As $z \to \infty$, we replace the z by $1/t$ to obtain

$$\sin\left(\frac{1}{t}\right) = \sum_{n=0}^{\infty} \frac{(-1)^n}{(2n+1)!\, t^{2n+1}}. \tag{7.3}$$

Clearly, from the definition, $\sin z$ has an essential singularity at infinity. This result could be anticipated from Exercise 6.1.7 since

$$\begin{aligned} \sin z &= \sin iy, \qquad \text{when } x = 0, \\ &= i \sinh y, \end{aligned}$$

which approaches infinity exponentially as $y \to \infty$.

Branch points. There is another sort of singularity that will be important in the later sections of this chapter. Consider

$$f(z) = z^a,$$

in which a is not an integer.[1] As z moves around the unit circle from e^0 to $e^{2\pi i}$,

$$f(z) \to e^{2\pi a i} \neq e^{0i},$$

for nonintegral a. As in Section 6.6, we have a branch point. The points e^{0i} and $e^{2\pi i}$ in the z-plane coincide but these coincident points lead to *different* values of $f(z)$; that is, $f(z)$ is a multivalued function. The problem is resolved by constructing a cut line so that $f(z)$ will be uniquely specified for a given point in the z-plane.

Note carefully that a function with a branch point and a required cut line will not be continuous across the cut line. In general, there will be a phase difference on opposite sides of this cut line. Hence, line integrals on opposite sides of this branch point cut line will not generally cancel each other. Numerous examples of this appear in the Exercises.

The cut line used to convert a multiply connected region into a simply connected region (Section 6.3) is completely different. Our function is continuous across the cut line, and no phase difference exists.

[1] $z = 0$ is technically a singular point, for z^a has only a finite number of derivatives, whereas an analytic function is guaranteed an infinite number of derivatives (Section 6.4).

EXAMPLE 7.1.1

Consider the function

$$f(z) = (z^2 - 1)^{1/2} = (z + 1)^{1/2}(z - 1)^{1/2}. \tag{7.4}$$

The first factor on the right-hand side, $(z + 1)^{1/2}$, has a branch point at $z = -1$. The second factor has a branch point at $z = +1$. To check on the possibility of taking the line segment joining $z = +1$ and $z = -1$ as a cut line let us follow the phases of these two factors as we move along the contour shown in Fig. 7.1.

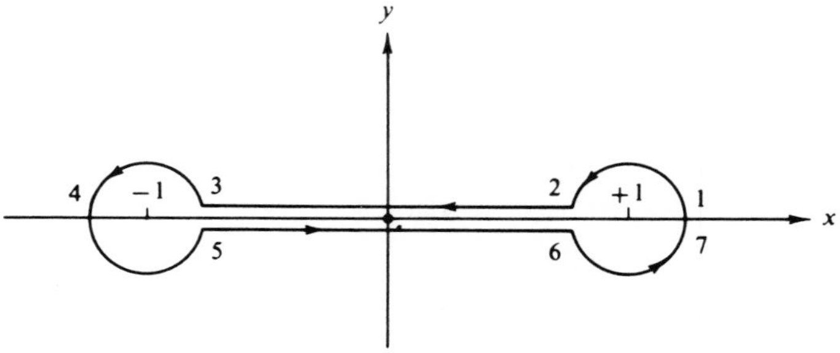

FIG. 7.1

Starting at point 1 where both $(z + 1)$ and $(z - 1)$ have a phase of zero, we can build up Table 7.1.

TABLE 7.1

	Phase Angle		
Point	$(z+1)$	$(z-1)$	$(z+1)^{1/2}(z-1)^{1/2}$
1	0	0	0
2	0	π	$\pi/2$
3	0	π	$\pi/2$
4	π	π	π
5	2π	π	$3\pi/2$
6	2π	π	$3\pi/2$
7	2π	2π	2π

Moving from point 1 to point 2 the phase of $(z - 1)$ increases by π, that is, $z - 1$ becomes negative. The phase of $z - 1$ then stays constant until the circle is completed, moving from 6 to 7. The phase of $(z + 1)$ shows a similar behavior increasing by 2π as we move from 3 to 5. The phase of $f(z) = (z + 1)^{1/2}(z - 1)^{1/2}$ tabulated in the final column is one half the sum of the two preceding columns.

Two features emerge;

1. The phase at points 5 and 6 is not the same as the phase at points 2 and 3. This behavior can be expected at a cut line.

2. The phase at point 7 exceeds that at point 1 by 2π and the function $f(z) = (z^2 - 1)^{1/2}$ is therefore *single-valued* for the contour shown, encircling *both* branch points.

If we take the x-axis $-1 \leqslant x \leqslant 1$ as a cut line, $f(z)$ is uniquely specified. Alternatively, the positive x-axis for $x > 1$, and the negative x-axis for $x < -1$ may be taken as cut lines. The branch points can not be encircled, and the function remains single-valued.

Generalizing from this example, the phase of a function

$$f(z) = f_1(z) \cdot f_2(z) \cdot f_3(z) \cdots.$$

is the algebraic sum of the phase of its individual factors:

$$\arg f(z) = \arg f_1(z) + \arg f_2(z) + \arg f_3(z) + \cdots.$$

The phase of an individual factor may be taken as the arctangent of the ratio of its imaginary part to its real part,

$$\arg f_i(z) = \tan^{-1}(v_i/u_i).$$

For the case of a factor of the form

$$f_i(z) = (z - z_0),$$

the phase corresponds to the phase angle of a two-dimensional vector from $+z_0$ to z, the phase increasing by 2π as the point $+z_0$ is encircled. Conversely, the traversal of any closed loop not encircling z_0 does *not* change the phase of $z - z_0$.

As a final note on singularities, Liouville's theorem (Exercise 6.4.6) states "*A function which is everywhere finite (bounded) and analytic must be a constant.*"

This is readily proved by the use of Cauchy's integral formula. Conversely, the slightest deviation of an analytic function from a constant value implies that there must be at least one singularity somewhere in the infinite complex plane. Apart from the trivial constant functions, then, singularities are a fact of life, and we must learn to live with them. But we shall do more than that. We shall use singularities to develop the powerful and useful calculus of residues.

EXERCISES

7.1.1 The function $f(z)$ expanded in a Laurent series exhibits a pole of order m at $z = z_0$. Show that the coefficient of $(z - z_0)^{-1}$, a_{-1}, is given by

$$a_{-1} = \frac{1}{(m-1)!} \frac{d^{m-1}}{dz^{m-1}} \left[(z - z_0)^m f(z)\right]_{z = z_0},$$

with

$$a_{-1} = [(z - z_0)f(z)]_{z=z_0},$$

when the pole is a simple pole ($m = 1$).

7.1.2 A function $f(z)$ can be represented by

$$f(z) = \frac{f_1(z)}{f_2(z)},$$

in which $f_1(z)$ and $f_2(z)$ are analytic. The denominator $f_2(z)$ vanishes at $z = z_0$ showing that $f(z)$ has a pole at $z = z_0$. However, $f_1(z_0) \neq 0$, $f_2'(z_0) \neq 0$. Show that a_{-1}, the coefficient of $(z - z_0)^{-1}$ in a Laurent expansion of $f(z)$ at $z = z_0$, is given by

$$a_{-1} = \frac{f_1(z_0)}{f_2'(z_0)}.$$

7.1.3 In analogy with Example 7.1.1 consider in detail the phase of each factor and the resultant overall phase of $f(z) = (z^2 + 1)^{1/2}$ following a contour similar to that of Fig. 7.1, but encircling the new branch points.

7.1.4 The Legendre function of the second kind, $Q_\nu(z)$, has branch points at $z = \pm 1$. The branch points are joined by a cut line along the real (x) axis.

(a) Show that $Q_0(z) = \frac{1}{2} \ln((z + 1)/(z - 1))$ is single valued (with the real axis $-1 \leq x \leq 1$ taken as a cut line).

(b) For real argument x and $|x| < 1$ it is convenient to take

$$Q_0(x) = \tfrac{1}{2} \ln((1 + x)/(1 - x)).$$

Show that

$$Q_0(x) = \tfrac{1}{2}[Q_0(x + i0) + Q_0(x - i0)].$$

Here $x + i0$ indicates z approaches the real axis from above, $x - i0$ indicates an approach from below.

7.1.5 As an example of an essential singularity consider $e^{1/z}$ as z approaches zero. For any complex number z_0, $z_0 \neq 0$, show that

$$e^{1/z} = z_0$$

has an infinite number of solutions.

7.2 Calculus of Residues

Residue theorem. If the Laurent expansion of a function is integrated term by term by using a closed contour that encircles one isolated singular point z_0 once in a counterclockwise sense, we obtain

$$a_n \oint (z - z_0)^n \, dz = a_n \frac{(z - z_0)^{n+1}}{n + 1}\bigg|_{z_1}^{z_1}$$

$$= 0 \quad \text{for} \quad n \neq -1. \tag{7.5}$$

However, if $n = -1$,

$$a_{-1} \oint (z - z_0)^{-1} \, dz = a_{-1} \oint \frac{ire^{i\theta} \, d\theta}{re^{i\theta}} = 2\pi i a_{-1}. \tag{7.6}$$

Summarizing Eqs. 7.5 and 7.6, we have

$$\frac{1}{2\pi i}\oint f(z)\,dz = a_{-1}. \tag{7.7}$$

The constant a_{-1}, the coefficient of $(z - z_0)^{-1}$ in the Laurent expansion, is called the *residue* of $f(z)$ at $z = z_0$.

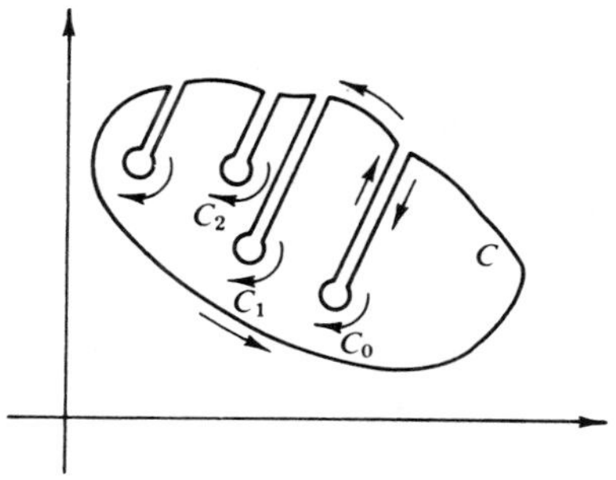

FIG. 7.2 Excluding isolated singularities

A set of isolated singularities can be handled very nicely by deforming our contour as shown in Fig. 7.2. Cauchy's integral theorem (Section 6.3) leads to

$$\oint_C f(z)\,dz + \oint_{C_0} f(z)\,dz + \oint_{C_1} f(z)\,dz + \oint_{C_2} f(z)\,dz + \cdots = 0. \tag{7.8}$$

The circular integral around any given singular point is given by Eq. 7.7.

$$\oint_{C_i} f(z)\,dz = -2\pi i\, a_{-1}\,(z = z_i), \tag{7.9}$$

assuming a Laurent expansion about the singular point, $z = z_i$. The negative sign comes from the clockwise integration as shown in Fig. 7.2. Combining Eqs. 7.8 and 7.9, we have

$$\oint_C f(z)\,dz = 2\pi i(a_{-1z_0} + a_{-1z_1} + a_{-1z_2} + \cdots)$$

$$= 2\pi i \quad \text{(sum of enclosed residues)}. \tag{7.10}$$

This is the *residue theorem.*

Cauchy principal value. Occasionally an isolated first-order pole will be directly on the contour of integration. In this case we may deform the contour to include or exclude the residue as desired by including a semicircular detour of

FIG. 7.3 By-passing singular points

infinitesimal radius. This is shown in Fig. 7.3. The integration over the semicircle then gives

$$\pi i a_{-1} \qquad \text{if counterclockwise,}$$

$$-\pi i a_{-1} \qquad \text{if clockwise.}$$

This contribution, $+$ or $-$, appears on the left-hand side of Eq. 7.10. If our detour were clockwise, the residue would not be enclosed and there would be no corresponding term on the right-hand side of Eq. 7.10. However, if our detour were counterclockwise, this residue would be enclosed by the contour C and a term $2\pi i a_{-1}$ would appear on the right-hand side of Eq. 7.10. The net result for either clockwise or counterclockwise detour is that a simple pole on the contour is counted as one half what it would be if it were within the contour. This corresponds to taking the Cauchy principal value.

For instance, let us suppose that $f(z)$ with a singularity at $z = x_0$ is integrated over the entire real axis. The contour is closed with an infinite semicircle in the upper half-plane. Then

$$\begin{aligned} \oint f(z)\,dz &= \int_{-\infty}^{x_0-\delta} f(x)\,dx + \int_{Cx_0} f(z)\,dz \\ &\quad + \int_{x_0+\delta}^{\infty} f(x)\,dx + \int_C \text{infinite semicircle} \qquad (7.11) \\ &= 2\pi i \sum \textit{enclosed} \text{ residues.} \end{aligned}$$

If the small semicircle C_{x_0} includes x_0 (by going below the x axis, counterclockwise), x_0 is enclosed, and its contribution appears *twice*—as $\pi i a_{-1}$ in $\int_{Cx_0}$ and as $2\pi i a_{-1}$ in the term $2\pi i \sum$ enclosed residues—for a net contribution of $\pi i a_{-1}$. If the upper small semicircle is elected, x_0 is excluded. The only contribution is from the *clockwise* integration over C_{x_0} which yields $-\pi i a_{-1}$. Moving this to the extreme right of Eq. 7.11, we have $+\pi i a_{-1}$, as before.

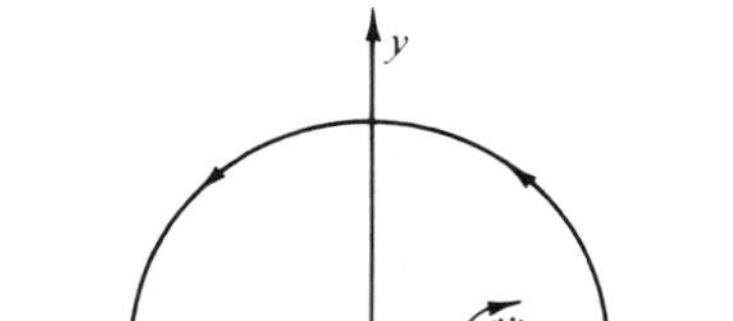

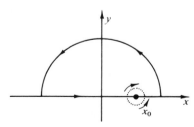

FIG. 7.4 Closing the contour with an infinite radius semicircle

The integrals along the x-axis may be combined and the semicircle radius permitted to approach zero. We have

$$\lim_{\delta \to 0} \left\{ \int_{-\infty}^{x_0-\delta} f(x)\,dx + \int_{x_0+\delta}^{\infty} f(x)\,dx \right\} = P \int_{-\infty}^{\infty} f(x)\,dx. \qquad (7.12)$$

P indicates Cauchy principal value and represents the above limiting process. Note carefully that the Cauchy principal value is a balancing or cancelling process. In the vicinity of our singularity at $z = x_0$,

$$f(x) \approx \frac{a_{-1}}{x - x_0}. \tag{7.13}$$

This is odd, relative to x_0. The symmetric or even interval (relative to x_0) provides cancellation of the shaded areas, Fig. 7.5. The contribution of the singularity is in the integration about the semicircle.

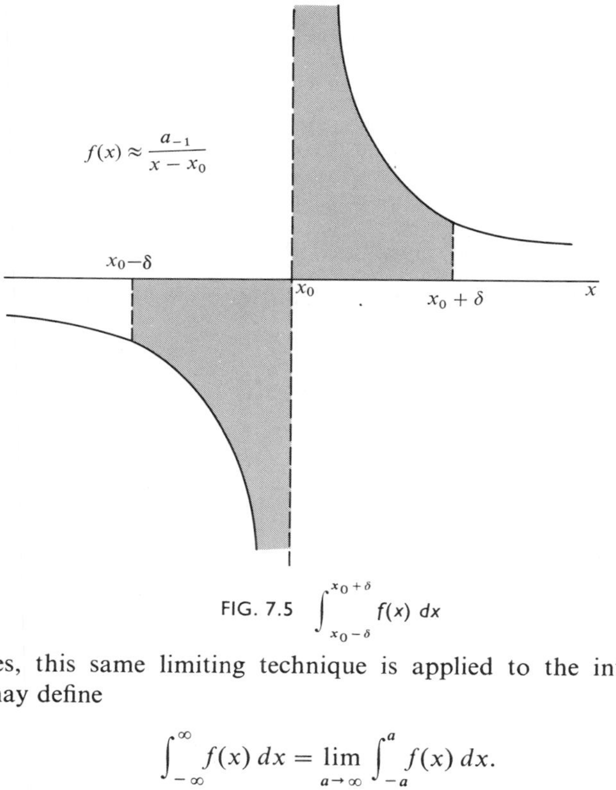

FIG. 7.5 $\int_{x_0-\delta}^{x_0+\delta} f(x)\,dx$

Sometimes, this same limiting technique is applied to the integration limits $\pm\infty$. We may define

$$\int_{-\infty}^{\infty} f(x)\,dx = \lim_{a\to\infty} \int_{-a}^{a} f(x)\,dx. \tag{7.14}$$

Example 2.9.1 is an illustration of this form of the Cauchy principal value.

An alternate treatment moves the pole off the contour and then considers the limiting behavior as it is brought back. This technique is illustrated in Example 7.2.3, in which the singular points are moved off the contour in such a way that the solution is forced into the form desired to satisfy the boundary conditions of the physical problem.

Evaluation of definite integrals—$\int_0^{2\pi} f(\sin\theta, \cos\theta)\,d\theta$. The calculus of residues is useful in evaluating a wide variety of definite integrals in both physical and purely mathematical problems. We consider, first, integrals of the form

$$I = \int_0^{2\pi} f(\sin\theta, \cos\theta)\, d\theta, \tag{7.15}$$

where f is finite for all values of θ. We also require f to be a rational function of $\sin\theta$ and $\cos\theta$ so that it will be single-valued. Let

$$z = e^{i\theta}, \qquad dz = ie^{i\theta}\, d\theta.$$

From this,

$$d\theta = -i\,\frac{dz}{z}, \qquad \sin\theta = \frac{z - z^{-1}}{2i}, \qquad \cos\theta = \frac{z + z^{-1}}{2}. \tag{7.16}$$

Our integral becomes

$$I = -i \oint f\left(\frac{z - z^{-1}}{2i}, \frac{z + z^{-1}}{2}\right) \frac{dz}{z}, \tag{7.17}$$

with the path of integration the unit circle. By the residue theorem, Eq. 7.10,

$$I = (-i)2\pi i \sum \text{residues within the unit circle.} \tag{7.18}$$

Note that we are after the residues of $f(z)/z$. Illustrations of integrals of this type are provided by Ex. 7.2.4–7.

Evaluation of definite integrals—$\int_{-\infty}^{\infty} f(x)\, dx$. Suppose that our definite integral has the form

$$I = \int_{-\infty}^{\infty} f(x)\, dx \tag{7.19}$$

and satisfies the two conditions:

(a) $f(z)$ is analytic in the upper half-plane except for a finite number of poles. (It will be assumed that there are no poles on the real axis. If poles are present on the real axis, they may be included or excluded as discussed earlier in this section.)

(b) $f(z)$ vanishes as strongly [1] as $1/z^2$ for $|z| \to \infty$, $0 \leqslant \arg z \leqslant \pi$.

With these conditions, we may take as a contour of integration the real axis and a semicircle in the upper half-plane as shown in Fig. 7.6. We let the radius R of the semicircle become infinitely large. Then

$$\begin{aligned} I &= \lim_{R\to\infty} \int_{-R}^{R} f(x)\, dx + \lim_{R\to\infty} \int_0^{\pi} f(Re^{i\theta}) iRe^{i\theta} d\theta \\ &= 2\pi i \sum \text{residues (upper half-plane).} \end{aligned} \tag{7.20}$$

From the second condition, the second integral (over the semicircle) vanishes, and

$$\int_{-\infty}^{\infty} f(x)\, dx = 2\pi i \sum \text{residues (upper half-plane).} \tag{7.21}$$

[1] We could use $f(z)$ vanishes faster than $1/z$, but we wish to have $f(z)$ single-valued.

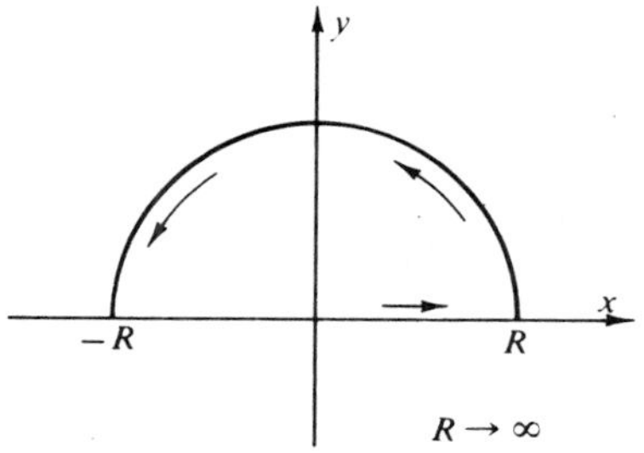

FIG. 7.6

EXAMPLE 7.2.1

Evaluate

$$I = \int_{-\infty}^{\infty} \frac{dx}{1 + x^2}. \tag{7.22}$$

From Eq. 7.21

$$\int_{-\infty}^{\infty} \frac{dx}{1 + x^2} = 2\pi i \sum \text{residues (upper half-plane)}.$$

Here and in every other similar problem we have the question—where are the poles? Rewriting the integrand as

$$\frac{1}{z^2 + 1} = \frac{1}{z + i} \cdot \frac{1}{z - i}, \tag{7.23}$$

we see that there are simple poles (order 1) at $z = i$ and $z = -i$.

A simple pole at $z = z_0$ indicates (and is indicated by) a Laurent expansion of the form

$$f(z) = \frac{a_{-1}}{z - z_0} + a_0 + \sum_{n=1}^{\infty} a_n(z - z_0)^n. \tag{7.24}$$

The residue a_{-1} is easily isolated as

$$a_{-1} = (z - z_0)f(z)|_{z=z_0}. \tag{7.25}$$

Using Eq. 7.25, we find that the residue at $z = i$ is $1/2i$, whereas that at $z = -i$ is $-1/2i$.

Then

$$\int_{-\infty}^{\infty} \frac{dx}{1 + x^2} = 2\pi i \cdot \frac{1}{2i} = \pi. \tag{7.26}$$

Here we have used $a_{-1} = 1/2i$ for the residue of the one included pole at $z = i$. The reader should satisfy himself that it is possible to use the lower semicircle and that this choice will lead to the same result, $I = \pi$. A somewhat more delicate problem is provided by the next example.

Evaluation of definite integrals—$\int_{-\infty}^{\infty} f(x)e^{iax}\,dx$. Consider the definite integral

$$I = \int_{-\infty}^{\infty} f(x)e^{iax}\,dx \tag{7.27}$$

with a real and positive. This is a Fourier transform, Chapter 15. We shall assume the condition

$$\lim_{|z|\to\infty} f(z) = 0, \qquad 0 \leqslant \arg z \leqslant \pi. \tag{7.28}$$

We employ the contour shown in Fig. 7.6. The application of the calculus of residues is the same as on the preceding page, but here we have to work a little harder to show that the integral over the (infinite) semicircle goes to zero. This integral becomes

$$I_R = \int_0^{\pi} f(Re^{i\theta})e^{iaR\cos\theta - aR\sin\theta}\, iRe^{i\theta}\,d\theta. \tag{7.29}$$

Let R be so large that $|f(z)| = |f(Re^{i\theta})| < \varepsilon$. Then

$$\begin{aligned} |I_R| &\leqslant \varepsilon R \int_0^{\pi} e^{-aR\sin\theta}\,d\theta \\ &= 2\varepsilon R \int_0^{\pi/2} e^{-aR\sin\theta}\,d\theta. \end{aligned} \tag{7.30}$$

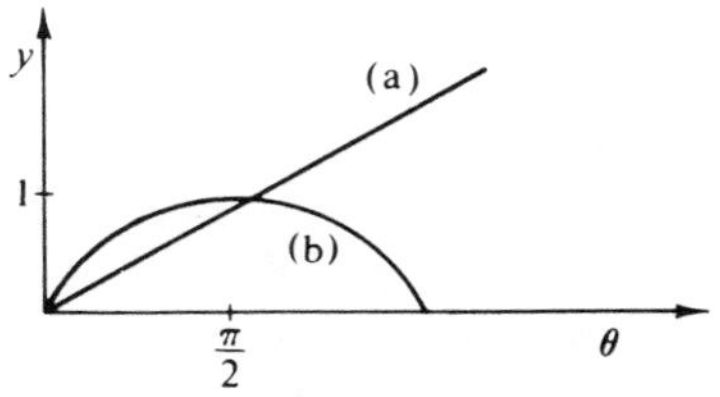

FIG. 7.7 (a) $y = (2/\pi)\theta$, (b) $y = \sin\theta$

In the range $[0, \pi/2]$,

$$\frac{2}{\pi}\theta \leqslant \sin\theta.$$

Therefore,

$$|I_R| \leqslant 2\varepsilon R \int_0^{\pi/2} e^{-aR2\theta/\pi}\,d\theta. \tag{7.31}$$

Now, integrating by inspection,

$$|I_R| \leqslant 2\varepsilon R\,\frac{1 - e^{-aR}}{aR2/\pi}.$$

Finally,

$$\lim_{R\to\infty} |I_R| \leqslant \frac{\pi}{a}\,\varepsilon. \tag{7.32}$$

From Eq. 7.28, $\varepsilon \to 0$ as $R \to \infty$, and

$$\lim_{R\to\infty} |I_R| = 0. \tag{7.33}$$

This useful result is sometimes called *Jordan's lemma.* With it, we are prepared to tackle Fourier integrals of the form shown in Eq. 7.27.

Using the contour shown in Fig. 7.6, we have

$$\int_{-\infty}^{\infty} f(x)e^{iax}\,dx + \lim_{R\to\infty} I_R = 2\pi i \sum \text{residues (upper half-plane)}.$$

Since the integral over the upper semicircle I_R vanishes as $R \to \infty$, (Jordan's lemma),

$$\int_{-\infty}^{\infty} f(x)e^{iax}\,dx = 2\pi i \sum \text{residues (upper half-plane)}. \tag{7.34}$$

Example 7.2.2—Singularity on Contour of Integration

Evaluate

$$I = \int_0^{\infty} \frac{\sin x}{x}\,dx. \tag{7.35}$$

This may be taken as the imaginary part[1] of

$$I_z = \int_{-\infty}^{\infty} \frac{e^{iz}\,dz}{z}. \tag{7.36}$$

Now the only pole is a simple pole at $z = 0$ and the residue there by Eq. 7.25 is $a_{-1} = 1$. We choose the contour shown in Fig. 7.8(a) to avoid the pole, (b) to

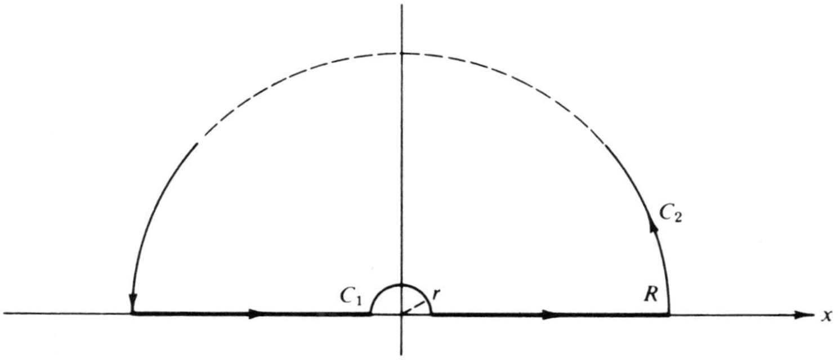

FIG. 7.8

include the real axis, and (c) to yield a vanishingly small integrand for $z = iy$, $y \to \infty$. Note that in this case a semicircle in the lower half-plane would be disastrous. We have

$$\oint \frac{e^{iz}\,dz}{z} = \int_{-R}^{-r} e^{ix}\frac{dx}{x} + \int_{C_1} \frac{e^{iz}\,dz}{z} + \int_r^R \frac{e^{ix}\,dx}{x} + \int_{C_2} \frac{e^{iz}\,dz}{z} = 0, \tag{7.37}$$

[1] One can use $\int[(e^{iz} - e^{-iz})/2iz]dz$, but then two different contours will be needed for the two exponentials (cf. Example 7.2.3).

the final zero coming from the residue theorem (Eq. 7.10). By Jordan's lemma above,

$$\int_{C_2} \frac{e^{iz}\,dz}{z} = 0, \tag{7.38}$$

and

$$\oint \frac{e^{iz}\,dz}{z} = \int_{C_1} \frac{e^{iz}\,dz}{z} + P\int_{-\infty}^{\infty} \frac{e^{ix}\,dx}{x} = 0. \tag{7.39}$$

The integral over the small semicircle yields $(-)\,\pi i$ times the residue of 1, minus as a result of going clockwise. Taking the imaginary part,[1]

$$\int_{-\infty}^{\infty} \frac{\sin x}{x}\,dx = \pi, \tag{7.40}$$

or

$$\int_{0}^{\infty} \frac{\sin x}{x}\,dx = \frac{\pi}{2}. \tag{7.41}$$

The contour of Fig. 7.8, although convenient, is not at all unique. Another choice of contour for evaluating Eq. 7.35 is presented as Exercise 7.2.11.

Example 7.2.3—Quantum Mechanical Scattering

The quantum mechanical analysis of scattering leads to the function

$$I(\sigma) = \int_{-\infty}^{\infty} \frac{x \sin x\,\mathrm{dx}}{x^2 - \sigma^2}, \tag{7.42}$$

where σ is real and positive. From the physical conditions of the problem there is a further requirement: $I(\sigma)$ is to have the form $e^{i\sigma}$ so that it will represent an outgoing scattered wave.

Using

$$\begin{aligned}\sin z &= \frac{1}{i}\sinh iz \\ &= \frac{1}{2i}e^{iz} - \frac{1}{2i}e^{-iz},\end{aligned} \tag{7.43}$$

we write Eq. 7.42 in the complex plane as

$$I(\sigma) = I_1 + I_2, \tag{7.44}$$

[1] Alternatively, we may combine the integrals of Eq. 7.37 as

$$\int_{-R}^{-r} e^{ix}\frac{dx}{x} + \int_{r}^{R} e^{ix}\frac{dx}{x} = \int_{r}^{R}(e^{ix} - e^{-ix})\frac{dx}{x} = i2\int_{r}^{R}\frac{\sin x}{x}\,dx.$$

with

$$I_1 = \frac{1}{2i}\int_{-\infty}^{\infty} \frac{ze^{iz}}{z^2 - \sigma^2}\, dz,$$
$$I_2 = -\frac{1}{2i}\int_{-\infty}^{\infty} \frac{ze^{-iz}}{z^2 - \sigma^2}\, dz. \tag{7.45}$$

Integral I_1 is similar to Example 7.2.2 and, as in that case, we may complete the contour by an infinite semicircle in the upper half plane. For I_2 the exponential is negative and we complete the contour by an infinite semicircle in the lower half plane, as shown in Fig. 7.9. As in Example 7.2.2, neither semicircle contributes anything to the integral—Jordan's lemma.

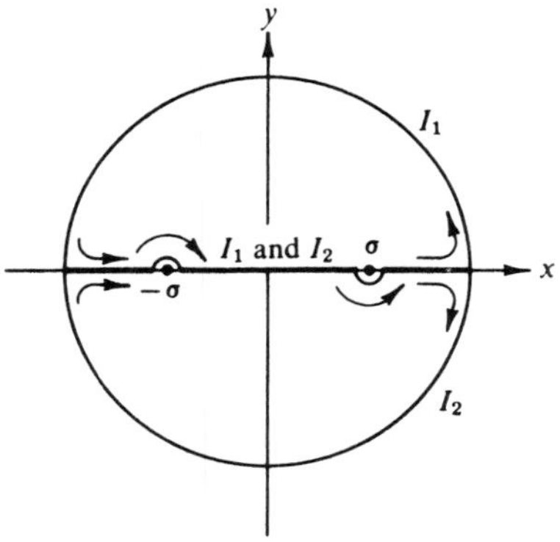

FIG. 7.9

There is still the problem of locating the poles and evaluating the residues. We find poles at $z = +\sigma$ and $z = -\sigma$ *on the contour of integration.* The residues are

	$z = \sigma$	$z = -\sigma$
I_1	$\frac{e^{i\sigma}}{2}$	$\frac{e^{-i\sigma}}{2}$
I_2	$\frac{e^{-i\sigma}}{2}$	$\frac{e^{i\sigma}}{2}$

Detouring around the poles, as shown in Fig. 7.9 (it matters little whether we go above or below), the residue theorem leads to

$$I_1 - \pi i\left(\frac{1}{2i}\right)\frac{e^{-i\sigma}}{2} + \pi i\left(\frac{1}{2i}\right)\frac{e^{i\sigma}}{2} = 2\pi i\left(\frac{1}{2i}\right)\frac{e^{i\sigma}}{2}, \tag{7.46}$$

for we have enclosed the singularity at $z = \sigma$ but excluded the one at $z = -\sigma$. In similar fashion, but noting that the contour for I_2 is *clockwise*,

$$I_2 - \pi i\left(\frac{-1}{2i}\right)\frac{e^{i\sigma}}{2} + \pi i\left(\frac{-1}{2i}\right)\frac{e^{-i\sigma}}{2} = -2\pi i\left(\frac{-1}{2i}\right)\frac{e^{i\sigma}}{2}. \tag{7.47}$$

Adding Eqs. 7.46 and 7.47, we have

$$I(\sigma) = I_1 + I_2 = \frac{\pi}{2}(e^{i\sigma} + e^{-i\sigma}) = \pi \cosh i\sigma$$
$$= \pi \cos \sigma. \tag{7.48}$$

This is a perfectly good evaluation of Eq. 7.42, but unfortunately the cosine dependence is appropriate for a standing wave and not for the outgoing scattered wave as specified.

To obtain the desired form we try a different technique. Instead of dodging around the singular points, let us move them off the real axis. Specifically, let $\sigma \to \sigma + i\gamma$, $-\sigma \to -\sigma - i\gamma$, where γ is positive but small and will eventually be made to approach zero, that is,

$$I(\sigma) = \lim_{\gamma \to 0} I(\sigma + i\gamma). \tag{7.49}$$

With this simple substitution, the first integral I_1 becomes

$$I_1(\sigma + i\gamma) = 2\pi i \left(\frac{1}{2i}\right) \frac{e^{i(\sigma + i\gamma)}}{2} \tag{7.50}$$

by direct application of the residue theorem. Also

$$I_2(\sigma + i\gamma) = -2\pi i \left(\frac{-1}{2i}\right) \frac{e^{i(\sigma + i\gamma)}}{2}. \tag{7.51}$$

Adding Eqs. 7.50 and 7.51 and then letting $\gamma \to 0$, we obtain

$$\begin{aligned} I(\sigma) &= \lim_{\gamma \to 0} [I_1(\sigma + i\gamma) + I_2(\sigma + i\gamma)] \\ &= \lim_{\gamma \to 0} \pi e^{i(\sigma + i\gamma)} = \pi e^{i\sigma}, \end{aligned} \tag{7.52}$$

a result that does fit the boundary conditions of our scattering problem.

It is interesting to note that the substitution $\sigma \to \sigma - i\gamma$ would have led to

$$I(\sigma) = \pi e^{-i\sigma}, \tag{7.53}$$

which could represent an incoming wave. Our earlier result (Eq. 7.48) is seen to be the arithmetic average of Eqs. 7.52 and 7.53. This average is the Cauchy principal value of the integral. Note that we have these possibilities (Eqs. 7.48, 7.52 and 7.53) because our integral is an improper integral. It is not uniquely defined until we specify the particular limiting process (or average) to be used.

Evaluation of definite integrals—exponential forms. With exponential or hyperbolic functions present in the integrand, life gets somewhat more complicated than before. Instead of a general overall prescription, the contour must be chosen to fit the specific integral. These cases are also opportunities to illustrate the versatility and power of contour integration.

As an example, we consider an integral that will be quite useful in developing a relation between $z!$ and $(-z)!$. Notice how the periodicity along the imaginary axis is exploited.

Example 7.2.4

Evaluate

$$I = \int_{-\infty}^{\infty} \frac{e^{ax}}{1 + e^x}\, dx, \qquad 0 < a < 1. \tag{7.54}$$

The limits on a are necessary (and sufficient) to prevent the integral from diverging

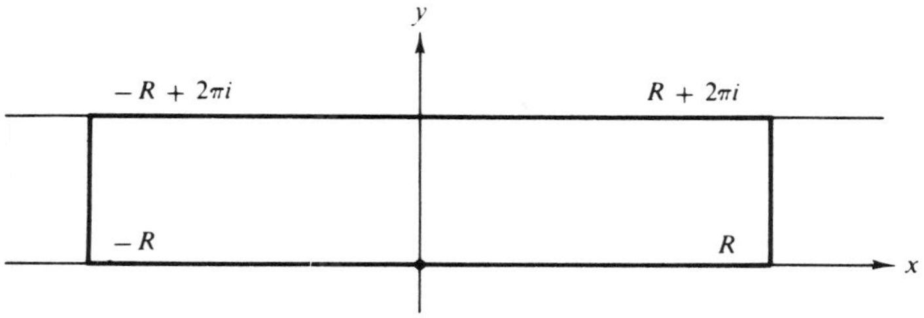

FIG. 7.10

as $x \rightarrow \pm\infty$. This integral (Eq. 7.54) may be handled by replacing the real variable x by the complex variable z and integrating around the contour shown in Fig. 7.10. If we take the limit as $R \rightarrow \infty$, the real axis, of course, leads to the integral we want. The return path along $y = 2\pi$ is chosen to leave the denominator of the integral invariant, at the same time introducing a constant factor $e^{i2\pi a}$ in the numerator. We have, in the complex plane,

$$\oint \frac{e^{az}}{1+e^z}\,dz = \lim_{R\rightarrow\infty}\left(\int_{-R}^{R}\frac{e^{ax}}{1+e^x}\,dx - e^{i2\pi a}\int_{-R}^{R}\frac{e^{ax}}{1+e^x}\,dx\right)$$
$$= (1 - e^{i2\pi a})\int_{-\infty}^{\infty}\frac{e^{ax}}{1+e^x}\,dx. \tag{7.55}$$

In addition there are two vertical sections ($0 \leqslant y \leqslant 2\pi$), which vanish (exponentially) as $R \rightarrow \infty$.

Now where are the poles and what are the residues? We have a pole when

$$e^z = e^x e^{iy} = -1. \tag{7.56}$$

Equation 7.56 is satisfied at $z = 0 + i\pi$. By a Laurent expansion in powers of $(z - i\pi)$ the pole is seen to be a simple pole with a residue of $-e^{i\pi a}$. Then, applying the residue theorem once more,

$$(1 - e^{i2\pi a})\int_{-\infty}^{\infty}\frac{e^{ax}}{1+e^x}\,dx = 2\pi i(-e^{i\pi a}). \tag{7.57}$$

This quickly reduces to

$$\int_{-\infty}^{\infty}\frac{e^{ax}}{1+e^x}\,dx = \frac{\pi}{\sin a\pi}, \qquad 0 < a < 1. \tag{7.58}$$

Using the beta function (Section 10.4), the integral can be shown to be equal to the product $(a-1)!\,(-a)!$. This results in the interesting and useful factorial function relation

$$a!(-a)! = \frac{\pi a}{\sin \pi a}. \tag{7.59}$$

Although Eq. 7.58 holds for real a, $0 < a < 1$, Eq. 7.59 may be extended by analytic continuation to all values of a, real and complex, excluding only real integral values.

As a final example of contour integrals of exponential functions we consider Bernoulli numbers again.

EXAMPLE 7.2.5—BERNOULLI NUMBERS

In Section 5.9 the Bernoulli numbers were defined by the expansion

$$\frac{x}{e^x - 1} = \sum_{n=0}^{\infty} \frac{B_n}{n!} x^n. \tag{7.60}$$

Replacing x with z, we have a Taylor series with

$$B_n = \frac{n!}{2\pi i} \oint_{C_0} \frac{z}{e^z - 1} \frac{dz}{z^{n+1}}, \tag{7.61}$$

where the contour C_0 is around the origin counterclockwise with $|z| < 2\pi$ to avoid the poles at $\pm 2\pi i$.

For $n = 0$ we have a simple pole at $z = 0$ with a residue of $+1$. Hence by Eq. 7.10

$$B_0 = \frac{0!}{2\pi i} \cdot 2\pi i(1) = 1. \tag{7.62}$$

For $n = 1$ the singularity at $z = 0$ becomes a second-order pole. The residue may be shown to be $-\frac{1}{2}$ by series expansion of the exponential, followed by a binomial expansion. This results in

$$B_1 = \frac{1!}{2\pi i} \cdot 2\pi i \left(-\frac{1}{2}\right) = -\frac{1}{2}. \tag{7.63}$$

For $n \geqslant 2$ this procedure becomes rather tedious, and we resort to a different means of evaluating Eq. 7.61. The contour is deformed, as shown in Fig. 7.11.

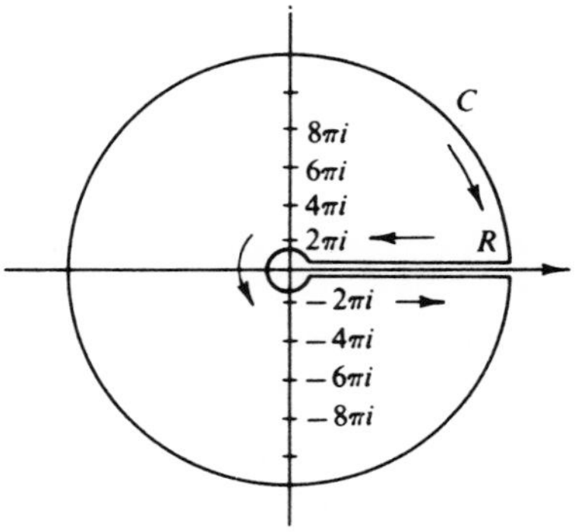

FIG. 7.11 Contour of integration for Bernoulli numbers

The new contour C still encircles the origin, as required, but now it also encircles (in a negative direction) an infinite series of singular points along the imaginary axis at $z = \pm p2\pi i$, $p = 1, 2, 3, \ldots$. The integration back and forth along the x-axis cancels out and for $R \to \infty$, the integration over the infinite circle, yields zero. Remember that $n \geqslant 2$. Therefore

$$\oint_{C_0} \frac{z}{e^z - 1}\frac{dz}{z^{n+1}} = -2\pi i \sum_{p=1}^{\infty} \text{residues} \qquad (z = \pm p2\pi i). \tag{7.64}$$

At $z = p2\pi i$ we have a simple pole with a residue $(p2\pi i)^{-n}$. When n is odd, the residue from $z = p2\pi i$ exactly cancels that from $z = -p2\pi i$ and $B_{n\,\text{odd}} = 0$, $n = 3, 5, 7$, etc. For n even the residues add, giving

$$\begin{aligned} B_n &= \frac{n!}{2\pi i}(-2\pi i)2\sum_{p=1}^{\infty}\frac{1}{p^n(2\pi i)^n} \\ &= -\frac{(-1)^{n/2}2\,n!}{(2\pi)^n}\sum_{p=1}^{\infty}p^{-n} \\ &= -\frac{(-1)^{n/2}2\,n!}{(2\pi)^n}\zeta(n), \end{aligned} \tag{7.65}$$

where $\zeta(n)$ is the Riemann zeta function introduced in Section 5.9. Equation 7.65 corresponds to Eq. 5.151 of Section 5.9.

EXERCISES

7.2.1 Determine the nature of the singularities of each of the following functions and evaluate the residues ($a > 0$).

(a) $\dfrac{1}{z^2 + a^2}$. (b) $\dfrac{1}{(z^2 + a^2)^2}$.

(c) $\dfrac{z^2}{(z^2 + a^2)^2}$. (d) $\dfrac{\sin 1/z}{z^2 + a^2}$.

(e) $\dfrac{ze^{+iz}}{z^2 + a^2}$. (f) $\dfrac{ze^{+iz}}{z^2 - a^2}$.

(g) $\dfrac{e^{+iz}}{z^2 - a^2}$. (h) $\dfrac{z^{-k}}{z + 1}$, $0 < k < 1$.

7.2.2 Locate the singularities and evaluate the residues of each of the following functions

(a) $z^{-n}(e^z - 1)^{-1}$, $z \neq 0$,

(b) $\dfrac{z^2 e^z}{1 + e^{2z}}$.

7.2.3 The statement that the integral half way around a singular point is equal to one half the integral all the way around was limited to simple poles. Show, by a specific example, that

$$\int_{\text{Semicircle}} f(z)\,dz = \frac{1}{2}\oint_{\text{Circle}} f(z)\,dz$$

does not necessarily hold if the integral encircles a pole of higher order. *Hint*. Try $f(z) = z^{-2}$.

7.2.4. Apply the calculus of residues to show that

$$\int_0^{2\pi} \frac{d\theta}{1+\varepsilon\cos\theta} = \frac{2\pi}{\sqrt{1-\varepsilon^2}},$$

where $-1 < \varepsilon < 1$.

7.2.5 Generalizing Ex. 7.2.4 show that

$$\int_0^{2\pi} \frac{d\theta}{a \pm b\cos\theta} = \int_0^{2\pi} \frac{d\theta}{a \pm b\sin\theta} = \frac{2\pi}{(a^2-b^2)^{1/2}}, \qquad \text{for} \quad a > |b|.$$

What happens if $|b| > |a|$?

7.2.6 Show that

$$\int_0^{2\pi} \frac{d\theta}{1 - 2t\cos\theta + t^2} = \frac{2\pi}{1-t^2}.$$

What limits are placed on the parameter t? Why?

7.2.7 With the calculus of residues show that

$$\int_0^{\pi} \cos^{2n}\theta\, d\theta = \pi\frac{(2n)!}{2^{2n}(n!)^2} = \pi\frac{(2n-1)!!}{(2n)!!}, \qquad n = 0, 1, 2, \ldots.$$

(The double factorial notation is defined in Section 10.1).

Hint: $\cos\theta = \frac{1}{2}(e^{i\theta} + e^{-i\theta}) = \frac{1}{2}(z + z^{-1}), \qquad |z| = 1.$

7.2.8 Prove that

$$\int_0^{\infty} \frac{\sin^2 x}{x^2}\, dx = \frac{\pi}{2}.$$

Hint. $\sin^2 x = \frac{1}{2}(1 - \cos 2x)$.

7.2.9 A quantum mechanical calculation of a transition probability leads to the function $f(t, \omega) = 2(1 - \cos\omega t)/\omega^2$. Show that

$$\int_{-\infty}^{\infty} f(t, \omega)\, d\omega = 2\pi t.$$

7.2.10 Show that ($a > 0$)

(a) $\displaystyle\int_{-\infty}^{\infty} \frac{\cos x}{x^2 + a^2}\, dx = \frac{\pi}{a} e^{-a}.$

How is the right side modified if $\cos x$ is replaced by $\cos kx$?

(b) $\displaystyle\int_{-\infty}^{\infty} \frac{x \sin x}{x^2 + a^2}\, dx = \pi e^{-a}.$

How is the right side modified if $\sin x$ is replaced by $\sin kx$?

7.2.11 Use the contour shown (Fig. 7.12) with $R \to \infty$ to prove that

$$\int_{-\infty}^{\infty} \frac{\sin x}{x}\, dx = \pi.$$

7.2.12 In the quantum theory of atomic collisions we encounter the integral

$$I = \int_{-\infty}^{\infty} \frac{\sin t}{t} e^{ipt}\, dt$$

in which p is real. Show that

$$I = 0, \qquad |p| > 1$$
$$I = \pi, \qquad |p| < 1.$$

What happens if $p = \pm 1$?

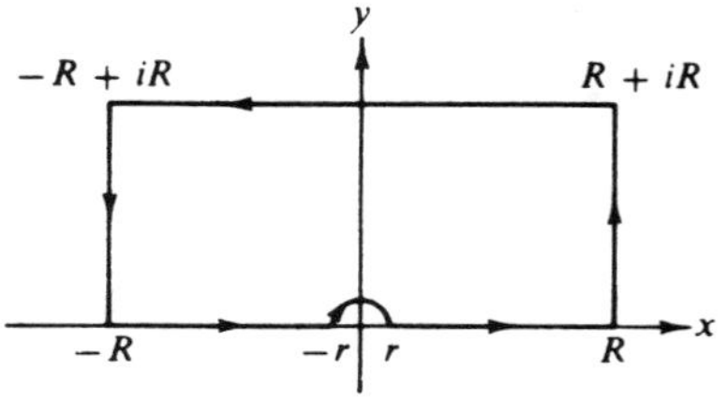

FIG. 7.12

7.2.13 Evaluate

$$\int_0^{\infty} \frac{(\ln x)^2}{1 + x^2}\, dx.$$

(a) by appropriate series expansion of the integrand to obtain

$$4 \sum_{n=0}^{\infty} (-1)^n (2n+1)^{-3},$$

(b) by contour integration to obtain

$$\frac{\pi^3}{8}.$$

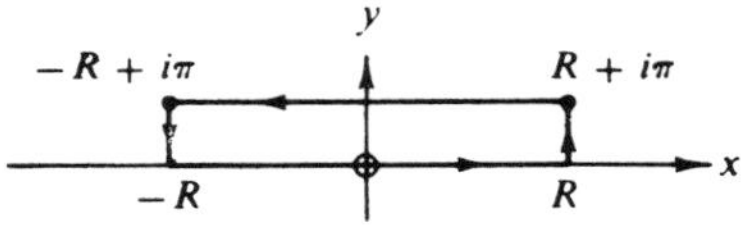

FIG. 7.13

Hint. $x \to z = e^t$. Try the contour shown in Fig. 7.13, letting $R \to \infty$.

7.2.14 Show that

$$\int_0^{\infty} \frac{x^a}{(x+1)^2}\, dx = \frac{\pi a}{\sin \pi a},$$

where $-1 < a < 1$. Here is still another way of deriving Eq. 7.59.
Hint. Use the contour shown in Fig. 7.14, noting that $z = 0$ is a branch point and the positive x-axis is a cut line. Note also the comments on phases following Example 7.1.1.

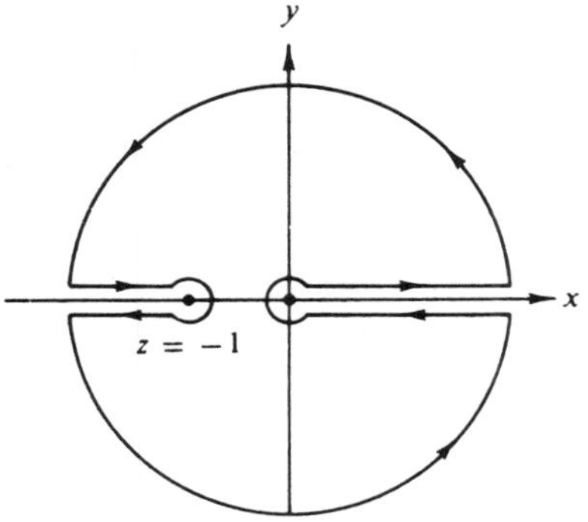

FIG. 7.14

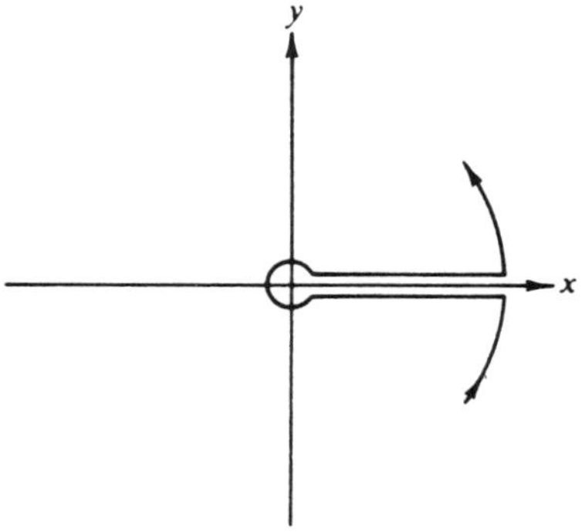

FIG. 7.15

7.2.15 Show that

$$\int_0^{\infty} \frac{x^{-a}}{x+1}\, dx = \frac{\pi}{\sin a\pi},$$

where $0 < a < 1$. This opens up another way of deriving the factorial function relation given by Eq. 7.59.

Hint. You have a branch point and you will need a cut line. Recall that $z^{-a} = w$ in polar form is

$$[re^{i(\theta+2\pi n)}]^{-a} = \rho e^{i\varphi},$$

which leads to $-a\theta - 2an\pi = \varphi$.
You must restrict n to zero (or any other single integer) in order that φ may be uniquely specified. Try the contour shown in Fig. 7.15.

7.2.16 Several of the Bromwich integrals, Section 15.11, involve a portion that may be approximated by

$$I(y) = \int_{0+iy}^{a+iy} \frac{e^{zt}}{z^{1/2}}\, dz .$$

Here a and t are positive and finite. Show that

$$\lim_{y\to\infty} I(y) = 0.$$

7.3 Applications of the Calculus of Residues

In this section, we consider three applications of the calculus of residues. First, two from mathematical analysis, and then a problem of the stability of an amplifier.

Inversion of power series. In Section 5.7 the problem of taking a known series expansion

$$w = a_1 z + a_2 z^2 + a_3 z^3 + \cdots \tag{7.66}$$

and rewriting it as

$$z = b_1 w + b_2 w^2 + b_3 w^3 + \cdots \tag{7.67}$$

was investigated. The first few coefficients b_i were found in terms of the known a_i by a direct substitution procedure. We can now rederive these coefficients by a more powerful and more elegant method.

We need the equality

$$z(w) = \frac{1}{2\pi i} \int_C \frac{t(dw/dt)}{w(t) - w(z)}\, dt, \tag{7.68}$$

where t is a (complex) variable of integration. To verify Eq. 7.68 we calculate the residue at $t = z$. Using Exercise 7.1.1 with $m = 1$ and a Taylor expansion of $w(t)$ about the point $t = z$ we find that the residue of the integrand is

$$\begin{aligned} a_{-1} &= \lim_{t\to z} \frac{(t-z)t(dw/dt)}{w(t) - w(z)} \\ &= \lim_{t\to z} \frac{(t-z)z(dw/dt)}{w(z) + (t-z)(dw/dt) + [(t-z)^2/2!](d^2w/dt^2) + \cdots - w(z)}. \end{aligned} \tag{7.69}$$

In the limit as $t \to z$, Eq. 7.69 reduces to $a_{-1} = z$ verifying Eq. 7.68.

The contour C in Eq. 7.68 is chosen so that $w(t) \neq 0$ on and within the contour except for the zero of $w(t)$ at the origin. Then z is restricted to keep $|w(z)| < |w(t)|$ over the entire contour. With this restriction the integrand may be expanded by the binomial theorem to yield

$$z(w) = \frac{1}{2\pi i}\oint \frac{t(dw/dt)}{w(t)} \sum_{n=0}^{\infty}\left[\frac{w(z)}{w(t)}\right]^n dt$$

$$= \sum_{n=0}^{\infty}[w(z)]^n \frac{1}{2\pi i}\oint \frac{t(dw/dt)}{[w(t)]^{n+1}}\, dt. \tag{7.70}$$

Comparing Eqs. 7.67 and 7.70, we equate coefficients of equal powers of $w(z)$ to obtain

$$b_n = \frac{1}{2\pi i}\oint \frac{t(dw/dt)}{[w(t)]^{n+1}}\, dt \tag{7.71}$$

Now the problem reduces to an evaluation of the integral in Eq. 7.71. From Eq. 7.66 the integrand I of Eq. 7.71 becomes

$$I = \frac{t(a_1 + 2a_2 t + \cdots)}{t^{n+1}(a_1 + a_2 t + \cdots)^{n+1}}, \tag{7.72}$$

showing a pole of order n at $t = 0$. Since $t = 0$ is the only zero of $w(t)$ on and within contour C,

$$b_n = \frac{1}{(n-1)!}\frac{d^{n-1}}{dt^{n-1}}\left[\frac{t^{n+1}(dw/dt)}{[w(t)]^{n+1}}\right]_{t=0} \tag{7.73}$$

by Exercise 7.1.1.

In an alternate procedure we integrate Eq. 7.71 by parts to obtain

$$b_n = \frac{1}{2\pi i}\oint \frac{t(dw/dt)}{w^{n+1}}\, dt = -\frac{1}{2\pi n i}\oint t\frac{d}{dt}[w^{-n}]\, dt$$

$$= \frac{1}{2\pi n i}\oint \frac{dt}{w^n}. \tag{7.74}$$

There is no "integrated part" because we have integrated a single-valued function over a closed loop. Now, applying the residue theorem, we have

$$b_n = \frac{1}{n!}\left[\frac{d^{n-1}}{dt^{n-1}}\left(\frac{t}{w(t)}\right)^n\right]_{t=0}. \tag{7.75}$$

This method of inverting a series is particularly useful when $w(t)$ can be expressed in closed form so that Eqs. 7.73 or 7.75 can be differentiated directly.

If the series expansion of $w(t)$ is used, we get the new coefficients b_i in terms of the original known a_j. The first few coefficients are

$$\begin{aligned} b_1 &= a_1^{-1}, \\ b_2 &= -a_2 a_1^{-3}, \\ b_3 &= a_1^{-3}(2a_2^2 a_1^{-2} - a_3 a_1^{-1}), \\ b_4 &= a_1^{-4}(-5a_2^3 a_1^{-3} + 5a_2 a_3 a_1^{-2} - a_4 a_1^{-1}), \end{aligned} \tag{7.76}$$

in agreement with the results of Section 5.7.

Infinite products. By Cauchy's integral formula (Section 6.4) and the residue theorem

$$\frac{1}{2\pi i}\oint_{C_m}\frac{f(z)}{z-z_0}\,dz = f(z_0) + \sum_{k=1}^{m} a_{-1}(z=z_k). \tag{7.77}$$

Here $f(z)$ is a function that is analytic everywhere except at isolated (first-order) poles, $z = z_k$, $k = 1, 2, 3, \ldots, m$. The index k is ordered so that $|z_1| \leqslant |z_2| \leqslant |z_3| \leqslant \cdots$. The contour C_m is a circle centered at the origin and enclosing poles at $z = z_1, z_2, \ldots, z_m$. Equation 7.77 follows by deforming C_m, as shown in Fig. 7.16.

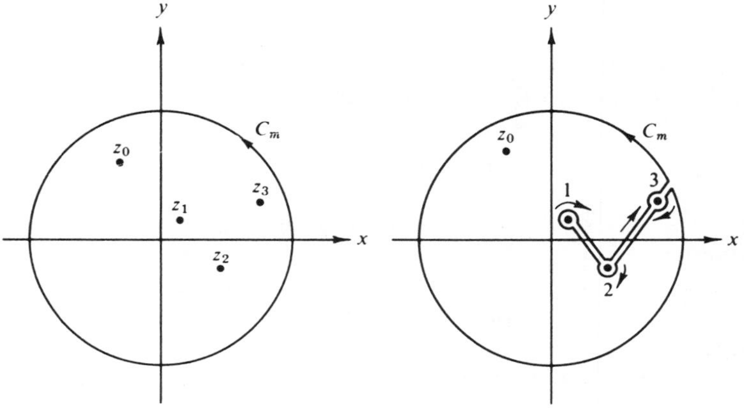

FIG. 7.16 Excluding isolated singularities

The residue of $f(z)/(z - z_0)$ at $z = z_k$ is just

$$a_{-1} = \lim_{z\to z_k}\frac{(z-z_k)f(z)}{z-z_0} = \frac{b_{-1}(z=z_k)}{z_k - z_0}, \tag{7.78}$$

where b_{-1} is the residue of $f(z)$ at $z = z_k$. Therefore Eq. 7.77 may be rewritten as

$$f(z_0) = \frac{1}{2\pi i}\oint_{C_m}\frac{f(z)}{z-z_0}\,dz + \sum_{k=1}^{m}\frac{b_{-1}(z=z_k)}{z_0 - z_k}. \tag{7.79}$$

Returning to our original integral, we have the identity

$$\frac{1}{2\pi i}\oint_{C_m}\frac{f(z)}{z-z_0}\,dz = \frac{1}{2\pi i}\oint_{C_m}\frac{f(z)\,dz}{z} + \frac{1}{2\pi i}\oint_{C_m}\frac{z_0 f(z)\,dz}{z(z-z_0)}, \tag{7.80}$$

Applying Eq. 7.79 to the first integral on the right side of Eq. 7.80, $z_0 = 0$, we obtain

$$\frac{1}{2\pi i}\oint_{C_m}\frac{f(z)}{z}\,dz = f(0) + \sum_{k=1}^{m}\frac{b_{-1}(z=z_k)}{z_k}. \tag{7.81}$$

Now if $f(z)/z^2$ falls off faster than $|z|^{-1}$ as $|z| \to \infty$, the second integral on the right-hand side of Eq. 7.80 will vanish as the radius of C_m is extended to infinity ($m \to \infty$). Then, substituting Eq. 7.81 into Eq. 7.80 and the result into Eq. 7.79, we have

$$f(z_0) = f(0) + \sum_{k=1}^{\infty} b_{-1}\left(\frac{1}{z_0 - z_k} + \frac{1}{z_k}\right). \tag{7.82}$$

This is a form of the Mittag–Leffler expansion.

The next step is to replace $f(z)$ by a function that is more convenient to handle. Suppose $g(z)$ is a function that is everywhere analytic in the finite complex plane and has simple zeros at isolated points $z = z_k$. We construct

$$f(z) = \frac{dg(z)/dz}{g(z)} = \frac{g'}{g}, \tag{7.83}$$

and this function $f(z)$ will satisfy the conditions of Eq. 7.77. With b_{-1}, the residue of g'/g; equal to unity at all poles, z_k (Eq. 7.82), becomes

$$\frac{g'(z)}{g(z)} = \frac{g'(0)}{g(0)} + \sum_{k=1}^{\infty}\left(\frac{1}{z - z_k} + \frac{1}{z_k}\right). \tag{7.84}$$

The subscript zero of the z_0 has been dropped as no longer necessary. By integrating Eq. 7.84 from 0 to z we obtain

$$\ln g(z) = \ln g(0) + \frac{g'(0)}{g(0)} z + \sum_{k=1}^{\infty}\left[\ln\left(1 - \frac{z}{z_k}\right) + \frac{z}{z_k}\right]. \tag{7.85}$$

Converting to exponential form, we have

$$\begin{aligned} g(z) &= g(0) \exp\left\{\frac{g'(0)}{g(0)} z + \sum_{k=1}^{\infty}\left[\ln\left(1 - \frac{z}{z_k}\right) + \frac{z}{z_k}\right]\right\} \\ &= g(0) \exp\left(\frac{g'(0)}{g(0)} z\right) \prod_{k=1}^{\infty}\left[\left(1 - \frac{z}{z_k}\right)e^{z/z_k}\right], \end{aligned} \tag{7.86}$$

giving $g(z)$ as an infinite product.

This final result (Eq. 7.86) is sometimes called the Weierstrass factor theorem. In using this equation it should be remembered that $g(z)$ is required to be analytic in the entire (finite) complex plane and that it has (simple) zeros at $z = z_k$. Also $g'(z)/zg(z)$ must vanish as $|z| \to \infty$.

EXAMPLE 7.3.1—$(\sin z)/z$

Expand $(\sin z)/z$ as an infinite product. We note that $g(z) = (\sin z)/z$ has no singularities in the finite plane and that it has simple zeros at $z = \pm k\pi$, $k = 1, 2, 3, \ldots$. Therefore $z_k = \pm k\pi$. Since

$$g(0) = 1 \quad \text{and} \quad g'(0) = 0, \tag{7.87}$$

Eq. 7.86 yields

$$\begin{aligned} \frac{\sin z}{z} &= \prod_{k=1}^{\infty}\left(1 - \frac{z}{k\pi}\right)e^{z/k\pi}\left(1 + \frac{z}{k\pi}\right)e^{-z/k\pi} \\ &= \prod_{k=1}^{\infty}\left(1 - \frac{z^2}{k^2\pi^2}\right). \end{aligned} \tag{7.88}$$

This is the origin of Eq. 5.187 of Section 5.10.

Stability of amplifying circuits. An amplifier with feedback can be represented by a diagram (Fig. 7.17). The original signal is E_s and the final output E_0.

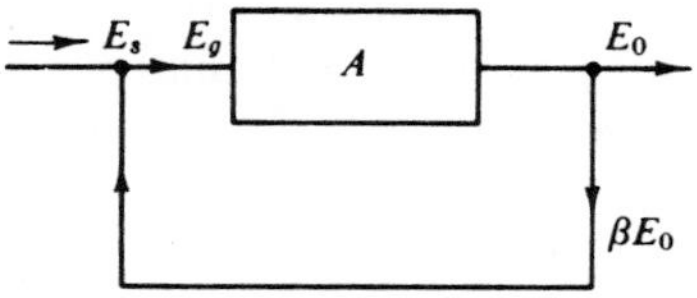

FIG. 7.17 Feedback circuit

Some of the output, βE_0, in which β may be a complex quantity, is fed back to join the signal E_s. The sum E_g is the actual input to the amplifying stage. We have

$$E_g = E_s + \beta E_0 . \tag{7.89}$$

Also

$$E_0 = AE_g , \tag{7.90}$$

where A is the amplification. Like β, A may be complex. Both A and β may depend on the angular frequency ω of the input (and output) signals. Eliminating E_g, we have

$$E_0 = AE_s + A\beta E_0 . \tag{7.91}$$

Instability (oscillation, etc.) is characterized by an output without any input, that is, $E_s = 0$, $E_0 \neq 0$. From Eq. 7.91 the condition for instability becomes

$$1 - A(\omega)\, \beta(\omega) = 0, \tag{7.92}$$

in which the real part of ω is greater than zero to correspond to a physically attainable frequency.

If $f(z) = 1 - A(z)\beta(z) = 0$ has a solution on the positive real axis, we have an exponentially increasing instability. If there is a complex solution in the right half of the complex plane, an oscillatory instability will result. So we try to determine whether $1 - A\beta$ has any solutions or roots anywhere in the right half of the complex plane.

Now the number of zeros (N), each multiplied by its multiplicity, minus the number of poles (P), each multiplied by its order or multiplicity, of a function $f(z)$ is given by[1]

$$\frac{1}{2\pi i} \oint \frac{f'(z)}{f(z)}\, dz = N - P. \tag{7.93}$$

Here N and P include the zeros and poles within the contour C. To prove Eq. 7.93 consider a zero of order n of $f(z)$ at $z = z_1$. Then

$$f(z) = (z - z_1)^n\, g(z). \tag{7.94}$$

[1] This is a special case of a more general integral sometimes known as Cauchy's integral.

Differentiating, we obtain

$$f'(z) = n(z - z_1)^{n-1} g(z) + (z - z_1)^n g'(z) \tag{7.95}$$

and the integrand of Eq. 7.93 becomes

$$\frac{f'(z)}{f(z)} = \frac{n}{z - z_1} + \frac{g'(z)}{g(z)}. \tag{7.96}$$

Integrating around z_1, we have

$$\frac{1}{2\pi i} \oint_{C_1} \frac{f'(z)}{f(z)} dz = n, \tag{7.97}$$

the multiplicity of the zero.

For a pole of order m at $z = z_2$

$$f(z) = (z - z_2)^{-m} h(z). \tag{7.98}$$

This time the integrand becomes

$$\frac{f'(z)}{f(z)} = \frac{-m}{z - z_2} + \frac{h'(z)}{h(z)}, \tag{7.99}$$

and integrating around the pole at z_2

$$\frac{1}{2\pi i} \oint_{C_2} \frac{f'(z)}{f(z)} dz = -m, \tag{7.100}$$

the negative of the order of the pole. Combining Eqs. 7.97 and 7.100, we obtain Eq. 7.93.

We shall assume that our function $f(z) = 1 - A\beta$ has no poles for $\mathscr{R}(\omega) \geqslant 0$. Then, if we integrate $f'(z)/f(z)$ around the right side of the complex plane, Eq. 7.93 will indicate whether we have included any zeros. If

$$\oint_C \frac{f'(z)}{f(z)} dz = 0, \tag{7.101}$$

then $N = 0$ and our system is stable. Here C is the imaginary axis and an (infinite) semicircle enclosing the right side of the z-plane.

To evaluate Eq. 7.93, that is, to check on Eq. 7.101, let

$$\begin{aligned} f(z) &= re^{i\theta}, \\ \ln f(z) &= \ln \mathrm{r} + i\theta. \end{aligned} \tag{7.102}$$

Then

$$\frac{f'(z)}{f(z)} = \frac{d}{dz} \ln f(z) = \frac{1}{r}\frac{dr}{dz} + i\frac{d\theta}{dz}. \tag{7.103}$$

Substituting into Eq. 7.93, we obtain

$$\oint_C \frac{f'(z)}{f(z)} dz = \oint \frac{dr}{r} + i \oint d\theta. \tag{7.104}$$

Since C is a closed contour, the first integral on the right (which yields $\ln r$) vanishes. The second integral yields 2π or zero, depending on whether the origin is included or excluded. We therefore conclude that the stability of our amplifier depends on whether a polar plot of $f(z) = 1 - A(z)\,\beta(z)$ excludes the origin as z ranges along the imaginary axis from $-i\infty$ to $+i\infty$.[1] If the origin is *not* included,

$$\oint d\theta = 0, \tag{7.105}$$

$$\frac{1}{2\pi i}\oint_C \frac{f'(z)}{f(z)}\,dz = N = 0, \tag{7.106}$$

and there are no zeros in the right half plane. Hence the amplifier is stable. This is the Nyquist stability criterion.

EXERCISES

7.3.1 (Inversion of a power series.) With the behavior of Eq. 7.71 for $n = 0$ ($b_0 = 0$), explain why the series inverted (Eq. 7.66) was adjusted to have *no constant* term.

7.3.2 Derive the infinite product representation

$$\cos x = \prod_{n=1}^{\infty}\left[1 - \frac{4x^2}{(2n-1)^2\pi^2}\right].$$

7.3.3 The product z times the gamma function, $z\Gamma(z)$ has simple poles at $z = -1, -2, -3, \ldots, -n, \ldots$. Develop an infinite product representation of the reciprocal of the gamma function.
Hint. You will encounter the Euler-Mascheroni constant.

7.3.4 (a) The function $g(z)$ is analytic in the entire finite complex plane and has simple zeros at $z = z_k$. Further, $g'(z)/z\,g(z)$ vanishes as $|z| \to \infty$. If, finally, $g(z)$ is an even function of z, show that

$$g(z) = g(0)\prod_{k=1}^{\infty}\left(1 - \frac{z^2}{z_k^2}\right),$$

where only one of the two zeros z_i and $-z_i$ is included in the infinite product.

(b) If $g(0) = 1$, show that

$$g''(0) = -2\sum_{k=1}^{\infty}\frac{1}{z_k^2},$$

$$g^{(4)}(0) = 3[g''(0)]^2 - 12\sum_{k=1}^{\infty}\frac{1}{z_k^4}.$$

Hint. One way is to consider $g''(0)$ as the integral of $g'(z)/z^2$ and $g^{(4)}(0)$ as the integral of $g''(z)/z^3$.

7.3.5 From the results of Exercise 7.3.4 show that

$$\sum_{n=1}^{\infty} n^{-2} = \frac{\pi^2}{6},$$

$$\sum_{n=1}^{\infty} n^{-4} = \frac{\pi^4}{90},$$

in agreement with Section 5.9.
Hint. Try $g(z) = (\sin z)/z$.

[1] For $\omega \to \infty$, $A \to 0$. Hence the infinite semicircle maps into the point $f(z) = 1$.

7.3.6 A simple series *RLC* circuit (Fig. 7.18) obeys the equation

$$L\ddot{I} + R\dot{I} + C^{-1}I = 0$$

for $t > 0$; the switch is closed at $t = 0$. Assume a solution of the form

$$I = I_0 e^{pt}.$$

Show that if

(a) $R = 0$, then p is on the imaginary axis (steady oscillation),

(b) $0 < \dfrac{R}{2L} < \dfrac{1}{\sqrt{LC}}$, then p is in the left half of the complex plane (damped oscillations),

(c) $\dfrac{R}{2L} = \dfrac{1}{\sqrt{LC}}$, then p is on the negative real axis, double root (critical damping),

(d) $\dfrac{R}{2L} > \dfrac{1}{\sqrt{LC}}$, then p is on the negative real axis (overdamping).

Note that if R is made negative in these last three cases p goes into the right half of the complex plane and we have instability.

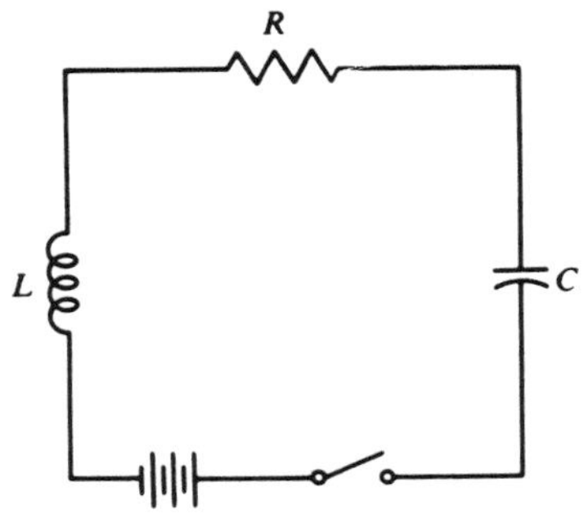

FIG. 7.18 *RLC* Circuit

7.4 The Method of Steepest Descents

In analyzing problems in mathematical physics it is often desirable to know the behavior of a function for large values of the variable, that is, the asymptotic behavior of the function. Specific examples are furnished by the various Bessel functions (Chapter 11) and by the gamma function (Chapter 10). The method of steepest descents is one of determining such asymptotic behavior when the function can be expressed as an integral of the general form

$$I(s) = \int_C g(z)\, e^{sf(z)}\, dz. \tag{7.107}$$

For the present, let us take s to be real. The contour of integration C is then chosen so that the real part of $f(z)$ approaches minus infinity at both limits and that the integrand will vanish at the limits, or is chosen as a closed contour. It is further assumed that the factor $g(z)$ in the integrand is dominated by the exponential in the region of interest.

If the parameter s is large and positive, the value of the integrand will become large when the real part of $f(z)$ is large and small when the real part of $f(z)$ is small or negative. In particular, as s is permitted to increase indefinitely (leading to the asymptotic dependence), the entire contribution of the integrand to the integral will come from the region in which the real part of $f(z)$ takes on a positive maximum value. Away from this positive maximum the integrand will become negligibly small in comparison. This is seen by expressing $f(z)$ as

$$f(z) = u(x, y) + i\, v(x, y). \tag{7.108}$$

Then the integral may be written

$$I(s) = \int_C g(z)\, e^{su(x,y)} e^{isv(x,y)}\, dz. \tag{7.109}$$

If now, in addition, we impose the condition that the imaginary part of the exponent, $i\,v(x, y)$, be constant in the region in which the real part takes on its maximum value, that is, $v(x, y) = v(x_0, y_0) = v_0$, we may approximate the integral by

$$I(s) \approx e^{isv_0} \int_C g(z)\, e^{su(x,y)}\, dz. \tag{7.110}$$

Away from the maximum of the real part, the imaginary part may be permitted to oscillate as it wishes, for the integrand is negligibly small and the varying phase factor is therefore irrelevant.

The real part of $s\,f(z)$ is a maximum for a given s when the real part of $f(z)$, $u(x, y)$, is a maximum. This implies that

$$\frac{\partial u}{\partial x} = \frac{\partial u}{\partial y} = 0 \tag{7.111}$$

and therefore, by use of the Cauchy Riemann conditions of Section 6.2, that

$$\frac{df(z)}{dz} = 0. \tag{7.112}$$

We proceed to search for such zeros of the derivative.

It is essential to note that the maximum value of $u(x, y)$ is the maximum only along a given contour. In the finite plane neither the real nor the imaginary part of our analytic function possesses an absolute maximum. This may be seen by recalling that both u and v satisfy Laplace's equation

$$\frac{\partial^2 u}{\partial x^2} + \frac{\partial^2 u}{\partial y^2} = 0. \tag{7.113}$$

From this, if the second derivative with respect to x is positive, the second derivative with respect to y must be negative, and therefore neither u nor v can possess an absolute maximum or minimum. Since the function $f(z)$ was taken to be analytic, singular points are clearly excluded. The vanishing of the derivative (Eq. 7.112) then implies that we have a saddle point, a stationary value, which may be a maximum of $u(x, y)$ for one contour and a minimum for another.

Our problem, then, is to choose the contour of integration to satisfy two conditions. (a) The contour must be chosen so that $u(x, y)$ has a maximum at the saddle point. (b) The contour must pass through the saddle in such a way that the imaginary part, $v(x, y)$, is a constant. This second condition leads to the path of steepest descent and gives the method its name. From Section 6.2, especially Exercise 6.2.1, and Section 6.7 we know that the curves corresponding to u = constant and v = constant form an orthogonal system. This means that a curve $v = c_i$, constant, is everywhere tangential to the gradient of u, ∇u. Hence the curve v = constant is the curve that gives the line of steepest descent from the saddle point.[1]

[1] The line of steepest ascent is also characterized by constant v. The saddle point must be inspected carefully to distinguish the line of steepest descent from the line of steepest ascent. This is discussed later in two examples.

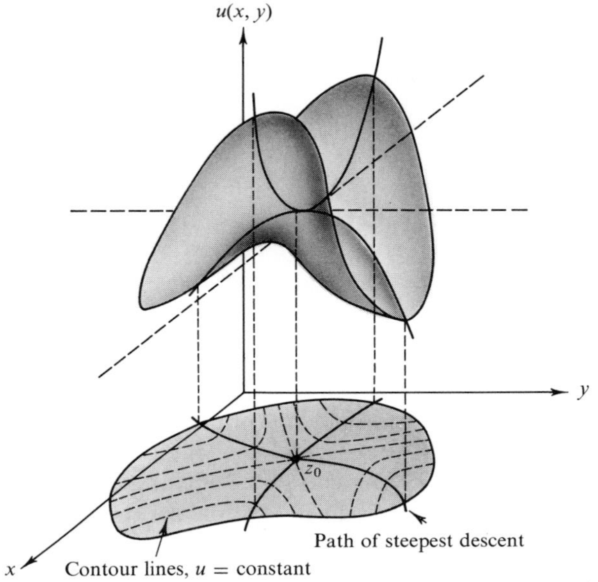

FIG. 7.19 A saddle point

At the saddle point the function $f(z)$ can be expanded in a Taylor series to give

$$f(z) = f(z_0) + \tfrac{1}{2}(z - z_0)^2 f''(z_0) + \cdots . \tag{7.114}$$

The first derivative is absent, since obviously Eq. 7.112 is satisfied. The first correction term, $\frac{1}{2}(z - z_0)^2 f''(z_0)$, is real and negative. It is real, for we have specified that the imaginary part shall be constant along our contour and negative because we are moving down from the saddle point or mountain pass. Then, assuming that $f''(z_0) \neq 0$,

$$f(z) - f(z_0) \approx \tfrac{1}{2}(z - z_0)^2 f''(z_0) = -\frac{1}{2s} t^2, \tag{7.115}$$

which serves to define a new variable t. If $(z - z_0)$ is written in polar form

$$(z - z_0) = \delta e^{i\alpha}, \tag{7.116}$$

(with the phase α held constant), we have

$$t^2 = -s f''(z_0)\, \delta^2 e^{2i\alpha}. \tag{7.117}$$

Since t is real,[1] it may be written

$$t = \pm \delta |s f''(z_0)|^{1/2}. \tag{7.118}$$

Substituting Eq. 7.115 into Eq. 7.107, we obtain

$$I(s) \approx g(z_0)\, e^{sf(z_0)} \int_{-\infty}^{\infty} e^{-t^2/2} \frac{dz}{dt}\, dt. \tag{7.119}$$

[1] The phase of the contour (specified by α) at the saddle point is chosen so that $\mathscr{I}[f(z) - f(z_0)] = 0$, that is, $\frac{1}{2}(z - z_0)^2 f''(z_0)$ must be real.

Since

$$\frac{dz}{dt} = \left(\frac{dt}{dz}\right)^{-1} = \left(\frac{dt}{d\delta}\frac{d\delta}{dz}\right)^{-1} = |s f''(z_0)|^{-1/2} e^{i\alpha}, \tag{7.120}$$

Eq. 7.119 becomes

$$I(s) \approx \frac{g(z_0)\, e^{sf(z_0)} e^{i\alpha}}{|s f''(z_0)|^{1/2}} \int_{-\infty}^{\infty} e^{-t^2/2}\, dt. \tag{7.121}$$

It will be noted that the limits have been set as minus infinity to plus infinity. This is permissible, for the integrand is essentially zero when t departs appreciably from the origin. Noting that the remaining integral is just a Gauss error integral, we finally obtain

$$I(s) \approx \frac{\sqrt{2\pi}\, g(z_0)\, e^{sf(z_0)} e^{i\alpha}}{|s f''(z_0)|^{1/2}}. \tag{7.122}$$

The phase α was introduced in Eq. 7.116 as the phase of the contour as it passed through the saddle point. It is chosen so that the two conditions given [α = constant; $\mathcal{R}f(z)$ = maximum] are satisfied. It sometimes happens that the contour passes through two or more saddle points in succession. If this is the case, we need only add the contribution made by Eq. 7.122 from each of the saddle points in order to get an approximation for the total integral.

One note of warning: we assumed that the only significant contribution to the integral came from the immediate vicinity of the saddle point(s) $z = z_0$, that is,

$$\mathcal{R}[f(z)] = u(x, y) \ll u(x_0, y_0)$$

over the entire contour away from $z_0 = x_0 + iy_0$. This condition must be checked for each new problem (Ex. 7.4.5.)

Example 7.4.1—Asymptotic Form of the Hankel Function, $H_\nu^{(1)}(s)$

In Section 11.4 it is shown that the Hankel functions, which satisfy Bessel's equation, may be defined by

$$H_\nu^{(1)}(s) = \frac{1}{\pi i} \int_{0C_1}^{-\infty} e^{(s/2)(z-1/z)} \frac{dz}{z^{\nu+1}}, \tag{7.123}$$

$$H_\nu^{(2)}(s) = \frac{1}{\pi i} \int_{-\infty C_2}^{0} e^{(s/2)(z-1/z)} \frac{dz}{z^{\nu+1}}. \tag{7.124}$$

The contour C_1 is the curve in the upper half plane of Fig. 7.20. The contour C_2 is in the lower half plane. We apply the method of steepest descents to the first Hankel function, $H_\nu^{(1)}(s)$, which is conveniently in the form specified by Eq. 7.107, with $f(z)$ given by

$$f(z) = \frac{1}{2}\left(z - \frac{1}{z}\right). \tag{7.125}$$

By differentiating we obtain

$$f'(z) = \frac{1}{2} + \frac{1}{2z^2}. \tag{7.126}$$

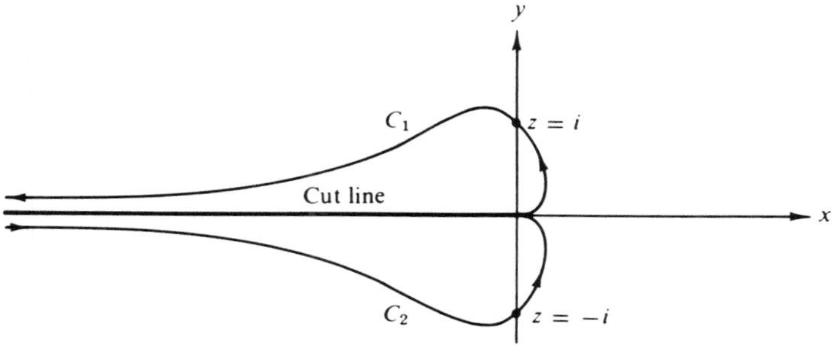

FIG. 7.20 Hankel function contours

Setting $f'(z) = 0$ in accordance with Eq. 7.112,

$$z = i, \ -i. \tag{7.127}$$

Hence there are saddle points at $z = +i$ and $z = -i$. The integral for $H_\nu^{(1)}(s)$ is chosen so that it starts at the origin, moves out tangentially to the positive real axis, then around through the saddle point at $z = +i$, and on out to minus infinity, asymptotic with the negative real axis. We must choose the contour through the point $z = +i$ so that the real part of $(z - 1/z)$ will be a maximum and the phase will be constant in the vicinity of the saddle point.

In the vicinity of the saddle point at $z_0 = +i$ we have

$$z - i = \delta e^{i\alpha}, \tag{7.128}$$

where δ is a small number. Then

$$\begin{aligned} 2f(z) = z - \frac{1}{z} &= \delta e^{i\alpha} + i - \frac{1}{\delta e^{i\alpha} + i} \\ &= \delta \cos \alpha + i(\delta \sin \alpha + 1) - \frac{1}{\delta \cos \alpha + i(\delta \sin \alpha + 1)} \\ &= \delta \cos \alpha + i(\delta \sin \alpha + 1) - \frac{\delta \cos \alpha - i(\delta \sin \alpha + 1)}{1 + 2\delta \sin \alpha + \delta^2}. \end{aligned} \tag{7.129}$$

Therefore our real part becomes

$$\mathscr{R}\left(z - \frac{1}{z}\right) = \delta \cos \alpha - \delta \cos \alpha(1 + 2\delta \sin \alpha + \delta^2)^{-1}. \tag{7.130}$$

Recalling that δ is small, we expand by the binomial theorem and neglect terms of order δ^3 and higher.

$$\mathscr{R}\left(z - \frac{1}{z}\right) = 2\delta^2 \cos \alpha \sin \alpha + O(\delta^3) \approx \delta^2 \sin 2\alpha. \tag{7.131}$$

We see that the real part of $(z - 1/z)$ will take on an extreme value if $\sin 2\alpha$ is an extremum, that is, if 2α is $\pi/2$ or $3\pi/2$. Hence the phase of the contour α should be chosen to be $\pi/4$ or $3\pi/4$. One choice will represent the path of steepest descent we want. The other choice will represent a path of steepest ascent that we must avoid. We distinguish the two possibilities by substituting in the specific values of α. For $\alpha = \pi/4$

$$\mathscr{R}\left(z - \frac{1}{z}\right) = \delta^2, \tag{7.132}$$

whereas, for $\alpha = 3\pi/4$

$$\mathscr{R}\left(z - \frac{1}{z}\right) = -\delta^2. \tag{7.133}$$

Direct substitution into Eq. 7.122 now yields

$$H_\nu^{(1)}(s) = \frac{1}{\pi i} \frac{\sqrt{2\pi} i^{-\nu-1} e^{(s/2)(i-1/i)} e^{i(3\pi/4)}}{|(s/2)(-2/i^3)|^{1/2}}$$

$$= \sqrt{\frac{2}{\pi s}}\, e^{(i\pi/2)(-\nu-2)} e^{is} e^{i(3\pi/4)}. \tag{7.134}$$

By combining terms we finally obtain

$$H_\nu^{(1)}(s) \approx \sqrt{\frac{2}{\pi s}}\, e^{i(s-\nu(\pi/2)-\pi/4)} \tag{7.135}$$

as the leading term of the asymptotic expansion of the Hankel function $H_\nu^{(1)}(s)$. Additional terms, if desired, may be picked up by the Stokes method, described in Section 11.6.

EXAMPLE 7.4.2—ASYMPTOTIC FORM OF THE FACTORIAL FUNCTION, $s!$

In many physical problems, particularly in the field of statistical mechanics, it is desirable to have an accurate approximation of the gamma or factorial function of very large numbers. As developed in Section 10.1, the factorial function may be defined by the integral

$$s! = \int_0^\infty \rho^s e^{-\rho}\, d\rho = s^{s+1} \int_{0C}^\infty e^{s(\ln z - z)}\, dz. \tag{7.136}$$

Here we have made the substitution $\rho = zs$ in order to throw the integral into the form required by Eq. 7.107. As before, we assume that s is real and positive, from which it follows that the integrand vanishes at the limits 0 and ∞. By differentiating the z-dependence appearing in the exponent, we obtain

$$\frac{df(z)}{dz} = \frac{d}{dz}(\ln z - z) = \frac{1}{z} - 1, \tag{7.137}$$

which shows that the point $z = 1$ is a saddle point. We let

$$z - 1 = \delta e^{i\alpha}, \tag{7.138}$$

with δ small to describe the contour in the vicinity of the saddle point. Substituting into $f(z)$, we develop a series expansion

$$\begin{aligned} f(z) &= \ln(1 + \delta e^{i\alpha}) - (1 + \delta e^{i\alpha}) \\ &= \delta e^{i\alpha} - \tfrac{1}{2}\delta^2 e^{2i\alpha} + \cdots - 1 - \delta e^{i\alpha} \\ &= -1 - \tfrac{1}{2}\delta^2 e^{2i\alpha}. \end{aligned} \tag{7.139}$$

From this we see that the integrand takes on a maximum value (e^{-s}) at the saddle point if we choose our contour C to follow the real axis, a conclusion that the reader may well have reached more or less intuitively.

Direct substitution into Eq. 7.122 with $\alpha = 0$ now gives

$$s! \approx \frac{\sqrt{2\pi s^{s+1}} e^{-s}}{|s(-1^{-2})|^{1/2}} \tag{7.140}$$

Thus the first term in the asymptotic expansion of the factorial function is

$$s! \approx \sqrt{2\pi s}\, s^s e^{-s}. \tag{7.141}$$

Additional terms in the asymptotic series are developed in Section 10.3. Equation 7.141 is Stirling's formula.

In the foregoing example the calculation was carried out by assuming s to be real. This assumption is not necessary. The student may show (Exercise 7.4.6) that Eq. 7.141 also holds when s is replaced by the complex variable w, provided only that the real part of w is required to be large and positive.

EXERCISES

7.4.1 Using the method of steepest descents, evaluate the second Hankel function given by

$$H_\nu^{(2)}(s) = \frac{1}{\pi i}\int_{-\infty C_2}^{0} e^{(s/2)(z-1/z)} \frac{dz}{z^{\nu+1}},$$

with contour C_2 as shown in Fig. 7.20.

Ans. $H_\nu^{(2)}(s) \approx \sqrt{\dfrac{2}{\pi s}}\, e^{-i(s-\pi/4-\nu\pi/2)}.$

7.4.2 The negative square root in Eq. 7.118 does not appear in Eq. 7.121. What is the justification for dropping it? Illustrate your argument by detailed reference to $H_\nu^{(1)}(z)$, Example 7.4.1.

7.4.3 (a) In applying the method of steepest descent to the Hankel function $H_\nu^{(1)}(s)$, show that

$$\mathscr{R}[f(z)] < \mathscr{R}[f(z_0)] = 0$$

for z on the contour C_1 but away from the point $z = z_0 = i$.

(b) Show that $\mathscr{R}[f(z)] > 0$ for $0 < r < 1$, $\begin{cases} \dfrac{\pi}{2} < \theta \leqslant \pi \\ -\pi \leqslant \theta < -\dfrac{\pi}{2} \end{cases}$

and for $r > 1$, $-\dfrac{\pi}{2} < \theta < \dfrac{\pi}{2}$.

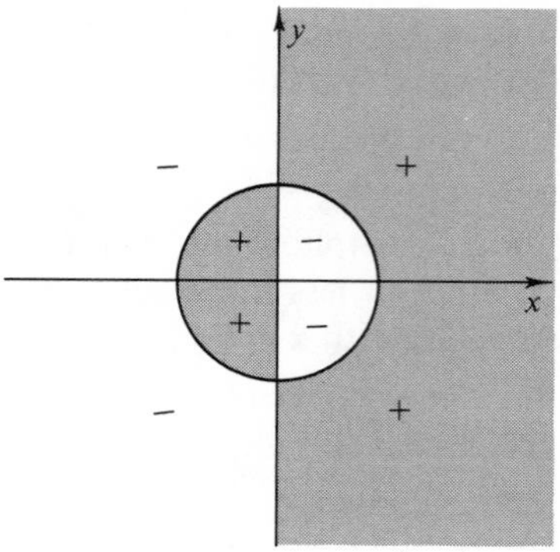

FIG. 7.21

This is why C_1 may *not* be deformed to pass through the second saddle point $z = -i$.

7.4.4 Determine the asymptotic dependence of the modified Bessel functions $I_\nu(x)$, given

$$I_\nu(x) = \frac{1}{2\pi i}\int_C e^{(x/2)(t+1/t)}\frac{dt}{t^{\nu+1}}.$$

The contour starts and ends at $t = -\infty$, encircling the origin in a positive sense and passing through *two* saddle points.

7.4.5 Determine the asymptotic dependence of the modified Bessel function of the second kind, $K_\nu(x)$, by using

$$K_\nu(x) = \frac{1}{2}\int_0^\infty e^{(-x/2)(s+1/s)}\frac{ds}{s^{1-\nu}}.$$

7.4.6 Show that Stirling's formula

$$s! \approx \sqrt{2\pi s}\, s^s e^{-s}$$

holds for complex values of s (with $\mathscr{R}(s)$ large and positive). *Hint.* This involves assigning a phase to s and then demanding that $\mathscr{I}[sf(z)] =$ constant in the vicinity of the saddle point.

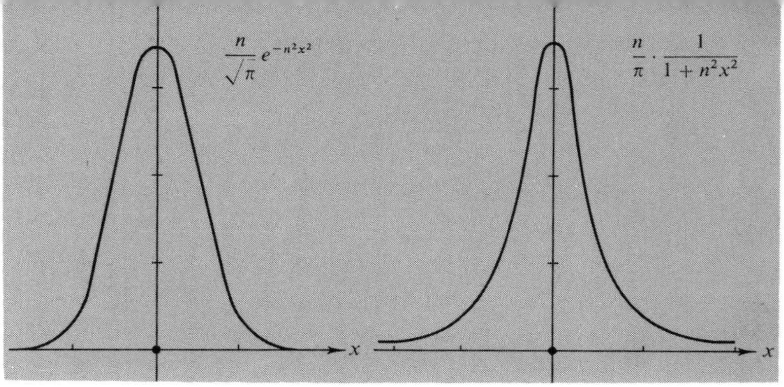

CHAPTER 8

SECOND-ORDER DIFFERENTIAL EQUATIONS

8.1 Partial Differential Equations of Theoretical Physics

Almost all of the elementary and a great many of the advanced parts of theoretical physics are formulated in terms of differential equations, often partial differential equations. Among the most frequently encountered are the following:

1. Laplace's equation, $\nabla^2\psi = 0$.
   This very common and very important equation occurs in studies of
   a. electromagnetic phenomena including electrostatics, dielectrics, steady currents, and magnetostatics,
   b. hydrodynamics (irrotational flow of perfect fluid and surface waves),
   c. heat flow,
   d. gravitation.
2. Poisson's equation, $\nabla^2\psi = -\rho/\varepsilon_0$.
3. The wave (Helmholtz) and time-independent diffusion equations, $\nabla^2\psi \pm k^2\psi = 0$.
   These equations appear in such diverse phenomena as
   a. elastic waves in solids including vibrating strings, bars, membranes,
   b. sound or acoustics,
   c. electromagnetic waves,
   d. nuclear reactors.
4. The time-dependent diffusion equation

$$\nabla^2\psi = \frac{1}{a^2}\frac{\partial\psi}{\partial t}$$

and the corresponding four-dimensional forms involving the d'Alembertian, a four-dimensional analogue of the Laplacian in Minkowski space,

$$\Box^2 = \nabla^2 + \frac{\partial^2}{\partial x_4^2} = \frac{\partial^2}{\partial x^2} + \frac{\partial^2}{\partial y^2} + \frac{\partial^2}{\partial z^2} + \frac{\partial^2}{(ic)^2\partial t^2}.$$

5. The time-dependent wave equation, $\Box^2\psi = 0$.

6. The scalar potential equation, $\Box^2\psi = -\rho/\varepsilon_0$.

7. The Klein-Gordon equation, $\Box^2\psi = \mu^2\psi$, and the corresponding vector equations in which the scalar function ψ is replaced by a vector function.

Other more complicated forms are common.

8. The Schrödinger wave equation,

$$-\frac{\hbar^2}{2m}\nabla^2\psi + V\psi = i\hbar\frac{\partial\psi}{\partial t}$$

and

$$-\frac{\hbar^2}{2m}\nabla^2\psi + V\psi = E\psi$$

for the time-independent case.

9. The equations for elastic waves and for viscous fluids and the telegraphy equation.

10. Maxwell's coupled partial differential equations for electric and magnetic fields and those of Dirac for relativistic electron wave functions.

All of these equations can be written in the form

$$H\psi = F,$$

in which H is a differential operator,

$$H\left(\frac{\partial}{\partial x}, \frac{\partial}{\partial y}, \frac{\partial}{\partial z}, \frac{\partial}{\partial t}, x, y, z\right),$$

F is a known function, and ψ is the unknown scalar (or vector) function.

Two characteristics are particularly important:

1. All of these equations are linear[1] in the unknown function ψ. As the easier physical and mathematical problems are being solved, nonlinear differential equations such as those describing shock wave phenomena are receiving more and more attention. However, both the nonlinear differential equations themselves and the numerical techniques that we often resort to for determining solutions are beyond the scope of this book.

2. These equations are all second-order differential equations. [Maxwell's and Dirac's equations are first-order but involve two unknown functions. Eliminating one unknown yields a second-order differential equation for the other (cf. Section 1.9).]

Occasionally we do encounter equations of higher order. In both the theory of the slow motion of a viscous fluid and the theory of an elastic body we find the equation

$$(\nabla^2)^2\psi = \left(\frac{\partial^4}{\partial x^4} + 2\frac{\partial^4}{\partial x^2\,\partial y^2} + \frac{\partial^4}{\partial y^4}\right)\psi = 0.$$

[1] Cf. Section 2.5 for definition of linearity.

Fortunately for introductory treatments such as this one, these higher-order differential equations are relatively rare.

Some general techniques for solving the partial differential equations are discussed in this section.

1. Separation of variables. The partial differential equation is split into ordinary differential equations which may be attacked by Frobenius' method, Section 8.4. This separation technique is introduced in Section 2.5 and is discussed further in Section 8.2. It does not always work, but when it does it is often the simplest method.

2. Integral solutions employing a Green's function. An introduction to the Green's function technique is given in Section 8.6. A more detailed treatment appears in Chapter 16.

3. Other analytical methods such as the use of integral transforms. Some of the techniques in this class are developed and applied in Chapter 15.

4. Numerical calculations. The development of modern high-speed calculating machines has opened up a wealth of possibilities based on the calculus of finite differences. Two of these, the Runge-Kutta method and a predictor-corrector method are discussed in Section 8.7. For further details of these methods and other methods the reader could start with R. W. Hamming's *Numerical Methods for Scientists and Engineers* (McGraw-Hill, 1962) and proceed to specialized references.

8.2 Separation of Variables—Ordinary Differential Equations

The equations of mathematical physics listed in Section 8.1 are all partial differential equations. Our first technique for their solution splits the partial differential equation of n variables into n ordinary differential equations. Each separation introduces an arbitrary constant of separation. If we have n variables, we shall have to introduce $n - 1$ constants, determined by the conditions imposed in the problem being solved.

In Section 2.5 the technique of separation of variables was illustrated for the wave equation in cartesian and spherical polar coordinates. In the spherical polar coordinate system the wave equation

$$\nabla^2\psi + k^2\psi = 0 \tag{8.1}$$

led to an azimuthal equation

$$\frac{d^2\Phi(\varphi)}{d\varphi^2} + m^2\,\Phi(\varphi) = 0, \tag{8.2}$$

in which $-m^2$ is a separation constant. As an illustration of how the constant is restricted, we note that φ in spherical polar coordinates is an azimuth angle. If this is a classical problem, we shall certainly require that the azimuthal solution $\Phi(\varphi)$ be single-valued, that is,

$$\Phi(\varphi + 2\pi) = \Phi(\varphi). \tag{8.3}$$

This is equivalent to requiring the azimuthal solution to have a period of 2π or some integral multiple of it.[1] Therefore m must be an integer. Which integer it is depends on the details of the problem. This is taken up in Chapter 9.

The coordinate systems described in Chapter 2 were selected from the infinitely many possible orthogonal coordinate systems, because in these systems[2] the wave and Helmholtz equations would separate into ordinary differential equations. Going through the fourteen coordinate systems one by one, the separated ordinary differential equations show considerable duplication. Whenever a coordinate corresponds to an axis of translation or to an azimuth angle, the separated equation always has the form

$$\frac{d^2\Phi(\varphi)}{d\varphi^2} = -m^2\,\Phi(\varphi)$$

for φ, the azimuth angle, and

$$\frac{d^2Z(z)}{dz^2} = \pm a^2\,Z(z) \tag{8.4}$$

for z, an axis of translation in one of the cylindrical coordinate systems. The solutions, of course, are $\sin az$ and $\cos az$ for $-a^2$ and the corresponding hyperbolic function (or exponentials) $\sinh az$ and $\cosh az$ for $+a^2$.

The Legendre equation,

$$\begin{gathered}\frac{1}{\sin\theta}\frac{d}{d\theta}\left(\sin\theta\,\frac{d\Theta}{d\theta}\right) + l(l+1)\Theta = 0,\\ (1-x^2)\frac{d^2y}{dx^2} - 2x\frac{dy}{dx} + l(l+1)y = 0,\end{gathered} \tag{8.5}$$

and the associated Legendre equation

$$\begin{gathered}\frac{1}{\sin\theta}\frac{d}{d\theta}\left(\sin\theta\,\frac{d\Theta}{d\theta}\right) + l(l+1)\Theta - \frac{m^2}{\sin^2\theta}\,\Theta = 0,\\ (1-x^2)\frac{d^2y}{dx^2} - 2x\frac{dy}{dx} + l(l+1)y - \frac{m^2}{1-x^2}\,y = 0,\end{gathered} \tag{8.6}$$

[3]

also appear frequently. As noted in Section 2.5 these equations appear when ∇^2 is used in spherical polar coordinates. Prolate and oblate spheroidal coordinates also give rise to the Legendre and associated Legendre equations.

A third equation frequently encountered is Bessel's differential equation,

$$x^2\frac{d^2y}{dx^2} + x\frac{dy}{dx} + (x^2 - n^2)y = 0. \tag{8.7}$$

[1] This also applies in most quantum mechanical problems, but the argument is much more involved.

[2] Bipolar and its rotational forms, toroidal and bispherical coordinates, are exceptions. They were included to illustrate how special coordinate systems can be used to advantage in special problems.

[3] These are equivalent algebraic forms in which $x = \cos\theta$.

In Sections 2.5 and 2.6 spherical polar and circular cylindrical coordinates yielded varieties of Bessel's equation. The separation of variables of Laplace's equation in parabolic coordinates also gives rise to Bessel's equation. It may be noted that the Bessel equation is notorious for the variety of disguises it may assume. For an extensive tabulation of possible forms, the reader is referred to *Tables of Functions* by Jahnke and Emde.[1]

Other occasionally encountered ordinary differential equations include the Laguerre and associated Laguerre equations from the supremely important hydrogen atom problem in quantum mechanics:

$$x \frac{d^2y}{dx^2} + (1 - x)\frac{dy}{dx} + \alpha y = 0, \tag{8.8}$$

$$x \frac{d^2y}{dx^2} + (1 + k - x)\frac{dy}{dx} + \alpha y = 0. \tag{8.9}$$

From the quantum mechanical theory of the linear oscillator we have Hermite's equation,

$$\frac{d^2y}{dx^2} - 2x\frac{dy}{dx} + 2\alpha y = 0. \tag{8.10}$$

Finally, from time to time we find the Chebyshev differential equation

$$(1 - x^2)\frac{d^2y}{dx^2} - x\frac{dy}{dx} + n^2 y = 0. \tag{8.11}$$

For convenient reference, the forms of the solutions of Laplace's equation' Helmholtz' equation, and the diffusion equation for spherical polar coordinates are collected in Table 8.1. The solutions of Laplace's equation in circular cylindrical coordinates are presented in Table 8.2.

For the Helmholtz and the diffusion equation, the constant $\pm k^2$ is added to the separation constant $\pm\alpha^2$ to define a new parameter γ^2 or $-\gamma^2$. For the choice $+\gamma^2$ (with $\gamma^2 > 0$), we get $J_m(\gamma\rho)$ and $N_m(\gamma\rho)$. For the choice $-\gamma^2$ (with $\gamma^2 > 0$), we get $I_m(\gamma\rho)$ and $K_m(\gamma\rho)$ as above. An illustration of the former case (with k^2 and α^2 interchanged) appears in Example 2.6.1.

These ordinary differential equations and two generalizations of them will be examined and systematized in the next section. General properties following from the form of the differential equations are taken up in Chapter 9. The individual solutions are developed and applied in Chapters 10 to 13.

The practicing physicist may and probably will meet other second-order ordinary differential equations, some of which may possibly be transformed into the examples studied here. Some he may have to solve by the techniques of Sections 8.4 and 8.5. Others may drive him to a calculating machine for a numerical solution.

[1] Fourth revised edition. New York: Dover (1945), p. 146. Also, *Tables of Higher Functions* by E. Jahnke, F. Emde, and F. Lösch, 6th Ed. New York: McGraw-Hill (1960).

TABLE 8.1

Solutions in Spherical Polar Coordinates[a]

$$\psi = \sum_{l,m} a_{lm}\psi_{lm}$$

1.	$\nabla^2\psi = 0$	$\psi_{lm} = \begin{Bmatrix} r^l \\ r^{-l-1} \end{Bmatrix} \begin{Bmatrix} P_l{}^m(\cos\theta) \\ Q_l{}^m(\cos\theta) \end{Bmatrix} \begin{Bmatrix} \cos m\varphi \\ \sin m\varphi \end{Bmatrix}^b$
2.	$\nabla^2\psi + k^2\psi = 0$	$\psi_{lm} = \begin{Bmatrix} j_l(kr) \\ n_l(kr) \end{Bmatrix} \begin{Bmatrix} P_l{}^m(\cos\theta) \\ Q_l{}^m(\cos\theta) \end{Bmatrix} \begin{Bmatrix} \cos m\varphi \\ \sin m\varphi \end{Bmatrix}^b$
3.	$\nabla^2\psi - k^2\psi = 0$	$\psi_{lm} = \begin{Bmatrix} i_l(kr) \\ k_l(kr) \end{Bmatrix} \begin{Bmatrix} P_l{}^m(\cos\theta) \\ Q_l{}^m(\cos\theta) \end{Bmatrix} \begin{Bmatrix} \cos m\varphi \\ \sin m\varphi \end{Bmatrix}^b$

[a] References for some of the functions are $P_l^m(\cos\theta)$, $m = 0$, Section 12.1; $m \neq 0$, Section 12.5; $Q_l^m(\cos\theta)$, Section 12.10; $j_l(kr)$, $n_l(kr)$, $i_l(kr)$, and $k_l(kr)$, Section 11.7.

[b] $\cos m\varphi$ and $\sin m\varphi$ may be replaced by $e^{\pm im\varphi}$.

TABLE 8.2

Solutions in Circular Cylindrical Coordinates[a]

$$\psi = \sum_{m,\alpha} a_{m\alpha}\psi_{m\alpha}, \qquad \nabla^2\psi = 0$$

(a)	$\psi_{m\alpha} = \begin{Bmatrix} J_m(\alpha\rho) \\ N_m(\alpha\rho) \end{Bmatrix} \begin{Bmatrix} \cos m\varphi \\ \sin m\varphi \end{Bmatrix} \begin{Bmatrix} e^{-\alpha z} \\ e^{\alpha z} \end{Bmatrix}$
(b)	$\psi_{m\alpha} = \begin{Bmatrix} I_m(\alpha\rho) \\ K_m(\alpha\rho) \end{Bmatrix} \begin{Bmatrix} \cos m\varphi \\ \sin m\varphi \end{Bmatrix} \begin{Bmatrix} \cos \alpha z \\ \sin \alpha z \end{Bmatrix}$
(c) If $\alpha = 0$ (no z-dependence)	$\psi_m = \begin{Bmatrix} \rho^m \\ \rho^{-m} \end{Bmatrix} \begin{Bmatrix} \cos m\varphi \\ \sin m\varphi \end{Bmatrix}$

[a] References for the radial functions are $J_m(\alpha\rho)$, Section 11.1; $N_m(\alpha\rho)$, Section 11.3; $I_m(\alpha\rho)$ and $K_m(\alpha\rho)$, Section 11.6.

EXERCISES

8.2.1 Show that when Laplace's equation is separated in prolate spheroidal coordinates (Section 2.10) the equations for the ξ_1- and ξ_2-dependence are the associated Legendre equation.

8.2.2 Show that when Laplace's equation is separated in oblate spheroidal coordinates (Sections 2.11 and 12.10), the equation for the v-dependence is Legendre's associated equation, and the equation for the u-dependence is the associated Legendre equation of imaginary argument.
Note. $\cosh ix = \cos x$, $\sinh ix = i \sin x$, Ex. 6.1.7.

8.2.3 The quantum mechanical angular momentum operator is given by $\mathbf{L} = -i(\mathbf{r} \times \nabla)$. Show that

$$\mathbf{L} \cdot \mathbf{L}\psi = l(l+1)\psi$$

leads to the associated Legendre equation.

Hint. Exercises 1.9.6 and 2.4.14 may be helpful.

8.2.4 The one-dimensional Schrödinger wave equation for a particle in a potential field $V = \frac{1}{2}kx^2$ is

$$-\frac{\hbar^2}{2m}\frac{d^2\psi}{dx^2} + \frac{1}{2}kx^2\psi = E\psi(x)$$

(a) Using $\xi = ax$ and a constant λ,

$$a = \left(\frac{mk}{\hbar^2}\right)^{1/4}$$

$$\lambda = \frac{2E}{\hbar}\left(\frac{m}{k}\right)^{1/2}$$

show that

$$\frac{d^2\psi(\xi)}{d\xi^2} + (\lambda - \xi^2)\,\psi(\xi) = 0.$$

(b) Substituting

$$\psi(\xi) = y(\xi)\,e^{-\xi^2/2}$$

show that $y(\xi)$ satisfies the Hermite differential equation.

8.2.5 Verify that the following are solutions of Laplace's equation:

(a) $\psi_1 = 1/r$,

(b) $\psi_2 = \dfrac{1}{2r}\ln\dfrac{r+z}{r-z}$

8.2.6 If Ψ is a solution of Laplace's equation, $\nabla^2\Psi = 0$, show that $\partial\Psi/\partial z$ is also a solution. *Note*. The z derivatives of $1/r$ generate the Legendre polynomials, $P_n(\cos\theta)$, Ex. 12.1.6. The z derivatives of $(1/2r)\ln[(r+z)/(r-z)]$ generate the Legendre functions, $Q_n(\cos\theta)$.

8.3 Singular Points

In this section the concept of singular point or singularity (as applied to a differential equation) is introduced. The interest in this concept stems from its usefulness in classifying differential equations and in investigating the feasibility of a series solution (Section 8.4). First, a definition.

All of the ordinary differential equations listed in Section 8.2 may be solved for d^2y/dx^2. Using the notation $d^2y/dx^2 = y''$, we have[1]

$$y'' = f(x, y, y'). \tag{8.12}$$

Now, if in Eq. 8.12 y and y' can take on all finite values at $x = x_0$ and y'' remains finite, point $x = x_0$ is an ordinary point. On the other hand, if y'' becomes infinite for *any* finite choice of y and y', point $x = x_0$ is labeled a singular point.

[1] This prime notation, $y' = dy/dx$, was introduced by Lagrange in the late 18th century as an abbreviation for Leibnitz's more explicit but more cumbersome dy/dx.

Another way of presenting this definition of singular point is to write our differential equation as

$$y'' + P(x)y' + Q(x)y = 0. \tag{8.13}$$

Now, if the functions $P(x)$ and $Q(x)$ remain finite at $x = x_0$, point $x = x_0$ is an ordinary point. However, if either $P(x)$ or $Q(x)$ (or both) diverges as $x \to x_0$, point x_0 is a singular point.

Using Eq. 8.13, we may distinguish between two kinds of singular points.

1. If either $P(x)$ or $Q(x)$ diverges as $x \to x_0$ but $(x - x_0) P(x)$ and $(x - x_0)^2 Q(x)$ remain finite as $x \to x_0$, then $x = x_0$ is called a regular or nonessential singular point.
2. If $P(x)$ diverges faster than $1/(x - x_0)$, so that $(x - x_0) P(x)$ goes to infinity as $x \to x_0$, or $Q(x)$ diverges faster than $1/(x - x_0)^2$ so that $(x - x_0)^2 Q(x)$ goes to infinity as $x \to x_0$, then point $x = x_0$ is labeled an irregular or essential singularity.

These definitions hold for all finite values of x_0. The analysis of point $x \to \infty$ is similar to the treatment of functions of a complex variable (Section 6.6). We set $x = 1/z$, substitute into the differential equation, and then let $z \to 0$. By changing variables in the derivatives, we have,

$$\frac{dy(x)}{dx} = \frac{dy(z^{-1})}{dz}\frac{dz}{dx} = -\frac{1}{x^2}\frac{dy(z^{-1})}{dz} = -z^2\frac{dy(z^{-1})}{dz} \tag{8.14}$$

$$\frac{d^2y(x)}{dx^2} = \frac{d}{dz}\left[\frac{dy(x)}{dx}\right]\frac{dz}{dx} = (-z^2)\left[-2z\frac{dy(z^{-1})}{dz} - z^2\frac{d^2y(z^{-1})}{dz^2}\right]$$

$$= 2z^3\frac{dy(z^{-1})}{dz} + z^4\frac{d^2y(z^{-1})}{dz^2}. \tag{8.15}$$

Using these results, we transform Eq. 8.13 into

$$z^4\frac{d^2y}{dz^2} + [2z^3 - z^2P(z^{-1})]\frac{dy}{dz} + Q(z^{-1})y = 0. \tag{8.16}$$

The behavior at $x = \infty$ ($z = 0$) then depends on the behavior of the new coefficients

$$\frac{2z^3 - z^2P(z^{-1})}{z^4} \quad \text{and} \quad \frac{Q(z^{-1})}{z^4}$$

as $z \to 0$. If these two expressions remain finite, point $x = \infty$ is an ordinary point. If they diverge no more rapidly than $1/z$ and $1/z^2$, respectively, point $x = \infty$ is a regular singular point, otherwise an irregular singular point (an essential singularity).

EXAMPLE 8.3.1

Bessel's equation is

$$x^2y'' + xy' + (x^2 - n^2)y = 0. \tag{8.17}$$

Comparing it with Eq. 8.13, we have

$$P(x) = \frac{1}{x}, \qquad Q(x) = 1 - \frac{n^2}{x^2},$$

which shows that point $x = 0$ is a regular singularity. By inspection we see that there are no other singular points in the finite range. As $x \to \infty$ ($z \to 0$), from Eq. 8.16 we have the coefficients

$$\frac{2z^3 - z^2 \cdot z}{z^4} \quad \text{and} \quad \frac{1 - n^2 z^2}{z^4}.$$

Since the latter expression diverges as z^4, point $x = \infty$ is an irregular or essential singularity.

The ordinary differential equations of Section 8.2, plus two others, the hypergeometric and the confluent hypergeometric, have singular points, as shown in Table 8.3.

TABLE 8.3

Equation	Regular singularity $x=$	Irregular singularity $x=$
1. Hypergeometric $x(x-1)y'' + [(1+a+b)x - c]y' + aby = 0.$	$0, 1, \infty$	—
2. Legendre[a] $(1-x^2)y'' - 2xy' + l(l+1)y = 0.$	$-1, 1, \infty$	—
3. Chebyshev $(1-x^2)y'' - xy' + n^2y = 0.$	$-1, 1, \infty$	—
4. Confluent hypergeometric $xy'' + (c-x)y' - ay = 0.$	0	∞
5. Bessel $x^2y'' + xy' + (x^2 - n^2)y = 0.$	0	∞
6. Laguerre[a] $xy'' + (1-x)y' + ay = 0.$	0	∞
7. Simple harmonic oscillator $y'' + \omega^2 y = 0.$	—	∞
8. Hermite $y'' - 2xy' + 2\alpha y = 0.$	—	∞

[a] The associated equations have the same singular points.

It will be seen that the first three equations in the preceding tabulation, hypergeometric, Legendre, and Chebyshev, all have three regular singular points. The hypergeometric equation with regular singularities at 0, 1, and ∞ is taken as the standard, the canonical form. The solutions of the other two may then be expressed in terms of its solutions, the hypergeometric functions. This is done in Chapter 13.

In a similar manner the confluent hypergeometric equation is taken as the canonical form of a linear second-order differential equation with one regular and one irregular singular point.

In a sense the confluent hypergeometric equation can be "derived" from the hypergeometric equation:

$$x(x-1)y'' + [(1+a+b)x - c]y' + aby = 0. \tag{8.18}$$

If we let $bx = z$, we obtain

$$\frac{z}{b}\left(\frac{z}{b} - 1\right)b^2 \frac{d^2y}{dz^2} + \left[(1+a+b)\frac{z}{b} - c\right]b\frac{dy}{dz} + aby = 0.$$

Dividing by b,

$$z\left(\frac{z}{b} - 1\right)\frac{d^2y}{dz^2} + \left[\frac{(1+a)z}{b} + z - c\right]\frac{dy}{dz} + ay = 0.$$

Finally by letting $b \to \infty$, our equation becomes

$$-z\frac{d^2y}{dz^2} - (c - z)\frac{dy}{dz} + ay = 0, \tag{8.19}$$

which, within a common factor of -1, is just the confluent hypergeometric equation. Our transformation has the effect of merging (confluence) the two regular singularities at $x = 0$ and 1 into one regular singularity at $z = 0$ and converting the singularity at infinity from a regular to an irregular singularity.

EXERCISES

8.3.1 Show that Legendre's equation has regular singularities at $x = -1$, 1, and ∞.

8.3.2 Show that Laguerre's equation, like the Bessel equation, has a regular singularity at $x = 0$ and an irregular singularity at $x = \infty$.

8.3.3 Show that the substitution

$$x \to \frac{1-x}{2}, \qquad a = -l, \qquad b = l + 1, \qquad c = 1$$

converts the hypergeometric equation into Legendre's equation.

8.4 Series Solutions—Frobenius' Method

In this section we develop a method of obtaining one solution of the linear, second-order, homogeneous differential equation. The method, a series expansion, will always work, provided the point of expansion is no worse than a regular singular point. In physics, this very gentle condition is almost always satisfied.

A *linear*, *second-order*, *homogeneous* differential equation may be put in the form

$$\frac{d^2y}{dx^2} + P(x)\frac{dy}{dx} + Q(x)y = 0. \tag{8.20}$$

The equation is *homogeneous* because each term contains $y(x)$ or a derivative; *linear* because each y, dy/dx, or d^2y/dx^2 appears as the first power—and no products. In this section we shall develop (at least) one solution of Eq. 8.20. In Section 8.5 we shall develop a second, independent solution and prove that no third, independent solution exists. Therefore, the most general solution of Eq. 8.20 may be written

$$y(x) = c_1y_1(x) + c_2y_2(x). \tag{8.20a}$$

Our physical problem may lead to a *nonhomogeneous*, linear, second-order differential equation

$$\frac{d^2y}{dx^2} + P(x)\frac{dy}{dx} + Q(x)y = F(x). \tag{8.20b}$$

The function on the right, $F(x)$, represents a source (such as electrostatic charge) or a driving force (as in a driven oscillator). Specific solutions of this nonhomogeneous equation are touched on in Ex. 8.5.19. They are explored in some detail, using Green's function techniques, in Sections 8.6, 16.5, and 16.6, and with a Laplace transform technique in Section 15.10. Calling this solution y_p, we may add to it any solution of the corresponding homogeneous equation (Eq. 8.20). Hence, the most general solution of Eq. 8.20*b* is

$$y(x) = c_1y_1(x) + c_2y_2(x) + y_p(x). \tag{8.20c}$$

The constants c_1 and c_2 will eventually be fixed by boundary conditions.

For the present, we assume that $F(x) = 0$, that our differential equation is homogeneous. We shall attempt to develop a solution of our linear, second-order, homogeneous differential equation, Eq. 8.20, by substituting in a power series with undetermined coefficients. Also available as a parameter is the power of the lowest nonvanishing term of the series. To illustrate, we apply the method to two important differential equations. First the linear oscillator equation

$$\frac{d^2y}{dx^2} + \omega^2y = 0, \tag{8.21}$$

with known solutions $y = \sin \omega x$, $\cos \omega x$.

We try

$$\begin{aligned} y(x) &= x^k(a_0 + a_1x + a_2x^2 + a_3x^3 + \cdots) \\ &= \sum_{\lambda=0}^{\infty} a_\lambda x^{k+\lambda}, \qquad a_0 \neq 0, \end{aligned} \tag{8.22}$$

with the exponent k and all the coefficients a_λ still undetermined. By differentiating twice, we obtain

$$\frac{dy}{dx} = \sum_{\lambda=0}^{\infty} a_\lambda (k+\lambda) x^{k+\lambda-1},$$

$$\frac{d^2y}{dx^2} = \sum_{\lambda=0}^{\infty} a_\lambda (k+\lambda)(k+\lambda-1) x^{k+\lambda-2}.$$

By substituting into Eq. 8.21, we have

$$\sum_{\lambda=0}^{\infty} a_\lambda (k+\lambda)(k+\lambda-1) x^{k+\lambda-2} + \omega^2 \sum_{\lambda=0}^{\infty} a_\lambda x^{k+\lambda} = 0. \tag{8.23}$$

From our analysis of the uniqueness of power series (Chapter 5) the coefficients of each power of x on the left-hand side of Eq. 8.23 must vanish individually.

The lowest power of x appearing in Eq. 8.23 is x^{k-2}, for $\lambda = 0$ in the first summation. The requirement that the coefficient vanish[1] yields

$$a_0 k(k-1) = 0.$$

We had chosen a_0 as the coefficient of the lowest nonvanishing terms of the series (Eq. 8.22), hence, by definition, $a_0 \neq 0$. Therefore we have

$$k(k-1) = 0. \tag{8.24}$$

This equation, coming from the coefficient of the lowest power of x, we call the *indicial equation.* The indicial equation and its roots are of critical importance to our analysis. Clearly, in this example we must require either that $k = 0$ or $k = 1$.

Before considering these two possibilities for k, we return to Eq. 8.23 and demand that the remaining net coefficients, say the coefficient of x^{k+j} ($j \geqslant 0$), vanish. We set $\lambda = j + 2$ in the first summation and $\lambda = j$ in the second. (They are independent summations and λ is a dummy index.) This results in

$$a_{j+2}(k+j+2)(k+j+1) + \omega^2 a_j = 0$$

or

$$a_{j+2} = -a_j \frac{\omega^2}{(k+j+2)(k+j+1)}. \tag{8.25}$$

This is a two-term *recurrence relation.* Given a_j, we may compute a_{j+2} and then a_{j+4}, a_{j+6}, and so on up as far as desired. The reader will note that for this example, if we start with a_0, Eq. 8.25 leads to the even coefficients a_2, a_4, etc., and ignores a_1, a_3, a_5, etc. Since a_1 is arbitrary, let us set it equal to zero (cf. Ex. 8.4.2 and 8.4.3) and then by Eq. 8.25

$$a_3 = a_5 = a_7 = \cdots = 0,$$

and all the odd power coefficients vanish. Do not worry about the lost odd powers; the object here is to get a solution. The rejected odd powers will actually reappear when the *second* root of the indicial equation is used.

Returning to Eq. 8.24, our indicial equation, we first try the solution $k = 0$.

[1] Uniqueness of power series, Section 5.7.

The recurrence relation (Eq. 8.25) becomes

$$a_{j+2} = -a_j \frac{\omega^2}{(j+2)(j+1)}, \tag{8.26}$$

which leads to

$$a_2 = -a_0 \frac{\omega^2}{1 \cdot 2} = -\frac{\omega^2}{2!} a_0,$$

$$a_4 = -a_2 \frac{\omega^2}{3 \cdot 4} = +\frac{\omega^4}{4!} a_0,$$

$$a_6 = -a_4 \frac{\omega^2}{5 \cdot 6} = -\frac{\omega^6}{6!} a_0, \text{ etc.}$$

By inspection (and mathematical induction)

$$a_{2n} = (-1)^n \frac{\omega^{2n}}{(2n)!} a_0, \tag{8.27}$$

and our solution is

$$y(x)_{k=0} = a_0\left[1 - \frac{(\omega x)^2}{2!} + \frac{(\omega x)^4}{4!} - \frac{(\omega x)^6}{6!} + \cdots\right]$$

$$= a_0 \cos \omega x. \tag{8.28}$$

If we choose the indicial equation root $k = 1$ (Eq. 8.25) the recurrence relation becomes

$$a_{j+2} = -a_j \frac{\omega^2}{(j+3)(j+2)}. \tag{8.29}$$

Substituting in $j = 0, 2, 4$, successively, we obtain

$$a_2 = -a_0 \frac{\omega^2}{2 \cdot 3} = -\frac{\omega^2}{3!} a_0,$$

$$a_4 = -a_2 \frac{\omega^2}{4 \cdot 5} = +\frac{\omega^4}{5!} a_0,$$

$$a_6 = -a_4 \frac{\omega^2}{6 \cdot 7} = -\frac{\omega^6}{7!} a_0, \text{ etc.}$$

Again, by inspection and mathematical induction,

$$a_{2n} = (-1)^n \frac{\omega^{2n}}{(2n+1)!} a_0. \tag{8.30}$$

For this choice, $k = 1$, we obtain

$$y(x)_{k=1} = a_0 x\left[1 - \frac{(\omega x)^2}{3!} + \frac{(\omega x)^4}{5!} - \frac{(\omega x)^6}{7!} + \cdots\right]$$

$$= \frac{a_0}{\omega}\left[(\omega x) - \frac{(\omega x)^3}{3!} + \frac{(\omega x)^5}{5!} - \frac{(\omega x)^7}{7!} + \cdots\right]$$

$$= \frac{a_0}{\omega} \sin \omega x. \tag{8.31}$$

To summarize this approach, Eq. 8.23 may be written schematically as shown in Fig. 8.1. From the uniqueness of power series (Section 5.7), the total coefficient of each power of x must vanish—all by itself. The requirement that the first coefficient (I) vanish leads to the indicial equation, Eq. 8.24. The second coefficient is handled by setting $a_1 = 0$. The vanishing of the coefficient of x^k (and higher powers, taken one at a time) leads to the recurrence relation Eq. 8.25.

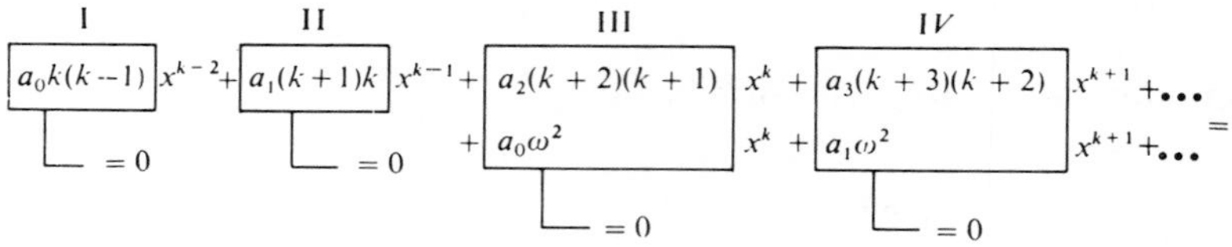

FIG. 8.1

This series substitution, known as Frobenius' method, has given us two series solutions of the linear oscillator equation. However, there are two points about such series solutions that must be strongly emphasized:

1. The series solution should always be substituted back into the differential equation, to see if it works, as a precaution against algebraic and logical errors. Conversely, if it works, it is a solution.

2. The acceptability of a series solution depends on its convergence (including asymptotic convergence). It is quite possible for Frobenius' method to give a series solution which satisfies the original differential equation when substituted in but which does *not* converge over the region of interest. Legendre's differential equation illustrates this situation.

The alert reader will note that we obtained one solution of even symmetry, $y_1(x) = y_1(-x)$, and one of odd symmetry, $y_2(x) = -y_2(-x)$. This is not just an accident but a direct consequence of the form of the differential equation. Writing a general differential equation as

$$\mathscr{L}(x)\, y(x) = 0, \tag{8.32}$$

in which $\mathscr{L}(x)$ is the differential operator, we see that for the linear oscillator equation (Eq. 8.21) $\mathscr{L}(x)$ is even, that is,

$$\mathscr{L}(x) = \mathscr{L}(-x). \tag{8.33}$$

Often this is described as even parity.

Whenever the differential operator has a specific parity or symmetry, either even or odd, we may interchange $+x$ and $-x$, and Eq. 8.32 becomes

$$\pm\mathscr{L}(x)\, y(-x) = 0, \tag{8.34}$$

+ if $\mathscr{L}(x)$ is even, − if $\mathscr{L}(x)$ is odd. Clearly, if $y(x)$ is a solution of the differential equation, $y(-x)$ is also a solution. Then any solution may be resolved into even and odd parts,

$$y(x) = \tfrac{1}{2}[y(x) + y(-x)] + \tfrac{1}{2}[y(x) - y(-x)], \tag{8.35}$$

the first bracket on the right giving an even solution, the second an odd solution.

If we refer back to Section 8.3, we can see that Legendre, Chebyshev, Bessel, simple harmonic oscillator, and Hermite equations (or differential operators) all exhibit this even parity. Solutions of all of them may be presented as series of even powers of x and separate series of odd powers of x. The Laguerre differential operator has neither even nor odd symmetry; hence its solutions cannot be expected to exhibit even or odd parity.

Limitations of series approach—Bessel's equation. This attack on the linear oscillator equation was perhaps a bit too easy. By substituting the power series (Eq. 8.22) into the differential equation (Eq. 8.21) we obtained two independent solutions with no trouble at all.

To get some idea of what can happen we try to solve Bessel's equation,

$$x^2 y'' + xy' + (x^2 - n^2)y = 0 \tag{8.36}$$

using y' for dy/dx and y'' for d^2y/dx^2. Again assuming a solution of the form

$$y(x) = \sum_{\lambda=0}^{\infty} a_\lambda x^{k+\lambda},$$

we differentiate and substitute into Eq. 8.36. The result is

$$\begin{aligned}\sum_{\lambda=0}^{\infty} a_\lambda (k+\lambda)(k+\lambda-1)x^{k+\lambda} + \sum_{\lambda=0}^{\infty} a_\lambda (k+\lambda)x^{k+\lambda}\\ + \sum_{\lambda=0}^{\infty} a_\lambda x^{k+\lambda+2} - \sum_{\lambda=0}^{\infty} a_\lambda n^2 x^{k+\lambda} = 0.\end{aligned} \tag{8.37}$$

By setting $\lambda = 0$, the coefficient of x^k, the lowest power of x appearing on the left-hand side, is

$$a_0[k(k-1) + k - n^2] = 0, \tag{8.38}$$

and again $a_0 \neq 0$ by definition. Equation 8.38 therefore yields the *indicial equation*

$$k^2 - n^2 = 0 \tag{8.39}$$

with solutions $k = \pm n$.

It is of some interest to examine the coefficient of x^{k+1} also. Here we obtain

$$a_1[(k+1)k + k + 1 - n^2] = 0$$

or

$$a_1(k+1-n)(k+1+n) = 0. \tag{8.40}$$

For $k = \pm n$ neither $k + 1 - n$ nor $k + 1 + n$ vanishes and we *must* require $a_1 = 0$.[1]

Proceeding to the coefficient of x^{k+j} for $k = n$, we set $\lambda = j$ in the first, second, and fourth terms of Eq. 8.37 and $\lambda = j - 2$ in the third term. By requiring the resultant

[1] $k = \pm n = -\tfrac{1}{2}$ are exceptions.

coefficient of x^{k+j} to vanish we obtain

$$a_j[(n+j)(n+j-1)+(n+j)-n^2]+a_{j-2}=0.$$

When j is replaced by $j+2$, this can be rewritten

$$a_{j+2}=-a_j\frac{1}{(j+2)(2n+j+2)}, \tag{8.41}$$

which is the desired recurrence relation. Repeated application of this recurrence relation leads to

$$a_2=-a_0\frac{1}{2(2n+2)}=-\frac{a_0 n!}{2^2 1!(n+1)!},$$

$$a_4=-a_2\frac{1}{4(2n+4)}=+\frac{a_0 n!}{2^4 2!(n+2)!},$$

$$a_6=-a_4\frac{1}{6(2n+6)}=-\frac{a_0 n!}{2^6 3!(n+3)!},\text{ etc.,}$$

and in general

$$a_{2p}=(-1)^p\frac{a_0\,n!}{2^{2p}p!(n+p)!}. \tag{8.42}$$

Inserting these coefficients in our assumed series solution, we have

$$y(x)=a_0x^n\left[1-\frac{n!x^2}{2^2 1!(n+1)!}+\frac{n!x^4}{2^4 2!(n+2)!}-\cdots\right]. \tag{8.43}$$

In summation form

$$\begin{aligned} y(x)&=a_0\sum_{j=0}^{\infty}(-1)^j\frac{n!x^{n+2j}}{2^{2j}j!(n+j)!}\\ &=a_0 2^n\sum_{j=0}^{\infty}(-1)^j\frac{n!}{j!(n+j)!}\left(\frac{x}{2}\right)^{n+2j} \end{aligned} \tag{8.44}$$

In Chapter 11 the final summation is identified as the Bessel function $J_n(x)$. Notice that this solution $J_n(x)$ has either even or odd symmetry[1] as might be expected from the form of Bessel's equation.

When $k=-n$ and n is not an integer, we may generate a second distinct series to be labeled $J_{-n}(x)$. However, when $-n$ is a negative integer, trouble develops. The recurrence relation for the coefficients a_j is still given by Eq. 8.41, but with $2n$ replaced by $-2n$. Then, when $j+2=2n$ or $j=2(n-1)$, the coefficient a_{j+2} blows up and we have no series solution. This catastrophe can be remedied in Eq. 8.44, as it is done in Chapter 11, with the result that

$$J_{-n}(x)=(-1)^nJ_n(x),\qquad n\text{ an integer.} \tag{8.45}$$

[1] $J_n(x)$ is an even function if n is an even integer, an odd function if n is an odd integer. For nonintegral n the x^n has no such simple symmetry.

The second solution simply reproduces the first. We have failed to construct a second independent solution for Bessel's equation by this series technique when n is an integer.

By substituting in an infinite series, we have obtained two solutions for the linear oscillator equation and one for Bessel's equation (two if n is not an integer). To the questions "Can we always do this? Will this method always work?", the answer is no, we cannot always do this. This method of series solution will not always work.

The success of the series substitution method depends on the roots of the indicial equation and the degree of singularity of the coefficients in the differential equation. To understand better the effect of the equation coefficients on this naïve series substitution approach consider four simple equations

$$y'' - \frac{6}{x^2} y = 0, \tag{8.46a}$$

$$y'' - \frac{6}{x^3} y = 0, \tag{8.46b}$$

$$y'' + \frac{1}{x} y' - \frac{a^2}{x^2} y = 0, \tag{8.46c}$$

$$y'' + \frac{1}{x^2} y' - \frac{a^2}{x^2} y = 0. \tag{8.46d}$$

The reader may show easily that for Eq. 8.46*a* the indicial equation is

$$k^2 - k - 6 = 0,$$

giving $k = 3, -2$. Since the equation is homogeneous in x (counting d^2/dx^2 as x^{-2}), there is no recurrence relation; $a_i = 0$ for $i > 0$. However, we are left with two perfectly good solutions, x^3 and x^{-2}.

Equation 8.46*b* differs from 8.46*a* by only one power of x, but this sends the indicial equation to

$$-6a_0 = 0,$$

with no solution at all, for we have agreed that $a_0 \neq 0$. Our series substitution worked for Eq. 8.46*a*, which had only a regular singularity, but broke down at Eq. 8.46*b*, which has an irregular singular point at the origin.

Continuing with Eq. 8.46*c*, we have added a term y'/x. The indicial equation is

$$k^2 - a^2 = 0,$$

but again there is no recurrence relation. The solutions are $y = x^a$, x^{-a}, both perfectly acceptable one term series.

When we change the power of x in the coefficient of y' from -1 to -2, Eq. 8.46*d*, there is a drastic change in the solution. The indicial equation (with only the y' term contributing) becomes

$$k = 0.$$

There is a recurrence relation

$$a_{j+1} = -a_j \frac{a^2 - j(j-1)}{j+1}.$$

Unless the parameter a is selected to make the series terminate, we have

$$\lim_{j\to\infty}\left|\frac{a_{j+1}}{a_j}\right| = \lim_{j\to\infty}\frac{j(j-1)}{j+1}$$

$$= \lim_{j\to\infty}\frac{j^2}{j} = \infty .$$

Hence our series solution diverges for all $x \neq 0$. Again our method worked for Eq. 8.46*c* with a regular singularity but failed when we had the irregular singularity of 8.46*d*.

Fuchs's theorem. The answer to the basic question when the method of series substitution can be expected to work is given by Fuchs's theorem, which asserts that we can always obtain at least one power series solution, provided we are expanding about a point that is an ordinary point or at worst a regular singular point. If we attempt an expansion about an irregular or essential singularity, our method may fail as it did for Eqs. 8.46*b* and 8.46*d*. Fortunately, the more important equations of mathematical physics listed in Section 8.3 have no irregular singularities in the finite plane. Further discussion of Fuchs's theorem appears in Section 8.5.

From Table 8.3, Section 8.3, infinity is seen to be a singular point for all of the equations considered. As a further illustration of Fuchs's theorem, Legendre's equation (with infinity as a regular singularity) has a convergent series solution in negative powers of the argument (Section 12.10). In contrast, Bessel's equation (with an irregular singularity at infinity) yields asymptotic series (Sections 5.11 and 11.6). While extremely useful, these asymptotic solutions are technically divergent.

Summary. If we are expanding about an ordinary point or at worst about a regular singularity, the series substitution approach will yield at least one solution (Fuchs's theorem).

Whether we get one or two distinct solutions depends on the roots of the indicial equation.

1. If the two roots of the indicial equation are equal, we can obtain only one solution by this series substitution method.
2. If the two roots differ by a nonintegral number, two independent solutions may be obtained.
3. If the two roots differ by an integer, the larger of the two will yield a solution. The smaller may or may not give a solution, depending on the behavior of the coefficients. In the linear oscillator equation we obtain two solutions; for Bessel's equation, only one solution.

EXERCISES

8.4.1 A series solution of Eq. 8.20 is attempted, expanding about the point $x = x_0$. If x_0 is an ordinary point show that the indicial equation has roots $k = 0, 1$.

8.4.2 In the development of a series solution of the simple harmonic oscillator equation the second series coefficient a_1 was neglected except to set it equal to zero. From the coefficient of the next to the lowest power of x, x^{k-1}, develop a second indicial type equation.

(a) (SHO equation with $k = 0$). Show that a_1 may be assigned any finite value (including zero).

(b) (SHO equation with $k = 1$). Show that a_1 *must* be set equal to zero.

8.4.3 Analyze the series solutions of the following differential equations to see when a_1 *may* be set equal to zero without irrevocably losing anything and when a_1 *must* be set equal to zero.

(a) Legendre, (b) Chebyshev, (c) Bessel, (d) Hermite.

Ans. (a) Legendre, (b) Chebyshev, and (d) Hermite: For $k = 0$, a_1 *may* be set equal to zero; For $k = 1$, a_1 *must* be set equal to zero.

(c) Bessel: a_1 *must* be set equal to zero (exception noted in footnote, p. 395)

8.4.4 Solve the Legendre equation

$$(1 - x^2)y'' - 2xy' + n(n + 1)y = 0$$

by direct series substitution.

(a) Verify that the indicial equation is

$$k(k - 1) = 0.$$

(b) Using $k = 0$, obtain a series of even powers of x, ($a_1 = 0$).

$$y_{\text{even}} = a_0\left[1 - \frac{n(n+1)}{2!}x^2 + \frac{n(n-2)(n+1)(n+3)}{4!}x^4 + \cdots\right]$$

where

$$a_{j+2} = \frac{j(j+1) - n(n+1)}{(j+1)(j+2)}a_j.$$

(c) Using $k = 1$ develop a series of odd powers of x ($a_1 = 0$).

$$y_{\text{odd}} = a_0\left[x - \frac{(n-1)(n+2)}{3!}x^3 + \frac{(n-1)(n-3)(n+2)(n+4)}{5!}x^5 + \cdots\right]$$

where

$$a_{j+2} = \frac{(j+1)(j+2) - n(n+1)}{(j+2)(j+3)}a_j.$$

(d) Show that both solutions, y_{even} and y_{odd}, diverge for $x = \pm 1$ *if the series continue to infinity*.

(e) Finally, show that by an appropriate choice of n, one series at a time may be converted into a polynomial, thereby avoiding the divergence catastrophe.

8.4.5 (a) Develop series solutions for Hermite's differential equation

$$y'' - 2xy' + 2\alpha y = 0.$$

Ans. $k(k - 1) = 0$, indicial equation.

For $k = 0$

$$a_{j+2} = 2a_j\frac{j - \alpha}{(j+1)(j+2)}, \qquad (j \text{ even}),$$

$$y_{\text{even}} = a_0\left[1 + \frac{2(-\alpha)x^2}{2!} + \frac{2^2(-\alpha)(2-\alpha)x^4}{4!} + \cdots\right].$$

For $k = 1$

$$a_{j+2} = 2a_j\frac{j + 1 - \alpha}{(j+2)(j+3)}, \qquad (j \text{ even}),$$

$$y_{\text{odd}} = a_0\left[x + \frac{2(1-\alpha)x^3}{3!} + \frac{2^2(1-\alpha)(3-\alpha)x^5}{5!} + \cdots\right].$$

(b) Show that both series solutions are convergent, the series behaving like the series expansion of e^{2x^2}.

(c) Show that by appropriate choice of α the series solutions may be cut off and converted to finite polynomials. (These polynomials, properly normalized, become the Hermite polynomials in Section 13.1.)

8.4.6 Laguerre's differential equation is

$$x\, L_n''(x) + (1 - x)\, L_n'(x) + n\, L_n(x) = 0.$$

Develop a series solution selecting the parameter n to make your series a polynomial.

8.4.7 Solve the Chebyshev equation

$$(1 - x^2)\, T_n'' - xT_n' + n^2 T_n = 0,$$

by series substitution. What restrictions are imposed on n if you demand that the series solution converge for $x = \pm 1$?

Ans. The infinite series does converge for $x = \pm 1$. Therefore, there is no restriction on n (cf. Ex. 5.2.16).

8.4.8 Solve

$$(1 - x^2) U_n''(x) - 3x\, U_n'(x) + n(n + 2)\, U_n(x) = 0,$$

choosing the root of the indicial equation to obtain a series of *odd* powers of x. Since the series will diverge for $x = 1$, choose n to convert it into a polynomial.

$$k(k - 1) = 0.$$

For $k = 1$

$$a_{j+2} = \frac{(j + 1)(j + 3) - n(n + 2)}{(j + 2)(j + 3)}\, a_j.$$

8.4.9 Obtain a series solution of the hypergeometric equation

$$x(x - 1)y'' + [(1 + a + b)x - c]y' + aby = 0.$$

Test your solution for convergence.

8.4.10 Obtain two series solutions of the confluent hypergeometric equation

$$xy'' + (c - x)y' - ay = 0.$$

Test your solutions for convergence.

8.4.11 A quantum mechanical analysis of the Stark effect (parabolic coordinates) leads to the differential equation

$$\frac{d}{d\xi}\left(\xi \frac{du}{d\xi}\right) + \left(\frac{1}{2} E\xi + \alpha - \frac{m^2}{4\xi} - \frac{1}{4} F\xi^2\right) u = 0.$$

Here α is a separation constant, E is the total energy, and F is a constant where Fz is the potential energy added to the system by the introduction of an electric field.

Using the larger root of the indicial equation, develop a power series solution about $\xi = 0$. Evaluate the first three coefficients in terms of a_0.

$$\text{Indicial equation} \qquad k^2 - \frac{m^2}{4} = 0,$$

$$u(\xi) = a_0 \xi^{m/2} \left\{1 - \frac{\alpha}{m + 1}\xi + \left[\frac{\alpha^2}{2(m + 1)(m + 2)} - \frac{E}{4(m + 2)}\right]\xi^2 + \cdots\right\}$$

Note that the perturbation E does not appear until a_3 is included. (This is not Eq. 2.120 of Section 2.12 because here $\xi = (\xi_{2.12})^{1/2}$).

8.4.12 For the special case of no azimuthal dependence, the quantum mechanical analysis of the hydrogen molecular ion leads to the equation (Section 2.10)

$$\frac{d}{d\eta}\left[(1 - \eta^2)\frac{du}{d\eta}\right] + \alpha u + \beta \eta^2 u = 0.$$

Develop a power-series solution for $u(\eta)$. Evaluate the first three nonvanishing coefficients in terms of a_0.

Indicial equation $\quad k(k-1)=0,$

$$u_{k=1}=a_0\eta\left\{1+\frac{2-\alpha}{6}\eta^2+\left[\frac{(2-\alpha)(12-\alpha)}{120}-\frac{\beta}{20}\right]\eta^4+\cdots\right\}.$$

8.4.13 To a good approximation, the interaction of two nucleons may be described by a meson potential

$$V=\frac{Ae^{-ax}}{x},$$

attractive for A negative. Develop a series solution of the resultant Schrödinger wave equation

$$\frac{\hbar^2}{2m}\frac{d^2\psi}{dx^2}+(E-V)\psi=0,$$

through the first three nonvanishing coefficients.

$$\psi_{k=1}=a_0\{x+\tfrac{1}{2}A'x^2+\tfrac{1}{6}[\tfrac{1}{2}A'^2-E'-\alpha A']x^3+\cdots\},$$

where the prime indicates multiplication by $2m/\hbar^2$.

8.4.14 Near the nucleus of a complex atom the potential energy of one electron is given by

$$V=\frac{Ze^2}{r}(1+b_1r+b_2r^2),$$

where the coefficients b_1 and b_2 arise from screening effects. For the case of zero angular momentum show that the first three terms of the solution of the Schrödinger equation have the same form as those of Exercise 8.4.13. By appropriate translation of coefficients or parameters, write out the first three terms in a series expansion of the wave function.

8.5 A Second Solution

In Section 8.4 a solution of a second-order homogeneous differential equation was developed by substituting in a power series. By Fuchs's theorem this is possible, provided the power series is an expansion about an ordinary point or a nonessential singularity.[1] There is no guarantee that this approach will yield the two independent solutions we expect from a linear second-order differential equation. Indeed, the technique gave only one solution for Bessel's equation (n an integer). In this section we develop two methods of obtaining a second independent solution: an integral method and a power series containing a logarithmic term. First, however, we consider the question of independence of a set of functions.

Linear independence of solutions. Given a set of functions, φ_λ, the criterion for linear *dependence* is the existence of a relation of the form

$$\sum_\lambda k_\lambda \varphi_\lambda = 0, \tag{8.47}$$

in which not all of the coefficients k_λ are zero. On the other hand, if the only solution of Eq. 8.47 is $k_\lambda = 0$ for all λ, the set of functions φ_λ is said to be linearly *independent*.

[1] This is why the classification of singularities in Section 8.3 is of vital importance to us.

It may be helpful to think of linear dependence of vectors. Consider **A**, **B**, and **C** in three dimensional space with $\mathbf{A} \cdot \mathbf{B} \times \mathbf{C} \neq 0$. Then no relation of the form

$$a\mathbf{A} + b\mathbf{B} + c\mathbf{C} = 0 \tag{8.48a}$$

exists. **A**, **B**, and **C** are linearly independent. On the other hand any fourth vector **D** may be expressed as a linear combination of **A**, **B**, and **C**. See Section 4.4. We can always write an equation of the form

$$\mathbf{D} - a\mathbf{A} - b\mathbf{B} - c\mathbf{C} = 0, \tag{8.48b}$$

and the four vectors are *not* linearly independent. The three noncoplanar vectors **A**, **B**, and **C** span our real three-dimensional space.

Let us assume that the functions φ_λ are differentiable as needed. Then, differentiating Eq. 8.47 repeatedly, we generate a set of equations

$$\sum_\lambda k_\lambda \varphi'_\lambda = 0 \tag{8.49a}$$

$$\sum_\lambda k_\lambda \varphi''_\lambda = 0, \text{ etc.} \tag{8.49b}$$

This gives us a set of homogeneous linear equations in which k_λ are the unknown quantities. By Section 4.1 there is a solution $k_\lambda \neq 0$ only if the determinant of the coefficients of the k_λ's vanishes. This means

$$\begin{vmatrix} \varphi_1 & \varphi_2 & \cdots & \varphi_n \\ \varphi'_1 & \varphi'_2 & \cdots & \varphi'_n \\ \cdots & \cdots & \cdots & \cdots \\ \varphi_1^{(n-1)} & \varphi_2^{(n-1)} & \cdots & \varphi_n^{(n-1)} \end{vmatrix} = 0. \tag{8.50}$$

This determinant is called the Wronskian.

1. If the Wronskian is not equal to zero, then Eq. 8.47 has no solution other than $k_\lambda = 0$. The set of functions φ_λ is therefore independent.

2. If the Wronskian vanishes at isolated values of the argument, this does not necessarily prove linear dependence. However, if the Wronskian is zero over the entire range of the variable, the functions φ_λ are linearly dependent over this range[1] (cf. Exercise 8.5.2 for the simple case of two functions).

EXAMPLE 8.5.1—LINEAR INDEPENDENCE

The solutions of the linear oscillator equation 8.21 are $\varphi_1 = \sin \omega x$, $\varphi_2 = \cos \omega x$. The Wronskian becomes

$$\begin{vmatrix} \sin \omega x & \cos \omega x \\ \omega \cos \omega x & -\omega \sin \omega x \end{vmatrix} = -\omega \neq 0.$$

[1] Cf. p. 187 of H. LASS, *Elements of Pure and Applied Mathematics*. New York: McGraw-Hill (1957) for proof of this assertion. It is assumed that the functions have continuous derivatives and that at least one of the minors of the bottom row of Eq. 8.50 (Laplace expansion) does not vanish in $[a, b]$, the interval under consideration.

These two solutions, φ_1 and φ_2, are therefore linearly independent. For just two functions this means that one is not a multiple of the other, which is obviously true in this case.

EXAMPLE 8.5.2—LINEAR DEPENDENCE

For an illustration of linear dependence, consider the solutions of the one-dimensional diffusion equation. We have $\varphi_1 = e^x$, $\varphi_2 = e^{-x}$, and we add $\varphi_3 = \cosh x$, also a solution. The Wronskian is

$$\begin{vmatrix} e^x & e^{-x} & \cosh x \\ e^x & -e^{-x} & \sinh x \\ e^x & e^{-x} & \cosh x \end{vmatrix} = 0.$$

The determinant vanishes for all x, because the first and third rows are identical. Hence e^x, e^{-x}, and $\cosh x$ are linearly dependent, and indeed we have a relation of the form of Eq. 8.47:

$$e^x + e^{-x} - 2\cosh x = 0 \qquad \text{with} \quad k_\lambda \neq 0.$$

A second solution. Returning to our linear, second-order, homogeneous, differential equation of the general form

$$y'' + P(x)y' + Q(x)y = 0, \tag{8.51}$$

let y_1 and y_2 be two independent solutions. Then the Wronskian, by definition, is

$$W = y_1 y_2' - y_1' y_2 \tag{8.52}$$

By differentiating the Wronskian, we obtain

$$\begin{aligned} W' &= y_1' y_2' + y_1 y_2'' - y_1'' y_2 - y_1' y_2' \\ &= y_1[-P(x)y_2' - Q(x)y_2] - y_2[-P(x)y_1' - Q(x)y_1] \\ &= -P(x)(y_1 y_2' - y_1' y_2). \end{aligned} \tag{8.53}$$

The expression in parentheses is just W, the Wronskian, and we have

$$W' = -P(x)W. \tag{8.54}$$

If $P(x) = 0$, that is,

$$y'' + Q(x)y = 0, \tag{8.55}$$

the Wronskian

$$W = y_1 y_2' - y_1' y_2 = \text{constant}. \tag{8.56}$$

Since our original differential equation is homogeneous, we may multiply the solutions y_1 and y_2 by whatever constants we wish and arrange to have the Wronskian equal to unity (or -1). This case, $P(x) = 0$, appears more frequently than might be expected. The reader will recall that ∇^2 in cartesian coordinates contains no first derivative. Similarly the radial dependence of $\nabla^2(r\psi)$ in spherical polar coordinates lacks a first derivative. Finally, every linear second-order differential equation can be transformed into an equation of the form of Eq. 8.55 (cf. Exercise 8.5.7).

Let us now assume that we have one solution of Eq. 8.51 by a series substitution (or by guessing). We now proceed to develop a second, independent solution. Rewriting Eq. 8.54 as

$$\frac{dW}{W} = -P\,dx_1,$$

we integrate, from $x_1 = a$ to $x_1 = x$ to obtain

$$\ln \frac{W(x)}{W(a)} = -\int_a^x P(x_1)\,dx_1,$$

or

$$W(x) = W(a)\exp\left[-\int_a^x P(x_1)\,dx_1\right].^{1} \tag{8.57}$$

But

$$\begin{aligned} W(x) &= y_1 y_2' - y_1' y_2 \\ &= y_1^2 \frac{d}{dx}\left(\frac{y_2}{y_1}\right). \end{aligned} \tag{8.58}$$

By combining Eqs. 8.57 and 8.58, we have

$$\frac{d}{dx}\left(\frac{y_2}{y_1}\right) = W(a)\frac{\exp[-\int_a^x P(x_1)\,dx_1]}{y_1^2} \tag{8.59}$$

Finally, by integrating Eq. 8.59 from $x_2 = b$ to $x_2 = x$ we get

$$y_2(x) = y_1(x)\,W(a)\int_b^x \frac{\exp[-\int_a^{x_2} P(x_1)\,dx_1]}{[y_1(x_2)]^2}\,dx_2. \tag{8.60}$$

Here a and b are arbitrary constants and a term $y_1(x)\,y_2(b)/y_1(b)$ has been dropped, for it leads to nothing new. Since $W(a)$, the Wronskian evaluated at $x = a$, is a constant and our solutions for the homogeneous differential equation always contain an unknown normalizing factor, we set $W(a) = 1$ and write

$$y_2(x) = y_1(x)\int^x \frac{\exp[-\int^{x_2} P(x_1)\,dx_1]}{[y_1(x_2)]^2}\,dx_2. \tag{8.61}$$

Note that the lower limits $x_1 = a$ and $x_2 = b$ have been omitted. If they are retained, they simply make a contribution equal to a constant times the known first solution, $y_1(x)$, hence add nothing new.

If we have the important special case of $P(x) = 0$, Eq. 8.61 reduces to

$$y_2(x) = y_1(x)\int^x \frac{dx_2}{[y_1(x_2)]^2}. \tag{8.62}$$

This means that by using either Eq. 8.61 or 8.62 we can take one known solution and by integrating can generate a second independent solution of Eq. 8.51. This technique is used in Section 12.10 to generate a second solution of Legendre's differential equation.

[1] If $P(x_1)$ remains finite, $a \leqslant x_1 \leqslant x$, $W(x) \neq 0$ unless $W(a) = 0$. That is, the Wronskian *of our two solutions* is either identically zero or never zero.

EXAMPLE 8.5.3

From Eq. 8.21 with $P(x) = 0$, let one solution be $y_1 = \sin x$. By applying Eq. 8.62 we obtain

$$y_2(x) = \sin x \int^x \frac{dx_2}{\sin^2 x_2}$$
$$= \sin x(-\cot x) = -\cos x,$$

which is clearly independent (not a linear multiple) of $\sin x$.

Series form of the second solution Further insight into the nature of the second solution of our differential equation may be obtained by the following sequence of operations:

1. Express $P(x)$ and $Q(x)$ in Eq. 8.51 as

$$P(x) = \sum_{i=-1}^{\infty} p_i x^i, \qquad Q(x) = \sum_{j=-2}^{\infty} q_j x^j. \tag{8.63}$$

The lower limits of the summations are selected to just satisfy Fuchs's theorem and because of this we may gain a better understanding of Fuchs's theorem.

2. Develop the first few terms of a power-series solution, as in Section 8.4.

3. Using this solution as y_1, obtain a second series type solution, y_2, with Eq. 8.61, integrating term by term.

Proceeding with Step 1, we have

$$y'' + (p_{-1}x^{-1} + p_0 + p_1 x + \cdots)y' + (q_{-2}x^{-2} + q_{-1}x^{-1} + \cdots)y = 0, \tag{8.64}$$

in which point $x = 0$ is at worst a regular singular point. If $p_{-1} = q_{-1} = q_{-2} = 0$, it reduces to an ordinary point. Substituting

$$y = \sum_{\lambda=0}^{\infty} a_\lambda x^{k+\lambda}$$

(Step 2), we obtain

$$\sum_{\lambda=0}^{\infty} (k+\lambda)(k+\lambda-1)a_\lambda x^{k+\lambda-2} + \sum_{i=-1}^{\infty} p_i x^i \sum_{\lambda=0}^{\infty} (k+\lambda)a_\lambda x^{k+\lambda-1}$$
$$+ \sum_{j=-2}^{\infty} q_j x^j \sum_{\lambda=0}^{\infty} a_\lambda x^{k+\lambda} = 0. \tag{8.65}$$

Assuming that $p_{-1} \neq 0$, $q_{-2} \neq 0$, our indicial equation is

$$k(k-1) + p_{-1}k + q_{-2} = 0,$$

which sets the net coefficient of x^{k-2} equal to zero. This reduces to

$$k^2 + (p_{-1} - 1)k + q_{-2} = 0. \tag{8.66}$$

We denote the two roots of this indicial equation by $k = \alpha$ and $k = \alpha - n$ where n is zero or a positive integer. (If n is not an integer, we expect two independent series solutions by the methods of Section 8.4 and there is no problem.) Then

$$(k-\alpha)(k-\alpha+n) = 0, \tag{8.67}$$

and, equating coefficients of k in Eqs. 8.66 and 8.67,

$$p_{-1} - 1 = n - 2\alpha. \tag{8.68}$$

The series solution corresponding to $k = \alpha$ may be written

$$y_1 = x^\alpha \sum_{\lambda=0}^{\infty} a_\lambda x^\lambda.$$

Substituting this series solution into Eq. 8.61 (Step 3), we are faced with

$$y_2(x) = y_1(x) \int^x \frac{\exp(-\int_a^{x_2} \sum_{i=-1}^{\infty} p_i x_1^i \, dx_1)}{x_2^{2\alpha}(\sum_{\lambda=0}^{\infty} a_\lambda x_2^\lambda)^2} \, dx_2, \tag{8.69}$$

where the solutions y_1 and y_2 have been normalized so that the Wronskian, $W(a) = 1$. Tackling the exponential factor first,

$$\int_a^{x_2} \sum_{i=-1}^{\infty} p_i x_1^i \, dx_1 = p_{-1} \ln x_2 + \sum_{k=0}^{\infty} \frac{p_k}{k+1} x_2^{k+1} + f(a) \tag{8.70}$$

Hence

$$\exp\left(-\int_a^{x_2} \sum_i p_i x_1^i \, dx_1\right) = \exp[-f(a)] x_2^{-p_{-1}} \exp\left(-\sum_{k=0}^{\infty} \frac{p_k}{k+1} x_2^{k+1}\right)$$

$$= \exp[-f(a)] x_2^{-p_{-1}} \left[1 - \sum_{k=0}^{\infty} \frac{p_k}{k+1} x_2^{k+1} + \frac{1}{2!}\left(\sum_{k=0}^{\infty} \frac{p_k}{k+1} x_2^{k+1}\right)^2 - \cdots\right] \tag{8.71}$$

This final series expansion of the exponential is certainly convergent if the original expansion of the coefficient $P(x)$ was convergent.

The denominator in Eq. 8.69 may be handled by writing

$$\begin{aligned} \left[x_2^{2\alpha}\left(\sum_{\lambda=0}^{\infty} a_\lambda x_2^\lambda\right)^2\right]^{-1} &= x_2^{-2\alpha}\left(\sum_{\lambda=0}^{\infty} a_\lambda x_2^\lambda\right)^{-2} \\ &= x_2^{-2\alpha} \sum_{\lambda=0}^{\infty} b_\lambda x_2^\lambda. \end{aligned} \tag{8.72}$$

Neglecting constant factors that will be picked up anyway by the requirement that $W(a) = 1$, we obtain

$$y_2(x) = y_1(x) \int^x x_2^{-p_{-1}-2\alpha} \left(\sum_{\lambda=0}^{\infty} c_\lambda x_2^\lambda\right) dx_2 \tag{8.73}$$

By Eq. 8.68

$$x_2^{-p_{-1}-2\alpha} = x_2^{-n-1}, \tag{8.74}$$

and we have assumed here that n is an integer. Therefore the integration indicated in Eq. 8.73 leads to a coefficient of $y_1(x)$ consisting of two parts:

1. A power series starting with x^{-n}.
2. A logarithmic term from the integration of x^{-1} (when $\lambda = n$). This term always appears when n is an integer *unless* c_n fortuitously happens to vanish.

EXAMPLE 8.5.4—A SECOND SOLUTION OF BESSEL'S EQUATION

From Bessel's equation, Eq. 8.36 (divided by x^2 to agree with Eq. 8.51), we have

$$P(x) = x^{-1} \qquad Q(x) = 1 \qquad \text{(for the case} \quad n = 0).$$

Hence, $p_{-1} = 1, q_0 = 1$; all other p_i's and q_j's vanish. The Bessel indicial equation is

$$k^2 = 0,$$

(Eq. 8.39 with $n = 0$). Hence we verify Eqs. 8.66–8.68 with n and $\alpha = 0$.

Our first solution is available from Eq. 8.44. Relabeling it to agree with Chapter 11 (and using $a_0 = 1$),

$$y_1(x) = J_0(x) = 1 - \frac{x^2}{4} + \frac{x^4}{64} - O(x^6). \tag{8.75a}$$

Now, substituting all this into Eq. 8.61 we have the specific case corresponding to Eq. 8.69

$$y_2(x) = J_0(x) \int^x \frac{\exp[-\int^{x_2} x_1^{-1}\, dx_1]}{[1 - x_2^2/4 + x_2^4/64 - \cdots]^2}\, dx_2. \tag{8.75b}$$

From the numerator of the integrand

$$\exp\left[-\int^{x_2} \frac{dx_1}{x_1}\right] = \exp[-\ln x_2] = \frac{1}{x_2}.$$

This corresponds to the $x_2^{-p_{-1}}$ in Eq. 8.71. From the denominator of the integrand

$$\left[1 - \frac{x_2^2}{4} + \frac{x_2^4}{64}\right]^{-2} = 1 + \frac{x_2^2}{2} + \frac{5x_2^4}{32} + \cdots.$$

Corresponding to Eq. 8.73,

$$\begin{aligned} y_2(x) &= J_0(x) \int^x \frac{1}{x_2}\left[1 + \frac{x_2^2}{2} + \frac{5x_2^4}{32} + \cdots\right] dx_2 \\ &= J_0(x)\left\{\ln x + \frac{x^2}{4} + \frac{5x^4}{128} + \cdots\right\}. \end{aligned} \tag{8.75c}$$

Let us check this result. From Eq. 11.63, which gives the standard form of the second solution,

$$\begin{aligned} N_0(x) &= \frac{2}{\pi}[\ln x - \ln 2 + \gamma]J_0(x) \\ &\quad + \frac{2}{\pi}\left\{\frac{x^2}{4} - \frac{3x^4}{128} + \cdots\right\}. \end{aligned} \tag{8.75d}$$

Two points arise: (1) Since Bessel's equation is homogeneous, we may multiply $y_2(x)$ by any constant. To match $N_0(x)$, we multiply our $y_2(x)$ by $2/\pi$. (2) To our second solution $(2/\pi)y_2(x)$, we may add any constant multiple of the first solution. Again to match $N_0(x)$ we add

$$\frac{2}{\pi}[-\ln 2 + \gamma]J_0(x),$$

where γ is the usual Euler-Mascheroni constant (Section 5.2).[1] Our new, modified second solution is

$$y_2(x) = \frac{2}{\pi}[\ln x - \ln 2 + \gamma]J_0(x) + \frac{2}{\pi}J_0(x)\left\{\frac{x^2}{4} + \frac{5x^4}{128} + \cdots\right\}. \tag{8.75e}$$

Now the comparison with $N_0(x)$ becomes a simple multiplication of $J_0(x)$ from Eq. 8.75*a* and the curly bracket of Eq. 8.75*c*. The multiplication checks —through terms of order x^2 and x^4, which is all we carried. Our second solution from Eqs. 8.61 and 8.69 agrees with the standard second solution, the Neumann function, $N_0(x)$.

From the preceding analysis, the second solution of Eq. 8.51, $y_2(x)$, may be written

$$y_2(x) = y_1(x)\ln x + \sum_{j=-n}^{\infty} d_j x^{j+\alpha}, \tag{8.75f}$$

the first solution times $\ln x$ and another power series, this one starting with $x^{\alpha-n}$, which means that we may look for a logarithmic term when the indicial equation of Section 8.4 gives only one series solution. With the form of the second solution specified by Eq. 8.75*f*, we can substitute Eq. 8.75*f* into the original differential equation and determine the coefficients d_j exactly as in Section 8.4.

The second solution will usually diverge at the origin because of the logarithmic factor and the negative powers of x in the series. For this reason $y_2(x)$ is often referred to as the irregular solution. The first series solution, $y_1(x)$, which usually converges at the origin, is called the regular solution. The question of behavior at the origin is discussed in more detail in Chapters 11 and 12 in which we take up Bessel functions, modified Bessel functions, and Legendre functions.

Summary. These two sections (together with the Exercises) provide a complete solution of our linear, homogeneous, second-order differential equation—assuming that the point of expansion is no worse than a regular singularity. At least one solution can always be obtained by series substitution (Section 8.4). A second, linearly independent solution can be constructed by the Wronskian double integral, Eq. 8.61. This is all there are: no third, linearly independent solution exists (cf. Ex. 8.5.6).

The *nonhomogeneous*, linear, second-order differential equation will have an

[1] The Neumann function N_0 is defined as it is in order to achieve convenient *asymptotic* properties, Section 11.6.

additional solution: the particular solution. This particular solution may be obtained by the method of variation of parameters, Ex. 8.5.19, or by techniques such as Green's functions, Section 8.6.

EXERCISES

8.5.1 The criterion for the linear *independence* of three vectors **A**, **B**, and **C** is that the equation

$$a\mathbf{A} + b\mathbf{B} + c\mathbf{C} = 0$$

(analogous to Eq. 8.47) has no solution other than the trivial $a = b = c = 0$. Using components $\mathbf{A} = (A_1, A_2, A_3)$, etc., set up the determinant criterion for the existence or nonexistence of a nontrivial solution for the coefficients a, b, and c. Show that your criterion is equivalent to the scalar product $\mathbf{A} \cdot \mathbf{B} \times \mathbf{C}$.

8.5.2 If the Wronskian of two functions y_1 and y_2 is identically zero, show by direct integration that

$$y_1 = cy_2,$$

that is, that y_1 and y_2 are dependent. Assume the functions have continuous derivatives and that at least one of the functions does not vanish in the interval under consideration.

8.5.3 Consider two functions $\varphi_1 = x$ and $\varphi_2 = |x| = x \operatorname{sgn} x$. The function $\operatorname{sgn} x$ is just the *sign* of x. Since $\varphi_1' = 1$ and $\varphi_2' = \operatorname{sgn} x$, $W(\varphi_1, \varphi_2) = 0$ for any interval including $[-1, +1]$. Does the vanishing of the Wronskian over $[-1, +1]$ prove that φ_1 and φ_2 are linearly dependent? Clearly they are not. What is wrong?

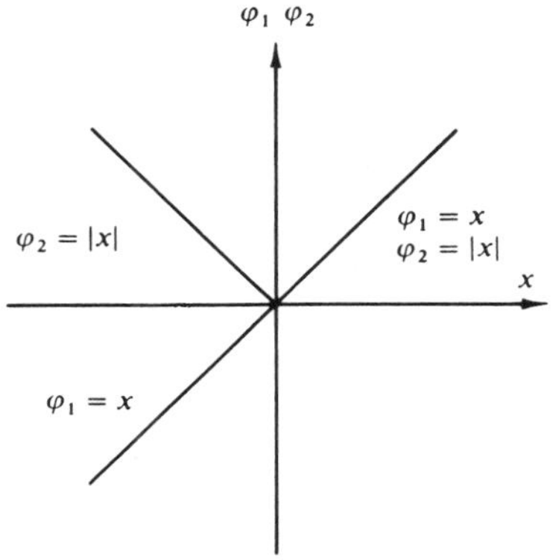

FIG. 8.2 x and $|x|$

8.5.4 Explain that *linear independence* does not mean the absence of any dependence. Illustrate your argument with cosh x and e^x.

8.5.5 Legendre's differential equation

$$(1 - x^2)y'' - 2xy' + n(n+1)y = 0$$

has a regular solution $P_n(x)$ and an irregular solution $Q_n(x)$. Show that the Wronskian of P_n and Q_n is given by

$$P_n(x)\, Q_n'(x) - P_n'(x)\, Q_n(x) = \frac{A_n}{1 - x^2}$$

with A_n *independent* of x.

8.5.6 Show, by means of the Wronskian, that a linear, second-order, homogeneous, differential equation of the form

$$y''(x) + P(x)\, y'(x) + Q(x)\, y(x) = 0$$

cannot have three independent solutions. (Assume a third solution and show that the Wronskian vanishes for all x.)

8.5.7 Transform our linear, second-order, differential equation

$$y'' + P(x)y' + Q(x)y = 0$$

by the substitution

$$y = z \exp[-\tfrac{1}{2}\textstyle\int^x P(t)dt]$$

and show that the resulting differential equation for z is

$$z'' + q(x)z = 0,$$

where

$$q(x) = Q(x) - \tfrac{1}{2}P'(x) - \tfrac{1}{4}P^2(x).$$

Note: This substitution can be *derived* by the technique of Ex. 8.5.18.

8.5.8 Use the result of Exercise 8.5.7 to show that the replacement of $\varphi(r)$ by $r\,\varphi(r)$ may be expected to eliminate the first derivative from the Laplacian in spherical polar coordinates.

8.5.9 By direct differentiation and substitution show that

$$y_2(x) = y_1(x) \int^x \frac{\exp[-\int^s P(t)dt]}{[y_1(s)]^2} ds$$

satisfies

$$y_2''(x) + P(x)\, y_2'(x) + Q(x)\, y_2(x) = 0.$$

Note: The **Leibnitz** formula for the derivative of an integral is

$$\frac{d}{d\alpha}\int_{g(\alpha)}^{h(\alpha)} f(x, \alpha)\, dx = \int_{g(\alpha)}^{h(\alpha)} \frac{\partial f(x, \alpha)}{\partial \alpha} dx + f[h(\alpha), \alpha]\frac{dh(\alpha)}{d\alpha} - f[g(\alpha), \alpha]\frac{dg(\alpha)}{d\alpha}.$$

8.5.10 In the equation

$$y_2(x) = y_1(x) \int^x \frac{\exp[-\int^s P(t)dt]}{[y_1(s)]^2} ds$$

$y_1(x)$ satisfies

$$y_1'' + P(x)y_1' + Q(x)y_1 = 0.$$

The function $y_2(x)$ is a linearly *independent* second-solution of the same equation. Show that the inclusion of lower limits on the two integrals leads to nothing new; that is, it throws in only over-all factors and/or a multiple of the known solution $y_1(x)$.

8.5.11 Given that one solution of

$$R'' + \frac{1}{r}R' - \frac{m^2}{r^2}R = 0$$

is $R = r^m$, show that Eq. 8.61 predicts a second solution, $R = r^{-m}$.

8.5.12 Using $y_1(x) = \sum_{n=0}^{\infty} (-1)^n x^{2n+1}/(2n+1)!$ as a solution of the linear oscillator equation, follow the analysis culminating in Eq. 8.75*f* and show that $c_1 = 0$ so that the second solution does not, in this case, contain a logarithmic term.

8.5.13 Show that when n is *not* an integer the second solution of Bessel's equation, obtained from Eq. 8.61, does *not* contain a logarithmic term.

8.5.14 (a) One solution of Hermite's differential equation

$$y'' - 2xy' + 2\alpha y = 0$$

for $\alpha = 0$ is $y_1(x) = 1$. Find a second solution $y_2(x)$, using Eq. 8.61. Show that your second solution is equivalent to y_{odd} (Exercise 8.4.5).

(b) Find a second solution for $\alpha = 1$, where $y_1(x) = x$, using Eq. 8.61. Show that your second solution is equivalent to y_{even} (Exercise 8.4.5).

8.5.15 One solution of Laguerre's differential equation

$$xy'' + (1 - x)y' + ny = 0$$

for $n = 0$ is $y_1(x) = 1$. Using Eq. 8.61, develop a second, linearly independent solution. Exhibit the logarithmic term explicitly.

8.5.16 One solution of the Chebyshev equation

$$(1 - x^2)y'' - xy' + n^2y = 0$$

for $n = 0$ is $y_1 = 1$.

(a) Using Eq. 8.61, develop a second, linearly independent solution.

(b) Find a second solution by direct integration of the Chebyshev equation.

Hint: Let $v = y'$ and integrate. Compare your result with the second solution given in Section 13.3.

Ans. (a) $y_2 = \sin^{-1}x$.

(b) The second solution, $V_n(x)$, is not defined for $n = 0$.

8.5.17 The radial Schrödinger wave equation has the form

$$\left[-\frac{\hbar^2}{2m}\frac{d^2}{dr^2} + l(l+1)\frac{\hbar^2}{2mr^2} + V(r)\right]y(r) = Ey(r).$$

The potential energy $V(r)$ may be expanded about the origin as

$$V(r) = \frac{b_{-1}}{r} + b_0 + b_1 r + \dots.$$

(a) Show that there is one (regular) solution starting with r^{l+1}.

(b) From Eq. 8.62 show that the irregular solution diverges at the origin as r^{-l}.

8.5.18 Show that if a second solution, y_2, is assumed to have the form $y_2(x) = y_1(x)f(x)$ substitution back into the original equation

$$y_2'' + P(x)y_2' + Q(x)y_2 = 0$$

leads to

$$f(x) = \int^x \frac{\exp[-\int^s P(t)dt]}{[y_1(s)]^2}ds$$

in agreement with Eq. 8.61.

8.5.19 If our linear, second-order differential equation is nonhomogeneous, that is, of the form of Eq. 8.20*b*, the most general solution is

$$y(x) = y_1(x) + y_2(x) + y_p(x).$$

(y_1 and y_2 are solutions of the homogeneous equation.)
Show that

$$y_p(x) = y_2(x) \int^x \frac{y_1(s)F(s)\,ds}{W\{y_1(s), y_2(s)\}} - y_1(x) \int^x \frac{y_2(s)F(s)\,ds}{W\{y_1(s), y_2(s)\}}$$

with $W\{y_1(s), y_2(s)\}$ the Wronskian of $y_1(s)$ and $y_2(s)$.
Hint: As in Ex. 8.5.18 let $y_p(x) = y_1(x)\,v(x)$ and develop a first order differential equation for $v'(x)$.

8.6 Nonhomogeneous Equation—Green's Function

The series substitution of Section 8.4 and the Wronskian double integral of Section 8.5 provide the most general solution of the *homogeneous*, linear, second-order differential equation. The specific solution, y_p, linearly dependent on the source term ($F(x)$ of Eq. 8.20*b*) may be cranked out by the variation of parameters method, Ex. 8.5.19. In this section, we turn to a different method of solution—Green's functions.

For a brief introduction to Green's function method, as applied to the solution of a nonhomogeneous partial differential equation, it is helpful to use the electrostatic analog. In the presence of charges the electrostatic potential ψ satisfies Poisson's nonhomogeneous equation (cf. Section 1.14)

$$\nabla^2\psi = -\frac{\rho}{\varepsilon_0}, \qquad \text{(mks units)} \tag{8.76}$$

and Laplace's homogeneous equation,

$$\nabla^2\psi = 0, \tag{8.77}$$

in the absence of electric charge ($\rho = 0$). If the charges are point charges q_i, we know that the solution is

$$\psi = \frac{1}{4\pi\varepsilon_0} \sum_i \frac{q_i}{r_i}, \tag{8.78}$$

a superposition of single-point charge solutions obtained from Coulomb's law for the force between two point charges q_1 and q_2,

$$\mathbf{F} = \frac{q_1 q_2 \mathbf{r}_0}{4\pi\varepsilon_0 r^2}. \tag{8.79}$$

By replacement of the discrete point charges with a smeared out distributed charge, charge density ρ, Eq. 8.78 becomes

$$\psi(r = 0) = \frac{1}{4\pi\varepsilon_0} \int \frac{\rho(\mathbf{r})}{r}\,d\tau \tag{8.80}$$

or, for the potential at $\mathbf{r} = \mathbf{r}_1$ away from the origin and the charge at $\mathbf{r} = \mathbf{r}_2$,

$$\psi(\mathbf{r}_1) = \frac{1}{4\pi\varepsilon_0} \int \frac{\rho(\mathbf{r}_2)}{|\mathbf{r}_1 - \mathbf{r}_2|}\,d\tau_2 \tag{8.81}$$

Dirac delta function. A formal derivation and generalization of this result is facilitated by using $\delta(x)$, the Dirac delta function, as in Section 1.15. For the one-dimensional case, the Dirac delta function is often defined by the following properties:

$$\delta(x) = 0, \qquad x \neq 0, \tag{8.82a}$$

$$\int_{-\infty}^{\infty} \delta(x)\, dx = 1, \tag{8.82b}$$

and

$$\int_{-\infty}^{\infty} f(x)\delta(x)\, dx = f(0). \tag{8.82c}$$

Here it is assumed that $f(x)$ is continuous at $x = 0$.

From these defining equations $\delta(x)$ must be an infinitely high, infinitely thin spike—as in the description of an impulsive force (Section 15.8) or charge density for a point charge.[1] The problem is that *no such function exists* in the usual sense of function. It is possible to approximate the delta function by a variety of functions, Eqs. 8.83*a*–8.83*d* and Figs. 8.3*a*–8.3*d*:

$$\delta_n(x) = \begin{cases} 0, & x < -\dfrac{1}{2n} \\ n, & -\dfrac{1}{2n} < x < \dfrac{1}{2n} \\ 0, & x > \dfrac{1}{2n} \end{cases} \tag{8.83a}$$

$$\delta_n(x) = \frac{n}{\sqrt{\pi}} \exp(-n^2x^2) \tag{8.83b}$$

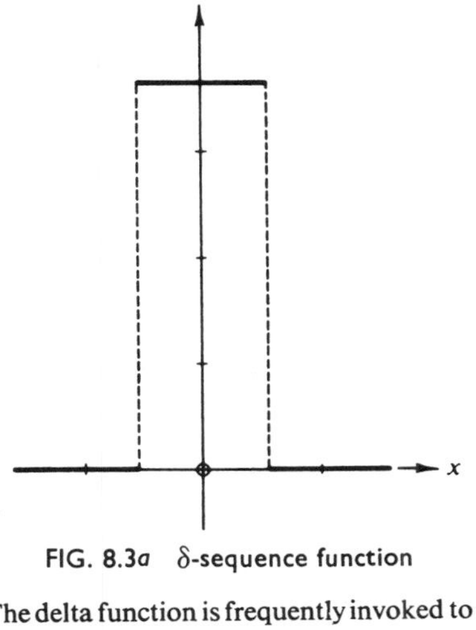

FIG. 8.3*a* δ-sequence function

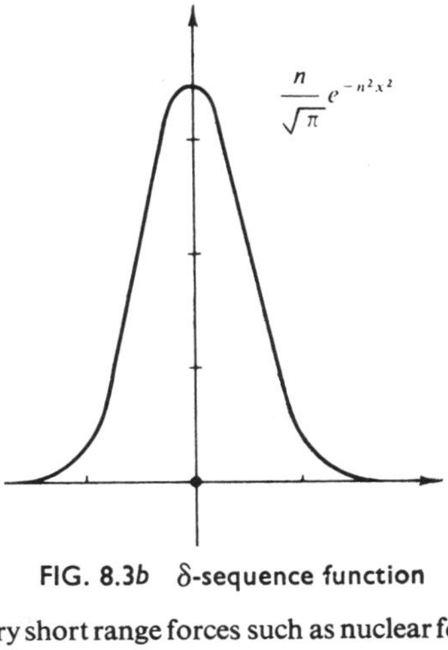

FIG. 8.3*b* δ-sequence function

[1] The delta function is frequently invoked to describe very short range forces such as nuclear forces.

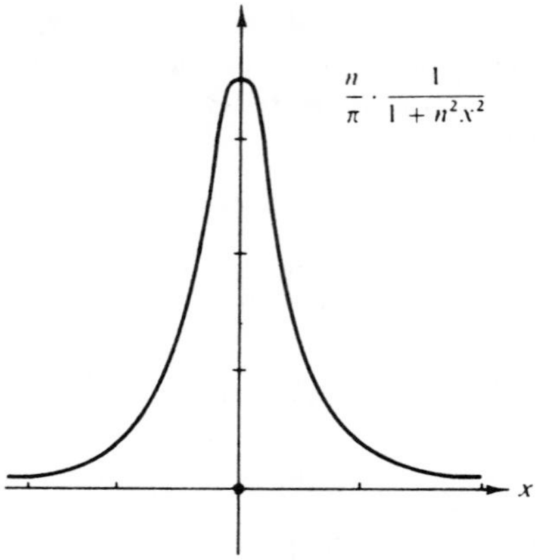

FIG. 8.3*c* δ-sequence function

$$\delta_n(x) = \frac{n}{\pi} \cdot \frac{1}{1+n^2x^2} \tag{8.83c}$$

$$\delta_n(x) = \frac{\sin nx}{\pi x} = \frac{1}{2\pi} \int_{-n}^{n} e^{ixt}\, dt \tag{8.83d}$$

These approximations have varying degrees of usefulness. Equation 8.83*a* is useful in providing a simple derivation of the integral property, Eq. 8.82*c*. Equation 8.83*b* is convenient to differentiate. Its derivatives lead to the Hermite polynomials, Eq. 13.7. Equation 8.83*d* is particularly useful in Fourier analysis and in its applications to quantum mechanics.

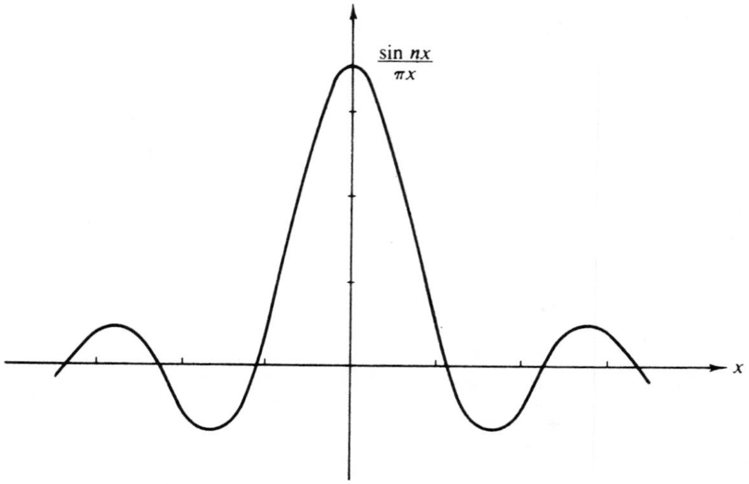

FIG. 8.3*d* δ-sequence function

For most physical purposes, such approximations are quite adequate. From a mathematical point of view, the situation is still unsatisfactory: the limits

$$\lim_{n\to\infty} \delta_n(x)$$

do not exist.

A way out of this difficulty is provided by the theory of distributions. Recognizing that Eq. 8.82c is the fundamental property, we focus our attention on it rather than on $\delta(x)$ itself. Equations 8.83a–8.83d with $n = 1, 2, 3, \ldots$ may be interpreted as *sequences* of normalized functions:

$$\int_{-\infty}^{\infty} \delta_n(x)\, dx = 1. \tag{8.84}$$

The sequence of integrals has the limit

$$\lim_{n\to\infty} \int_{-\infty}^{\infty} \delta_n(x) f(x)\, dx = f(0). \tag{8.85a}$$

Note carefully that Eq. 8.85a is the limit of a sequence of integrals. Again, the limit of $\delta_n(x)$, $n \to \infty$, does not exist. (The limits for all four forms of $\delta_n(x)$ diverge at $x = 0$).

We may treat $\delta(x)$ consistently in the form

$$\int_{-\infty}^{\infty} \delta(x) f(x)\, dx = \lim_{n\to\infty} \int_{-\infty}^{\infty} \delta_n(x) f(x)\, dx. \tag{8.85b}$$

$\delta(x)$ is labeled a distribution (not a function) defined by the sequences $\delta_n(x)$ as indicated in Eq. 8.85b. We might emphasize that the integral on the left-hand side of Eq. 8.85b is not a Riemann integral.[1] It is a limit.

This distribution $\delta(x)$ is only one of an infinity of possible distributions, but it is the one we are interested in because of Eq. 8.82c.

We shall use $\delta(x)$ frequently and shall call it the Dirac delta function—for historical reasons. Remember that it is not really a function. It is essentially a shorthand notation, defined implicitly as the limit of integrals of a sequence, $\delta_n(x)$, according to Eq. 8.85b. It should be understood that our Dirac delta function has significance only as part of an integrand and never as an end result.

Shifting our singularity to the point $x = x'$, the Dirac delta function is written $\delta(x - x')$. Equation 8.82c becomes

$$\int_{-\infty}^{\infty} f(x)\, \delta(x - x')\, dx = f(x'), \tag{8.86}$$

As a description of a singularity at $x = x'$, the Dirac delta function may be written as $\delta(x - x')$ or as $\delta(x' - x)$. Going to three dimensions and using spherical polar coordinates, we obtain

$$\int_0^{2\pi}\int_0^{\pi}\int_0^{\infty} \delta(\mathbf{r}) r^2\, dr \sin\theta\, d\theta\, d\varphi = \iiint_{-\infty}^{\infty} \delta(x)\, \delta(y) \delta(z)\, dx\, dy\, dz = 1. \tag{8.87}$$

[1] It can be treated as a Stieltjes integral if desired. $\delta(x)\, dx$ is replaced by $dS(x)$, where $S(x)$ is the Heaviside step function (cf. Ex. 8.6.3).

This corresponds to a singularity (or source) at the origin. Again, if our source is at $\mathbf{r} = \mathbf{r}_1$, Eq. 8.87 becomes

$$\int\int\int \delta(\mathbf{r}_2 - \mathbf{r}_1) r_2^2 \, dr_2 \sin\theta_2 \, d\theta_2 \, d\varphi_2 = 1. \tag{8.88}$$

As already mentioned,

$$\delta(\mathbf{r}_2 - \mathbf{r}_1) = \delta(\mathbf{r}_1 - \mathbf{r}_2). \tag{8.89}$$

Poisson's equation—Green's function solution. Returning to our electrostatic problem, we use ψ as the potential corresponding to the given distribution of charge and therefore satisfying Poisson's equation

$$\nabla^2 \psi = -\frac{\rho}{\varepsilon_0}, \tag{8.90}$$

whereas a function φ, which we label a Green's function, is required to satisfy Poisson's equation with a point source at the point defined by $\mathbf{r}_2$:

$$\nabla^2 \varphi = -\delta(\mathbf{r}_1 - \mathbf{r}_2). \tag{8.91}$$

Physically, then, φ is the potential at $\mathbf{r}_1$ corresponding to a unit source (ε_0) at $\mathbf{r}_2$. By Green's theorem (Section 1.11)

$$\int (\psi \, \nabla^2 \varphi - \varphi \, \nabla^2 \psi) \, d\tau_2 = \int (\psi \, \nabla \varphi - \varphi \, \nabla \psi) \cdot d\boldsymbol{\sigma}. \tag{8.92}$$

Assuming that the integrand falls off faster than r^{-2}, we may simplify our problem by taking the volume so large that the surface integral vanishes, leaving

$$\int \psi \, \nabla^2 \varphi \, d\tau_2 = \int \varphi \, \nabla^2 \psi \, d\tau_2 \tag{8.93}$$

or by substituting in Eqs. 8.90 and 8.91,

$$-\int \psi(\mathbf{r}_2) \, \delta(\mathbf{r}_1 - \mathbf{r}_2) \, d\tau_2 = -\int \frac{\varphi(\mathbf{r}_1, \mathbf{r}_2) \, \rho(\mathbf{r}_2)}{\varepsilon_0} \, d\tau_2. \tag{8.94}$$

Integration by employing the defining property of the Dirac delta function (Eq. 8.82*c*), produces

$$\psi(\mathbf{r}_1) = \frac{1}{\varepsilon_0} \int \varphi(\mathbf{r}_1, \mathbf{r}_2) \, \rho(\mathbf{r}_2) \, d\tau_2. \tag{8.95}$$

Note that we have used Eq. 8.91 to eliminate $\nabla^2 \varphi$ but that the function φ itself is still unknown. In Section 1.14, Gauss's law, we found that

$$\int \nabla^2 \left(\frac{1}{r}\right) d\tau = \begin{cases} 0, \\ -4\pi, \end{cases} \tag{8.96}$$

0 if the volume did not include the origin and -4π if the origin were included. This result from Section 1.14 may be rewritten as

$$\nabla^2 \left(\frac{1}{4\pi r}\right) = -\delta(\mathbf{r}), \quad \text{or} \quad \nabla^2 \left(\frac{1}{4\pi r_{12}}\right) = -\delta(\mathbf{r}_1 - \mathbf{r}_2) \tag{8.97}$$

corresponding to a shift of the electrostatic charge from the origin to the position $\mathbf{r} = \mathbf{r}_2$. Here $r_{12} = |\mathbf{r}_1 - \mathbf{r}_2|$, and the Dirac delta function $\delta(\mathbf{r}_1 - \mathbf{r}_2)$ vanishes unless $\mathbf{r}_1 = \mathbf{r}_2$. Therefore, in a comparison of Eqs. 8.91 and 8.97 the function φ (Green's function) is given by

$$\varphi(\mathbf{r}_1, \mathbf{r}_2) = \frac{1}{4\pi|\mathbf{r}_1 - \mathbf{r}_2|}. \tag{8.98}$$

The solution of our differential equation (Poisson's equation) is

$$\psi(\mathbf{r}_1) = \frac{1}{4\pi\varepsilon_0}\int \frac{\rho(\mathbf{r}_2)}{|\mathbf{r}_1 - \mathbf{r}_2|}\,d\tau_2 \tag{8.99}$$

in complete agreement with Eq. 8.81.

In summary, Green's function, $\varphi(\mathbf{r}_1, \mathbf{r}_2)$, often written $G(\mathbf{r}_1, \mathbf{r}_2)$ as a reminder of the name, is a solution of Eq. 8.91. It enters in an integral solution of our differential equation, as in Eq. 8.81. For the simple, but important electrostatic case we obtain Green's function $G(\mathbf{r}_1, \mathbf{r}_2)$ by Gauss's law, comparing Eqs. 8.91 and 8.97. Finally, from the final solution (Eq. 8.99), it is possible to develop a physical interpretation of Green's function. It occurs as a weighting function or influence function which enhances or reduces the effect of the charge element $\rho(\mathbf{r}_2)\,d\tau_2$ according to its distance from the field point $\mathbf{r}_1$. Green's function, $G(\mathbf{r}_1, \mathbf{r}_2)$, gives the effect of a unit point source at $\mathbf{r}_2$ in producing a potential at $\mathbf{r}_1$. This is how it was introduced in Eq. 8.91; this is how it appears in Eq. 8.99.

An important property of Green's function is the symmetry of its two variables, that is,

$$G(\mathbf{r}_1, \mathbf{r}_2) = G(\mathbf{r}_2, \mathbf{r}_1). \tag{8.100}$$

Although this is obvious in the electrostatic case just considered, it can be proved under much more general conditions. In place of Eq. 8.91, let us require that $G(\mathbf{r}, \mathbf{r}_1)$ satisfy[1]

$$\nabla \cdot [p(\mathbf{r})\,\nabla G(\mathbf{r}, \mathbf{r}_1)] + \lambda q(\mathbf{r})\,G(\mathbf{r}, \mathbf{r}_1) = -\delta(\mathbf{r} - \mathbf{r}_1), \tag{8.101}$$

corresponding to a mathematical point source at $\mathbf{r} = \mathbf{r}_1$. Here the functions $p(\mathbf{r})$ and $q(\mathbf{r})$ are well-behaved but otherwise arbitrary functions of $\mathbf{r}$. Green's function, $G(\mathbf{r}, \mathbf{r}_2)$, satisfies the same equation but the subscript 1 is replaced by subscript 2.

$$\nabla \cdot [p(\mathbf{r})\,\nabla G(\mathbf{r}, \mathbf{r}_2)] + \lambda\, q(\mathbf{r})\,G(\mathbf{r}, \mathbf{r}_2) = -\delta(\mathbf{r} - \mathbf{r}_2). \tag{8.102}$$

Then $G(\mathbf{r}, \mathbf{r}_2)$ is a sort of potential at $\mathbf{r}$, created by a unit point source at $\mathbf{r}_2$. We multiply the equation for $G(\mathbf{r}, \mathbf{r}_1)$ by $G(\mathbf{r}, \mathbf{r}_2)$ and the equation for $G(\mathbf{r}, \mathbf{r}_2)$ by $G(\mathbf{r}, \mathbf{r}_1)$ and then subtract the two:

$$\begin{aligned} G(\mathbf{r}, \mathbf{r}_2)\nabla \cdot [p(\mathbf{r})\,\nabla G(\mathbf{r}, \mathbf{r}_1)] &- G(\mathbf{r}, \mathbf{r}_1)\,\nabla \cdot [p(\mathbf{r})\,\nabla\, G(\mathbf{r}, \mathbf{r}_2)] \\ &= -G(\mathbf{r}, \mathbf{r}_2)\,\delta(\mathbf{r} - \mathbf{r}_1) + G(\mathbf{r}, \mathbf{r}_1)\,\delta(\mathbf{r} - \mathbf{r}_2). \end{aligned} \tag{8.103}$$

By integrating over whatever volume is involved, we obtain a surface integral by Green's theorem:

[1] Equation 8.101 is a three-dimensional version of the *self-adjoint* eigenvalue equation, Eq. 9.4.

$$\int_S [G(\mathbf{r}, \mathbf{r}_2)\, p(\mathbf{r})\, \nabla G(\mathbf{r}, \mathbf{r}_1) - G(\mathbf{r}, \mathbf{r}_1)\, p(\mathbf{r})\, \nabla G(\mathbf{r}, \mathbf{r}_2)] \cdot d\boldsymbol{\sigma}$$
$$= -G(\mathbf{r}_1, \mathbf{r}_2) + G(\mathbf{r}_2, \mathbf{r}_1). \quad (8.104)$$

The terms on the right-hand side appear when we use the Dirac delta functions and carry out the volume integration. Under the requirement that Green's functions, $G(\mathbf{r}, \mathbf{r}_1)$ and $G(\mathbf{r}, \mathbf{r}_2)$, have the same values over the surface S and that their normal derivatives have the same values over the surfaces S, or that the Green's functions vanish (Dirichlet boundary conditions, Section 9.1)[1] over the surface S, the surface integral vanishes and

$$G(\mathbf{r}_1, \mathbf{r}_2) = G(\mathbf{r}_2, \mathbf{r}_1), \quad (8.105)$$

which shows that Green's function is symmetric. If the eigenfunctions are complex, boundary conditions corresponding to Eqs. 9.20–9.22 are appropriate. Equation 8.105 becomes

$$G(\mathbf{r}_1, \mathbf{r}_2) = G^*(\mathbf{r}_2, \mathbf{r}_1). \quad (8.106)$$

Note that this symmetry property holds for Green's function in every equation in the form of Eq. 8.101. In Chapter 9 we shall call equations in this form self-adjoint. The symmetry is the basis of various reciprocity theorems; the effect of a charge at $\mathbf{r}_2$ on the potential at $\mathbf{r}_1$ is the same as the effect of a charge at $\mathbf{r}_1$ on the potential at $\mathbf{r}_2$.

This use of Green's functions is a powerful technique for solving many of the more difficult problems of mathematical physics. We shall return to it when we take up integral equations in Chapter 16.

EXERCISES

8.6.1 Let

$$\delta_n(x) = \begin{cases} 0, & x < -\dfrac{1}{2n}, \\ n, & -\dfrac{1}{2n} < x < \dfrac{1}{2n}, \\ 0, & \dfrac{1}{2n} < x. \end{cases}$$

Show that

$$\lim_{n\to\infty} \int_{-\infty}^{\infty} f(x)\, \delta_n(x)\, dx = f(0)$$

assuming that $f(x)$ is continuous at $x = 0$.

[1] Any attempt to demand that the normal derivatives vanish at the surface (Neumann's conditions, Section 9.1) leads to trouble with Gauss's Law. It is like demanding that $\int \mathbf{E} \cdot d\boldsymbol{\sigma} = 0$ when you know perfectly well that there is some electric charge inside the surface.

8.6.2 Verify that the sequence $\delta_n(x)$, based on the function

$$\delta_n = \begin{cases} 0, & x < 0 \\ ne^{-nx}, & x > 0, \end{cases}$$

is a delta sequence (satisfying Eq. 8.85*a*). Note that the singularity is at $+0$, the positive side of the origin.

8.6.3 (a) If we define a sequence $\delta_n(x) = n/(2\cosh^2 nx)$, show that

$$\int_{-\infty}^{\infty} \delta_n(x)\,dx = 1, \qquad \text{independent of } n.$$

(b) Continuing this analysis, show that

$$\int_{-\infty}^{x} \delta_n(x)\,dx = \frac{1}{2}\,[1 + \tanh nx] \equiv S_n(x)$$

and

$$\lim_{n\to\infty} S_n(x) = \begin{cases} 0, & x < 0, \\ 1, & x > 0. \end{cases}$$

This is the Heaviside unit step function.

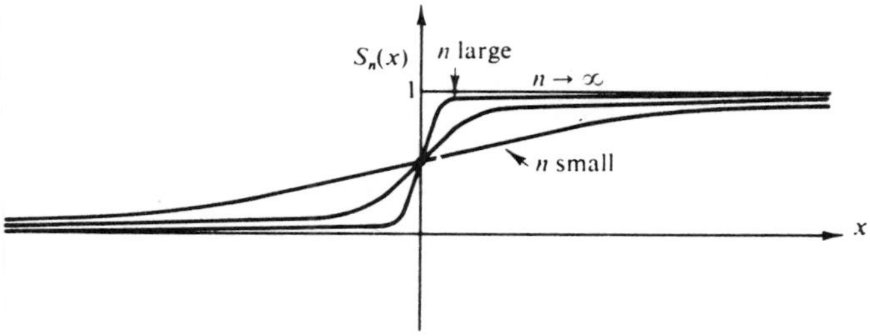

FIG. 8.4 $\frac{1}{2}[1 + \tanh nx]$ and the Heaviside unit step function

8.6.4 Using the Gauss error curve delta sequence (δ_n) show that

$$x\,\frac{d}{dx}\,\delta(x) = -\delta(x)$$

treating $\delta(x)$ and its derivative as in Eq. 8.85*b*.

8.6.5 Show that

$$\int_{-\infty}^{\infty} \delta'(x) f(x)\,dx = -f'(0).$$

Here we assume that $f'(x)$ is continuous at $x = 0$.

8.6.6 Prove that

$$\delta(f(x)) = \left|\frac{df(x)}{dx}\right|^{-1} \delta(x - x_0),$$

where x_0 is chosen so that $f(x_0) = 0$. *Hint*. Note that $\delta(f)\,df = \delta(x)\,dx$.

8.6.7 Show that in spherical polar coordinates $(r, \cos\theta, \varphi)$ the delta function $\delta(\mathbf{r}_1 - \mathbf{r}_2)$ becomes

$$\frac{1}{r_1^2}\delta(r_1 - r_2)\,\delta(\cos\theta_1 - \cos\theta_2)\,\delta(\varphi_1 - \varphi_2).$$

Generalize this to the curvilinear coordinates (q_1, q_2, q_3) of Section 2.1 with scale factors h_1, h_2, and h_3.

8.6.8 A rigorous development of Fourier transforms (Sneddon, *Fourier Transforms*) includes as a theorem the relations

$$\lim_{a\to\infty}\frac{2}{\pi}\int_{x_1}^{x_2} f(u+x)\frac{\sin ax}{x}\,dx = \begin{cases} f(u+0)+f(u-0), & x_1 < 0 < x_2 \\ f(u+0), & x_1 = 0 < x_2 \\ f(u-0), & x_1 < 0 = x_2 \\ 0 & x_1 < x_2 < 0 \ \text{ or } \ 0 < x_1 < x_2 \end{cases}$$

Verify these results using the Dirac delta function.

8.6.9 Show that

$$G(\mathbf{r}_1, \mathbf{r}_2) = \frac{e^{ik|\mathbf{r}_1-\mathbf{r}_2|}}{4\pi|\mathbf{r}_1 - \mathbf{r}_2|}$$

is a Green's function satisfying the differential equation

$$(\nabla_1^2 + k^2)\,G(\mathbf{r}_1, \mathbf{r}_2) = -\delta(\mathbf{r}_1 - \mathbf{r}_2).$$

This involves two parts:

(a) Show that $G(\mathbf{r}_1, \mathbf{r}_2)$ satisfies the homogeneous differential equation away from $\mathbf{r}_1 = \mathbf{r}_2$

(b) Show that for $|\mathbf{r}_1 - \mathbf{r}_2|$ small enough

$$\int(\nabla_1^2 + k^2)\,G(\mathbf{r}_1, \mathbf{r}_2)\,d\tau_1 = \begin{cases} 0, & \mathbf{r}_2 \neq \mathbf{r}_1, \\ -1, & \mathbf{r}_2 = \mathbf{r}_1. \end{cases}$$

8.7 Numerical Solutions

The analytic solutions and approximate solutions to differential equations in this chapter and in succeeding chapters may suffice to solve the problem at hand—particularly if there is some symmetry present. The power series solutions will tell how the solution behaves at small values of x. The asymptotic solutions (cf. Sections 11.6 and 12.10) will tell how the solution behaves at large values of x. These limiting cases and also the possible resemblance of our differential equation to the standard forms with known solutions (Chapters 11–13) are invaluable in helping us gain an understanding of the general behavior of our solution.

However, the usual situation is that we have a *different* equation, perhaps a different potential in the Schrödinger wave equation, and we want a reasonably exact solution. So we turn to numerical techniques.

First-Order Differential Equations. The differential equation involves a continuity of points. The independent variable x is continuous. The (unknown) dependent variable $y(x)$ is assumed continuous. The concept of differentiation demands continuity. Our numerical processes replace these continua by discrete sets. We consider x at

$$x_0, \qquad x_0 + h, \qquad x_0 + 2h, \qquad x_0 + 3h, \qquad \text{etc.},$$

where h is some small interval. The smaller h is, the better the approximation—in principle. But if h is made too small, the demands on machine time will be excessive, and accuracy may actually decline because of accumulated round-off errors. We refer to the successive discrete values of x as x_n, x_{n+1}, etc., and the corresponding values of $y(x)$ as $y(x_n) = y_n$. If x_0 and y_0 are given, the problem is to find y_1, then to find y_2, and so on.

Runge-Kutta Method. Consider the ordinary first-order differential equation

$$\frac{d}{dx} y(x) = f(x, y) \tag{8.107}$$

with the initial condition $y(x_0) = y_0$. Using a Taylor expansion, we could approximate y_1 as

$$\begin{aligned} y_1 &= y_0 + h\, y_0' \\ &= y_0 + h\, f(x_0, y_0), \end{aligned} \tag{8.108}$$

neglecting the terms of order h^2.

The Runge-Kutta method is a refinement of this, with an error of order h^5. The relevant formulas are

$$y_{n+1} = y_n + [k_0 + 2k_1 + 2k_2 + k_3]/6, \tag{8.109}$$

where

$$\begin{aligned} k_0 &= h\, f(x_n, y_n), \\ k_1 &= h\, f(x_n + \tfrac{1}{2}h, y_n + \tfrac{1}{2}k_0), \\ k_2 &= h\, f(x_n + \tfrac{1}{2}h, y_n + \tfrac{1}{2}k_1), \\ k_3 &= h\, f(x_n + h, y_n + k_2). \end{aligned} \tag{8.110}$$

A derivation of these equations appears in Ralston and Wilf[1] (Chapter 9 by M. J. Romanelli).

The Runge-Kutta method is stable, meaning that small errors do not get amplified. It is self-starting, meaning that we just take the x_0 and y_0 and away we go. But it has disadvantages. Four separate calculations of $f(x, y)$ are required at each step. The errors, while of order h^5 per step, are not known. One checks the numerical solution by cutting h in half and repeating the calculation. If the second result agrees with the first, then h was small enough.

Finally the Runge-Kutta method can be extended to a set of *coupled* first order equations:

$$\begin{aligned} \frac{du}{dx} &= f_1(x, u, v) \\ \frac{dv}{dx} &= f_2(x, u, v), \qquad \text{etc.,} \end{aligned} \tag{8.111}$$

[1] *Mathematical Methods for Digital Computers*, A. Ralston, H. S. Wilf, eds. New York: Wiley (1960).

with as many *dependent* variables as desired.

Predictor-Corrector Methods. As an alternate attack on Eq. 8.107, we might estimate or *predict* a tentative value of y_{n+1} by

$$\begin{aligned}\bar{y}_{n+1} &= y_{n-1} + 2h\, y_n' \\ &= y_{n-1} + 2h f(x_n, y_n).\end{aligned} \tag{8.112}$$

This is not quite the same as Eq. 8.108. Rather, it may be interpreted as

$$y_n' \approx \frac{\Delta y}{\Delta x} = \frac{y_{n+1} - y_{n-1}}{2h}, \tag{8.113}$$

the derivative as a tangent being replaced by a chord. Next we *calculate*

$$y_{n+1}' = f(x_{n+1}, \bar{y}_{n+1}). \tag{8.114}$$

Then to *correct* for the crudeness of Eq. 8.112, we take

$$y_{n+1} = y_n + \frac{h}{2}(\bar{y}_{n+1}' + y_n'). \tag{8.115}$$

Here, the finite difference ratio $\Delta y/h$ is approximated by the average of the two derivatives. This technique—a prediction followed by a correction (and iteration until agreement is reached)—is the heart of the predictor-corrector method. It should be emphasized that the above set of equations is intended only to illustrate the predictor-corrector method. The accuracy of this set (to order h^3) is usually inadequate.

The iteration (substituting y_{n+1} from Eq. 8.115 back into Eq. 8.114 and recycling until y_{n+1} settles down to some limit) is time-consuming in a computing machine operation. Consequently, the iteration is usually replaced by an intermediate step (the *modifier*) between Eqs. 8.112 and 8.114.

This modified predictor-corrector method has the major advantage over the Runge-Kutta method of requiring only two computations of $f(x, y)$ per step, instead of four. Unfortunately, the method as originally developed was unstable—small errors (round-off and truncation) tended to propagate and beome amplified.

This very serious problem of instability has been overcome in a version of the predictor-corrector method devised by Hamming. The formulas (which are moderately involved), a partial derivation, and detailed instructions for starting the solution are all given by Ralston (Chapter 8 of Ralston and Wilf). Hamming's method is accurate to order h^4. It is stable for all reasonable values of h, and it provides an estimate of the error. Unlike the Runge-Kutta method, it is *not* self-starting. For example, Eq. 8.112 requires both y_{n-1} and y_n. Starting values (y_0, y_1, y_2, y_3) for the Hamming predictor-corrector method may be computed by series solution (power series for small x, asymptotic series for large x) or by the Runge-Kutta method.

The Hamming predictor-corrector method may be extended to cover a set of coupled first-order differential equations, i.e., Eq. 8.111.

Second-order differential equations. Any second-order differential equation

$$y''(x) + P(x)\,y'(x) + Q(x)\,y(x) = F(x), \tag{8.116}$$

may be split into two first-order differential equations by writing

$$y'(x) = z(x), \tag{8.117}$$

and then

$$z'(x) + P(x)\,z(x) + Q(x)\,y(x) = F(x). \tag{8.118}$$

These coupled first-order differential equations may be solved by either the Runge-Kutta or Hamming predictor-corrector techniques described above.

When the second-order differential equation may be put in the form[1]

$$y''(x) + f(x)\,y(x) = g(x), \tag{8.119}$$

a host of special methods are available for numerical integration. The Numerov method is particularly useful.[2]

REFERENCES

BATEMAN, H., *Partial Differential Equations of Mathematical Physics*. New York: Dover (1944; first edition, 1932). A wealth of applications of various partial differential equations in classical physics. Excellent examples of the use of different coordinate systems—ellipsoidal, paraboloidal, toroidal coordinates, and so on.

HAMMING, R. W., *Numerical Methods for Scientists and Engineers*. New York: McGraw-Hill (1962).

INCE, E. L., *Ordinary Differential Equations*. New York: Dover (1926). The classic work in the theory of ordinary differential equations.

MURPHY, G. M., *Ordinary Differential Equations and Their Solutions*. Princeton, New Jersey: Van Nostrand (1960). A thorough, relatively readable treatment of ordinary differential equations, both linear and nonlinear.

RALSTON, A., and H. WILF, Eds., *Mathematical Methods for Digital Computers*. New York: Wiley (1960).

[1]The Schrödinger wave equation may be put in this form (cf. Ex. 8.5.7).

[2]D. R. Hartree, *The Calculation of Atomic Structures*, Chapter 4. New York: Wiley (1957).

$$\overline{\mathscr{L}}u = \mathscr{L}u = \frac{d}{dx}\left[p(x)\frac{du(x)}{dx}\right] + q(x)\,u(x)$$

$$\mathscr{L}u(x) + \lambda\, w(x)\, u(x) = 0.$$

CHAPTER 9

STURM-LIOUVILLE THEORY—ORTHOGONAL FUNCTIONS

9.1 Self-Adjoint Differential Equations

In Chapter 8 we studied, classified, and solved linear, second-order, differential equations corresponding to linear, second-order, differential operators of the general form

$$\mathscr{L}\,u(x) = p_0(x)\frac{d^2}{dx^2}u(x) + p_1(x)\frac{d}{dx}u(x) + p_2(x)\,u(x). \tag{9.1}$$

The *functions* $p_0(x)$, $p_1(x)$, and $p_2(x)$ are not to be confused with the constants p_i of Section 8.5. Reference to Eq. 8.51 shows that $P(x) = p_1(x)/p_0(x)$ and $Q(x) = p_2(x)/p_0(x)$.

These coefficients, $p_0(x)$, $p_1(x)$, and $p_2(x)$ are real functions of x and over the region of interest, $a \le x \le b$, the first $2 - i$ derivatives of $p_i(x)$ are continuous. Further, $p_0(x)$ does not vanish for $a < x < b$. Now, the zeros of $p_0(x)$ are singular points (Section 8.3) and the preceding statement simply means that we choose our interval $[a, b]$ so that there are no singular points in the interior of the interval. There may be and often are singular points on the boundaries.

It is convenient in the mathematical theory of differential equations to define an *adjoint*[1] *operator* $\overline{\mathscr{L}}$ by

$$\begin{aligned}\overline{\mathscr{L}}u &= \frac{d^2}{dx^2}[p_0 u] - \frac{d}{dx}[p_1 u] + p_2 u \\ &= p_0\frac{d^2u}{dx^2} + (2p_0' - p_1)\frac{du}{dx} + (p_0'' - p_1' + p_2)u.\end{aligned} \tag{9.2}$$

[1] The *adjoint* operator bears a somewhat forced relationship to the *adjoint* matrix. A better justification for the nomenclature is found in a comparison of the *self-adjoint* operator (plus appropriate boundary conditions) with the *self-adjoint* matrix. The significant properties are developed in Section 9.2. It is just because of these properties that we are interested in *self-adjoint* operators.

In a comparison of Eqs. 9.1 and 9.2 the necessary and sufficient condition that $\mathscr{L} = \bar{\mathscr{L}}$ is that

$$p_0'(x) = \frac{dp_0(x)}{dx} = p_1(x). \tag{9.3}$$

When this condition is satisfied,

$$\bar{\mathscr{L}}u = \mathscr{L}u = \frac{d}{dx}\left[p(x)\frac{du(x)}{dx}\right] + q(x)\,u(x) \tag{9.4}$$

and the operator $\mathscr{L}$ is said to be *self-adjoint*. Here, for the self-adjoint case, $p_0(x)$ is replaced by $p(x)$ and $p_2(x)$ by $q(x)$ to avoid unnecessary subscripts.

In a survey of the differential equations introduced in Section 8.2, Legendre's equation and the linear oscillator equation are self-adjoint, but others, such as the Laguerre and Hermite equations, are not. However, the theory of linear, second-order, self-adjoint differential equations is perfectly general because we can *always* transform the non-self-adjoint operator into the required self-adjoint form. Consider Eq. 9.1 with $p_0' \neq p_1$. If we multiply $\mathscr{L}$ by[1]

$$\frac{1}{p_0(x)}\exp\left[\int^x \frac{p_1(t)}{p_0(t)}\,dt\right]$$

we obtain

$$\frac{1}{p_0(x)}\exp\left[\int^x \frac{p_1(t)}{p_0(t)}\,dt\right]\mathscr{L}\,u(x) = \frac{d}{dx}\left\{\exp\left[\int^x \frac{p_1(t)}{p_0(t)}\,dt\right]\frac{du(x)}{dx}\right\}$$
$$+ \frac{p_2(x)}{p_0(x)}\cdot\exp\left[\int^x \frac{p_1(t)}{p_0(t)}\,dt\right]u, \tag{9.5}$$

which is clearly self-adjoint. Notice the $p_0(x)$ in the denominator. This is why we require $p_0(x) \neq 0$, $a < x < b$. In the following development we assume that $\mathscr{L}$ has been put in self-adjoint form.

Eigenfunctions, eigenvalues. Let us form the differential equation

$$\mathscr{L}u(x) + \lambda\, w(x)\, u(x) = 0. \tag{9.6}$$

Here λ is a constant and $w(x)$ is a known function of x, called a density or weighting function. The significance of these labels will appear in subsequent sections. We require that $w(x) > 0$ except, possibly, at isolated points at which $w(x) = 0$. For a given choice of the parameter λ, a function $u_\lambda(x)$, which satisfies Eq. 9.6 *and the imposed boundary conditions*, is called an *eigenfunction* corresponding to λ. The constant λ is then called an *eigenvalue*. There is no guarantee that an eigenfunction

[1] If we multiply $\mathscr{L}$ by $f(x)/p_0(x)$ and then demand that

$$f'(x) = \frac{fp_1}{p_0},$$

so that the new operator will be self-adjoint, we obtain

$$f(x) = \exp\left[\int^x \frac{p_1(t)}{p_0(t)}\,dt\right]$$

$u_\lambda(x)$ will exist for any arbitrary choice of parameter λ. Indeed, the requirement that there be an eigenfunction often restricts the acceptable values of λ to a discrete set. Here we have one mathematical approach to the process of quantization in quantum mechanics.

EXAMPLE 9.1.1—LEGENDRE'S EQUATION

Legendre's equation is given by

$$(1 - x^2)y'' - 2xy' + n(n + 1)y = 0. \tag{9.7}$$

From Eqs. 9.1 and 9.6

$$\begin{aligned} p_0(x) &= 1 - x^2 = p \qquad & w(x) &= 1, \\ p_1(x) &= -2x = p' \qquad & \lambda &= n(n + 1), \\ p_2(x) &= 0 = q. \end{aligned} \tag{9.8}$$

The reader will recall that our series solutions of Legendre's equation (Section 8.4)[1] diverged unless n was restricted to one of the integers. This represents a quantization of the parameter λ.

When the equations of Chapter 8 are transformed into self-adjoint form, we find the following values of the coefficients and parameters (Table 9.1).

TABLE 9.1

Equation	$p(x)$	$q(x)$	λ	$w(x)$
Legendre	$1 - x^2$	0	$l(l + 1)$	1
Associated Legendre	$1 - x^2$	$-m^2/(1 - x^2)$	$l(l + 1)$	1
Chebyshev I	$(1 - x^2)^{1/2}$	0	n^2	$(1 - x^2)^{-1/2}$
Chebyshev II	$(1 - x^2)^{3/2}$	0	$n(n + 2)$	$(1 - x^2)^{1/2}$
Bessel	x	$-\dfrac{n^2}{x}$	a^2	x
Laguerre	xe^{-x}	0	α	e^{-x}
Associated Laguerre	$x^{k+1}e^{-x}$	0	$\alpha - k$	$x^k e^{-x}$
Hermite	e^{-x^2}	0	2α	e^{-x^2}
Simple harmonic oscillator[a]	1	0	ω^2	1

[a] This will form the basis for Chapter 14, Fourier series.

[1] Cf. also Sections 5.2 and 12.10.

The coefficient $p(x)$ is the coefficient of the second derivative of the eigenfunction and hopefully can be identified with no difficulty. The eigenvalue λ is the parameter or function of the parameter that is available (in a term of the form $\lambda\, w(x)\, y(x)$). Any x dependence apart from the eigenfunction becomes the weighting function $w(x)$. If there is another term containing the eigenfunction (not the derivatives), the coefficient of the eigenfunction in this additional term is identified as $q(x)$. If no such term is present, $q(x)$ is simply zero.

EXAMPLE 9.1.2—DEUTERON

Further insight into the concepts of eigenfunction and eigenvalue may be provided by an extremely simple model of the deuteron. The neutron–proton nuclear interaction is represented by a square well potential: $V = V_0 < 0$ for $0 \leqslant r < a$, $V = 0$ for $r > a$. The Schrödinger wave equation is

$$-\frac{\hbar^2}{2M}\nabla^2\psi + V\psi = E\psi. \tag{9.9}$$

With $\psi = \psi(r)$, we may write $u(r) = r\psi(r)$, and using Exercise 2.4.15, the wave equation becomes

$$\frac{d^2u}{dr^2} + k_1^2 u = 0, \tag{9.10}$$

with

$$k_1^2 = \frac{2M}{\hbar^2}(E - V_0) > 0 \tag{9.11}$$

for the interior range, $0 \leqslant r < a$. For $a < r < \infty$, we have

$$\frac{d^2u}{dr^2} - k_2^2 u = 0 \tag{9.12}$$

with

$$k_2^2 = -\frac{2ME}{\hbar^2} > 0. \tag{9.13}$$

From the boundary condition that ψ remain finite, $u(0) = 0$ and

$$u_1(r) = \sin k_1 r, \qquad 0 \leqslant r < a. \tag{9.14}$$

In the range outside the potential well, we have a linear combination of the two exponentials,

$$u_2(r) = A \exp k_2 r + B \exp(-k_2 r), \qquad a < r < \infty. \tag{9.15}$$

Continuity of particle density and current demand that $u_1(a) = u_2(a)$ and that $u_1'(a) = u_2'(a)$. These *joining conditions* give

$$\begin{aligned} \sin k_1 a &= A \exp k_2 a + B \exp(-k_2 a), \\ k_1 \cos k_1 a &= k_2 A \exp k_2 a - k_2 B \exp(-k_2 a). \end{aligned} \tag{9.16}$$

The condition that we actually have one proton–neutron combination is that $\int \psi^*\psi \, d\tau = 1$. This constraint can be met if we impose a boundary condition that $\psi(r)$ remain finite as $r \to \infty$. And this, in turn, means that $A = 0$. Dividing the above pair of equations,

$$\tan k_1 a = -\frac{k_1}{k_2} = -\sqrt{\frac{E - V_0}{-E}}, \tag{9.17}$$

a transcendental equation for the energy E with only certain *discrete* solutions. If E is such that Eq. 9.17 can be satisfied, our solutions $u_1(r)$ and $u_2(r)$ can satisfy the boundary conditions. If Eq. 9.17 is not satisfied, *no acceptable solution exists.* The values of E for which Eq. 9.17 is satisfied are the eigenvalues; the corresponding functions u_1 and u_2 (or ψ) are the eigenfunctions. For the actual deuteron problem, there is one (and only one) negative value of E satisfying Eq. 9.17, i.e., the deuteron has one and only one bound state.

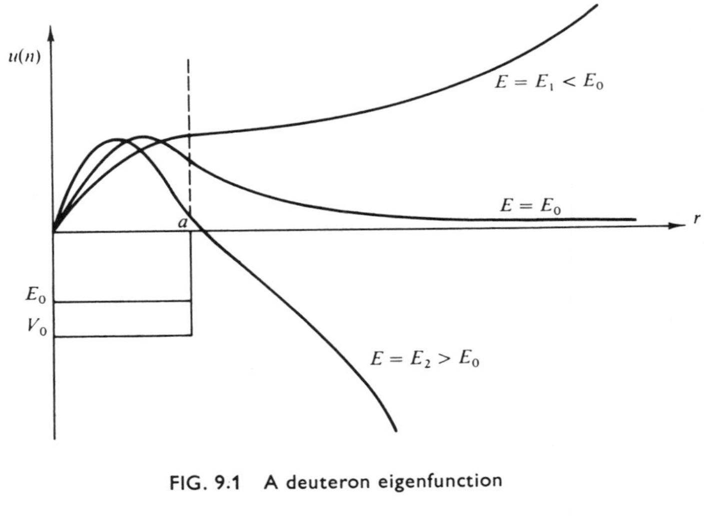

FIG. 9.1 A deuteron eigenfunction

Now, what happens if E does *not* satisfy Eq. 9.17, if E is *not* an eigenvalue? In graphical form, imagine that E and therefore k_1 are varied slightly.

For $E = E_1 < E_0$, k_1 is reduced, and $\sin k_1 a$ has not turned down as much. The joining conditions, Eqs. 9.16, require $A > 0$ and the wave function goes to $+\infty$, exponentially. For $E = E_2 > E_0$, k_1 is larger, $\sin k_1 a$ peaks sooner and is descending more rapidly at $r = a$. The joining conditions demand $A < 0$, and the wave function goes to $-\infty$, exponentially. Only for $E = E_0$, an eigenvalue, will the wave function have the required negative exponential asymptotic behavior.

Boundary conditions. In the foregoing definition of eigenfunction, it was noted that the eigenfunction $u_\lambda(x)$ was required to satisfy certain imposed boundary conditions. These boundary conditions may take three forms:

1. Cauchy boundary conditions. The value of a function and normal derivative specified on the boundary. In electrostatics this would mean φ, the potential, and E_n the normal components of the electric field.
2. Dirichlet boundary conditions. The value of a function specified on the boundary.
3. Neumann boundary conditions. The normal derivative (normal gradient) of a function specified on the boundary. In the electrostatic case this would be E_n and therefore σ, the surface charge density.

A summary of the relation of these three types of boundary condition to the three types of two-dimensional partial differential equation is given in Table 9.2. For extended discussions of these partial differential equations the reader may consult Sommerfeld, Chapter 2, or Morse and Feshbach, Chapter 6.

TABLE 9.2

	Type of partial differential equation		
Boundary conditions	Elliptic	Hyperbolic	Parabolic
	Laplace, Poisson in (x, y)	Wave equation in (x, t)	Diffusion equation in (x, t)
Cauchy			
Open surface	Unphysical results (instability)	*Unique, stable solution*	Too restrictive
Closed surface	Too restrictive	Too restrictive	Too restrictive
Dirichlet			
Open surface	Insufficient	Insufficient	*Unique, stable solution in* one direction
Closed surface	*Unique, stable solution*	Solution not unique	Too restrictive
Neumann			
Open surface	Insufficient	Insufficient	*Unique, stable solution* in one direction
Closed surface	*Unique, stable solution*	Solution not unique	Too restrictive

Parts of Table 9.2 are simply a matter of maintaining internal consistency, of common sense. For instance, for Poisson's equation with a closed surface, Dirichlet conditions lead to a unique, stable solution. Neumann conditions, independent of the Dirichlet conditions, likewise lead to a unique stable solution independent of the Dirichlet solution. Therefore, Cauchy boundary conditions (meaning Dirichlet plus Neumann) could lead to an inconsistency.

The term boundary conditions includes as a special case the concept of initial conditions. For instance, specifying the initial position x_0 and the initial velocity v_0

in some dynamical problem would correspond to the Cauchy boundary conditions. The only difference in the present usage of boundary conditions in these one-dimensional problems is that we are going to apply the conditions on *both* ends of the allowed range of the variable.

Usually the boundary conditions in one of the three forms will guarantee that at the ends of our interval (that is, at the boundary) the following products will vanish:

$$p(x)\,u(x)\,\frac{du(x)}{dx}\bigg|_{x=a} = 0$$

and (9.18)

$$p(x)\,u(x)\frac{du(x)}{dx}\bigg|_{x=b} = 0.$$

Here $u(x)$ is a solution of the particular differential equation (9.6) being considered. We can however, work with a somewhat less restrictive set of boundary conditions,

$$vpu'|_{x=a} = vpu'|_{x=b}, \tag{9.19}$$

in which $u(x)$ and $v(x)$ are solutions of the differential equation corresponding to the same or to different eigenvalues.

In anticipation of applications in quantum mechanics let us consider the possibility that Eq. 9.6 may have complex solutions, $u(x)$ and $v(x)$. We replace Eq. 9.19 with the complex boundary condition

$$v^*pu'|_{x=a} = v^*pu'|_{x=b}, \tag{9.20}$$

in which v^* is the complex conjugate of $v(x)$. Clearly, if $v(x)$ is real, $v = v^*$ and Eq. 9.20 will reduce to Eq. 9.19. Because we take the coefficient $p_0(x)$ to be real, the boundary condition may equally well be written as the complex conjugate of Eq. 9.20,

$$vpu^{*\prime}|_{x=a} = vpu^{*\prime}|_{x=b}. \tag{9.21}$$

Finally, since $u(x)$ and $v(x)$ are to be any two solutions, we may interchange u and v to obtain

$$v^{*\prime}pu|_{x=a} = v^{*\prime}pu|_{x=b} \tag{9.22}$$

and its complex conjugate.

Hermitian operators. We now prove an important property of the combination self-adjoint, second-order differential operator (Eq. 9.6), plus solutions that satisfy boundary conditions given by Eqs. 9.20 and 9.22.

By integrating over the range $a \leqslant x \leqslant b$ we obtain

$$\int_a^b v^*\mathscr{L}u\,dx = \int_a^b v^*(pu')'\,dx + \int_a^b v^*qu\,dx \tag{9.23}$$

using Eq. 9.4. Integrating by parts, we have

$$\int_a^b v^*(pu')'\,dx = v^*pu'\bigg|_a^b - \int_a^b v^{*\prime}pu'\,dx. \tag{9.24}$$

The integrated part vanishes on application of the boundary conditions (Eq. 9.20). Integrating the remaining integral by parts a second time, we have

$$-\int_a^b v^{*\prime} p u' \, dx = -v^{*\prime} p u \Big|_a^b + \int_a^b u(p v^{*\prime})' \, dx. \tag{9.25}$$

Again the integrated part vanishes in an application of Eq. 9.22. A combination of Eqs. 9.23–9.25 gives us

$$\int_a^b v^* \mathscr{L} u \, dx = \int_a^b u \mathscr{L} v^* \, dx. \tag{9.26}$$

This property, given by Eq. 9.26, is expressed by saying that the operator $\mathscr{L}$ is Hermitian with respect to the functions $u(x)$ and $v(x)$ which satisfy the boundary conditions specified by Eq. 9.20.

EXAMPLE 9.1.3—CHOICE OF INTEGRATION INTERVAL, $[a,b]$

For $\mathscr{L} = d^2/dx^2$ a possible eigenvalue equation is

$$\frac{d^2}{dx^2} y(x) + n^2 y(x) = 0, \tag{9.27}$$

with eigenfunctions

$$u_n = \cos nx$$
$$v_m = \sin mx.$$

Equation 9.19 becomes

$$-n \sin mx \sin nx \,|_a^b = 0$$

or

$$m \cos mx \cos nx \,|_a^b = 0,$$

interchanging u_n and v_m. Since $\sin mx$ and $\cos nx$ are periodic with period 2π (for n and m integral), Eq. 9.19 is clearly satisfied if $a = x_0$ and $b = x_0 + 2\pi$.

EXERCISES

9.1.1 Show that Laguerre's equation may be put into self-adjoint form by multiplying by e^{-x} and that $w(x) = e^{-x}$ is the weighting function.

9.1.2 Show that the Hermite equation may be put into self-adjoint form by multiplying by e^{-x^2} and that this gives $w(x) = e^{-x^2}$ as the appropriate density function.

9.1.3 Show that the Chebyshev equation (type I) may be put into self-adjoint form by multiplying by $(1-x^2)^{-1/2}$ and that this gives $w(x) = (1-x^2)^{-1/2}$ as the appropriate density function.

9.1.4 Show the following when the linear second-order differential equation is expressed in self-adjoint form:

(a) The Wronskian is equal to a constant divided by the initial coefficient p.

$$W = \frac{C}{p}$$

(b) A second solution is given by

$$y_2(x) = C\, y_1(x) \int^x \frac{dt}{p[y_1(t)]^2}\,.$$

9.1.5 $U_n(x)$, the Chebyshev polynomial (type II) satisfies the differential equation

$$(1-x^2)\, U_n''(x) - 3x\, U_n'(x) + n(n+2)\, U_n(x) = 0.$$

(a) Locate the singular points that appear in the *finite* plane and show whether they are regular or irregular.
(b) Put this equation in self-adjoint form.
(c) Identify the complete eigenvalue.
(d) Identify the weighting function.
(e) Write out an orthogonality integral for $U_n(x)$ and $U_m(x)$, $n \neq m$.

9.1.6 For the very special case $\lambda = 0$ and $q(x) = 0$ the self-adjoint eigenvalue equation becomes

$$\frac{d}{dx}\left[p(x)\frac{du(x)}{dx}\right] = 0,$$

satisfied by

$$\frac{du}{dx} = \frac{1}{p(x)}\,.$$

Use this to obtain a second solution of the following:
(a) Legendre's equation,
(b) Laguerre's equation,
(c) Hermite's equation.

Ans. (a) $u_2(x) = \dfrac{1}{2}\ln\dfrac{1+x}{1-x}$,

(b) $u_2(x) - u_2(x_0) = \displaystyle\int_{x_0}^{x} e^t \frac{dt}{t}$,

(c) $u_2(x) = \displaystyle\int_0^x e^{t^2}\,dt.$

These second solutions illustrate the divergent behavior usually found in a second solution.

9.1.7 Given that $\mathscr{L}u = 0$ and $g\mathscr{L}u$ is self-adjoint, show that for the adjoint operator $\bar{\mathscr{L}}$, $\bar{\mathscr{L}}(gu) = 0$.

9.1.8 Show that if a function ψ is required to satisfy Laplace's equation in a finite region of space and to satisfy Dirichlet boundary conditions over the entire closed bounding surface, then ψ is unique.
Hint. One of the forms of Green's theorem, Section 1.11 will be helpful.

9.2 Hermitian (Self-Adjoint) Operators

Hermitian or self-adjoint operators have three properties that are of extreme importance in physics, both classical and quantum.

1. The eigenvalues of an Hermitian operator are real.
2. The eigenfunctions of an Hermitian operator are orthogonal.
3. The eigenfunctions of an Hermitian operator form a complete set.[1]

Real eigenvalues. We proceed to prove the first two of these three properties. Let

$$\mathcal{L}u_i + \lambda_i w u_i = 0, \tag{9.28}$$

$$\mathcal{L}u_j + \lambda_j w u_j = 0, \tag{9.29}$$

$$\mathcal{L}u_j^* + \lambda_j^* w u_j^* = 0. \tag{9.30}$$

Here $\mathcal{L}$ is a real operator (p and q are real functions of x) and $w(x)$ is a real function. But we permit λ_k, the eigenvalues, and u_k, the eigenfunctions, to be complex. Multiplying Eq. 9.28 by u_j^* and Eq. 9.30 by u_i and then subtracting, we have

$$u_j^* \mathcal{L} u_i - u_i \mathcal{L} u_j^* = (\lambda_j^* - \lambda_i) w u_i u_j^* \tag{9.31}$$

We integrate over the range $a \leqslant x \leqslant b$,

$$\int_a^b u_j^* \mathcal{L} u_i \, dx - \int_a^b u_i \mathcal{L} u_j^* \, dx = (\lambda_j^* - \lambda_i) \int_a^b u_i u_j^* w \, dx. \tag{9.32}$$

Since $\mathcal{L}$ is Hermitian, the left-hand side vanishes by Eq. 9.26 and

$$(\lambda_j^* - \lambda_i) \int_a^b u_i u_j^* w \, dx = 0. \tag{9.33}$$

If $i = j$, the integral cannot vanish [$w(x) > 0$, apart from isolated points], except in the trivial case $u_i = 0$. Hence the coefficient $(\lambda_i^* - \lambda_i)$ must be zero,

$$\lambda_i^* = \lambda_i, \tag{9.34}$$

which is a mathematical statement that the eigenvalue is real. Since λ_i can represent any one of the eigenvalues, this proves the first property. This is an exact analog of the nature of the eigenvalues of real symmetric (and of Hermitian) matrices (cf. Section 4.6).

This reality of the eigenvalues of Hermitian operators has a fundamental significance in quantum mechanics. In quantum mechanics the eigenvalues correspond to precisely measurable quantities, such as energy and angular momentum. With the theory formulated in terms of Hermitian operators, this proof of the reality of the eigenvalues guarantees that the theory will predict real numbers for these measurable physical quantities.

Orthogonal eigenfunctions. If we now take $i \neq j$ and if $\lambda_i \neq \lambda_j$, the integral of the product of the two different eigenfunctions must vanish.

$$\int_a^b u_i u_j^* w \, dx = 0. \tag{9.35}$$

[1] This third property is not universal. It *does* hold for our linear, second-order differential operators in Sturm-Liouville (self-adjoint) form.

This condition, called orthogonality, is the continuum analog of the vanishing of a scalar product of two vectors.[1] We say that the eigenfunctions $u_i(x)$ and $u_j(x)$ are orthogonal with respect to the weighting function $w(x)$ over the interval $[a, b]$. Equation 9.35 constitutes a partial proof of the second property of our Hermitian operators. Again the precise analogy with matrix analysis should be noted. Indeed, we can establish a one-to-one correspondence between this Sturm-Liouville theory of differential equations and the treatment of Hermitian matrices. Historically, this correspondence has been significant in establishing the mathematical equivalence of matrix mechanics developed by Heisenberg and wave mechanics developed by Schrödinger. Today, the two diverse approaches are merged into the theory of quantum mechanics and the mathematical formulation which is more convenient for a particular problem is used for that problem. Actually, the mathematical alternatives do not end here. Integral equations, Chapter 16, form a third equivalent and sometimes more convenient or more powerful approach.

This proof of orthogonality is not quite complete. There is a loophole, because we may have $i \neq j$ but still have $\lambda_i = \lambda_j$. Such a case is labeled degenerate. Illustrations of degeneracy are given at the end of this section. If $\lambda_i = \lambda_j$, the integral in Eq. 9.33 need not vanish. This means that linearly independent eigenfunctions corresponding to the same eigenvalue are not automatically orthogonal and that some other method must be sought to obtain an orthogonal set. Although the eigenfunctions in this degenerate case may not be orthogonal, they can always be made orthogonal. One method is developed in the next section.

It will be seen in succeeding chapters that it is just as desirable to have a given set of functions orthogonal as it is to have an orthogonal coordinate system. We can work with nonorthogonal functions, but they are likely to prove as messy as an oblique coordinate system.

The third property, completeness, is defined and discussed in Section 9.4. The formal proof, which involves the calculus of variations (cf. Chapter 17), is not given here. For details the reader is referred to Courant and Hilbert, Vol. I, or to Titchmarsh.

EXAMPLE 9.2.1. FOURIER SERIES: ORTHOGONALITY

Continuing Example 9.1.3, the eigenvalue equation, Eq. 9.27,

$$\frac{d^2}{dx^2}\, y(x) + n^2\, y(x) = 0,$$

[1] From the definition of Riemann integral

$$\int_a^b f(x)\, g(x)\, dx = \lim_{N\to\infty} \left(\sum_{i=1}^{N} f(x_i)\, g(x_i)\right) \Delta x,$$

where $x_0 = a$, $x_N = b$, and $x_i - x_{i-1} = \Delta x$. If we interpret $f(x_i)$ and $g(x_i)$ as the ith components of an N component vector, then this sum (and therefore this integral) corresponds directly to a scalar product of vectors, Eq. 1.22. The vanishing of the scalar product is the condition for *orthogonality* of the vectors—or functions.

perhaps describes a quantum mechanical particle in a box, perhaps a vibrating violin string with (degenerate) eigenfunctions—$\cos nx$, $\sin nx$.

With n real (here taken to be integral), the orthogonality integrals become

$$\text{(a)} \qquad \int_{x_0}^{x_0+2\pi} \sin mx \sin nx \, dx = C_n \, \delta_{nm},$$

$$\text{(b)} \qquad \int_{x_0}^{x_0+2\pi} \cos mx \cos nx \, dx = D_n \, \delta_{nm},$$

$$\text{(c)} \qquad \int_{x_0}^{x_0+2\pi} \sin mx \cos nx \, dx = 0.$$

For an interval of 2π the preceding analysis guarantees the Kronecker delta in (a) and (b) but not the zero in (c) because (c) involves degenerate eigenfunctions. However, inspection shows that (c) always vanishes for all integral m and n.

Our Sturm-Liouville theory says nothing about the values of C_n and D_n. Actual calculation yields

$$C_n = \begin{cases} \pi, & n \neq 0, \\ 0, & n = 0, \end{cases}$$

$$D_n = \begin{cases} \pi, & n \neq 0, \\ 2\pi, & n = 0. \end{cases}$$

These orthogonality integrals form the basis of the Fourier series developed in Chapter 14.

Example 9.2.2. Expansion in Orthogonal Eigenfunctions: Square Wave

The property of completeness means that certain classes of functions (that is, sectionally or piecewise continuous) may be represented by a series of orthogonal eigenfunctions to any desired degree of accuracy. Consider the square wave

$$f(x) = \begin{cases} \dfrac{h}{2}, & 0 < x < \pi, \\ -\dfrac{h}{2}, & -\pi < x < 0. \end{cases} \tag{9.36}$$

This function may be expanded in any of a variety of eigenfunctions—Legendre, Hermite, Chebyshev, etc. The choice of eigenfunction is made on the basis of convenience. To illustrate the expansion technique, let us choose the eigenfunctions of Example 9.2.1 above, $\cos nx$ and $\sin nx$.

The eigenfunction series is conveniently (and conventionally) written

$$f(x) = \frac{a_0}{2} + \sum_{n=1}^{\infty} (a_n \cos nx + b_n \sin nx).$$

From the orthogonality integrals of Example 9.2.1 the coefficients are given by

$$a_n = \frac{1}{\pi}\int_{-\pi}^{\pi} f(t)\cos nt\,dt,$$

$$b_n = \frac{1}{\pi}\int_{-\pi}^{\pi} f(t)\sin nt\,dt, \qquad n = 0, 1, 2, \ldots.$$

Direct substitution of $\pm h/2$ for $f(t)$ yields

$$a_n = 0,$$

which is expected here because of the antisymmetry, and

$$b_n = \frac{h}{n\pi}(1 - \cos n\pi) = \begin{cases} 0, & n \text{ even}, \\ \dfrac{2h}{n\pi}, & n \text{ odd}. \end{cases}$$

Hence the eigenfunction (Fourier) expansion of the square wave is

$$f(x) = \frac{2h}{\pi}\sum_{n=0}^{\infty}\frac{\sin(2n+1)x}{(2n+1)}. \tag{9.37}$$

Additional examples, using other eigenfunctions, appear in Chapters 11 and 12.

The concept of degeneracy was introduced earlier. If N linearly independent eigenfunctions correspond to the same eigenvalue, the eigenvalue is said to be N-fold degenerate. A particularly simple illustration is provided by the eigenvalues and eigenfunctions of the linear oscillator equation, Example 9.2.1. For each value of the eigenvalue n, there are two possible solutions: $\sin nx$ and $\cos nx$ (and any linear combination). We may say the eigenfunctions are degenerate or the eigenvalue is degenerate.

A more involved example is furnished by the physical system of an electron in an atom (nonrelativistic treatment, spin neglected). From the Schrödinger equation, Eq. 13.53 for hydrogen, the total energy of the electron is our eigenvalue. We may label it E_{nLM} using the quantum numbers n, L, and M as subscripts. For each distinct set of quantum numbers (n, L, M) there is a distinct, linearly independent eigenfunction $\psi_{nLM}(r, \theta, \varphi)$. For hydrogen, the energy E_{nLM} is independent of L and M. With $0 \leqslant L \leqslant n - 1$ and $-L \leqslant M \leqslant L$, the eigenvalue is n^2-fold degenerate. (Including the electron spin would raise this to $2n^2$). In atoms with more than one electron, the electrostatic potential is no longer a simple r^{-1} potential. The energy depends upon L as well as n, although *not* upon M. E_{nLM} is still $(2L + 1)$-fold degenerate. This degeneracy may be removed by applying an external magnetic field, giving rise to the Zeeman effect.

EXERCISES

9.2.1 The functions $u_1(x)$ and $u_2(x)$ are eigenfunctions of the same Hermitian operator but for distinct eigenvalues λ_1 and λ_2. Prove that $u_1(x)$ and $u_2(x)$ are linearly independent.

9.2.2 (a) Show that the first derivatives of the Legendre polynomials satisfy a self-adjoint differential equation with eigenvalue $\lambda = n(n+1) - 2$.
(b) Show that these Legendre polynomial derivatives satisfy an orthogonality relation

$$\int_{-1}^{1} P_m'(x)P_n'(x)(1-x^2)\,dx = 0, \qquad m \neq n.$$

Note: In Section 12.5 $(1-x^2)^{1/2}P_n'(x)$ will be labelled an associated Legendre polynomial, $P_n^1(x)$.

9.2.3 A set of functions $u_n(x)$ satisfy the Sturm-Liouville equation

$$\frac{d}{dx}\left[p(x)\frac{d}{dx}u_n(x)\right] + \lambda_n w(x)u_n(x) = 0.$$

They also satisfy appropriate boundary conditions and have distinct eigenvalues. Hence, $u_m(x)$ and $u_n(x)$ are orthogonal. Prove that $du_m(x)/dx$ and $du_n(x)/dx$ are also orthogonal with $p(x)$ as a weighting function.

9.2.4 One solution of Legendre's equation, Eq. 9.7, is the polynomial $P_n(x)$. A second solution, developed in Section 12.10, is denoted by $Q_n(x)$. The integral of the product $P_n(x)Q_l(x)(n \neq l)$ over the orthogonality interval $[-1, 1]$ does *not* vanish. Explain why the Sturm-Liouville theory does not guarantee a value of zero in this case.

9.2.5 With $\mathscr{L}$ *not* self-adjoint,

$$\mathscr{L}u_i + \lambda_i w u_i = 0$$

and

$$\bar{\mathscr{L}}v_j + \lambda_j w v_j = 0.$$

(a) Show that

$$\int_a^b v_j \mathscr{L}u_i\,dx = \int_a^b u_i \bar{\mathscr{L}}v_j\,dx$$

provided

$$u_i p_0 v_j'\Big|_a^b = v_j p_0 u_i'\Big|_a^b$$

and

$$u_i(p_1 - p_0')v_j\Big|_a^b = 0.$$

(b) Show that the orthogonality integral for the eigenfunctions u_i and v_j becomes

$$\int_a^b u_i v_j w\,dx = 0 \qquad (\lambda_i \neq \lambda_j).$$

9.3 Schmidt Orthogonalization

This is a method of taking a nonorthogonal set of linearly independent functions[1] and literally constructing an orthogonal set over an arbitrary interval and with respect to an arbitrary weight or density factor. Here, for convenience, the functions are assumed to be real. The generalization to the complex case should offer little difficulty.

[1] Such a set of functions might well arise from the solutions of a (partial) differential equation in which the eigenvalue was independent of one or more of the constants of separation. As an example, we have the hydrogen atom problem (Sections 9.2 and 13.2). The eigenvalue (energy) is independent of both the electron orbital angular momentum and its projection on the z-axis, m. The student should note, however, that the origin of the set of functions is irrelevant to the Schmidt orthogonalization procedure.

So far no normalization has been specified. This means that

$$\int_a^b \varphi_i^2 w\, dx = N_i^2,$$

but no attention has been paid to the value of N_i. Since our basic equation, (9.6), is linear and homogeneous, we may multiply our solution by any constant and it will still be a solution. We now demand that each solution $\varphi_i(x)$ be multiplied by N_i^{-1} so that the new (normalized) φ_i will satisfy

$$\int_a^b \varphi_i^2(x)\, w(x)\, dx = 1 \tag{9.38}$$

or

$$\int_a^b \varphi_i(x)\, \varphi_j(x)\, w(x)\, dx = \delta_{ij}. \tag{9.39}$$

Equation 9.38 says that we have normalized to unity. Including the property of orthogonality, we have Eq. 9.39. Functions satisfying this equation are said to be orthonormal (orthogonal plus unit normalization). It should be emphasized that other normalizations are possible, and indeed, by historical convention, each of the special functions of mathematical physics treated in Chapters 12 and 13 will be normalized differently!

We consider three sets of functions: an original, given set $u_n(x)$, $n = 0, 1, 2, \ldots$; an orthogonalized set $\psi_n(x)$ to be constructed; and a final set of functions $\varphi_n(x)$ which are the normalized ψ_n's. The original u_n's may be degenerate eigenfunctions, but this is not necessary. We will have

$u_n(x)$	$\psi_n(x)$	$\varphi_n(x)$
linearly independent	*linearly independent*	*linearly independent*
nonorthogonal	*orthogonal*	*orthogonal*
unnormalized	unnormalized	*normalized*
		(orthonormal)

The Schmidt procedure is to take the nth ψ function (ψ_n) to be $u_n(x)$ plus an unknown linear combination of the previous φ's. The presence of the new $u_n(x)$ will guarantee linear independence. The requirement that $\psi_n(x)$ be orthogonal to each of the previous φ's yields just enough constraints to determine each of the unknown coefficients. Then the fully determined ψ_n will be normalized to unity, yielding $\varphi_n(x)$. Then the sequence of steps is repeated for $\psi_{n+1}(x)$.

Starting with $n = 0$, let

$$\psi_0(x) = u_0(x) \tag{9.40}$$

with no "previous" φ's to worry about. Normalizing,

$$\varphi_0(x) = \frac{\psi_0(x)}{[\int \psi_0^2 w\, dx]^{1/2}}. \tag{9.41}$$

For $n = 1$, let

$$\psi_1(x) = u_1(x) + a_{10}\varphi_0(x). \tag{9.42}$$

We demand that $\psi_1(x)$ be orthogonal to $\varphi_0(x)$. (At this stage, the normalization of $\psi_1(x)$ is irrelevant.) This demand leads to

$$\int \psi_1 \varphi_0 w\, dx = \int u_1 \varphi_0 w\, dx + a_{10} \int \varphi_0^2 w\, dx \qquad (9.43)$$
$$= 0.$$

Since φ_0 is normalized to unity (Eq. 9.41), we have

$$a_{10} = -\int u_1 \varphi_0 w\, dx, \qquad (9.44)$$

fixing the value of a_{10}. Normalizing, we define

$$\varphi_1(x) = \frac{\psi_1(x)}{(\int \psi_1^2 w\, dx)^{1/2}}. \qquad (9.45)$$

Generalizing, we have

$$\varphi_i(x) = \frac{\psi_i(x)}{(\int \psi_i^2(x)\, w(x)\, dx)^{1/2}} \qquad (9.46)$$

where

$$\psi_i(x) = u_i + a_{i0}\varphi_0 + a_{i,1}\varphi_1 + \cdots + a_{i,i-1}\varphi_{i-1}. \qquad (9.47)$$

The coefficients a_{ij} are given by

$$a_{ij} = -\int u_i \varphi_j w\, dx. \qquad (9.48)$$

It will be noticed that although this is one possible way of constructing an orthogonal or orthonormal set the functions $\varphi_i(x)$ are not unique. There is an infinite number of possible orthonormal sets for a given interval and a given density function. As an illustration of the freedom involved, consider two (nonparallel) vectors **A** and **B** in the xy-plane. We may normalize **A** to unit magnitude and then form $\mathbf{B}' = a\mathbf{A} + \mathbf{B}$ so that $\mathbf{B}'$ is perpendicular to **A**. By normalizing $\mathbf{B}'$ we have completed the Schmidt orthogonalization for two vectors. But any two perpendicular unit vectors such as **i** and **j** could have been chosen as our orthonormal set. Again, with an infinite number of possible rotations of **i** and **j** about the z-axis, we have an infinite number of possible orthonormal sets.

Example 9.3.1 Legendre Polynomials by Schmidt Orthogonalization

Let us form an orthonormal set from the set of functions $u_n(x) = x^n$, $n = 0, 1, 2, \ldots$. The interval is $-1 \leqslant x \leqslant 1$ and the density function is $w(x) = 1$.

In accordance with the Schmidt orthogonalization process described,

$$u_0 = 1 \quad \text{and} \quad \varphi_0 = \frac{1}{\sqrt{2}}. \qquad (9.49)$$

Then

$$\psi_1(x) = x + a_{10}\frac{1}{\sqrt{2}} \qquad (9.50)$$

and

$$a_{10} = -\int_{-1}^{1} \frac{x}{\sqrt{2}}\,dx = 0 \tag{9.51}$$

by symmetry. Normalizing, we obtain

$$\varphi_1(x) = \sqrt{\frac{3}{2}}\,x. \tag{9.52}$$

Continuing the Schmidt process, we define

$$\psi_2(x) = x^2 + a_{20}\frac{1}{\sqrt{2}} + a_{21}\sqrt{\frac{3}{2}}\,x \tag{9.53}$$

where

$$a_{20} = -\int_{-1}^{1} \frac{x^2}{\sqrt{2}}\,dx = -\frac{\sqrt{2}}{3}, \tag{9.54}$$

$$a_{21} = -\int_{-1}^{1} \sqrt{\frac{3}{2}}\,x^3\,dx = 0, \tag{9.55}$$

again by symmetry. Therefore

$$\psi_2(x) = x^2 - \tfrac{1}{3}, \tag{9.56}$$

and, on normalizing to unity,

$$\varphi_2(x) = \sqrt{\tfrac{5}{2}}\cdot\tfrac{1}{2}(3x^2 - 1). \tag{9.57}$$

The next function $\varphi_3(x)$ is

$$\varphi_3(x) = \sqrt{\tfrac{7}{2}}\cdot\tfrac{1}{2}(5x^3 - 3x). \tag{9.58}$$

Reference to Chapter 12 will show that

$$\varphi_n(x) = \sqrt{\frac{2n+1}{2}}\,P_n(x), \tag{9.59}$$

where $P_n(x)$ is the nth-order Legendre polynomial. Our Schmidt process provides a possible but very cumbersome method of generating the Legendre polynomials.

Orthogonal polynomials. This particular example has been chosen strictly to illustrate the Schmidt procedure. Although it has the advantage of introducing the Legendre polynomials, the initial functions $u_n = x^n$ are not degenerate eigenfunctions and they are not solutions of Legendre's equation. They are simply a set of functions which we have here rearranged to create an orthonormal set for the given interval and given weighting function. That we obtained the Legendre polynomials is not quite black magic but a direct consequence of the choice of interval and weighting function. The use of $u_n(x) = x^n$ but with other choices of interval and weighting function leads to other sets of orthogonal polynomials as shown in Table 9.3. We consider these polynomials in detail in Chapters 12 and 13 as solutions of particular differential equations.

An examination of this orthogonalization process will reveal two arbitrary features. First, as emphasized before, it is not necessary to normalize the functions

TABLE 9.3

Orthogonal Polynomials Generated by Schmidt Orthogonalization of $u_n(x)=x^n$, $n=0,1,2,\ldots$

Polynomials	Interval	Weighting function $w(x)$	Standard normalization
Legendre	$-1\leqslant x\leqslant 1$	1	$\int_{-1}^{1}[P_n(x)]^2\,dx=\frac{2}{2n+1}$
Chebyshev I	$-1\leqslant x\leqslant 1$	$(1-x^2)^{-1/2}$	$\int_{-1}^{1}[T_n(x)]^2(1-x^2)^{-1/2}\,dx=\begin{cases}\pi/2, & n\neq 0\\ \pi, & n=0\end{cases}$
Chebyshev II	$-1\leqslant x\leqslant 1$	$(1-x^2)^{1/2}$	$\int_{-1}^{1}[U_n(x)]^2(1-x^2)^{1/2}\,dx=\frac{\pi}{2}$
Laguerre	$0\leqslant x<\infty$	e^{-x}	$\int_{0}^{\infty}[L_n(x)]^2e^{-x}\,dx=1$
Associated Laguerre	$0\leqslant x<\infty$	x^ke^{-x}	$\int_{0}^{\infty}[L_n^k(x)]^2x^ke^{-x}\,dx=\frac{(n+k)!}{n!}$
Hermite	$-\infty<x<\infty$	e^{-x^2}	$\int_{-\infty}^{\infty}[H_n(x)]^2e^{-x^2}\,dx=2^n\pi^{1/2}n!$

to unity. In the example just given we could have required

$$\int_{-1}^{1}\varphi_n(x)\,\varphi_m(x)\,dx=\frac{2}{2n+1}\,\delta_{nm}, \tag{9.60}$$

and the resulting set would have been the actual Legendre polynomials. Second, the sign of φ_n is always indeterminant. In the example we chose the sign by requiring the coefficient of the highest power of x in the polynomial to be positive. For the Laguerre polynomials, on the other hand, we would require the coefficient of the highest power to be $(-1)^n/n!$

EXERCISES

9.3.1 Apply the Schmidt procedure to form the first three Laguerre polynomials

$$u_n(x)=x^n,\qquad n=0,1,2,\ldots,$$
$$0\leqslant x<\infty,$$
$$w(x)=e^{-x}.$$

The conventional normalization is

$$\int_0^{\infty}L_m(x)\,L_n(x)\,e^{-x}\,dx=\delta_{mn}$$

Ans. $L_0=1,$
$L_1=1-x,$
$L_2=\dfrac{(2-4x+x^2)}{2}.$

9.3.2 Using the Schmidt orthogonalization procedure, construct the lowest three Hermite polynomials:

$$u_n(x) = x^n, \qquad n = 0, 1, 2, \cdots \qquad -\infty < x < \infty, \qquad w(x) = e^{-x^2}.$$

For this set of polynomials the usual normalization is

$$\int_{-\infty}^{\infty} H_m(x)\, H_n(x)\, w(x)\, dx = \delta_{mn} 2^m m! \pi^{1/2}.$$

Ans. $H_0 = 1,$
$H_1 = 2x,$
$H_2 = 4x^2 - 2.$

9.3.3 Use the Schmidt orthogonalization scheme to construct the first three Chebyshev polynomials (type I).

$$u_n(x) = x^n, \qquad n = 0, 1, 2, \cdots \qquad -1 \leqslant x \leqslant 1, \qquad w(x) = (1 - x^2)^{-1/2}.$$

Take the normalization

$$\int_{-1}^{1} T_m(x)\, T_n(x)\, w(x)\, dx = \delta_{mn} \begin{cases} \pi, & m = n = 0 \\ \dfrac{\pi}{2}, & m = n \geqslant 1. \end{cases}$$

Hint. The needed integrals are given in Ex. 10.4.3.

Ans. $T_0 = 1,$
$T_1 = x,$
$T_2 = 2x^2 - 1,$
$(T_3 = 4x^3 - 3x).$

9.3.4 Use the Schmidt orthogonalization scheme to construct the first three Chebyshev polynomials (type II).

$$u_n(x) = x^n, \qquad n = 0, 1, 2, \cdots \qquad -1 \leqslant x \leqslant 1, \qquad w(x) = (1 - x^2)^{+1/2}.$$

Take the normalization to be

$$\int_{-1}^{1} U_m(x)\, U_n(x)\, w(x)\, dx = \delta_{mn} \frac{\pi}{2}.$$

Hint.

$$\int_{-1}^{1} (1 - x^2)^{1/2} x^{2n}\, dx = \frac{\pi}{2} \times \frac{1 \cdot 3 \cdot 5 \cdots (2n-1)}{4 \cdot 6 \cdot 8 \cdots (2n+2)}, \qquad n = 1, 2, 3 \cdots$$

$$= \frac{\pi}{2}, \qquad n = 0.$$

Ans. $U_0 = 1, \qquad U_1 = 2x, \qquad U_2 = 4x^2 - 1.$

9.3.5 As a modification of Ex. 9.3.2 apply the Schmidt orthogonalization procedure to the set $u_n(x) = x^n$, $n = 0, 1, 2, \ldots$, $0 \leq x < \infty$. Take $w(x)$ to be $\exp[-x^2]$. Find the first two nonvanishing polynomials. Normalize so that the coefficient of the highest power of x is unity. In Ex. 9.3.2, the interval $(-\infty, \infty)$ led to the Hermite polynomials. These are certainly not the Hermite polynomials.

Ans. $\varphi_0 = 1.$
$\varphi_1 = x - \pi^{-1/2}$

9.3.6 Form an orthogonal set over the interval $0 \leqslant x < \infty$, using $u_n(x) = e^{-nx}$, $n = 1, 2, 3, \ldots$. Take the weighting factor, $w(x)$, to be unity. These functions are solutions of $u_n'' - n^2 u_n = 0$, which is clearly already in Sturm-Liouville (self-adjoint) form. Why doesn't the Sturm-Liouville theory guarantee the orthogonality of these functions?

9.4 Completeness of Eigenfunctions

The third important property of an Hermitian operator is that its eigenfunctions form a complete set. This completeness means that any well-behaved (at least piecewise continuous) function $F(x)$ can be approximated by a series

$$F(x) = \sum_{n=0}^{\infty} a_n \varphi_n(x) \tag{9.61}$$

to any desired degree of accuracy.[1] More precisely, the set $\varphi_n(x)$ is called complete[2] if the limit of the mean square error vanishes;

$$\lim_{m\to\infty} \int_a^b \left[F(x) - \sum_{n=0}^{m} a_n \varphi_n(x)\right]^2 w(x)\,dx = 0. \tag{9.62}$$

Proof that this can indeed be done with the Hermitian operator eigenfunctions is given by Titchmarsh and by Courant and Hilbert (Vol. I). It is worth noting that our Hermitian operators here are linear, second-order differential operators in self-adjoint form.

In the language of linear algebra, we have a linear space, a function space. The linearly independent, orthonormal functions $\varphi_n(x)$, form the basis for this (infinite-dimensional) space. Equation 9.61 is a statement that the functions $\varphi_n(x)$ span this linear space. With an inner product defined by Eq. 9.64, our linear space is a Hilbert space.

The question of completeness of a set of functions is often determined by comparison with a Laurent series, Section 6.5. In Section 14.1 this is done for Fourier series, thus establishing the completeness of Fourier series. For all of the orthogonal polynomials mentioned in Section 9.3, it is possible to find a polynomial expansion of each power of z,

$$z^n = \sum_{i=0}^{n} a_i P_i(z), \tag{9.63}$$

where $P_i(z)$ is the ith polynomial. Exs. 12.4.6, 13.1.6, 13.2.5, and 13.3.15 are specific examples of Eq. 9.63. Using Eq. 9.63, the Laurent expansion of $f(z)$ may be reexpressed in terms of the polynomials showing that the polynomial expansion exists. (And existing, it is unique, Ex. 9.4.1.)

In Eq. 9.61 the expansion coefficients a_m may be determined by

$$a_m = \int_a^b F(x)\varphi_m(x)w(x)\,dx. \tag{9.64}$$

This follows from multiplying Eq. 9.61 by $\varphi_m(x)\,w(x)$ and integrating. From the orthogonality of the eigenfunctions, $\varphi_n(x)$, only the mth term survives. Here we see the value of orthogonality. Equation 9.64 may be compared with the dot or inner product of vectors, Section 1.3, and a_m interpreted as the mth projection of the function $F(x)$.

For a known function, $F(x)$, Eq. 9.64 gives a_m as a *definite* integral which can always be evaluated, by machine if not analytically.

For examples of particular eigenfunction expansions, see the following: Fourier

[1] If we have a finite set, as with vectors, the summation is over the number of linearly independent members of the set.

[2] Many authors use the term closed here.

series, Section 9.2 and Chapter 14; Bessel and Fourier-Bessel expansions, Section 11.2; Legendre series, Section 12.3; Laplace series, Section 12.6; Hermite series, Section 13.1; Laguerre series, Section 13.2; and Chebyshev series, Section 13.3.

Bessel's inequality. If the set of functions $\varphi_n(x)$ does not form a complete set, possibly because we simply have not included the required infinite number of members of an infinite set, we are led to Bessel's inequality. First, consider the finite case. Let A be an n component vector,

$$\mathbf{A} = \mathbf{e}_1 a_1 + \mathbf{e}_2 a_2 + \cdots + \mathbf{e}_n a_n, \tag{9.65}$$

in which $\mathbf{e}_i$ is a unit vector and a_i is the corresponding component (projection) of $\mathbf{A}$, that is,

$$a_i = \mathbf{A} \cdot \mathbf{e}_i. \tag{9.66}$$

Then

$$\left(\mathbf{A} - \sum_i \mathbf{e}_i a_i\right)^2 \geqslant 0. \tag{9.67}$$

If we sum over all n components, clearly the summation equals $\mathbf{A}$ by Eq. 9.65 and the equality holds. If, however, the summation does not include all n components, the inequality results. By expanding Eq. 9.67 and remembering that the unit vectors satisfy an orthogonality relation,

$$\mathbf{e}_i \cdot \mathbf{e}_j = \delta_{ij}, \tag{9.68}$$

we have

$$A^2 \geqslant \sum_i a_i^2. \tag{9.69}$$

This is Bessel's inequality.

For functions we consider the integral

$$\int_a^b \left[f(x) - \sum_i a_i\, \varphi_i(x)\right]^2 w(x)\, dx \geqslant 0. \tag{9.70}$$

This is the continuum analog of Eq. 9.67, letting $n \to \infty$ and replacing the summation by an integration. Again, with the weighting factor $w(x) > 0$ the integrand is nonnegative. The integral vanishes by Eq. 9.61 if we have a complete set. Otherwise, it is positive. Expanding the squared term, we obtain

$$\int_a^b [f(x)]^2\, w(x)\, dx - 2 \sum_i a_i \int_a^b f(x)\, \varphi_i(x)\, w(x)\, dx + \sum_i a_i^2 \geqslant 0. \tag{9.71}$$

Applying Eq. 9.64, we have

$$\int_a^b [f(x)]^2\, w(x)\, dx \geqslant \sum_i a_i^2. \tag{9.72}$$

Hence the sum of the squares of the expansion coefficients a_i is less than or equal to the weighted integral of $[f(x)]^2$, the equality holding if and only if the expansion is exact, that is, if the set of functions $\varphi_n(x)$ is a complete set.

Bessel's inequality has a variety of uses, including proof of convergence of Fourier series.

Schwarz inequality. The frequently used Schwarz inequality is similar to the Bessel inequality. Consider the quadratic equation

$$\sum_{i=1}^{n}(a_i x + b_i)^2 = \sum_{i=1}^{n} a_i^2(x + b_i/a_i)^2 = 0. \tag{9.73}$$

If b_i/a_i = constant, c, then the solution is $x = -c$. If b_i/a_i is *not* a constant, all terms cannot vanish simultaneously for real x. So the solution must be complex. Expanding, we find that

$$x^2 \sum_{i}^{n} a_i^2 + 2x \sum_{i}^{n} a_i b_i + \sum_{i}^{n} b_i^2 = 0, \tag{9.74}$$

and since x is complex (or $= -b_i/a_i$) the quadratic formula for x[1] leads to

$$\left(\sum_{i=1}^{n} a_i b_i\right)^2 \leqslant \left(\sum_{i=1}^{n} a_i^2\right)\left(\sum_{i=1}^{n} b_i^2\right), \tag{9.75}$$

the equality holding when b_i/a_i equals a constant.

Once more, in terms of vectors, we have

$$(\mathbf{a} \cdot \mathbf{b})^2 = a^2 b^2 \cos^2 \theta \leqslant a^2 b^2, \tag{9.76}$$

where θ is the included angle.

The Schwarz inequality for functions has the form

$$\left[\int_a^b f(x)\, g(x)\, dx\right]^2 \leqslant \int_a^b [f(x)]^2\, dx \int_a^b [g(x)]^2\, dx, \tag{9.77}$$

the equality holding if and only if $g(x) = \alpha f(x)$, α being a constant.

In quantum mechanics, $f(x)$ and $g(x)$ might each represent a state or configuration of a physical system. Then the Schwarz inequality guarantees that the inner product $\int f(x)\, g(x)\, dx$ *exists.*

In the advanced physics literature, $\int f(x)\, g(x)\, dx$ is usually written $\langle f, g\rangle$. Then Eq. 9.77 appears as

$$\langle f, g\rangle^2 \leqslant \langle f, f\rangle\langle g, g\rangle.$$

Equation 9.77 may be derived by considering the equation

$$\int [\lambda f(x) + g(x)]^2\, dx = 0 \tag{9.78}$$

and proceeding as in Eq. 9.73 for an n-component vector.

If $g(x)$ is a normalized eigenfunction, $\varphi_i(x)$, Eq. 9.77 yields [here $w(x) = 1$]

$$a_i^2 \leqslant \int_a^b [f(x)]^2\, dx, \tag{9.79}$$

a result that also follows from Eq. 9.72.

[1] With discriminant $b^2 - 4ac$ negative (or zero).

Dirac delta function. Our orthonormal set of eigenfunctions, $\varphi_n(x)$, provides another interesting representation of the Dirac delta function. Consider the sum

$$K(x, t) = K(t, x) = \sum_{n=0}^{\infty} \varphi_n(x)\varphi_n(t). \tag{9.80}$$

(For convenience we assume that $\varphi_n(x)$ has been redefined to include $[w(x)]^{1/2}$ if $w(x) \neq 1$.) This series in Eq. 9.80 is assuredly not uniformly convergent, but it may be used as part of an integrand in which the ensuing integration will make it convergent (cf. Section 5.5).

Suppose we form the integral

$$\int F(t)K(x, t)\, dt,$$

where it is assumed that $F(t)$ can be expanded in a series of eigenfunctions, $\varphi_p(t)$. We obtain

$$\int F(t)K(x, t)\, dt = \int \sum_{p=0}^{\infty} a_p\, \varphi_p(t) \sum_{n=0}^{\infty} \varphi_n(x)\, \varphi_n(t)\, dt$$

$$= \sum_{p=0}^{\infty} a_p\, \varphi_p(x) = F(x), \tag{9.81}$$

the cross products $\varphi_p \varphi_n$ $(n \neq p)$ vanishing by orthogonality (Eq. 9.39). Referring back to the definition of the Dirac delta function (Sections 1.15 and 8.6) the result of Eq. 9.81 means that

$$K(x, t) = \delta(x - t) = \sum_{n=0}^{\infty} \varphi_n(x)\, \varphi_n(t). \tag{9.82}$$

We may easily show that

$$\int K(x, t)\, dt = 1 \tag{9.83}$$

by letting $F(t) = \varphi_0$, a constant. Confirmation of the behavior at $x = t$ may be found by using the Bessel inequality. We have

$$K(t, t) = \sum_{n=0}^{\infty} [\varphi_n(t)]^2. \tag{9.84}$$

Using Eq. 9.72, we obtain

$$\int [K(t, t)]^2\, dt \ngeq \sum_{n=0}^{\infty} a_n^2 = \sum_{n=0}^{\infty} 1 = \infty. \tag{9.85}$$

Hence, $K(x, t)$ diverges for $x = t$, as expected.

Green's function. A series somewhat similar to that representing $K(x, t)$ results when we expand the Green's function in the eigenfunctions of the corresponding homogeneous equation. In the inhomogeneous Helmholtz equation we have

$$\nabla^2\psi(\mathbf{r}) + k^2\psi(\mathbf{r}) = -\rho(\mathbf{r}). \tag{9.86}$$

The homogeneous Helmholtz equation is satisfied by its eigenfunctions φ_n.

$$\nabla^2 \varphi_n(\mathbf{r}) + k_n^2 \varphi_n(\mathbf{r}) = 0. \tag{9.87}$$

As outlined in Section 8.6, the Green's function $G(\mathbf{r}_1, \mathbf{r}_2)$ satisfies the point source equation

$$\nabla^2 G(\mathbf{r}_1, \mathbf{r}_2) + k^2 G(\mathbf{r}_1, \mathbf{r}_2) = -\delta(\mathbf{r}_1 - \mathbf{r}_2). \tag{9.88}$$

We expand the Green's function in a series of eigenfunctions of the homogeneous equation (9.87), that is,

$$G(\mathbf{r}_1, \mathbf{r}_2) = \sum_{n=0}^{\infty} a_n(\mathbf{r}_2)\, \varphi_n(\mathbf{r}_1), \tag{9.89}$$

and by substituting into Eq. 9.88 obtain

$$-\sum_{n=0}^{\infty} a_n(\mathbf{r}_2)\, k_n^2\, \varphi_n(\mathbf{r}_1) + k^2 \sum_{n=0}^{\infty} a_n(\mathbf{r}_2)\, \varphi_n(\mathbf{r}_1) = -\sum_{n=0}^{\infty} \varphi_n(\mathbf{r}_1)\, \varphi_n(\mathbf{r}_2). \tag{9.90}$$

Here $\delta(\mathbf{r}_1 - \mathbf{r}_2)$ has been replaced by its eigenfunction expansion, Eq. 9.82. When we employ the orthogonality of $\varphi_n(\mathbf{r}_1)$ to isolate a_n and then substitute into Eq. 9.89, the Green's function becomes

$$G(\mathbf{r}_1, \mathbf{r}_2) = \sum_{n=0}^{\infty} \frac{\varphi_n(\mathbf{r}_1)\, \varphi_n(\mathbf{r}_2)}{k_n^2 - k^2}, \tag{9.91}$$

a bilinear expansion, symmetric with respect to $\mathbf{r}_1$ and $\mathbf{r}_2$ as expected. Finally $\psi(\mathbf{r}_1)$, the desired solution of the inhomogeneous equation, is given by

$$\psi(\mathbf{r}_1) = \int G(\mathbf{r}_1, \mathbf{r}_2) \rho(\mathbf{r}_2)\, d\tau_2 \tag{9.92}$$

If we generalize our inhomogeneous differential equation to

$$\mathscr{L}\psi + \lambda\psi = -\rho, \tag{9.93}$$

where $\mathscr{L}$ is an Hermitian operator, we find that

$$G(\mathbf{r}_1, \mathbf{r}_2) = \sum_{n=0}^{\infty} \frac{\varphi_n(\mathbf{r}_1)\, \varphi_n(\mathbf{r}_2)}{\lambda_n - \lambda}, \tag{9.94}$$

where λ_n is the nth eigenvalue and φ_n, the corresponding orthonormal eigenfunction of the homogeneous differential equation

$$\mathscr{L}\psi + \lambda\psi = 0. \tag{9.95}$$

The Green's function will be encountered again in Section 16.5, in which we investigate it in more detail and relate it to integral equations.

EXERCISES

9.4.1 A function $f(x)$ is expanded in a series of orthonormal eigenfunctions

$$f(x) = \sum_{n=0}^{\infty} a_n\, \varphi_n(x).$$

Show that the series expansion is unique for a given set of $\varphi_n(x)$. The functions $\varphi_n(x)$ are being taken here as the *basis* vectors in an infinite dimensional Hilbert space.

9.4.2 In place of the expansion of a function $F(x)$ given by

$$F(x) = \sum_{n=0}^{\infty} a_n \varphi_n(x)$$

with

$$a_n = \int_a^b F(x)\, \varphi_n(x)\, w(x)\, dx,$$

take the *finite* series approximation

$$F(x) \approx \sum_{n=0}^{m} c_n \varphi_n(x).$$

Show that the mean square error

$$\int_a^b \left[F(x) - \sum_{n=0}^{m} c_n \varphi_n(x)\right]^2 w(x)\, dx$$

is minimized by taking $c_n = a_n$.

9.4.3 Derive the Schwarz inequality from the identity

$$\left[\int_a^b f(x)\, g(x)\, dx\right]^2 = \int_a^b [f(x)]^2\, dx \int_a^b [g(x)]^2\, dx - \frac{1}{2}\int_a^b\int_a^b [f(x)\, g(y) - f(y)\, g(x)]^2\, dx\, dy.$$

9.4.4 If the functions $f(x)$ and $g(x)$ of the Schwarz inequality, Eq. 9.77, may be expanded in a series of eigenfunctions $\varphi_i(x)$, show that Eq. 9.77 reduces to Eq. 9.75 (with n possibly infinite).
Note the description of $f(x)$ as a vector in a function space in which $\varphi_i(x)$ corresponds to the unit vector $\mathbf{e}_i$.

9.4.5 The operator H is Hermitian and positive definite, that is,

$$\int_a^b f^* H f\, dx > 0.$$

Prove the generalized Schwarz inequality:

$$\left|\int_a^b f^* H g\, dx\right|^2 \leqslant \int_a^b f^* H f\, dx \int_a^b g^* H g\, dx.$$

9.4.6 The Dirac delta function representation given by Eq. 9.82

$$\delta(x - t) = \sum_{n=0}^{\infty} \varphi_n(x)\, \varphi_n(t)$$

is often called the *closure relation*. For an orthonormal set of functions, φ_n, show that closure implies completeness, that is, that Eq. 9.61 follows from Eq. 9.82.
Hint. One can take

$$F(x) = \int F(t)\, \delta(x - t)\, dt.$$

9.4.7 Substitute Eq. 9.91, the eigenfunction expansion of Green's function, into Eq. 9.92 and then show that Eq. 9.92 is indeed a solution of the nonhomogeneous Helmholtz equation (9.86).

9.4.8 (a) Starting with a one-dimensional nonhomogeneous differential equation, (Eq. 9.93), assume that $\psi(x)$ and $\rho(x)$ may be represented by eigenfunction expansions. Without any use of the Dirac delta function or its representations show that

$$\psi(x) = \sum_{n=0}^{\infty} \frac{\int_a^b \rho(t)\varphi_n(t)\, dt}{\lambda_n - \lambda}\, \varphi_n(x).$$

Note that (1) if $\rho = 0$, no solution exists unless $\lambda = \lambda_n$ and (2) if $\lambda = \lambda_n$, no solution exists unless ρ is orthogonal to φ_n. This same behavior will reappear with integral equations in Section 16.4.

(b) Interchanging summation and integration show that you have constructed the Green's function corresponding to Eq. 9.94.

9.4.9 The eigenfunctions of the Schrödinger equation are often complex. In this case the orthogonality integral, Eq. 9.39, is replaced by

$$\int_a^b \varphi_i^*(x)\varphi_j(x)w(x)\,dx = \delta_{ij}.$$

Instead of Eq. 9.82 we have

$$\delta(\mathbf{r}_1 - \mathbf{r}_2) = \sum_{n=0}^{\infty} \varphi_n(\mathbf{r}_1)\varphi_n^*(\mathbf{r}_2).$$

Show that the Green's function, Eq. 9.91, becomes

$$G(\mathbf{r}_1, \mathbf{r}_2) = \sum_{n=0}^{\infty} \frac{\varphi_n(\mathbf{r}_1)\varphi_n^*(\mathbf{r}_2)}{k_n^2 - k^2}$$

$$= G^*(\mathbf{r}_2, \mathbf{r}_1).$$

9.4.10 A normalized wave function $\psi(x) = \sum_{n=0}^{\infty} a_n\,\varphi_n(x)$. The expansion coefficients a_n are known as probability amplitudes. We may define a density matrix ρ with elements $\rho_{ij} = a_i a_j^*$. Show that

$$(\rho^2)_{ij} = \rho_{ij}$$

or

$$\rho^2 = \rho.$$

This result, by definition, makes ρ a projection operator.

$$\textit{Hint.} \int \psi^*\psi\,dx = 1.$$

REFERENCES

MILLER, K. S., *Linear Differential Equations in the Real Domain.* New York: Norton (1963).

TITCHMARSH, E. C., *Eigenfunction Expansions Associated with Second Order Differential Equations.* London: Oxford University Press, Vol. I, Second Edition (1962), Vol. II (1958).

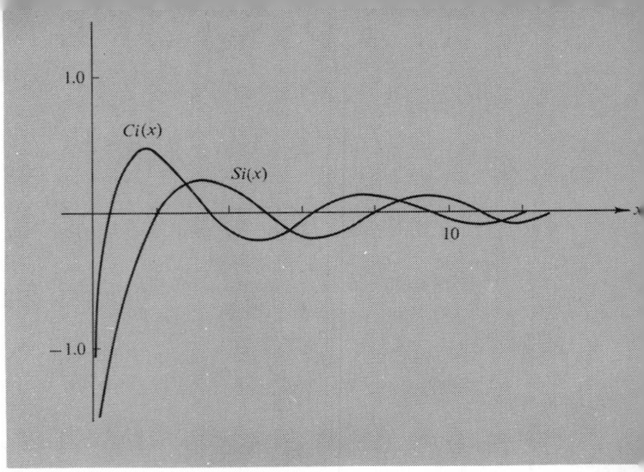

CHAPTER 10

THE GAMMA FUNCTION (FACTORIAL FUNCTION)

The Gamma Function (Factorial Function)

The gamma function appears occasionally in physical problems such as the normalization of Coulomb wave functions and the computation of probabilities in statistical mechanics. In general, however, it has less direct physical application and interpretation than, say, the Legendre and Bessel functions of Chapters 11 and 12. Rather its importance stems from its usefulness in developing other functions which have direct physical application. The gamma function, therefore, is included here. A discussion of the numerical evaluation of the gamma function appears in Section 10.3.

10.1 Definitions, Simple Properties

At least three different, convenient definitions of the gamma function are in common use. Our first task is to state these definitions, to develop some simple, direct consequences, and to show the equivalence of the three forms.

Infinite limit (Euler). The first definition, named after Euler is

$$\Gamma(z) \equiv \lim_{n\to\infty} \frac{1 \cdot 2 \cdot 3 \cdots n}{z(z+1)(z+2)\cdots(z+n)} n^z. \tag{10.1}$$

Here and elsewhere in this chapter z may be either real or complex. Replacing z with $z+1$, we have

$$\begin{aligned}\Gamma(z+1) &= \lim_{n\to\infty} \frac{1 \cdot 2 \cdot 3 \cdots n}{(z+1)(z+2)(z+3)\cdots(z+n+1)} n^{z+1} \\ &= \lim_{n\to\infty} \frac{nz}{z+n+1} \cdot \frac{1 \cdot 2 \cdot 3 \cdots n}{z(z+1)(z+2)\cdots(z+n)} n^z \\ &= z\,\Gamma(z).\end{aligned} \tag{10.2}$$

This is the basic functional relation for the gamma function. It should be noted that it is a *difference* equation. It has been shown that the gamma function is one of a general class of functions which do not satisfy any differential equation with rational coefficients. Also from the definition

$$\Gamma(1) = \lim_{n\to\infty} \frac{1 \cdot 2 \cdot 3 \cdots n}{1 \cdot 2 \cdot 3 \cdots n(n+1)} n$$
$$= 1. \tag{10.3}$$

Now, application of Eq. 10.2 gives

$$\begin{aligned} \Gamma(2) &= 1, \\ \Gamma(3) &= 2\,\Gamma(2) = 2, \\ \Gamma(n) &= 1 \cdot 2 \cdot 3 \cdots (n-1) \end{aligned} \tag{10.4}$$

Definite integral (Euler). A second definition, also frequently called Euler's form, is

$$\Gamma(z) \equiv \int_0^\infty e^{-t} t^{z-1}\, dt, \qquad \mathscr{R}(z) > 0. \tag{10.5}$$

The restriction on z is necessary to avoid divergence of the integral. When the gamma function does appear in physical problems, it is often in this form or some variation such as

$$\Gamma(z) = 2 \int_0^\infty e^{-t^2} t^{2z-1}\, dt, \qquad \mathscr{R}(z) > 0, \tag{10.6}$$

$$\Gamma(z) = \int_0^1 \left[\ln\left(\frac{1}{t}\right)\right]^{z-1} dt, \qquad \mathscr{R}(z) > 0. \tag{10.7}$$

When $z = \frac{1}{2}$, Eq. 10.6 is just the Gauss error function, and we have the interesting result

$$\Gamma(\tfrac{1}{2}) = \sqrt{\pi}. \tag{10.8}$$

To show the equivalence of these two definitions, Eqs. 10.1 and 10.5, consider the function of two variables

$$F(z, n) = \int_0^n \left(1 - \frac{t}{n}\right)^n t^{z-1}\, dt, \qquad \mathscr{R}(z) > 0 \tag{10.9}$$

with n a positive integer.[1] Since

$$\lim_{n\to\infty} \left(1 - \frac{t}{n}\right)^n \equiv e^{-t}, \tag{10.10}$$

from the definition of the exponential,

$$\lim_{n\to\infty} F(z, n) = F(z, \infty) = \int_0^\infty e^{-t} t^{z-1}\, dt$$
$$\equiv \Gamma(z) \tag{10.11}$$

by Eq. 10.5.

[1] The form of $F(z, n)$ is suggested by the beta function (cf. Eq. 10.60).

Returning to $F(z, n)$, we evaluate it in successive integrations by parts. For convenience let $u = t/n$. Then

$$F(z, n) = n^z \int_0^1 (1 - u)^n u^{z-1}\, du. \tag{10.12}$$

Integrating by parts,

$$\frac{F(z, n)}{n^z} = (1 - u)^n \frac{u^z}{z}\Big|_0^1 + \frac{n}{z}\int_0^1 (1 - u)^{n-1} u^z\, du. \tag{10.13}$$

Repeating this with the integrated part vanishing at both end points each time, we finally get

$$\begin{aligned} F(z, n) &= n^z \frac{n(n-1)\cdots 1}{z(z+1)\cdots(z+n-1)} \int_0^1 u^{z+n-1}\, du \\ &= \frac{1 \cdot 2 \cdot 3 \cdots n}{z(z+1)(z+2)\cdots(z+n)}\, n^z \end{aligned} \tag{10.14}$$

This is identical with the expression on the right side of Eq. 10.1. Hence

$$\lim_{n\to\infty} F(z, n) = F(z, \infty) \equiv \Gamma(z), \tag{10.15}$$

by Eq. 10.1, completing the proof.

Infinite product (Weierstrass). The third definition (Weierstrass' form) is

$$\frac{1}{\Gamma(z)} \equiv z e^{\gamma z} \prod_{n=1}^{\infty} \left(1 + \frac{z}{n}\right) e^{-z/n}, \tag{10.16}$$

where γ is the usual Euler-Mascheroni constant,

$$\gamma = 0.577\,216 \cdots. \tag{10.17}$$

This form can be derived from the original definition (Eq. 10.1) by rewriting it as

$$\begin{aligned} \Gamma(z) &= \lim_{n\to\infty} \frac{1 \cdot 2 \cdot 3 \cdots n}{z(z+1)\cdots(z+n)}\, n^z \\ &= \lim_{n\to\infty} \frac{1}{z} \prod_{m=1}^{n} \left(1 + \frac{z}{m}\right)^{-1} n^z. \end{aligned} \tag{10.18}$$

Inverting and using

$$n^{-z} = e^{(-\ln n)z}, \tag{10.19}$$

we obtain

$$\frac{1}{\Gamma(z)} = z \lim_{n\to\infty} e^{(-\ln n)z} \prod_{m=1}^{n} \left(1 + \frac{z}{m}\right). \tag{10.20}$$

Multiplying and dividing by

$$\exp\left[\left(1 + \frac{1}{2} + \frac{1}{3} + \cdots + \frac{1}{n}\right)z\right] = \prod_{m=1}^{n} e^{z/m}, \tag{10.21}$$

$$\begin{aligned} \frac{1}{\Gamma(z)} &= z\left\{\lim_{n\to\infty} \exp\left[\left(1 + \frac{1}{2} + \frac{1}{3} + \cdots + \frac{1}{n} - \ln n\right)z\right]\right\} \\ &\quad \times \left[\lim_{n\to\infty} \prod_{m=1}^{n} \left(1 + \frac{z}{m}\right) e^{-z/m}\right]. \end{aligned} \tag{10.22}$$

As shown in Section 5.2, the infinite series in the exponent converges and defines γ, the Euler-Mascheroni constant. Hence Eq. 10.16 follows.

It was also shown in Section 5.10 that the Weierstrass infinite product definition of $\Gamma(z)$ led directly to an important identity,

$$\Gamma(z)\,\Gamma(1-z) = \frac{\pi}{\sin \pi z}. \tag{10.23}$$

Again, letting $z = \frac{1}{2}$, we obtain

$$\Gamma(\tfrac{1}{2}) = \sqrt{\pi} \tag{10.24}$$

(taking the positive square root) in agreement with Eq. 10.8.

The Weierstrass definition shows immediately that $\Gamma(z)$ has simple poles at $z = 0, -1, -2, -3 \cdots$ and that $[\Gamma(z)]^{-1}$ has no poles in the finite complex plane, which means that $\Gamma(z)$ has no zeros. This behavior may also be seen in Eq. 10.23, in which we note that $\pi/(\sin \pi z)$ is never equal to zero.

Factorial notation. So far this discussion has been presented in terms of the classical notation. As pointed out by Jeffreys and others, the -1 of the $z-1$ exponent in our second definition (Eq. 10.5) is a continual nuisance. Accordingly, Eq. 10.5 is rewritten as

$$\int_0^\infty e^{-t} t^z \, dt \equiv z!, \qquad \mathscr{R}(z) > -1, \tag{10.25}$$

to *define* a factorial function $z!$. Occasionally we may still encounter Gauss's notation, $\Pi(z)$, for the factorial function

$$\prod(z) = z!. \tag{10.26}$$

The Γ notation is due to Legendre. The factorial function of Eq. 10.25 is, of course, related to the gamma function by

$$\Gamma(z) = (z-1)!$$

or

$$\Gamma(z+1) = z! \tag{10.27}$$

If $z = n$, a positive integer (Eq. 10.4) shows that

$$z! = n! = 1 \cdot 2 \cdot 3 \cdots n, \tag{10.28}$$

the familiar factorial. However, it should be noted carefully that since $z!$ is now defined by Eq. 10.25 (or equivalently by Eq. 10.27) *the factorial function is no longer limited to positive integral values of the argument*. The difference relation (Eq. 10.2) becomes

$$(z-1)! = \frac{z!}{z}. \tag{10.29}$$

This shows immediately that

$$0! = 1 \tag{10.30}$$

and

$$n! = \pm\infty \quad \text{for} \quad n, \text{ a } \textit{negative} \text{ integer}. \tag{10.31}$$

In terms of the factorial function (Eq. 10.23) becomes

$$z!(-z)! = \frac{\pi z}{\sin \pi z}. \tag{10.32}$$

By restricting ourselves to the real values of the argument $x!$ defines the curve shown in Fig. 10.1. The minimum of the curve is

$$x! = (0.461,63 \cdots)! = 0.885,60 \cdots. \tag{10.33}$$

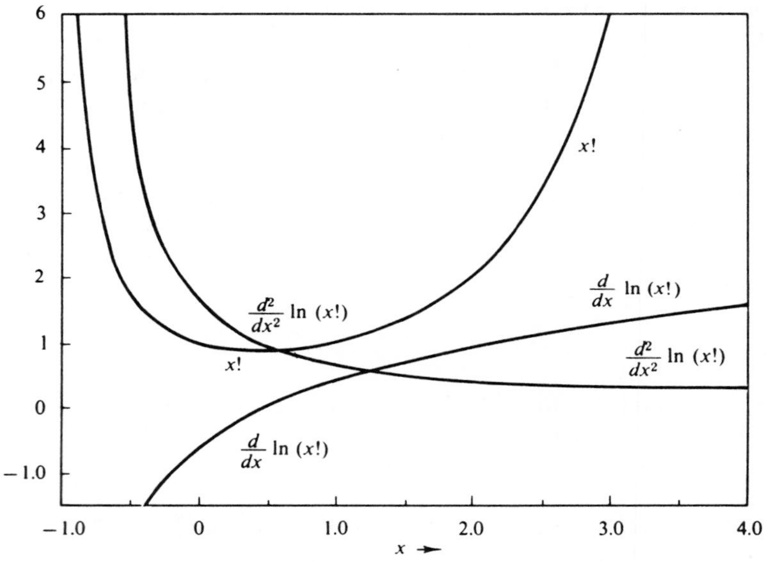

FIG. 10.1 The factorial function and the first two derivatives of ln(x!)

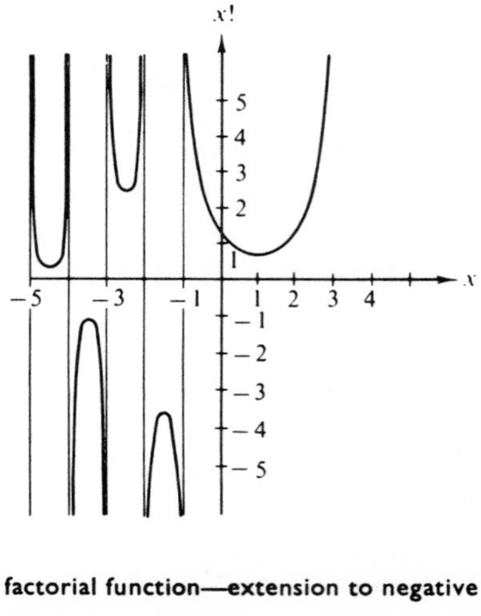

FIG. 10.2 The factorial function—extension to negative arguments

Integral representation. An integral representation that is useful in developing asymptotic series for the Bessel functions is

$$\int_C e^{-z} z^{\nu}\, dz = (e^{2\pi i\nu} - 1)\nu! \tag{10.34}$$

where C is the contour shown in Figure 10.3. This contour integral representation is particularly useful when ν is not an integer, $z = 0$ then being a *branch point*.

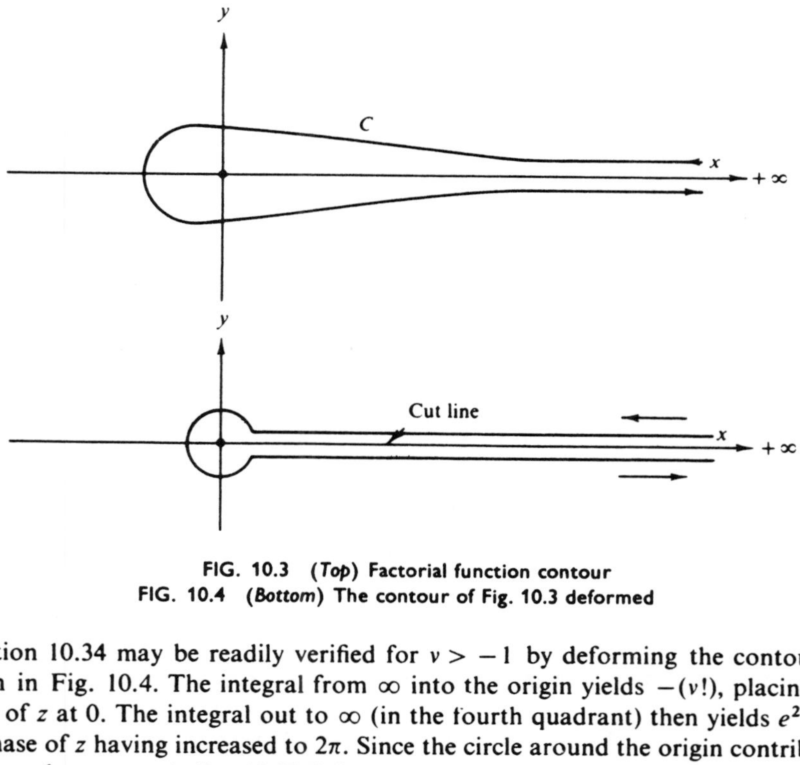

FIG. 10.3 **(*Top*) Factorial function contour**
FIG. 10.4 **(*Bottom*) The contour of Fig. 10.3 deformed**

Equation 10.34 may be readily verified for $\nu > -1$ by deforming the contour as shown in Fig. 10.4. The integral from ∞ into the origin yields $-(\nu!)$, placing the phase of z at 0. The integral out to ∞ (in the fourth quadrant) then yields $e^{2\pi i\nu}\nu!$, the phase of z having increased to 2π. Since the circle around the origin contributes nothing when $\nu > -1$, Eq. 10.34 follows.

It is often convenient to throw this result into a more symmetrical form

$$\int_C e^{-z}(-z)^{\nu}\, dz = 2i \sin \nu\pi\ \nu! \tag{10.35}$$

This corresponds to choosing the phase of z to have a range of $-\pi$ to $+\pi$ in Eq. 10.34.

This analysis establishes Eqs. 10.34 and 10.35 for $\nu > -1$. It is relatively simple to extend the range to include all nonintegral ν. First, we note that the integral exists for $\nu < -1$ as long as we stay away from the origin. Second, integrating by parts we find that Eq. 10.35 yields the familiar difference relation (Eq. 10.29). If we take the difference relation to define the factorial function of $\nu < -1$, then Eqs. 10.34 and 10.35 are verified for all ν (except negative integers).

EXERCISES

10.1.1 Derive the recurrence relations

$$\Gamma(z+1) = z\,\Gamma(z)$$

from the Euler integral form (Eq. 10.5),

$$\Gamma(z) = \int_0^\infty e^{-t} t^{z-1}\, dt.$$

10.1.2 In a power series solution for the Legendre functions of the second kind we encounter the expression

$$\frac{(n+1)(n+2)(n+3)\cdots(n+2s-1)(n+2s)}{2\cdot 4\cdot 6\cdot 8\cdots(2s-2)(2s)\cdot(2n+3)(2n+5)(2n+7)\cdots(2n+2s+1)},$$

in which s is a positive integer. Rewrite this expression in terms of factorials.

10.1.3 Show that

$$\frac{(s-n)!}{(2s-2n)!} = \frac{(-1)^{n-s}(2n-2s)!}{(n-s)!}.$$

Here s and n are integers with $s < n$. This result can be used to avoid negative factorials such as in the series representations of the spherical Neumann functions and the Legendre functions of the second kind.

10.1.4 Show that $\Gamma(z)$ may be written

$$\Gamma(z) = 2\int_0^\infty e^{-t^2} t^{2z-1}\, dt, \qquad \mathscr{R}(z) > 0,$$

$$\Gamma(z) = \int_0^1 \left[\ln\left(\frac{1}{t}\right)\right]^{z-1} dt,\ \mathscr{R}(z) > 0.$$

10.1.5 In a Maxwellian distribution, the fraction of particles between the speed v and $v + dv$ is

$$\frac{dN}{N} = 4\pi\left(\frac{m}{2\pi kT}\right)^{3/2} \exp\left(-mv^2/2kT\right)v^2\, dv,$$

N being the total number of particles. The average or expectation value of v^n is defined as $\langle v^n\rangle = N^{-1}\int v^n\, dN$. Show that

$$\langle v^n\rangle = \left(\frac{2kT}{m}\right)^{n/2}\left(\frac{n+1}{2}\right)!\Big/\frac{1}{2}!.$$

10.1.6 By transforming the integral into a gamma function show that

$$-\int_0^1 x^k \ln x\, dx = \frac{1}{(k+1)^2} \qquad k > -1.$$

10.1.7 Show that

$$\int_0^\infty e^{-x^4}\, dx = (\tfrac{1}{4})!$$

10.1.8 Show that

$$\lim_{x \to 0} \frac{(ax-1)!}{(x-1)!} = \frac{1}{a}.$$

10.1.9 Verify

(a) $\int_0^\infty e^{-r} \ln r \, dr = -\gamma.$

(b) $\int_0^\infty re^{-r} \ln r \, dr = 1 - \gamma.$

(c) $\int_0^\infty r^n e^{-r} \ln r \, dr = (n-1)! + n\int_0^\infty r^{n-1} e^{-r} \ln r \, dr, \qquad n = 1, 2, 3, \cdots.$

Hint. These may be verified by integration by parts, three parts, or differentiating the integral form of $n!$ with respect to n.

10.1.10 Locate the poles of $\Gamma(z)$. Show that they are simple poles and determine the residues.

10.1.11 Show that the equation $x! = k$, $k \neq 0$, has an infinite number of real roots.

10.1.12 In many problems in electromagnetic theory we encounter the products

$$2n(2n-2)\cdots 6 \cdot 4 \cdot 2 \equiv (2n)!!$$

$$(2n+1)(2n-1)\cdots 5 \cdot 3 \cdot 1 \equiv (2n+1)!!$$

Show that

$$(2n)!! = 2^n n!$$

$$(2n+1)!! = \frac{(2n+1)!}{2^n n!}.$$

10.1.13 (a) Develop a recurrence relation for $(2n+1)!!$.
(b) Show that $(-1)!! = 1$ is consistent with part *a*.

Ans. $(2n+1)!! = (2n+1)(2n-1)!!$.

10.1.14 For s a non-negative integer, show that

$$(-2s-1)!! = \frac{(-1)^s}{(2s-1)!!} = \frac{(-1)^s 2^s s!}{(2s)!}.$$

10.1.15 Express the coefficient of the nth term of the expansion of $(1+x)^{1/2}$
(a) in terms of factorials of integers.
(b) in terms of the double factorial (!!) functions.

Ans.

$$a_n = (-1)^{n+1} \frac{(2n-3)!}{2^{2n-2} n!(n-2)!} = (-1)^{n+1} \frac{(2n-3)!!}{(2n)!!} \qquad n = 2, 3, 4, \ldots.$$

10.1.16 Express the coefficient of the nth term of the expansion of $(1+x)^{-1/2}$
(a) in terms of the factorials of integers.
(b) in terms of the double factorial (!!) functions.

Ans. $a_n = (-1)^n \dfrac{(2n)!}{2^{2n}(n!)^2} = (-1)^n \dfrac{(2n-1)!!}{(2n)!!}, \; n = 1, 2, 3, \ldots.$

10.1.17 (a) Show that

$$\Gamma(\tfrac{1}{2} - n)\Gamma(\tfrac{1}{2} + n) = (-1)^n \pi$$

where n is an integer.

(b) Express $\Gamma(\frac{1}{2}+n)$ and $\Gamma(\frac{1}{2}-n)$ separately in terms of $\pi^{1/2}$ and a !! function (Exercise 10.1.12)

$$\textit{Ans.}\ \Gamma(\tfrac{1}{2}+n) = \frac{(2n-1)!!}{2^n}\pi^{1/2}.$$

10.1.18 From one of the definitions of the factorial or gamma function, show that

$$|(ix)!|^2 = \frac{\pi x}{\sinh \pi x}.$$

10.1.19 Prove that

$$|\Gamma(\alpha + i\beta)| = \frac{\Gamma(1+\alpha)}{\alpha}\prod_{n=0}^{\infty}\left[1 + \frac{\beta^2}{(\alpha+n)^2}\right]^{-1/2}.$$

This equation has been useful in calculations in the theory of beta decay.

10.1.20 Show that

$$|(n+ib)!| = \left(\frac{\pi b}{\sinh \pi b}\right)^{1/2}\prod_{s=1}^{n}(s^2+b^2)^{1/2}$$

for n, a positive integer.

10.1.21 Show that

$$|x!| \geq |(x+iy)!|$$

for all x. The variables x and y are real.

10.1.22 Show that

$$|(-\tfrac{1}{2}+iy)!|^2 = \frac{\pi}{\cosh \pi y}.$$

10.1.23 The wave function of a particle scattered by a pure Coulomb potential is $\psi(r, \theta)$. At the origin the wave function becomes

$$\psi(0) = e^{-\pi\gamma/2}\Gamma(1+i\gamma)$$

where $\gamma = Z_1 Z_2 e^2/\hbar v$. Show that

$$\psi^*(0)\psi(0) = \frac{2\pi\gamma}{e^{2\pi\gamma}-1}.$$

10.1.24 Derive the contour integral representation

$$2i \sin \nu\pi\ \nu! = \int_c e^{-z}(-z)^\nu\, dz.$$

10.2 Digamma and Polygamma Functions

Digamma functions. As may be noted from the three definitions in Section 10.1, it is inconvenient to deal with the derivatives of the gamma or factorial function directly. Instead, it is customary to take the natural logarithm of the factorial function, (Eq. 10.1), convert the product to a sum, and then differentiate, that is,

$$z! = \lim_{n\to\infty} \frac{n!}{(z+1)(z+2)\cdots(z+n)} n^z \tag{10.36}$$

and

$$\ln(z!) = \lim_{n\to\infty} [\ln(n!) + z \ln n - \ln(z+1) - \ln(z+2) - \cdots - \ln(z+n)], \tag{10.37}$$

in which the logarithm of the limit is equal to the limit of the logarithm. Differentiating with respect to z, we obtain

$$\frac{d}{dz}\ln(z!) \equiv \mathsf{F}(z) = \lim_{n\to\infty}\left(\ln n - \frac{1}{z+1} - \frac{1}{z+2} - \cdots - \frac{1}{z+n}\right), \tag{10.38}$$

which defines $\mathsf{F}(z)$, the digamma function. From the definition of the Euler-Mascheroni constant[1] Eq. 10.38 may be rewritten

$$\mathsf{F}(z) = -\gamma - \sum_{n=1}^{\infty}\left(\frac{1}{z+n} - \frac{1}{n}\right). \tag{10.39}$$

One application of Eq. 10.39 is in the derivation of the series form of the Neumann function (Section 11.3). Clearly

$$\mathsf{F}(0) = -\gamma = -0.577\,215\,664\,901\cdots.^2 \tag{10.40}$$

Another, perhaps more useful expression for $\mathsf{F}(z)$ is derived in Section 10.3.

Polygamma function. The digamma function may be differentiated repeatedly, giving rise to the polygamma function:

$$\begin{aligned}\mathsf{F}^{(m)}(z) &\equiv \frac{d^{m+1}}{dz^{m+1}}\ln(z!) \\ &= (-1)^{m+1} m! \sum_{n=1}^{\infty}\frac{1}{(z+n)^{m+1}}, \qquad m = 1, 2, 3, \ldots.\end{aligned} \tag{10.41}$$

A plot of $\mathsf{F}(x)$ and $\mathsf{F}'(x)$ is included in Fig. 10.1. Since the series in Eq. 10.41 defines the Riemann zeta function[3] (with $z = 0$),

$$\zeta(m) \equiv \sum_{n=1}^{\infty}\frac{1}{n^m}, \tag{10.42}$$

we have

$$\mathsf{F}^{(m)}(0) = (-1)^{m+1} m!\, \zeta(m+1). \tag{10.43}$$

The values of the polygamma functions of positive integral argument, $\mathsf{F}^{(m)}(n)$, may be calculated using Ex. 10.2.4.

In terms of the perhaps more common Γ notation,

$$\frac{d^{n+1}}{dz^{n+1}}\ln\Gamma(z) = \frac{d^n}{dz^n}\psi(z) = \psi^{(n)}(z). \tag{10.44a}$$

[1] Cf. Sections 5.2 and 5.6.

[2] γ has been computed to 1271 places by D. E. Knuth, *Math. Comp.* **16**, 275 (1962) and to 3566 decimal places by D. W. Sweeney, *Math. Comp.* **17**, 170 (1963). It may be of interest that the fraction 228/395 gives γ accurate to six places.

[3] Section 5.9. For $z \neq 0$ this series may be used to define a generalized zeta function.

From Eq. 10.27,

$$\psi^{(n)}(z) = \mathsf{F}^{(n)}(z-1). \tag{10.44b}$$

Riemann zeta function. For convenience, the Riemann zeta function is given in Table 10.1. The general behavior is described in Fig. 5.9.

TABLE 10.1

Riemann Zeta Function

s	$\zeta(s)$
2	1.64493 40668
3	1.20205 69032
4	1.08232 32337
5	1.03692 77551
6	1.01734 30620
7	1.00834 92774
8	1.00407 73562
9	1.00200 83928
10	1.00099 45751

It is now possible to write a Maclaurin expansion for $\ln(z!)$.

$$\ln(z!) = -\gamma z + \frac{z^2}{2}\zeta(2) - \frac{z^3}{3}\zeta(3) + \cdots + (-1)^n \frac{z^n}{n}\zeta(n) + \cdots, \tag{10.44c}$$

convergent for $|z| < 1$; for $z = x$, the range is $-1 < x \leqslant 1$. Alternate forms of this series appear in Ex. 5.9.13. Eq. 10.44c is a possible means of computing $z!$ for real or complex z, but Stirling's series (Section 10.3) is usually better, and, in addition, an excellent table of values of the gamma function for complex arguments based on the use of Stirling's series and the recurrence relation (Eq. 10.29) is now available.[1]

The digamma and polygamma functions may also be used in summing series. If the general term of the series has the form of a rational fraction (with the highest power of the index in the numerator at least two less than the highest power of the index in the denominator), it may be transformed by the method of partial fractions (cf. Section 15.7). The infinite series may then be expressed as a finite sum of digamma and polygamma functions. The usefulness of this method depends on the availability of tables of digamma and polygamma functions. Such tables and examples of series summation are given in AMS-55, Chapter 6.

[1] *Table of the Gamma Function for Complex Arguments*, National Bureau of Standards, Applied Mathematics Series No. 34.

EXERCISES

10.2.1 Verify that the following two forms of the digamma function,

$$F(x) = \sum_{r=1}^{x} \frac{1}{r} - \gamma$$

and

$$F(x) = \sum_{r=1}^{\infty} \frac{x}{r(r+x)} - \gamma,$$

are equal to each other (for x a positive integer).

10.2.2 Show that $F(z)$ has the series expansion

$$F(z) = -\gamma + \sum_{n=2}^{\infty} (-1)^n \zeta(n) z^{n-1}.$$

10.2.3 For a power series expansion of $\ln (z!)$ AMS-55 lists

$$\ln (z!) = -\ln (1+z) + z(1-\gamma) + \sum_{n=2}^{\infty} (-1)^n [\zeta(n) - 1] z^n/n.$$

(a) Show that this agrees with Eq. 10.44*c* for $|z| < 1$.
(b) What is the range of convergence of this new expression?

10.2.4 Derive the difference relation for the polygamma function

$$F^{(m)}(z+1) = F^{(m)}(z) + (-1)^m \frac{m!}{(z+1)^{m+1}}, \qquad m = 0, 1, 2, \ldots .$$

10.2.5 Show that if

$$\Gamma(x+iy) = u + iv$$

then

$$\Gamma(x-iy) = u - iv.$$

This is a special case of the Schwarz reflection principle, Section 6.5.

10.2.6 The Pochhammer symbol $(a)_n$ is defined as

$$(a)_n = a(a+1) \ldots (a+n-1)$$
$$(a)_0 = 1$$

(for integral n).
(a) Express $(a)_n$ in terms of factorials.
(b) Find $(d/da)(a)_n$ in terms of $(a)_n$ and digamma functions.

Ans. $\frac{d}{da}(a)_n = (a)_n[F(a+n-1) - F(a-1)].$

10.2.7 Dirac relativistic wave functions for hydrogen involve factors such as $[2(1-\alpha^2 Z^2)^{1/2}]!$ where α, the fine structure constant, is $\frac{1}{137}$ and Z is the atomic number. Expand $[2(1-\alpha^2 Z^2)^{1/2}]!$ in a series of powers of $\alpha^2 Z^2$.

10.2.8 The quantum mechanical description of a particle in a coulomb field requires a knowledge of the phase of the complex factorial function. Determine the phase of $(1 + ib)!$ for small b.

10.2.9 The total energy radiated by a black body is given by

$$u = \frac{8\pi k^4 T^4}{c^3 h^3} \int_0^\infty \frac{x^3}{e^x - 1}\, dx.$$

Show that the integral in this expression is equal to $3!\ \zeta(4)$. [$\zeta(4) = \pi^4/90 = 1.0823\ldots$.] The final result is the Stefan-Boltzmann law.

10.2.10 As a generalization of the result in Exercise 10.2.9, show that

$$\int_0^\infty \frac{x^s\, dx}{e^x - 1} = s!\ \zeta(s+1), \qquad \mathscr{R}(s) > 0.$$

10.2.11 Prove that

$$\int_0^\infty \frac{x^s\, dx}{e^x + 1} = s!(1 - 2^{-s})\zeta(s+1), \qquad \mathscr{R}(s) > 0.$$

Ex. 10.2.10 and 10.2.11 actually constitute Mellin integral transforms (cf. Section 15.1).

10.2.12 Prove that

$$\psi^{(n)}(z) = (-1)^{n+1} \int_0^\infty \frac{t^n e^{-zt}}{1 - e^{-t}}\, dt, \qquad \mathscr{R}(z) > 0.$$

10.2.13 Using di- and polygamma functions sum the series
(a)

$$\sum_{n=1}^{\infty} \frac{1}{n(n+1)}$$

(b)

$$\sum_{n=2}^{\infty} \frac{1}{n^2 - 1}.$$

Note: You can use Exercise 10.2.4 to calculate the needed digamma functions.

10.2.14 Verify the contour integral representation of $\zeta(s)$,

$$\zeta(s) = -\frac{(-s)!}{2\pi i} \int_C \frac{(-z)^{s-1}}{e^z - 1}\, dz.$$

The contour C is the same as that for Eq. 10.35. The points $z = \pm 2n\pi i$, $n = 1, 2, 3 \ldots$ are all excluded.

10.2.15 Show that $\zeta(s)$ is analytic in the entire finite complex plane except at $s = 1$ where it has a simple pole with a residue of $+1$.
Hint: The contour integral representation will be useful.

10.3 Stirling's Series

For computation of $\ln(z!)$ for very large z (statistical mechanics) and for numerical computations at nonintegral values of z a series expansion of $\ln(z!)$ in negative powers of z is desirable. Perhaps the most elegant way of deriving such an expansion is by the method of steepest descents (Section 7.4). The following method, starting with a numerical integration formula, does not require knowledge of contour integration and is particularly direct.

Derivation from Euler-Maclaurin integration formula. The Euler-Maclaurin formula for evaluating a definite integral[1] is

$$\int_0^n f(x)\,dx = \tfrac{1}{2}f(0) + f(1) + f(2) + \cdots + \tfrac{1}{2}f(n)$$

$$- b_2[f'(n) - f'(0)] - b_4[f'''(n) - f'''(0)] - \cdots, \qquad (10.45)$$

in which the b_{2n} are related to the Bernoulli numbers B_{2n} (cf., Section 5.9) by

$$(2n)!b_{2n} = B_{2n}, \qquad (10.46)$$

$$\begin{aligned} B_0 &= 1, & B_6 &= \tfrac{1}{42}, \\ B_2 &= \tfrac{1}{6}, & B_8 &= -\tfrac{1}{30}, \\ B_4 &= -\tfrac{1}{30}, & B_{10} &= \tfrac{5}{66}, \text{ etc.} \end{aligned} \qquad (10.47)$$

By applying Eq. 10.45 to the definite integral

$$\int_0^\infty \frac{dx}{(z+x)^2} = \frac{1}{z} \qquad (10.48)$$

(for z not on the negative real axis) we obtain

$$\frac{1}{z} = \frac{1}{2z^2} + \mathsf{F}'(z) - \frac{2!b_2}{z^3} - \frac{4!b_4}{z^5} - \cdots. \qquad (10.49)$$

Using Eq. 10.46 and solving for $\mathsf{F}'(z)$, we have

$$\begin{aligned} \mathsf{F}'(z) &= \frac{1}{z} - \frac{1}{2z^2} + \frac{B_2}{z^3} + \frac{B_4}{z^5} + \cdots \\ &= \frac{1}{z} - \frac{1}{2z^2} + \sum_{n=1}^{\infty} \frac{B_{2n}}{z^{2n+1}}. \end{aligned} \qquad (10.50)$$

Since the Bernoulli numbers diverge strongly, this series does not converge! It is a semiconvergent or asymptotic series, useful for computation despite its divergence (cf. Section 5.11).

Integrating once, we get the digamma function

$$\begin{aligned} \mathsf{F}(z) &= C_1 + \ln z + \frac{1}{2z} - \frac{B_2}{2z^2} - \frac{B_4}{4z^4} - \cdots \\ &= C_1 + \ln z + \frac{1}{2z} - \sum_{n=1}^{\infty} \frac{B_{2n}}{2nz^{2n}}. \end{aligned} \qquad (10.51)$$

Integrating Eq. 10.51 with respect to z from $z - 1$ to z and then letting z approach infinity, C_1, the constant of integration may be shown to vanish. This gives us a second expression for the digamma function, often more useful than Eq. 10.38.

[1] Obtained by repeated integration by parts, Section 5.9.

Stirling's series. The indefinite integral of the digamma function (Eq. 10.51) is

$$\ln(z!) = C_2 + \left(z + \frac{1}{2}\right) \ln z - z + \frac{B_2}{2z} + \cdots + \frac{B_{2n}}{2n(2n-1)z^{2n-1}} + \cdots, \quad (10.52)$$

in which C_2 is another constant of integration. To fix C_2 it is convenient to use the doubling or Legendre duplication formula derived in Section 10.4,

$$z!(z - \tfrac{1}{2})! = 2^{-2z}\pi^{1/2}(2z)! \quad (10.53)$$

This may be proved directly when z is a positive integer by writing $(2z)!$ as a product of even terms times a product of odd terms and extracting a factor of two from each term (Exercise 10.3.5). Substituting Eq. 10.52 into the logarithm of the doubling formula, we find that C_2 is

$$C_2 = \tfrac{1}{2} \ln 2\pi, \quad (10.54)$$

giving

$$\ln(z!) = \frac{1}{2}\ln 2\pi + \left(z + \frac{1}{2}\right) \ln z - z + \frac{1}{12z} - \frac{1}{360z^3} + \frac{1}{1260z^5} - \cdots. \quad (10.55)$$

This is Stirling's series, an asymptotic expansion. The absolute value of the error is less than the absolute value of the first term neglected.

The constants of integration C_1 and C_2 may also be evaluated by comparison with the first term of the series expansion obtained by the method of "steepest descent". This is carried out in Section 7.4.

To help convey a feeling of the remarkable precision of Stirling's series for $s!$ the ratio of the first term of Stirling's approximation to $s!$ is plotted in Fig. 10.5.

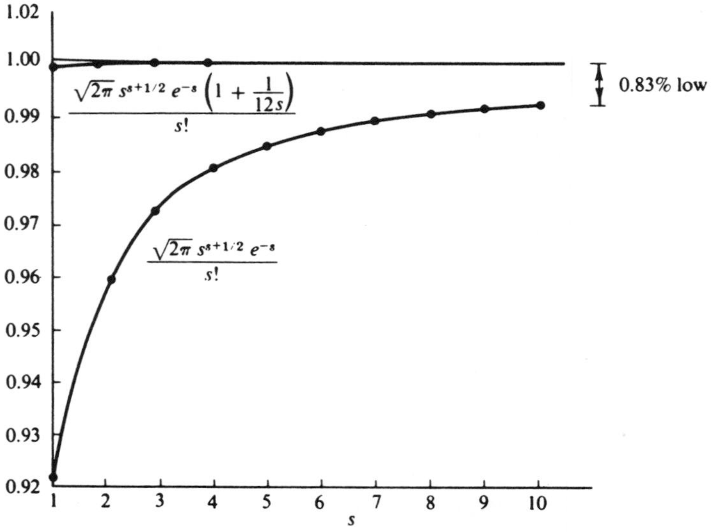

FIG. 10.5 Accuracy of Stirling's formula

A tabulation gives the ratio of the first term in the expansion to $s!$ and the ratio of the first two terms in the expansion to $s!$. The derivation of these forms is Exercise 10.3.1.

TABLE 10.2

s	$\dfrac{\sqrt{2\pi}\, s^{s+1/2} e^{-s}}{s!}$	$\dfrac{\sqrt{2\pi}\, s^{s+1/2} e^{-s}[1 + \frac{1}{12s}]}{s!}$
1	0.92213	0.99898
2	0.95950	0.99949
3	0.97270	0.99972
4	0.97942	0.99983
5	0.98349	0.99988
6	0.98621	0.99992
7	0.98817	0.99994
8	0.98964	0.99995
9	0.99078	0.99996
10	0.99170	0.99998

Numerical Computation. The possibility of using the Maclaurin expansion, Eq. 10.44c, for the numerical evaluation of the factorial function is mentioned in Section 10.2. However, for large x, Stirling's series, Eq. 10.55, gives much more rapid convergence. The *Table of the Gamma Function for Complex Arguments*, National Bureau of Standards, Applied Mathematics Series No. 34, is based on the use of Stirling's series for $z = x + iy$, $9 \leqslant x \leqslant 10$. Lower values of x are reached with the recurrence relation, Eq. 10.29. Now, suppose the numerical value of $x!$ is needed for some particular value of x in a program in a large, highspeed digital computer. How shall we instruct the computer to compute $x!$? Stirling's series followed by the recurrence relation is a good possibility. An even better possibility is to fit $x!$, $0 \leqslant x \leqslant 1$, by a short power series (polynomial) and then calculate $x!$ directly from this empirical fit. Presumably, you have told the computing machine the values of the coefficients of your polynomial. Such polynomial fits have been made by Hastings[1] for various accuracy requirements. For example,

$$x! = 1 + \sum_{n=1}^{8} b_n x^n + \varepsilon(x), \tag{10.56a}$$

with

$$\begin{aligned} b_1 &= -0.57719\,1652 & b_5 &= -0.75670\,4078 \\ b_2 &= 0.98820\,5891 & b_6 &= 0.48219\,9394 \\ b_3 &= -0.89705\,6937 & b_7 &= -0.19352\,7818 \\ b_4 &= 0.91820\,6857 & b_8 &= 0.03586\,8343 \end{aligned} \tag{10.56b}$$

with the magnitude of the error $|\varepsilon(x)| < 3 \times 10^{-7}$, $0 \leqslant x \leqslant 1$.

[1] *Approximations for Digital Computers*, C. Hastings, Jr. Princeton, New Jersey: Princeton Univ. Press (1955).

This is *not* a least squares fit. Hastings employed a Chebyshev polynomial technique similar to that described in Section 13.3 to minimize the *maximum* value of $|\varepsilon(x)|$.

EXERCISES

10.3.1 Rewrite Stirling's series to give $z!$ instead of $\ln(z!)$.

$$\textit{Ans}:\ z! = \sqrt{2\pi}\, z^{z+1/2}e^{-z}\left(1 + \frac{1}{12z} + \frac{1}{288z^2} - \frac{139}{51{,}840z^3} + \cdots\right).$$

10.3.2 Use Stirling's formula to estimate 52!, the number of possible rearrangements of cards in a standard deck of playing cards.

10.3.3 By integrating Eq. 10.51 from $z-1$ to z and then letting $z \to \infty$, evaluate the constant C_1 in the asymptotic series for the digamma function $F(z)$.

10.3.4 Show that the constant C_2 in Stirling's formula equals $\frac{1}{2}\ln 2\pi$ by using the logarithm of the doubling formula.

10.3.5 By direct expansion verify the doubling formula for $z = n + \frac{1}{2}$; n is an integer.

10.3.6 Without using Stirling's series show that

(a) $\ln(n!) < \int_1^{n+1} \ln x\, dx,$

(b) $\ln(n!) > \int_1^{n} \ln x\, dx; \qquad n$ is an integer $\geqslant 2$.

Notice that the arithmetic mean of these two integrals gives a good approximation for Stirling's series.

10.3.7 Test for convergence

$$\sum_{p=0}^{\infty}\left[\frac{(p-\frac{1}{2})!}{p!}\right]^2 \times \frac{2p+1}{2p+2} = \pi \sum_{p=0}^{\infty} \frac{(2p-1)!!(2p+1)!!}{(2p)!!(2p+2)!!}.$$

This series arises in an attempt to describe the magnetic field created by and enclosed by a current loop.

10.3.8 Show that

$$\lim_{x\to\infty} x^{b-a}\frac{(x+a)!}{(x+b)!} = 1.$$

10.3.9 Show that

$$\lim_{n\to\infty} \frac{(2n-1)!!}{(2n)!!}\, n^{1/2} = \pi^{-1/2}.$$

10.4 The Beta Function

Using the integral definition (Eq. 10.25), we write the product of two factorials as the product of two integrals. To facilitate a change in variables, the integrals are taken over a finite range.

$$m!n! = \lim_{a^2\to\infty} \int_0^{a^2} e^{-u}u^m\,du \int_0^{a^2} e^{-v}v^n\,dv, \qquad \begin{array}{l}\mathcal{R}(m) > -1,\\ \mathcal{R}(n) > -1.\end{array} \tag{10.57a}$$

Replacing u with x^2 and v with y^2, we obtain

$$m!n! = \lim_{a\to\infty} 4\int_0^a e^{-x^2}x^{2m+1}\,dx \int_0^a e^{-y^2}y^{2n+1}\,dy. \tag{10.57b}$$

Transforming to polar coordinates gives us

$$\begin{aligned} m!n! &= \lim_{a\to\infty} 4\int_0^a e^{-r^2}r^{2m+2n+3}\,dr \int_0^{\pi/2} \cos^{2m+1}\theta \sin^{2n+1}\theta\,d\theta \\ &= (m+n+1)!\,2\int_0^{\pi/2} \cos^{2m+1}\theta \sin^{2n+1}\theta\,d\theta. \end{aligned} \tag{10.58}$$

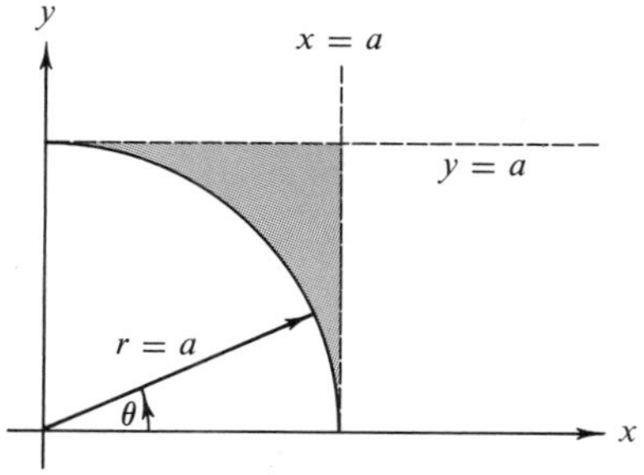

FIG. 10.6 Transformation from cartesian to polar coordinates

The definite integral, together with the factor 2, has been named the beta function

$$\begin{aligned} B(m+1, n+1) &\equiv 2\int_0^{\pi/2} \cos^{2m+1}\theta \sin^{2n+1}\theta\,d\theta \\ &= \frac{m!n!}{(m+n+1)!} = B(n+1, m+1). \end{aligned} \tag{10.59a}$$

Equivalently, in terms of the gamma function

$$B(p, q) = \frac{\Gamma(p)\Gamma(q)}{\Gamma(p+q)}. \tag{10.59b}$$

The only reason for choosing $m + 1$ and $n + 1$, rather than m and n, as the arguments of B is to be in agreement with the conventional, historical beta function. Clearly the extra 1's are about as useful as the human appendix.

In this manipulation the transformation from cartesian to polar coordinates needs some justification. As seen in Fig. 10.6 the shaded area is being neglected. However, the maximum value of the integrand in this region is $e^{-a^2}a^{2m+2n+3}$ which vanishes so strongly as a approaches infinity that the integral over the neglected region vanishes.

Definite integrals, alternate forms. One immediate application of the beta function is in the evaluation of a wide variety of definite integrals having the form of Eq. 10.59*a*. The substitution $t = \cos^2\theta$ converts the beta function to

$$B(m + 1, n + 1) = \int_0^1 t^m(1 - t)^n \, dt, \tag{10.60}$$

which is almost identical to the integral in Eq. 10.9[1] The further substitution $t = u/(1 + u)$ yields still another useful form,

$$B(m + 1, n + 1) = \int_0^\infty \frac{u^m}{(1 + u)^{m+n+2}} \, du. \tag{10.61}$$

The beta function as a definite integral is useful in establishing integral representations of the Bessel function (Ex. 11.1.18) and the hypergeometric function (Ex. 13.4.7).

Verification of $\pi a/\sin \pi a$ relation. If we take $m = a$, $n = -a$, $-1 < a < 1$, then

$$\int_0^\infty \frac{u^a}{(1 + u)^2} \, du = a!(-a)! \tag{10.62}$$

By contour integration this integral may be shown to be equal to $\pi a/\sin \pi a$ (Ex. 7.2.14) thus providing another method of obtaining Eq. 10.32.

Derivation of Legendre duplication formula. The form of Eq. 10.59 suggests that the beta function may be useful in deriving the doubling formula used in the preceding section. For $\mathscr{R}(z) > -1$,

$$\frac{z!z!}{(2z + 1)!} = \int_0^1 t^z(1 - t)^z \, dt, \tag{10.63}$$

and by substituting $t = (1 + s)/2$

$$\begin{aligned} \frac{z!z!}{(2z + 1)!} &= 2^{-2z-1} \int_{-1}^1 (1 - s^2)^z \, ds \\ &= 2^{-2z} \int_0^1 (1 - s^2)^z \, ds; \end{aligned} \tag{10.64}$$

the last equality holds because the integrand is even. To throw this integral back into the form of Eq. 10.60 we substitute $u = s^2$ and

[1] The Laplace transform convolution theorem provides an alternate derivation of Eq. 10.60, cf. Ex. 15.10.2.

$$\frac{z!z!}{(2z+1)!} = 2^{-2z-1}\int_0^1 (1-u)^z u^{-1/2}\,du$$
$$= 2^{-2z-1}\frac{z!(-\frac{1}{2})!}{(z+\frac{1}{2})!} \tag{10.65}$$

Rearranging terms and recalling that $(-\frac{1}{2})! = \pi^{1/2}$, we quickly reduce these equations to one form of the Legendre duplication formula,

$$z!(z+\tfrac{1}{2})! = 2^{-2z-1}\pi^{1/2}(2z+1)! \tag{10.66a}$$

Dividing by $(z+\frac{1}{2})$, we obtain an alternate form of the duplication formula.

$$z!(z-\tfrac{1}{2})! = 2^{-2z}\pi^{1/2}(2z)! \tag{10.66b}$$

Although the integrals used in this derivation are defined only for $\mathscr{R}(z) > -1$, the results (Eqs. 10.66*a* and 10.66*b*) hold for all z by analytic continuation.[1]

Using the result of Ex. 10.1.12, Eq. 10.66*a* may be rewritten (with $z = n$, an integer) as

$$(n+\tfrac{1}{2})! = \pi^{1/2}(2n+1)!!/2^{n+1}. \tag{10.66c}$$

This is often convenient for eliminating factorials of fractions.

Incomplete beta function. Just as there is an incomplete gamma function (Section 10.5), so is there an incomplete beta function,

$$B_x(p,q) = \int_0^x t^{p-1}(1-t)^{q-1}\,dt, \qquad \begin{array}{l} 0 \leqslant x \leqslant 1 \\ p > 0 \\ q > 0 \text{ (if } x = 1). \end{array} \tag{10.67}$$

Clearly $B_{x=1}(p,q)$ becomes the regular (complete) beta function, Eq. 10.60. A power series expansion of $B_x(p,q)$ is the subject of Ex. 5.2.18 and 5.7.9. The relation to hypergeometric functions appears in Section 13.4.

EXERCISES

10.4.1 Derive the doubling formula for the factorial function by integrating $(\sin 2\theta)^{2n+1} = (2\sin\theta\cos\theta)^{2n+1}$ (and using the beta function).

[1] If $2z$ is a negative integer, we get the valid but unilluminating result $\infty = \infty$.

10.4.2 Verify the following beta function identities:

(a) $B(a, b) = B(a+1, b) + B(a, b+1)$,

(b) $B(a, b) = \dfrac{a+b}{b} B(a, b+1)$,

(c) $B(a, b) = \dfrac{b-1}{a} B(a+1, b-1)$,

(d) $B(a, b)\, B(a+b, c) = B(b, c)\, B(a, b+c)$.

10.4.3 (a) Show that

$$\int_{-1}^{1} (1-x^2)^{1/2} x^{2n}\, dx = \begin{cases} \pi/2, & n = 0 \\ \pi \dfrac{(2n-1)!!}{(2n+2)!!}, & n = 1, 2, 3 \ldots. \end{cases}$$

(b) Show that

$$\int_{-1}^{1} (1-x^2)^{-1/2} x^{2n}\, dx = \begin{cases} \pi, & n = 0 \\ \pi \dfrac{(2n-1)!!}{(2n)!!}, & n = 1, 2, 3 \ldots. \end{cases}$$

10.4.4 Show that

$$\int_{-1}^{1} (1-x^2)^n\, dx = \begin{cases} 2^{2n+1} \dfrac{n!\, n!}{(2n+1)!}, & n > -1 \\ 2 \dfrac{(2n)!!}{(2n+1)!!}, & n = 0, 1, 2 \ldots. \end{cases}$$

10.4.5 Evaluate $\int_{-1}^{1} (1+x)^a (1-x)^b\, dx$ in terms of the beta function.

Ans. $2^{a+b+1} B(a+1, b+1)$.

10.4.6 Show, by means of the beta function, that

$$\int_t^z \frac{dx}{(z-x)^{1-\alpha}(x-t)^\alpha} = \frac{\pi}{\sin \pi\alpha}, \qquad 0 < \alpha < 1.$$

This result is used in Section 16.2 to solve Abel's generalized integral equation.

10.4.7 Show that the Dirichlet integral

$$\iint x^p y^q\, dA = \frac{p!q!}{(p+q+2)!} = \frac{B(p+1, q+1)}{p+q+2},$$

where the range of integration is the triangle bounded by the positive x- and y-axes and the line $x + y = 1$.

10.4.8 Show that

$$\int_0^\infty \int_0^\infty e^{-(x^2+y^2+2xy\cos\theta)}\, dx\, dy = \frac{\theta}{2\sin\theta}.$$

What are the limits on θ?

Hint. Consider oblique xy coordinates.

Ans. $-\pi < \theta < \pi$.

10.4.9 Evaluate (using the beta function)

(a) $\displaystyle\int_0^{\pi/2} \cos^{1/2}\theta\, d\theta = \frac{(2\pi)^{3/2}}{16[(\frac{1}{4})!]^2}$,

(b) $$\int_0^{\pi/2} \cos^n \theta \, d\theta = \int_0^{\pi/2} \sin^n \theta \, d\theta = \frac{\sqrt{\pi}[(n-1)/2]!}{2(n/2)!} = \begin{cases} \dfrac{(n-1)!!}{n!!} & \text{for } n \text{ odd}, \\[2ex] \dfrac{\pi}{2} \cdot \dfrac{(n-1)!!}{n!!} & \text{for } n \text{ even}. \end{cases}$$

10.4.10 From

$$J_\nu(z) = \sum_{s=0}^{\infty} (-1)^s \frac{1}{s!(s+\nu)!} \left(\frac{z}{2}\right)^{2s+\nu}$$

and

$$B(m+1, n+1) = 2 \int_0^{\pi/2} \sin^{2m+1} \theta \cos^{2n+1} \theta \, d\theta,$$

show that

$$J_\nu(z) = \frac{2}{\pi^{1/2}(\nu - \frac{1}{2})!} \left(\frac{z}{2}\right)^\nu \int_0^{\pi/2} \sin^{2\nu} \theta \cos(z \cos \theta) \, d\theta, \qquad \mathscr{R}(\nu) > -\tfrac{1}{2}.$$

Since $J_\nu(z)$ is a Bessel function (Section 11.1), a useful integral representation of the Bessel function is established.

10.4.11 Given that the associated Legendre polynomial $P_m^m(x) = (2m-1)!!\,(1-x^2)^{m/2}$, Section 12.5, show that

(a) $$\int_{-1}^{1} [P_m^m(x)]^2 \, dx = \frac{2}{(2m+1)} (2m)!, \qquad m = 0, 1, 2, \ldots.$$

(b) $$\int_{-1}^{1} [P_m^m(x)]^2 \frac{dx}{1-x^2} = 2 \cdot (2m-1)!, \qquad m = 1, 2, 3, \ldots.$$

10.4.12 Show that

(a) $$\int_0^1 (x^2)^{s+\frac{1}{2}} (1-x^2)^{-1/2} \, dx = \frac{(2s)!!}{(2s+1)!!},$$

(b) $$\int_0^1 (x^2)^p (1-x^2)^q \, dx = \frac{1}{2} \frac{(p-\frac{1}{2})!\, q!}{(p+q+\frac{1}{2})!}.$$

10.5 The Incomplete Gamma Functions and Related Functions

Generalizing the Euler definition of the gamma function (Eq. 10.5), we define the incomplete gamma functions by the *variable* limit integrals

$$\gamma(a, x) = \int_0^x e^{-t} t^{a-1} \, dt, \qquad \mathscr{R}(a) > 0$$

and

$$\Gamma(a, x) = \int_x^\infty e^{-t} t^{a-1} \, dt. \tag{10.68}$$

Clearly, the two functions are related, for

$$\gamma(a, x) + \Gamma(a, x) = \Gamma(a). \tag{10.69}$$

The choice of employing $\gamma(a, x)$ or $\Gamma(a, x)$ is purely a matter of convenience. If the parameter a is a positive integer, Eqs. 10.68 may be integrated completely to yield

$$\begin{aligned} \gamma(n, x) &= (n-1)!\left(1 - e^{-x}\sum_{s=0}^{n-1}\frac{x^s}{s!}\right) \\ \Gamma(n, x) &= (n-1)!\, e^{-x}\sum_{s=0}^{n-1}\frac{x^s}{s!}, \qquad n = 1, 2, \ldots. \end{aligned} \tag{10.70}$$

For nonintegral a a power series expansion of $\gamma(a, x)$ for small x and an asymptotic expansion of $\Gamma(a, x)$ are developed in Sections 5.6 and 5.11.

$$\begin{aligned} \gamma(a, x) &= x^a \sum_{n=0}^{\infty}(-1)^n \frac{x^n}{n!(a+n)}, \\ \Gamma(a, x) &= x^{a-1}e^{-x}\sum_{n=0}^{\infty}\frac{(a-1)!}{(a-1-n)!}\cdot\frac{1}{x^n} \\ &= x^{a-1}e^{-x}\sum_{n=0}^{\infty}(-1)^n\frac{(n-a)!}{(-a)!}\cdot\frac{1}{x^n}. \end{aligned} \tag{10.71}$$

These incomplete gamma functions may also be expressed quite elegantly in terms of confluent hypergeometric functions (cf. Section 13.5).

Exponential integral. Although the incomplete gamma function $\Gamma(a, x)$ in its general form (Eq. 10.68) is only infrequently encountered in physical problems, a special case is quite common and very useful. We define the exponential integral by[1]

$$-Ei(-x) \equiv \int_x^{\infty}\frac{e^{-t}}{t}\,dt = E_1(x). \tag{10.72}$$

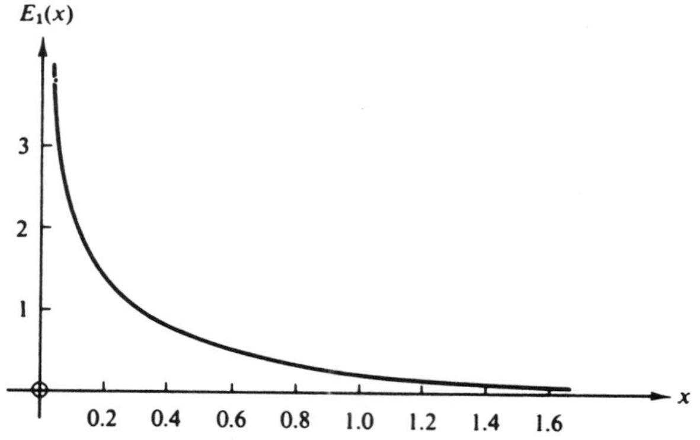

FIG. 10.7 The exponential integral, $E_1(x) = -Ei(-x)$

[1] The appearance of the two minus signs in $-Ei(-x)$ is an historical monstrosity. AMS-55 denotes this integral as $E_1(x)$.

Then

$$E_1(x) = \Gamma(0, x)$$
$$= \lim_{a\to 0}[\Gamma(a) - \gamma(a, x)]. \tag{10.73}$$

Caution is needed here, for the integral in Eq. 10.72 diverges logarithmically as $x \to 0$. We may split the divergent term in the series expansion for $\gamma(a, x)$,

$$E_1(x) = \lim_{a\to 0} \left[\frac{a\Gamma(a) - x^a}{a}\right] - \sum_{n=1}^{\infty} \frac{(-1)^n x^n}{n \cdot n!}, \tag{10.74}$$

Using l'Hospital's rule and

$$\frac{d}{da}\{a\Gamma(a)\} = \frac{d}{da} a! = a!\,F(a) \tag{10.74a}$$

and then Eq. 10.40,

$$E_1(x) = -\gamma - \ln x - \sum_{n=1}^{\infty} \frac{(-1)^n x^n}{n \cdot n!}, \tag{10.75}$$

useful for small x. An asymptotic expansion is given in Section 5.11.

Further special forms related to the exponential integral are the sine integral, cosine integral, and logarithmic integral defined by[1]

$$si(x) = -\int_x^{\infty} \frac{\sin t}{t}\,dt$$
$$Ci(x) = -\int_x^{\infty} \frac{\cos t}{t}\,dt \tag{10.76}$$
$$li(x) = \int_0^x \frac{du}{\ln u} = Ei(\ln x).$$

By transforming from real to imaginary argument, we can show that

$$si(x) = \frac{1}{2i}[Ei(ix) - Ei(-ix)] = \frac{1}{2i}[E_1(ix) - E_1(-ix)], \tag{10.77}$$

whereas

$$Ci(x) = \tfrac{1}{2}[Ei(ix) + Ei(-ix)] = -\tfrac{1}{2}[E_1(ix) + E_1(-ix)], \quad |\arg x| < \frac{\pi}{2}. \tag{10.78}$$

Adding these two relations, we obtain

$$Ei(ix) = Ci(x) + i\,si(x), \tag{10.79}$$

to show that the relation among these integrals is exactly analogous to that among e^{ix}, $\cos x$, and $\sin x$.

[1] Another sine integral is given by $Si(x) = si(x) + \pi/2$.

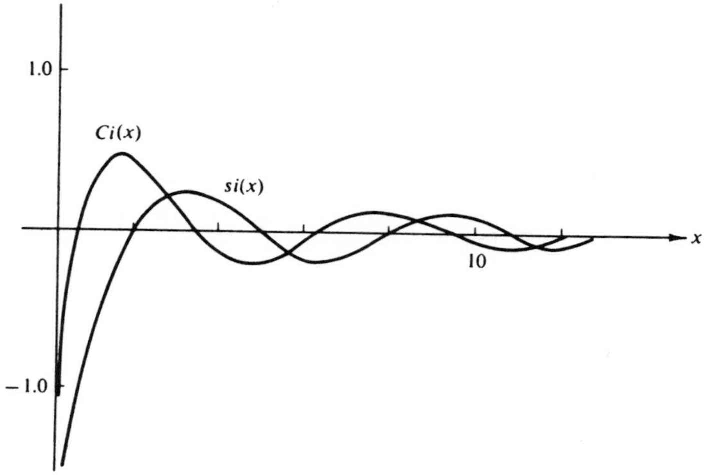

FIG. 10.8 Sine and cosine integrals

Asymptotic expansions of $Ci(x)$ and $si(x)$ are developed in Section 5.11. Power series expansions about the origin for $Ci(x)$, $si(x)$, and $li(x)$ may be obtained from those for the exponential integral, $E_1(x)$ or by direct integration, Ex. 10.5.10.

Error integrals. The error integrals

$$\operatorname{erf} z = \frac{2}{\sqrt{\pi}} \int_0^z e^{-t^2}\, dt,$$
$$\operatorname{erfc} z = \frac{2}{\sqrt{\pi}} \int_z^\infty e^{-t^2}\, dt, \tag{10.80a}$$

are introduced in Section 5.11. Asymptotic forms are developed there. From the general form of the integrands and Eq. 10.6, we expect that erf z and erfc z may be written as incomplete gamma functions with $a = \frac{1}{2}$. The relations are

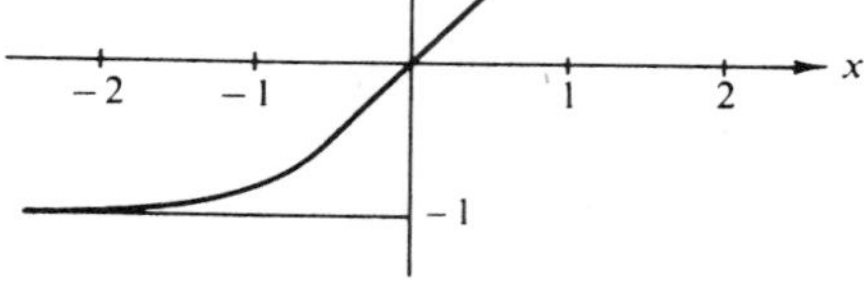

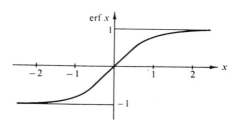

FIG. 10.9 Error function, erf x

$$\begin{aligned} \operatorname{erf} z &= \pi^{-1/2}\gamma(\tfrac{1}{2}, z^2) \\ \operatorname{erfc} z &= \pi^{-1/2}\Gamma(\tfrac{1}{2}, z^2). \end{aligned} \tag{10.80b}$$

The power series expansion of erf z follows directly from Eq. 10.71.

EXERCISES

10.5.1 Show that

$$\gamma(a, x) = e^{-x} \sum_{n=0}^{\infty} \frac{(a-1)!}{(a+n)!} x^{a+n}.$$

(a) By repeatedly integrating by parts.
(b) Demonstrate this relation by transforming it into Eq. 10.71.

10.5.2 Show that

(a) $\dfrac{d^m}{dx^m}[x^{-a}\gamma(a, x)] = (-1)^m x^{-a-m}\gamma(a+m, x),$

(b) $\dfrac{d^m}{dx^m}[e^x \gamma(a, x)] = e^x \dfrac{\Gamma(a)}{\Gamma(a-m)}\gamma(a-m, x).$

10.5.3 Show that $\gamma(a, x)$ and $\Gamma(a, x)$ satisfy the recurrence relations

(a) $\gamma(a+1, x) = a\,\gamma(a, x) - x^a e^{-x},$
(b) $\Gamma(a+1, x) = a\,\Gamma(a, x) + x^a e^{-x}.$

10.5.4 The potential produced by a 1s hydrogen electron is (Ex. 12.8.6) given by

$$V(\mathbf{r}) = \frac{q}{4\pi\varepsilon_0 a_0}\left\{\frac{1}{2r}\gamma(3, 2r) + \Gamma(2, 2r)\right\}.$$

(a) For $r \ll 1$ show that

$$V(\mathbf{r}) = \frac{q}{4\pi\varepsilon_0 a_0}\left\{1 - \frac{2}{3}r^2 + \cdots\right\}.$$

(b) For $r \gg 1$ show that

$$V(\mathbf{r}) = \frac{q}{4\pi\varepsilon_0 a_0}\cdot\frac{1}{r}.$$

Note. For computation at intermediate values of r, Eqs. 10.70 are convenient.

10.5.5 The potential of a 2p hydrogen electron is found to be (Ex. 12.8.7)

$$\begin{aligned} V(\mathbf{r}) = {} & \frac{1}{4\pi\varepsilon_0}\cdot\frac{q}{24a_0}\left\{\frac{1}{r}\gamma(5, r) + \Gamma(4, r)\right\} \\ & - \frac{1}{4\pi\varepsilon_0}\cdot\frac{q}{120a_0}\left\{\frac{1}{r^3}\gamma(7, r) + r^2\Gamma(2, r)\right\} P_2(\cos\theta). \end{aligned}$$

Here r is expressed in units of a_0, the Bohr radius. $P_2(\cos\theta)$ is a Legendre polynomial (Section 12.1).

(a) For $r \ll 1$, show that

$$V(\mathbf{r}) = \frac{1}{4\pi\varepsilon_0} \cdot \frac{q}{a_0} \left\{ \frac{1}{4} - \frac{1}{120} r^2 P_2(\cos\theta) + \cdots \right\}.$$

(b) For $r \gg 1$, show that

$$V(\mathbf{r}) = \frac{1}{4\pi\varepsilon_0} \cdot \frac{q}{a_0 r} \left\{ 1 - \frac{6}{r^2} P_2(\cos\theta) + \cdots \right\}.$$

10.5.6 Prove that the exponential integral

$$\int_x^\infty \frac{e^{-t}}{t}\, dt = -\gamma - \ln x - \sum_{n=1}^\infty \frac{(-1)^n x^n}{n \cdot n!}.$$

γ is the Euler-Mascheroni constant.

10.5.7 Show that $E_1(z)$ may be written

$$E_1(z) = e^{-z} \int_0^\infty \frac{e^{-zt}}{1+t}\, dt.$$

Show also that we must impose the condition $|\arg z| \le \pi/2$.

10.5.8 Related to the exponential integral (Eq. 10.72) by a simple change of variable is the function

$$E_n(x) = \int_1^\infty \frac{e^{-xt}}{t^n}\, dt.$$

Show that $E_n(x)$ satisfies the recurrence relation

$$E_{n+1}(x) = \frac{1}{n} e^{-x} - \frac{x}{n} E_n(x), \qquad n = 1, 2, 3, \ldots.$$

10.5.9 With $E_n(x)$ defined in Exercise 10.5.8 show that $E_n(0) = 1/(n-1)$, $n > 1$.

10.5.10 Develop the following power series expansions

(a) $$si(x) = -\frac{\pi}{2} + \sum_{n=0}^\infty \frac{(-1)^n x^{2n+1}}{(2n+1)(2n+1)!}$$

(b) $$Ci(x) = \gamma + \ln x + \sum_{n=1}^\infty \frac{(-1)^n x^{2n}}{2n(2n)!}.$$

10.5.11 An analysis of a center-fed linear antenna leads to the expression

$$\int_0^x \frac{1 - \cos t}{t}\, dt.$$

Show that this is equal to

$$\gamma + \ln x - Ci(x).$$

10.5.12 Using the relation

$$\Gamma(a) = \gamma(a, x) + \Gamma(a, x)$$

show that if $\gamma(a, x)$ satisfies the relations of Ex. 10.5.2, then $\Gamma(a, x)$ must satisfy the same relations.

REFERENCES

AMS-55, Handbook of Mathematical Functions with Formulas, Graphs, and Mathematical Tables, U.S. Department of Commerce, National Bureau of Standards, Applied Mathematics Series-55, M. Abramowitz and I. A. Stegun, Eds. Contains a wealth of information about gamma functions, incomplete gamma functions, exponential integrals, error functions, and related functions—Chapters 4–6.

ARTIN, Emil, *The Gamma Function.* (Translated by Michael Butler). New York: Holt, Rinehart and Winston (1964). Demonstrates that if a function $f(x)$ is smooth (log convex) and equal to $(n - 1)!$ when $x = n$, it is the gamma function.

DAVIS, H. T., *Tables of the Higher Mathematical Functions.* Bloomington, Indiana: Principia Press (1933). Volume I contains extensive information on the gamma function and the polygamma functions.

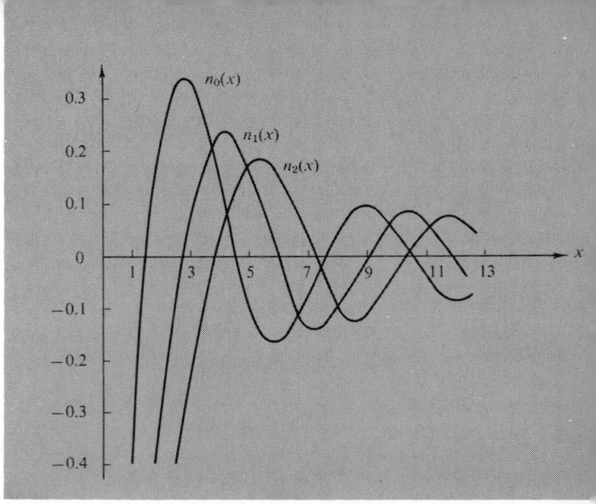

CHAPTER 11

BESSEL FUNCTIONS

11.1 Bessel Functions of the First Kind, $J_\nu(x)$

Generating function, integral order, $J_n(x)$. Bessel functions appear in a wide variety of physical problems. In Section 2.6, separation of the Helmholtz or wave equation in circular cylindrical coordinates led to Bessel's equation. In Section 11.7 it will be seen that the Helmholtz equation in spherical polar coordinates also leads to a form of Bessel's equation. Although Bessel functions are of interest primarily as solutions of a differential equation, it is instructive and convenient to develop them from a completely different approach, that of the generating function. This approach also has the advantage of focusing attention on the functions themselves rather than on the differential equations they happen to satisfy. Let us introduce a function of two variables,

$$g(x, t) = e^{(x/2)(t-1/t)}. \tag{11.1}$$

Series form. Expanding this generating function in a Laurent series (Section 6.5), we obtain

$$e^{(x/2)(t-1/t)} = \sum_{n=-\infty}^{\infty} J_n(x)t^n, \tag{11.2}$$

and $J_n(x)$, which is the coefficient of t^n, is labeled a Bessel function of the first kind of integral order n.

Expanding the exponentials we have a product of Maclaurin series in $xt/2$ and $-x/2t$, respectively,

$$e^{xt/2} \cdot e^{-x/2t} = \sum_{r=0}^{\infty} \left(\frac{x}{2}\right)^r \frac{t^r}{r!} \sum_{s=0}^{\infty} (-1)^s \left(\frac{x}{2}\right)^s \frac{t^{-s}}{s!}. \tag{11.3}$$

For a given s we get t^n ($n \geqslant 0$) from

$$\left(\frac{x}{2}\right)^{n+s} \frac{t^{n+s}}{(n+s)!} (-1)^s \left(\frac{x}{2}\right)^s \frac{t^{-s}}{s!} \tag{11.4}$$

The coefficient of t^n is then[1]

$$J_n(x) = \sum_{s=0}^{\infty} \frac{(-1)^s}{s!(n+s)!}\left(\frac{x}{2}\right)^{n+2s} = \frac{x^n}{2^n n!} - \cdots. \tag{11.5}$$

This series form exhibits the behavior of the Bessel function $J_n(x)$ for small x and permits numerical evaluation of $J_n(x)$. The results for J_0, J_1, and J_2 are shown in Fig. 11.1. From Section 5.3, the error in using only a finite number of terms in numerical evaluation is less than the first term omitted. For instance, if we want $J_n(x)$ to $\pm 1\%$ accuracy, the first term alone of Eq. 11.5 will suffice, provided the ratio of the second term to the first is less than 1% (in magnitude) or $x < 0.2(n+1)^{1/2}$. The Bessel functions oscillate but are *not* periodic—except in the limit as $x \to \infty$ (Section 11.6). The amplitude of $J_n(x)$ is not constant but decreases asymptotically as $x^{-1/2}$.

Equation 11.5 actually holds for $n < 0$, also giving

$$J_{-n}(x) = \sum_{s=0}^{\infty} \frac{(-1)^s}{s!(s-n)!}\left(\frac{x}{2}\right)^{2s-n}, \tag{11.6}$$

which amounts to replacing n by $-n$ in Eq. 11.5. Since n is an integer (here),

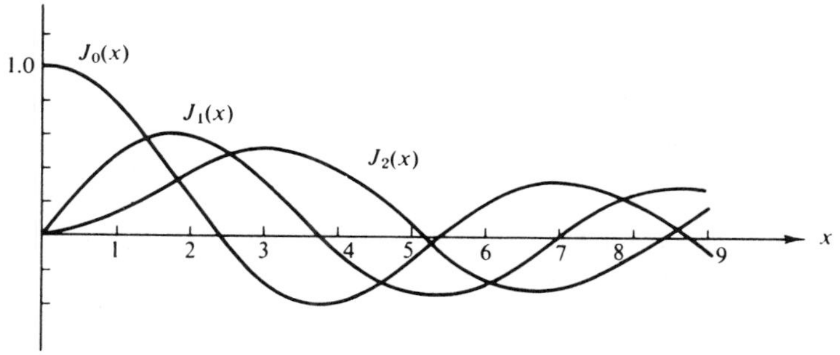

FIG. 11.1 Bessel functions, $J_0(x)$, $J_1(x)$, and $J_2(x)$

$(s-n)! \to \infty$ for $s = 0, \ldots, (n-1)$. Hence the series may be considered to start with $s = n$. Replacing s by $s+n$, we obtain

$$J_{-n}(x) = \sum_{s=0}^{\infty} \frac{(-1)^{s+n}}{s!(n+s)!}\left(\frac{x}{2}\right)^{n+2s}, \tag{11.7}$$

showing immediately that $J_n(x)$ and $J_{-n}(x)$ are not independent but are related by

$$J_{-n}(x) = (-1)^n J_n(x), \qquad \text{(integral } n\text{)}. \tag{11.8}$$

These series expressions (Eqs. 11.5 and 11.6) may be used with n replaced by ν to *define* $J_\nu(x)$ and $J_{-\nu}(x)$ for nonintegral ν (cf. Exercise 11.1.7).

[1] From the steps leading to this series and from its convergence characteristics it should be clear that this series may be used with x replaced by z and with z any point in the finite complex plane.

Recurrence relations. The recurrence relations for $J_n(x)$ and its derivatives may all be obtained by operating on the series, Eq. 11.5, though this requires a bit of clairvoyance (or a lot of trial and error). Verification of the known recurrence relations is straightforward, Ex. 11.1.7. Here it is convenient to obtain them from the generating function, $g(x, t)$. Differentiating Eq. 11.1 partially with respect to t, we find that

$$\frac{\partial}{\partial t} g(x, t) = \frac{1}{2} x\left(1 + \frac{1}{t^2}\right) e^{(x/2)(t - 1/t)} \tag{11.9}$$

$$= \sum_{n=-\infty}^{\infty} n\, J_n(x) t^{n-1}$$

and substituting Eq. 11.2 for the exponential and equating the coefficients of like powers of t[1] we obtain

$$J_{n-1}(x) + J_{n+1}(x) = \frac{2n}{x} J_n(x). \tag{11.10}$$

This is a three-term recurrence relation. Given J_0 and J_1, for example, J_2 (and any other integral order J_n) may be computed.

With the opportunities offered by modern digital computers (and the demands they levy) Eq. 11.10 has acquired an interesting new application. In computing a numerical value of $J_N(x_0)$ for a given x_0, one could use Eq. 11.5 for small x, or the asymptotic form, Eq. 11.144 of Section 11.6 for large x. A better way, in terms of accuracy and machine utilization, is to use the recurrence relation, Eq. 11.10, and work *down*.[2] With $n \gg N$ and $n \gg x_0$, assume

$$J_{n+1}(x_0) = 0 \qquad \text{and} \qquad J_n(x_0) = \alpha,$$

where α is some *small* number. Then Eq. 11.10 leads to $J_{n-1}(x_0)$, $J_{n-2}(x_0)$, etc., and finally to $J_0(x_0)$. Since α is arbitrary, the J_n's are all off by a common factor. This factor is determined by the condition

$$J_0(x_0) + 2 \sum_{m=1}^{\infty} J_{2m}(x_0) = 1. \tag{11.10a}$$

(Set $t = 1$ in Eq. 11.2.) The accuracy of this calculation is checked by trying again at $n' = n + 3$. This technique yields the desired $J_N(x_0)$ and all the lower integral index J's down to J_0.

High-speed, high-precision numerical computation is more or less an art. Modifications and refinements of this and other numerical techniques are being proposed year by year. For information on the current "state of the art" the student will have to go to the literature, and this means primarily to the journal *Mathematics of Computation*.

[1] This depends on the fact that the power series representation is unique (Sections 5.7, 6.5).

[2] I. A. Stegun, M. Abramowitz, MTAC-11, 255–257 (1957).

Differentiating Eq. 11.1 partially with respect to x, we have

$$\frac{\partial}{\partial x} g(x, t) = \frac{1}{2}\left(t - \frac{1}{t}\right) e^{(x/2)(t - 1/t)}$$
$$= \sum_{n=-\infty}^{\infty} J_n'(x)t^n. \tag{11.11}$$

Again, substituting in Eq. 11.2 and equating the coefficients of like powers of t, we obtain the result

$$J_{n-1}(x) - J_{n+1}(x) = 2J_n'(x). \tag{11.12}$$

As a special case of this general recurrence relation,

$$J_0'(x) = -J_1(x). \tag{11.13}$$

Adding Eqs. 11.10 and 11.12 and dividing by 2, we have

$$J_{n-1}(x) = \frac{n}{x} J_n(x) + J_n'(x). \tag{11.14}$$

Multiplying by x^n and rearranging terms produces

$$\frac{d}{dx}[x^n J_n(x)] = x^n J_{n-1}(x). \tag{11.15}$$

Subtracting Eq. 11.12 from 11.10 and dividing by 2 leads to

$$J_{n+1}(x) = \frac{n}{x} J_n(x) - J_n'(x). \tag{11.16}$$

Multiplying by x^{-n} and rearranging terms, we obtain

$$\frac{d}{dx}[x^{-n} J_n(x)] = -x^{-n} J_{n+1}(x). \tag{11.17}$$

Bessel's differential equation. Suppose we consider a set of functions $Z_\nu(x)$ which satisfies the basic recurrence relations (Eqs. 11.10 and 11.12), but with ν not necessarily an integer and Z_ν not necessarily given by the series (Eq. 11.5). Equation 11.14 may be rewritten ($n \to \nu$)

$$x Z_\nu'(x) = x Z_{\nu-1}(x) - \nu Z_\nu(x). \tag{11.18}$$

On differentiating with respect to x, we have

$$x Z_\nu''(x) + (\nu + 1)Z_\nu' - x Z_{\nu-1}' - Z_{\nu-1} = 0. \tag{11.19}$$

Multiplying by x and then subtracting Eq. 11.18 multiplied by ν gives us

$$x^2 Z_\nu'' + xZ_\nu' - \nu^2 Z_\nu + (\nu - 1)xZ_{\nu-1} - x^2 Z_{\nu-1}' = 0. \tag{11.20}$$

Now we rewrite Eq. 11.16 and replace n by $\nu - 1$.

$$xZ_{\nu-1}' = (\nu - 1)Z_{\nu-1} - xZ_\nu. \tag{11.21}$$

Using this to eliminate $Z_{\nu-1}$ and $Z'_{\nu-1}$ from Eq. 11.20, we finally get

$$x^2 Z''_\nu + xZ'_\nu + (x^2 - \nu^2)Z_\nu = 0. \tag{11.22}$$

This is just Bessel's equation. Hence *any* functions, $Z_\nu(x)$, that satisfy the recurrence relations (Eqs. 11.10 and 11.12, 11.14 and 11.16, or 11.15 and 11.17) satisfy Bessel's equation; that is, the unknown Z_ν are Bessel functions. In particular, we have shown that the functions $J_n(x)$, defined by our generating function, satisfy Bessel's equation.

If the argument is kx rather than x, Eq. 11.22 becomes

$$x^2 \frac{d^2}{dx^2} Z_\nu(kx) + x\frac{d}{dx} Z_\nu(kx) + (k^2x^2 - \nu^2)Z_\nu(kx) = 0. \tag{11.22a}$$

Integral representation. A particularly useful and powerful way of treating Bessel functions employs integral representations. If we return to the generating function (Eq. 11.2), and substitute $t = e^{i\theta}$,

$$e^{ix\sin\theta} = J_0(x) + 2(J_2(x)\cos 2\theta + J_4(x)\cos 4\theta + \cdots) + 2i(J_1(x)\sin\theta + J_3(x)\sin 3\theta + \cdots), \tag{11.23}$$

in which we have used the relations

$$\begin{aligned} J_1(x)e^{i\theta} + J_{-1}(x)e^{-i\theta} &= J_1(x)(e^{i\theta} - e^{-i\theta}) \\ &= 2iJ_1(x)\sin\theta, \\ J_2(x)e^{2i\theta} + J_{-2}(x)e^{-2i\theta} &= 2J_2(x)\cos 2\theta, \end{aligned} \tag{11.24}$$

and so on.

In summation notation,

$$\begin{aligned} \cos(x\sin\theta) &= J_0(x) + 2\sum_{n=1}^{\infty} J_{2n}(x)\cos(2n\theta), \\ \sin(x\sin\theta) &= 2\sum_{n=1}^{\infty} J_{2n-1}(x)\sin[(2n-1)\theta], \end{aligned} \tag{11.25}$$

equating real and imaginary parts, respectively. By employing the orthogonality properties of cosine and sine,[1]

$$\begin{aligned} \int_0^\pi \cos n\theta \cos m\theta \, d\theta &= \frac{\pi}{2}\delta_{nm} \\ \int_0^\pi \sin n\theta \sin m\theta \, d\theta &= \frac{\pi}{2}\delta_{nm}, \end{aligned} \tag{11.26}$$

in which n and m are *positive* integers (zero is excluded),[2] we obtain

[1] They are eigenfunctions of a self-adjoint equation (linear oscillator equation) and satisfy appropriate boundary conditions cf. Sections 9.2 and 14.1.

[2] Equation 11.26 holds for either $n = 0$ or $m = 0$ but not for both equal to zero.

$$\frac{1}{\pi}\int_0^\pi \cos(x\sin\theta)\cos n\theta\,d\theta = \begin{cases} J_n(x), & n \text{ even}, \\ 0, & n \text{ odd}, \end{cases} \tag{11.27}$$

$$\frac{1}{\pi}\int_0^\pi \sin(x\sin\theta)\sin n\theta\,d\theta = \begin{cases} 0, & n \text{ even}, \\ J_n(x), & n \text{ odd}. \end{cases} \tag{11.28}$$

If these two equations are added together,

$$\begin{aligned} J_n(x) &= \frac{1}{\pi}\int_0^\pi [\cos(x\sin\theta)\cos n\theta + \sin(x\sin\theta)\sin n\theta]\,d\theta \\ &= \frac{1}{\pi}\int_0^\pi \cos(n\theta - x\sin\theta)\,d\theta, \qquad n = 0, 1, 2, 3, \ldots. \end{aligned} \tag{11.29}$$

As a special case

$$J_0(x) = \frac{1}{\pi}\int_0^\pi \cos(x\sin\theta)\,d\theta. \tag{11.30}$$

This integral representation (Eq. 11.29) may be obtained somewhat more directly by employing contour integration (cf. Exercise 11.1.16). Many other integral representations exist (cf. Exercise 11.1.18).

Example 11.1.1. Fraunhofer Diffraction, Circular Aperture

In the theory of diffraction through a circular aperture we encounter the integral

$$\Phi \sim \int_0^a \int_0^{2\pi} e^{ibr\sin\theta}\,d\theta r\,dr \tag{11.31}$$

for Φ, the amplitude of the diffracted wave. Here θ is an azimuth angle in the plane of the circular aperture of radius a, and α is the angle defined by a point on a screen below the circular aperture relative to the normal through the center point. The parameter b is given by

$$b = \frac{2\pi}{\lambda}\sin\alpha, \tag{11.32}$$

with λ the wavelength of the incident wave. The other symbols are defined by Fig. 11.2. Expressing the exponential in trigonometric form and using Eq. 11.30, we obtain[1]

$$\Phi \sim 2\pi\int_0^a J_0(br)r\,dr. \tag{11.33}$$

Equation 11.15 enables us to integrate Eq. 11.33 immediately to obtain

$$\Phi \sim \frac{2\pi ab}{b^2}J_1(ab) \sim \frac{\lambda a}{\sin\alpha}J_1\left(\frac{2\pi a}{\lambda}\sin\alpha\right). \tag{11.34}$$

[1] We could also refer to Ex. 11.1.16(b).

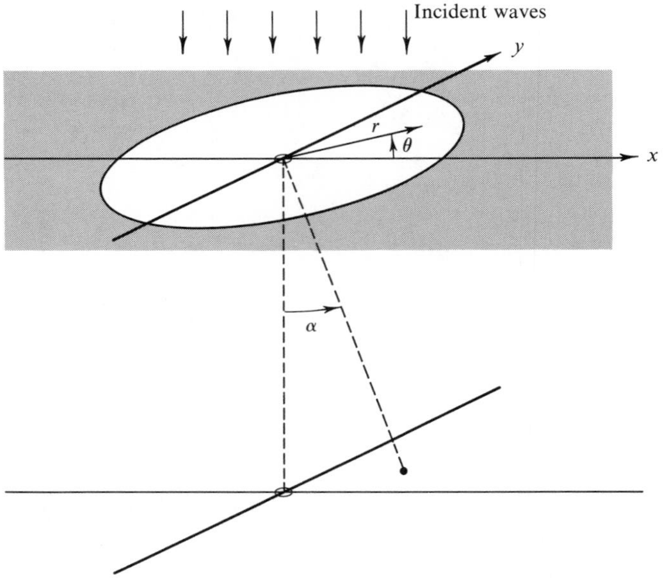

FIG. 11.2 Fraunhofer diffraction—circular aperture

The intensity of the light in the diffraction pattern is proportional to Φ^2 and

$$\Phi^2 \sim \left\{\frac{J_1[(2\pi a/\lambda)\sin\alpha]}{\sin\alpha}\right\}^2. \tag{11.35}$$

From Table 11.1, which lists the zeros of the Bessel functions and their first derivatives, this will have a zero at

$$\frac{2\pi a}{\lambda}\sin\alpha = 3.8317\cdots \tag{11.36}$$

or

$$\sin\alpha = \frac{3.8317\lambda}{2\pi a}. \tag{11.37}$$

For green light $\lambda = 5.5 \times 10^{-5}$ cm. Hence, if $a = 0.5$ cm,

$$\begin{aligned}\alpha \approx \sin\alpha &= 6.7 \times 10^{-5} \text{ (radian)}\\ &\approx 14 \text{ seconds of arc,}\end{aligned} \tag{11.38}$$

which shows that the bending or spreading of the light ray is extremely small. Had this analysis been known in the seventeenth century, the arguments against the wave theory of light would have collapsed.

In mid-twentieth century this same diffraction pattern appears in the scattering of nuclear particles by atomic nuclei—a striking demonstration of the wave properties of the nuclear particles.

Additional roots of the Bessel functions and their first derivatives may be found

TABLE 11.1

Zeros of the Bessel Functions and Their First Derivatives

Number of zero	$J_0(x)$	$J_1(x)$	$J_2(x)$	$J_3(x)$	$J_4(x)$	$J_5(x)$
1	2.4048	3.8317	5.1356	6.3802	7.5883	8.7715
2	5.5201	7.0156	8.4172	9.7610	11.0647	12.3386
3	8.6537	10.1735	11.6198	13.0152	14.3725	15.7002
4	11.7915	13.3237	14.7960	16.2235	17.6160	18.9801
5	14.9309	16.4706	17.9598	19.4094	20.8269	22.2178
	$J_0'(x)$	$J_1'(x)$	$J_2'(x)$	$J_3'(x)$		
1	3.8317	1.8412	3.0542	4.2012		
2	7.0156	5.3314	6.7061	8.0152		
3	10.1735	8.5363	9.9695	11.3459		

in C. L. Beattie, "Table of First 700 Zeros of Bessel Functions," *Bell Tech. J.* **37**, 689 (1958) and Bell Monograph 3055. *Note*. $J_0'(x) = -J_1(x)$.

A further example of the use of Bessel functions and their roots is provided by the electromagnetic resonant cavity.

EXAMPLE 11.1.2 CYLINDRICAL RESONANT CAVITY

In the interior of our resonant cavity, electromagnetic waves oscillate with a time dependence $e^{-i\omega t}$. From Example 2.6.1, the z component (E_z, space part only) satisfies the scalar Helmholtz equation

$$\nabla^2 E_z + \alpha^2 E_z = 0, \tag{11.39}$$

where $\alpha^2 = \omega^2 \varepsilon_0 \mu_0$. Further from Example 2.6.1,

$$(E_z)_{mnk} = \sum_{m,n} J_m(\gamma_{mn}\rho)e^{\pm im\varphi}[a_{mn}\sin kz + b_{mn}\cos kz]. \tag{11.40}$$

For the end surfaces at $z = 0$ and $z = l$ (as in Fig. 11.3), let us set $a_{mn} = 0$, and

$$k = \frac{p\pi}{l}, \quad p = 0, 1, 2, \ldots. \tag{11.41}$$

Maxwell's equations then guarantee that the tangential electric fields E_ρ and E_φ will vanish at $z = 0$ and l. This is the transverse magnetic or TM mode of oscillation. We have (Example 2.6.1)

$$\begin{aligned}\gamma^2 &= \omega^2\varepsilon_0\mu_0 - k^2 \\ &= \omega^2\varepsilon_0\mu_0 - \frac{p^2\pi^2}{l^2}.\end{aligned} \tag{11.42}$$

But there is the usual boundary condition that $E_z(\rho = a) = 0$. Hence, we must set

$$\gamma_{mn} = \frac{\alpha_{mn}}{a} \tag{11.43}$$

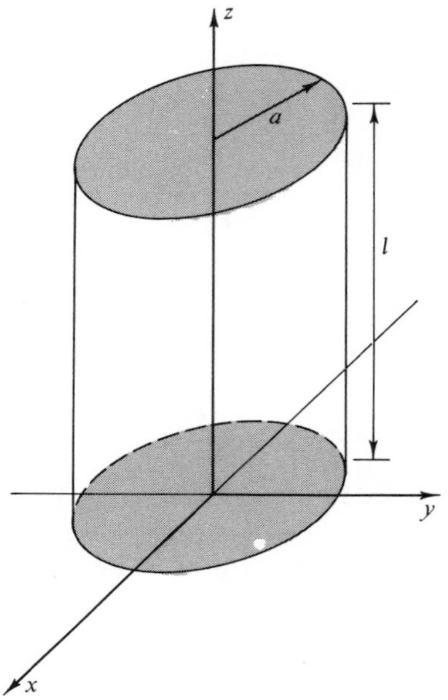

FIG. 11.3 Cylindrical resonant cavity

where α_{mn} is the nth zero of J_m.

The result of the two boundary conditions and the separation constant m^2 is that the angular frequency of our oscillation depends upon three discrete parameters

$$\omega_{mnp} = \frac{1}{\sqrt{\varepsilon_0 \mu_0}} \sqrt{\frac{\alpha_{mn}^2}{a^2} + \frac{p^2\pi^2}{l^2}}, \qquad \begin{cases} m = 0, 1, 2 \cdots \\ n = 1, 2, 3 \cdots \\ p = 0, 1, 2 \cdots. \end{cases} \tag{11.44}$$

These are the allowable resonant frequencies.

Alternate approaches. Bessel functions are introduced here by means of a generating function, Eq. 11.2. Other approaches are possible. Listing the various possibilities, we have:

1. Generating function (black magic), Eq. 11.2.
2. Series solution of Bessel's differential equation, Section 8.4.
3. Contour integrals: Some writers prefer to start with contour integral definitions of the Hankel functions, Sections 7.4 and 11.4, and develop the Bessel function $J_\nu(x)$ from the Hankel functions.
4. Direct solution of physical problems: Example 11.1.1, Fraunhofer diffraction with a circular aperture, illustrates this. Incidentally, Eq. 11.31 can be treated by series expansion, if desired. Feynman[1] develops Bessel functions from a consideration of cavity resonators.

[1] *R. P. Feynman, R. B. Leighton, and M. Sands, The Feynman Lectures on Physics, Vol. II, Chap.* 23. Addison-Wesley, Reading, Massachusetts (1964).

In case the generating function seems too arbitrary, it can be derived from a contour integral, Ex. 11.1.16, or from the Bessel function recurrence relations, Ex. 11.1.6.

Bessel functions of nonintegral order. These different approaches are not exactly equivalent. The generating function approach is very convenient for deriving two recurrence relations, Bessel's differential equation, integral representations, addition theorems (Ex. 11.1.2), and upper and lower bounds (Ex. 11.1.1). However the reader will probably have noticed that the generating function defined only Bessel functions of integral order, J_0, J_1, J_2, etc. This is a great limitation of the generating function approach. However, the Bessel function of the first kind $J_\nu(x)$, may easily be defined for nonintegral ν by using the series (Eq. 11.5) as a new definition.

The recurrence relations may be verified by substituting in the series form of $J_\nu(x)$ (Ex. 11.1.7). From these relations Bessel's equation follows. In fact, if ν is not an integer, there is actually an important simplification. It is found that J_ν and $J_{-\nu}$ are independent, for no relation of the form of Eq. 11.8 exists. On the other hand, for $\nu = n$, an integer, we will need another solution. The development of this second solution and an investigation of its properties form the subject of Section 11.3.

EXERCISES

11.1.1 From the product of the generating functions $g(x, t) \cdot g(-x, t)$ show that

$$1 = [J_0(x)]^2 + 2[J_1(x)]^2 + 2[J_2(x)]^2 + \cdots$$

and therefore that $|J_0(x)| \leqslant 1$ and $|J_n(x)| \leqslant 1/\sqrt{2}$, $n = 1, 2, 3, \ldots$.

11.1.2 Using a generating function $g(x, t) = g(u + v, t) = g(u, t) \cdot g(v, t)$, show that

(a) $J_n(u + v) = \sum_{s=-\infty}^{\infty} J_s(u) \,.\, J_{n-s}(v)$,

(b) $J_0(u + v) = J_0(u)\, J_0(v) + 2 \sum_{s=1}^{\infty} J_s(u)\, J_{-s}(v)$.

These are addition theorems for the Bessel functions.

11.1.3 Using only the generating function

$$e^{(x/2)(t-1/t)} = \sum_{n=-\infty}^{\infty} J_n(x) t^n$$

and not the explicit series form of $J_n(x)$, show that $J_n(x)$ has odd or even parity according to whether n is odd or even, that is,[1]

$$J_n(x) = (-1)^n J_n(-x).$$

[1]This is easily seen from the series form (Eq. 11.5)

11.1.4 Derive the Jacobi-Anger expansion

$$e^{iz\cos\theta} = \sum_{m=-\infty}^{\infty} i^m J_m(z) e^{im\theta}.$$

This is an expansion of a plane wave in a series of cylindrical waves.

11.1.5 Show that

(a) $\cos x = J_0(x) + 2\sum_{n=1}^{\infty}(-1)^n J_{2n}(x)$,

(b) $\sin x = 2\sum_{n=1}^{\infty}(-1)^{n+1} J_{2n-1}(x)$.

11.1.6 To help remove the generating function from the realm of black magic, show that it can be *derived* from the recurrence relation, Eq. 11.10.
Hint: 1. Assume a generating function of the form

$$g(x, t) = \sum_{m=-\infty}^{\infty} J_m(x) t^m.$$

2. Multiply Eq. 11.10 by t^n and sum over n.
3. Rewrite the above result as

$$\left(t + \frac{1}{t}\right) g(x, t) = \frac{2t}{x} \frac{\partial g(x, t)}{\partial t}.$$

4. Integrate and adjust the constant of integration (a function of x) so that the coefficient of t^0 is $J_0(x)$ as given by Eq. 11.5.

11.1.7 Show, by direct differentiation, that

$$J_\nu(x) = \sum_{s=0}^{\infty} \frac{(-1)^s}{s!(s+\nu)!}\left(\frac{x}{2}\right)^{\nu+2s}$$

satisfies the two recurrence relations

$$J_{\nu-1}(x) + J_{\nu+1}(x) = \frac{2\nu}{x} J_\nu(x),$$

$$J_{\nu-1}(x) - J_{\nu+1}(x) = 2J_\nu'(x),$$

and Bessel's differential equation

$$x^2 J_\nu''(x) + x J_\nu'(x) + (x^2 - \nu^2) J_\nu(x) = 0.$$

11.1.8 Prove that

(a) $\dfrac{\sin x}{x} = \displaystyle\int_0^{\pi/2} J_0(x\cos\theta)\cos\theta\, d\theta$,

(b) $\dfrac{1 - \cos x}{x} = \displaystyle\int_0^{\pi/2} J_1(x\cos\theta)\, d\theta$.

Hint. The definite integral

$$\int_0^{\pi/2} \cos^{2s+1}\theta\, d\theta = \frac{2\cdot 4\cdot 6\cdots(2s)}{1\cdot 3\cdot 5\cdots(2s+1)}$$

may be useful.

11.1.9 Show that

$$J_0(x) = \frac{2}{\pi}\int_0^1 \frac{\cos xt}{\sqrt{1-t^2}}\, dt,$$

This is a Fourier cosine transform (cf. Section 15.3).

11.1.10 Derive

$$J_n(x) = (-1)^n x^n \left(\frac{d}{x\,dx}\right)^n J_0(x).$$

Hint. Try mathematical induction.

11.1.11 Show that between any two consecutive zeros of $J_n(x)$ there is one and only one zero of $J_{n+1}(x)$. *Hint.* Eqs. 11.15 and 11.17 may be useful.

11.1.12 An analysis of antenna radiation patterns for a system with a circular aperture involves the equation

$$g(u) = \int_0^1 f(r)\, J_0(ur) r\, dr.$$

If $f(r) = 1 - r^2$, show that

$$g(u) = \frac{2}{u^2} J_2(u).$$

11.1.13 The differential cross section in a nuclear scattering experiment is given by $d\sigma/d\Omega = |f(\theta)|^2$. An approximate treatment leads to

$$f(\theta) = \frac{-ik}{2\pi} \int_0^{2\pi} \int_0^R \exp\,[ik\rho \sin\theta \sin\varphi]\rho\, d\rho\, d\varphi.$$

Here θ is angle through which the scattered particle is scattered. R is the nuclear radius. Show that

$$\frac{d\sigma}{d\Omega} = (\pi R^2)\,\frac{1}{\pi}\left[\frac{J_1(kR\sin\theta)}{\sin\theta}\right]^2$$

11.1.14 A set of functions $C_n(x)$ satisfies the recurrence relations

$$C_{n-1}(x) - C_{n+1}(x) = \frac{2n}{x} C_n(x),$$

$$C_{n-1}(x) + C_{n+1}(x) = 2C_n'(x).$$

(a) What linear second-order differential equation do the $C_n(x)$ satisfy?

(b) By a change of variable transform your differential equation into Bessel's equation. This suggests that $C_n(x)$ may be expressed in terms of Bessel functions of transformed argument.

11.1.15 A particle (mass m) is contained in a right circular cylinder (pillbox) of radius R and height H. The particle is described by a wave function satisfying the Schrödinger wave equation

$$-\frac{\hbar^2}{2m}\nabla^2\psi(\rho, \varphi, z) = E\psi(\rho, \varphi, z)$$

and the condition that the wave function go to zero over the surface of the pillbox. Find the lowest (zero point) permitted energy.

$$\textit{Ans.}\ E = \frac{\hbar^2}{2m}\left[\left(\frac{z_{pq}}{R}\right)^2 + \left(\frac{n\pi}{H}\right)^2\right],$$

where z_{pq} is the qth zero of J_p, the index p fixed by the azimuthal dependence.

$$E_{\min} = \frac{\hbar^2}{2m}\left[\left(\frac{2.405}{R}\right)^2 + \left(\frac{\pi}{H}\right)^2\right].$$

11.1.16 (a) Show by direct differentiation and substitution that

$$J_\nu(x) = \frac{1}{2\pi i}\int_C e^{(x/2)(t-1/t)} t^{-\nu-1}\, dt$$

or that the equivalent equation

$$J_\nu(x) = \frac{1}{2\pi i}\left(\frac{x}{2}\right)^\nu \int e^{s-x^2/4s} s^{-\nu-1}\, ds$$

satisfies Bessel's equation. C is the contour shown in Fig. 11.4. The negative real axis is cut line.

Hint. Show that the total integrand (after substituting in Bessel's differential equation) may be written as a total derivative:

$$\frac{d}{dt}\left\{\exp\left[\frac{x}{2}\left(t-\frac{1}{t}\right)\right] t^{-\nu-1}\left[\nu+\frac{x}{2}\left(t+\frac{1}{t}\right)\right]\right\}.$$

(b) Show that the first integral (with n an integer) may be transformed into

$$\begin{aligned} J_n(x) &= \frac{1}{2\pi}\int_0^{2\pi} e^{i(x\sin\theta-n\theta)}\, d\theta \\ &= \frac{i^{-n}}{2\pi}\int_0^{2\pi} e^{i(x\cos\theta+n\theta)}\, d\theta. \end{aligned}$$

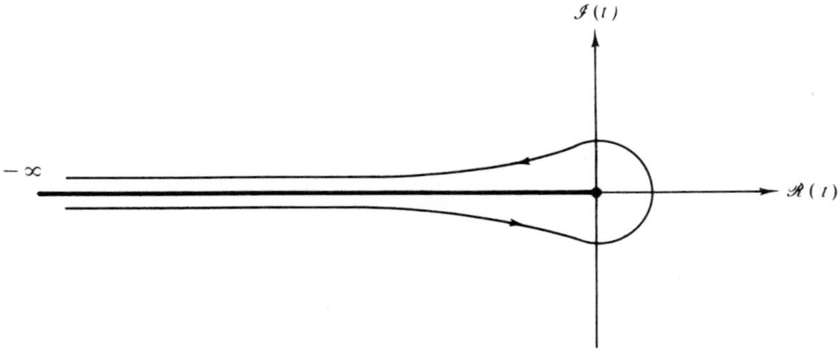

FIG. 11.4 Bessel function contour

11.1.17 The contour C in Exercise 11.1.16 is deformed to the path $-\infty$ to -1, unit circle $e^{-i\pi}$ to $e^{i\pi}$, and finally -1 to $-\infty$. Show that

$$\begin{aligned} J_\nu(x) &= \frac{1}{\pi}\int_0^\pi \cos(\nu\theta - x\sin\theta)\, d\theta \\ &\quad - \frac{\sin\nu\pi}{\pi}\int_0^\infty e^{(-\nu\theta - x\sinh\theta)}\, d\theta. \end{aligned}$$

This is Bessel's integral. *Hint.* The negative values of the variable of integration u may be handled by using

$$u = te^{\pm i\pi}.$$

11.1.18 (a) Show that

$$J_\nu(x) = \frac{2}{\pi^{1/2}(\nu-\frac{1}{2})!}\left(\frac{x}{2}\right)^\nu \int_0^{\pi/2} \cos(x\sin\theta)\cos^{2\nu}\theta\, d\theta,$$

where $\nu > -\frac{1}{2}$. *Hint.* Here is a chance to use series expansion and term-by-term integration. The formulas of Section 10.4 will prove useful.

(b) Transform the integral in part (a) into

$$J_\nu(x) = \frac{1}{\pi^{1/2}(\nu-\frac{1}{2})!}\left(\frac{x}{2}\right)^\nu \int_0^\pi \cos(x\cos\theta)\sin^{2\nu}\theta\, d\theta$$

$$= \frac{1}{\pi^{1/2}(\nu - \frac{1}{2})!}\left(\frac{x}{2}\right)^\nu \int_0^\pi e^{\pm ix\cos\theta} \sin^{2\nu}\theta \, d\theta$$

$$= \frac{1}{\pi^{1/2}(\nu - \frac{1}{2})!}\left(\frac{x}{2}\right)^\nu \int_{-1}^1 e^{\pm ipx}(1 - p^2)^{\nu - 1/2} \, dp.$$

These are alternate integral representations of $J_\nu(x)$.

11.1.19 (a) From

$$J_\nu(x) = \frac{1}{2\pi i}\left(\frac{x}{2}\right)^\nu \int t^{-\nu-1} e^{t - x^2/4t} \, dt$$

derive the recurrence relation

$$J'_\nu(x) = \frac{\nu}{x} J_\nu(x) - J_{\nu+1}(x).$$

(b) From

$$J_\nu(x) = \frac{1}{2\pi i} \int t^{-\nu-1} e^{(x/2)(t - 1/t)} \, dt$$

derive the recurrence relation

$$J'_\nu(x) = \tfrac{1}{2} J_{\nu-1}(x) - \tfrac{1}{2} J_{\nu+1}(x).$$

11.1.20 Show that the recurrence relation

$$J'_n(x) = \tfrac{1}{2}[J_{n-1}(x) - J_{n+1}(x)]$$

follows directly from differentiation of

$$J_n(x) = \frac{1}{\pi} \int_0^\pi \cos(n\theta - x\sin\theta) \, d\theta.$$

11.1.21 Evaluate

$$\int_0^\infty e^{-ax} J_0(bx) \, dx, \qquad a, b > 0.$$

Actually the results holds for $a \geqslant 0$, $-\infty < b < \infty$. This is a Laplace transform of J_0.
Hint. Either an integral representation of J_0 or a series expansion will be helpful.

11.1.22 Using trigonometric forms, verify that

$$J_0(br) = \frac{1}{2\pi} \int_0^{2\pi} e^{i\,br\sin\theta} \, d\theta.$$

11.1.23 (*a*) Plot the intensity (Φ^2 of Eq. 11.35) as a function of ($\sin\alpha/\lambda$) along a diameter of the circular diffraction pattern. Locate the first two minima.
(*b*) What fraction of the total light *intensity* falls within the central maximum?
Hint. $[J_1(x)]^2/x$ may be written as a derivative and the area integral of the intensity integrated by inspection.

11.1.24 The fraction of light incident on a circular aperture (normal incidence) that is transmitted is given by

$$T = 2\int_0^{2ka} J_2(x) \frac{dx}{x} - \frac{1}{2ka}\int_0^{2ka} J_2(x) \, dx.$$

Here a is the radius of the aperture, and k is the wave number, $2\pi/\lambda$. Show that

(a) $$T = 1 - \frac{1}{ka}\sum_{n=0}^{\infty} J_{2n+1}(2ka),$$

(b) $$T = 1 - \frac{1}{2ka}\int_0^{2ka} J_0(x) \, dx$$

11.1.25 The amplitude $U(\rho, \varphi, t)$ of a vibrating circular membrane of radius a satisfies the wave equation

$$\nabla^2 U - \frac{1}{v^2}\frac{\partial^2 U}{\partial t^2} = 0.$$

Here v is the phase velocity of the wave fixed by the elastic constants and whatever damping is imposed.

(a) Show that a solution is

$$U(\rho, \varphi, t) = J_m(k\rho)(a_1 e^{im\varphi} + a_2 e^{-im\varphi})(b_1 e^{i\omega t} + b_2 e^{-i\omega t}).$$

(b) From the Dirichlet boundary condition, $J_m(ka) = 0$, find the allowable values of the wavelength λ. $(k = 2\pi/\lambda)$.

Note. There are other Bessel functions besides J_m but they all diverge at $\rho = 0$. This is shown explicitly in Section 11.3. The divergent behavior is actually implicit in Eq. 11.6.

11.1.26 Example 11.1.2 describes the TM modes of electromagnetic cavity oscillation. The transverse electric (TE) modes differ in that we work from the z component of the magnetic induction $\mathbf{B}$:

$$\nabla^2 B_z + \alpha^2 B_z = 0$$

with boundary conditions

$$B_z(0) = B_z(l) = 0 \quad \text{and} \quad \left.\frac{\partial B_z}{\partial \rho}\right|_{\rho=a} = 0.$$

Show that the TE resonant frequencies are given by

$$\omega_{mnp} = \frac{1}{\sqrt{\varepsilon_0 \mu_0}}\sqrt{\frac{\beta_{mn}^2}{a^2} + \frac{p^2\pi^2}{l^2}}, \qquad p = 1, 2, 3, \ldots.$$

11.1.27 Plot the three lowest TM and the three lowest TE angular resonant frequencies, ω_{mnp}, as a function of the radius/length (a/l) ratio for $0 \leqslant a/l \leqslant 1.5$.

Hint. Try plotting ω^2 (in units of $(\varepsilon_0 \mu_0 a^2)^{-1}$) vs. $(a/l)^2$. Why this choice?

11.1.28 A thin conducting disk of radius a carries a charge q. Show that the potential is described by

$$\varphi(r, z) = \frac{q}{4\pi\varepsilon_0 a}\int_0^\infty e^{-k|z|} J_0(kr)\frac{\sin ka}{k}\,dk,$$

where J_0 is the usual Bessel function and r and z are the familiar cylindrical coordinates.

Note: This is a difficult problem. One approach is through Fourier transforms such as Ex. 15.3.6. For a discussion of the physical problem see Jackson (*Classical Electrodynamics*, pp. 90–93).

11.2 Orthogonality

If Bessel's equation, Eq.11.22a, is divided by x, we see that it becomes self adjoint, and, therefore, by the Sturm–Liouville theory, Section 9.2, the solutions are expected to be orthogonal—if we can arrange to have appropriate boundary conditions satisfied. To take care of the boundary conditions, we introduce parameters a and $\alpha_{\nu m}$ into the argument of J_ν to get $J_\nu(\alpha_{\nu m}\rho/a)$. Here, a is the upper limit of the cylindrical radial coordinate ρ. From Eq.11.22a,

$$\rho \frac{d^2}{d\rho^2} J_\nu\left(\alpha_{\nu m}\frac{\rho}{a}\right) + \frac{d}{d\rho} J_\nu\left(\alpha_{\nu m}\frac{\rho}{a}\right) + \left(\frac{\alpha_{\nu m}^2 \rho}{a^2} - \frac{\nu^2}{\rho}\right) J_\nu\left(\alpha_{\nu m}\frac{\rho}{a}\right) = 0. \tag{11.45}$$

changing the parameter $\alpha_{\nu m}$ to $\alpha_{\nu n}$, $J_\nu(\alpha_{\nu n}\rho/a)$ satisfies

$$\rho \frac{d^2}{d\rho^2} J_\nu\left(\alpha_{\nu n}\frac{\rho}{a}\right) + \frac{d}{d\rho} J_\nu\left(\alpha_{\nu n}\frac{\rho}{a}\right) + \left(\frac{\alpha_{\nu n}^2 \rho}{a^2} - \frac{\nu^2}{\rho}\right) J_\nu\left(\alpha_{\nu n}\frac{\rho}{a}\right) = 0. \tag{11.45a}$$

Proceeding as in Section 9.2, we multiply Eq. 11.45 by $J_\nu(\alpha_{\nu n}\rho/a)$ and Eq. 11.45a by $J_\nu(\alpha_{\nu m}\rho/a)$ and subtract, obtaining

$$J_\nu\left(\alpha_{\nu n}\frac{\rho}{a}\right) \frac{d}{d\rho}\left[\rho \frac{d}{d\rho} J_\nu\left(\alpha_{\nu m}\frac{\rho}{a}\right)\right] - J_\nu\left(\alpha_{\nu m}\frac{\rho}{a}\right) \frac{d}{d\rho}\left[\rho \frac{d}{d\rho} J_\nu\left(\alpha_{\nu n}\frac{\rho}{a}\right)\right] = \frac{\alpha_{\nu n}^2 - \alpha_{\nu m}^2}{a^2} \rho J_\nu\left(\alpha_{\nu m}\frac{\rho}{a}\right) J_\nu\left(\alpha_{\nu n}\frac{\rho}{a}\right). \tag{11.46}$$

Integrating from $\rho = 0$ to $\rho = a$,

$$\int_0^a J_\nu\left(\alpha_{\nu n}\frac{\rho}{a}\right) \frac{d}{d\rho}\left[\rho \frac{d}{d\rho} J_\nu\left(\alpha_{\nu m}\frac{\rho}{a}\right)\right] d\rho - \int_0^a J_\nu\left(\alpha_{\nu m}\frac{\rho}{a}\right) \frac{d}{d\rho}\left[\rho \frac{d}{d\rho} J_\nu\left(\alpha_{\nu n}\frac{\rho}{a}\right)\right] d\rho = \frac{\alpha_{\nu n}^2 - \alpha_{\nu m}^2}{a^2} \int_0^a J_\nu\left(\alpha_{\nu m}\frac{\rho}{a}\right) J_\nu\left(\alpha_{\nu n}\frac{\rho}{a}\right) \rho \, d\rho. \tag{11.47}$$

Upon integrating by parts, the left-hand side of Eq. 11.47 becomes

$$\left|\rho J_\nu\left(\alpha_{\nu n}\frac{\rho}{a}\right) \frac{d}{d\rho} J_\nu\left(\alpha_{\nu m}\frac{\rho}{a}\right)\right|_0^a - \left|\rho J_\nu\left(\alpha_{\nu m}\frac{\rho}{a}\right) \frac{d}{d\rho} J_\nu\left(\alpha_{\nu n}\frac{\rho}{a}\right)\right|_0^a \tag{11.48}$$

For $\nu \geqslant 0$, the factor ρ guarantees a zero at the lower limit, $\rho = 0$. Actually, the lower limit on the index ν may be extended down to $\nu > -1$, Ex. 11.2.4.[1] At $\rho = a$, each expression vanishes if we choose the parameters $\alpha_{\nu n}$ and $\alpha_{\nu m}$ to be zeros or roots of J_ν, i.e., $J_\nu(\alpha_{\nu m}) = 0$. The subscripts now become meaningful: $\alpha_{\nu m}$ is the mth zero of J_ν.

With this choice of parameters, the left-hand side vanishes (the Sturm-Liouville boundary conditions are satisfied) and for $m \neq n$,

$$\int_0^a J_\nu\left(\alpha_{\nu m}\frac{\rho}{a}\right) J_\nu\left(\alpha_{\nu n}\frac{\rho}{a}\right) \rho \, d\rho = 0. \tag{11.49}$$

This gives us orthogonality over the interval $[0, a]$.

[1] The case $\nu = -1$ reverts to $\nu = +1$, Eq. 11.8.

Normalization. The normalization integral may be developed by returning to Eq. 11.48, setting $\alpha_{\nu n} = \alpha_{\nu m} + \varepsilon$, and taking the limit $\varepsilon \to 0$ (cf. Ex. 11.2.2). With the aid of the recurrence relation, Eq. 11.16, the result may be written

$$\int_0^a \left[J_\nu\left(\alpha_{\nu m}\frac{\rho}{a}\right)\right]^2 \rho\, d\rho = \frac{a^2}{2}\,[J_{\nu+1}(\alpha_{\nu m})]^2. \tag{11.50}$$

Bessel series. If we assume that the set of Bessel functions $J_\nu(\alpha_{\nu m}\rho/a)$ (ν fixed, $m = 1, 2, 3, \ldots$) is complete, then any well-behaved but otherwise arbitrary function $f(\rho)$ may be expanded in a Bessel series (Bessel–Fourier or Fourier–Bessel)

$$f(\rho) = \sum_{m=1}^{\infty} c_{\nu m} J_\nu\left(\alpha_{\nu m}\frac{\rho}{a}\right), \qquad 0 \leqslant \rho \leqslant a, \quad \nu > -1. \tag{11.51}$$

The coefficients $c_{\nu m}$ are determined by using Eq. 11.50,

$$c_{\nu m} = \frac{2}{a^2[J_{\nu+1}(\alpha_{\nu m})]^2}\int_0^a f(\rho) J_\nu\left(\alpha_{\nu m}\frac{\rho}{a}\right)\rho\, d\rho. \tag{11.52}$$

A similar series expansion involving $J_\nu(\beta_{\nu m}\rho/a)$ with $(d/d\rho)J_\nu(\beta_{\nu m}\rho/a)\,|_{\rho=a} = 0$ is included in Exercises 11.2.3 and 11.2.6(b).

Example 11.2.1 Electrostatic Potential in a Hollow Cylinder

From Example 2.6.1, the solution of Laplace's equation is a linear combination of

$$\begin{aligned}\psi_{km}(\rho, \varphi, z) &= P_{km}(\rho)\Phi_m(\varphi)Z_k(z)\\ &= J_m(k\rho)\cdot[a_m \sin m\varphi + b_m \cos m\varphi]\\ &\quad\cdot[c_1\, e^{kz} + c_2\, e^{-kz}].\end{aligned} \tag{11.53}$$

The particular linear combination is determined by the boundary conditions to be satisfied.

Our cylinder here has a radius a and a height l. The top end section has a potential distribution $\psi(\rho, \varphi)$. Elsewhere on the surface the potential is zero.[1] The problem is to find the electrostatic potential

$$\psi(\rho, \varphi, z) = \sum_{k,m} \psi_{km}(\rho, \varphi, z) \tag{11.54}$$

everywhere in the interior.

For convenience, the circular cylindrical coordinates are placed as shown in Fig. 11.3. Since $\psi(\rho, \varphi, 0) = 0$, we take $c_1 = -c_2 = \frac{1}{2}$. The z dependence becomes $\sinh kz$, vanishing at $z = 0$. The requirement that $\psi = 0$ on the cylindrical sides is met by requiring the separation constant k to be

$$k = k_{mn} = \alpha_{mn}/a, \tag{11.55}$$

[1] If $\psi = 0$ at $z = 0, l$, but $\psi \neq 0$ for $\rho = a$, the modified Bessel functions, Section 11.5, are involved.

where the first subscript m gives the index of the Bessel function, while the second subscript identifies the particular zero of J_m.

The electrostatic potential becomes

$$\psi(\rho, \varphi, z) = \sum_{m=0}^{\infty} \sum_{n=1}^{\infty} J_m\left(\alpha_{mn}\frac{\rho}{a}\right) \cdot [a_{mn} \sin m\varphi + b_{mn} \cos m\varphi] \cdot \sinh\left(\alpha_{mn}\frac{z}{a}\right). \tag{11.56}$$

Equation 11.56 is a double series: a Bessel series in ρ and a Fourier series in φ.

At $z = l$, $\psi = \psi(\rho, \varphi)$, a known function of ρ and φ. Therefore,

$$\psi(\rho, \varphi) = \sum_{m=0}^{\infty} \sum_{n=1}^{\infty} J_m\left(\alpha_{mn}\frac{\rho}{a}\right) \cdot [a_{mn} \sin m\varphi + b_{mn} \cos m\varphi] \cdot \sinh\left(\alpha_{mn}\frac{l}{a}\right). \tag{11.57}$$

The constants a_{mn} and b_{mn} are evaluated by using Eqs. 11.49 and 11.50 and the corresponding equations for $\sin \varphi$ and $\cos \varphi$ (Example 9.2.1 and Eqs. 14.7–14.9) We find[1]

$$\left.\begin{matrix} a_{mn} \\ b_{mn} \end{matrix}\right\} = 2\left[\pi a^2 \sinh\left(\alpha_{mn}\frac{l}{a}\right) J_{m+1}^2(\alpha_{mn})\right]^{-1} \cdot \int_0^{2\pi}\int_0^a \psi(\rho, \varphi) J_m\left(\alpha_{mn}\frac{\rho}{a}\right) \left\{\begin{matrix} \sin m\varphi \\ \cos m\varphi \end{matrix}\right\} \rho \, d\rho \, d\varphi. \tag{11.58}$$

These are definite integrals, i.e., numbers. Substituting back into Eq. 11.56 the series is specified and the potential $\psi(\rho, \varphi, z)$ determined. The problem is solved.

Continuum form. The Bessel series, Eq. 11.51 and Exercise 11.2.6 apply to expansions over the finite interval $[0, a]$. If $a \to \infty$ then the series forms may be expected to go over into integrals. The discrete roots $\alpha_{\nu m}$ become a continuous variable α. A similar situation is encountered in Fourier series, Section 14.2. The development of the Bessel integral from the Bessel series is left as Exercise 11.2.8.

For operations with a continuum of Bessel functions, $J_\nu(\alpha\rho)$, a key relation is the Bessel function closure equation

$$\int_0^\infty J_\nu(\alpha\rho) J_\nu(\alpha'\rho)\rho \, d\rho = \frac{1}{\alpha}\,\delta(\alpha - \alpha'). \tag{11.59}$$

This may be proved by the use of Hankel transforms, Section 15.1. An alternate approach, starting from a relation similar to Eq. 9.82, is given by Morse and Feshbach, Section 6.3.

[1] If $m = 0$, the factor 2 is omitted (cf. Eq. 14.8).

EXERCISES

11.2.1 (a) Show that

$$(a^2 - b^2)\int_0^P J_\nu(ax)\, J_\nu(bx)\, x\, dx = P[b\, J_\nu(aP)\, J'_\nu(bP) - aJ'_\nu(aP)\, J_\nu(bP)]$$

with

$$J'_\nu(aP) = \frac{d}{d(ax)} J_\nu(ax)\Big|_{x=P}.$$

(b) $$\int_0^P [J_\nu(ax)]^2 x\, dx = \frac{P^2}{2}\left\{[J'_\nu(aP)]^2 + \left(1 - \frac{\nu^2}{a^2P^2}\right)[J_\nu(aP)]^2\right\}, \quad \nu > -1.$$

These two integrals are usually called the first and second Lommel integrals.
Hint. We have the development of the orthogonality of the Bessel functions as an analogy.

11.2.2 Show that

$$\int_0^a \left[J_\nu\left(\alpha_{\nu m}\frac{\rho}{a}\right)\right]^2 \rho\, d\rho = \frac{a^2}{2}[J_{\nu+1}(\alpha_{\nu m})]^2, \qquad \nu > -1.$$

Here $\alpha_{\nu m}$ is the mth zero of J_ν.
Hint. With $\alpha_{\nu n} = \alpha_{\nu m} + \varepsilon$, expand $J_\nu[(\alpha_{\nu m} + \varepsilon)\rho/a]$ about $\alpha_{\nu m}\rho/a$ by a Taylor expansion.

11.2.3 (a) If $\beta_{\nu m}$ is the mth zero of $(d/d\rho)J_\nu(\beta_{\nu m}\,\rho/a)$, show that the Bessel functions are orthogonal over the interval $[0, a]$ with an orthogonality integral

$$\int_0^a J_\nu\left(\beta_{\nu m}\frac{\rho}{a}\right) J_\nu\left(\beta_{\nu n}\frac{\rho}{a}\right)\rho\, d\rho = 0, \qquad m \neq n, \quad \nu > -1.$$

(b) Derive the corresponding normalization integral ($m = n$).

$$\textit{Ans.} \quad \frac{a^2}{2}\left(1 - \frac{\nu^2}{\beta_{\nu m}^2}\right)[J_\nu(\beta_{\nu m})]^2 \quad \nu > -1.$$

11.2.4 Verify that the orthogonality equation, Eq. 11.49 and the normalization equation, Eq. 11.50 hold for $\nu > -1$.
Hint. Using power series expansions, examine the behavior of Eq. 11.48 as $\rho \to 0$.

11.2.5 From Eq. 11.49 develop a proof that $J_\nu(z), \nu > -1$, has no complex roots. *Hint:* (a) Use the series form of $J_\nu(z)$ to exclude pure imaginary roots. (b) Assume $\alpha_{\nu m}$ to be complex and take $\alpha_{\nu n}$ to be $\alpha_{\nu m}^*$.

11.2.6 (a) In the series expansion

$$f(\rho) = \sum_{m=1}^{\infty} c_{\nu m} J_\nu\left(\alpha_{\nu m}\frac{\rho}{a}\right), \qquad 0 \leqslant \rho \leqslant a, \quad \nu > -1,$$

with $J_\nu(\alpha_{\nu m}) = 0$, show that the coefficients are given by

$$c_{\nu m} = \frac{2}{a^2[J_{\nu+1}(\alpha_{\nu m})]^2}\int_0^a f(\rho)\, J_\nu\left(\alpha_{\nu m}\frac{\rho}{a}\right)\rho\, d\rho.$$

(b) In the series expansion

$$f(\rho) = \sum_{m=1}^{\infty} d_{\nu m} J_\nu\left(\beta_{\nu m}\frac{\rho}{a}\right), \qquad 0 \leqslant \rho \leqslant a, \quad \nu > -1,$$

with $(d/d\rho)J_\nu(\beta_{\nu m}\rho/a)|_{\rho=a} = 0$, show that the coefficients are given by

$$d_{\nu m} = \frac{2}{a^2(1 - \nu^2/\beta_{\nu m}^2)[J_\nu(\beta_{\nu m})]^2} \int_0^a f(\rho) J_\nu\left(\beta_{\nu m}\frac{\rho}{a}\right) \rho \, d\rho.$$

11.2.7 A right circular cylinder has an electrostatic potential of $\psi(\rho, \varphi)$ on both ends. The potential on the curved cylindrical surface is zero. Find the potential at all interior points.
Hint. Choose your coordinate system and adjust your z dependence to exploit the symmetry of your potential.

11.2.8 For the continuum case, show that Eqs. 11.51 and 11.52 are replaced by

$$f(\rho) = \int_0^\infty a(\alpha) \, J_\nu(\alpha\rho) \, d\alpha,$$

$$a(\alpha) = \alpha \int_0^\infty f(\rho) \, J_\nu(\alpha\rho)\rho \, d\rho.$$

Hint. The corresponding case for sines and cosines is worked out in Section 15.2.

11.2.9 Analogous to $\cos x$ and $\sin x$ of Section 5.10, $J_0(x)$ and $J_1(x)$ may be expressed as infinite products. Show that

(a) $J_0(x) = \prod_{m=1}^{\infty}\left(1 - \frac{x^2}{\alpha_{0m}^2}\right),$

(b) $J_1(x) = x \prod_{m=1}^{\infty}\left(1 - \frac{x^2}{\alpha_{1m}^2}\right).$

11.3 Neumann Functions, Bessel Functions of the Second Kind, $N_\nu(x)$

From the theory of differential equations it is known that Bessel's equation has two independent solutions. Indeed, for nonintegral order ν we have already found two solutions and labeled them $J_\nu(x)$ and $J_{-\nu}(x)$, using the infinite series (Eq. 11.5). The trouble is that when ν is integral Eq. 11.8 holds and we have but one independent solution. A second solution may be developed by the methods of Section 8.5. This yields a perfectly good second solution of Bessel's equation, but it is not the usual standard form.

Definition. As an alternate approach we take the particular linear combination of $J_\nu(x)$ and $J_{-\nu}(x)$

$$N_\nu(x) = \frac{\cos \nu\pi \, J_\nu(x) - J_{-\nu}(x)}{\sin \nu\pi}. \tag{11.60}$$

This is the Neumann function.[1] For nonintegral ν, $N_\nu(x)$ clearly satisfies Bessel's equation, for it is a linear combination of known solutions, $J_\nu(x)$ and $J_{-\nu}(x)$. However, for integral ν, $\nu = n$, Eq. 11.8 applies and Eq. 11.60 becomes indeterminate. The definition of $N_\nu(x)$ was chosen deliberately for this indeterminate

[1] In AMS-55 and in most mathematics tables, this is labelled $Y_\nu(x)$.

property. Evaluating $N_n(x)$ by L'Hospital's rule for indeterminate forms, we obtain

$$\begin{aligned} N_n(x) &= \left.\frac{(d/d\nu)[\cos \nu\pi J_\nu(x) - J_{-\nu}(x)]}{(d/d\nu)\sin \nu\pi}\right|_{\nu=n} \\ &= \left.\frac{-\pi \sin n\pi J_n(x) + [\cos n\pi\, \partial J_\nu/\partial\nu - \partial J_{-\nu}/\partial\nu]}{\pi \cos n\pi}\right|_{\nu=n} \\ &= \frac{1}{\pi}\left[\frac{\partial J_\nu(x)}{\partial\nu} - (-1)^n \frac{\partial J_{-\nu}(x)}{\partial\nu}\right]\Bigg|_{\nu=n} \end{aligned} \tag{11.61}$$

Series form. A series expansion[1] gives the horrible result

$$\begin{aligned} N_n(x) &= \frac{2}{\pi} J_n(x) \ln\left(\frac{x}{2}\right) \\ &- \frac{1}{\pi}\sum_{r=0}^{\infty}(-1)^r \frac{1}{r!(n+r)!}\left(\frac{x}{2}\right)^{n+2r} [\mathrm{F}(r) + \mathrm{F}(n+r)] \\ &- \frac{1}{\pi}\sum_{r=0}^{n-1}\frac{(n-r-1)!}{r!}\left(\frac{x}{2}\right)^{-n+2r} \end{aligned} \tag{11.62}$$

which exhibits the logarithmic dependence that was to be expected. This, of course, verifies the independence of J_n and N_n. $\mathrm{F}(r)$ is the digamma function which arises from differentiating the factorials in the denominator of $J_\nu(x)$ (cf. Section 10.2 and especially Eq. 10.39). Using the properties of the digamma function, we rewrite Eq. 11.62 in the only slightly less horrible form

$$\begin{aligned} N_n(x) &= \frac{2}{\pi}\left[\ln\left(\frac{x}{2}\right) + \gamma - \frac{1}{2}\sum_{p=1}^{n} p^{-1}\right] J_n(x) \\ &- \frac{1}{\pi}\sum_{r=0}^{\infty}(-1)^r \frac{(x/2)^{n+2r}}{r!(n+r)!}\sum_{p=1}^{r}\left[\frac{1}{p} + \frac{1}{p+n}\right] \\ &- \frac{1}{\pi}\sum_{r=0}^{n-1}\frac{(n-r-1)!}{r!}\left(\frac{x}{2}\right)^{-n+2r} \end{aligned} \tag{11.63}$$

For $n = 0$ we have the limiting value

$$N_0(x) = \frac{2}{\pi}(\ln x + \gamma - \ln 2) + O(x^2) \tag{11.64}$$

and for $\nu > 0$

$$N_\nu(x) = -\frac{(\nu-1)!}{\pi}\left(\frac{2}{x}\right)^\nu + \cdots .\text{[2]} \tag{11.65}$$

To verify that $N_\nu(x)$, our Neumann function or Bessel function of the second kind, actually does satisfy Bessel's equation for integral n, we may proceed as follows. Differentiating Bessel's equation for $J_{\pm\nu}(x)$ with respect to ν, we have

$$x^2\frac{d^2}{dx^2}\left(\frac{\partial J_{\pm\nu}}{\partial\nu}\right) + x\frac{d}{dx}\left(\frac{\partial J_{\pm\nu}}{\partial\nu}\right) + (x^2 - \nu^2)\frac{\partial J_{\pm\nu}}{\partial\nu} = 2\nu J_{\pm\nu}. \tag{11.66}$$

[1] Using $(d/d\nu)x^\nu = x^\nu \ln x$.

[2] Note that this limiting form applies to *both* integral and nonintegral values of the index ν.

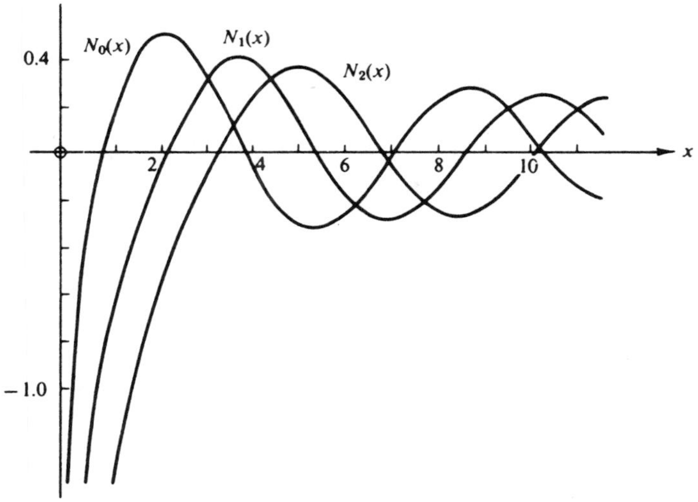

FIG. 11.5 Neumann functions, $N_0(x)$, $N_1(x)$, and $N_2(x)$

Multiplying the equation for $J_{-\nu}$ by $(-1)^\nu$, subtracting from the equation for J_ν (as suggested by Eq. 11.61), and taking the limit $\nu \to n$, we obtain

$$x^2 \frac{d^2}{dx^2} N_n + x \frac{d}{dx} N_n + (x^2 - n^2) N_n = \frac{2n}{\pi} [J_n - (-1)^n J_{-n}]. \tag{11.67}$$

For $\nu = n$, an integer, the right-hand side vanishes by Eq. 11.8 and $N_n(x)$ is seen to be a solution of Bessel's equation. The most general solution for any ν can therefore be written

$$y(x) = A\, J_\nu(x) + B\, N_\nu(x). \tag{11.68}$$

It is seen from Eq. 11.62 that N_n diverges at least logarithmically. Any boundary condition that requires the solution to be finite at the origin [as in our vibrating circular membrane (section 11.1)] automatically excludes $N_n(x)$. Conversely, in the absence of such a requirement $N_n(x)$ must be considered.

To a certain extent, the definition of the Neumann function $N_n(x)$ is arbitrary. Equation 11.63 contains terms of the form $a_n J_n(x)$. Clearly, any finite value of the constant a_n would still give us a second solution of Bessel's equation. Why should a_n have the particular value shown in Eq. 11.63? The answer involves the asymptotic dependence developed in Section 11.6. If J_n corresponds to a cosine wave, then N_n corresponds to a sine wave. This simple and convenient asymptotic phase relationship is a consequence of the particular admixture of J_n in N_n.

Recurrence relations. Substituting Eq. 11.60 for $N_\nu(x)$ (nonintegral ν) or Eq. 11.61 (integral ν) into the recurrence relations (Eqs. 11.10 and 11.12) for $J_n(x)$, we see immediately that $N_\nu(x)$ satisfies these same recurrence relations. This actually constitutes another proof that N_ν is a solution. Note carefully that the converse is not necessarily true. All solutions need not satisfy the same recurrence relations. An example of this sort of trouble appears in Section 11.5.

Wronskian formulas. From Section 8.5 we have the Wronskian formula for solutions of the Bessel equation

$$u_\nu(x)\, v_\nu'(x) - u_\nu'(x)\, v_\nu(x) = \frac{A_\nu}{x}, \tag{11.69}$$

in which A_ν is a parameter that depends on the particular Bessel functions $u_\nu(x)$ and $v_\nu(x)$ being considered. It is a constant in the sense that it is independent of x. Consider the special case

$$u_\nu(x) = J_\nu(x), \qquad v_\nu(x) = J_{-\nu}(x), \tag{11.70}$$

$$J_\nu J_{-\nu}' - J_\nu' J_{-\nu} = \frac{A_\nu}{x}. \tag{11.71}$$

Since A_ν is a constant, it may be identified at any convenient point such as $x = 0$. Using the first terms in the series expansions (Eqs. 11.5 and 11.6), we obtain

$$\begin{aligned} J_\nu &\to \frac{x^\nu}{2^\nu \nu!}, & J_{-\nu} &\to \frac{2^\nu x^{-\nu}}{(-\nu)!}, \\ J_\nu' &\to \frac{\nu x^{\nu-1}}{2^\nu \nu!}, & J_{-\nu}' &\to \frac{-\nu 2^\nu x^{-\nu-1}}{(-\nu)!}. \end{aligned} \tag{11.72}$$

Substitution into Eq. 11.69 yields

$$\begin{aligned} J_\nu(x)\, J_{-\nu}'(x) - J_\nu'(x)\, J_{-\nu}(x) &= \frac{-2\nu}{x\nu!(-\nu)!} \\ &= -\frac{2 \sin \nu\pi}{\pi x} \end{aligned} \tag{11.73}$$

using Eq. 10.32

$$\nu!(-\nu)! = \frac{\pi\nu}{\sin \pi\nu}.$$

Note that A_ν vanishes for integral ν, as it must, since the nonvanishing of the Wronskian is a test of the independence of the two solutions. By Eq. 11.73 J_n and J_{-n} are clearly linearly dependent.

Using our recurrence relations, we may readily develop a large number of alternate forms, among which are

$$J_\nu J_{-\nu+1} + J_{-\nu} J_{\nu-1} = \frac{2 \sin \nu\pi}{\pi x}, \tag{11.74}$$

$$J_\nu J_{-\nu-1} + J_{-\nu} J_{\nu+1} = -\frac{2 \sin \nu\pi}{\pi x}, \tag{11.75}$$

$$J_\nu N_\nu' - J_\nu' N_\nu = \frac{2}{\pi x}, \tag{11.76}$$

$$J_\nu N_{\nu+1} - J_{\nu+1} N_\nu = -\frac{2}{\pi x}. \tag{11.77}$$

Many more will be found in the references given.

The reader will recall that in Chapter 8 Wronskians were of great value in two respects: (1) in establishing the linear independence or linear dependence of solutions of differential equations and (2) in developing an integral form of a second solution. Here, the specific forms of the Wronskians and Wronskian-derived combinations of Bessel functions are useful primarily to illustrate the general behavior of the various Bessel functions. Wronskians are of great use in checking tables of Bessel functions. In Chapter 16, Wronskians reappear in connection with Green's functions.

EXAMPLE 11.3.1 COAXIAL WAVE GUIDES

We are interested in an electromagnetic wave confined between the concentric, conducting cylindrical surfaces $\rho = a$ and $\rho = b$. Most of the mathematics is worked out in Examples 2.6.1 and 11.1.2. To go from the standing wave of these examples to the traveling wave here, we let $a_{mn} = ib_{mn}$ in Eq. 11.40 and obtain

$$E_z = \sum_{m,n} b_{mn} J_m(\gamma\rho) e^{\pm im\varphi} e^{i(kz-\omega t)}. \tag{11.78}$$

One generalization is needed. The origin $\rho = 0$ is now excluded $(0 < a \leqslant \rho \leqslant b)$. Hence, the Neumann function $N_m(\gamma\rho)$ may *not* be excluded. $E_z(\rho, \varphi, z, t)$ becomes

$$E_z = \sum_{m,n} [b_{mn} J_m(\gamma\rho) + c_{mn} N_m(\gamma\rho)] e^{\pm im\varphi} e^{i(kz-\omega t)}. \tag{11.79}$$

With the condition

$$H_z = 0, \tag{11.80}$$

we have the basic equations for a TM (transverse magnetic) wave.

The (tangential) electric field must vanish at the conducting surfaces (Dirichlet boundary condition) or

$$b_{mn} J_m(\gamma a) + c_{mn} N_m(\gamma a) = 0, \tag{11.81}$$

$$b_{mn} J_m(\gamma b) + c_{mn} N_m(\gamma b) = 0. \tag{11.82}$$

These transcendental equations may be solved for γ (γ_{mn}) and the ratio c_{mn}/b_{mn}. From Example 2.6.1,

$$k^2 = \omega^2 \mu_0 \varepsilon_0 - \gamma^2. \tag{11.83}$$

Since k^2 must be positive for a real wave, the minimum frequency that will be propagated (in this TM mode) is

$$\omega = \frac{\gamma}{\sqrt{\mu_0 \varepsilon_0}}, \tag{11.84}$$

with γ fixed by the boundary conditions, Eqs. 11.81 and 11.82. This is the cutoff frequency of the wave guide.

There is also a TE (transverse electric) mode with $E_z = 0$, and H_z given by Eq. 11.79. Then we have Neumann boundary conditions in place of Eqs. 11.81 and 11.82. Finally, for the coaxial guide (*not* for the plain cylindrical guide, $a = 0$), a TEM (transverse electron agnetic) mode, $E_z = H_z = 0$, is possible. This corresponds to a plane wave as in free space. Ex. 2.9.5 is an example of this for two parallel wires.

The simpler cases (no Neumann functions, simpler boundary conditions) of a circular wave guide are included as Exs. 11.3.7 and 11.3.8.

To conclude this discussion of Neumann functions, the Neumann function, $N_\nu(x)$, is introduced for the following reasons:

1. It is a second, independent solution of Bessel's equation, which completes the general solution.
2. It is required for specific physical problems such as electromagnetic waves in coaxial cables.
3. It leads to a Green's function for the Bessel equation (Section 11.5).
4. It leads directly to the two Hankel functions (Section 11.4).

EXERCISES

11.3.1 Verify the expansions (leading term only)

$$N_0(x) \to \frac{2}{\pi}(\ln x + \gamma - \ln 2)$$

$$N_\nu(x) \to -\frac{(\nu-1)!}{\pi}\left(\frac{2}{x}\right)^\nu, \qquad \nu > 0$$

$$x \ll 1.$$

by direct differentiation of the definition of the Neumann function as indicated in Eq. 11.61.

11.3.2 Prove that the Neumann functions N_n (with n an integer) satisfy the recurrence relations

$$N_{n-1}(x) + N_{n+1}(x) = \frac{2n}{x} N_n(x)$$

$$N_{n-1}(x) - N_{n+1}(x) = 2N_n'(x).$$

Hint. These relations may be proved by differentiating the recurrence relations for J_ν or by using the limit form of N_ν but *not* dividing everything by zero.

11.3.3 Show that

$$N_{-n}(x) = (-1)^n N_n(x).$$

11.3.4 Show that

$$N_0'(x) = -N_1(x).$$

11.3.5 If Y and Z are any two solutions of Bessel's equation, show that

$$Y_\nu(x)Z_\nu'(x) - Y_\nu'(x)Z_\nu(x) = \frac{A_\nu}{x},$$

in which A_ν may depend on ν but is independent of x. This is really a special case of Ex. 9.1.4.

11.3.6 Verify the Wronskian formulas

$$J_\nu(x)\, J_{-\nu+1}(x) + J_{-\nu}(x)\, J_{\nu-1}(x) = \frac{2 \sin \nu\pi}{\pi x},$$

$$J_\nu(x)\, N_\nu'(x) - J_\nu'(x)\, N_\nu(x) = \frac{2}{\pi x}.$$

11.3.7 A cylindrical waveguide has radius r_0. Find the nonvanishing components of the electric and magnetic fields for
(a) TM_{01}, transverse magnetic wave ($H_z = H_\rho = E_\theta = 0$),
(b) TE_{01}, transverse electric wave ($E_z = E_\rho = H_\theta = 0$).
The subscripts 01 indicate that the longitudinal component (E_z or H_z) involves J_0 and the boundary condition is satisfied by the *first* zero of J_0 or J_0'.

11.3.8 For a given mode of oscillation the *minimum* frequency that will be passed by a circular cylindrical wave guide is

$$\nu_{\min} = \frac{c}{\lambda_c},$$

in which λ_c is fixed by the boundary condition

$$J_n\left(\frac{2\pi r_0}{\lambda_c}\right) = 0 \quad \text{for } TM_{nm} \text{ mode},$$

$$J_n'\left(\frac{2\pi r_0}{\lambda_c}\right) = 0 \quad \text{for } TE_{nm} \text{ mode}.$$

The subscript n denotes the order of the Bessel function and m indicates the zero used. Find this cut-off wavelength, λ_c for the three TM and three TE modes with the longest cut-off wavelengths. Explain your results in terms of the graph of J_0, J_1, and J_2, (Fig. 11.1).

11.4 Hankel Functions

Many authors prefer to introduce the Hankel functions by means of integral representations and then use them to define the Neumann function, $N_\nu(z)$. An outline of this approach is given at the end of this section.

Definitions. As we have already obtained the Neumann function by more elementary (and less powerful) techniques, we may use it to define the Hankel functions, $H_\nu^{(1)}(x)$ and $H_\nu^{(2)}(x)$:

$$H_\nu^{(1)}(x) = J_\nu(x) + i\, N_\nu(x) \tag{11.85}$$

and

$$H_\nu^{(2)}(x) = J_\nu(x) - i\, N_\nu(x). \tag{11.86}$$

This is exactly analogous to taking

$$e^{\pm i\theta} = \cos\theta \pm i \sin\theta. \tag{11.87}$$

The extent of the analogy will be seen even better when the asymptotic forms are considered (Section 11.6). Indeed, it is their asymptotic behavior that makes the Hankel functions useful.

Series expansion of $H_\nu^{(1)}(x)$ and $H_\nu^{(2)}(x)$ may be obtained by combining Eqs. 11.5 and 11.63. Often only the first term is of interest; it is given by

$$H_0^{(1)}(x) \approx i\frac{2}{\pi}\ln x + 1 + i\frac{2}{\pi}(\gamma - \ln 2) + \cdots, \tag{11.88}$$

$$H_\nu^{(1)}(x) \approx -i\frac{(\nu-1)!}{\pi}\left(\frac{2}{x}\right)^\nu + \cdots, \qquad \nu > 0, \tag{11.89}$$

$$H_0^{(2)}(x) \approx -i\frac{2}{\pi}\ln x + 1 - i\frac{2}{\pi}(\gamma - \ln 2) + \cdots, \tag{11.90}$$

$$H_\nu^{(2)}(x) \approx i\frac{(\nu-1)!}{\pi}\left(\frac{2}{x}\right)^\nu + \cdots, \qquad \nu > 0. \tag{11.91}$$

Since the Hankel functions are linear combinations (with constant coefficients) of J_ν and N_ν, they satisfy the same recurrence relations (Eqs. 11.10 and 11.12).

$$H_{\nu-1}(x) + H_{\nu+1}(x) = \frac{2\nu}{x} H_\nu(x), \tag{11.92}$$

$$H_{\nu-1}(x) - H_{\nu+1}(x) = 2H_\nu'(x), \tag{11.93}$$

for both $H_\nu^{(1)}(x)$ and $H_\nu^{(2)}(x)$.

A variety of Wronskian formulas can be developed:

$$H_\nu^{(2)}H_{\nu+1}^{(1)} - H_\nu^{(1)}H_{\nu+1}^{(2)} = \frac{4}{i\pi x}, \tag{11.94}$$

$$J_{\nu-1}H_\nu^{(1)} - J_\nu H_{\nu-1}^{(1)} = \frac{2}{i\pi x}, \tag{11.95}$$

$$J_\nu H_{\nu-1}^{(2)} - J_{\nu-1}H_\nu^{(2)} = \frac{2}{i\pi x}. \tag{11.96}$$

EXAMPLE 11.4.1 CYLINDRICAL TRAVELING WAVES

As an illustration of the use of Hankel functions, consider a two-dimensional wave problem similar to the vibrating circular membrane of Ex. 11.1.25. Now imagine that the waves are generated at $r = 0$ and move outward to infinity. We replace our standing waves by traveling ones. The differential equation remains the same, but the boundary conditions change. We now demand that for large r the solution behave like

$$U \to e^{i(kr - \omega t)} \tag{11.97}$$

to describe an outgoing wave. As before, k is the wave number. This assumes, for simplicity, that there is no azimuthal dependence, that is, no angular momentum, or $m = 0$. In Sections 7.4 and 11.6, $H_0^{(1)}(kr)$ is shown to have the asymptotic behavior

$$H_0^{(1)}(kr) \to e^{ikr}. \tag{11.98}$$

This boundary condition at infinity then determines our wave solution as

$$U(r, t) = H_0^{(1)}(kr)\, e^{-i\omega t}. \tag{11.99}$$

This solution diverges as $r \to 0$, which is just the behavior to be expected with a source at the origin.

The choice of a *two*-dimensional wave problem to illustrate the Hankel function $H_0^{(1)}(z)$ is not accidental. Bessel functions may appear in a variety of ways, such as in the separation of conical coordinates. However, they enter most commonly from the radial equations from the separation of variables in the Helmholtz equation in cylindrical and in spherical polar coordinates. We have taken a degenerate form of cylindrical coordinates for this illustration. Had we used spherical polar coordinates (spherical waves) we should have encountered index $\nu = n + \frac{1}{2}$, n an integer. These special values yield the spherical Bessel functions to be discussed in Section 11.7.

Contour integral representation of the Hankel functions. The integral representation (Schlaefli integral)

$$J_\nu(x) = \frac{1}{2\pi i} \int e^{(x/2)(t - 1/t)} \frac{dt}{t^{\nu+1}} \tag{11.100}$$

may easily be established for $\nu = n$, an integer [recognizing that the numerator is the generating function (Eq. 11.1) and integrating around the origin]. If ν is not an integer, the integrand is not single-valued and a cut line is needed in our complex plane. Choosing the negative real axis as the cut line and using the contour shown in Fig. 11.6, Eq. 11.100 can be extended to nonintegral ν. Substituting Eq. 11.100

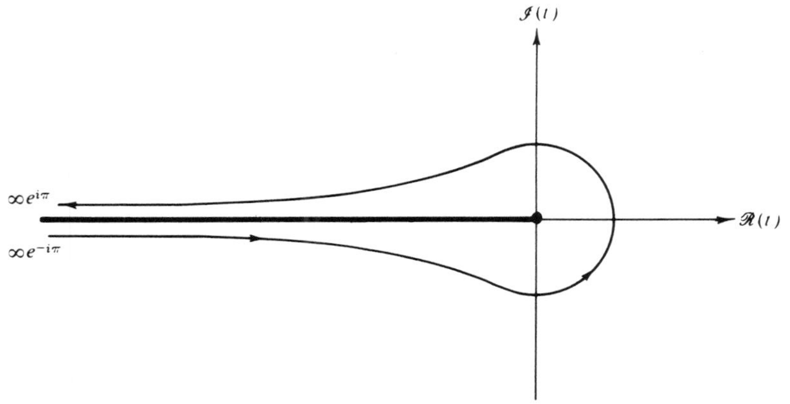

FIG. 11.6 Bessel function contour

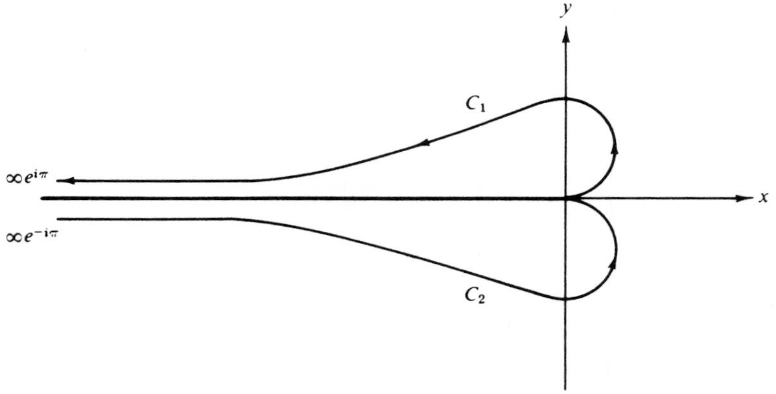

FIG. 11.7 Hankel function contours

into Bessel's differential equation, the combined integrand can be represented by an exact differential which vanishes as $t \to \infty e^{\pm i\pi}$ (cf. Ex. 11.1.16).

We now deform the contour so that it approaches the origin along the positive real axis, as shown in Fig. 11.7. This particular approach guarantees that the exact differential mentioned will vanish as $t \to 0$ because of the $e^{-x/2t}$ factor. Hence each of the separate portions $\infty\, e^{-i\pi}$ to 0 and 0 to $\infty\, e^{i\pi}$ is a solution of Bessel's equation. We define

$$H_\nu^{(1)}(x) = \frac{1}{\pi i} \int_0^{\infty e^{i\pi}} e^{(x/2)(t-1/t)} \frac{dt}{t^{\nu+1}} \tag{11.101}$$

$$H_\nu^{(2)}(x) = \frac{1}{\pi i} \int_{\infty e^{-i\pi}}^{0} e^{(x/2)(t-1/t)} \frac{dt}{t^{\nu+1}}. \tag{11.102}$$

These expressions are particularly convenient because they may be handled by the method of steepest descents (Section 7.4). $H_\nu^{(1)}(x)$ has a saddle point at $t = +i$, whereas $H_\nu^{(2)}(x)$ has a saddle point at $t = -i$.

The problem of relating Eqs. 11.101 and 11.102 to our earlier definition of the Hankel function (Eqs. 11.85 and 11.86) remains. Since Eqs. 11.100–11.102 combined yield

$$J_\nu(x) = \tfrac{1}{2}[H_\nu^{(1)}(x) + H_\nu^{(2)}(x)] \tag{11.103}$$

by inspection, we need only show that

$$N_\nu(x) = \frac{1}{2i}[H_\nu^{(1)}(x) - H_\nu^{(2)}(x)] \tag{11.104}$$

This may be accomplished by the following steps:

1. With the substitutions $t = e^{i\pi}/s$ for $H_\nu^{(1)}$ and $t = e^{-i\pi}/s$ for $H_\nu^{(2)}$, we obtain

$$H_\nu^{(1)}(x) = e^{-i\nu\pi} H_{-\nu}^{(1)}(x), \tag{11.105}$$

$$H_\nu^{(2)}(x) = e^{i\nu\pi} H_{-\nu}^{(2)}(x). \tag{11.106}$$

2. From Eqs. 11.103 ($\nu \to -\nu$), and 11.105 and 11.106,

$$J_{-\nu}(x) = \tfrac{1}{2}[e^{i\nu\pi} H_\nu^{(1)}(x) + e^{-i\nu\pi} H_\nu^{(2)}(x)]. \tag{11.107}$$

3. Finally, substitute J_ν (Eq. 11.103) and $J_{-\nu}$ (Eq. 11.107) into the defining equation for N_ν, Eq. 11.60. This leads to Eq. 11.104 and establishes the contour integrals Eqs. 11.101 and 11.102 as the Hankel functions.

Integral representations have appeared before: Eq. 10.35 for $\Gamma(z)$ and various representations of $J_\nu(z)$ in Section 11.1. With these integral representations of the Hankel functions, it is perhaps appropriate to ask why we are interested in integral representations. There are at least four reasons. The first is simply esthetic appeal —some people find them attractive. Second, the integral representations help to distinguish between two linearly independent solutions. In Fig. 11.7, the contours C_1 and C_2 cross *different* saddle points (Section 7.4). For the Legendre functions the contour for $P_n(z)$ (Fig. 12.9) and that for $Q_n(z)$ (Fig. 12.18) encircle *different* singular points.

Third, the integral representations facilitate manipulations, analysis, the development of relations among the various special functions. Fourth, and probably most important of all, the integral representations are extremely useful in developing asymptotic expansions. One approach, the method of steepest descents, appears in Section 7.4. A second approach, the direct expansion of an integral representation is given in Section 11.6 for the modified Bessel function $K_\nu(z)$. This same technique may be used to obtain asymptotic expansions of the confluent hypergeometric functions, M and U—Ex. 13.5.13.

In conclusion, the Hankel functions are introduced here for the following reasons:

1. As analogs of $e^{\pm ix}$ they are useful for describing traveling waves.
2. They offer an alternate (contour integral) and a rather elegant definition of Bessel functions.
3. $H_\nu^{(1)}$ is used to define the modified Bessel function K_ν of Section 11.5.

EXERCISES

11.4.1 Verify the Wronskian formulas

(a) $J_\nu(x)\, H_\nu^{(1)\prime}(x) - J_\nu'(x)\, H_\nu^{(1)}(x) = \dfrac{i2}{\pi x}$,

(b) $J_\nu(x)\, H_\nu^{(2)\prime}(x) - J_\nu'(x)\, H_\nu^{(2)}(x) = \dfrac{-i2}{\pi x}$,

(c) $N_\nu(x)\, H_\nu^{(1)\prime}(x) - N_\nu'(x)\, H_\nu^{(1)}(x) = \dfrac{-2}{\pi x}$,

(d) $N_\nu(x)\, H_\nu^{(2)\prime}(x) - N_\nu'(x)\, H_\nu^{(2)}(x) = \dfrac{-2}{\pi x}$,

(e) $H_\nu^{(1)}(x)\, H_\nu^{(2)\prime}(x) - H_\nu^{(1)\prime}(x)\, H_\nu^{(2)}(x) = \dfrac{-i4}{\pi x}$,

$$(f)\ H_\nu^{(2)}(x)\, H_{\nu+1}^{(1)}(x) - H_\nu^{(1)}(x)\, H_{\nu+1}^{(2)}(x) = \frac{4}{i\pi x},$$

$$(g)\ J_{\nu-1}(x)\, H_\nu^{(1)}(x) - J_\nu(x)\, H_{\nu-1}^{(1)}(x) = \frac{2}{i\pi x}.$$

11.4.2 Show that the integral forms

$$(a)\ \frac{1}{i\pi}\int_{0C_1}^{\infty e^{i\pi}} e^{(x/2)(t-1/t)}\frac{dt}{t^{\nu+1}} = H_\nu^{(1)}(x),$$

$$(b)\ \frac{1}{i\pi}\int_{\infty e^{-i\pi}C_2}^{0} e^{(x/2)(t-1/t)}\frac{dt}{t^{\nu+1}} = H_\nu^{(2)}(x),$$

satisfy Bessel's differential equation. The contours C_1 and C_2 are shown in Fig. 11.7.

11.4.3 Using the integrals and contours given in problem 11.4.2, show that

$$\frac{1}{2i}[H_\nu^{(1)}(x) - H_\nu^{(2)}(x)] = N_\nu(x).$$

11.4.4 Show that the integrals in Exercise 11.4.2 may be transformed to yield

$$(a)\ H_\nu^{(1)}(x) = \frac{1}{\pi i}\int_{C_3} e^{x\sinh\gamma - \nu\gamma}\, d\gamma, \qquad (b)\ H_\nu^{(2)}(x) = \frac{1}{\pi i}\int_{C_4} e^{x\sinh\gamma-\nu\gamma}\, d\gamma.$$

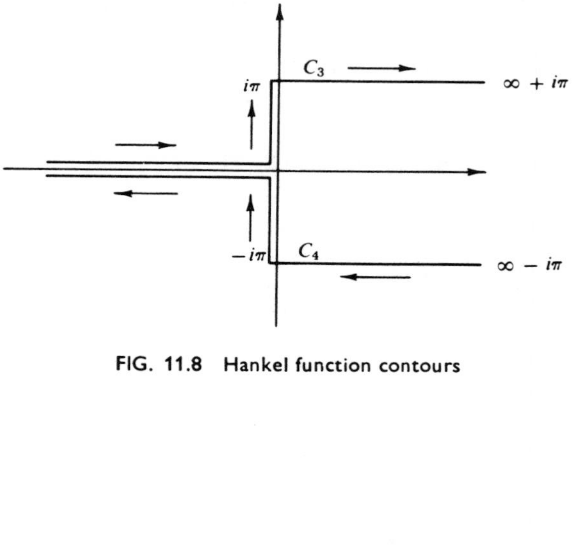

FIG. 11.8 Hankel function contours

11.5 Modified Bessel Functions, $I_\nu(x)$ and $K_\nu(x)$

Frequently in physical problems the equation

$$x^2\frac{d^2}{dx^2}y(x) + x\frac{d}{dx}y(x) - (x^2+\nu^2)\,y(x) = 0 \tag{11.108}$$

occurs. This is not quite Bessel's equation, but it may be put in the form of Bessel's equation by replacing x by $-it$

$$x\frac{d}{dx} \to t\frac{d}{dt},$$
$$x^2\frac{d^2}{dx^2} \to t^2\frac{d^2}{dt^2}.$$

Then

$$t^2\frac{d^2}{dt^2}y(-it) + t\frac{d}{dt}y(-it) + (t^2 - \nu^2)\,y(-it) = 0, \tag{11.109}$$

or $y(-it)$ is a Bessel function. It is customary (and convenient) to choose the normalization so that

$$y(x) = I_\nu(x) = i^{-\nu}J_\nu(ix). \tag{11.110}$$

Often this is written

$$I_\nu(x) = e^{-\nu\pi i/2}J_\nu(xe^{i\pi/2}). \tag{11.111}$$

Series form. In terms of infinite series, this is equivalent to removing the $(-1)^s$ sign in Eq. 11.5 and writing

$$I_\nu(x) = \sum_{s=0}^{\infty}\frac{1}{s!(s+\nu)!}\left(\frac{x}{2}\right)^{2s+\nu},$$
$$I_{-\nu}(x) = \sum_{s=0}^{\infty}\frac{1}{s!(s-\nu)!}\left(\frac{x}{2}\right)^{2s-\nu} \tag{11.112}$$

For integral ν this yields

$$I_n(x) = I_{-n}(x). \tag{11.113}$$

Recurrence relations. The recurrence relations satisfied by $I_\nu(x)$ may be developed from the series expansions, but it is perhaps easier to work from the existing recurrence relations for $J_\nu(x)$. Let us replace x by $-ix$ and rewrite Eq. 11.110 as

$$J_\nu(x) = i^\nu I_\nu(-ix). \tag{11.114}$$

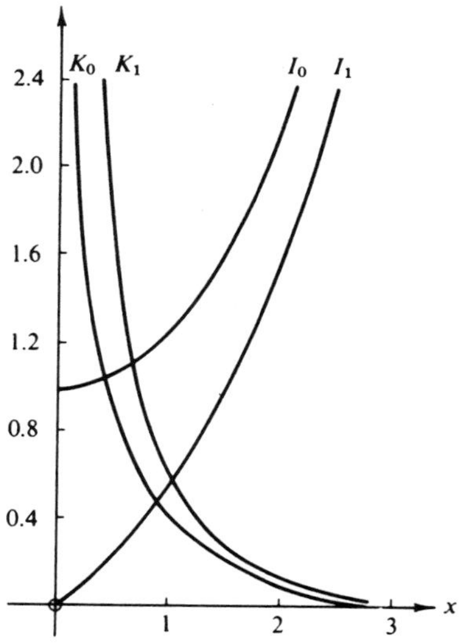

FIG. 11.9 Modified Bessel functions

Then Eq. 11.10 becomes

$$i^{\nu-1}I_{\nu-1}(-ix) + i^{\nu+1}I_{\nu+1}(-ix) = \frac{2\nu}{x}i^\nu I_\nu(-ix).$$

Replacing x by ix, we have a recurrence relation for $I_\nu(x)$,

$$I_{\nu-1}(x) - I_{\nu+1}(x) = \frac{2\nu}{x}I_\nu(x). \tag{11.115}$$

Equation 11.12 transforms to

$$I_{\nu-1}(x) + I_{\nu+1}(x) = 2I'_\nu(x). \tag{11.116}$$

These are the recurrence relations used in Exercise 11.1.14.

From Eq. 11.113 it is seen that we have but one independent solution when ν is an integer, exactly as in the Bessel functions J_ν. The choice of a second, independent solution of Eq. 11.108 is essentially a matter of convenience. The second solution given here is selected on the basis of its asymptotic behavior—as shown in the next section. The confusion of choice and notation for this solution is perhaps greater than anywhere else in this field.[1] Many authors[2] choose to define a second solution in terms of the Hankel function $H_\nu^{(1)}(x)$ by

$$\begin{aligned} K_\nu(x) &\equiv \frac{\pi}{2} i^{\nu+1} H_\nu^{(1)}(ix) \\ &= \frac{\pi}{2} i^{\nu+1}[J_\nu(ix) + i\, N_\nu(ix)]. \end{aligned} \tag{11.117}$$

Using Eqs. 11.60 and 11.110 we may transform Eq. 11.117 to[3]

$$K_\nu(x) = \frac{\pi}{2} \frac{I_{-\nu}(x) - I_\nu(x)}{\sin \nu\pi}, \tag{11.118}$$

analogous to Eq. 11.60 for $N_\nu(x)$. The choice of Eq. 11.117 as a definition is somewhat unfortunate in that the function $K_\nu(x)$ does *not* satisfy the same recurrence relations as $I_\nu(x)$ (cf. Exercises 11.5.7 and 11.5.8). To avoid this annoyance other authors[4] have included an additional factor of cosine $n\pi$. This permits K_ν to satisfy the same recurrence relations as I_ν, but it has the disadvantage of making $K_\nu = 0$ for $\nu = \frac{1}{2}, \frac{3}{2}, \frac{5}{2}, \ldots$.

The series expansion of $K_\nu(x)$ follows directly from the series form of $H_\nu^{(1)}(ix)$. The lowest order terms are

$$\begin{aligned} K_0(x) &= -\ln x - \gamma + \ln 2 + \cdots, \\ K_\nu(x) &= 2^{\nu-1}(\nu-1)!x^{-\nu} + \cdots. \end{aligned} \tag{11.119}$$

Because the modified Bessel function I_ν is related to the Bessel function J_ν, much as sinh is related to sine, I_ν and the second solution K_ν are sometimes referred to as hyperbolic Bessel functions.

Example 11.5.1 Neutron Diffusion Theory

An illustration of the occurrence of modified Bessel functions is provided by the theory of diffusion of thermal neutrons. In the steady state the continuity equation is

$$D\,\nabla^2\varphi - \sum_a \varphi + S = 0, \tag{11.120}$$

the first term representing diffusion, the second term giving the absorption losses,

[1] A discussion and comparison of notations will be found in MTOAC **1**, 207–308 (1944).
[2] Watson, Morse and Feshbach, Jeffreys and Jeffreys (without the $\pi/2$).
[3] For integral index n we take the limit as $\nu \to n$.
[4] Whittaker and Watson.

and the third representing the source strength. The parameter D is given by

$$D = \frac{\lambda_s}{3(1 - 2/3A)}$$

in which λ_s is the mean free path between scattering collisions. A is the atomic number of the scattering nucleus and enters here as a correction for the anisotropy of scattering in the laboratory system. The neutron flux φ is the product of neutron density times average velocity. $\sum_a$ is the macroscopic absorption cross section, the product of absorption probability per atom (σ_a) and the number of atoms per unit volume. It is assumed that the absorption is small compared to the scattering.

In this illustration let us consider the neutron source as an infinitely long line source embedded in an infinite diffusing medium. This line source is the z-axis. For the emission of S_0 neutrons per unit length per unit time

$$S = S_0\,\delta(\boldsymbol{\rho}),$$

in which $\delta(\boldsymbol{\rho})$ is a cylindrical Dirac delta function. The symmetry of the problem demands cylindrical coordinates. Because there is no dependence on z or θ the diffusion equation quickly reduces to

$$\rho^2 \frac{d^2\varphi}{d\rho^2} + \rho \frac{d\varphi}{d\rho} - \rho^2 \frac{\sum_a}{D} \varphi = 0 \tag{11.121}$$

away from the line source ($\rho \neq 0$). Using

$$k^2 = \frac{\sum_a}{D},$$

where k^{-1} is called "diffusion length,"

$$\varphi = a_1\, I_0(k\rho) + a_2\, K_0(k\rho). \tag{11.122}$$

Since $I_0(k\rho)$ increases exponentially for large $k\rho$ [cf., the series expansion (Eq. 11.112) and Section 11.6], its coefficient a_1 must be set equal to zero; then

$$\varphi = a_2\, K_0(k\rho). \tag{11.123}$$

The net flow of neutrons through an area $d\boldsymbol{\sigma}$ is given by

$$-D\, \nabla\varphi \cdot d\boldsymbol{\sigma}.$$

The constant of integration a_2 may be determined by requiring that D times the integral of the negative gradient of the neutron flux around the edge of a small pillbox of unit height be equal to the production within the pillbox S_0. The box is small ($\rho \to 0$) to eliminate absorption and gives

$$S_0 = \lim_{\rho \to 0} D a_2 \int -\nabla K_0(k\rho) \cdot \boldsymbol{\rho}_0\, \rho\, d\theta, \tag{11.124}$$

which is actually a two-dimensional form of Gauss's law (Section 1.14). Using the series form of $K_0(k\rho)$ this becomes

$$S_0 = D a_2 \lim_{\rho \to 0} \frac{2\pi\rho}{\rho} \tag{11.125}$$

or

$$a_2 = \frac{S_0}{2\pi D}.$$

By substituting into Eq. 11.123, we have

$$\varphi = \frac{S_0}{2\pi D} K_0(k\rho) \tag{11.126}$$

for the complete solution of this diffusion problem.

Green's function. A comparison with Section 8.6 will show that we have, in effect, determined a Green's function for the diffusion equation (11.120). This may be readily verified by rewriting Eq. 11.120 as

$$D\,\nabla^2 G(k\rho) - Dk^2\,G(k\rho) = -\delta(\boldsymbol{\rho}), \tag{11.127}$$

for a unit line source (per unit length) along the z-axis. Integrating over the pillbox volume, we have

$$\begin{aligned}\int \nabla^2 G(k\rho)\,d\tau - k^2 \int G(k\rho)\,d\tau &= -\frac{1}{D}\int \delta(\boldsymbol{\rho})\,d\tau \\ &= -\frac{1}{D}.\end{aligned}$$

The second integral on the left vanishes as $\rho \to 0$. The evaluation of the first integral on the left may be carried out by Gauss's theorem and corresponds exactly to our previous solution. The result is

$$G(k\rho) = \frac{1}{2\pi D} K_0(k\rho). \tag{11.128}$$

This is the Green's function.

If our source is now generalized to be a continuous distribution of parallel line sources described by the radial vector $\boldsymbol{\rho}_2$, we find the Green's function is

$$G(\boldsymbol{\rho}_1, \boldsymbol{\rho}_2) = \frac{1}{2\pi D} K_0(k\,|\boldsymbol{\rho}_1 - \boldsymbol{\rho}_2|). \tag{11.129}$$

The resulting neutron flux is

$$\begin{aligned}\varphi(\boldsymbol{\rho}_1) &= \int G(\boldsymbol{\rho}_1, \boldsymbol{\rho}_2) S(\boldsymbol{\rho}_2)\,d\tau_2 \\ &= \frac{1}{2\pi D}\int K_0(k\,|\boldsymbol{\rho}_1 - \boldsymbol{\rho}_2|) S(\boldsymbol{\rho}_2)\,d\tau_2,\end{aligned} \tag{11.130}$$

where the integration is over the plane perpendicular to the line sources and $S(\boldsymbol{\rho}_2)$ is the source strength (per unit volume per second).

To put the modified Bessel functions $I_\nu(x)$ and $K_\nu(x)$ in proper perspective, they are introduced here.

1. These functions are solutions of the frequently encountered modified Bessel equation

2. They are needed for specific physical problems such as diffusion problems.
3. $K_\nu(x)$ provides a Green's function.
4. $K_\nu(x)$ leads to a convenient determination of asymptotic behavior (Section 11.6).

EXERCISES

11.5.1 Show that

$$e^{(x/2)(t+1/t)} = \sum_{n=-\infty}^{\infty} I_n(x)t^n,$$

thus generating modified Bessel functions, $I_n(x)$.

11.5.2 Verify the following identities

(a) $1 = I_0(x) + 2\sum_{n=1}^{\infty}(-1)^n I_{2n}(x)$,

(b) $e^x = I_0(x) + 2\sum_{n=1}^{\infty} I_n(x)$,

(c) $e^{-x} = I_0(x) + 2\sum_{n=1}^{\infty}(-1)^n I_n(x)$,

(d) $\cosh x = I_0(x) + 2\sum_{n=1}^{\infty} I_{2n}(x)$,

(e) $\sinh x = 2\sum_{n=1}^{\infty} I_{2n-1}(x)$.

11.5.3 (a) From the generating function of Ex. 11.5.1 show that

$$I_n(x) = \frac{1}{2\pi i}\oint \exp\,[(x/2)(t + 1/t)]\,\frac{dt}{t^{n+1}}.$$

(b) For $n = \nu$, not an integer, show that the above integral representation may be generalized to

$$I_\nu(x) = \frac{1}{2\pi i}\int_C \exp\,[(x/2)(t + 1/t)]\,\frac{dt}{t^{\nu+1}}.$$

The contour C is the same as that for $J_\nu(x)$, Fig. 11.6.

11.5.4 For $\nu > -\frac{1}{2}$ show that $I_\nu(z)$ may be represented by

$$\begin{aligned} I_\nu(z) &= \frac{1}{\pi^{1/2}(\nu - \frac{1}{2})!}\left(\frac{z}{2}\right)^\nu \int_0^\pi e^{\pm z\cos\theta}\sin^{2\nu}\theta\,d\theta \\ &= \frac{1}{\pi^{1/2}(\nu - \frac{1}{2})!}\left(\frac{z}{2}\right)^\nu \int_{-1}^1 e^{\pm zp}(1 - p^2)^{\nu-1/2}\,dp \\ &= \frac{2}{\pi^{1/2}(\nu - \frac{1}{2})!}\left(\frac{z}{2}\right)^\nu \int_0^{\pi/2} \cosh(z\cos\theta)\sin^{2\nu}\theta\,d\theta. \end{aligned}$$

11.5.5 A cylindrical cavity has a radius a and a height l, Fig. 11.3. The ends, $z = 0$ and l are at zero potential. The cylindrical walls, $\rho = a$, have a potential $V = V(\varphi, z)$.
(a) Show that the electrostatic potential $\Phi(\rho, \varphi, z)$ has the functional form

$$\Phi(\rho, \varphi, z) = \sum_{m=0}^{\infty} \sum_{n=1}^{\infty} I_m(k_n \rho) \sin k_n z \cdot (a_{mn} \sin m\varphi + b_{mn} \cos m\varphi), \qquad \text{where} \quad k_n = \frac{n\pi}{l}$$

(b) Show that the coefficients a_{mn} and b_{mn} are given by

$$\left.\begin{matrix} a_{mn} \\ b_{mn} \end{matrix}\right\} = \frac{2}{\pi l I_m(k_n a)} \int_0^{2\pi} \int_0^l V(\varphi, z) \sin k_n z . \left\{\begin{matrix} \sin m\varphi \\ \cos m\varphi \end{matrix}\right\} dz\, d\varphi.$$

Hint. Expand $V(\varphi, z)$ as a double series and use the orthogonality of the trigonometric functions.

11.5.6 Verify that $K_\nu(x)$ is given by

$$K_\nu(x) = \frac{\pi}{2} \frac{I_{-\nu}(x) - I_\nu(x)}{\sin \nu\pi}$$

and from this show that

$$K_\nu(x) = K_{-\nu}(x).$$

11.5.7 Show that $K_\nu(x)$ satisfies the recurrence relations

$$K_{\nu-1}(x) - K_{\nu+1}(x) = -\frac{2\nu}{x} K_\nu(x),$$

$$K_{\nu-1}(x) + K_{\nu+1}(x) = -2K_\nu'(x).$$

11.5.8 If $\mathscr{K}_\nu = e^{\nu\pi i} K_\nu$, show that $\mathscr{K}_\nu$ satisfies the *same* recurrence relations as I_ν.

11.5.9 For $\nu > -\frac{1}{2}$ show that $K_\nu(z)$ may be represented by

$$K_\nu(z) = \frac{\pi^{1/2}}{(\nu - \frac{1}{2})!} \left(\frac{z}{2}\right)^\nu \int_0^\infty e^{-z\cosh t} \sinh^{2\nu} t\, dt \qquad -\frac{\pi}{2} < \arg z < \frac{\pi}{2}$$

$$= \frac{\pi^{1/2}}{(\nu - \frac{1}{2})!} \left(\frac{z}{2}\right)^\nu \int_1^\infty e^{-zp} (p^2 - 1)^{\nu - 1/2}\, dp.$$

11.5.10 Show that $I_\nu(x)$ and $K_\nu(x)$ satisfy the Wronskian relation

$$I_\nu(x) K_\nu'(x) - I_\nu'(x) K_\nu(x) = -\frac{1}{x}.$$

This result is quoted in Section 16.6 in the development of a Green's function.

11.5.11 If $r = (x^2 + y^2)^{1/2}$, prove that

$$\frac{1}{r} = \frac{2}{\pi} \int_0^\infty \cos(xt) K_0(yt)\, dt.$$

11.5.12 Show that $-\frac{1}{4} N_0(k\rho)$ is a possible Green's function for the Helmholtz equation in circular cylindrical coordinates

$$\nabla^2 \psi + k^2 \psi = 0$$

(no axial or azimuthal dependence).

11.5.13 The Green's functions satisfying

$$(\nabla^2 - k^2) G(\mathbf{r}) = -\delta(\mathbf{r})$$

are

$$G_1(x) = \frac{1}{2k} e^{-kx},$$

$$G_2(\rho) = \frac{1}{2\pi} K_0(k\rho),$$

$$G_3(r) = \frac{k}{4\pi} \frac{e^{-kr}}{kr},$$

corresponding to a plane source, a line source, and a point source respectively.
(a) By integrating the point source to form a line source show that

$$\int_\rho^\infty \frac{e^{-kr}}{(r^2-\rho^2)^{1/2}}\,dr = K_0(k\rho)$$

or

$$\int_1^\infty \frac{e^{-\alpha u}}{(u^2-1)^{1/2}}\,du = \int_0^\infty e^{-\alpha\cosh t}\,dt = K_0(\alpha).$$

Hint. Show that $G_2(\rho) = \int_{-\infty}^{\infty} G_3(r)\,dx$ where $x^2 = r^2 - \rho^2$.

(b) By integrating the line source to form a plane source show that

$$\int_x^\infty \frac{K_0(k\rho)\rho\,d\rho}{(\rho^2-x^2)^{1/2}} = \frac{\pi}{2}\frac{e^{-kx}}{k}.$$

or

$$\int_1^\infty \frac{K_0(\alpha u)u\,du}{(u^2-1)^{1/2}} = \frac{\pi}{2}\frac{e^{-\alpha}}{\alpha}.$$

Hint. Show that $G_1(x) = \int_{-\infty}^{\infty} G_2(\rho)\,dy$ where $y^2 = \rho^2 - x^2$.

11.6 Asymptotic Expansions

Stokes's method. If $Z_\nu(x)$, a solution of Bessel's equation is replaced by $x^{-1/2}\,y(x)$, then $y(x)$ satisfies the equation

$$y'' + \left(1 - \frac{\nu^2 - \frac{1}{4}}{x^2}\right)y = 0. \tag{11.131}$$

For large values of $x(x \gg \nu)$, this becomes

$$y_1'' + y_1 = 0, \tag{11.132}$$

with the familiar solutions

$$y_1 = a_0 \sin x + b_0 \cos x. \tag{11.133}$$

We may improve the approximation (still for large x) by replacing the constants a_0 and b_0 by a series of negative powers of x,

$$a_0 \to \sum_{n=0}^{\infty} a_n x^{-n}, \qquad b_0 \to \sum_{n=0}^{\infty} b_n x^{-n}.$$

This means that we must look for a solution of the form

$$y(x) = \sum_{n=0}^{\infty} a_n x^{-n} \sin x + \sum_{n=0}^{\infty} b_n x^{-n} \cos x. \tag{11.134}$$

By substituting Eq. 11.134 into the original differential equation (11.131), we may obtain

$$\begin{aligned} y(x) = &\left[a_0 - \frac{\nu^2 - \frac{1}{4}}{2x} b_0 - \frac{(\nu^2 - \frac{1}{4})(\nu^2 - \frac{9}{4})}{2\cdot(2x)^2} a_0 + \cdots\right]\sin x \\ &+ \left[b_0 + \frac{\nu^2 - \frac{1}{4}}{2x} a_0 - \frac{(\nu^2 - \frac{1}{4})(\nu^2 - \frac{9}{4})}{2\cdot(2x)^2} b_0 + \cdots\right]\cos x. \end{aligned} \tag{11.135}$$

This process may be continued as far as desired. It is worth noting that this form terminates and yields an exact result for $\nu = \pm\frac{1}{2}, \pm\frac{3}{2}, \pm\frac{5}{2}, \ldots$, etc. These exact results are the spherical Bessel functions of Section 11.7.

With the two constants a_0 and b_0 Stokes's method yields a general solution for Bessel's equation for large values of the variable x. However, the usefulness of this solution is limited in that it has not yet been related or joined to the standard solutions $J_\nu(x)$ and $N_\nu(x)$. In particular, the relative contributions of $\sin x$ and $\cos x$ are undetermined. This joining or determination of the phase may be carried out by using the method of steepest descent (Section 7.4) to determine the leading terms in the asymptotic expansions of $J_\nu(x)$ and $N_\nu(x)$.

Expansion of an integral representation, $K_\nu(z)$. As an alternate and direct approach, consider the integral representation (Exercise 11.5.9)

$$K_\nu(z) = \frac{\pi^{1/2}}{(\nu - \frac{1}{2})!}\left(\frac{z}{2}\right)^\nu \int_1^\infty e^{-zx}(x^2 - 1)^{\nu - 1/2}dx, \qquad \nu > -\frac{1}{2}. \tag{11.136}$$

For the present let us take z to be real, although Eq. 11.136 may be established for $-\pi/2 < \arg z < \pi/2$. We have two problems: first, to show that K_ν as given in Eq. 11.136, actually satisfies the modified Bessel equation (11.108); second, to show that it has the proper normalization.

1. That Eq. 11.136 is a solution of the modified Bessel equation may be verified by direct substitution. We obtain

$$z^{\nu+1}\int_1^\infty \frac{d}{dx}[e^{-zx}(x^2 - 1)^{\nu + 1/2}]\,dx = 0,$$

which transforms the combined integrand into the derivative of a function that vanishes at both end points. The rejection of the possibility that this solution contains I_ν constitutes Exercise 11.6.1.

2. The normalization may be verified by substituting $x = 1 + t/z$.

$$\begin{aligned}\frac{\pi^{1/2}}{(\nu - \frac{1}{2})!}\left(\frac{z}{2}\right)^\nu &\int_1^\infty e^{-zx}(x^2 - 1)^{\nu - 1/2}\,dx \\ &= \frac{\pi^{1/2}}{(\nu - \frac{1}{2})!}\left(\frac{z}{2}\right)^\nu e^{-z}\int_0^\infty e^{-t}\left(\frac{t^2}{z^2} + \frac{2t}{z}\right)^{\nu - 1/2}\frac{dt}{z} \\ &= \frac{\pi^{1/2}}{(\nu - \frac{1}{2})!}\frac{e^{-z}}{2^\nu z^\nu}\int_0^\infty e^{-t}t^{2\nu - 1}\left(1 + \frac{2z}{t}\right)^{\nu - 1/2}dt.\end{aligned} \tag{11.137}$$

This substitution has changed the limits of integration to a more convenient range and has isolated the negative exponential dependence, e^{-z}. The last integral in Eq. 11.137 may be evaluated for $z = 0$ to yield $(2\nu - 1)!$. Then, using the duplication formula (Section 10.4),

$$\lim_{z\to 0} K_\nu(z) = \frac{(\nu - 1)!2^{\nu - 1}}{z^\nu}, \qquad \nu > 0, \tag{11.138}$$

in agreement with Eq. 11.119, thus checking the normalization.[1]

Now to develop an asymptotic series for $K_\nu(z)$. Equation 11.136 may be rewritten (for large z) as

$$K_\nu(z) = \sqrt{\frac{\pi}{2z}} \frac{e^{-z}}{(\nu - \frac{1}{2})!} \int_0^\infty e^{-t} t^{\nu - 1/2} \left(1 + \frac{t}{2z}\right)^{\nu - 1/2} dt.$$

We expand $(1 + t/2z)^{\nu - 1/2}$ by the binomial theorem to obtain

$$K_\nu(z) = \sqrt{\frac{\pi}{2z}} \frac{e^{-z}}{(\nu - \frac{1}{2})!} \sum_{r=0}^{\infty} \frac{(\nu - \frac{1}{2})!}{r!(\nu - r - \frac{1}{2})!} (2z)^{-r} \int_0^\infty e^{-t} t^{\nu + r - 1/2} dt. \tag{11.139}$$

Term-by-term integration (valid for asymptotic series) yields the desired asymptotic expansion of $K_\nu(z)$.

$$K_\nu(z) \sim \sqrt{\frac{\pi}{2z}} e^{-z} \left[1 + \frac{(4\nu^2 - 1^2)}{1!\,8z} + \frac{(4\nu^2 - 1^2)(4\nu^2 - 3^2)}{2!(8z)^2} + \cdots\right]. \tag{11.140}$$

Although the integral of Eq. 11.136, integrating along the real axis, was convergent only for $-\pi/2 < \arg z < +\pi/2$, Eq. 11.140 may be extended to $-3\pi/2 < \arg z < 3\pi/2$. Considered as an infinite series, Eq. 11.140 is actually divergent. However, this series is asymptotic in the sense that for large enough z, $K_\nu(z)$ may be approximated to any fixed degree of accuracy. (Cf. Section 5.11 for a definition and discussion of asymptotic series.)

It is convenient to rewrite Eq. 11.140 as

$$K_\nu(z) = \sqrt{\frac{\pi}{2z}} e^{-z} [P_\nu(iz) + iQ_\nu(iz)] \tag{11.140a}$$

where

$$P_\nu(z) \sim 1 - \frac{(\mu - 1)(\mu - 9)}{2!(8z)^2} + \frac{(\mu - 1)(\mu - 9)(\mu - 25)(\mu - 49)}{4!(8z)^4} - \cdots, \tag{11.140b}$$

$$Q_\nu(z) \sim \frac{\mu - 1}{1!(8z)} - \frac{(\mu - 1)(\mu - 9)(\mu - 25)}{3!(8z)^3} + \cdots, \tag{11.140c}$$

and

$$\mu = 4\nu^2.$$

Then with the asymptotic form of $K_\nu(z)$, Eq. 11.140*a*, we can obtain expansions for all the other Bessel and hyperbolic Bessel functions by the defining relations:

1. From

$$\frac{\pi}{2} i^{\nu + 1} H_\nu^{(1)}(iz) = K_\nu(z) \tag{11.141}$$

[1] For $\nu = 0$ the integral diverges logarithmically in agreement with the logarithmic divergence of $K_0(z)$ (Section 11.5).

we have

$$H_\nu^{(1)}(z) = \sqrt{\frac{2}{\pi z}} \exp i\left[z - \left(\nu + \frac{1}{2}\right)\frac{\pi}{2}\right] \cdot [P_\nu(z) + iQ_\nu(z)],$$
$$-\pi < \arg z < 2\pi. \qquad (11.142)$$

2. The second Hankel function is just the complex conjugate of the first (for real argument),

$$H_\nu^{(2)}(z) = \sqrt{\frac{2}{\pi z}} \exp -i\left[z - \left(\nu + \frac{1}{2}\right)\frac{\pi}{2}\right] \cdot [P_\nu(z) - iQ_\nu(z)],$$
$$-2\pi < \arg z < \pi. \qquad (11.143)$$

An alternate derivation of the asymptotic behavior of the Hankel functions appears in Section 7.4 as an application of the method of steepest descents.

3. Since $J_\nu(z)$ is the real part of $H_\nu^{(1)}(z)$,

$$J_\nu(z) = \sqrt{\frac{2}{\pi z}}\left\{P_\nu(z) \cos\left[z - \left(\nu + \frac{1}{2}\right)\frac{\pi}{2}\right] - Q_\nu(z) \sin\left[z - \left(\nu + \frac{1}{2}\right)\frac{\pi}{2}\right]\right\},$$
$$-\pi < \arg z < \pi.$$
$$(11.144)$$

4. The Neumann function is the imaginary part of $H_\nu^{(1)}(z)$, or

$$N_\nu(z) = \sqrt{\frac{2}{\pi z}}\left\{P_\nu(z) \sin\left[z - \left(\nu + \frac{1}{2}\right)\frac{\pi}{2}\right] + Q_\nu(z) \cos\left[z - \left(\nu + \frac{1}{2}\right)\frac{\pi}{2}\right]\right\},$$
$$-\pi < \arg z < \pi.$$
$$(11.145)$$

5. Finally, the regular hyperbolic or modified Bessel function $I_\nu(z)$ is given by

$$I_\nu(z) = i^{-\nu} J_\nu(iz) \qquad (11.146)$$

or

$$I_\nu(z) = \frac{e^z}{\sqrt{2\pi z}} [P_\nu(iz) - iQ_\nu(iz)], \qquad -\frac{\pi}{2} < \arg z < \frac{\pi}{2}. \qquad (11.146a)$$

This completes our determination of the asymptotic expansions. However, it is perhaps worth noting the primary characteristics. Apart from the ubiquitous $z^{-1/2}$, J_ν and N_ν behave as cosine and sine, respectively. The zeros are *almost* evenly spaced at intervals of π; the spacing becomes exactly π in the limit as $z \to \infty$. The Hankel functions have been defined to behave like the imaginary exponentials, and the modified Bessel functions, I_ν and K_ν, go into the positive and negative exponentials. This asymptotic behavior may be sufficient to eliminate immediately one of these functions as a solution for a physical problem. Indeed, in Section 11.5 this was the basis for eliminating I_0 as a solution in the infinite medium neutron diffusion problem.

It is of some interest to consider the accuracy of the asymptotic forms taking just the first term, for example,

$$J_n(x) \approx \sqrt{\frac{2}{\pi x}} \cos\left[x - \left(n + \frac{1}{2}\right)\left(\frac{\pi}{2}\right)\right] \tag{11.147}$$

A comparison with the exact values of $J_0(x)$ appears in Fig. 11.10.

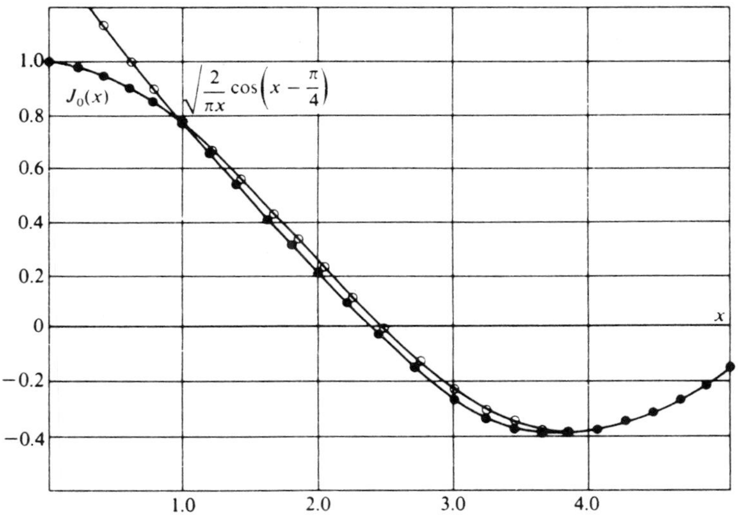

FIG. 11.10 Asymptotic approximation of $J_0(x)$

As another possible measure of the accuracy of Eq. 11.147, we might compare the locations of the zeros given by Eq. 11.147 with the exact zeros or roots, α_{ns} for the sth root of $J_n(x)$, from Table 11.1. The discrepancy

$$\delta_{ns} = \left[s + \frac{n}{2} - \frac{1}{4}\right]\pi - \alpha_{ns}, \tag{11.147a}$$

is -0.0486 for α_{01}, the first zero of $J_0(x)$, and drops rapidly for $s > 1$. Moving to $J_1(x)$, $\delta_{11} = 0.0953$, $\delta_{12} = 0.0530$, etc. For larger n, the accuracy drops rapidly even for relatively large values of s (and x). Clearly the condition for the validity of Eq. 11.147 is that the sine term be negligible, i.e., that

$$8x \gg 4n^2 - 1. \tag{11.147b}$$

For n or $\nu > 1$, the asymptotic region may be far out.

As pointed out in Section 11.3, the asymptotic forms may be used to evaluate the various Wronskian formulas (cf. Exercise 11.6.3).

Numerical Evaluation. When a program in a large highspeed computing machine calls for one of the Bessel or modified Bessel functions, the programmer has two alternatives. He can store all the Bessel functions and tell the computer how to locate the required value, or he can instruct the computer to simply calculate the needed value. The first alternative would be fairly slow and would place unreasonable demands on the storage capacity. So our programmer adopts the "compute it yourself" alternative.

The computation of $J_n(x)$ using the recurrence relation, Eq. 11.10 is discussed in Section 11.1. For N_n, I_n, and K_n, the preferred methods are the series if x is small and the asymptotic forms (with many terms in the series of negative powers) if x is large. The criteria of large and small may vary as shown in Table 11.2.

TABLE 11.2

Equations for the Computation of Neumann and the Modified Bessel Functions

	Power series	Asymptotic series
$N_n(x)$	Eq. 11.63, $x \leqslant 4$	Eq. 11.145, $x > 4$
$I_n(x)$	Eq. 11.112, $x \leqslant 12$ or $\leqslant n$	Eq. 11.146*a*, $x > 12$ and $> n$
$K_n(x)$	Eq. 11.119, $x \leqslant 1$	Eq. 11.140*a*, $x > 1$

In actual practice, it is found convenient to limit the series (power or asymptotic) computation of $N_n(x)$ and $K_n(x)$ to $n = 0, 1$. Then $N_n(x)$, $n \geqslant 2$ is computed using the recurrence relation, Eq. 11.10. $K_n(x)$, $n \geqslant 2$ is computed using the recurrence relations of Ex. 11.5.7. $I_n(x)$ could be handled this way, if desired, but direct application of the power series or asymptotic series is feasible for all values of n and x.

EXERCISES

11.6.1 In checking the normalization of the integral representation of $K_\nu(z)$ (Eq. 11.136), we assumed that $I_\nu(z)$ was not present. How do we know that the integral representation (Eq. 11.136) does not yield $K_\nu(z) + \varepsilon I_\nu(z)$ with $\varepsilon \neq 0$?

11.6.2 (a) Show that

$$y(z) = z^\nu \int e^{-zt}(t^2 - 1)^{\nu-1/2}\, dt$$

satisfies the modified Bessel equation, provided the contour is chosen so that

$$e^{-zt}(t^2 - 1)^{\nu+1/2}$$

has the same value at the initial and final points of the contour.

(b) Verify that the contours shown in Fig. 11.11 are suitable for this problem.

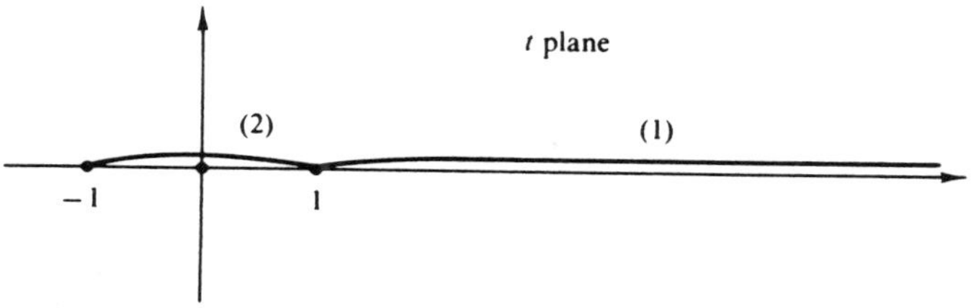

FIG. 11.11 Modified Bessel function contours

11.6.3 Use the asymptotic expansions to verify the following Wronskian formulas:

(a) $J_\nu(x)\, J_{-\nu-1}(x) + J_{-\nu}(x)\, J_{\nu+1}(x) = -\dfrac{2 \sin \nu\pi}{\pi x}$,

(b) $J_\nu(x)\, N_{\nu+1}(x) - J_{\nu+1}(x)\, N_\nu(x) = -\dfrac{2}{\pi x}$,

(c) $J_\nu(x)\, H^{(2)}_{\nu-1}(x) - J_{\nu-1}(x)\, H^{(2)}_\nu(x) = \dfrac{2}{i\pi x}$,

(d) $I_\nu(x)\, K'_\nu(x) - I'_\nu(x)\, K_\nu(x) = -\dfrac{1}{x}$,

(e) $I_\nu(x)\, K_{\nu+1}(x) + I_{\nu+1}(x)\, K_\nu(x) = \dfrac{1}{x}$.

11.6.4 Show that

$$K_0(x) = \int_0^\infty J_0(xt) \frac{t}{1+t^2}\, dt.$$

Hint: (a) Show that the K_0 defined here satisfies the modified Bessel equation. (b) Show that in the limit as $x \to \infty$ this $K_0(x) \to 0$ thus excluding $I_0(x)$.

11.7 Spherical Bessel Functions

When the Helmholtz equation in spherical polar (or conical) coordinates is separated, the radial equation has the form

$$r^2 \frac{d^2R}{dr^2} + 2r \frac{dR}{dr} + [k^2r^2 - n(n+1)]R = 0; \tag{11.148}$$

k enters from the Helmholtz equation and the integer n is a separation constant which often has the physical interpretation of angular momentum. Clearly, this is *not* Bessel's equation, but if we substitute

$$R(kr) = \frac{Z(kr)}{(kr)^{1/2}}$$

Eq. 11.148 becomes

$$r^2 \frac{d^2Z}{dr^2} + r \frac{dZ}{dr} + \left[k^2r^2 - \left(n + \frac{1}{2}\right)^2\right]Z = 0, \tag{11.149}$$

which *is* Bessel's equation. Z is a Bessel function of order $n + \frac{1}{2}$ (n an integer). Because of the importance of spherical coordinates this combination,

$$\frac{Z_{n+1/2}(kr)}{(kr)^{1/2}},$$

occurs quite often.

Definitions. It is convenient to label these functions spherical Bessel functions with the following defining equations:

$$
\begin{aligned}
j_n(x) &= \sqrt{\frac{\pi}{2x}}\, J_{n+1/2}(x), \\
n_n(x) &= \sqrt{\frac{\pi}{2x}}\, N_{n+1/2}(x) = (-1)^{n+1}\sqrt{\frac{\pi}{2x}}\, J_{-n-1/2}(x),\ ^{1} \\
h_n^{(1)}(x) &= \sqrt{\frac{\pi}{2x}}\, H_{n+1/2}^{(1)}(x) = j_n(x) + i\, n_n(x), \\
h_n^{(2)}(x) &= \sqrt{\frac{\pi}{2x}}\, H_{n+1/2}^{(2)}(x) = j_n(x) - i\, n_n(x).
\end{aligned}
\tag{11.150}
$$

These spherical Bessel functions can be expressed in series form by using the series (Eq. 11.5) for J_n, replacing n with $n + \frac{1}{2}$.

$$J_{n+1/2}(x) = \sum_{s=0}^{\infty} \frac{(-1)^s}{s!(s+n+\frac{1}{2})!}\left(\frac{x}{2}\right)^{2s+n+1/2} \tag{11.151}$$

Using the Legendre duplication formula,

$$z!(z+\tfrac{1}{2})! = 2^{-2z-1}\pi^{1/2}(2z+1)!, \tag{11.152}$$

we have

$$
\begin{aligned}
j_n(x) &= \sqrt{\frac{\pi}{2x}} \sum_{s=0}^{\infty} \frac{(-1)^s 2^{2s+2n+1}(s+n)!}{\pi^{1/2}(2s+2n+1)!s!}\left(\frac{x}{2}\right)^{2s+n+1/2} \\
&= 2^n x^n \sum_{s=0}^{\infty} \frac{(-1)^s (s+n)!}{s!(2s+2n+1)!} x^{2s}.
\end{aligned}
\tag{11.153}
$$

Now $N_{n+1/2}(x) = (-1)^{n+1} J_{-n-1/2}(x)$ and from Eq. 11.5 we find that

$$J_{-n-1/2}(x) = \sum_{s=0}^{\infty} \frac{(-1)^s}{s!(s-n-\frac{1}{2})!}\left(\frac{x}{2}\right)^{2s-n-1/2} \tag{11.154}$$

This yields

$$n_n(x) = (-1)^{n+1}\frac{2^n \pi^{1/2}}{x^{n+1}} \sum_{s=0}^{\infty} \frac{(-1)^s}{s!(s-n-\frac{1}{2})!}\left(\frac{x}{2}\right)^{2s} \tag{11.155}$$

The Legendre duplication formula can be used again to give

$$n_n(x) = \frac{(-1)^{n+1}}{2^n x^{n+1}} \sum_{s=0}^{\infty} \frac{(-1)^s (s-n)!}{s!(2s-2n)!} x^{2s}. \tag{11.156}$$

However, this is awkward to handle for n, a positive integer, because of the factorials in both numerator and denominator. These spherical Bessel functions are closely related to the trigonometric functions, as may be seen by considering the special case, $n = 0$. We find

[1] This is possible because $\cos(n+\frac{1}{2})\pi = 0$.

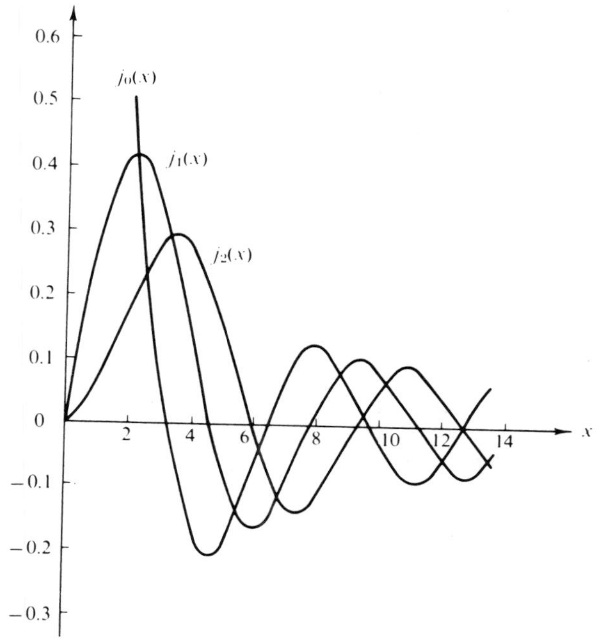

FIG. 11.12 Spherical Bessel functions

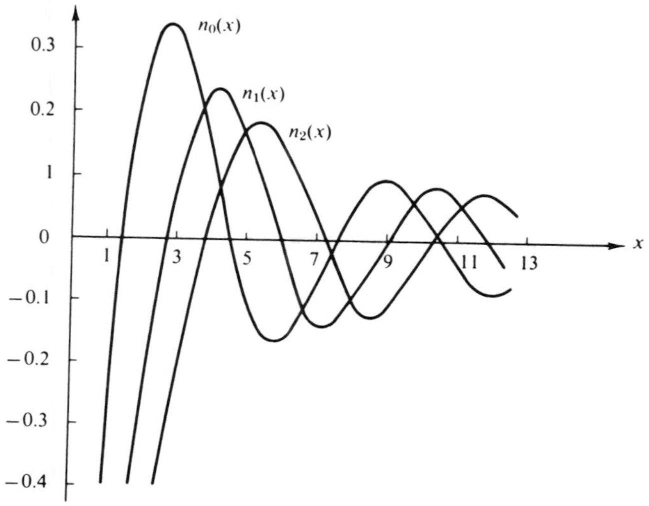

FIG. 11.13 Spherical Neumann functions

$$j_0(x) = \sum_{s=0}^{\infty} \frac{(-1)^s}{(2s+1)!} x^{2s}$$

$$= \frac{\sin x}{x}, \tag{11.157}$$

whereas for n_0 Eq. 11.156 yields

$$n_0(x) = -\frac{\cos x}{x}. \tag{11.158}$$

From the definition of the spherical Hankel functions (Eq. 11.150),

$$\begin{aligned} h_0^{(1)}(x) &= \frac{1}{x}(\sin x - i\cos x) = -\frac{i}{x}e^{ix} \\ h_0^{(2)}(x) &= \frac{1}{x}(\sin x + i\cos x) = \frac{i}{x}e^{-ix} \end{aligned} \tag{11.159}$$

The relationship to the trigonometric functions may perhaps best be seen by considering the asymptotic series of Section 11.6. For $\nu = \pm\frac{1}{2}, \pm\frac{3}{2}, \pm\frac{5}{2}$, etc., the asymptotic series terminate, giving finite combinations of trigonometric functions.

Limiting values. For $x \ll 1$,[1] Eqs. 11.153 and 11.156 yield

$$j_n(x) \approx \frac{2^n n!}{(2n+1)!} x^n = \frac{x^n}{(2n+1)!!} \tag{11.160a}$$

$$n_n(x) \approx \frac{(-1)^{n+1}}{2^n} \cdot \frac{(-n)!}{(-2n)!} x^{-n-1} \tag{11.160b}$$

$$= -\frac{(2n)!}{2^n n!} x^{-n-1} = -(2n-1)!!\, x^{-n-1}.$$

The transformation of factorials in the expressions for $n_n(x)$ employs Ex. 10.1.3. The limiting values of the spherical Hankel functions go as $\pm i n_n(x)$.

The asymptotic values of j_n, n_n, $h_n^{(2)}$, and $h_n^{(1)}$ may be obtained from the Bessel asymptotic forms, Section 11.6,

$$j_n(x) \sim \frac{1}{x}\sin\left(x - \frac{n\pi}{2}\right), \tag{11.161a}$$

$$n_n(x) \sim -\frac{1}{x}\cos\left(x - \frac{n\pi}{2}\right), \tag{11.161b}$$

$$h_n^{(1)}(x) \sim (-i)^{n+1}\frac{e^{ix}}{x}, \tag{11.161c}$$

$$h_n^{(2)}(x) \sim i^{n+1}\frac{e^{-ix}}{x}. \tag{11.161d}$$

The condition for these spherical Bessel forms is that $x \gg n(n+1)/2$.

[1] The condition that the second term in the series be negligible compared to the first is actually $x \ll 2[(2n+2)(2n+3)/(n+1)]^{1/2}$ for $j_n(x)$.

Recurrence relations. The recurrence relations to which we now turn provide a convenient way of developing the higher-order spherical Bessel functions. These recurrence relations may be derived from the series, but as with the modified Bessel functions it is easier to substitute into the known recurrence relations (Eqs. 11.10 and 11.12). This gives

$$f_{n-1}(x) + f_{n+1}(x) = \frac{2n+1}{x} f_n(x), \tag{11.162a}$$

$$n f_{n-1}(x) - (n+1) f_{n+1}(x) = (2n+1) f_n'(x). \tag{11.162b}$$

Rearranging these relations (or substituting into Eqs. 11.15 and 11.17), we obtain

$$\frac{d}{dx}[x^{n+1} f_n(x)] = x^{n+1} f_{n-1}(x), \tag{11.163a}$$

$$\frac{d}{dx}[x^{-n} f_n(x)] = -x^{-n} f_{n+1}(x). \tag{11.163b}$$

Here f_n may represent j_n, n_n, $h_n^{(1)}$, or $h_n^{(2)}$.

Applying Eq. 11.163*b*, we quickly get

$$\begin{aligned} j_1(x) &= \frac{\sin x}{x^2} - \frac{\cos x}{x}, \\ j_2(x) &= \left(\frac{3}{x^3} - \frac{1}{x}\right)\sin x - \frac{3}{x^2}\cos x, \end{aligned} \tag{11.164}$$

$$\begin{aligned} n_1(x) &= -\frac{\cos x}{x^2} - \frac{\sin x}{x}, \\ n_2(x) &= -\left(\frac{3}{x^3} - \frac{1}{x}\right)\cos x - \frac{3}{x^2}\sin x, \text{ etc.} \end{aligned} \tag{11.165}$$

By mathematical induction we may establish the Rayleigh formulas

$$j_n(x) = (-1)^n x^n \left(\frac{d}{x\,dx}\right)^n \left(\frac{\sin x}{x}\right), \tag{11.166}$$

$$n_n(x) = -(-1)^n x^n \left(\frac{d}{x\,dx}\right)^n \left(\frac{\cos x}{x}\right). \tag{11.167}$$

We note that the spherical Bessel functions $j_n(x)$ and $n_n(x)$ can always be expressed as $\sin x$ and $\cos x$ with coefficients that are polynomials involving negative powers of x. For the spherical Hankel functions

$$\begin{aligned} h_n^{(1)}(x) &= -i(-1)^n x^n \left(\frac{d}{x\,dx}\right)^n \left(\frac{e^{ix}}{x}\right) \\ h_n^{(2)}(x) &= i(-1)^n x^n \left(\frac{d}{x\,dx}\right)^n \left(\frac{e^{-ix}}{x}\right). \end{aligned} \tag{11.168}$$

Numerical computation. The spherical Bessel and modified Bessel functions may be computed using the same techniques described in Sections 11.1 and 11.6 for evaluating the Bessel functions. For $j_n(x)$ and $i_n(x)$[1] it is convenient to use Eq. 11.162*a* and Ex. 11.7.19 and work *downward*, as is done for $J_n(x)$. Normalization is accomplished by comparing with the known forms of $j_0(x)$ and $i_0(x)$, Eq. 11.157 and Ex. 11.7.16. For $n_n(x)$ and $k_n(x)$, Eq. 11.162*a* and Ex. 11.7.20 are used again, but this time working *upward*, starting with the known forms of $n_0(x)$, $n_1(x)$, $k_0(x)$, and $k_1(x)$, Eq. 11.165 and Ex. 11.7.18.

Orthogonality. We may take the orthogonality integral for the ordinary Bessel functions (Eq. 11.50),

$$\int_0^a J_\nu\left(\alpha_{\nu p}\frac{\rho}{a}\right)J_\nu\left(\alpha_{\nu q}\frac{\rho}{a}\right)\rho\,d\rho = \frac{a^2}{2}[J_{\nu+1}(\alpha_{\nu p})]^2\,\delta_{pq} \tag{11.169}$$

and substitute in the expression for j_n to obtain

$$\int_0^a j_n\left(\alpha_{np}\frac{\rho}{a}\right)j_n\left(\alpha_{nq}\frac{\rho}{a}\right)\rho^2\,d\rho = \frac{a^3}{2}[j_{n+1}(\alpha_{np})]^2\,\delta_{pq}. \tag{11.170}$$

Here α_{np} and α_{nq} are roots of j_n.

This represents orthogonality with respect to the roots of the Bessel functions. An illustration of this sort of orthogonality is provided later in this section by the problem of a particle in a sphere. Equation 11.170 guarantees orthogonality of the wave functions $j_n(r)$ for fixed n. (If n varies, the spherical harmonic will provide orthogonality.) Another form, orthogonality with respect to the indices, may be developed as follows. From Eq. 11.148, with $x = kr$,

$$\begin{aligned} x^2 j_n'' + 2xj_n' + [x^2 - n(n+1)]j_n &= 0, \\ x^2 j_m'' + 2xj_m' + [x^2 - m(m+1)]j_m &= 0. \end{aligned} \tag{11.171}$$

We multiply the first equation by j_m, the second by j_n, and subtract:

$$\begin{aligned} [n(n+1) - m(m+1)]j_n j_m &= x^2(j_n'' j_m - j_n j_m'') + 2x(j_n' j_m - j_n j_m') \\ &= \frac{d}{dx}[x^2(j_n' j_m - j_n j_m')]. \end{aligned} \tag{11.172}$$

The last equation is an expression of the Sturm-Liouville form of Eq. 11.148 (cf. Section 9.1). Integrating from zero to infinity, we obtain

$$[n(n+1) - m(m+1)]\int_0^\infty j_m(x)\,j_n(x)\,dx = \left|x^2(j_n' j_m - j_n j_m')\right|_0^\infty. \tag{11.173}$$

The right-hand side clearly vanishes for $x = 0$ (m, n are assumed to be non-negative integers).

As $x \to \infty$, it will also vanish provided $(n - m)$ is even, as shown by examination of the asymptotic forms

[1] The spherical modified Bessel functions, $i_n(x)$ and $k_n(x)$, are defined in Ex. 11.7.16.

$$j_n(x) \to \frac{1}{x} \sin\left(x - \frac{n\pi}{2}\right),$$

$$j_n'(x) \to \frac{1}{x} \cos\left(x - \frac{n\pi}{2}\right).$$

We find

$$x^2(j_n' j_m - j_n j_m') \to \cos\left(x - \frac{n\pi}{2}\right)\sin\left(x - \frac{m\pi}{2}\right) - \sin\left(x - \frac{n\pi}{2}\right)\cos\left(x - \frac{m\pi}{2}\right) + O(x^{-1})$$

$$= \sin\frac{\pi}{2}(n - m) + O(x^{-1}).$$

Therefore

$$\int_0^\infty j_m(x)\, j_n(x)\, dx = \frac{\sin[(n-m)\pi/2]}{n(n+1) - m(m+1)}. \tag{11.174}$$

The right-hand side of Eq. 11.174 vanishes for $(n - m)$ even. If $(n - m)$ is odd, one of the spherical Bessel functions, say j_m, will be odd, whereas the other will be even,

$$j_p(x) = (-1)^p\, j_p(-x). \tag{11.175}$$

Hence the integral will vanish if we extend the lower limit to $-\infty$. We then have

$$\int_{-\infty}^\infty j_m(x)\, j_n(x)\, dx = 0, \qquad m \neq n, \qquad m, n \geqslant 0. \tag{11.176}$$

If $m = n$ (cf. Exercise 11.7.11), we have

$$\int_{-\infty}^\infty [j_n(x)]^2\, dx = \frac{\pi}{2n+1}. \tag{11.177}$$

Example 11.7.1. Particle in a Sphere

An illustration of the use of the spherical Bessel functions is provided by the problem of a quantum mechanical particle in a sphere of radius a. Quantum theory requires that the wave function ψ, describing our particle, satisfy

$$-\frac{\hbar^2}{2m}\nabla^2\psi = E\psi, \tag{11.178}$$

and the boundary conditions (a) $\psi(r \leqslant a)$ remains finite, (b) $\psi(a) = 0$. This corresponds to a potential $V = 0$, $r \leqslant a$, and $V = \infty$, $r > a$. Here $\hbar$ is Planck's constant (divided by 2π), m, the mass of our particle, and E, its energy. Let us determine the *minimum* value of the energy for which our wave equation has an acceptable solution. Equation 11.178 is just Helmholtz' equation with a radial part (cf. Section 2.5 for separation of variables)

$$\frac{d^2R}{dr^2} + \frac{2}{r}\frac{dR}{dr} + \left[\frac{2mE}{\hbar^2} - \frac{n(n+1)}{r^2}\right]R = 0. \tag{11.179}$$

Hence by Eq. 11.148, with $n = 0$,

$$R = Aj_0\left(\frac{\sqrt{2mE}}{\hbar}r\right) + Bn_0\left(\frac{\sqrt{2mE}}{\hbar}r\right) \tag{11.180}$$

We choose the index $n = 0$, for any angular dependence would raise the energy. The spherical Neumann function is rejected because of its divergent behavior at the origin. Technically, the spherical Neumann function n_0 is a Green's function satisfying Green's equation and *not* satisfying the Schrödinger wave equation *at the origin.* To satisfy the second boundary condition (for all angles), we require

$$\frac{\sqrt{2mE}}{\hbar}a = \alpha,$$

where α is a root of j_0, that is, $j_0(\alpha) = 0$. This has the effect of limiting the allowable energies to a certain discrete set or, in other words, application of boundary condition (b) quantizes the energy E. The smallest α is the first zero of j_0,

$$\alpha = \pi$$

and

$$E_{\min} = \frac{\pi^2\hbar^2}{2ma^2} = \frac{h^2}{8ma^2}, \tag{11.181}$$

which means that for any finite sphere the particle will have a positive minimum or zero-point energy. This is an illustration of the Heisenberg uncertainty principle.

In solid state physics, astrophysics, and other areas of physics, we may wish to know how many different solutions (energy states) there are corresponding to energies less than or equal to some fixed energy E_0. For a cubic volume (Ex. 2.5.3) the problem is fairly simple. The considerably more difficult spherical case is worked out by R. H. Lambert, *Am. J. Phys.* **36**, 417 and 1169 (1968).

The spherical Bessel functions will enter again in connection with spherical waves, but further consideration is postponed until the corresponding angular functions, the Legendre functions, have been introduced.

EXERCISES

11.7.1 Show that if

$$n_n(x) = \sqrt{\frac{\pi}{2x}}N_{n+1/2}(x),$$

it automatically equals

$$(-1)^{n+1}\sqrt{\frac{\pi}{2x}}J_{-n-1/2}(x).$$

11.7.2 Verify that, to within the factor $\sqrt{\pi/2x}$, j_0, j_1, n_0, and n_1 are given by the asymptotic forms of J_ν and N_ν; $\nu = \frac{1}{2}, \frac{3}{2}$ (Eqs. 11.144 and 11.145).

11.7.3 Use the integral representation of $J_\nu(x)$,

$$J_\nu(x) = \frac{1}{\pi^{1/2}(\nu - \frac{1}{2})!}\left(\frac{x}{2}\right)^\nu \int_{-1}^{1} e^{\pm ixp}(1 - p^2)^{\nu-1/2}\,dp,$$

to show that the spherical Bessel functions $j_n(x)$ are expressible in terms of trigonometric functions; that is,

$$j_0(x) = \frac{\sin x}{x},$$

$$j_1(x) = \frac{\sin x}{x^2} - \frac{\cos x}{x}.$$

11.7.4 Split the series form of $n_n(x)$, Eq. 11.156, into two parts, one with $0 \leq s \leq n$, and a second with $s > n$. Rewrite the general term in the first part in terms of factorials of nonnegative argument. As a partial check on your series verify that $n_1(x) = -(\cos x)/x^2 - (\sin x)/x$.

$$\textit{Ans. } n_n(x) = \frac{-1}{2^n x^{n+1}} \sum_{s=0}^{n} \frac{(2n-2s)!}{(n-s)!} \frac{x^{2s}}{s!}$$

$$+ \frac{(-1)^{n+1}}{2^n x^{n+1}} \sum_{s=n+1}^{\infty} (-1)^s \frac{(s-n)!}{(2s-2n)!} \frac{x^{2s}}{s!}.$$

Hint. Cf. Ex. 10.1.3.

11.7.5 (a) Derive the recurrence relations

$$f_{n-1}(x) + f_{n+1}(x) = \frac{2n+1}{x} f_n(x),$$

$$n f_{n-1}(x) - (n+1) f_{n+1}(x) = (2n+1) f_n'(x),$$

satisfied by the spherical Bessel functions, $j_n(x)$, $n_n(x)$, $h_n^{(1)}(x)$, and $h_n^{(2)}(x)$.

(b) Show, from these two recurrence relations, that the spherical Bessel function $f_n(x)$ satisfies the differential equation

$$x^2 f_n''(x) + 2x f_n'(x) + [x^2 - n(n+1)] f_n(x) = 0.$$

11.7.6 Prove by mathematical induction that

$$j_n(x) = (-1)^n x^n \left(\frac{d}{xdx}\right)^n \left(\frac{\sin x}{x}\right)$$

for n an arbitrary non-negative integer.

11.7.7 From the discussion of orthogonality of the spherical Bessel functions, show that a Wronskian relation for $j_n(x)$ and $n_n(x)$ is

$$j_n(x)\, n_n'(x) - j_n'(x)\, n_n(x) = \frac{1}{x^2}.$$

11.7.8 Verify

$$h_n^{(1)}(x)\, h_n^{(2)\prime}(x) - h_n^{(1)\prime}(x)\, h_n^{(2)}(x) = -\frac{2i}{x^2}.$$

11.7.9 Verify Poisson's integral representation of the spherical Bessel function,

$$j_n(z) = \frac{z^n}{2^{n+1} n!} \int_0^{\pi} \cos(z \cos\theta) \sin^{2n+1}\theta \, d\theta.$$

11.7.10 From Eq. 11.173 show that Eq. 11.174 follows and also that

$$\int_0^{\infty} J_\mu(x)\, J_\nu(x) \frac{dx}{x} = \frac{2}{\pi} \frac{\sin[(\mu-\nu)\pi/2]}{\mu^2 - \nu^2}, \qquad \mu + \nu > -1.$$

11.7.11 Derive Eq. 11.177.

$$\int_{-\infty}^{\infty} [j_n(x)]^2\, dx = \frac{\pi}{2n+1}.$$

11.7.12 Set up the orthogonality integral for $j_L(kr)$ in a sphere of radius R with the boundary condition

$$j_L(kR) = 0.$$

The result is used in classifying electromagnetic radiation according to its angular momentum.

11.7.13 The Fresnel integrals occurring in diffraction theory are given by

$$x(t) = \int_0^t \cos(v^2)\, dv,$$

$$y(t) = \int_0^t \sin(v^2)\, dv.$$

Show that these integrals may be expanded in series of spherical Bessel functions,

$$x(s) = \frac{1}{2}\int_0^s j_{-1}(u)u^{1/2}\, du = s^{1/2}\sum_{n=0}^{\infty} j_{2n}(s),$$

$$y(s) = \frac{1}{2}\int_0^s j_0(u)u^{1/2} du = s^{1/2}\sum_{n=0}^{\infty} j_{2n+1}(s).$$

Hint. To establish the equality of the integral and the sum, you may wish to work with their derivatives. The spherical Bessel analogs of Eqs. 11.12 and 11.14 are helpful.

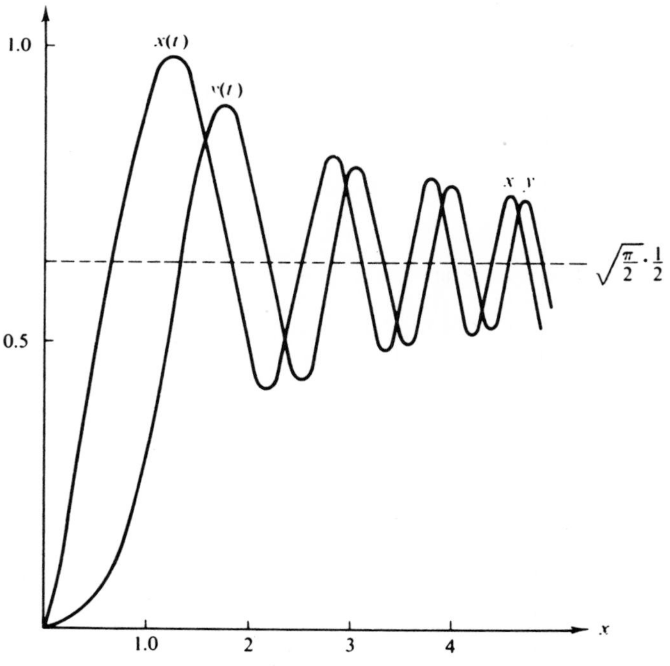

FIG. 11.14 **Fresnel integrals**

11.7.14 A hollow sphere of radius a (Helmholtz resonator) contains standing sound waves. Find the minimum frequency of oscillation in terms of the radius a and the velocity of sound v. The sound waves satisfy the wave equation

$$\nabla^2\psi = \frac{1}{v^2}\frac{\partial^2\psi}{\partial t^2}$$

and the boundary condition

$$\frac{\partial \psi}{\partial r} = 0, \qquad r = a.$$

This is a Neumann boundary condition. Example 11.7.1 has the same differential equation but with a Dirichlet boundary condition.

Ans. $\nu_{\min} = 0.3313\, v/a$,
$\lambda_{\max} = 3.018a$.

11.7.15 Show that Green's functions for the radial part of the three dimensional Helmholtz equation (singularity or source at the origin) are

(a) $-\dfrac{k}{4\pi} n_0(kr)$, standing wave,

(b) $\dfrac{ik}{4\pi} h_0^{(1)}(kr)$, outgoing wave (assume time dependence $e^{-i\omega t}$).

The generalization to a source away from the origin is covered in Section 16.6.

11.7.16 Defining the spherical modified Bessel functions by

$$i_n(x) = \sqrt{\frac{\pi}{2x}}\, I_{n+1/2}(x),$$

$$k_n(x) = \sqrt{\frac{2}{\pi x}}\, K_{n+1/2}(x),$$

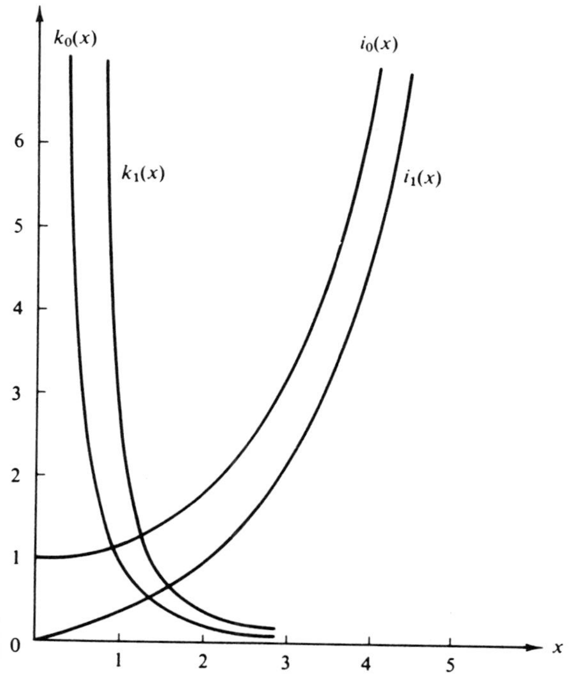

FIG. 11.15 Spherical modified Bessel functions

show that

$$i_0(x) = \frac{\sinh x}{x}$$

$$k_0(x) = \frac{e^{-x}}{x}.$$

Note that the numerical factors in the definitions of i_n and k_n are *not* identical.

11.7.17 (a) Show that the parity of $i_n(x)$ is $(-1)^n$.
(b) Show that $k_n(x)$ has no definite parity.

11.7.18 Show that the spherical modified Bessel functions satisfy the following relations:

(a) $i_n(x) = i^{-n} j_n(ix)$,

$k_n(x) = -i^n h_n^{(1)}(ix)$,

(b) $i_{n+1}(x) = x^n \dfrac{d}{dx}(x^{-n} i_n)$,

$k_{n+1}(x) = -x^n \dfrac{d}{dx}(x^{-n} k_n)$,

(c) $i_n(x) = x^n \left(\dfrac{d}{x\,dx}\right)^n \dfrac{\sinh x}{x}$,

$k_n(x) = (-1)^n x^n \left(\dfrac{d}{x\,dx}\right)^n \dfrac{e^{-x}}{x}$.

11.7.19 Show that the recurrence relations for $i_n(x)$ and $k_n(x)$ are

(a) $$i_{n-1}(x) - i_{n+1}(x) = \frac{2n+1}{x} i_n(x),$$

$$n\, i_{n-1}(x) + (n+1)\, i_{n+1}(x) = (2n+1)\, i_n'(x),$$

(b) $$k_{n-1}(x) - k_{n+1}(x) = -\frac{2n+1}{x} k_n(x),$$

$$n\, k_{n-1}(x) + (n+1)\, k_{n+1}(x) = -(2n+1)\, k_n'(x).$$

11.7.20 Derive the limiting values for the spherical modified Bessel functions

(a) $i_n(x) \approx \dfrac{x^n}{(2n+1)!!}$

$k_n(x) \approx \dfrac{(2n-1)!!}{x^{n+1}}, \qquad x \ll 1.$

(b) $i_n(x) \sim \dfrac{e^x}{2x}$

$k_n(x) \sim \dfrac{e^{-x}}{x}, \qquad x \gg n(n+1)/2.$

11.7.21 Show that the Wronskian of the spherical modified Bessel functions is given by

$$i_n(x)\, k_n'(x) - i_n'(x)\, k_n(x) = -\frac{1}{x^2}.$$

11.7.22 Show that a Green's function for the radial part of the three dimensional modified Helmholtz equation, $(\nabla^2 - k^2)\psi = 0$, is

$$\frac{k}{4\pi} k_0(kr).$$

(Source at origin.)

REFERENCES

WATSON, G. N., *A Treatise on the Theory of Bessel Functions*. 2nd Ed. Cambridge: Cambridge University Press (1952). This is the definitive text on Bessel functions and their properties. Although difficult reading, it is invaluable as the ultimate reference.

RELTON, F. E., *Applied Bessel Functions*. London: Blackie and Son (1946). A very readable short volume. Perhaps the best reference for the beginning student.

See also the references listed at the end of Chapter 13.

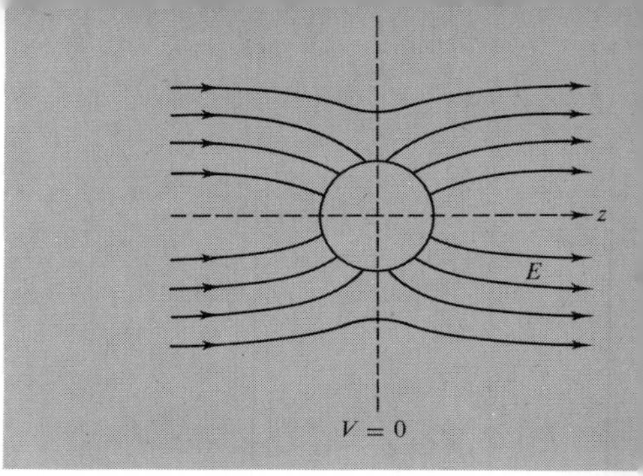

CHAPTER 12

LEGENDRE FUNCTIONS

12.1 Generating Function

The Legendre differential equation has already been encountered in the separation of variables for Laplace's equation, Helmholtz' equation, and similar differential equations in spherical polar coordinates and also in oblate and prolate spheroidal coordinates. In Section 9.1 the Legendre equation was shown to be self-adjoint, and, as a consequence, the (undetermined) solutions were orthogonal.

The polynomial solutions were actually constructed in Section 9.3 by the Schmidt process. However, it was not proved that this process would generate all of the Legendre polynomials nor that these polynomials were actually solutions of Legendre's equation.

Physical basis—electrostatics. As with Bessel functions, it is convenient to introduce the Legendre polynomials by means of a generating function. However, a direct physical interpretation is possible. Consider an electric charge q placed on the z-axis at $z = a$. As shown in Fig. 12.1, the electrostatic potential of charge q is

$$\varphi = \frac{1}{4\pi\varepsilon_0} \cdot \frac{q}{r_1} \text{ (mks units)} \tag{12.1}$$

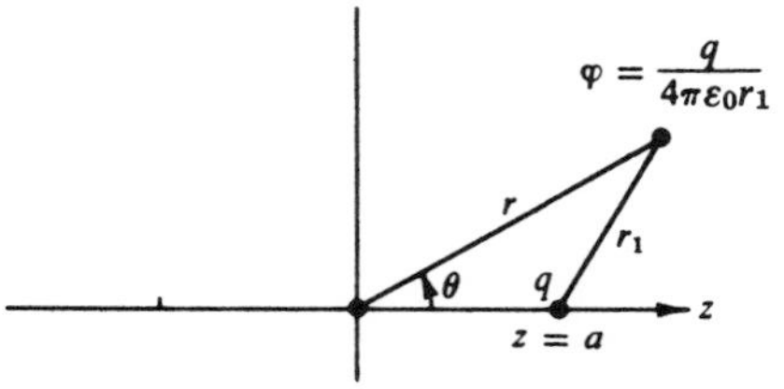

FIG. 12.1 Electrostatic potential. Charge q displaced from origin

Our problem is to express the electrostatic potential in terms of the spherical polar coordinates r and θ (the coordinate φ is absent because of symmetry about the z-axis). Using the law of cosines, we obtain

$$\varphi = \frac{q}{4\pi\varepsilon_0}(r^2 + a^2 - 2ar\cos\theta)^{-1/2}. \tag{12.2}$$

Legendre polynomials. Consider the case of $r > a$ or, more precisely, $r^2 > |a^2 - 2ar\cos\theta|$. The radical may be expanded by the binomial series to give

$$\varphi = \frac{q}{4\pi\varepsilon_0 r}\sum_{n=0}^{\infty} P_n(\cos\theta)\left(\frac{a}{r}\right)^n, \tag{12.3}$$

a series of powers of (a/r) with the coefficient of the nth power denoted by $P_n(\cos\theta)$. The P_n are the Legendre polynomials (Fig. 12.2) and may be defined by

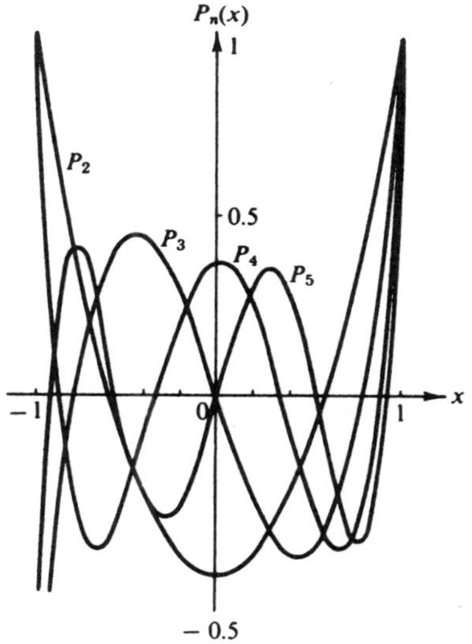

FIG. 12.2 Legendre polynomials, $P_2(x)$, $P_3(x)$, $P_4(x)$ and $P_5(x)$

$$g(t, x) = (1 - 2xt + t^2)^{-1/2} = \sum_{n=0}^{\infty} P_n(x)t^n, \qquad |t| < 1. \tag{12.4}$$

This is our generating function. In the next section it is shown that $|P_n(\cos\theta)| \leqslant 1$, which means that the series expansion (Eq. 12.4) is convergent for $|t| < 1$.[1] Indeed the series is convergent for $|t| = 1$ except for $|x| = 1$.

Actually since Eq. 12.4 defines the Legendre polynomials, $P_n(x)$, convergence of the series is not necessary. We can still obtain the explicit values of the polynomials and develop useful relations between them even when the series diverges. However,

[1] Note that the series in Eq. 12.3 is convergent for $r > a$ even though the binomial expansion involved is valid only for $r > (a^2 + 2ar)^{1/2}$, $\cos\theta = -1$.

the property of convergence is convenient in order to be able to exploit the properties of power series (Section 5.7).

In physical applications, Eq. 12.4 often appears in the vector form

$$\frac{1}{|\mathbf{r}_1 - \mathbf{r}_2|} = \frac{1}{r_>} \sum_{n=0}^{\infty} \left(\frac{r_<}{r_>}\right)^n P_n(\cos\theta) \tag{12.4a}$$

where

$$\left.\begin{aligned} r_> &= |\mathbf{r}_1| \\ r_< &= |\mathbf{r}_2| \end{aligned}\right\} \quad \text{for} \quad |\mathbf{r}_1| > |\mathbf{r}_2|,$$

and

$$\left.\begin{aligned} r_> &= |\mathbf{r}_2| \\ r_< &= |\mathbf{r}_1| \end{aligned}\right\} \quad \text{for} \quad |\mathbf{r}_2| > |\mathbf{r}_1|.$$

Using the binomial theorem (Section 5.6) and Exercise 10.1.16, we expand the generating function as follows:

$$\begin{aligned} (1 - 2xt + t^2)^{-1/2} &= \sum_{n=0}^{\infty} \frac{(2n)!}{2^{2n}(n!)^2} (2xt - t^2)^n \qquad (12.5) \\ &= \sum_{n=0}^{\infty} \frac{(2n-1)!!}{(2n)!!} (2xt - t^2)^n. \end{aligned}$$

Binomial expansion of the $(2xt - t^2)^n$ factor yields the double series

$$\begin{aligned} (1 - 2xt + t^2)^{-1/2} &= \sum_{n=0}^{\infty} \frac{(2n)!}{2^{2n}(n!)^2} t^n \sum_{k=0}^{n} (-1)^k \frac{n!}{k!(n-k)!} (2x)^{n-k} t^k \qquad (12.6) \\ &= \sum_{n=0}^{\infty} \sum_{k=0}^{n} (-1)^k \frac{(2n)!}{2^{2n} n!\,k!(n-k)!} (2x)^{n-k} t^{n+k}. \end{aligned}$$

From Eq. 5.64 of Section 5.4 (rearranging the order of summation), Eq. 12.6 becomes

$$(1 - 2xt + t^2)^{-1/2} = \sum_{n=0}^{\infty} \sum_{k=0}^{[n/2]} (-1)^k \frac{(2n-2k)!}{2^{2n-2k} k!(n-k)!(n-2k)!} \cdot (2x)^{n-2k} t^n. \tag{12.7}$$

with the variable t independent of the index k.[1] Now, equating our two power series (Eqs. 12.4 and 12.7) term by term, we have

$$P_n(x) = \sum_{k=0}^{[n/2]} (-1)^k \frac{(2n-2k)!}{2^n k!(n-k)!(n-2k)!} x^{n-2k}. \tag{12.8}$$

Returning to the electric charge on the z-axis, we demonstrate the usefulness and power of the generating function by adding a charge $-q$ at $z = -a$, as shown in Fig. 12.3. The potential becomes

[1] $[n/2] = n/2$ for n even, $(n-1)/2$ for n odd.

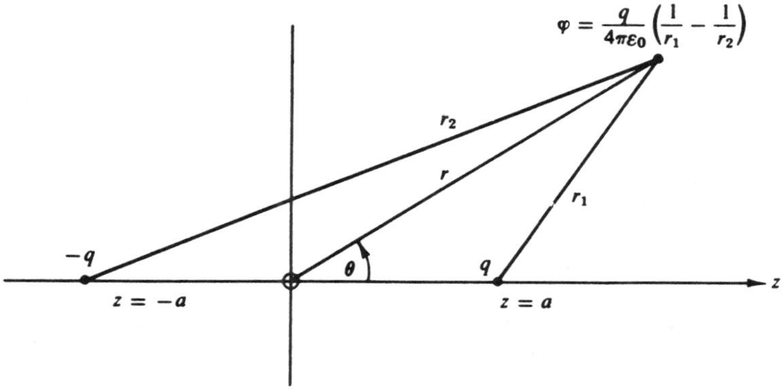

FIG. 12.3 Electric dipole

$$\varphi = \frac{q}{4\pi\varepsilon_0}\left(\frac{1}{r_1} - \frac{1}{r_2}\right), \tag{12.9}$$

and by using the law of cosines we have

$$\varphi = \frac{q}{4\pi\varepsilon_0 r}\left\{\left[1 - 2\left(\frac{a}{r}\right)\cos\theta + \left(\frac{a}{r}\right)^2\right]^{-1/2} - \left[1 + 2\left(\frac{a}{r}\right)\cos\theta + \left(\frac{a}{r}\right)^2\right]^{-1/2}\right\}. \tag{12.10}$$

$(r > a)$

Clearly, the second radical is like the first, except that a has been replaced by $-a$. Then, using Eq. 12.4, we obtain

$$\begin{aligned}\varphi &= \frac{q}{4\pi\varepsilon_0 r}\left[\sum_{n=0}^{\infty} P_n(\cos\theta)\left(\frac{a}{r}\right)^n - \sum_{n=0}^{\infty} P_n(\cos\theta)(-1)^n\left(\frac{a}{r}\right)^n\right] \\ &= \frac{2q}{4\pi\varepsilon_0 r}\left[P_1(\cos\theta)\left(\frac{a}{r}\right) + P_3(\cos\theta)\left(\frac{a}{r}\right)^3 + \cdots\right].\end{aligned} \tag{12.11}$$

The first term (and dominant term for $r \gg a$) is

$$\varphi = \frac{2aq}{4\pi\varepsilon_0} \cdot \frac{P_1(\cos\theta)}{r^2}, \tag{12.12}$$

which is the usual electric dipole potential. Here $2aq$ is the dipole moment (Fig. 12.3).

Linear electric multipoles. This analysis may be extended by placing additional charges on the z-axis so that the P_1 term, as well as the P_0 (monopole) term, is cancelled out. For instance, charges of q at $z = a$ and $z = -a$, $-2q$ at $z = 0$ give rise to a potential whose series expansion starts with $P_2(\cos\theta)$. This is a linear electric quadrupole. Two linear quadrupoles may be placed so that the quadrupole term is cancelled, but the P_3, the octupole term, survives. Other configurations are possible. For instance (Fig. 12.4), alternate positive and negative charges at the vertices of

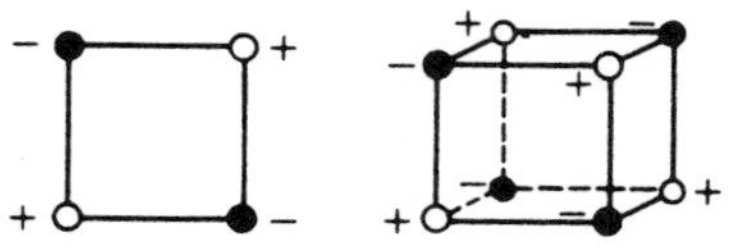

FIG. 12.4 Electric quadrupole, electric octupole

a square (or parallelogram) yield a quadrupole field, alternate charges on the vertices of a cube, an octupole field. Further information about electric multipoles and other methods of treating them can be found in the texts on electromagnetic theory listed in the introduction.

Before leaving multipole fields, perhaps two points should be emphasized. First, an electric (or magnetic) multipole has an absolute significance only if all lower-order terms vanish. For instance, the potential of one charge q at $z = a$ was expanded in a series of Legendre polynomials. Although we may refer to the $P_1(\cos\theta)$ term in this expansion as a dipole term, it should be remembered that this term exists only because of our choice of coordinates. We actually have a monopole, $P_0(\cos\theta)$.

Second, in physical systems we do not encounter pure multipoles. As an example, the potential of the finite dipole (q at $z = a$, $-q$ at $z = -a$) contained a $P_3(\cos\theta)$ term. These higher-order terms may be eliminated by shrinking the multipole to a point multipole, in this case keeping the product qa constant ($a \to 0$, $q \to \infty$) to maintain the same dipole moment.

It might also be noted that a multipole expansion is actually a decomposition into the irreducible representations of the rotation group (Section 4.9).

Extension to Gegenbauer polynomials. The generating function $g(t, x)$ used here is actually a special case of a more general generating function,

$$\frac{2^m}{(1 - 2xt + t^2)^{m+1/2}} = \frac{\pi^{1/2}}{(m - \frac{1}{2})!} \sum_{n=0}^{\infty} T_n^m(x) t^n. \tag{12.13}$$

The coefficients $T_n^m(x)$ are often called Gegenbauer polynomials. For $m = 0$ this reduces to Eq. 12.4; that is, $T_n^0(x) = P_n(x)$. The cases $m = \pm\frac{1}{2}$ are considered in Chapter 13 in connection with the Chebyshev polynomials.

EXERCISES

12.1.1 Develop the electrostatic potential for the array of charges shown. This is a linear electric quadrupole (Fig. 12.5).

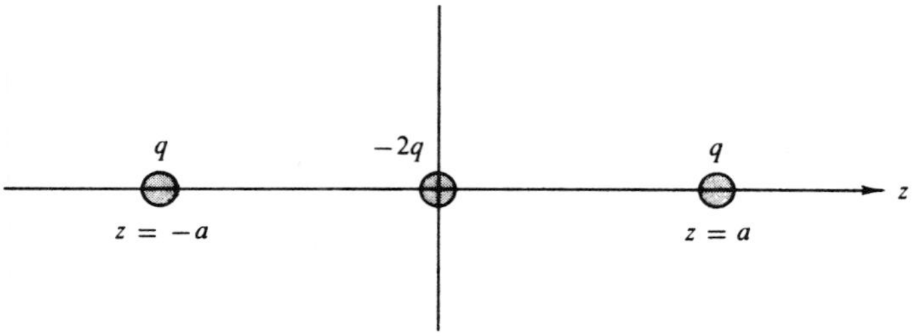

FIG. 12.5 Linear electric quadrupole

12.1.2 Show that the electrostatic potential produced by a charge q at $z = a$ for $r < a$ is

$$\varphi(\mathbf{r}) = \frac{q}{4\pi\varepsilon_0 a} \sum_{n=0}^{\infty} \left(\frac{r}{a}\right)^n P_n(\cos\theta).$$

12.1.3 Using $\mathbf{E} = -\nabla\varphi$, determine the components of the electric field corresponding to the (pure) electric dipole potential

$$\varphi(\mathbf{r}) = \frac{2aq\,P_1(\cos\theta)}{4\pi\varepsilon_0 r^2}.$$

Here it is assumed that $r \gg a$.

$$E_r = +\frac{4aq\cos\theta}{4\pi\varepsilon_0 r^3},$$

$$E_\theta = +\frac{2aq\sin\theta}{4\pi\varepsilon_0 r^3},$$

$$E_\varphi = 0.$$

12.1.4 A point electric dipole of strength $p^{(1)}$ is placed at $z = a$; a second point electric dipole of equal but opposite strength is at the origin. Keeping the product $p^{(1)}a$ constant, let $a \to 0$. Show that this results in a point electric quadrupole.

Hint. Ex. 12.2.4 (when proved) will be helpful.

12.1.5 A point charge q is in the interior of a hollow conducting sphere of radius r_0. The charge q is displaced a distance a from the center of the sphere. If the conducting sphere is grounded show that the potential in the interior produced by q and the distributed induced charge is the same as that produced by q and its image charge q'. The image charge is at a distance $a' = r_0^2/a$ from the center, collinear with q and the origin.

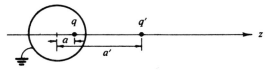

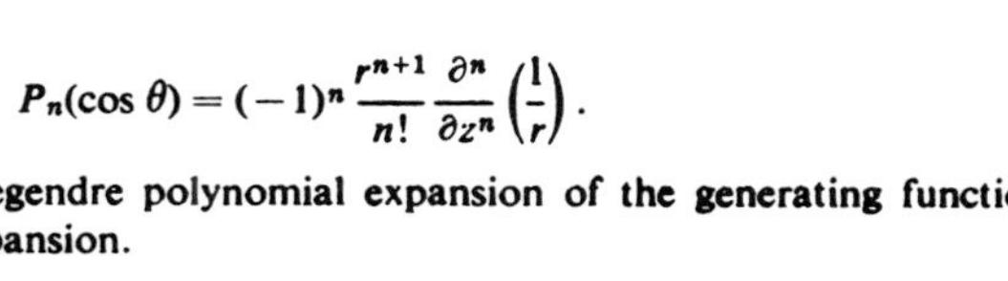

Hint: Calculate the electrostatic potential for $a < r_0 < a'$. Show that the potential vanishes for $r = r_0$ if we take $q' = -qr_0/a$.

12.1.6 Prove that

$$P_n(\cos\theta) = (-1)^n \frac{r^{n+1}}{n!} \frac{\partial^n}{\partial z^n}\left(\frac{1}{r}\right).$$

Hint. Compare the Legendre polynomial expansion of the generating function with a Taylor series expansion.

12.1.7 By differentiation and direct substitution of the series form, Eq. 12.8, show that $P_n(x)$ satisfies the Legendre differential equation. Note that there is no restriction upon x. We may have any x, $-\infty < x < \infty$ and indeed any z in the entire finite complex plane.

12.1.8 The Chebyshev polynomials (type II) are generated by (Eq. 13.64, Section 13.3)

$$\frac{1}{1-2xt+t^2} = \sum_{n=0}^{\infty} U_n(x)t^n.$$

Using the techniques of Section 5.4 for transforming series develop a series representation of $U_n(x)$.

$$Ans:\ U_n(x) = \sum_{k=0}^{[n/2]} (-1)^k \frac{(n-k)!}{k!(n-2k)!}(2x)^{n-2k}.$$

12.2 Recurrence Relations and Special Properties

Recurrence relations. The Legendre polynomial generating function provides a convenient way of deriving the recurrence relations[1] and some special properties. If our generating function (Eq. 12.4) is differentiated with respect to t, we obtain

$$\frac{\partial g(t,x)}{\partial t} = \frac{x-t}{(1-2xt+t^2)^{3/2}} = \sum_{n=0}^{\infty} n\,P_n(x)t^{n-1}. \tag{12.14}$$

By substituting Eq. 12.4 into this and rearranging terms, we have

$$(1-2xt+t^2)\sum_{n=0}^{\infty} n\,P_n(x)t^{n-1} + (t-x)\sum_{n=0}^{\infty} P_n(x)t^n = 0. \tag{12.15}$$

The left-hand side is a power series in t. Since this power series vanishes for all values of t, we may put the coefficient of each power of t equal to zero, that is, our power series is unique (Section 5.7). This may be done easily by separating the individual summations and using distinctive summation indices,

$$\begin{aligned}\sum_{m=0}^{\infty} m\,P_m(x)t^{m-1} - \sum_{n=0}^{\infty} 2nx\,P_n(x)t^n + \sum_{s=0}^{\infty} s\,P_s(x)t^{s+1}&\\ + \sum_{s=0}^{\infty} P_s(x)t^{s+1} - \sum_{n=0}^{\infty} x\,P_n(x)t^n = 0.&\end{aligned} \tag{12.16}$$

Now letting $m = n+1$, $s = n-1$, we find

$$(2n+1)x\,P_n(x) = (n+1)\,P_{n+1}(x) + n\,P_{n-1}(x), \qquad n = 1, 2, 3, \ldots. \tag{12.17}$$

This is another three-term recurrence relation similar to (but not identical with) the recurrence relation for Bessel functions. With this recurrence relation we may easily construct the higher Legendre polynomials. If we take $n = 1$ and insert the easily found values of $P_0(x)$ and $P_1(x)$ (Exercise 12.1.5 or Eq. 12.8),

$$3x\,P_1(x) = 2P_2(x) + P_0(x) \tag{12.18}$$

or

$$P_2(x) = \tfrac{1}{2}(3x^2 - 1). \tag{12.19}$$

[1] We can also apply the explicit series form (Eq. 12.8) directly.

TABLE 12.1

Legendre Polynomials

$P_0(x) = 1$
$P_1(x) = x$
$P_2(x) = \frac{1}{2}(3x^2 - 1)$
$P_3(x) = \frac{1}{2}(5x^3 - 3x)$
$P_4(x) = \frac{1}{8}(35x^4 - 30x^2 + 3)$
$P_5(x) = \frac{1}{8}(63x^5 - 70x^3 + 15x)$

This process may be continued indefinitely. The first few Legendre polynomials are listed in Table 12.1.

Cumbersome as it may appear at first, this technique is actually more efficient for a large digital computer than direct evaluation of the series (Eq. 12.8). For greater stability (to avoid undue accumulation and magnification of round off error), Eq. 12.17 is rewritten as

$$P_{n+1}(x) = 2xP_n(x) - P_{n-1}(x) - [xP_n(x) - P_{n-1}(x)]/(n + 1). \qquad (12.17a)$$

One starts with $P_0(x) = 1$, $P_1(x) = x$, and computes the *numerical* values of all the $P_n(x)$ for a given value of x up to the desired $P_N(x)$. The values of $P_n(x)$, $0 \leqslant n < N$ are available as a fringe benefit.

Differential equations. More information about the behavior of the Legendre polynomials can be obtained if we now differentiate Eq. 12.4 with respect to x. This gives

$$\frac{\partial g(t, x)}{\partial x} = \frac{t}{(1 - 2xt + t^2)^{3/2}} = \sum_{n=0}^{\infty} P'_n(x)t^n \qquad (12.20)$$

or

$$(1 - 2xt + t^2)\sum_{n=0}^{\infty} P'_n(x)t^n - t\sum_{n=0}^{\infty} P_n(x)t^n = 0. \qquad (12.21)$$

As before, the coefficient of each power of t is set equal to zero and we obtain

$$P'_{n+1}(x) + P'_{n-1}(x) = 2xP'_n(x) + P_n(x). \qquad (12.22)$$

A more useful relation may be found by differentiating Eq. 12.17 with respect to x and multiplying by two. To this we add $(2n + 1)$ times Eq. 12.22, cancelling the P'_n term. The result is

$$P'_{n+1}(x) - P'_{n-1}(x) = (2n + 1)\, P_n(x). \qquad (12.23)$$

From Eqs. 12.17 and 12.23 numerous additional equations may be developed, including

$$P'_{n+1}(x) = (n + 1)\, P_n(x) + x\, P'_n(x), \qquad (12.24)$$

$$P'_{n-1}(x) = -n\, P_n(x) + x\, P'_n(x), \qquad (12.25)$$

$$(1 - x^2)\, P'_n(x) = n\, P_{n-1}(x) - nx\, P_n(x), \qquad (12.26)$$

$$(1 - x^2)\, P'_n(x) = (n + 1)xP_n(x) - (n + 1)P_{n+1}(x). \qquad (12.27)$$

By differentiating Eq. 12.26 and using Eq. 12.25 to eliminate $P'_{n-1}(x)$ we find that $P_n(x)$ satisfies the linear, second-order differential equation

$$(1 - x^2)P''_n(x) - 2x\, P'_n(x) + n(n+1)\, P_n(x) = 0. \tag{12.28}$$

The previous equations, Eqs. 12.22–12.27, are all first-order differential equations, but with polynomials of two different indices. The price for having all indices alike is a second-order differential equation. Equation 12.28 is Legendre's differential equation. We now see that the polynomials $P_n(x)$ generated by the expansion of $(1 - 2xt + t^2)^{-1/2}$ satisfy Legendre's equation which, of course, is why they are called Legendre polynomials.

In Eq. 12.28 differentiation is with respect to x ($x = \cos\theta$). Frequently, we encounter Legendre's equation expressed in terms of differentiation with respect to θ,

$$\frac{1}{\sin\theta}\frac{d}{d\theta}\left(\sin\theta\,\frac{dP_n(\cos\theta)}{d\theta}\right) + n(n+1)P_n(\cos\theta) = 0. \tag{12.29}$$

Our generating function provides still more information about the Legendre polynomials. If we set $x = 1$, Eq. 12.4 becomes

$$\begin{aligned}\frac{1}{(1 - 2t + t^2)^{1/2}} &= \frac{1}{1-t}\\ &= \sum_{n=0}^{\infty} t^n.\end{aligned} \tag{12.30}$$

But

$$\frac{1}{(1 - 2tx + t^2)^{1/2}_{x=1}} = \sum_{n=0}^{\infty} P_n(1)\, t^n.$$

Comparing the two series expansions,

$$P_n(1) = 1. \tag{12.31}$$

If we let $x = -1$, the same sort of analysis shows that

$$P_n(-1) = (-1)^n. \tag{12.32}$$

For obtaining these results, the generating function is more convenient than the explicit series form.

If we take $x = 0$, using the expansion

$$(1 + t^2)^{-1/2} = 1 - \tfrac{1}{2}t^2 + \tfrac{3}{8}t^4 + \cdots + (-1)^n \frac{1\cdot 3\cdots(2n-1)}{2^n\, n!}\, t^{2n} + \cdots, \tag{12.33}$$

we have

$$P_{2n}(0) = (-1)^n \frac{1\cdot 3\cdots(2n-1)}{2^n n!} = (-1)^n \frac{(2n-1)!!}{(2n)!!} \tag{12.34}$$

$$P_{2n+1}(0) = 0, \qquad n = 0, 1, 2, \ldots. \tag{12.35}$$

These results also follow from Eq. 12.8, by inspection.

Parity. Some of these results are special cases of the parity property of the Legendre polynomials. We refer once more to Eq. 12.4. If we replace x by $-x$ and t by $-t$, the generating function is unchanged. Hence

$$\begin{aligned} g(t, x) &= g(-t, -x) \\ &= [1 - 2(-t)(-x) + (-t)^2]^{-1/2} \\ &= \sum_{n=0}^{\infty} P_n(-x)(-t)^n \\ &= \sum_{n=0}^{\infty} P_n(x)t^n. \end{aligned} \tag{12.36}$$

Comparing these two series, we have

$$P_n(-x) = (-1)^n P_n(x); \tag{12.37}$$

that is, the polynomial functions are odd or even (with respect to $x = 0$, $\theta = \pi/2$) according to whether the index n is odd or even. This is the parity or reflection property that plays such an important role in quantum mechanics. For central forces the index n is a measure of the orbital angular momentum, thus linking parity and orbital angular momentum.

The reader will see this parity property confirmed by the series solution and for the special values tabulated in Table 12.1. It might also be noted that Eq. 12.37 may be predicted by inspection of Eq. 12.17, the recurrence relation. Specifically, if $P_{n-1}(x)$ and $xP_n(x)$ are even, then $P_{n+1}(x)$ must be even.

Upper and lower bounds for $P_n(\cos\theta)$. Finally, in addition to these results, our generating function enables us to set an upper limit on $|P_n(\cos\theta)|$. We have

$$\begin{aligned} (1 - 2t\cos\theta + t^2)^{-1/2} &= (1 - te^{i\theta})^{-1/2}(1 - te^{-i\theta})^{-1/2} \\ &= (1 + \tfrac{1}{2}te^{i\theta} + \tfrac{3}{8}t^2e^{2i\theta} + \cdots) \\ &\quad \cdot (1 + \tfrac{1}{2}te^{-i\theta} + \tfrac{3}{8}t^2e^{-2i\theta} + \cdots), \end{aligned} \tag{12.38}$$

with all coefficients *positive*. Our Legendre polynomial, $P_n(\cos\theta)$, still the coefficient of t^n, may now be written as a sum of terms of the form

$$\begin{aligned} a_m(e^{im\theta} + e^{-im\theta})/2 &= a_m \cosh im\theta \\ &= a_m \cos m\theta \end{aligned} \tag{12.39a}$$

with all the a_m *positive*. Then

$$P_n(\cos\theta) = \sum_{m=0 \text{ or } 1}^{n} a_m \cos m\theta. \tag{12.39b}$$

This series is clearly a maximum when $\theta = 0$ and $\cos m\theta = 1$. But for $x = \cos\theta = 1$, Eq. 12.31 shows that $P_n(1) = 1$. Therefore,

$$|P_n(\cos\theta)| \leqslant P_n(1) = 1. \tag{12.39c}$$

In this section, various useful properties of the Legendre polynomials are derived from the generating function, Eq. 12.4. The explicit series representation, Eq. 12.8, offers an alternate and sometimes superior approach. Table 12.2 offers a comparison of the two approaches.

TABLE 12.2

Comparison of Generating Function plus Recurrence Relations and Series Expansion, Eq. 12.8

Application	Generating function recurrence relations Eqs. 12.4, 12.17, and 12.22	Series Eq. 12.8
Table 12.1 numerical value	computer choice	more direct
Derivation of differential equation, Eq. 12.27	moderately involved	verification easy, derivation requires clairvoyance
$P_n(1)$, Eq. 12.30	easy	awkward
$P_n(0)$, Eq. 12.34	easy	by inspection
Parity, Eq. 12.36	easy	by inspection
Bounds, Eq. 12.38	fairly easy	awkward

EXERCISES

12.2.1 Given the series

$$\alpha_0 + \alpha_2 \cos^2\theta + \alpha_4 \cos^4\theta + \alpha_6 \cos^6\theta = a_0 P_0 + a_2 P_2 + a_4 P_4 + a_6 P_6.$$

Express the coefficients α_i as a column vector $\boldsymbol{\alpha}$ and the coefficients a_i as a column vector $\mathbf{a}$ and determine the matrices A and B such that

$$\mathsf{A}\boldsymbol{\alpha} = \mathbf{a} \quad \text{and} \quad \mathsf{B}\mathbf{a} = \boldsymbol{\alpha}.$$

Check your computation by showing that $\mathsf{AB} = \mathbf{1}$ (unit matrix). Repeat for the odd case

$$\alpha_1 \cos\theta + \alpha_3 \cos^3\theta + \alpha_5 \cos^5\theta + \alpha_7 \cos^7\theta = a_1 P_1 + a_3 P_3 + a_5 P_5 + a_7 P_7.$$

12.2.2 By differentiating the generating function, $g(t, x)$, with respect to t, multiplying by $2t$, and then adding $g(t, x)$, show that

$$\frac{1 - t^2}{(1 - 2tx + t^2)^{3/2}} = \sum_{n=0}^{\infty} (2n + 1) P_n(x) t^n.$$

This result is useful in calculating the charge induced on a grounded metal sphere by a point charge q.

12.2.3 A point electric octupole may be constructed by placing a point electric quadrupole (pole strength $p^{(2)}$ in the z-direction) at $z = a$ and an equal but opposite point electric quadrupole at $z = 0$ and then letting $a \to 0$, subject to $p^{(2)}a = \text{constant}$. Find the electrostatic potential corresponding to a point electric octupole. Show from the construction of the point electric octupole that the corresponding potential may be obtained by differentiating the point quadrupole potential.

12.2.4 Operating in *spherical polar coordinates* show that

$$\frac{\partial}{\partial z}\left[\frac{P_n(\cos\theta)}{r^{n+1}}\right] = -(n + 1)\frac{P_{n+1}(\cos\theta)}{r^{n+2}}.$$

This is the key step in the mathematical argument that the derivative of one multipole leads to the next higher multipole.
Hint: Cf. Exercise 2.4.10

12.2.5 From

$$P_L(\cos\theta) = \frac{1}{L!}\frac{\partial^L}{\partial t^L}(1 - 2t\cos\theta + t^2)^{-1/2}\Big|_{t=0}$$

show that

$$P_L(1) = 1, \qquad P_L(-1) = (-1)^L.$$

12.2.6 Prove that

$$P_n'(1) = \frac{d}{dx}P_n(x)\Big|_{x=1} = \frac{1}{2}n(n + 1).$$

12.2.7 Show that $P_n(\cos\theta) = (-1)^n P_n(-\cos\theta)$ by use of the recurrence relation relating P_n, P_{n+1}, and P_{n-1} and your knowledge of P_0 and P_1.

12.2.8 From Eq. 12.38 write out the coefficient of t^2 in terms of $\cos n\theta$, $n \leqslant 2$. This coefficient is $P_2(\cos\theta)$.

12.3 Orthogonality

Legendre's differential equation (12.28) may be written in the form

$$\frac{d}{dx}[(1-x^2)\,P_n'(x)] + n(n+1)\,P_n(x) = 0, \tag{12.40}$$

showing clearly that it is self-adjoint. Subject to satisfying certain boundary conditions, then, it is known that the solutions $P_n(x)$ will be orthogonal. Repeating the Sturm-Liouville analysis (Section 9.2), we multiply Eq. 12.40 by $P_m(x)$ and subtract the corresponding equation with m and n interchanged. Integrating from -1 to $+1$, we get

$$\begin{aligned}\int_{-1}^{1}\left\{P_m(x)\frac{d}{dx}[(1-x^2)\,P_n'(x)] - P_n(x)\frac{d}{dx}[(1-x^2)\,P_m'(x)]\right\}dx \\ = [m(m+1)-n(n+1)]\int_{-1}^{1} P_n(x)\,P_m(x)\,dx.\end{aligned} \tag{12.41}$$

Integrating by parts, the integrated part vanishing because of the factor $(1-x^2)$,[1] we have

$$[m(m+1)-n(n+1)]\int_{-1}^{1} P_n(x)\,P_m(x)\,dx = 0. \tag{12.42}$$

Then for $m \neq n$

$$\begin{aligned}\int_{-1}^{1} P_n(x)\,P_m(x)\,dx &= 0, \\ \int_{0}^{\pi} P_n(\cos\theta)\,P_m(\cos\theta)\sin\theta\,d\theta &= 0,\end{aligned} \tag{12.43}$$

showing that $P_n(x)$ and $P_m(x)$ are orthogonal for the interval $[-1, 1]$. This orthogonality may also be demonstrated quite readily by using Rodrigues' definition of $P_n(x)$ (cf. Section 12.4, Exercise 12.4.2).

We still need to evaluate the integral (Eq. 12.42) when $n = m$. Certainly it is no longer zero. From our generating function

$$(1-2tx+t^2)^{-1} = \left[\sum_{n=0}^{\infty} P_n(x)t^n\right]^2. \tag{12.44}$$

Integrating from $x = -1$ to $x = +1$, we have

$$\int_{-1}^{1}\frac{dx}{1-2tx+t^2} = \sum_{n=0}^{\infty} t^{2n}\int_{-1}^{1}[P_n(x)]^2\,dx; \tag{12.45}$$

the cross terms in the series vanish by means of Eq. 12.43. Using $y = 1-2tx+t^2$, we obtain

$$\int_{-1}^{1}\frac{dx}{1-2tx+t^2} = \frac{1}{2t}\int_{(1-t)^2}^{(1+t)^2}\frac{dy}{y} = \frac{1}{t}\ln\left(\frac{1+t}{1-t}\right). \tag{12.46}$$

[1] This, of course, is why the limits were chosen as -1 and $+1$.

Expanding this in a power series (Ex. 5.4.1) gives us

$$\frac{1}{t}\ln\left(\frac{1+t}{1-t}\right) = 2\sum_{n=0}^{\infty}\frac{t^{2n}}{2n+1}. \tag{12.47}$$

Since our power series representation is known to be unique, we must have

$$\int_{-1}^{1}[P_n(x)]^2\,dx = \frac{2}{2n+1}. \tag{12.48}$$

We shall return to this result in Section 12.6 when we construct the orthonormal spherical harmonics.

Expansion of functions, Legendre series. In addition to orthogonality, the Sturm-Liouville theory shows that the Legendre polynomials form a complete set. Let us assume, then, that the series

$$\sum_{n=0}^{\infty} a_n P_n(x) = f(x), \tag{12.49}$$

converges uniformly to $f(x)$ in $[-1, 1]$. This demands that $f(x)$ and $f'(x)$ be at least sectionally continuous in this interval. The coefficients a_n are found by multiplying the series by $P_m(x)$ and integrating term by term. Using the orthogonality property expressed in Eqs. 12.43 and 12.48, we obtain

$$\frac{2}{2m+1}a_m = \int_{-1}^{1} f(x)\,P_m(x)\,dx \tag{12.50}$$

or

$$f(x) = \sum_{n=0}^{\infty}\frac{2n+1}{2}\left(\int_{-1}^{1} f(t)P_n(t)\,dt\right)P_n(x). \tag{12.51}$$

This expansion in a series of Legendre polynomials is usually referred to as a Legendre series.[1] Its properties are quite similar to the more familiar Fourier series (Chapter 14). In particular, we can use the orthogonality property, (Eq. 12.43) to show that the series is unique.

Equation 12.3, which leads directly to the generating function definition of Legendre polynomials, is a Legendre expansion of $1/r_1$. This Legendre expansion of $1/r_1$ or $1/r_{12}$ appears in several exercises of Section 12.8. Going beyond a simple Coulomb field, the $1/r_{12}$ is often replaced by a potential $V(|\mathbf{r}_1 - \mathbf{r}_2|)$ and the solution of the problem is again effected by a Legendre expansion. In nuclear physics calculations, the coefficients a_n may be computed (by a computing machine) up through a_{100}.

EXAMPLE 12.3.1. EARTH'S GRAVITATIONAL FIELD

An example of a Legendre series is provided by the description of the earth's

[1] Note that Eq. 12.50 gives a_m as a *definite* integral, i.e., a number for a given $f(x)$.

gravitational potential U (for exterior points), neglecting azimuthal effects. With

$$R = \text{equatorial radius}$$
$$= 6378.1 \pm 0.1 \text{ km}$$
$$\frac{GM}{R} = 62.494 \pm 0.001 \text{ km}^2/\text{sec}^2$$

we write

$$U(r, \theta) = \frac{GM}{R}\left[\frac{R}{r} - \sum_{n=2}^{\infty} a_n \left(\frac{R}{r}\right)^{n+1} P_n(\cos\theta)\right], \tag{12.52}$$

a Legendre series. Artificial satellite motions have shown that

$$a_2 = (1.08279 \pm 0.00015) \times 10^{-3}$$
$$a_3 = (-2.4 \pm 0.3) \times 10^{-6}.$$

This is the famous pear-shaped deformation of the earth.

$$a_4 = (-1.4 \pm 0.2) \times 10^{-6}.$$

The higher-order coefficients are approximately zero, within experimental accuracy. The reader might note that P_1 is omitted, since it would represent a displacement and not a deformation.

More recent satellite data permit a determination of the longitudinal dependence of the earth's gravitational field. Such dependence may be described by a Laplace series (Section 12.6).

EXAMPLE 12.3.2. SPHERE IN A UNIFORM FIELD

Another illustration of the use of Legendre polynomials is provided by the problem of a neutral conducting sphere (radius r_0) placed in a (previously) uniform electric field. The problem is to find the new, perturbed, electrostatic potential. Calling the electrostatic potential V[1],

$$\nabla^2 V = 0, \tag{12.53}$$

Laplace's equation. We select spherical polar coordinates because of the spherical shape of the conductor. (This will simplify the application of the boundary condition at the surface of the conductor.) Separating variables and glancing at Table 8.1 if necessary, we obtain

$$V = \sum_{n=0}^{\infty} a_n r^n P_n(\cos\theta) + \sum_{n=0}^{\infty} b_n \frac{P_n(\cos\theta)}{r^{n+1}}. \tag{12.54}$$

No φ-dependence appears because of the axial symmetry of our problem. (The center of the conducting sphere is taken as the origin and the z-axis is oriented parallel to the original uniform field.)

[1] It should be emphasized that this is not a presentation of a Legendre series expansion of a known $V(\cos\theta)$. Here we are back to *boundary value* problems.

It might be noted here that n is an integer, *because* only for integral n is the θ dependence well-behaved at $\cos\theta = \pm 1$. For nonintegral n, the solutions of Legendre's equation diverge at the ends of the interval $[-1, 1]$, the poles $\theta = 0, \pi$ of the sphere (cf. Example 5.2.4 and Exercises 5.2.15 and 8.4.4). It is for this same reason that the *second* solution of Legendre's equation, Q_n, is also excluded.

Now, our (Dirichlet) boundary conditions come in. If the original unperturbed electrostatic field is E_0, we require

$$\begin{aligned} V(r \to \infty) &= -E_0 z = -E_0 r \cos\theta \\ &= -E_0 r\, P_1(\cos\theta). \end{aligned} \tag{12.55}$$

Since our Legendre series is unique, we may equate coefficients of $P_n(\cos\theta)$ in Eq. 12.54 ($r \to \infty$) and Eq. 12.55 to obtain

$$\begin{aligned} a_n &= 0, \qquad n > 1, \\ a_1 &= -E_0 . \end{aligned} \tag{12.56}$$

If $a_n \neq 0$ for $n > 1$, these terms would dominate at large r and the boundary condition (Eq. 12.55) could not be satisfied. We may choose the conducting sphere and the plane $\theta = \pi/2$ to be at zero potential, which means that Eq. 12.54 now becomes

$$\begin{aligned} V(r = r_0) &= a_0 + \frac{b_0}{r_0} + \left(\frac{b_1}{r_0^2} - E_0 r_0\right) P_1(\cos\theta) + \sum_{n=2}^{\infty} b_n \frac{P_n(\cos\theta)}{r_0^{n+1}} \\ &= 0. \end{aligned} \tag{12.57}$$

In order that this may hold for all values of θ, each coefficient of $P_n(\cos\theta)$ must vanish.[1] Hence

$$\begin{aligned} a_0 &= b_0 = 0,^{2} \\ b_n &= 0, \qquad n \geqslant 2, \end{aligned} \tag{12.58}$$

whereas

$$b_1 = E_0 r_0^3 . \tag{12.59}$$

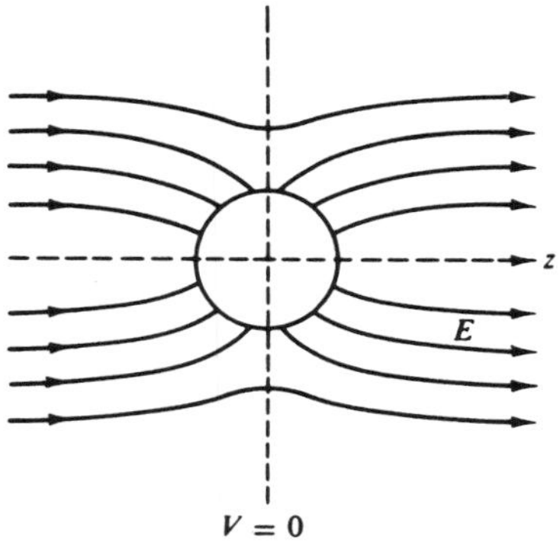

FIG. 12.6 Conducting sphere in a uniform field

The electrostatic potential (outside the sphere) is then

$$\begin{aligned} V &= -E_0 r\, P_1(\cos\theta) + \frac{E_0 r_0^3}{r^2} P_1(\cos\theta) \\ &= -E_0 r\, P_1(\cos\theta)\left(1 - \frac{r_0^3}{r^3}\right). \end{aligned} \tag{12.60}$$

[1] Again this is equivalent to saying that a series expansion in Legendre polynomials (or any complete orthogonal set) is unique.

[2] The coefficient of P_0 is $a_0 + b_0/r_0$. We set $b_0 = 0$ (and therefore $a_0 = 0$ also), since there is no net charge on the sphere. If there is a net charge q, then $b_0 \neq 0$.

In Section 1.15 it was shown that a solution of Laplace's equation which satisfied the boundary conditions over the entire boundary was unique. The electrostatic potential V, as given by Eq. 12.60, is a solution of Laplace's equation. It satisfies our boundary conditions and is therefore the solution of Laplace's equation for this problem.

It may further be shown (Exercise 12.3.9) that there is an induced surface charge density

$$\sigma = -\varepsilon_0 \frac{\partial V}{\partial r}\bigg|_{r=r_0} = 3\varepsilon_0 E_0 \cos\theta \tag{12.61}$$

on the surface of the sphere and an induced electric dipole moment (Ex. 12.3.9)

$$P = 4\pi r_0^3 \varepsilon_0 E_0 . \tag{12.62}$$

EXAMPLE 12.3.3 ELECTROSTATIC POTENTIAL OF A RING OF CHARGE

As a further example, consider the electrostatic potential produced by a conducting ring carrying a total electric charge q. From electrostatics (and Section 1.14), the potential ψ satisfies Laplace's equation. Separating variables in spherical polar coordinates (cf. Table 8.1),

$$\psi(r, \theta) = \sum_{n=0}^{\infty} a_n \frac{a^n}{r^{n+1}} P_n(\cos\theta), \qquad r > a. \tag{12.63a}$$

Here a is the radius of the ring which is assumed to be in the $\theta = \pi/2$ plane. There is no φ (azimuthal) dependence because of the cylindrical symmetry of the system.

The terms with positive exponent radial dependence are rejected since the potential must have an asymptotic behavior

$$\psi \sim \frac{q}{4\pi\varepsilon_0} \cdot \frac{1}{r}, \qquad r \gg a. \tag{12.63b}$$

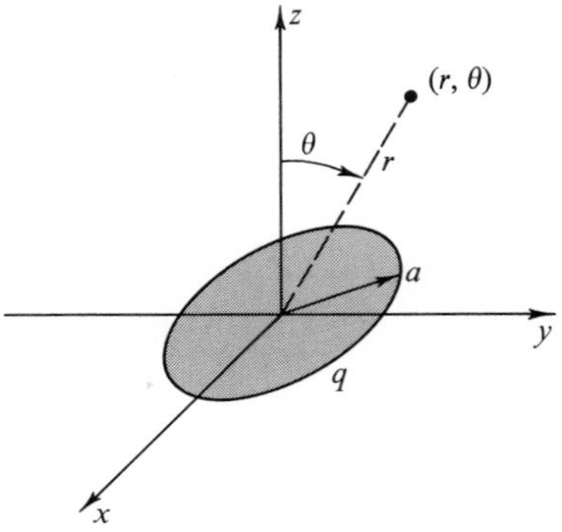

FIG. 12.7 Charged, conducting ring

The problem is to determine the coefficients a_n in Eq. 12.63*a*. This may be done by evaluating $\psi(r, \theta)$ at $\theta = 0$, $r = z$, and comparing with an independent calculation of the potential from Coulomb's law. In effect, we are using a boundary condition along the z-axis. From Coulomb's law (with all charge equidistant),

$$\psi(r, \theta) = \frac{q}{4\pi\varepsilon_0} \cdot \frac{1}{(z^2 + a^2)^{1/2}}, \qquad r = z,$$

$$= \frac{q}{4\pi\varepsilon_0 z} \sum_{s=0}^{\infty} (-1)^s \frac{(2s)!}{2^{2s}(s!)^2} \left(\frac{a}{z}\right)^{2s}, \qquad z > a. \tag{12.63c}$$

The last step uses the result of Exercise 10.1.16. Now, Eq. 12.63*a* evaluated at $\theta = 0$, $r = z$ (with $P_n(1) = 1$), yields

$$\psi(r, \theta) = \sum_{n=0}^{\infty} a_n \frac{a^n}{z^{n+1}}, \qquad r = z. \tag{12.63d}$$

Comparing Eqs. 12.63*c* and 12.63*d*, $a_n = 0$ for n odd. Setting $n = 2s$,

$$a_{2s} = \frac{q}{4\pi\varepsilon_0} (-1)^s \frac{(2s)!}{2^{2s}(s!)^2}, \tag{12.63e}$$

and our electrostatic potential $\psi(r, \theta)$ is given by

$$\psi(r, \theta) = \frac{q}{4\pi\varepsilon_0 r} \sum_{s=0}^{\infty} (-1)^s \frac{(2s)!}{2^{2s}(s!)^2} \left(\frac{a}{r}\right)^{2s} P_{2s}(\cos\theta), \qquad r > a. \tag{12.63f}$$

The magnetic analog of this problem appears in Section 12.5—Example 12.5.1.

EXERCISES

12.3.1 You have constructed a set of orthogonal functions by the Schmidt process (Section 9.3), taking $u_n(x) = x^n$, $n = 0, 1, 2, \ldots$, in increasing order with $w(x) = 1$ and an interval $-1 \leqslant x \leqslant 1$. Prove that the nth such function constructed is proportional to $P_n(x)$. *Hint*. Use mathematical induction.

12.3.2 Expand the Dirac delta function in a series of Legendre polynomials, using the interval $-1 \leqslant x \leqslant 1$.

12.3.3 A particular function $f(x)$ defined over the interval $[-1, 1]$ is expanded in a Legendre series over this same interval. Show that the expansion is unique.

12.3.4 Derive the recurrence relation

$$(1 - x^2)P_n'(x) = nP_{n-1}(x) - nxP_n(x)$$

from the Legendre polynomial generating function.

12.3.5 Evaluate $\int_0^1 P_n(x)\,dx$.

Ans. $n = 2s$; $\quad$ 1 for $s = 0$, $\quad$ 0 for $s > 0$,

$n = 2s + 1$; $\quad P_{2s}(0)/(2s+2) = (-1)^s(2s-1)!!/(2s+2)!!$

Hint. Use a recurrence relation to replace $P_n(x)$ by derivatives and then integrate by inspection!

12.3.6 Prove that

$$\int_{-1}^{1} x(1-x^2)\,P_n'\,P_m'\,dx = 0, \qquad \text{unless } m = n \pm 1.$$

12.3.7 The amplitude of a scattered wave is given by

$$f(\theta) = \lambda \sum_{l=0}^{\infty} (2l+1)\exp[i\delta_l]\sin\delta_l\,P_l(\cos\theta).$$

Here θ is the angle of scattering, l the angular momentum, and δ_l the phase shift produced by the central potential that is doing the scattering. The total cross section is $\sigma_{\text{tot}} = \int f^*(\theta)\,f(\theta)\,d\Omega$. Show that

$$\sigma_{\text{tot}} = 4\pi\lambda^2 \sum_{l=0}^{\infty}(2l+1)\sin^2\delta_l\,.$$

12.3.8 The coincidence counting rate, $W(\theta)$, in a gamma-gamma angular correlation experiment has the form

$$W(\theta) = \sum_{n=0}^{\infty} a_{2n}P_{2n}(\cos\theta).$$

Show that data in the range $\pi/2 \le \theta \le \pi$ can, in principle, define the function, $W(\theta)$, (and permit a determination of the coefficients a_{2n}). This means that while data in the range $0 \le \theta < \pi/2$ may be useful as a check, they are not essential.

12.3.9 A conducting sphere of radius r_0 is placed in an initially uniform electric field, $\mathbf{E}_0$. Show the following:

(a) The induced surface charge density is

$$\sigma = 3\varepsilon_0 E_0 \cos\theta.$$

(b) The induced electric dipole moment is

$$P = 4\pi r_0^3\,\varepsilon_0 E_0\,.$$

The induced electric dipole moment can be calculated either from the surface charge [part (a)], or by noting that the final electric field $\mathbf{E}$ is the result of superimposing a dipole field on the original uniform field.

12.3.10 A charge q is displaced a distance a along the z-axis from the center of a spherical cavity of radius R.

(a) Show that the electric field averaged over the volume $a \leqslant r \leqslant R$ is zero.

(b) Show that the electric field averaged over the volume $0 \leqslant r \leqslant a$ is

$$\mathbf{E} = \mathbf{k}E_z = -\mathbf{k}\frac{q}{4\pi\varepsilon_0 a^2}, \qquad \text{(mks units)}$$

$$= -\mathbf{k}\frac{nqa}{3\varepsilon_0},$$

where n is the number of such displaced charges per unit volume. This is a basic calculation in the polarization of a dielectric. *Hint.* $\mathbf{E} = -\nabla\varphi$.

12.3.11 As an extension of Example 12.3.3, find the potential $\psi(r, \theta)$ produced by a charged conducting disk, Fig. 12.8, for $r > a$, the radius of the disk.

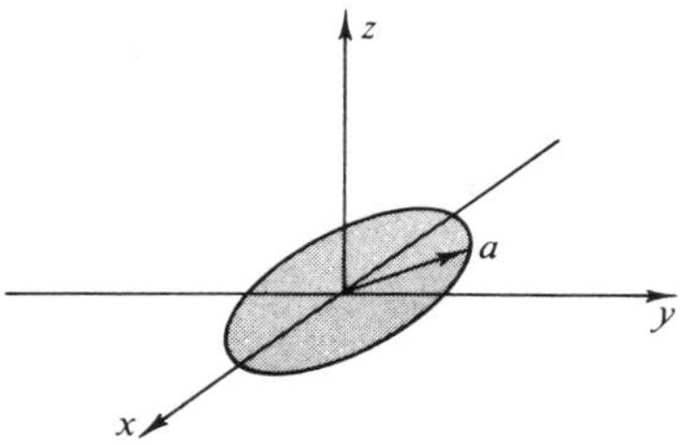

FIG. 12.8 Charged, conducting disk

From Ex. 2.11.2, the charge density σ (on each side of the disk) is

$$\sigma(\rho) = \frac{q}{4\pi a(a^2 - \rho^2)^{1/2}}, \qquad \rho^2 = x^2 + y^2.$$

$$Ans. \quad \psi(r, \theta) = \frac{q}{4\pi\varepsilon_0 r}\sum_{l=0}^{\infty}(-1)^l \frac{1}{2l+1}\left(\frac{a}{r}\right)^{2l} P_{2l}(\cos\theta).$$

12.3.12 (a) From the result of Ex. 12.3.11, calculate the potential of the disk. Since you are violating the condition $r > a$, justify your calculation carefully.
Hint. You may run into the series given in Ex. 5.2.19.
(b) Calculate the capacitance of the disk. Compare with Ex. 2.11.2.

12.3.13 The hemisphere defined by $r = a, 0 \le \theta < \pi/2$ has an electrostatic potential $+V_0$. The hemisphere $r = a, \pi/2 < \theta \le \pi$ has an electrostatic potential $-V_0$. Show that the potential at *interior* points is

$$V = V_0 \sum_{n=0}^{\infty} \frac{4n+3}{2n+2}\left(\frac{r}{a}\right)^{2n+1} P_{2n}(0)P_{2n+1}(\cos\theta)$$

$$= V_0 \sum_{n=0}^{\infty}(-1)^n \frac{(4n+3)(2n-1)!!}{(2n+2)!!}\left(\frac{r}{a}\right)^{2n+1} P_{2n+1}(\cos\theta).$$

Hint. You need Ex. 12.3.5.

12.3.14 A conducting sphere of radius a is divided into two electrically separate hemispheres by a thin insulating barrier at its equator. The top hemisphere is maintained at a potential V_0, the bottom hemisphere at $-V_0$.
(a) Show that the electrostatic potential *exterior* to the two hemispheres is

$$V(r, \theta) = V_0 \sum_{s=0}^{\infty}(-1)^s(4s+3)\frac{(2s-1)!!}{(2s+2)!!}\left(\frac{a}{r}\right)^{2s+2} P_{2s+1}(\cos\theta).$$

(b) Calculate the electric charge density σ on the outside surface. Note that your series diverges at $\cos\theta = \pm 1$ as you expected from the infinite capacitance of this system (zero thickness for the insulating barrier).

$$Ans. \quad \sigma = \varepsilon_0 E_n = -\varepsilon_0 \left.\frac{\partial V}{\partial r}\right|_{r=a} = \frac{\varepsilon_0 V_0}{a}\sum_{s=0}^{\infty}(-1)^s(4s+3)\frac{(2s-1)!!}{(2s)!!} P_{2s+1}(\cos\theta).$$

12.4 Alternate Definitions of Legendre Polynomials

Rodrigues' formula. The series form of the Legendre polynomials (Eq. 12.8) of Section 12.1 may be transformed as follows. From Eq. 12.8

$$P_n(x) = \sum_{r=0}^{[n/2]} (-1)^r \frac{(2n-2r)!}{2^n r!(n-r)!(n-2r)!} x^{n-2r} \tag{12.64}$$

For n an integer

$$P_n(x) = \sum_{r=0}^{[n/2]} (-1)^r \frac{1}{2^n r!(n-r)!}\left(\frac{d}{dx}\right)^n x^{2n-2r}$$

$$= \frac{1}{2^n n!}\left(\frac{d}{dx}\right)^n \sum_{r=0}^{n} \frac{(-1)^r n!}{r!(n-r)!} x^{2n-2r}. \tag{12.64a}$$

Note the extension of the upper limit. The reader is asked to show in Exercise 12.4.1 that the additional terms $[n/2]+1$ to n in the summation contribute nothing. However, the effect of these extra terms is to permit the replacement of the new summation by $(x^2-1)^n$ (binomial theorem once again) to obtain

$$P_n(x) = \frac{1}{2^n n!}\left(\frac{d}{dx}\right)^n (x^2-1)^n. \tag{12.65}$$

This is Rodrigues' formula. It is useful in proving many of the properties of the Legendre polynomials such as orthogonality. A related application is seen in Exercise 12.4.3.

Schlaefli Integral. Rodrigues' formula provides a means of developing an integral representation of $P_n(z)$. Using Cauchy's integral formula (Section 6.4)

$$f(z) = \frac{1}{2\pi i}\oint \frac{f(t)}{t-z}\,dt \tag{12.66}$$

with

$$f(z) = (z^2-1)^n, \tag{12.67}$$

we have

$$(z^2-1)^n = \frac{1}{2\pi i}\oint \frac{(t^2-1)^n}{t-z}\,dt. \tag{12.68}$$

Differentiating n times with respect to z and multiplying by $1/2^n n!$ gives us

$$P_n(z) = \frac{1}{2^n n!}\frac{d^n}{dz^n}(z^2-1)^n$$

$$= \frac{2^{-n}}{2\pi i}\oint \frac{(t^2-1)^n}{(t-z)^{n+1}}\,dt \tag{12.69}$$

with the contour enclosing the point $t = z$.

This is the Schlaefli integral. Margenau and Murphy[1] use this to derive the recurrence relations we obtained from the generating function.

[1] Margenau, H., and Murphy, G. M., *The Mathematics of Physics and Chemistry*, Second Ed., Section 3.5. Van Nostrand, Princeton, New Jersey (1956).

The Schlaefli integral may readily be shown to satisfy Legendre's equation by differentiation and direct substitution. We obtain

$$(1-z^2)\frac{d^2P_n}{dz^2} - 2z\frac{dP_n}{dz} + n(n+1)P_n = \frac{n+1}{2^n 2\pi i}\oint \frac{d}{dt}\left[\frac{(t^2-1)^{n+1}}{(t-z)^{n+2}}\right] dt. \tag{12.70}$$

For integral n our function $(t^2-1)^{n+1}/(t-z)^{n+2}$ is single-valued, and the integral around the closed path vanishes. The Schlaefli integral may also be used to define $P_\nu(z)$ for nonintegral ν integrating around the points $t = z$, $t = 1$, but not crossing the cut line -1 to $-\infty$. We could equally well encircle the points $t = z$ and $t = -1$, but this would lead to nothing new. A contour about $t = +1$ and $t = -1$ will lead to a second solution $Q_\nu(z)$ in Section 12.10.

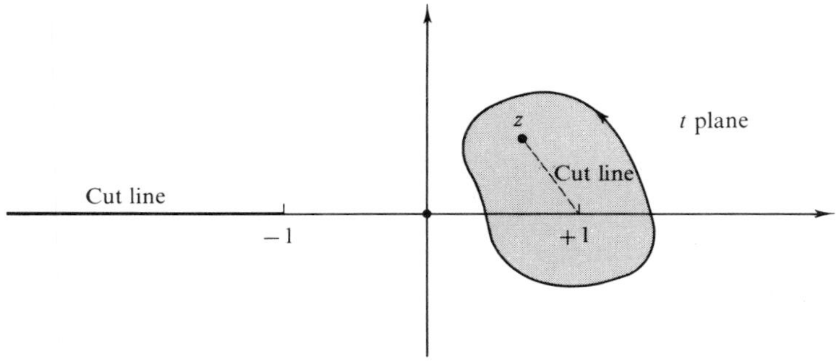

FIG. 12.9 Schlaefli integral contour

A change of variable

$$t = z + \sqrt{z^2-1}\, e^{i\varphi} \tag{12.71}$$

with $\mathscr{R}(z) > 0$ so that we still enclose the point $t = 1$ and not the point $t = -1$ yields

$$P_n(z) = \frac{1}{2\pi i}\oint \frac{2^n(z^2-1)^{n/2}e^{in\varphi}(z+\sqrt{z^2-1}\cos\varphi)^n i\sqrt{z^2-1}\,e^{i\varphi}}{2^n(\sqrt{z^2-1}\,e^{i\varphi})^{n+1}}\, d\varphi$$

$$= \frac{1}{2\pi}\int_0^{2\pi}(z+\sqrt{z^2-1}\cos\varphi)^n\, d\varphi. \tag{12.72}$$

This is Laplace's first integral representation. The contour is a circle of radius $|\sqrt{z^2-1}|$ about the point $t = z$. A second representation may be obtained by replacing n by $-n-1$. The differential equation is invariant to this substitution.

$$P_n(z) = \frac{1}{2\pi}\int_0^{2\pi}(z+\sqrt{z^2-1}\cos\varphi)^{-n-1}\, d\varphi. \tag{12.73}$$

The justification for calling this integral $P_n(z)$ rather than $P_{-n-1}(z)$ is the subject of Exercise 12.4.13.

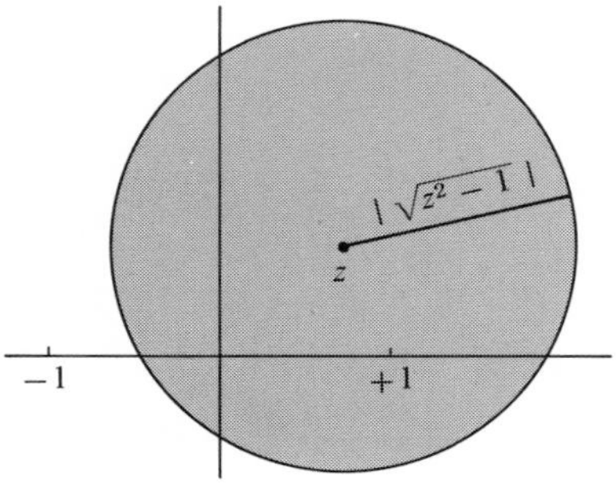

FIG. 12.10 Contour for Laplace's integral representation of the Legendre function

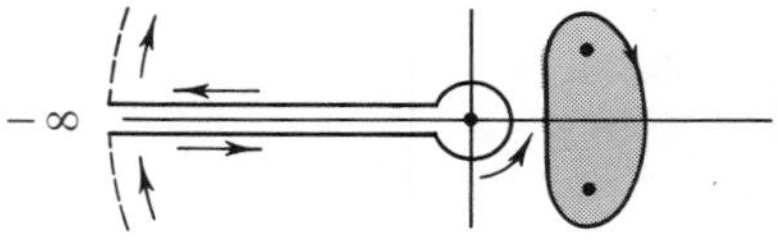

FIG. 12.11 Contour for $P_\nu(\xi)$, nonintegral ν

In Section 12.1 the generating function, despite its immediate and important physical application, was really pulled out of nowhere. Perhaps it may be made less mysterious by going just a little further. Using the substitution

$$t = z + \sqrt{z^2 - 1}\cos\varphi, \tag{12.74}$$

Eq. 12.73 becomes

$$P_n(z) = \frac{1}{2\pi i}\oint \frac{t^{-n-1}}{[1 - 2tz + t^2]^{1/2}}\,dt, \tag{12.75}$$

our contour enclosing the origin (integral n).[1] By the calculus of residues (Section 7.2) this is simply the coefficient of t^n in the expansion of $(1 - 2tz + t^2)^{-1/2}$, $|t^2 - 2tz| < 1$. Hence starting with the Schlaefi integral, a solution of **Legendre's** equation, we have developed a generating function for the Legendre polynomials.[2]

EXERCISES

12.4.1 Show that *each* term in the summation

$$\sum_{r=[n/2]+1}^{n} \left(\frac{d}{dx}\right)^n \frac{(-1)^r\, n!}{r!\,(n-r)!} x^{2n-2r}$$

vanishes (r and n integral).

[1] If n is not an integer, $t = 0$ is a branch point and the path of integration is a loop around the singular points $t = z \pm \sqrt{z^2 - 1}$, *clockwise*.

[2] In case this seems too much like going around in a circle, we could start with a power series solution of Legendre's equation, Ex. 8.4.4. Normalization could be achieved by demanding that $P_n(1) = 1$, or by developing Rodrigues' formula as in this section, and then using Ex. 12.4.2. The remainder of the derivation would follow as given in this section.

12.4.2 Using Rodrigues' formula, show that the $P_n(x)$ are orthogonal and that

$$\int_{-1}^{1} [P_n(x)]^2\, dx = \frac{2}{2n+1}.$$

12.4.3 Show that $\int_{-1}^{1} x^m P_n(x)\, dx = 0$ when $m < n$. *Hint*. Use Rodrigues' formula.

12.4.4 Show that

$$\int_{-1}^{1} x^n P_n(x)\, dx = \frac{2^{n+1}\, n!\, n!}{(2n+1)!}.$$

12.4.5 Show that

$$\int_{-1}^{1} x^{2r} P_{2n}(x)\, dx = \frac{2^{2n+1}(2r)!\,(r+n)!}{(2r+2n+1)!\,(r-n)!}.$$

12.4.6 As a generalization of Exercises 12.4.4 and 12.4.5, show that the Legendre expansions of x^s are

(a) $$x^{2r} = \sum_{n=0}^{r} \frac{2^{2n}(4n+1)(2r)!\,(r+n)!}{(2r+2n+1)!\,(r-n)!} P_{2n}(x), \qquad s = 2r,$$

(b) $$x^{2r+1} = \sum_{n=0}^{r} \frac{2^{2n+1}(4n+3)(2r+1)!\,(r+n+1)!}{(2r+2n+3)!\,(r-n)!} P_{2n+1}(x), \qquad s = 2r+1.$$

12.4.7 A plane wave may be expanded in a series of spherical waves by the Rayleigh equation

$$e^{ikr\cos\gamma} = \sum_{n=0}^{\infty} a_n j_n(kr) P_n(\cos\gamma).$$

Show that $a_n = i^n(2n+1)$.

Hint. 1. Use the orthogonality of the P_n to solve for $a_n j_n(kr)$.
2. Differentiate n times with respect to (kr) and set $r = 0$ to eliminate the r-dependence.
3. Evaluate the remaining integral by Exercise 12.4.4.

Note. This problem may also be treated by noting that both sides of the equation satisfy the Helmholtz equation. The equality can be established by showing that the solutions have the same behavior at the origin and also behave alike at large distances.

12.4.8 Verify the Rayleigh equation of Exercise 12.4.7 by starting with the following steps:
1. Differentiate with respect to (kr) to establish

$$\sum_n a_n j_n'(kr) P_n(\cos\gamma) = i \sum_n a_n j_n(kr) \cos\gamma P_n(\cos\gamma).$$

2. Use a recurrence relation to replace $\cos\gamma P_n(\cos\gamma)$ by a linear combination of P_{n-1} and P_{n+1}.
3. Use a recurrence relation to replace j_n' by a linear combination of j_{n-1} and j_{n+1}.

12.4.9 From Ex. 12.4.7, show that

$$j_n(kr) = \frac{1}{2i^n} \int_{-1}^{1} e^{ikr\mu} P_n(\mu)\, d\mu.$$

12.4.10 The Legendre polynomials and the spherical Bessel functions are related by

$$j_n(z) = \tfrac{1}{2}(-i)^n \int_0^{\pi} e^{iz\cos\theta} P_n(\cos\theta) \sin\theta\, d\theta, \qquad n = 0, 1, 2, \ldots.$$

Verify this relation by transforming the right-hand side into

$$\frac{z^n}{2^{n+1}n!}\int_0^\pi \cos(z\cos\theta)\sin^{2n+1}\theta\,d\theta$$

and using Exercise 11.7.9.

12.4.11 By direct evaluation of the Schlaefli integral show that $P_n(1) = 1$.

12.4.12 Explain why the contour of the Schlaefli integral, Eq. 12.69, is chosen to enclose the points $t = z$ *and* $t = 1$ when $n \to \nu$, not an integer.

12.4.13 Using the Laplace integral representations, show that $P_n(z) = P_{-n-1}(z)$. *Hint.* Since $P_{-n-1}(z)$ satisfies Legendre's equation, it must be a linear combination of $P_{+n}(z)$, regular at $z = \pm 1$ and $Q_{+n}(z)$ divergent, at $z = \pm 1$.

$$P_{-n-1}(z) = a_1 P_n(z) + a_2 Q_n(z).$$

Show that
(a) $a_2 = 0$ and (b) $a_1 = 1$.

12.5 Associated Legendre Functions

When Helmholtz' equation is separated in spherical polar, oblate spheroidal, or prolate spheroidal coordinates, one of the separated ordinary differential equations is the associated Legendre equation. Only if the azimuthal separation constant m^2 is zero do we have Legendre's equation, Eq. 12.28.

One way of developing the solution of the associated Legendre equation is to start with the regular Legendre equation and convert it into the associated Legendre equation by multiple differentiation. We take Legendre's equation

$$(1 - x^2)P_n'' - 2xP_n' + n(n+1)P_n = 0 \tag{12.76}$$

and with the help of Leibnitz' formula[1] differentiate m times. The result is,

$$(1 - x^2)u'' - 2x(m+1)u' + (n-m)(n+m+1)u = 0, \tag{12.77}$$

where

$$u \equiv \frac{d^m}{dx^m} P_n(x). \tag{12.78}$$

Now let

$$v(x) = (1 - x^2)^{m/2}u(x) = (1 - x^2)^{m/2}\frac{d^m P_n(x)}{dx^m}. \tag{12.79}$$

Solving for u and differentiating, we obtain

$$u' = \left(v' + \frac{mxv}{1 - x^2}\right)(1 - x^2)^{-m/2}, \tag{12.80}$$

[1] Leibnitz' formula for the nth derivative of a product is

$$\frac{d^n}{dx^n}[A(x)\,B(x)] = \sum_{s=0}^{n}\binom{n}{s}\frac{d^{n-s}}{dx^{n-s}}A(x)\frac{d^s}{dx^s}B(x), \qquad \binom{n}{s} = \frac{n!}{(n-s)!s!},$$

a binomial coefficient.

$$u'' = \left[v'' + \frac{2mxv'}{1-x^2} + \frac{mv}{1-x^2} + \frac{m(m+2)x^2v}{(1-x^2)^2}\right]\cdot(1-x^2)^{-m/2} \tag{12.81}$$

Substituting into Eq. 12.77, we find that the new function v satisfies the differential equation

$$(1-x^2)v'' - 2xv' + \left[n(n+1) - \frac{m^2}{1-x^2}\right]v = 0, \tag{12.82}$$

which is the associated Legendre equation reducing to Legendre's equation, as it must be when m is set equal to zero. Expressed in spherical polar coordinates, the associated Legendre equation is,

$$\frac{1}{\sin\theta}\frac{d}{d\theta}\left(\sin\theta\frac{dv}{d\theta}\right) + \left[n(n+1) - \frac{m^2}{\sin^2\theta}\right]v = 0. \tag{12.83}$$

Associated Legendre functions. The regular solutions, relabeled $P_n^m(x)$ are

$$v \equiv P_n^m(x) = (1-x^2)^{m/2}\frac{d^m}{dx^m}P_n(x). \tag{12.84}$$

These are the associated Legendre functions.[1]

From the form of Eq. 12.84 we might expect m to be nonnegative, differentiating a negative number of times not having been defined. However, if $P_n(x)$ is expressed by Rodrigues' formula, this limitation on m is relaxed, and we may have $-n \leqslant m \leqslant n$, negative as well as positive values of m being permitted. Using Leibnitz' differentiation formula once again, the reader may show (Exercise 12.5.1) that

TABLE 12.3

Associated Legendre Functions

$P_1^1(x) = (1-x^2)^{1/2} = \sin\theta$
$P_2^1(x) = 3x(1-x^2)^{1/2} = 3\cos\theta\sin\theta$
$P_2^2(x) = 3(1-x^2) = 3\sin^2\theta$
$P_3^1(x) = \frac{3}{2}(5x^2-1)(1-x^2)^{1/2} = \frac{3}{2}(5\cos^2\theta-1)\sin\theta$
$P_3^2(x) = 15x(1-x^2) = 15\cos\theta\sin^2\theta$
$P_3^3(x) = 15(1-x^2)^{3/2} = 15\sin^3\theta$
$P_4^1(x) = \frac{5}{2}(7x^3-3x)(1-x^2)^{1/2} = \frac{5}{2}(7\cos^3\theta-3\cos\theta)\sin\theta$
$P_4^2(x) = \frac{15}{2}(7x^2-1)(1-x^2) = \frac{15}{2}(7\cos^2\theta-1)\sin^2\theta$
$P_4^3(x) = 105x(1-x^2)^{3/2} = 105\cos\theta\sin^3\theta$
$P_4^4(x) = 105(1-x^2)^2 = 105\sin^4\theta$

[1] Occasionally (as in AMS–55), the reader will find the associated Legendre functions defined with an additional factor of $(-1)^m$. This $(-1)^m$ seems an unnecessary complication at this point. It will be included in the definition of the spherical harmonics $Y_n^m(\theta, \varphi)$ in Section 12.6.

$P_n^m(x)$ and $P_n^{-m}(x)$ are related by

$$P_n^{-m}(x) = (-1)^m \frac{(n-m)!}{(n+m)!} P_n^m(x). \tag{12.84a}$$

From our definition of the associated Legendre functions, $P_n^m(x)$,

$$P_n^0(x) = P_n(x). \tag{12.85}$$

In addition, we may develop Table 12.3.

As with the Legendre polynomials, a generating function for the associated Legendre functions does exist:

$$\frac{(2m)!(1-x^2)^{m/2}}{2^m m!(1-2tx+t^2)^{m+1/2}} = \sum_{s=0}^{\infty} P_{s+m}^m(x)t^s. \tag{12.86}$$

However, because of its more cumbersome form and lack of any direct physical application, it is seldom used.

Recurrence relations. As expected, the associated Legendre functions satisfy recurrence relations. Because of the existence of two indices instead of just one, we have a wide variety of recurrence relations:

$$P_n^{m+1} - \frac{2mx}{(1-x^2)^{1/2}} P_n^m + [n(n+1) - m(m-1)]P_n^{m-1} = 0, \tag{12.87}$$

$$(2n+1)xP_n^m = (n+m)P_{n-1}^m + (n-m+1)P_{n+1}^m, \tag{12.88}$$

$$\begin{aligned}(2n+1)(1-x^2)^{1/2}P_n^m &\\ &= P_{n+1}^{m+1} - P_{n-1}^{m+1} \\ &= (n+m)(n+m-1)P_{n-1}^{m-1} - (n-m+1)(n-m+2)P_{n+1}^{m-1},\end{aligned} \tag{12.89}$$

$$(1-x^2)^{1/2}P_n^{m\prime} = \tfrac{1}{2}P_n^{m+1} - \tfrac{1}{2}(n+m)(n-m+1)P_n^{m-1}. \tag{12.90}$$

These relations, and many other similar ones, may be verified by use of the generating function (Eq. 12.4), by substitution of the series solution of the associated Legendre equation (12.82), or by reduction to the Legendre polynomial recurrence relations, using Eq. 12.84. As an example of the last method, consider the third equation in the preceding set. It is similar to Eq. 12.23.

$$(2n+1)\, P_n(x) = P'_{n+1}(x) - P'_{n-1}(x). \tag{12.91}$$

Let us differentiate this Legendre polynomial recurrence relation m times to obtain

$$\begin{aligned}(2n+1)\frac{d^m}{dx^m}P_n(x) &= \frac{d^m}{dx^m}P'_{n+1}(x) - \frac{d^m}{dx^m}P'_{n-1}(x) \\ &= \frac{d^{m+1}}{dx^{m+1}}P_{n+1}(x) - \frac{d^{m+1}}{dx^{m+1}}P_{n-1}(x).\end{aligned} \tag{12.92}$$

Now multiplying by $(1-x^2)^{(m+1)/2}$ and using the definition of $P_n^m(x)$, we obtain Eq. 12.89.

Parity. The parity relation satisfied by the associated Legendre functions may be determined by examination of the defining equation (12.84). As $x \to -x$, we already know that $P_n(x)$ contributes a $(-1)^n$. The m-fold differentiation yields a factor of $(-1)^m$. Hence we have

$$P_n^m(-x) = (-1)^{n+m} P_n^m(x). \tag{12.93}$$

A glance at Table 12.3 verifies this for $m \leqslant 4$.

Also, from the definition in Eq. 12.84

$$P_n^m(\pm 1) = 0, \quad \text{for} \quad m > 0. \tag{12.94}$$

Orthogonality. The orthogonality of the $P_n^m(x)$ follows from the differential equation just as in $P_n(x)$ (Section 12.3); the term $-m^2/(1-x^2)$ cancels out, assuming m is the same in both cases. However, it is instructive to demonstrate the orthogonality by another method, a method that will also provide the normalization constant.

Using the definition in Eq. 12.84 and Rodrigues' formula (Eq. 12.65) for $P_n(x)$, we find

$$\int_{-1}^{1} P_p^m(x)\, P_q^m(x)\, dx = \frac{(-1)^m}{2^{p+q} p!\, q!} \int_{-1}^{1} X^m \frac{d^{p+m}}{dx^{p+m}} X^p \frac{d^{q+m}}{dx^{q+m}} X^q\, dx. \tag{12.95}$$

The function X is given by $X \equiv (x^2 - 1)$. If $p \neq q$, let us assume that $p < q$. Notice that the superscript m is the same for both functions. This is an essential condition. The technique is to integrate repeatedly by parts; all the integrated parts will vanish as long as there is a factor $X = x^2 - 1$. Let us integrate $q + m$ times to obtain

$$\int_{-1}^{1} P_p^m(x)\, P_q^m(x)\, dx = \frac{(-1)^m(-1)^{q+m}}{2^{p+q} p!\, q!} \int_{-1}^{1} \frac{d^{q+m}}{dx^{q+m}} \left(X^m \frac{d^{p+m}}{dx^{p+m}} X^p \right) X^q\, dx. \tag{12.96}$$

The integrand on the right-hand side is now expanded by Leibnitz' formula to give

$$X^q \frac{d^{q+m}}{dx^{q+m}} \left(X^m \frac{d^{p+m}}{dx^{p+m}} X^p \right) = X^q \sum_{i=0}^{i=q+m} \frac{(q+m)!}{i!(q+m-i)!} \frac{d^{q+m-i}}{dx^{q+m-i}} X^m \frac{d^{p+m+i}}{dx^{p+m+i}} X^p. \tag{12.97}$$

Since the term X^m contains no power of x greater than x^{2m}, we must have

$$q + m - i \leqslant 2m \tag{12.98}$$

or the derivative will vanish. Similarly,

$$p + m + i \leqslant 2p. \tag{12.99}$$

In the solution of these equations for the index i, the conditions for a nonzero result are

$$i \geqslant q - m, \quad i \leqslant p - m. \tag{12.100}$$

If $p < q$, as assumed, there is no solution and the integral vanishes. The same result obviously must follow if $p > q$.

For the remaining case, $p = q$, we may still have the single term corresponding to $i = q - m$. Putting Eq. 12.97 into Eq. 12.96, we have

$$\int_{-1}^{1} [P_q^m(x)]^2\, dx = \frac{(-1)^{q+2m}(q+m)!}{2^{2q}q!q!(2m)!(q-m)!} \int_{-1}^{1} X^q \left(\frac{d^{2m}}{dx^{2m}} X^m\right)\left(\frac{d^{2q}}{dx^{2q}} X^q\right) dx. \tag{12.101}$$

Since

$$X^m = (x^2 - 1)^m = x^{2m} - mx^{2m-2} + \cdots, \tag{12.102}$$

$$\frac{d^{2m}}{dx^{2m}} X^m = (2m)!, \tag{12.103}$$

Eq. 12.101 reduces to

$$\int_{-1}^{1} [P_q^m(x)]^2\, dx = \frac{(-1)^{q+2m}(2q)!(q+m)!}{2^{2q}q!q!(q-m)!} \int_{-1}^{1} X^q\, dx. \tag{12.104}$$

The integral on the right is just

$$(-1)^q \int_0^\pi \sin^{2q+1}\theta\, d\theta = \frac{(-1)^q 2^{2q+1} q!q!}{(2q+1)!} \tag{12.105}$$

(cf. Exercise 10.4.9). Combining Eqs. 12.104 and 12.105, we have the orthogonality integral

$$\int_{-1}^{1} P_p^m(x)\, P_q^m(x)\, dx = \frac{2}{2q+1} \cdot \frac{(q+m)!}{(q-m)!}\, \delta_{p,q} \tag{12.106}$$

or, in spherical polar coordinates,

$$\int_0^\pi P_p^m(\cos\theta)\, P_q^m(\cos\theta)\sin\theta\, d\theta = \frac{2}{2q+1} \cdot \frac{(q+m)!}{(q-m)!}\, \delta_{p,q}. \tag{12.107}$$

The orthogonality of the Legendre polynomials is actually a special case of this result, obtained by setting m equal to zero; that is, for $m = 0$, Eq. 12.106 reduces to Eqs. 12.43 and 12.48.

It is possible to develop an orthogonality relation for associated Legendre functions of the same lower index but different upper index. We find

$$\int_{-1}^{1} P_n^m(x) P_n^k(x)(1 - x^2)^{-1}\, dx = \frac{(n+m)!}{m(n-m)!}\, \delta_{m,k}. \tag{12.108}$$

Note that a new weighting factor, $(1 - x^2)^{-1}$, has been introduced. This form is essentially a mathematical curiosity. In physical problems orthogonality of the φ dependence ties the two upper indices together and leads to Eq. 12.107.

EXAMPLE 12.5.1. MAGNETIC INDUCTION FIELD OF A CURRENT LOOP.

Like the other differential equations of mathematical physics, the associated

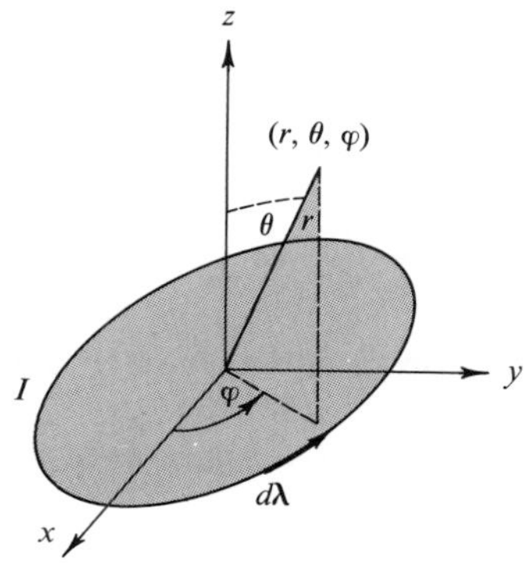

FIG. 12.12 Circular current loop

Legendre equation is likely to pop up quite unexpectedly. As an illustration, consider the magnetic induction field **B** and magnetic vector potential **A** created by a single circular current loop in the equatorial plane (Fig. 12.12).

We know from electromagnetic theory that the contribution of current element $I\,d\boldsymbol{\lambda}$ to the magnetic vector potential is

$$d\mathbf{A} = \frac{\mu_0}{4\pi}\frac{I\,d\boldsymbol{\lambda}}{r}. \tag{12.109}$$

This, plus the symmetry of our system, shows that **A** has only a $\boldsymbol{\varphi}_0$-component and that the component is independent of φ[1]

$$\mathbf{A} = \boldsymbol{\varphi}_0\, A_\varphi(r, \theta). \tag{12.110}$$

By Maxwell's equations

$$\nabla \times \mathbf{H} = \mathbf{J}, \qquad (\partial\mathbf{D}/\partial t = 0, \text{ mks units}). \tag{12.111}$$

Since

$$\mu_0\mathbf{H} = \mathbf{B} = \nabla \times \mathbf{A}, \tag{12.112}$$

we have

$$\nabla \times \nabla \times \mathbf{A} = \mu_0\mathbf{J}, \tag{12.113}$$

where **J** is the current density. In our problem **J** is zero everywhere except in the current loop. Therefore, away from the loop,

$$\nabla \times \nabla \times \boldsymbol{\varphi}_0\, A_\varphi(r, \theta) = 0, \tag{12.114}$$

which introduces Eq. 12.110.

Using the expression for the curl in spherical polar coordinates (Section 2.4), we obtain (Example 2.4.2)

$$\nabla \times \nabla \times \boldsymbol{\varphi}_0\, A_\varphi(r, \theta) = \boldsymbol{\varphi}_0\left[-\frac{\partial^2 A_\varphi}{\partial r^2} - \frac{2}{r}\frac{\partial A_\varphi}{\partial r} - \frac{1}{r^2}\frac{\partial^2 A_\varphi}{\partial \theta^2} - \frac{1}{r^2}\frac{\partial}{\partial\theta}(\cot\theta A_\varphi)\right]$$

$$= 0. \tag{12.115}$$

[1] Pair off corresponding current elements $I\,d\boldsymbol{\lambda}(\varphi_1)$ and $I\,d\boldsymbol{\lambda}(\varphi_2)$, where $\varphi - \varphi_1 = \varphi_2 - \varphi$.

Letting $A_\varphi(r, \theta) = R(r)\Theta(\theta)$ and separating variables, we have

$$r^2 \frac{d^2R}{dr^2} + 2r\frac{dR}{dr} - n(n+1)R = 0, \tag{12.116}$$

$$\frac{d^2\Theta}{d\theta^2} + \cot\theta \frac{d\Theta}{d\theta} + n(n+1)\Theta - \frac{\Theta}{\sin^2\theta} = 0. \tag{12.117}$$

The second equation is the associated Legendre equation (12.83) with $m = 1$, and we may immediately write

$$\Theta(\theta) = P_n^1(\cos\theta). \tag{12.118}$$

The separation constant $n(n+1)$ was chosen to keep this solution well behaved.

By trial, letting $R(r) = r^\alpha$, we find that $\alpha = n, -n-1$. The first possibility is discarded, for our solution must vanish as $r \to \infty$. Hence

$$A_{\varphi n} = \frac{b_n}{r^{n+1}} P_n^1(\cos\theta) = c_n\left(\frac{a}{r}\right)^{n+1} P_n^1(\cos\theta) \tag{12.119}$$

and

$$A_\varphi(r, \theta) = \sum_{n=1}^{\infty} c_n\left(\frac{a}{r}\right)^{n+1} P_n^1(\cos\theta), \qquad (r > a). \tag{12.120}$$

Here a is the radius of the current loop.

Since A_φ must be invariant to reflection in the equatorial plane by the symmetry of our problem,

$$A_\varphi(r, \cos\theta) = A_\varphi(r, -\cos\theta), \tag{12.121}$$

the parity property of $P_n^m(\cos\theta)$ (Eq. 12.93) shows that $c_n = 0$ for n even.

To complete the evaluation of the constants, we may use Eq. 12.120 to calculate B_z along the z-axis $[B_z = B_r(r, \theta = 0)]$ and compare with the expression obtained from the Biot and Savart law. This is the same technique that is used in Example 12.3.3. We have

$$\begin{aligned} B_r &= \nabla \times \mathbf{A}\big|_r \\ &= \frac{1}{r\sin\theta}\left[\frac{\partial}{\partial\theta}(\sin\theta\, A_\varphi)\right] \\ &= \frac{\cot\theta}{r} A_\varphi + \frac{1}{r}\frac{\partial A_\varphi}{\partial\theta}. \end{aligned} \tag{12.122}$$

Using

$$\begin{aligned} \frac{\partial P_n^1(\cos\theta)}{\partial\theta} &= -\sin\theta \frac{dP_n^1(\cos\theta)}{d(\cos\theta)} \\ &= -\frac{1}{2}P_n^2 + \frac{n(n+1)}{2} P_n^0 \end{aligned} \tag{12.123}$$

and Eq. 12.87 with $m = 1$,

$$P_n^2(\cos\theta) - \frac{2\cos\theta}{\sin\theta} P_n^1(\cos\theta) + n(n+1)\, P_n(\cos\theta) = 0, \tag{12.124}$$

we obtain

$$B_r(r, \theta) = \sum_{n=1}^{\infty} c_n n(n+1) \frac{a^{n+1}}{r^{n+2}} P_n(\cos\theta), \qquad r > a \tag{12.125}$$

(for all θ). In particular, for $\theta = 0$,

$$B_r(r, 0) = \sum_{n=1}^{\infty} c_n n(n+1) \frac{a^{n+1}}{r^{n+2}}. \tag{12.126}$$

We may also obtain

$$\begin{aligned} B_\theta(r, \theta) &= -\frac{1}{r}\frac{\partial(rA_\varphi)}{\partial r} \\ &= \sum_{n=1}^{\infty} c_n n \frac{a^{n+1}}{r^{n+2}} P_n^1(\cos\theta), \qquad r > a. \end{aligned} \tag{12.127}$$

The Biot and Savart law states that

$$d\mathbf{B} = \frac{\mu_0}{4\pi} I \frac{d\boldsymbol{\lambda} \times \mathbf{r}_0}{r^2} \qquad \text{(mks units).} \tag{12.128}$$

We now integrate over the perimeter of our loop (radius a); the magnetic induction field along the z-axis is $\mathbf{k}B_z$, where

$$\begin{aligned} B_z &= \frac{\mu_0 I}{2} a^2(a^2 + z^2)^{-3/2} \\ &= \frac{\mu_0 I}{2}\frac{a^2}{z^3}\left(1 + \frac{a^2}{z^2}\right)^{-3/2} \end{aligned} \tag{12.129}$$

Expanding by the binomial theorem

$$\begin{aligned} B_z &= \frac{\mu_0 I}{2}\frac{a^2}{z^3}\left[1 - \frac{3}{2}\left(\frac{a}{z}\right)^2 + \frac{15}{8}\left(\frac{a}{z}\right)^4 - \cdots\right] \\ &= \frac{\mu_0 I}{2}\frac{a^2}{z^3}\sum_{s=0}^{\infty}(-1)^s \frac{(2s+1)!!}{(2s)!!}\left(\frac{a}{z}\right)^{2s}, \qquad z > a. \end{aligned} \tag{12.130}$$

Equating Eqs. 12.126 and 12.130 term by term (with $r = z$),[1] we find

$$\begin{aligned} &c_1 = \frac{\mu_0 I}{4}, \qquad c_3 = -\frac{\mu_0 I}{16}, \qquad c_2 = c_4 = \cdots = 0. \\ &c_n = (-1)^{(n-1)/2} \frac{\mu_0 I}{2n(n+1)} \times \frac{(n/2)!}{[(n-1)/2]!(\frac{1}{2})!}, \qquad n \text{ odd.} \end{aligned} \tag{12.131}$$

Equivalently, we may write

$$c_{2n+1} = (-1)^n \frac{\mu_0 I}{2^{2n+2}} \cdot \frac{(2n)!}{n!(n+1)!} = (-1)^n \frac{\mu_0 I}{2} \cdot \frac{(2n-1)!!}{(2n+2)!!} \tag{12.132}$$

[1] The descending power series is also unique.

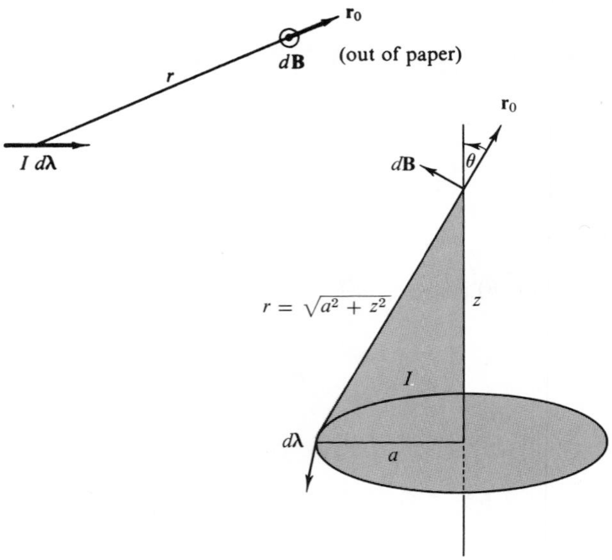

FIG. 12.13 Law of Biot and Savart applied to a circular loop

and

$$A_\varphi(r,\theta)=\left(\frac{a}{r}\right)^2\sum_{n=0}^{\infty}c_{2n+1}\left(\frac{a}{r}\right)^{2n}P_{2n+1}^{1}(\cos\theta), \tag{12.133}$$

$$B_r(r,\theta)=\frac{a^2}{r^3}\sum_{n=0}^{\infty}c_{2n+1}(2n+1)(2n+2)\left(\frac{a}{r}\right)^{2n}P_{2n+1}(\cos\theta), \tag{12.134}$$

$$B_\theta(r,\theta)=\frac{a^2}{r^3}\sum_{n=0}^{\infty}c_{2n+1}(2n+1)\left(\frac{a}{r}\right)^{2n}P_{2n+1}^{1}(\cos\theta). \tag{12.135}$$

These fields may be described in closed form by the use of elliptic integrals. Ex. 5.8.4 is an illustration of this approach. A third possibility is direct integration of Eq. 12.109 by expanding the factor $1/r$ as a Legendre polynomial generating function. The current is specified by Dirac delta functions. These methods have the advantage of yielding the constants c_n directly.

A comparison of magnetic current loop dipole fields and finite electric dipole fields may be of interest. For the magnetic current loop dipole the preceding analysis gives

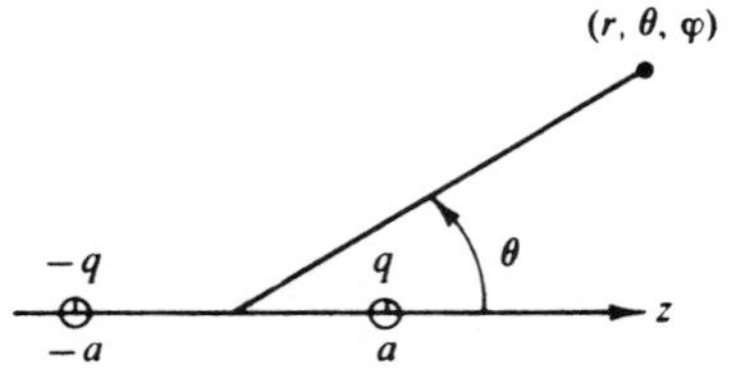

FIG. 12.14

$$B_r(r, \theta) = \frac{\mu_0 I}{2} \frac{a^2}{r^3}\left[P_1 - \frac{3}{2}\left(\frac{a}{r}\right)^2 P_3 + \cdots\right], \tag{12.136}$$

$$B_\theta(r, \theta) = \frac{\mu_0 I}{4} \frac{a^2}{r^3}\left[P_1^1 - \frac{3}{4}\left(\frac{a}{r}\right)^2 P_3^1 + \cdots\right]. \tag{12.137}$$

From the finite electric dipole potential of Section 12.1 we have

$$E_r(r, \theta) = \frac{qa}{\pi\varepsilon_0 r^3}\left[P_1 + 2P_3\left(\frac{a}{r}\right)^2 + \cdots\right], \tag{12.138}$$

$$E_\theta(r, \theta) = \frac{qa}{2\pi\varepsilon_0 r^3}\left[P_1^1 + \left(\frac{a}{r}\right)^2 P_3^1 + \cdots\right]. \tag{12.139}$$

The two fields agree in form as far as the leading term is concerned, and this is the basis for calling them both dipole fields.

As with electric multipoles, it is sometimes convenient to talk about *point* magnetic multipoles. For the dipole case, Eqs. 12.136 and 12.137, the point dipole is formed by taking the limit $a \to 0$, $I \to \infty$ with Ia^2 held constant. With $\mathbf{n}$ a unit vector normal to the current loop (positive sense by right hand rule, Section 1.10) the magnetic moment $\mathbf{m}$ is given by $\mathbf{m} = \mathbf{n}I\pi a^2$.

EXERCISES

12.5.1 Prove that

$$P_n^{-m}(x) = (-1)^m \frac{(n-m)!}{(n+m)!} P_n^m(x),$$

where $P_n^m(x)$ is defined by

$$P_n^m(x) = \frac{1}{2^n n!}(1-x^2)^{m/2} \frac{d^{n+m}}{dx^{n+m}}(x^2-1)^n.$$

Hint. One approach is to apply Leibnitz' formula to $(x+1)^n(x-1)^n$.

12.5.2 Show that

$$P_{2n}^1(0) = 0,$$

$$P_{2n+1}^1(0) = (-1)^n \frac{(2n+1)!}{(2^n n!)^2} = (-1)^n \frac{(2n+1)!!}{(2n)!!},$$

by each of the three methods:
(a) use of recurrence relations,
(b) expansion of the generating function, and
(c) Rodrigues' formula.

12.5.3 Evaluate $P_n^m(0)$.

$$Ans.\ P_n{}^m(0) = \begin{cases} (-1)^{(n-m)/2} \dfrac{(n+m)!}{2^n\left(\dfrac{n-m}{2}\right)!\left(\dfrac{n+m}{2}\right)!}, & n+m \text{ even}, \\ 0, & n+m \text{ odd}. \end{cases}$$

$$\text{Also, } P_n{}^m(0) = (-1)^{(n-m)/2}\frac{(n+m-1)!!}{(n-m)!!}, \qquad n+m \text{ even}.$$

12.5.4 Show that

$$P_n{}^n(\cos\theta) = (2n-1)!!\sin^n\theta, \qquad n = 0, 1, 2, \ldots .$$

12.5.5 Derive the associated Legendre recurrence relation

$$P_n^{m+1}(x) - \frac{2mx}{(1-x^2)^{1/2}}P_n^m(x) + [n(n+1) - m(m-1)]P_n^{m-1}(x) = 0.$$

12.5.6 Show that

$$\sin\theta\, P_n'(\cos\theta) = P_n^1(\cos\theta).$$

12.5.7 Show that

(a) $$\int_0^\pi \left(\frac{dP_n^m}{d\theta}\frac{dP_{n'}^m}{d\theta} + \frac{m^2 P_n^m P_{n'}^m}{\sin^2\theta}\right)\sin\theta\, d\theta = \frac{2n(n+1)}{2n+1}\frac{(n+m)!}{(n-m)!}\delta_{nn'},$$

(b) $$\int_0^\pi \left(\frac{P_n^1}{\sin\theta}\frac{dP_{n'}^1}{d\theta} + \frac{P_{n'}^1}{\sin\theta}\frac{dP_n^1}{d\theta}\right)\sin\theta\, d\theta = 0.$$

These integrals occur in the theory of scattering of electromagnetic waves by spheres.

12.5.8 As a repeat of Exercise 12.3.6, show, using associated Legendre functions, that

$$\int_{-1}^1 x(1-x^2)\,P_n'(x)\,P_m'(x)\,dx$$
$$= \frac{n+1}{2n+1}\cdot\frac{2}{2n-1}\cdot\frac{n!}{(n-2)!}\delta_{m,n-1} + \frac{n}{2n+1}\cdot\frac{2}{2n+3}\cdot\frac{(n+2)!}{n!}\delta_{m,n+1}.$$

12.5.9 Evaluate

$$\int_0^\pi \sin^2\theta\, P_n^1(\cos\theta)\,d\theta.$$

12.5.10 The associated Legendre polynomial $P_n^m(x)$ satisfies the self-adjoint differential equation

$$(1-x^2)P_n^{m\,\prime\prime}(x) - 2x\,P_n^{m\,\prime}(x) + \left[n(n+1) - \frac{m^2}{1-x^2}\right]P_n^m(x) = 0.$$

From the differential equations for $P_n^m(x)$ and $P_n^k(x)$ show that

$$\int_{-1}^1 P_n^m(x)\,P_n^k(x)\frac{dx}{1-x^2} = 0$$

for $k \neq m$.

12.5.11 Determine the vector potential of a magnetic quadrupole by differentiating the magnetic dipole potential.

$$Ans.\ \mathbf{A}_{MQ} = \frac{\mu_0}{2}(Ia^2)(dz)\,\boldsymbol{\varphi}_0\frac{P_2^1(\cos\theta)}{r^3} + \text{higher-order terms}.$$

$$\mathbf{B}_{MQ} = \mu_0(Ia^2)(dz)\left[\mathbf{r}_0\frac{3P_2(\cos\theta)}{r^4} + \boldsymbol{\theta}_0\frac{P_2^1(\cos\theta)}{r^4}\right]$$

This corresponds to placing a current loop of radius a at $z = dz$, an oppositely directed current loop at $z = -dz$, and letting $a \to 0$ subject to $(dz)x$ (dipole strength) equal constant.

Another approach to this problem would be to integrate dA (Eq. 12.109), to expand the denominator in a series of Legendre polynomials, and to use the Legendre polynomial addition theorem (Section 12.8).

12.5.12 A single loop of wire of radius a carries a current I.
(a) Find the magnetic induction $\mathbf{B}$ for $r < a$.
(b) Calculate the integral of the magnetic flux $(\mathbf{B}\cdot d\boldsymbol{\sigma})$ over the area of the current loop, that is,

$$\int_0^a \int_0^{2\pi} B_\theta\left(r, \theta = \frac{\pi}{2}\right) d\varphi\, r\, dr.$$

Ans. ∞.

The earth is within such a ring current in which I approximates millions of amperes arising from the drift of charged particles in the Van Allen belt.

12.5.13 A uniformly charged spherical shell is rotating with constant angular velocity.
(a) Calculate the magnetic induction $\mathbf{B}$ along the axis of rotation outside the sphere.
(b) Using the vector potential series of Section 12.5, find $\mathbf{A}$ and then $\mathbf{B}$ for all space outside the sphere.

12.5.14 In the liquid drop model of the nucleus, the spherical nucleus is subjected to small deformations. Consider a sphere of radius r_0 that is deformed so that its new surface is given by

$$r = r_0[1 + \alpha_2 P_2(\cos\theta)].$$

Find the area of the deformed sphere through terms of order α_2^2.
Hint.

$$dA = \left[r^2 + \left(\frac{dr}{d\theta}\right)^2\right]^{1/2} r\sin\theta\, d\theta\, d\varphi.$$

Ans. $A = 4\pi r_0^2[1 + \frac{4}{5}\alpha_2^2 + O(\alpha_2^3)]$.

12.5.15 A nuclear particle is in a potential $V(r, \theta, \varphi) = 0$ for $0 \leqslant r < a$ and ∞ for $r > a$. The particle is described by a wave function $\psi(r, \theta, \varphi)$ which satisfies the wave equation

$$-\frac{\hbar^2}{2M}\nabla^2\psi + V_0\psi = E\psi$$

and the boundary condition

$$\psi(r = a) = 0.$$

Show that for the energy E to be a minimum there must be no angular dependence in the wave function, that is, $\psi = \psi(r)$.
Hint. The problem centers on the boundary condition on the radial function.

12.6 Spherical Harmonics

In the separation of variables of (a) Laplace's equation, (b) Helmholtz's or the space-dependence of the classical wave equation, and (c) the Schrödinger wave equation for central force fields,

$$\nabla^2\psi + k^2 f(r)\psi = 0, \tag{12.140}$$

the angular dependence, coming entirely from the Laplacian operator, is

$$\frac{\Phi(\varphi)}{\sin\theta}\frac{d}{d\theta}\left(\sin\theta\frac{d\Theta}{d\theta}\right) + \frac{\Theta(\theta)}{\sin^2\theta}\frac{d^2\Phi(\varphi)}{d\varphi^2} + n(n+1)\,\Theta(\theta)\,\Phi(\varphi) = 0. \tag{12.141}$$

Azimuthal dependence—orthogonality. The separated azimuthal equation is

$$\frac{1}{\Phi(\varphi)}\frac{d^2\Phi(\varphi)}{d\varphi^2} = -m^2, \tag{12.142}$$

with solutions

$$\Phi(\varphi) = e^{-im\varphi}, e^{im\varphi}, \tag{12.143}$$

which readily satisfy the orthogonal condition

$$\int_0^{2\pi} e^{-im_1\varphi} e^{im_2\varphi}\, d\varphi = 2\pi\delta_{m_1,m_2}. \tag{12.144}$$

Notice that it is the product $\Phi^*_{m_1}(\varphi)\,\Phi_{m_2}(\varphi)$ that is taken and that * is used to indicate the complex conjugate function. This choice is not required, but it is convenient for quantum mechanical calculations. We could have used

$$\Phi = \sin m\varphi, \qquad \cos m\varphi \tag{12.145}$$

and the conditions of orthogonality that form the basis for Fourier series (Chapter 14).

In electrostatics and most other physical problems we require m to be an integer in order that $\Phi(\varphi)$ may be a single-valued function of the azimuth angle. In quantum mechanics the question is much more involved because the observable quantity that must be single-valued is the square of the magnitude of the wave function, $\Phi^*\Phi$. However, it can be shown that we must still have m integral or nonsensical currents will result.

By means of Eq. 12.144,

$$\Phi_m = \frac{1}{\sqrt{2\pi}} e^{im\varphi} \tag{12.146}$$

is orthonormal (orthogonal and normalized) with respect to integration over the azimuth angle φ.

Polar angle dependence. Splitting off the azimuthal dependence, the polar angle dependence (θ) leads to the associated Legendre equation (12.83), which is satisfied by the associated Legendre functions, that is, $\Theta(\theta) = P_n^m(\cos\theta)$. To include negative values of m, it is convenient to use Rodrigues' formula, Eq. 12.65 in the definition of $P_n^m(\cos\theta)$. This leads to

$$P_n^m(\cos\theta) = \frac{1}{2^n n!}(1 - x^2)^{m/2}\frac{d^{m+n}}{dx^{m+n}}(x^2 - 1)^n, \qquad -n \leqslant m \leqslant n. \tag{12.147}$$

$P_n^m(\cos\theta)$ and $P_n^{-m}(\cos\theta)$ are related as indicated in Exercise 12.5.1. An advantage of this approach over simply defining $P_n^m(\cos\theta)$ for $0 \leqslant m \leqslant n$ and requiring that $P_n^{-m} = P_n^m$ is that the recurrence relations valid for $0 \leqslant m \leqslant n$ remain valid for $-n \leqslant m < 0$.

Normalizing the associated Legendre function by Eq. 12.106, we obtain the orthonormal function

$$\mathscr{P}_n^m(\cos\theta) = \sqrt{\frac{2n+1}{2}\frac{(n-m)!}{(n+m)!}}\, P_n^m(\cos\theta), \qquad -n \leqslant m \leqslant n. \tag{12.148}$$

Spherical harmonics. The function $\Phi_m(\varphi)$ (Eq. 12.146) is orthonormal with respect to the azimuthal angle φ, whereas the function $\mathscr{P}_n^m(\cos\theta)$ (Eq. 12.148) is orthonormal with respect to the polar angle θ. We take the product of the two and define

$$Y_n^m(\theta, \varphi) \equiv (-1)^m \sqrt{\frac{2n+1}{4\pi}\frac{(n-m)!}{(n+m)!}}\, P_n^m(\cos\theta) e^{im\varphi} \tag{12.149}$$

to obtain functions of two angles (and two indices) which are orthonormal over the spherical surface. These $Y_n^m(\theta, \varphi)$ are spherical harmonics. The complete orthogonality integral becomes

Table 12.4

Spherical Harmonics (Condon-Shortley Phase)

$$Y_0^0(\theta, \varphi) = \frac{1}{\sqrt{4\pi}}$$

$$Y_1^1(\theta, \varphi) = -\sqrt{\frac{3}{8\pi}} \sin\theta e^{i\varphi}$$

$$Y_1^0(\theta, \varphi) = \sqrt{\frac{3}{4\pi}} \cos\theta$$

$$Y_1^{-1}(\theta, \varphi) = +\sqrt{\frac{3}{8\pi}} \sin\theta e^{-i\varphi}$$

$$Y_2^2(\theta, \varphi) = \sqrt{\frac{5}{96\pi}}\, 3\sin^2\theta e^{2i\varphi}$$

$$Y_2^1(\theta, \varphi) = -\sqrt{\frac{5}{24\pi}}\, 3\sin\theta\cos\theta e^{i\varphi}$$

$$Y_2^0(\theta, \varphi) = \sqrt{\frac{5}{4\pi}}\left(\frac{3}{2}\cos^2\theta - \frac{1}{2}\right)$$

$$Y_2^{-1}(\theta, \varphi) = +\sqrt{\frac{5}{24\pi}}\, 3\sin\theta\cos\theta e^{-i\varphi}$$

$$Y_2^{-2}(\theta, \varphi) = \sqrt{\frac{5}{96\pi}}\, 3\sin^2\theta\, e^{-2i\varphi}$$

$$\int_{\varphi=0}^{2\pi} \int_{\theta=0}^{\pi} Y_{n_1}^{m_1 *}(\theta, \varphi)\, Y_{n_2}^{m_2}(\theta, \varphi) \sin\theta \, d\theta \, d\varphi = \delta_{n_1, n_2} \delta_{m_1, m_2}. \qquad (12.150)$$

The extra $(-1)^m$ included in the defining equation of $Y_n^m(\theta, \varphi)$ deserves some comment. It is clearly legitimate, since Eq. 12.140 is linear and homogeneous. It is not necessary, but in moving on to certain quantum mechanical calculations, particularly in the quantum theory of angular momentum (Section 12.7), it is most convenient. The factor $(-1)^m$ is a phase factor, often called the Condon-Shortley phase, after the authors of a classic text on atomic spectroscopy. The effect of this $(-1)^m$ (Eq. 12.149) and the $(-1)^m$ of Eq. 12.84*a* for $P_n^{-m}(\cos\theta)$ is to introduce an alternation of sign among the *positive* m spherical harmonics. This is shown in Table 12.4.

Laplace series, fundamental expansion theorem. Part of the importance of spherical harmonics lies in the completeness property, a consequence of the Sturm-Liouville form of Laplace's equation. This property, in this case, means that any function $f(\theta, \varphi)$ (with sufficient continuity properties) evaluated over the surface of the sphere can be expanded in a uniformly convergent double series of spherical harmonics[1] (Laplace's series),

$$f(\theta, \varphi) = \sum_{m,n} a_{mn} Y_n^m(\theta, \varphi). \qquad (12.151)$$

If $f(\theta, \varphi)$ is known, the coefficients can be immediately found by the use of the orthogonality integral.

EXERCISES

12.6.1 Show that the parity of $Y_L^M(\theta, \varphi)$ is $(-1)^L$. Note the disappearance of any M dependence.

12.6.2 Prove that

$$Y_L^M(0, \varphi) = \left(\frac{2L+1}{4\pi}\right)^{1/2} \delta_{M0}.$$

12.6.3 In the theory of Coulomb excitation of nuclei we encounter $Y_L^M(\pi/2, 0)$. Show that

$$Y_L^M\left(\frac{\pi}{2}, 0\right) = \left(\frac{2L+1}{4\pi}\right)^{1/2} \frac{[(L-M)!(L+M)!]^{1/2}}{(L-M)!!(L+M)!!}(-1)^{(L+M)/2} \quad \text{for} \quad L+M \text{ even}$$

$$= 0 \quad \text{for} \quad L+M \text{ odd}.$$

Here

$$(2n)!! = 2n(2n-2)\cdots 6\cdot 4\cdot 2,$$

$$(2n+1)!! = (2n+1)(2n-1)\cdots 5\cdot 3\cdot 1.$$

[1] For a proof of this fundamental theorem see E. W. Hobson, *The Theory of Spherical and Ellipsoidal Harmonics*. New York: Chelsea (1955), Chapter VII.

12.6.4 (a) Express the elements of the quadrupole moment tensor $x_i x_j$ as a linear combination of the spherical harmonics Y_2^m (and Y_0^0).
Note: The tensor $x_i x_j$ is *reducible*. The Y_0^0 indicates the presence of a scalar component.
(*b*) The quadrupole moment tensor is usually defined as

$$\mathbf{Q}_{ij} = \int (3x_i x_j - r^2 \delta_{ij})\, \rho(\mathbf{r})\, d\tau,$$

with $\rho(\mathbf{r})$ the charge density. Express the components of $(3x_i x_j - r^2\delta_{ij})$ in terms of $r^2 Y_2^M$.
(c) What is the significance of the $-r^2\,\delta_{ij}$ term?
Hint. Cf. Section 3.4.

12.6.5 The orthogonal azimuthal functions yield a useful representation of the Dirac delta function. Show that

$$\delta(\varphi_1 - \varphi_2) = \frac{1}{2\pi} \sum_{m=-\infty}^{\infty} \exp[im(\varphi_1 - \varphi_2)].$$

12.6.6 Derive the spherical harmonic closure relation

$$\begin{aligned} \sum_{l=0}^{\infty} \sum_{m=-l}^{+l} Y_l^m(\theta_1, \varphi_1)\, Y_l^{m*}(\theta_2, \varphi_2) &= \frac{1}{\sin\theta_1}\, \delta(\theta_1 - \theta_2)\, \delta(\varphi_1 - \varphi_2) \\ &= \delta(\cos\theta_1 - \cos\theta_2)\, \delta(\varphi_1 - \varphi_2). \end{aligned}$$

12.6.7 The quantum mechanical angular momentum operators $L_x \pm iL_y$ are given by

$$L_x + iL_y = e^{i\varphi}\left(\frac{\partial}{\partial\theta} + i\cot\theta\,\frac{\partial}{\partial\varphi}\right),$$

$$L_x - iL_y = -e^{-i\varphi}\left(\frac{\partial}{\partial\theta} - i\cot\theta\,\frac{\partial}{\partial\varphi}\right).$$

Show that

(a) $(L_x + iL_y) Y_L^M(\theta, \varphi) = +\sqrt{(L-M)(L+M+1)}\; Y_L^{M+1}(\theta, \varphi)$

(b) $(L_x - iL_y)\; Y_L^M(\theta, \varphi) = +\sqrt{(L+M)(L-M+1)}\; Y_L^{M-1}(\theta, \varphi)$.

12.6.8 With $L_\pm$ given by

$$L_\pm = L_x \pm iL_y = \pm e^{\pm i\varphi}\left[\frac{\partial}{\partial\theta} \pm i\cot\theta\,\frac{\partial}{\partial\varphi}\right],$$

show that

(a) $Y_l^m = \sqrt{\dfrac{(l+m)!}{(2l)!\,(l-m)!}}\,(L_-)^{l-m}\, Y_l^l,$

(b) $Y_l^m = \sqrt{\dfrac{(l-m)!}{(2l)!\,(l+m)!}}\,(L_+)^{l+m}\, Y_l^{-l}.$

12.6.9 In some circumstances it is desirable to replace the imaginary exponential of our spherical harmonic by sine or cosine. Morse and Feshbach define

$$Y_{mn}^e = P_n^m(\cos\theta)\cos m\varphi,$$
$$Y_{mn}^0 = P_n^m(\cos\theta)\sin m\varphi,$$

where

$$\begin{aligned} \int_0^{2\pi}\!\int_0^{\pi} [Y_{mn}^{e\text{ or }0}(\theta, \varphi)]^2 \sin\theta\, d\theta\, d\varphi &= \frac{4\pi}{2(2n+1)}\frac{(n+m)!}{(n-m)!} \quad \text{for} \quad n = 1, 2, 3, .. \\ &= 4\pi \quad \text{for} \quad n = 0 \;(Y_{00}^0 \text{ does not exist}). \end{aligned}$$

These spherical harmonics are often named according to the patterns of their positive and negative regions on the surface of a sphere—zonal harmonics for $m = 0$, sectoral harmonics for $m = n$, and tesseral harmonics for $0 < m < n$. For Y^e_{mn}, $n = 4$, $m = 0, 2, 4$, indicate on a diagram of a hemisphere (one diagram for each spherical harmonic) the regions in which the spherical harmonic is positive.

12.6.10 A function $f(r, \theta, \varphi)$ may be expressed as a Laplace series

$$f(r, \theta, \varphi) = \sum_{l,m} a_{lm} r^l Y_l^m(\theta, \varphi).$$

With $\langle\ \rangle_{sphere}$ used to mean the average over a sphere (centered on the origin), show that

$$\langle f(r, \theta, \varphi)\rangle_{sphere} = f(0, 0, 0).$$

12.7 Angular Momentum and Ladder Operators

Orbital angular momentum. The classical concept of angular momentum $\mathbf{L}_{\text{classical}} = \mathbf{r} \times \mathbf{p}$ is presented in Section 1.4 to introduce the cross product. Following the usual Schrödinger representation of quantum mechanics, the classical linear momentum $\mathbf{p}$ is replaced by the operator $-i\nabla$. The quantum mechanical angular momentum *operator* becomes[1]

$$\mathbf{L}_{QM} = -i\mathbf{r} \times \nabla. \tag{12.152}$$

This is used repeatedly in Sections 1.8, 1.9, and 2.4 to illustrate vector differential operators. From Exercise 1.8.6, the angular momentum components satisfy a commutation relation

$$[L_i, L_j] = i\varepsilon_{ijk} L_k. \tag{12.153}$$

The ε_{ijk} is the Levi-Civita symbol of Section 3.4. A summation over the index k is understood.

From Exercises 2.4.10 and 2.4.11, we find

$$L_z = -i\frac{\partial}{\partial\varphi}, \tag{12.154}$$

in spherical polar coordinates. Hence

$$L_z Y_L^M(\theta, \varphi) = M Y_L^M(\theta, \varphi). \tag{12.155}$$

The differential operator corresponding to the square of the angular momentum

$$\mathbf{L}^2 = \mathbf{L}\cdot\mathbf{L} = L_x^2 + L_y^2 + L_z^2 \tag{12.156}$$

may be determined from

$$\mathbf{L}\cdot\mathbf{L} = -(\mathbf{r} \times \nabla)\cdot(\mathbf{r} \times \nabla) \tag{12.157}$$

which is the subject of Exercises 1.9.6 and 2.4.14 (b). From these, we find that $\mathbf{L}\cdot\mathbf{L}$ operating on a spherical harmonic yields[2]

[1] For simplicity, the $\hbar$ is dropped. This means that the angular momentum is measured in units of $\hbar$.

[2] In addition to these eigenvalue equations, the relation of $\mathbf{L}$ to rotations of coordinate systems and to rotations of functions is examined in Sections 4.9–4.11.

$$\mathbf{L}\cdot\mathbf{L}Y_L^M(\theta,\varphi) = -\left\{\frac{1}{\sin\theta}\frac{\partial}{\partial\theta}\left(\sin\theta\frac{\partial}{\partial\theta}\right) + \frac{1}{\sin^2\theta}\frac{\partial^2}{\partial\varphi^2}\right\}Y_L^M(\theta,\varphi), \tag{12.158}$$

or

$$\mathbf{L}\cdot\mathbf{L}Y_L^M(\theta,\varphi) = L(L+1)Y_L^M(\theta,\varphi). \tag{12.159}$$

This is Exercise 8.2.3.

Equation 12.153 presents the basic commutation relations of the components of the quantum mechanical angular momentum. Indeed, within the framework of quantum mechanics, these commutation relations *define* an angular momentum operator. From Eq. 12.155, our spherical harmonic $Y_L^M(\theta, \varphi)$ is an eigenfunction of L_z with eigenvalue M. Finally, from Eq. 12.159 $Y_L^M(\theta, \varphi)$ is also an eigenfunction of $\mathbf{L}^2$ with eigenvalue $L(L+1)$.

General operator approach. Apart from the replacement of $\mathbf{p}$ by $-i\nabla$, the analysis so far has been in terms of classical mathematics. Let us start anew with a more typical quantum mechanical analysis.

1. We assume an Hermitian operator $\mathbf{J}$ whose components satisfy the commutation relations

$$[J_i, J_j] = i\varepsilon_{ijk}J_k. \tag{12.160}$$

Otherwise $\mathbf{J}$ is arbitrary.

2. We assume that ψ_{JM} is simultaneously a normalized eigenfunction (or eigenvector) of J_z with eigenvalue M, and an eigenfunction of $\mathbf{J}^2$ with eigenvalue $J(J+1)$:[1]

$$J_z\psi_{JM} = M\psi_{JM}, \tag{12.161}$$

$$\mathbf{J}^2\psi_{JM} = J(J+1)\psi_{JM}. \tag{12.162}$$

Otherwise ψ_{JM} is assumed *unknown*.

Let us see what general conclusions we can develop. Then we shall let our general operators J_x, J_y, and J_z become the specific *orbital* angular momentum operators L_x, L_y, and L_z. ψ_{JM} will then become a function of the spherical polar coordinate angles θ and φ. We shall derive its form—in terms of Legendre polynomials and differential operators—and identify it with the spherical harmonic $Y_L^M(\theta, \varphi)$. This will illustrate the generality and power of operator techniques—particularly the use of *ladder operators*. It will also make clear the basis of the Condon-Shortley phase factor, the association of the $(-1)^M$ with the positive M spherical harmonics.

The ladder operators are defined as

$$\begin{aligned} J_+ &= J_x + iJ_y, \\ J_- &= J_x - iJ_y. \end{aligned} \tag{12.163}$$

[1] That ψ_{JM} is an eigenfunction of *both* J_z and $\mathbf{J}^2$ is a consequence of $[J_z, \mathbf{J}^2] = 0$.

In terms of these operators, $\mathbf{J}^2$ may be rewritten

$$\mathbf{J}^2 = \tfrac{1}{2}(J_+J_- + J_-J_+) + J_z^2. \tag{12.164}$$

From the commutation relations, Eq. 12.160, we find

$$[J_z, J_+] = +J_+, \qquad [J_z, J_-] = -J_-, \qquad [J_+, J_-] = 2J_z. \tag{12.165}$$

Since J_+ commutes with $\mathbf{J}^2$ (Exercise 12.7.1),

$$\mathbf{J}^2(J_+\psi_{JM}) = J_+(\mathbf{J}^2\psi_{JM}) = J(J+1)(J_+\psi_{JM}). \tag{12.166}$$

Therefore, $J_+\psi_{JM}$ is still an eigenfunction of $\mathbf{J}^2$ with eigenvalue $J(J+1)$. Similarly for $J_-\psi_{JM}$. But, from Eq. 12.165,

$$J_zJ_+ = J_+(J_z + 1), \tag{12.167}$$

or

$$J_z(J_+\psi_{JM}) = J_+(J_z + 1)\psi_{JM} = (M+1)(J_+\psi_{JM}). \tag{12.168}$$

Therefore, $J_+\psi_{JM}$ is still an eigenfunction of J_z but now with eigenvalue $M+1$. J_+ has *raised* the eigenvalue by 1 and so is often called a *raising operator*. Similarly J_- *lowers* the eigenvalue by 1 and so is often called a *lowering operator*.

Now what is the effect of letting first J_+ and then J_- operate on ψ_{JM}? The answer comes from expressing J_-J_+ (and J_+J_-) in terms of $\mathbf{J}^2$ and J_z. From Eqs. 12.160 and 12.164,

$$\begin{aligned} J_-J_+ &= \mathbf{J}^2 - J_z(J_z + 1), \\ J_+J_- &= \mathbf{J}^2 - J_z(J_z - 1). \end{aligned} \tag{12.169}$$

Then using Eqs. 12.161, 12.162, and 12.169,

$$\begin{aligned} J\ J_+\psi_{JM} &= [J(J+1) - M(M+1)]\psi_{JM} = (J-M)(J+M+1)\psi_{JM}, \\ J_+J_-\psi_{JM} &= [J(J+1) - M(M-1)]\psi_{JM} = (J+M)(J-M+1)\psi_{JM}. \end{aligned} \tag{12.170}$$

Now, multiply by ψ_{JM}^* and integrate (over all angles for the spherical harmonics). Since the ψ_{JM} have been assumed normalized,

$$\begin{aligned} \int \psi_{JM}^* J_- J_+ \psi_{JM}\, d\tau &= (J-M)(J+M+1) \geqslant 0, \\ \int \psi_{JM}^* J_+ J_- \psi_{JM}\, d\tau &= (J+M)(J-M+1) \geqslant 0. \end{aligned} \tag{12.171}$$

The $\geqslant 0$ part is worth a comment. In the language of quantum mechanics, J_+ and J_- are Hermitian conjugates,[1]

$$J_+^\dagger = J_-, \qquad J_-^\dagger = J_+. \tag{12.172}$$

Examples of this are provided by the matrices of Exercises 4.2.12 (spin $\frac{1}{2}$), 4.2.14 (spin 1), and 4.2.16 (spin 3/2). Therefore,

$$J_- J_+ = J_+^\dagger J_+, \qquad J_+ J_- = J_-^\dagger J_-, \tag{12.173}$$

[1] The Hermitian conjugation or *adjoint* operation is defined for matrices in Section 4.5, for (*real*) differential operators in Section 9.1.

and the expectation values, Eq. 12.171, must be positive or zero.[1] For our particular orbital angular momentum ladder operators, L_+ and L_-, explicit forms are given in Exercises 2.4.12 and 12.6.7. The reader can show (Exercise 12.7.2) that

$$\int Y_L^{M*} L_-(L_+ Y_L^M)\, d\Omega = \int (L_+ Y_L^M)^*(L_+ Y_L^M)\, d\Omega. \tag{12.174}$$

This is a sort of integration by parts (with the extra minus sign in L_- just cancelled by the minus sign in the integration by parts formula). Actually the equality is most easily verified by evaluating each side of Eq. 12.174, using Ex. 12.6.7.

From the right-hand side of Eq. 12.174, it is clear that the $\geqslant 0$ in Eq. 12.171 is valid. With the $\geqslant 0$ justified, we must have M restricted to the range $-J \leqslant M \leqslant J$.

Since J_+ raises the eigenvalue M to $M + 1$, we relabel the resultant eigenfunction $\psi_{J,M+1}$. The normalization is given by Eq. 12.171 as

$$J_+\psi_{JM} = \sqrt{(J - M)(J + M + 1)}\,\psi_{J,M+1}, \tag{12.175}$$

taking the positive square root and *not* introducing any phase factor. By the same arguments

$$J_-\psi_{JM} = \sqrt{(J + M)(J - M + 1)}\,\psi_{J,M-1}. \tag{12.176}$$

Both $\psi_{J,M+1}$ and $\psi_{J,M-1}$ remain normalized to unity. An explicit calculation of these results (using *known* ladder operators and *known* spherical harmonics) is the topic of Exercise 12.6.7. In Eqs. 12.175 and 12.176, the positive square root has been taken. Then the relative phase of $\psi_{J,M\pm1}$ and ψ_{JM} is determined by the ladder operators.

Repeated application of J_+ leads to

$$(J_+)^n\psi_{JM} = C_{JMn}\psi_{J,M+n}. \tag{12.177}$$

This operation *must* stop at $M' = M + n = J$, or else we would jump to $M' > J$ and be in contradiction with the conclusion from Eq. 12.171, $M \leqslant J$. Equivalently, we may say that whatever $M_{\max}$ is, since $J_+\psi_{JM_{\max}} = 0$, the left-hand side of Eq. 12.175 is zero, and, therefore, the right-hand side is zero. This yields $M_{\max} = J$. In the same fashion,

$$(J_-)^n\psi_{JM} = D_{JMn}\psi_{J,M-n} \tag{12.178}$$

must terminate at $M'' = M - n = -J$. We conclude from this first, that

$$J_+\psi_{J,J} = 0, \qquad J_-\psi_{J,-J} = 0. \tag{12.179}$$

Second, since M ranges from $+J$ to $-J$ in unit steps, $2J$ *must be an integer*. J is either an integer or half of an odd integer. As seen below, orbital angular momentum is described with integral J. But from the spins of some of the fundamental particles and of some nuclei, we do get $J = \frac{1}{2}, \frac{3}{2}, \frac{5}{2} \cdots$. Our angular momentum is quantized—essentially as a result of the commutation relations.

[1] For an excellent discussion of adjoint operators and Hilbert space, see A. Messiah, *Quantum Mechanics*, Chap. 7. New York: Wiley (1961).

Orbital angular momentum operators. Now, we return to our specific *orbital* angular momentum operators, L_x, L_y, and L_z. Equation 12.161 becomes

$$L_z\psi_{LM}(\theta, \varphi) = M\psi_{LM}(\theta, \varphi).$$

The explicit form of L_z indicates that $\psi_{LM}(\theta, \varphi)$ has a φ dependence of $e^{iM\varphi}$—with M an integer to keep ψ_{LM} single-valued. And if M is an integer, then L is an integer also.

To determine the θ dependence of $\psi_{LM}(\theta, \varphi)$, we proceed in two main steps: first the determination of $\psi_{LL}(\theta, \varphi)$, and second, the development of $\psi_{LM}(\theta, \varphi)$ in terms of ψ_{LL} with the phase fixed by ψ_{LO}.

Let

$$\psi_{LM}(\theta, \varphi) = \Theta_{LM}(\theta)e^{iM\varphi}. \tag{12.180}$$

From Eq. 12.179, using the form of L_+ given in Exercises 2.4.12 and 12.6.7,

$$e^{i(L+1)\varphi}\left[\frac{d}{d\theta} - L\cot\theta\right]\Theta_{LL}(\theta) = 0, \tag{12.181}$$

and

$$\psi_{LL}(\theta, \varphi) = c_L \sin^L \theta e^{iL\varphi}. \tag{12.182}$$

Normalizing,

$$c_L^* c_L \int_0^{2\pi}\int_0^{\pi} \sin^{2L+1}\theta\, d\theta\, d\varphi = 1. \tag{12.183}$$

The θ integral may be evaluated as a beta function (Exercise 10.4.9) and

$$|c_L| = \sqrt{\frac{(2L+1)!!}{4\pi(2L)!!}} = \frac{\sqrt{(2L)!}}{2^L L!}\sqrt{\frac{2L+1}{4\pi}}. \tag{12.184}$$

This completes our first step.

To obtain the ψ_{LM}, $M \neq \pm L$, we return to the ladder operators. From Eqs. 12.175 and 12.176 (J_+ replaced by L_+, and J_- replaced by L_-),

$$\begin{aligned}\psi_{LM}(\theta, \varphi) &= \sqrt{\frac{(L+M)!}{(2L)!(L-M)!}}(L_-)^{L-M}\psi_{LL}(\theta, \varphi),\\ \psi_{LM}(\theta, \varphi) &= \sqrt{\frac{(L-M)!}{(2L)!(L+M)!}}(L_+)^{L+M}\psi_{L,-L}(\theta, \varphi).\end{aligned} \tag{12.185}$$

Again, note that the relative phases are set by the ladder operators. L_+ and L_- operating on $\Theta_{LM}(\theta)e^{iM\varphi}$ may be written

$$\begin{aligned}L_+\Theta_{LM}(\theta)e^{iM\varphi} &= e^{i(M+1)\varphi}\left[\frac{d}{d\theta} - M\cot\theta\right]\Theta_{LM}(\theta)\\ &= -e^{i(M+1)\varphi}\sin^{1+M}\theta\frac{d}{d(\cos\theta)}\sin^{-M}\theta\,\Theta_{LM}(\theta),\end{aligned} \tag{12.186}$$

$$L_- \Theta_{LM}(\theta)e^{iM\varphi} = -e^{i(M-1)\varphi}\left[\frac{d}{d\theta} + M\cot\theta\right]\Theta_{LM}(\theta)$$

$$= e^{i(M-1)\varphi}\sin^{1-M}\theta \frac{d}{d(\cos\theta)}\sin^M\theta\,\Theta_{LM}(\theta).$$

Repeating these operations n times,

$$(L_+)^n\Theta_{LM}(\theta)e^{iM\varphi} = (-1)^n e^{i(M+n)\varphi}\sin^{n+M}\theta \frac{d^n}{d(\cos\theta)^n}\sin^{-M}\theta\,\Theta_{LM}(\theta),$$

$$(L_-)^n\Theta_{LM}(\theta)e^{iM\varphi} = e^{i(M-n)\varphi}\sin^{n-M}\theta \frac{d^n}{d(\cos\theta)^n}\sin^M\theta\,\Theta_{LM}(\theta). \tag{12.187}$$

From Eq. 12.185,

$$\psi_{LM}(\theta, \varphi) = c_L\sqrt{\frac{(L+M)!}{(2L)!(L-M)!}}\,e^{iM\varphi}\sin^{-M}\theta \frac{d^{L-M}}{d(\cos\theta)^{L-M}}\sin^{2L}\theta, \tag{12.188}$$

and for $M = -L$,

$$\begin{aligned}\psi_{L,-L}(\theta, \varphi) &= \frac{c_L}{(2L)!}e^{-iL\varphi}\sin^L\theta \frac{d^{2L}}{d(\cos\theta)^{2L}}\sin^{2L}\theta \\ &= (-1)^L c_L \sin^L\theta\, e^{-iL\varphi}.\end{aligned} \tag{12.189}$$

Note the characteristic $(-1)^L$ phase of $\psi_{L,-L}$ relative to $\psi_{L,L}$. This $(-1)^L$ enters from

$$\sin^{2L}\theta = (1-x^2)^L = (-1)^L(x^2-1)^L. \tag{12.190}$$

Combining Eqs. 12.185, 12.187, and 12.189,

$$\psi_{LM}(\theta, \varphi) = (-1)^L c_L\sqrt{\frac{(L-M)!}{(2L)!(L+M)!}}(-1)^{L+M}e^{iM\varphi}\sin^M\theta \frac{d^{L+M}}{d(\cos\theta)^{L+M}}\sin^{2L}\theta. \tag{12.191}$$

Eqs. 12.188 and 12.191 agree that

$$\psi_{Lo}(\theta, \varphi) = c_L\frac{1}{\sqrt{(2L)!}}\frac{d^L}{(d\cos\theta)^L}\sin^{2L}\theta. \tag{12.192}$$

Using Rodrigues' formula, Eq. 12.65,

$$\begin{aligned}\psi_{Lo}(\theta, \varphi) &= (-1)^L c_L\frac{2^L L!}{\sqrt{(2L)!}}P_L(\cos\theta) \\ &= (-1)^L\frac{c_L}{|c_L|}\sqrt{\frac{2L+1}{4\pi}}P_L(\cos\theta).\end{aligned} \tag{12.193}$$

The last equality follows from Eq. 12.184. We now demand that $\psi_{Lo}(0, 0)$ be real and positive. Therefore,

$$c_L = (-1)^L |c_L| = (-1)^L \frac{\sqrt{(2L)!}}{2^2 L!} \sqrt{\frac{2L+1}{4\pi}}. \tag{12.194}$$

With $(-1)^L c_L/|c_L| = 1$, $\psi_{LO}(\theta, \varphi)$ in Eq. 12.193 may be identified with the spherical harmonic $Y_L^O(\theta, \varphi)$ of Section 12.6.

When we substitute $(-1)^L c_L$ into Eq. 12.191,

$$\begin{aligned}\psi_{LM}(\theta, \varphi) &= \frac{\sqrt{(2L)!}}{2^L L!} \sqrt{\frac{2L+1}{4\pi}} \sqrt{\frac{(L-M)!}{(2L)!(L+M)!}} (-1)^{L+M} \\ &\quad \cdot e^{iM\varphi} \sin^M \theta \frac{d^{L+M}}{d(\cos\theta)^{L+M}} \sin^{2L}\theta \\ &= \sqrt{\frac{2L+1}{4\pi}} \sqrt{\frac{(L-M)!}{(L+M)!}} e^{iM\varphi}(-1)^M \\ &\quad \cdot \left\{\frac{1}{2^L L!}(1-x^2)^{M/2} \frac{d^{L+M}}{dx^{L+M}}(x^2-1)^L\right\}, \qquad x = \cos\theta, \quad M \geqslant 0.\end{aligned} \tag{12.195}$$

The expression in the curly bracket is identified as the associated Legendre function, (Eq. 12.147), and we have

$$\begin{aligned}\psi_{LM}(\theta, \varphi) = Y_L^M(\theta, \varphi) &= (-1)^M \sqrt{\frac{2L+1}{4\pi} \cdot \frac{(L-M)!}{(L+M)!}} \\ &\quad \cdot P_L^M(\cos\theta) e^{iM\varphi}, \qquad M \geqslant 0,\end{aligned} \tag{12.196}$$

in complete agreement with Section 12.6. Then by Eq. 12.84*a*, Y_L^M for negative superscript is given by

$$Y_L^{-M}(\theta, \varphi) = (-1)^M Y_L^{M*}(\theta, \varphi). \tag{12.197}$$

Our angular momentum eigenfunctions $\psi_{LM}(\theta, \varphi)$ are identified with the spherical harmonics. The phase factor $(-1)^M$ is associated with the positive values of M and is seen to be a consequence of the ladder operators.

Our development of spherical harmonics here may be considered a portion of Lie algebra—related to group theory—Section 4.9.

EXERCISES

12.7.1 Show that

(a) $[J_+, \mathbf{J}^2] = 0$,

(b) $[J_-, \mathbf{J}^2] = 0$.

12.7.2 Using the known forms of L_+ and L_- (Exercises 2.4.12 and 12.6.7), show that

$$\int Y_L^{M*} L_-(L_+ Y_L^M)\, d\Omega = \int (L_+ Y_L^M)^*(L_+ Y_L^M)\, d\Omega.$$

12.7.3 Derive the relations

(a) $\psi_{LM}(\theta, \varphi) = \sqrt{\dfrac{(L+M)!}{(2L)!(L-M)!}}\,(L_-)^{L-M}\psi_{LL}(\theta, \varphi),$

(b) $\psi_{LM}(\theta, \varphi) = \sqrt{\dfrac{(L-M)!}{(2L)!(L+M)!}}\,(L_+)^{L+M}\psi_{L,\,-L}(\theta, \varphi).$

12.7.4 Derive the multiple operator equations

(a) $(L_+)^n\, \Theta_{LM}(\theta)\, e^{iM\varphi} = (-1)^n e^{i(M+n)\varphi} \sin^{n+M}\theta \dfrac{d^n}{d(\cos\theta)^n} \sin^{-M}\theta\, \Theta_{LM}(\theta),$

(b) $(L_-)^n\, \Theta_{LM}(\theta)\, e^{iM\varphi} = e^{i(M-n)\varphi} \sin^{n-M}\theta \dfrac{d^n}{d(\cos\theta)^n} \sin^{M}\theta\, \Theta_{LM}(\theta).$

12.7.5 Show, using $(L_-)^n$, that

$$Y_L^{-M}(\theta, \varphi) = (-1)^M\, Y_L^{M*}(\theta, \varphi).$$

12.7.6 Verify by explicit calculation that

(a) $L_+\, Y_1^0(\theta, \varphi) = -\sqrt{\dfrac{3}{4\pi}} \sin\theta\, e^{i\varphi} = \sqrt{2}\, Y_1^1(\theta, \varphi),$

(b) $L_-\, Y_1^0(\theta, \varphi) = +\sqrt{\dfrac{3}{4\pi}} \sin\theta\, e^{-i\varphi} = \sqrt{2}\, Y_1^{-1}(\theta, \varphi).$

The signs (Condon-Shortley phase) are a consequence of the ladder operators, L_+ and L_-.

12.8 The Addition Theorem for Spherical Harmonics

Trigonometric identity. In the following discussion (θ_1, φ_1) and (θ_2, φ_2) denote two different directions in our spherical coordinate system, separated by an angle γ. These angles satisfy the trigonometric identity

$$\cos\gamma = \cos\theta_1 \cos\theta_2 + \sin\theta_1 \sin\theta_2 \cos(\varphi_1 - \varphi_2), \tag{12.198}$$

which is perhaps most easily proved by vector methods (cf. Chapter 1).

The addition theorem, then, asserts that

$$P_n(\cos\gamma) = \frac{4\pi}{2n+1} \sum_{m=-n}^{n} (-1)^m\, Y_n^m(\theta_1, \varphi_1)\, Y_n^{-m}(\theta_2, \varphi_2) \tag{12.199}$$

or equivalently

$$P_n(\cos\gamma) = \frac{4\pi}{2n+1} \sum_{m=-n}^{n} Y_n^m(\theta_1, \varphi_1) Y_n^{m*}(\theta_2, \varphi_2).^{1} \tag{12.200}$$

[1] The asterisk may go on *either* spherical harmonic.

In terms of the associated Legendre functions the addition theorem is

$$P_n(\cos\gamma) = P_n(\cos\theta_1)\,P_n(\cos\theta_2) + 2\sum_{m=1}^{n}\frac{(n-m)!}{(n+m)!}P_n^m(\cos\theta_1)\,P_n^m(\cos\theta_2)\cos m(\varphi_1-\varphi_2). \qquad (12.201)$$

Equation 12.198 is a special case of Eq. 12.201.

Derivation of addition theorem. We now derive Eq. 12.200. Let $g(\theta, \varphi)$ be a function that may be expanded in a Laplace series

$$\begin{aligned} g(\theta_1, \varphi_1) &= Y_n^m(\theta_1, \varphi_1) \qquad \text{relative to } x_1, y_1, z_1 \\ &= \sum_{m=-n}^{n} a_{nm} Y_n^m(\gamma, \psi) \qquad \text{relative to } x_2, y_2, z_2. \end{aligned} \qquad (12.202)$$

Actually, the choice of the 0 of the azimuth angle ψ is irrelevant. At $\gamma = 0$ we have

$$g(\theta_1, \varphi_1)\,|_{\gamma=0} = a_{n0}\left(\frac{2n+1}{4\pi}\right)^{1/2} \qquad (12.203)$$

since $P_n(1) = 1$, whereas $P_n^m(1) = 0$ $(m \neq 0)$. Multiplying Eq. (12.202) by $Y_n^{0*}(\gamma, \psi)$ and integrating over the sphere, we obtain

$$\int g(\theta_1, \varphi_1) Y_n^{0*}(\gamma, \psi)\, d\Omega_{\gamma,\psi} = a_{n0}. \qquad (12.204)$$

Now, using Eq. 12.202, we may rewrite Eq. 12.204

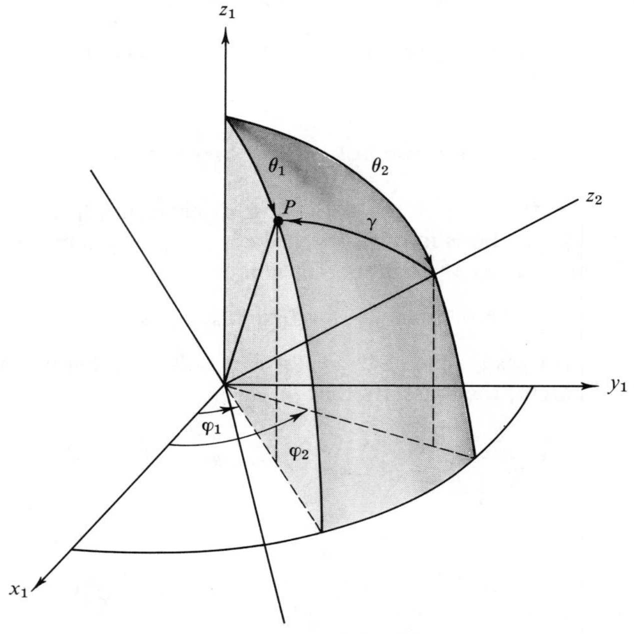

FIG. 12.15

$$\int Y_n^m(\theta_1, \varphi_1) Y_n^{0*}(\gamma, \psi)\, d\Omega = a_{n0}. \tag{12.205}$$

As for Eq. 12.202, we assume that $P_n(\cos\gamma)$ has an expansion of the form

$$P_n(\cos\gamma) = \sum_{m=-n}^{n} b_{nm}\, Y_n^m(\theta_1, \varphi_1), \tag{12.206}$$

where the b_{nm} will, of course, depend on θ_2, φ_2, that is, on the orientation of the z_2-axis. Multiplying by $Y_n^{m*}(\theta_1, \varphi_1)$ and integrating with respect to θ_1 and φ_1 over the sphere, we have

$$\int P_n(\cos\gamma) Y_n^{m*}(\theta_1, \varphi_1)\, d\Omega_{\theta_1, \varphi_1} = b_{nm}. \tag{12.207}$$

In terms of spherical harmonics, Eq. 12.207 becomes

$$\left(\frac{4\pi}{2n+1}\right)^{1/2} \int Y_n^0(\gamma, \psi) Y_n^{m*}(\theta_1, \varphi_1)\, d\Omega = b_{nm}. \tag{12.208}$$

Note that the subscripts have been dropped from the solid angle element $d\Omega$. Since the range of integration is over all solid angles, the choice of polar axis is irrelevant. Then in a comparison of Eqs. 12.205 and 12.208,

$$\begin{aligned} b_{nm}^* &= a_{n0}\left(\frac{4\pi}{2n+1}\right)^{1/2} \\ &= \frac{4\pi}{2n+1}\, g(\theta_1, \varphi_1)\big|_{\gamma=0} \quad \text{by Eq. 12.203} \\ &= \frac{4\pi}{2n+1}\, Y_n^m(\theta_2, \varphi_2) \quad \text{by Eq. 12.202.} \end{aligned} \tag{12.209}$$

The change in subscripts occurs because

$$\left.\begin{aligned} \theta_1 &\to \theta_2 \\ \varphi_1 &\to \varphi_2 \end{aligned}\right\} \quad \text{for} \quad \gamma \to 0.$$

Substituting back into Eq. 12.206, we obtain Eq. 12.200, thus proving our addition theorem.

The reader familiar with group theory will find a much more elegant proof of Eq. 12.200 by using the rotation group.[1] This is Ex. 4.9.10.

One application of the addition theorem is in the construction of a Green's function for the three-dimensional Laplace equation in spherical polar coordinates. If the source is on the polar axis at the point ($r = a$, $\theta = 0$, $\varphi = 0$), then by Eq. 12.4

$$\begin{aligned} \frac{1}{R} = \frac{1}{|\mathbf{r} - \mathbf{k}a|} &= \sum_{n=0}^{\infty} P_n(\cos\gamma)\frac{a^n}{r^{n+1}}, \qquad r > a \\ &= \sum_{n=0}^{\infty} P_n(\cos\gamma)\frac{r^n}{a^{n+1}}, \qquad r < a \end{aligned} \tag{12.210}$$

[1] Cf. M. E. Rose, *Elementary Theory of Angular Momentum.* New York: Wiley (1957).

Rotating our coordinate system to put the source at (a, θ_2, φ_2) and the point of observation at (r, θ_1, φ_1), we obtain

$$\begin{aligned} G(r, \theta_1, \varphi_1, a, \theta_2, \varphi_2) &= \frac{1}{R} \\ &= \sum_{n=0}^{\infty} \sum_{m=-n}^{n} \frac{4\pi}{2n+1} Y_n^{m*}(\theta_1, \varphi_1)\, Y_n^m(\theta_2, \varphi_2) \frac{a^n}{r^{n+1}}, \qquad r > a, \\ &= \sum_{n=0}^{\infty} \sum_{m=-n}^{n} \frac{4\pi}{2n+1} Y_n^{m*}(\theta_1, \varphi_1)\, Y_n^m(\theta_2, \varphi_2) \frac{r^n}{a^{n+1}}, \qquad r < a. \end{aligned} \tag{12.211}$$

In Section 16.6 this argument is reversed to provide another derivation of the Legendre polynomial addition theorem.

EXERCISES

12.8.1 In proving the addition theorem it was assumed that $Y_n^k(\theta_1, \varphi_1)$ could be expanded in a series of $Y_n^m(\theta_2, \varphi_2)$ in which m varied from $-n$ to $+n$ but n was held fixed. What arguments can you develop to justify summing only over the upper index m and *not* over the lower index n? *Hints.* One possibility is to examine the homogeneity of the Y_n^m, that is, Y_n^m may be expressed entirely in terms of the form $\cos^{n-p}\theta \sin^p\theta$ or $x^{n-p-s}y^p z^s/r^n$. Another possibility is to examine the behavior of the Legendre equation $[\nabla^2 + n(n+1)/r^2]\, P_n(\cos\theta) = 0$ under rotation of the coordinate system.

12.8.2 An atomic electron with angular momentum L and magnetic quantum number M has a wave function

$$\psi(r, \theta, \varphi) = f(r)\, Y_L^M(\theta, \varphi).$$

Show that the sum of the electron densities in a given complete shell is spherically symmetric, that is, that $\sum_{M=-L}^{L} \psi^*(r, \theta, \varphi)\, \psi(r, \theta, \varphi)$ is independent of θ and φ.

12.8.3 The potential of an electron at point $\mathbf{r}_e$ in the field of Z protons at $\mathbf{r}_p$ is

$$\varphi = -\frac{e^2}{4\pi\varepsilon_0} \sum_{p=1}^{Z} \frac{1}{|\mathbf{r}_e - \mathbf{r}_p|}.$$

Show that this may be written

$$\varphi = -\frac{e^2}{4\pi\varepsilon_0 r_e} \sum_{p=1}^{Z} \sum_{L,M} \left(\frac{r_p}{r_e}\right)^L \frac{4\pi}{2L+1} Y_L^{M*}(\theta_p, \varphi_p)\, Y_L^M(\theta_e, \varphi_e),$$

where $r_e > r_p$. How should φ be written for $r_e < r_p$?

12.8.4 Two protons are *uniformly* distributed within the same spherical volume. If the coordinates of one element of charge are $(r_1, \theta_1, \varphi_1)$ and the coordinates of the other are $(r_2, \theta_2, \varphi_2)$ and r_{12} is the distance between them, the element of energy of repulsion will be given by

$$d\psi = \rho^2 \frac{dv_1\, dv_2}{r_{12}} = \rho^2 \frac{r_1^2\, dr_1 \sin\theta_1\, d\theta_1\, d\varphi_1\, r_2^2\, dr_2 \sin\theta_2\, d\theta_2\, d\varphi_2}{r_{12}}$$

Here $\rho = \dfrac{\text{charge}}{\text{volume}} = \dfrac{3e}{4\pi R^3}$, charge density,

$r_{12}^2 = r_1^2 + r_2^2 - 2r_1r_2 \cos \gamma$.

Calculate the total electrostatic energy (of repulsion) of the two protons. This calculation is used in accounting for the mass difference in "mirror" nuclei, such as O^{15} and N^{15}.

$$\textit{Ans.}\ \text{For } r_2 > r_1 \quad \left.\begin{aligned} &\frac{3}{5}\frac{e^2}{R} \\ r_2 < r_1 \quad &\frac{3}{5}\frac{e^2}{R} \end{aligned}\right\} \frac{6}{5}\frac{e^2}{R} \text{ (total).}$$

This is *double* that required to create a uniformly charged sphere because we have two separate cloud charges interacting, *not* one charge interacting with itself (with permutation of pairs *not* considered).

12.8.5 Each of the two $1s$ electrons in helium may be described by a hydrogenic wave function

$$\psi(\mathbf{r}) = \left(\frac{Z^3}{\pi a_0^3}\right)^{1/2} e^{-Zr/a_0}$$

in the absence of the other electron. Here Z, the atomic number, is 2. The symbol a_0 is the Bohr radius, $\hbar^2/me^2$. Find the mutual potential energy of the two electrons given by

$$\int \psi^*(\mathbf{r}_1)\psi^*(\mathbf{r}_2)\frac{e^2}{r_{12}}\psi(\mathbf{r}_1)\psi(\mathbf{r}_2)\, d^3r_1\, d^3r_2. \qquad \textit{Ans.}\ \frac{5e^2Z}{8a_0}.$$

Note. $d^3r_1 = r_1^2 dr_1 \sin\theta_1 d\theta_1 d\varphi_1$,
$r_{12} = |\mathbf{r}_1 - \mathbf{r}_2|$.

12.8.6 The probability of finding a $1s$ hydrogen electron in a volume element $r^2\, dr \sin\theta\, d\theta\, d\varphi$ is

$$\frac{1}{\pi a_0^3} \exp\left[-2r/a_0\right] r^2\, dr \sin\theta\, d\theta\, d\varphi.$$

Find the corresponding electrostatic potential. Calculate the potential from

$$V(\mathbf{r}_1) = \frac{q}{4\pi\varepsilon_0}\int \frac{\rho(\mathbf{r}_2)}{r_{12}}\, d^3r_2$$

with $\mathbf{r}_1$ *not* on the z-axis. Expand r_{12}. Apply the Legendre polynomial addition theorem and show that the angular dependence of $V(\mathbf{r}_1)$ drops out.

$$\textit{Ans.}\ V(\mathbf{r}_1) = \frac{q}{4\pi\varepsilon_0}\left\{\frac{1}{2r_1}\gamma\left(3, \frac{2r_1}{a_0}\right) + \frac{1}{a_0}\Gamma\left(2, \frac{2r_1}{a_0}\right)\right\}.$$

12.8.7 A hydrogen electron in a $2p$ orbit has a charge distribution

$$\rho = \frac{q}{64\pi a_0^5} r^2 e^{-r/a_0} \sin^2\theta,$$

where a_0 is the Bohr radius, $\hbar^2/me^2$. Find the electrostatic potential corresponding to this charge distribution.

12.8.8 The electric current density produced by a $2p$ electron in a hydrogen atom is

$$\mathbf{J} = \boldsymbol{\varphi}_0 \frac{q\hbar}{32ma_0^5} e^{-r/a_0} r \sin\theta.$$

Using

$$\mathbf{A}(\mathbf{r}_1) = \frac{\mu_0}{4\pi}\int \frac{\mathbf{J}(\mathbf{r}_2)}{|\mathbf{r}_1 - \mathbf{r}_2|}\, d^3r_2$$

find the magnetic vector potential produced by this hydrogen electron. *Hint*. Resolve into cartesian components. Use the addition theorem to eliminate γ, the angle included between $\mathbf{r}_1$ and $\mathbf{r}_2$.

12.8.9 (a) As a Laplace series and as an example of Eq. 9.80 (now with complex functions), show that

$$\delta(\Omega_1 - \Omega_2) = \sum_{n,m} Y_n^{m*}(\theta_2, \varphi_2) Y_n^m(\theta_1, \varphi_1).$$

(b) Show also that this *same* Dirac delta function may be written

$$\delta(\Omega_1 - \Omega_2) = \sum_n \frac{2n+1}{4\pi} P_n(\cos\gamma).$$

Now, if you can justify equating the summations over *n term by term* you have an alternate derivation of the spherical harmonic addition theorem.

12.9 Integrals of the Product of Three Spherical Harmonics

Frequently in quantum mechanics we encounter integrals of the general form

$$\int Y_{L_1}^{M_1*} Y_{L_2}^{M_2} Y_{L_3}^{M_3}\, d\Omega \quad \text{or} \quad \int Y_{L_1}^{M_1*} P_{L_2} Y_{L_3}^{M_3}\, d\Omega,$$

in which the integration is over all solid angles. The first factor in the integrand may come from the wave function of a final state and the third factor from an initial state, whereas the middle factor may represent an operator that is being evaluated or whose "matrix element" is being determined.

By using group theoretical methods, as in the quantum theory of angular momentum, it is possible to give a general expression for the forms listed. The analysis involves the vector-addition or Clebsch-Gordan coefficients, which have been tabulated. Three general restrictions appear.[1] (1) The integral vanishes unless the *vector* sum of the L's (angular momentum) is zero, $|L_1 - L_3| \leqslant L_2 \leqslant L_1 + L_3$. (2) The integral vanishes unless $M_2 + M_3 = M_1$. Here we have the theoretical foundation of the vector model of atomic spectroscopy. (3) Finally, the integral vanishes unless the product $Y_{L_1}^{M_1*} Y_{L_2}^{M_2} Y_{L_3}^{M_3}$ is even, that is, unless $M_1 + M_2 + M_3 + L_1 + L_2 + L_3$ is an even integer. This is a parity conservation law.

Details of this general and powerful approach will be found in the references. The reader will note that the vector-addition coefficients are developed in terms of the Condon-Shortley phase convention in which the $(-1)^m$ of Eq. 12.149 is associated with the positive m.

It is possible to evaluate many of the commonly encountered integrals of this form with the techniques already developed. The integration over azimuth may be carried out by inspection.

$$\int_0^{2\pi} e^{-iM_1\varphi} e^{iM_2\varphi} e^{iM_3\varphi}\, d\varphi = 2\pi\delta_{M_2+M_3-M_1,0}. \tag{12.212}$$

[1] E. U. Condon and G. H. Shortley, *The Theory of Atomic Spectra*. Cambridge: Cambridge University Press (1951). M. E. Rose, *Elementary Theory of Angular Momentum*. New York: Wiley (1957). A. Edmonds, *Angular Momentum in Quantum Mechanics*. Princeton, New Jersey: Princeton University Press (1957). E. P. Wigner, *Group Theory and Its Applications to Quantum Mechanics* (translated by J. J. Griffin). New York: Academic Press (1959).

Physically this corresponds to the conservation of the z-component of angular momentum.

Application of recurrence relations. A glance at Table 12.4 will show that the θ-dependence of $Y_{L_2}^{M_2}$, that is, $P_{L_2}^{M_2}(\theta)$, can be expressed in terms of $\cos\theta$ and $\sin\theta$. However, a factor of $\cos\theta$ or $\sin\theta$ may be combined with the $Y_{L_3}^{M_3}$ factor by using the associated Legendre polynomial recurrence relations. For instance, from Eqs. 12.88 and 12.89 we get

$$\cos\theta Y_L^M = +\left[\frac{(L-M+1)(L+M+1)}{(2L+1)(2L+3)}\right]^{1/2} Y_{L+1}^M + \left[\frac{(L-M)(L+M)}{(2L-1)(2L+1)}\right]^{1/2} Y_{L-1}^M \tag{12.213}$$

$$e^{i\varphi}\sin\theta Y_L^M = -\left[\frac{(L+M+1)(L+M+2)}{(2L+1)(2L+3)}\right]^{1/2} Y_{L+1}^{M+1} + \left[\frac{(L-M)(L-M-1)}{(2L-1)(2L+1)}\right]^{1/2} Y_{L-1}^{M+1} \tag{12.214}$$

$$e^{-i\varphi}\sin\theta Y_L^M = +\left[\frac{(L-M+1)(L-M+2)}{(2L+1)(2L+3)}\right]^{1/2} Y_{L+1}^{M-1} - \left[\frac{(L+M)(L+M-1)}{(2L-1)(2L+1)}\right]^{1/2} Y_{L-1}^{M-1} \tag{12.215}$$

Using these equations, we obtain

$$\int Y_{L_1}^{M_1 *}\cos\theta Y_L^M\, d\Omega = \left[\frac{(L-M+1)(L+M+1)}{(2L+1)(2L+3)}\right]^{1/2} \delta_{M_1,M}\,\delta_{L_1,L+1} + \left[\frac{(L-M)(L+M)}{(2L-1)(2L+1)}\right]^{1/2} \delta_{M_1,M}\delta_{L_1,L-1}. \tag{12.216}$$

The occurence of the Kronecker delta $(L_1, L \pm 1)$ is an aspect of the conservation of angular momentum. Physically, this integral arises in a consideration of ordinary atomic electromagnetic radiation (electric dipole). It leads to the familiar selection rule that transitions to an atomic level with orbital angular momentum quantum number L_1 can originate only from atomic levels with quantum numbers $L_1 - 1$ or $L_1 + 1$. The application to expressions such as

$$\text{quadrupole moment} \sim \int Y_L^{M*} P_2(\cos\theta) Y_L^M\, d\Omega$$

is more involved but perfectly straightforward.

Slater coefficients. Integrals of the product of three spherical harmonics are also encountered in the evaluation of the interaction energy between two atomic electrons. We have integrals of the form

$$\int \psi_\alpha^*(1)\,\psi_\alpha(1)\,\frac{1}{r_{12}}\,\psi_\beta^*(2)\,\psi_\beta(2)\,d\tau_1\,d\tau_2.$$

If we expand $r_{12} = |\mathbf{r}_1 - \mathbf{r}_2|$ by the generating function expansion, expand the resulting $P_k(\cos\gamma)$ by the addition theorem, and assume the angular dependence of wave functions $\psi(1)$ and $\psi(2)$ to be $Y_{l_1}^{m_1}(\theta_1, \varphi_1)$ and $Y_{l_2}^{m_2}(\theta_2, \varphi_2)$, respectively, the foregoing integral will separate, giving us a radial integral and *two* integrals of the form

$$a_k'(l_1, m_1) \equiv \sqrt{\frac{4\pi}{2k+1}} \int Y_{l_1}^{m_1 *} Y_k^0 Y_{l_1}^{m_1}\, d\Omega. \tag{12.217}$$

From exchange terms in our total wave function $[\psi_\alpha(1) \rightarrow \psi_\alpha(2),\ \psi_\beta(2) \rightarrow \psi_\beta(1)]$ we get more integrals of the form

$$b_k'(l_1, m_1, l_2, m_2) \equiv \sqrt{\frac{4\pi}{2k+1}} \int Y_{l_2}^{m_2 *} Y_k^{m_2 - m_1} Y_{l_1}^{m_1}\, d\Omega. \tag{12.218}$$

The combinations needed for the computation of atomic wave functions are

$$a_k(l_1, m_1, l_2, m_2) = a_k'(l_1, m_1) \cdot a_k'(l_2, m_2) \tag{12.219}$$

$$b_k(l_1, m_1, l_2, m_2) = [b_k'(l_1, m_1, l_2, m_2)]^2. \tag{12.220}$$

These a_k and b_k, often called Slater coefficients, are tabulated in texts that deal with atomic structure.[1]

EXERCISES

12.9.1 Verify
(a) $a_0'(l, m) = 1,$
(b) $a_1'(1, m) = 0,$
(c) $a_2'(0, 0) = 0,$
(d) $a_2'(1, 0) = \frac{2}{5},$
(e) $a_2'(1, \pm 1) = -\frac{1}{5}.$

[1] D. R. Hartree, *The Calculation of Atomic Structures*. New York: Wiley (1953), Chapter 3. Condon-Shortley, *The Theory of Atomic Spectra*, §9[6], Tables 1[6], 2[6].

12.9.2 Verify

(a) $\int Y_L^M Y_0^0 Y_L^{M*}\, d\Omega = \dfrac{1}{\sqrt{4\pi}}$,

(b) $\int Y_L^M Y_1^0 Y_{L+1}^{M*}\, d\Omega = \sqrt{\dfrac{3}{4\pi}}\sqrt{\dfrac{(L+M+1)(L-M+1)}{(2L+1)(2L+3)}}$,

(c) $\int Y_L^M Y_1^1 Y_{L+1}^{M+1*}\, d\Omega = \sqrt{\dfrac{3}{8\pi}}\sqrt{\dfrac{(L+M+1)(L+M+2)}{(2L+1)(2L+3)}}$,

(d) $\int Y_L^M Y_1^1 Y_{L-1}^{M+1*}\, d\Omega = -\sqrt{\dfrac{3}{8\pi}}\sqrt{\dfrac{(L-M)(L-M-1)}{(2L-1)(2L+1)}}$.

These integrals were used in an investigation of the angular correlation of internal conversion electrons.

12.9.3 Show that

$$\text{(a)}\quad \int_{-1}^{1} x\, P_L(x)\, P_N(x)\, dx = \begin{cases} \dfrac{2(L+1)}{(2L+1)(2L+3)}, & N = L+1, \\[2ex] \dfrac{2L}{(2L-1)(2L+1)}, & N = L-1. \end{cases}$$

$$\text{(b)}\quad \int_{-1}^{1} x^2\, P_L(x)\, P_N(x)\, dx = \begin{cases} \dfrac{2(L+1)(L+2)}{(2L+1)(2L+3)(2L+5)}, & N = L+2, \\[2ex] \dfrac{2(2L^2+2L-1)}{(2L-1)(2L+1)(2L+3)}, & N = L, \\[2ex] \dfrac{2L(L-1)}{(2L-3)(2L-1)(2L+1)}, & N = L-2. \end{cases}$$

12.10 Legendre Functions of the Second Kind, $Q_n(z)$

In all the analysis so far in this chapter we have been dealing with one solution of Legendre's equation, the solution $P_n(\cos\theta)$, which is regular (finite) at the two singular points of the differential equation, $\cos\theta = \pm 1$. From the general theory of differential equations it is known that a second solution exists. We shall develop this second solution, Q_n, by a series solution of Legendre's equation. Later a closed form will be obtained.

Series solutions of Legendre's equation. To solve

$$\frac{d}{dx}\left[(1-x^2)\frac{dy}{dx}\right] + n(n+1)y = 0 \tag{12.221}$$

we proceed as in Chapter 8, letting

$$y = \sum_{\lambda=0}^{\infty} a_\lambda x^{k+\lambda}, \tag{12.222}$$

with footnote [1] attached to the series.

[1] Note that x may be replaced by the complex variable z.

with

$$y' = \sum_{\lambda=0}^{\infty} (k + \lambda)a_\lambda x^{k+\lambda-1}, \tag{12.223}$$

$$y'' = \sum_{\lambda=0}^{\infty} (k + \lambda)(k + \lambda - 1)a_\lambda x^{k+\lambda-2}. \tag{12.224}$$

Substitution into the original differential equation gives

$$\sum_{\lambda=0}^{\infty} (k + \lambda)(k + \lambda - 1)a_\lambda x^{k+\lambda-2} + \sum_{\lambda=0}^{\infty} [n(n + 1) - 2(k + \lambda) - (k + \lambda)(k + \lambda - 1)]a_\lambda x^{k+\lambda} = 0. \tag{12.225}$$

The *indicial equation is*

$$k(k - 1) = 0, \tag{12.226}$$

with solutions $k = 0, 1$. We try first $k = 0$ with $a_0 = 1$, $a_1 = 0$. Then our series is described by the recurrence relation

$$(\lambda + 2)(\lambda + 1)a_{\lambda+2} + [n(n + 1) - 2\lambda - \lambda(\lambda - 1)]a_\lambda = 0, \tag{12.227}$$

which becomes

$$a_{\lambda+2} = -\frac{(n + \lambda + 1)(n - \lambda)}{(\lambda + 1)(\lambda + 2)} a_\lambda. \tag{12.228}$$

Labeling this series p_n, we have

$$p_n(x) = 1 - \frac{n(n + 1)}{2!} x^2 + \frac{(n - 2)n(n + 1)(n + 3)}{4!} x^4 + \cdots. \tag{12.229}$$

The second solution of the indicial equation, $k = 1$, with $a_0 = 1$, $a_1 = 0$, leads to the recurrence relation

$$a_{\lambda+2} = -\frac{(n + \lambda + 2)(n - \lambda - 1)}{(\lambda + 2)(\lambda + 3)} a_\lambda. \tag{12.230}$$

Labeling this series q_n, we obtain

$$q_n(x) = x - \frac{(n - 1)(n + 2)}{3!} x^3 + \frac{(n - 3)(n - 1)(n + 2)(n + 4)}{5!} x^5 - \cdots. \tag{12.231}$$

Our general solution of Eq. 12.221, then, is

$$y_n(x) = A_n\, p_n(x) + B_n\, q_n(x), \tag{12.232}$$

provided we have convergence.

For n, a positive even integer (or zero), series p_n terminates, and with a proper choice of normalizing factor

$$P_n(x) = (-1)^{n/2} \frac{n!}{2^n[(n/2)!]^2}\, p_n(x)$$
$$= (-1)^s \frac{(2s)!}{2^{2s}(s!)^2}\, p_{2s}(x) = (-1)^s \frac{(2s - 1)!!}{(2s)!!}\, p_{2s}(x), \qquad \text{for} \quad n = 2s. \tag{12.233}$$

If n is a positive odd integer, series q_n terminates after a finite number of terms, and we may write

$$P_n(x) = (-1)^{(n-1)/2} \frac{n!}{2^{n-1}\{[(n-1)/2]!\}^2} q_n(x)$$

$$= (-1)^s \frac{(2s+1)!}{2^{2s}(s!)^2} q_{2s+1}(x) = (-1)^s \frac{(2s+1)!!}{(2s)!!} q_{2s+1}(x), \qquad \text{for } n = 2s+1.$$

(12.234)

Note that these expressions hold for all real values of x, $-\infty < x < \infty$, and for complex values in the finite complex plane. The constants that multiply p_n and q_n are chosen to make P_n agree with Legendre polynomials given by the generating function.

Equations 12.229 and 12.231 may still be used with $n = v$, not an integer, but now the series no longer terminates, and the range of convergence becomes $-1 < x < 1$. The end points, $x = \pm 1$ are *not* included.

It is sometimes convenient to reverse the order of the terms in the series. This may be done by putting

$$s = \frac{n}{2} - \lambda \qquad \text{in the first form of } P_n(x), \qquad n \text{ even},$$

$$s = \frac{n-1}{2} - \lambda \qquad \text{in the second form of } P_n(x), \qquad n \text{ odd},$$

so that Eqs. 12.233 and 12.234 become

$$P_n(x) = \sum_{s=0}^{[n/2]} (-1)^s \frac{(2n-2s)!}{2^n s!(n-s)!(n-2s)!} x^{n-2s}, \tag{12.235}$$

where the upper limit $s = n/2$ (for n even) or $(n-1)/2$ (for n odd). This reproduces Eq. 12.8 of Section 12.1, which is obtained directly from the generating function. This agreement with Eq. 12.8 is the reason for the particular choice of normalization in Eqs. 12.233 and 12.234.

We see immediately that for very large x

$$P_n(x) \to \frac{(2n)!}{2^n(n!)^2} x^n. \tag{12.236}$$

For use in Section 12.11 we need P_n of pure imaginary argument. If x in Eq. 12.235 is replaced by $i\zeta$, we have

$$P_n(i\zeta) = (-1)^{n/2} \sum_{s=0}^{[n/2]} \frac{(2n-2s)!}{2^n s!(n-s)!(n-2s)!} \zeta^{n-2s}. \tag{12.237}$$

$Q_n(x)$, functions of the second kind. It will be noticed that we have used only p_n for n even and q_n for n odd (because they terminated for this choice of n). We may now define a second solution of Legendre's equation by

$$Q_n(x) = (-1)^{n/2} \frac{[(n/2)!]^2 2^n}{n!} q_n(x) = (-1)^s \frac{(2s)!!}{(2s-1)!!} q_{2s}(x), \qquad \text{for } n \text{ even}, n = 2s,$$

(12.238)

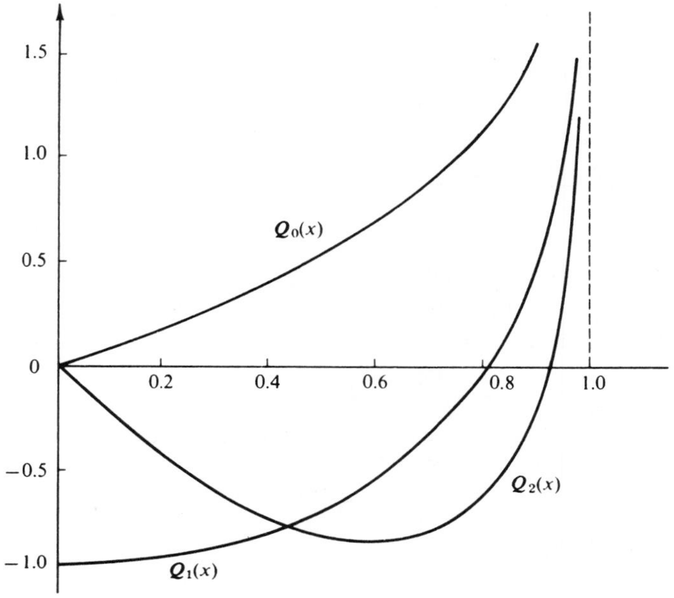

FIG. 12.16 Second Legendre function, $Q_n(x)$, $0 \leqslant x < 1$

$$Q_n(x) = (-1)^{(n+1)/2} \frac{\{[(n-1)/2]!\}^2 2^{n-1}}{n!} p_n(x)$$

$$= (-1)^{s+1} \frac{(2s)!!}{(2s+1)!!} p_{2s+1}(x), \qquad \text{for } n \text{ odd}, n = 2s+1. \qquad (12.239)$$

This choice of normalizing factors forces Q_n to satisfy the same recurrence relations as P_n. This may be verified by substituting Eqs. 12.238 and 12.239 into Eqs. 12.17 and 12.26. Inspection of the (series) recurrence relations (Eqs. 12.228 and 12.230), that is, by the Cauchy ratio test, shows that $Q_n(x)$ will converge for $-1 < x < 1$. If $|x| \geqslant 1$, our second solution *diverges.*

To develop a second solution in the range $x^2 > 1$, we seek a solution of Eq. 12.221 in the form of a *descending* power series. Let

$$y = \sum_{\lambda=0}^{\infty} b_{-\lambda} x^{k-\lambda} \qquad (12.240)$$

with

$$y' = \sum_{\lambda=0}^{\infty} (k-\lambda) b_{-\lambda} x^{k-\lambda-1} \qquad (12.241)$$

$$y'' = \sum_{\lambda=0}^{\infty} (k-\lambda)(k-\lambda-1) b_{-\lambda} x^{k-\lambda-2}. \qquad (12.242)$$

Substitution of these equations into Eq. 12.221 gives

$$\sum_{\lambda=0}^{\infty} (k-\lambda)(k-\lambda-1) b_{-\lambda} x^{k-\lambda-2} + \sum_{\lambda=0}^{\infty} [n(n+1) - 2(k-\lambda) - (k-\lambda)(k-\lambda-1)] b_{-\lambda} x^{k-\lambda} = 0. \qquad (12.243)$$

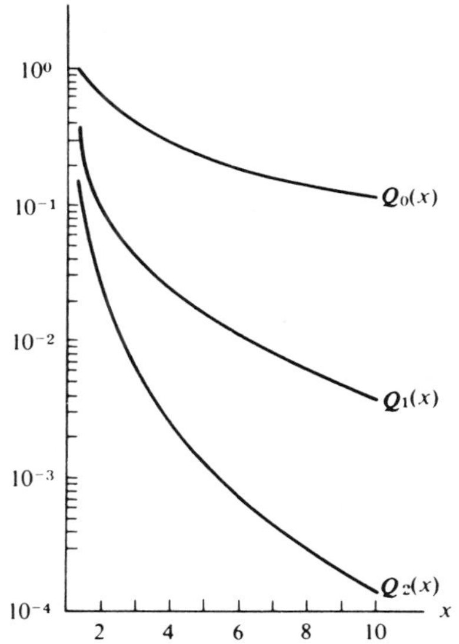

FIG. 12.17 Second Legendre function, $Q_n(x)$, $x > 1$

Requiring the coefficient of x^k to vanish leads to the indicial equation

$$k(k+1) - n(n+1) = 0, \tag{12.244}$$

with solutions

$$k = n, \qquad -n-1. \tag{12.245}$$

For integral n, the solution $k = n$ ($n \geqslant 0$) leads to the familiar Legendre polynomials $P_n(x)$. The second solution, $k = -n-1$, leads to a series with a recurrence relation

$$b_{-\lambda}(n+\lambda+1)(n+\lambda+2) + [n(n+1) - (n+\lambda+3)(n+\lambda+2)]b_{-\lambda-2} = 0 \tag{12.246}$$

or

$$b_{-\lambda-2} = \frac{(n+\lambda+1)(n+\lambda+2)}{(\lambda+2)(2n+\lambda+3)}\, b_{-\lambda}. \tag{12.247}$$

This gives

$$y(x) = b_0 x^{-n-1}\left[1 + \frac{(n+1)(n+2)}{2(2n+3)}x^{-2} + \frac{(n+1)(n+2)(n+3)(n+4)}{2\cdot 4\cdot(2n+3)(2n+5)}x^{-4} + \cdots\right] \tag{12.248}$$

with

$$b_{-2s} = \frac{(n+1)(n+2)\cdots(n+2s)}{2\cdot 4\cdots(2s)(2n+3)(2n+5)\cdots(2n+2s+1)}\, b_0. \tag{12.249}$$

This may also be written

$$y(x) = b_0 x^{-n-1} \sum_{s=0}^{\infty} \frac{(n+s)!(n+2s)!(2n+1)!}{s!(n!)^2(2n+2s+1)!} x^{-2s}. \tag{12.250}$$

One of the standard forms of the second solution is obtained by choosing

$$b_0 = \frac{2^n(n!)^2}{(2n+1)!}, \tag{12.251}$$

which leads to

$$Q_n(x) = 2^n x^{-n-1} \sum_{s=0}^{\infty} \frac{(n+s)!(n+2s)!}{s!(2n+2s+1)!} x^{-2s}, \qquad x^2 > 1. \tag{12.252}$$

By substituting in this series we see that $Q_n(x)$, $x^2 > 1$, still satisfies the same recurrence relations as $P_n(x)$.

Closed form solutions. Frequently, a closed form of the second solution, $Q_n(z)$, is desirable. This may be obtained by the method discussed in Section 8.5. We may write

$$Q_n(z) = P_n(z)\left\{A_n + B_n \int^z \frac{dx}{(1-x^2)[P_n(x)]^2}\right\}, \tag{12.253}$$

in which the constant A_n replaces the evaluation of the integral at the arbitrary lower limit. Both constants, A_n and B_n, may be determined for special cases.

For $n = 0$, Eq. 12.253 yields

$$\begin{aligned} Q_0(z) &= P_0(z)\left\{A_0 + B_0 \int^z \frac{dx}{(1-x^2)[P_0(x)]^2}\right\} \\ &= A_0 + B_0 \frac{1}{2} \ln \frac{1+z}{1-z} \\ &= A_0 + B_0\left(z + \frac{z^3}{3} + \frac{z^5}{5} + \cdots + \frac{z^{2s+1}}{2s+1} + \cdots\right), \end{aligned} \tag{12.254}$$

the last expression following from a Maclaurin expansion of the logarithm. Comparing this with the series solution (Eq. 12.231), we obtain

$$Q_0(z) = q_0(z) = z + \frac{z^3}{3} + \frac{z^5}{5} + \cdots + \frac{z^{2s+1}}{2s+1} + \cdots, \tag{12.255}$$

we have $A_0 = 0$, $B_0 = 1$. Similar results follow for $n = 1$. We obtain

$$\begin{aligned} Q_1(z) &= z\left[A_1 + B_1 \int^z \frac{dx}{(1-x^2)x^2}\right] \\ &= A_1 z + B_1 z\left(\frac{1}{2} \ln \frac{1+z}{1-z} - \frac{1}{z}\right). \end{aligned} \tag{12.256}$$

Expanding in a power series and comparing with $Q_1(z) = -p_1(z)$, we have $A_1 = 0$,

$B_1 = 1$. Therefore we may write

$$Q_0(z) = \frac{1}{2} \ln \frac{1+z}{1-z}$$

$$Q_1(z) = \frac{1}{2} z \ln \frac{1+z}{1-z} - 1, \qquad |z| < 1. \tag{12.257}$$

Perhaps the best way of determining the higher-order $Q_n(z)$ is to use the recurrence relation (Eq. 12.17), which may be verified for both $x^2 < 1$ and for $x^2 > 1$ by substituting in the series forms. This recurrence relation technique yields

$$Q_2(z) = \frac{1}{2} P_2(z) \ln \frac{1+z}{1-z} - \frac{3}{2} P_1(z). \tag{12.258}$$

Repeated application of the recurrence formula leads to

$$Q_n(z) = \frac{1}{2} P_n(z) \ln \frac{1+z}{1-z} - \frac{2n-1}{1 \cdot n} P_{n-1}(z) - \frac{2n-5}{3(n-1)} P_{n-3}(z) - \cdots. \tag{12.259}$$

From the form $\ln [(1+z)/(1-z)]$, it will be seen that for real z these expressions hold in the range $-1 < x < 1$. If we wish to have closed forms valid outside this range, we need only replace

$$\ln \frac{1+x}{1-x} \quad \text{by} \quad \ln \frac{z+1}{z-1}.$$

When using the latter form, valid for large z, the line interval $-1 \leqslant x \leqslant 1$ is taken as a cut line. Values of $Q_n(x)$, on the cut line, are customarily assigned by the relation

$$Q_n(x) = \frac{1}{2} [Q_n(x + i0) + Q_n(x - i0)], \tag{12.260}$$

the arithmetic average of approaches from the positive imaginary side and from the negative imaginary side. The reader will note that for $z \to x > 1, z - 1 \to (1-x)e^{\pm i\pi}$. The result of all this is that for *all* z, except on the real axis $-1 \leqslant x \leqslant 1$, we have

$$Q_0(z) = \tfrac{1}{2} \ln \frac{z+1}{z-1}, \tag{12.261}$$

$$Q_1(z) = \tfrac{1}{2} z \ln \frac{z+1}{z-1} - 1, \tag{12.262}$$

and so on.

The reader may verify that Eqs. 12.261 and 12.262 agree with Eq. 12.252. This was the basis for our choice of b_0.

In the next section we shall want $Q_n(y)$ on the imaginary axis. From the series form of $Q_n(y)$, $|z| > 1$, we may write

$$Q_n(iy) = (-i)^{n+1} 2^n y^{-n-1} \sum_{s=0}^{\infty} \frac{(n+s)!(n+2s)!(-i)^{2s}}{s!(2n+2s+1)!} y^{-2s}, \tag{12.263}$$

replacing y by iy. In the closed form it is convenient to take

$$\ln \frac{iy+1}{iy-1} = 2i\left(-\frac{1}{y} + \frac{1}{3y^3} - \frac{1}{5y^5} + \cdots\right)$$
$$= -2i \cot^{-1} y. \tag{12.264}$$

With this substitution, Eq. 12.259 becomes

$$Q_n(iy) = \frac{1}{2} P_n(iy)(-2i)\cot^{-1} y - \frac{2n-1}{n} P_{n-1}(iy) - \frac{2n-5}{3(n-1)} P_{n-3}(iy) - \cdots, \tag{12.265}$$

with special cases,

$$Q_0(iy) = -i \cot^{-1} y, \tag{12.266}$$

$$Q_1(iy) = y \cot^{-1} y - 1. \tag{12.267}$$

There is a discontinuity at $y = 0$ arising from the arc cotangent. Apart from this, Eq. 12.265 defines $Q_n(iy)$ for all real y, $0 < y < \infty$.

For convenient reference some special values of $Q_n(z)$ are given.

1. $Q_n(1) = \infty$, from the logarithmic term (Eq. 12.259).
2. $Q_n(\infty) = 0$, from the series expansion, (Eq. 12.252).
3. $Q_n(-z) = (-1)^{n+1} Q_n(z)$. This follows from the series form. It may also be derived by using $Q_0(z)$, $Q_1(z)$ and the recurrence relation (Eq. 12.17).
4. $Q_n(0) = 0$, for n even, by (3) above.
5. $Q_n(0) = (-1)^{(n+1)/2} \dfrac{\{[(n-1)/2]!\}^2}{n!} 2^{n-1} = (-1)^{s+1} \dfrac{(2s)!!}{(2s+1)!!}$, for n odd, $n = 2s + 1$

This last result comes from the series form (Eq. 12.239) with $p_n(0) = 1$.

As with the Bessel functions, a Wronskian relation can be developed.[1] We find

$$(1 - z^2)[P_n(z)\, Q_n'(z) - P_n'(z)\, Q_n(z)] = 1, \tag{12.268}$$

the constant being independent of n, $n = 0, 1, 2, 3, \ldots$. If we use Eq. 12.26 to eliminate the derivatives P_n' and Q_n', Eq. 12.268 becomes

$$P_n(z) Q_{n-1}(z) - P_{n-1}(z)\, Q_n(z) = \frac{1}{n}. \tag{12.269}$$

Associated Legendre functions of the second kind, $Q_n^m(x)$, may be developed in analogy to the associated Legendre functions of Section 12.5. Analogous to Eq. 12.84, we may define

$$Q_n^m(x) = (-1)^m (1 - x^2)^{m/2} \frac{d^m}{dx^m} Q_n(x) \tag{12.270}$$

[1] This Wronskian relation was used to get $Q_n(z)$ in closed form (Eq. 12.253).

and

$$Q_n^{-m}(x) = (-1)^m \frac{(n-m)!}{(n+m)!} Q_n^m(x), \qquad -1 < x < 1,\ m \geqslant 0. \tag{12.271}$$

Some authors omit either or both of the $(-1)^m$ factors. Using these definitions, we readily obtain

$$\begin{aligned} Q_1^1(x) &= -(1-x^2)^{1/2}\left(\frac{x}{1-x^2} + \frac{1}{2}\ln\frac{1+x}{1-x}\right), \\ Q_1^{-1}(x) &= -\tfrac{1}{2}\, Q^1{}_1(x). \end{aligned} \tag{12.272}$$

The $\ln[(1+x)/(1-x)]$ is characteristic of these $Q_n^m(x)$ for $m \neq 0$ as well as $m = 0$. For large x ($x^2 > 1$) or for the entire complex plane with $-1 < x < 1$ as a cut line AMS-55 gives

$$Q_n^m(z) = (z^2 - 1)^{m/2} \frac{d^m}{dz^m} Q_n(z), \qquad m \geqslant 0 \tag{12.273}$$

and Eq. 12.271 for negative superscript.

Integral representations. From the Schlaefi integral analysis (Section 12.4) a second, independent solution of Legendre's equation is

$$\int_C \frac{(t^2-1)^\nu}{(t-z)^{\nu+1}}\, dt,$$

in which the figure-eight contour (a) keeps the integrated function single-valued and (b) develops a solution distinct from $P_\nu(z)$. With the normalizing factor $(2^{\nu+2}\, i \sin \nu\pi)^{-1}$, we have

$$Q_\nu(z) = \frac{1}{2^{\nu+2}\, i \sin \pi\nu} \int_C \frac{(t^2-1)^\nu}{(z-t)^{\nu+1}}\, dt. \tag{12.274}$$

Here z is *not* on the real axis, $-1 \leqslant x \leqslant 1$. By deforming the contour, as shown in Fig. 12.19, we obtain

$$Q_\nu(z) = \frac{1}{2^{\nu+1}} \int_{-1}^{1} \frac{(1-t^2)^\nu}{(z-t)^{\nu+1}}\, dt, \qquad \nu > -1. \tag{12.275}$$

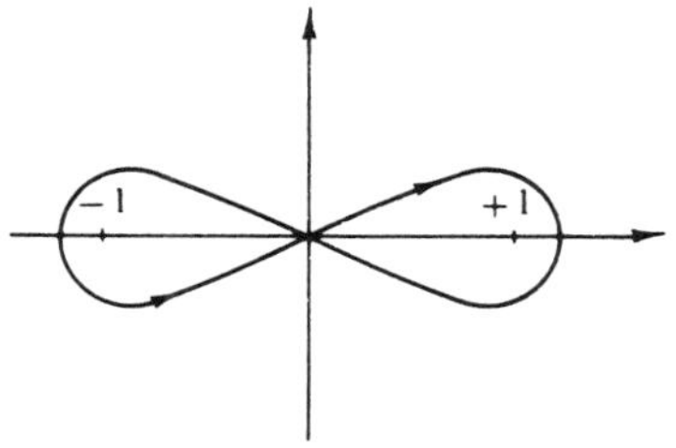

FIG. 12.18 Schlaefi integral contour for $Q_\nu(x)$

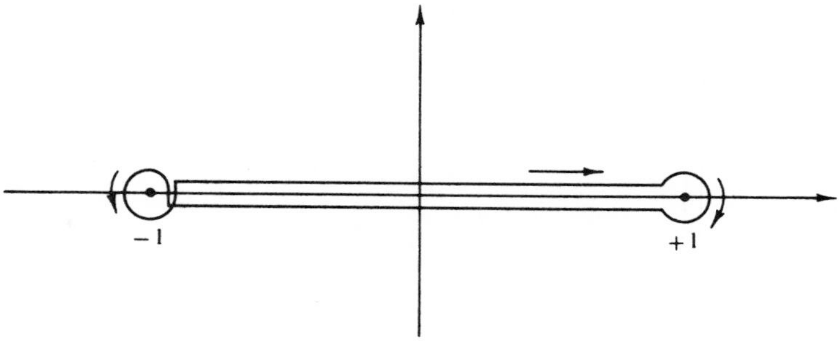

FIG. 12.19 Deformed Schlaefi contour

The identification of this integral as $Q_\nu(z)$ is left as Exercise 12.10.8.

The substitution

$$t = \frac{e^{\varphi}\sqrt{z+1} - \sqrt{z-1}}{e^{\varphi}\sqrt{z+1} + \sqrt{z-1}}$$

leads (after considerable manipulation) to

$$Q_\nu(z) = \int_0^\infty \frac{d\varphi}{[z + \sqrt{z^2 - 1}\cosh\varphi]^{\nu+1}}, \qquad \nu > -1, \qquad (12.276)$$

analogous to Eq. 12.73 for $P_n(z)$. To hold $Q_\nu(z)$ analytic and single-valued a cut line is run from $+1$ to $-\infty$ (in the z-plane).

EXERCISES

12.10.1 Derive the parity relation for $Q_n(x)$.

12.10.2 From Eqs. 12.233 and 12.234 show that

(a)

$$P_{2n}(x) = \frac{(-1)^n}{2^{2n-1}} \sum_{s=0}^{n} (-1)^s \frac{(2n+2s-1)!}{(2s)!(n+s-1)!(n-s)!} x^{2s}.$$

(b)

$$P_{2n+1}(x) = \frac{(-1)^n}{2^{2n}} \sum_{s=0}^{n} (-1)^s \frac{(2n+2s+1)!}{(2s+1)!(n+s)!(n-s)!} x^{2s+1}.$$

Check the normalization by showing that one term of each series agrees with the corresponding term of Eq. 12.8.

12.10.3 Show that

(a)

$$Q_{2n}(x) = (-1)^n 2^{2n} \sum_{s=0}^{n} (-1)^s \frac{(n+s)!(n-s)!}{(2s+1)!(2n-2s)!} x^{2s+1}$$

$$+ 2^{2n} \sum_{s=n+1}^{\infty} \frac{(n+s)!(2s-2n)!}{(2s+1)!(s-n)!} x^{2s+1}, \qquad |x| < 1.$$

(b) $$Q_{2n+1}(x) = (-1)^{n+1}2^{2n}\sum_{s=0}^{n}(-1)^s \frac{(n+s)!(n-s)!}{(2s)!(2n-2s+1)!}x^{2s}$$

$$+ 2^{2n+1}\sum_{s=n+1}^{\infty}\frac{(n+s)!(2s-2n-2)!}{(2s)!(s-n-1)!}x^{2s}, \qquad |x|<1.$$

12.10.4 Verify that the Legendre functions of the second kind, $Q_n(x)$, satisfy the same recurrence relations as $P_n(x)$, both for $|x| < 1$ and for $|x| > 1$.

$$(2n+1)x\,Q_n(x) = (n+1)\,Q_{n+1}(x) + n\,Q_{n-1}(x),$$

$$(2n+1)\,Q_n(x) = Q'_{n+1}(x) - Q'_{n-1}(x).$$

12.10.5 (a) Using the recurrence relations, prove (independently of the Wronskian relation) that

$$n[P_n(x)\,Q_{n-1}(x) - P_{n-1}(x)\,Q_n(x)] = P_1(x)\,Q_0(x) - P_0(x)\,Q_1(x).$$

(b) By direct substitution show that the right-hand side of this equation equals 1.

12.10.6 Evaluate $Q_n^m(0)$.

Ans. $$Q_n^m(0) = \begin{cases} 0, & n+m \text{ even} \\ (-1)^{(n+m+1)/2}2^n \dfrac{\left(\dfrac{m+m-1}{2}\right)!\left(\dfrac{n-m+1}{2}\right)!}{(n-m+1)!}, & n+m \text{ odd}, \quad m<n. \end{cases}$$

12.10.7 Assuming that

$$\int_{-1}^{1}\frac{(1-t^2)^\nu}{(z-t)^{\nu+1}}\,dt$$

satisfies Legendre's equation ($\nu > -1$), show from the parity that it cannot involve any $P_\nu(z)$.

12.10.8 Assuming that

$$\int_{-1}^{1}\frac{(1-t^2)^\nu}{(z-t)^{\nu+1}}\,dt$$

satisfies Legendre's equation ($\nu > -1$), show that the integral is $2^{\nu+1}\,Q_\nu(z)$. *Hint*. We can expand the denominator in powers of t/z (for $|z| > t$) and integrate by using the beta function (Section 10.4).

12.11 Application to Spheroidal Coordinate Systems

Let us consider the physical problems of dielectric and conducting spheroids placed in a uniform electric field. For simplicity the axis of symmetry of the spheroid will be taken parallel to the direction of the original electric field. This gives us complete axial symmetry and eliminates all azimuthal dependence.

Oblate spheroidal coordinates. If our spheroid is oblate, it will be convenient to choose oblate spheroidal coordinates so that one coordinate surface describes the spheroid. We can use the system defined by the transformation equations[1]

$$x = a\cosh u\sin v\cos\varphi,$$

$$y = a\cosh u\sin v\sin\varphi, \tag{12.277}$$

$$z = a\sinh u\cos v,$$

[1] This the same coordinate system as the oblate spheroidal system defined in Section 2.11, except that $v_{12.11} = \pi/2 - v_{2.11}$.

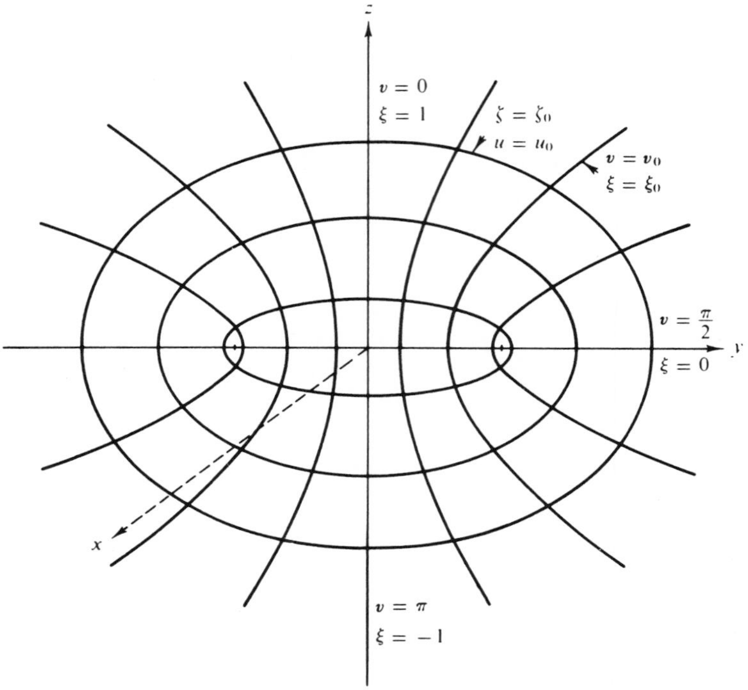

FIG. 12.20 Oblate spherical coordinates

the surface $u = u_0$ being an oblate spheroid, the surface $v = v_0$, a hyperboloid of one sheet, and the surface $\varphi = \varphi_0$, a half plane through the z-axis.

Here it is somewhat more convenient to replace this set (u, v, φ) by the coordinates (ξ, ζ, φ), defined by

$$\begin{aligned} \xi &= \cos v, \qquad -1 \leqslant \xi \leqslant 1, \\ \zeta &= \sinh u, \qquad 0 \leqslant \zeta < \infty. \end{aligned} \tag{12.278}$$

In this new system the surface $\zeta = 0$ is a circular disk, whereas $\zeta = \zeta_0 > 0$ is an ellipsoid of revolution (oblate spheroid). The surface $\xi = 0$ is a plane with a circular hole, whereas $\xi = 1$ is a line, perpendicular to the hole, $\xi = 0$, and the disk $\zeta = 0$ and passing through their common center. Eqs. 12.277 become

$$\begin{aligned} x &= \rho \cos \varphi \\ y &= \rho \sin \varphi \\ z &= a\xi\zeta, \end{aligned} \tag{12.279}$$

in which

$$\rho = a[(1 - \xi^2)(1 + \zeta^2)]^{1/2}. \tag{12.280}$$

The triad (ξ, ζ, φ) forms a right-handed set, $\boldsymbol{\xi}_0 \times \boldsymbol{\zeta}_0 = \boldsymbol{\varphi}_0$.

We seek solutions of Laplace's equation, which has become

$$\frac{\partial}{\partial\xi}\left[(1-\xi^2)\frac{\partial V}{\partial\xi}\right]+\frac{\partial}{\partial\zeta}\left[(1+\zeta^2)\frac{\partial V}{\partial\zeta}\right]+\frac{\xi^2+\zeta^2}{(1-\xi^2)(1+\zeta^2)}\frac{\partial^2 V}{\partial\varphi^2}=0. \tag{12.281}$$

Here we have used the scale factors

$$\begin{aligned} h_\xi &= a\left(\frac{\xi^2+\zeta^2}{1-\xi^2}\right)^{1/2}, \\ h_\zeta &= a\left(\frac{\xi^2+\zeta^2}{1+\zeta^2}\right)^{1/2}, \\ h_\varphi &= a[(1-\xi^2)(1+\zeta^2)]^{1/2}=\rho. \end{aligned} \tag{12.282}$$

We assume a solution of the form $X(\xi)\,Z(\zeta)\,\Phi(\varphi)$. Then, separating variables (cf. Section 2.5), we find that $X(\xi)$ must satisfy the associated Legendre equation, $Z(\zeta)$ the associated Legendre equation of pure imaginary argument, and $\Phi(\varphi)$ the usual simple harmonic oscillator equation. The potential V may be written

$$V(\xi,\zeta,\varphi)=\sum_{m,n}X_{mn}(\xi)\,Z_{mn}(\zeta)\,\Phi_m(\varphi), \tag{12.283}$$

where

$$\begin{aligned} X_{mn}(\xi) &= A\,P_n^m(\xi)+B\,Q_n^m(\xi) \\ Z_{mn}(\zeta) &= A'\,P_n^m(i\zeta)+B'\,Q_n^m(i\zeta) \\ \Phi_m(\varphi) &= C\cos m\varphi+D\sin m\varphi. \end{aligned} \tag{12.284}$$

Boundary conditions—conducting oblate spheroid. By applying our condition of axial symmetry, we set $m=0$, eliminating all the associated Legendre functions. However, we still need P_n and Q_n for both real and imaginary arguments.

Our boundary conditions, for the simpler case of a *conducting* oblate spheroid, are

$$\begin{aligned} V &= 0 \quad \text{for} \quad \zeta=\zeta_0 \qquad \text{(surface of spheroid)}, \\ V &\to -E_0 a\xi\zeta \quad \text{for} \quad \zeta\to\infty, \end{aligned}$$

where E_0 is the original, unperturbed electric field. To construct the proper ξ dependence we must have $n=1$ and $B=0$. (If $B\neq 0$, there will be trouble at $\xi=1$, $Q_n(1)=\infty$.) This yields

$$\begin{aligned} V(\xi,\zeta) &= P_1(\xi)[A'P_1(i\zeta)+B'Q_1(i\zeta)] \\ &= \xi[A'i\zeta+B'(\zeta\cot^{-1}\zeta-1)]. \end{aligned} \tag{12.285}$$

As $\zeta\to\infty$, $Q_1(i\zeta)\to 0$. Applying the second boundary condition,

$$A'=iE_0a, \tag{12.286}$$

we obtain

$$V(\xi,\zeta)=\xi[-E_0a\zeta+B'(\zeta\cot^{-1}\zeta-1)]. \tag{12.287}$$

By the first boundary condition the potential V vanishes on the spheroidal surface, $\zeta=\zeta_0$, or

$$0=\xi[-E_0a\zeta_0+B'(\zeta_0\cot^{-1}\zeta_0-1)], \tag{12.288}$$

and

$$B' = \frac{E_0 a\zeta_0}{\zeta_0 \cot^{-1}\zeta_0 - 1}. \tag{12.289}$$

The final result for $V(\xi, \zeta)$ is

$$V(\xi, \zeta) = -E_0 a\xi\zeta_0\left(\frac{\zeta}{\zeta_0} - \frac{\zeta\cot^{-1}\zeta - 1}{\zeta_0\cot^{-1}\zeta_0 - 1}\right), \zeta \geqslant \zeta_0. \tag{12.290}$$

To recapitulate, we have taken a summation of solutions of Laplace's equation to describe the potential. The arbitrary constants have been determined by demanding that the solution fit the boundary conditions. This gives Eq. 12.290. Finally, from Section 1.15, we know that our solution is unique.

The corresponding spherical problem (Section 12.3) is a limiting case. The limit, $\zeta_0 \to \infty$, $a \to 0$, with $a\zeta_0 = r_0$, a constant, will carry the oblate spheroidal system into the spherical system.

Boundary conditions—dielectric oblate spheroid. If our oblate spheroid is a *dielectric* spheroid, permittivity ε, the first boundary condition is replaced by

(a') V is continuous at $\zeta = \zeta_0$,[1]

(a'') $D_n = -\dfrac{\varepsilon}{h_\zeta}\dfrac{\partial V}{\partial\zeta}$ is continuous at $\zeta = \zeta_0$.

Let us *assume* that the electric field inside the dielectric is constant and parallel to the z(symmetry)-axis.[2] The justification of this assumption is that it works! Then, applying the boundary condition at infinity, we have

$$V_{\text{outside}} = \xi[-E_0 a\zeta + B'(\zeta\cot^{-1}\zeta - 1)] \tag{12.291}$$

$$V_{\text{inside}} = -E_{\text{in}} a\xi\zeta. \tag{12.292}$$

At $\zeta = \zeta_0$ conditions (a′) and (a″) give

$$-E_{\text{in}} a\zeta_0 = -E_0 a\zeta_0 + B'(\zeta_0\cot^{-1}\zeta_0 - 1) \tag{12.293}$$

$$\varepsilon E_{\text{in}} a = \varepsilon_0 E_0 a - \varepsilon_0 B'\left(\cot^{-1}\zeta_0 - \frac{\zeta_0}{1+\zeta_0^2}\right), \tag{12.294}$$

the factor h_ζ cancelling out. Eliminating B', we find that the ratio of the original field E_0 to the field in the interior of the dielectric E_{in} is

$$\frac{E_0}{E_{\text{in}}} = 1 + \left(\frac{\varepsilon}{\varepsilon_0} - 1\right)(1+\zeta_0^2)(1 - \zeta_0\cot^{-1}\zeta_0). \tag{12.295}$$

It is convenient to separate the geometric effect from the physical effect ($\varepsilon \neq \varepsilon_0$). We may *define* a depolarizing factor, L, by

$$E_{\text{in}} = E_0 - LP, \tag{12.296}$$

[1] This is equivalent to requiring the tangential component of **E** to be continuous.

[2] The alternative is to set up the interior potential as a summation of solutions of Laplace's equation analogous to Eq. 12.283.

where P, the polarization, is $E_{in}(\varepsilon - \varepsilon_0)$. Solving for L, we find

$$L = \frac{1}{\varepsilon_0}(1 + \zeta_0^2)(1 - \zeta_0 \cot^{-1} \zeta_0), \tag{12.297}$$

a function of the geometry of our oblate spheroid.

In the limit as $\zeta_0 \to \infty$

$$\lim_{\zeta_0 \to \infty} L = \frac{1}{3\varepsilon_0}, \tag{12.298}$$

which is just the value for a dielectric sphere. At the other extreme

$$\lim_{\zeta_0 \to 0} L = \frac{1}{\varepsilon_0}, \tag{12.299}$$

and this is just the value for a thin dielectric slab perpendicular to the uniform electric field.

Prolate spheroidal coordinates. The analysis of a dielectric *prolate* spheroid in a uniform electric field proceeds along the same lines, except that prolate spheroidal coordinates are used. We shall use[1]

$$\begin{aligned} x &= \rho \cos \varphi, \\ y &= \rho \sin \varphi, \\ z &= a\xi\eta, \end{aligned} \tag{12.300}$$

$$\begin{aligned} \eta &= \cosh u, \qquad 1 \leqslant \eta < \infty, \\ \xi &= \cos v, \qquad -1 \leqslant \xi \leqslant 1, \end{aligned} \tag{12.301}$$

and

$$\rho = a[(1 - \xi^2)(\eta^2 - 1)]^{1/2}.$$

If we express the Laplacian in terms of (ξ, η, φ) (cf. Eq. 2.106, Section 2.10) the variables will separate. Both the ξ- and η-dependence lead to the associated Legendre equation. The φ-dependence once again yields the simple harmonic oscillator equation. Hence $V(\xi, \eta, \varphi)$ is the same as $V(\xi, \zeta, \varphi)$ (Eqs. 12.283 and 12.284) but with the arguments $i\zeta$ replaced by real η. If $m = 0$ is set by azimuthal symmetry and boundary conditions are applied, the external potential becomes

$$V_{\text{outside}} = \xi\left[A'\eta + B'\left(\frac{1}{2}\eta \ln \frac{\eta + 1}{\eta - 1} - 1\right)\right]. \tag{12.302}$$

Leaving the intervening steps for the reader to supply by analogy with the oblate spheroid, we finally obtain a depolarizing factor L,

$$L = \frac{1}{\varepsilon_0}\left(\eta_0^2 - 1\right)\left(\frac{1}{2}\eta_0 \ln \frac{\eta_0 + 1}{\eta_0 - 1} - 1\right), \tag{12.303}$$

in which $\eta = \eta_0$ describes the prolate spheroidal surface. In the limit as $\eta_0 \to 1$

[1] This follows Section 2.10 exactly.

the prolate spheroid becomes an infinitely long thin needle. As expected,

$$\lim_{\eta_0 \to 1} L = 0. \tag{12.304}$$

On the other hand,

$$\lim_{\eta_0 \to \infty} L = \frac{1}{3\varepsilon_0}, \tag{12.305}$$

again corresponding to the spherical case.

It might perhaps be mentioned that although these electrostatic cases have only a little application the corresponding magnetic problems are rather important. The analysis for paramagnetic and diamagnetic spheroids in a uniform magnetic field is just like that of the electrostatic cases. Both electrostatic and magnetostatic problems involve solutions of Laplace's equation.

EXERCISES

12.11.1 (a) A charged isolated oblate spheroidal shell ($\zeta = \zeta_0$) is at a potential $V = V_0$. Show that the electrostatic potential outside the shell is

$$V = V_0 \frac{Q_0(i\zeta)}{Q_0(i\zeta_0)} = V_0 \frac{\cot^{-1}\zeta}{\cot^{-1}\zeta_0}.$$

(b) By letting $\zeta_0 \to 0$ we obtain a disk of radius a. Find the potential of this disk at large distances and from it calculate the total charge on the disk. Use this result to show that the capacitance of the disk is $C = 8\varepsilon_0 a$ (mks units).

12.11.2 For the charged conducting isolated oblate spheroid of Exercise 12.11.1 find the charge q on the spheroid and the capacitance by each of the following two methods:
(a) A consideration of the behavior of the potential at large distances.
(b) Application of Gauss's law (Section 1.14), integrating over the surface of the spheroid.
(c) Show that the capacitance reduces to the value for a sphere in the limit as the oblate spheroid approaches a spherical shape.

Ans. (a) and (b) $q = \dfrac{4\pi\varepsilon_0 a V_0}{\cot^{-1}\zeta_0}$.

12.11.3 A charged conducting isolated prolate spheroidal shell ($\eta = \eta_0$) is at a potential V_0.
(a) Show that the electrostatic potential outside the shell is

$$V = V_0 \frac{Q_0(\eta)}{Q_0(\eta_0)} = V_0 \frac{\ln[(\eta + 1)/(\eta - 1)]}{\ln[(\eta_0 + 1)/(\eta_0 - 1)]}.$$

(b) What is the capacitance of the spheroid?
Hint. Investigate the behavior of this potential at large distances from the shell.

12.11.4 (a) For the charged conducting isolated prolate spheroid of Exercise 12.11.3 find the charge q and the capacitance by use of Gauss's law (Section 1.14), integrating over the surface of the spheroid.

(b) Show that the capacitance reduces to the value for a sphere in the limit as the prolate spheroid approaches a spherical shape.

Ans. (a) $q = 4\pi\varepsilon_0 a V_0 / Q_0(\eta_0)$
$C = 4\pi\varepsilon_0 a / Q_0(\eta_0)$.

12.12 Vector Spherical Harmonics

Most of our attention in this chapter has been directed toward solving the equations of scalar fields such as the electrostatic field. This was done primarily because the scalar fields are easier to handle than vector fields! However, with scalar field problems under firm control, more and more attention is being paid to vector field problems.

Magnetic field of a current loop. To illustrate the difficulties let us consider the equation[1]

$$\nabla \times \nabla \times \mathbf{A} = \mu_0 \mathbf{J} \tag{12.306}$$

for the magnetic vector potential. Let us further suppose that the boundary conditions are best expressed in spherical polar coordinates. In the example of a current loop (Section 12.5) it was possible to handle this equation because the form of **A** was highly restricted. In general, this equation will yield three scalar equations, each involving all three components of A, A_r, A_θ, and A_φ. Such coupled differential equations can be solved, but the complexities are formidable.

Setting $\nabla \cdot \mathbf{A} = 0$, we can convert our equation into the vector Laplacian $\nabla^2 \mathbf{A}$. This will separate into one equation for each component in *cartesian* coordinates. Unfortunately, our boundary conditions (for the current loop) are in spherical coordinates. To satisfy them we would still have to mix the cartesian components A_x, A_y, and A_z in a form that would probably be both awkward and difficult to handle.

To facilitate the solution of Eq. 12.306 and other equations, such as the vector Helmholtz and the vector wave equation, various combinations of the (scalar) spherical harmonics have been used to construct vectors in spherical polar coordinates. One set, useful in quantum mechanics, has been described by Hill.[2] His three vector spherical harmonics are

$$\begin{aligned}\mathbf{V}_{LM} = \mathbf{r}_0\left[-\left(\frac{L+1}{2L+1}\right)^{1/2} Y_L^M\right] + \boldsymbol{\theta}_0\left\{\frac{1}{[(L+1)(2L+1)]^{1/2}}\frac{\partial Y_L^M}{\partial\theta}\right\} \\ + \boldsymbol{\varphi}_0\left\{\frac{iM}{[(L+1)(2L+1)]^{1/2}\sin\theta}\, Y_L^M\right\},\end{aligned} \tag{12.307}$$

$$\begin{aligned}\mathbf{W}_{LM} = \mathbf{r}_0\left[\left(\frac{L}{2L+1}\right)^{1/2} Y_L^M\right] + \boldsymbol{\theta}_0\left\{\frac{1}{[L(2L+1)]^{1/2}}\frac{\partial Y_L^M}{\partial\theta}\right\} \\ + \boldsymbol{\varphi}_0\left\{\frac{iM}{[L(2L+1)]^{1/2}\sin\theta}\, Y_L^M\right\},\end{aligned} \tag{12.308}$$

$$\mathbf{X}_{LM} = \boldsymbol{\theta}_0\left\{\frac{-M}{[L(L+1)]^{1/2}\sin\theta}\, Y_L^M\right\} + \boldsymbol{\varphi}_0\left\{\frac{-i}{[L(L+1)]^{1/2}}\frac{\partial Y_L^M}{\partial\theta}\right\}. \tag{12.309}$$

These functions satisfy a general orthogonality relation

$$\int \mathbf{A}_{LM}\cdot\mathbf{B}^*_{L'M'}\, d\Omega = \delta_{AB}\delta_{LL'}\delta_{MM'}, \tag{12.310}$$

[1] Cf. Exercise 1.14.5 for a derivation from Maxwell's equations.

[2] E. H. Hill, Theory of Vector Spherical Harmonics, *Am. J. Phys.* **22**, 211 (1954); also J. M. Blatt and V. Weisskopf, *Theoretical Nuclear Physics*. New York: Wiley (1952). Note that Hill assigns phases in accordance with the Condon-Shortley phase convention (Section 12.6).

where **A** and **B** may be **V**, **X**, or **W**. This may be verified by using the definitions of **V**, **X**, and **W** and reducing the integral to one of ordinary orthonormal spherical harmonics, $Y_L^M(\theta, \varphi)$.

Under the parity operations (coordinate inversion) the vector spherical harmonics transform as

$$\begin{aligned}\mathbf{V}_{LM}(\theta', \varphi') &= (-1)^{L+1}\mathbf{V}_{LM}(\theta, \varphi),\\ \mathbf{W}_{LM}(\theta', \varphi') &= (-1)^{L+1}\mathbf{W}_{LM}(\theta, \varphi),\\ \mathbf{X}_{LM}(\theta', \varphi') &= (-1)^{L}\mathbf{X}_{LM}(\theta, \varphi),\end{aligned} \tag{12.311}$$

where

$$\begin{aligned}\theta' &= \pi - \theta\\ \varphi' &= \pi + \varphi.\end{aligned} \tag{12.312}$$

In verifying these relations the reader should remember that the spherical polar coordinate unit vectors $\mathbf{r}_0$ and $\boldsymbol{\varphi}_0$ are odd and $\boldsymbol{\theta}_0$ is even. These properties may be verified by expressing the unit vectors $\mathbf{r}_0$, $\boldsymbol{\theta}_0$, and $\boldsymbol{\varphi}_0$ in terms of the cartesian unit vectors **i**, **j**, and **k** and spherical polar coordinates.

To demonstrate the use of the vector spherical harmonics consider Eq. 12.306 again. From Hill's table of differential relations

$$\nabla \cdot [F(r)\,\mathbf{V}_{LM}(\theta, \varphi)] = -\left(\frac{L+1}{2L+1}\right)^{1/2}\left[\frac{dF}{dr} + \frac{L+2}{r}F\right] Y_L^M(\theta, \varphi), \tag{12.313}$$

$$\nabla \cdot [F(r)\,\mathbf{W}_{LM}(\theta, \varphi)] = \left(\frac{L}{2L+1}\right)^{1/2}\left[\frac{dF}{dr} - \frac{L-1}{r}F\right] Y_L^M(\theta, \varphi), \tag{12.314}$$

$$\nabla \cdot [F(r)\,\mathbf{X}_{LM}(\theta, \varphi)] = 0. \tag{12.315}$$

The condition

$$\nabla \cdot \mathbf{A} = 0 \tag{12.316}$$

eliminates $\mathbf{V}_{LM}$ and $\mathbf{W}_{LM}$, leaving only $\mathbf{X}_{LM}$. In the absence of current ($\mathbf{J} = 0$), that is, away from the current loop, Eq. 12.306, subject to Eq. 12.316, becomes

$$\nabla^2\mathbf{A} = 0. \tag{12.317}$$

Using another of Hill's differential relations with $\mathbf{A}_{LM} = R(r)\,\mathbf{X}_{LM}(\theta, \varphi)$, we obtain

$$\nabla^2[R(r)\,\mathbf{X}_{LM}(\theta, \varphi)] = \left[\frac{d^2R}{dr^2} + \frac{2}{r}\frac{dR}{dr} - \frac{L(L+1)}{r^2}R\right]\mathbf{X}_{LM} = 0, \tag{12.318}$$

in agreement with our Eq. 12.116. We have

$$\mathbf{A}_{LM} = a_{LM}r^{-L-1}\,\mathbf{X}_{LM}(\theta, \varphi). \tag{12.319}$$

We note that there can be no azimuthal dependence because of the symmetry of our loop, $M = 0$, and our solution reduces to

$$\mathbf{A}_L = a_L r^{-L-1}\left\{\frac{-i}{[L(L+1)]^{1/2}}\frac{\partial Y_L^0}{\partial\theta}\right\}\boldsymbol{\varphi}_0. \tag{12.320}$$

This is equivalent to Eq. 12.119. The constants a_L are determined by fitting boundary conditions, as done in Section 12.5 for c_n. The magnetic field may be found from

$$\nabla \times [F(r)\mathbf{X}_{LM}] = i\left(\frac{L}{2L+1}\right)^{1/2}\left[\frac{dF}{dr} - \frac{L}{r}F\right]\mathbf{V}_{LM} + i\left(\frac{L+1}{2L+1}\right)^{1/2}\left[\frac{dF}{dr} + \frac{(L+1)}{r}F\right]\mathbf{W}_{LM}, \tag{12.321}$$

which corresponds to Eq. 12.122. [Here $F(r) = a_L r^{-L-1}$.]

The definitions of the vector spherical harmonics given here are dictated by convenience, primarily in quantum mechanical calculations, in which the angular momentum is a significant parameter. Morse and Feshbach describe another set of vector spherical harmonics, **B**, **C**, and **P**, in which the radial dependence is entirely in **P** and the angular dependence entirely in **B** and **C**. This set offers advantages in treating the wave equation when we want to separate the longitudinal and transverse parts of the wave.

Further examples of the usefulness and power of the vector spherical harmonics will be found in Blatt and Weisskopf, in Morse and Feshbach, and in Jackson's *Classical Electrodynamics*, which uses vector spherical harmonics in a description of multipole radiation and related electromagnetic problems.

Vector spherical harmonics may be developed as the result of coupling L units of orbital angular momentum and 1 unit of spin angular momentum. An extension, coupling L units of orbital angular momentum and 2 units of spin angular momentum to form *tensor* spherical harmonics, is presented by Mathews.[1] The major application of tensor spherical harmonics is in the investigation of gravitational radiation.

EXERCISES

12.12.1 Construct the $l = 0$, $m = 0$ and $l = 1$, $m = 0$ vector spherical harmonics.

Ans.
$$V_{00} = -\mathbf{r}_0(4\pi)^{-1/2}$$
$$X_{00} = 0$$
$$W_{00} = 0$$
$$V_{10} = -\mathbf{r}_0(2\pi)^{-1/2}\cos\theta - \boldsymbol{\theta}_0(8\pi)^{-1/2}\sin\theta$$
$$X_{10} = \boldsymbol{\varphi}_0 i(3/8\pi)^{1/2}\sin\theta$$
$$W_{10} = \mathbf{r}_0(4\pi)^{-1/2}\cos\theta - \boldsymbol{\theta}_0(8\pi)^{-1/2}\sin\theta.$$

12.12.2 Verify that the parity of $\mathbf{V}_{LM}$ is $(-1)^{L+1}$, the parity of $\mathbf{X}_{LM}$ is $(-1)^L$, and that of $\mathbf{W}_{LM}$ is $(-1)^{L+1}$. What happened to the M-dependence of the parity? *Hint.* $\mathbf{r}_0$ and $\boldsymbol{\varphi}_0$ have odd parity; $\boldsymbol{\theta}_0$ has even parity (cf. Exercise 2.4.6).

[1] J. Mathews, "Gravitational Multipole Radiation," in *H. P. Robertson, In Memoriam*. Philadelphia, Pennsylvania: Society for Industrial and Applied Mathematics (1963).

12.12.3 Verify the orthonormality of the vector spherical harmonics $\mathbf{V}_{LM}$, $\mathbf{X}_{LM}$, and $\mathbf{W}_{LM}$.

12.12.4 In *Classical Electrodynamics*, p. 545, Jackson defines $\mathbf{X}_{LM}$ by the equation

$$\mathbf{X}_{LM}(\theta, \varphi) = \frac{1}{\sqrt{L(L+1)}} \mathbf{L} Y_L^M(\theta, \varphi),$$

in which the angular momentum operator $\mathbf{L}$ is given by

$$\mathbf{L} = -i(\mathbf{r} \times \boldsymbol{\nabla}).$$

Show that this definition agrees with Eq. 12.309.

12.12.5 Show that

$$\sum_{M=-L}^{L} \mathbf{X}_{LM}^{*}(\theta, \varphi) \cdot \mathbf{X}_{LM}(\theta, \varphi) = \frac{2L+1}{4\pi}.$$

Hint. One way is to use Ex. 12.12.4 with $\mathbf{L}$ expanded in cartesian coordinates using the raising and lowering operators of Section 12.7.

12.12.6 Show that

$$\int \mathbf{X}_{LM} \cdot (\mathbf{r}_0 \times \mathbf{X}_{LM})\, d\Omega = 0.$$

The integrand represents an interference term in electromagnetic radiation which contributes to angular distributions but not to total intensity.

REFERENCES

Hobson, E. W., *The Theory of Spherical and Ellipsoidal Harmonics*. New York: Chelsea (1955). A most complete reference. This is the classic text on Legendre polynomials and all related functions.

See also the references listed at the end of Chapter 13.

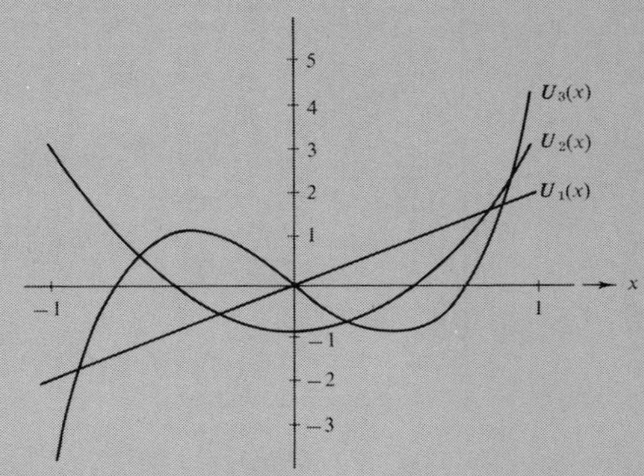

CHAPTER 13

SPECIAL FUNCTIONS

In this chapter we shall study four sets of orthogonal polynomials, Hermite, Laguerre, and Chebyshev[1] of first and second kinds. Although these four sets are of less importance in mathematical physics than the Bessel and Legendre functions of Chapters 11 and 12, they do occur. They are of occasional use and therefore deserve at least a little attention. Because the general mathematical techniques duplicate those of the preceding two chapters, the development of these functions is only outlined. Detailed proofs, along the lines of Chapters 11 and 12, are left to the reader. To conclude the chapter, these polynomials and other functions are expressed in terms of hypergeometric and confluent hypergeometric functions.

13.1 Hermite Functions

Generating functions—Hermite polynomials. The Hermite polynomials, $H_n(x)$, may be defined by the generating function

$$g(x, t) = e^{-t^2 + 2tx} = \sum_{n=0}^{\infty} H_n(x) \frac{t^n}{n!}. \tag{13.1}$$

Recurrence relations. Note the absence of a superscript, which distinguishes it from the unrelated Hankel functions. From the generating function we find that the Hermite polynomials satisfy the recurrence relations

$$H_{n+1}(x) = 2x\, H_n(x) - 2n\, H_{n-1}(x) \tag{13.2}$$

and

$$H_n'(x) = 2n\, H_{n-1}(x). \tag{13.3}$$

[1] This is the spelling choice of AMS-55. However a variety of forms, such as Tschebyscheff is encountered.

Equation 13.2 may be obtained by differentiating the generating function with respect to t; differentiation with respect to x leads to Eq. 13.3.

Direct expansion of the generating function easily gives $H_0(x) = 1$ and $H_1(x) = 2x$. Then Eq. 13.2 permits the construction of any $H_n(x)$ desired (integral n). For convenient reference the first several Hermite polynomials are listed in Table 13.1.

TABLE 13.1

Hermite Polynomials

$H_0(x) = 1$
$H_1(x) = 2x$
$H_2(x) = 4x^2 - 2$
$H_3(x) = 8x^3 - 12x$
$H_4(x) = 16x^4 - 48x^2 + 12$
$H_5(x) = 32x^5 - 160x^3 + 120x$
$H_6(x) = 64x^6 - 480x^4 + 720x^2 - 120$

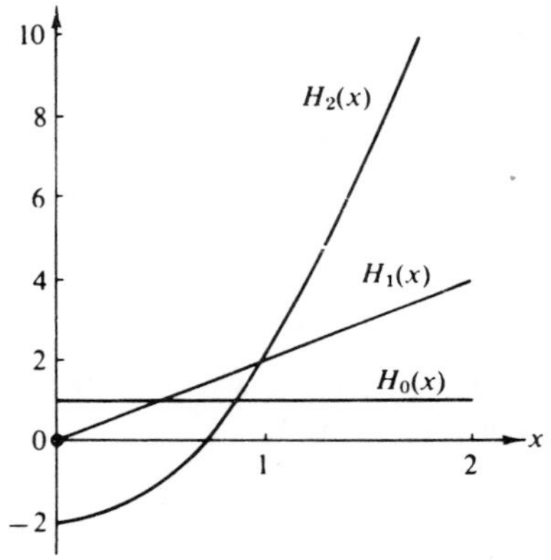

FIG. 13.1 Hermite polynomials

Special values of the Hermite polynomials follow from the generating function, that is,

$$H_{2n}(0) = (-1)^n \frac{(2n)!}{n!} \tag{13.4}$$

$$H_{2n+1}(0) = 0. \tag{13.5}$$

We also obtain from the generating function the important parity relation

$$H_n(x) = (-1)^n H_n(-x). \tag{13.6}$$

Alternate representations. Differentiation of the generating function[1] n times with respect to t and then setting t equal to zero yields

$$H_n(x) = (-1)^n e^{x^2} \frac{d^n}{dx^n} (e^{-x^2}). \tag{13.7}$$

This gives us a Rodrigues representation of $H_n(x)$. A second representation may be obtained by using the calculus of residues (Chapter 7). If we multiply Eq. 13.1 by t^{-m-1} and integrate around the origin, only the term with $H_m(x)$ will survive.

$$H_m(x) = \frac{m!}{2\pi i} \oint t^{-m-1} e^{-t^2+2tx}\, dt. \tag{13.8}$$

Also from Eq. 13.1, we may write our Hermite polynomial $H_n(x)$ in series form.

$$\begin{aligned} H_n(x) &= (2x)^n - \frac{2\,n!}{(n-2)!2!}(2x)^{n-2} + \frac{4\,n!}{(n-4)!4!}(2x)^{n-4} 1 \cdot 3 \cdots \\ &= \sum_{s=0}^{[n/2]} (-2)^s (2x)^{n-2s} \binom{n}{2s} 1 \cdot 3 \cdot 5 \cdots (2s-1) \\ &= \sum_{s=0}^{[n/2]} (-1)^s (2x)^{n-2s} \frac{n!}{(n-2s)!s!}. \end{aligned} \tag{13.9}$$

This terminates for integral n and yields our Hermite polynomial.

Orthogonality. The recurrence relations (Eqs. 13.2 and 13.3) lead to the second-order linear differential equation

$$H_n''(x) - 2x\, H_n'(x) + 2n\, H_n(x) = 0 \tag{13.10}$$

which is clearly *not* self-adjoint. To develop the orthogonal properties of the Hermite polynomials it is convenient to introduce a set of (unnormalized) functions φ_n by

$$\varphi_n(x) = e^{-x^2/2}\, H_n(x). \tag{13.11}$$

Substitution into Eq. 13.10 yields the differential equation for $\varphi_n(x)$,

$$\varphi_n''(x) + (2n + 1 - x^2)\, \varphi_n(x) = 0. \tag{13.12}$$

This is the differential equation for a quantum mechanical, simple harmonic oscillator which is perhaps the most important single application of the Hermite polynomials. Equation 13.12 is self-adjoint and the solutions $\varphi_n(x)$ are orthogonal for the interval $(-\infty < x < \infty)$.

The problem of normalizing these functions remains. Proceeding as in Section 12.3, we multiply Eq. 13.1 by itself and then by e^{-x^2}. This yields

$$e^{-x^2} e^{-s^2+2sx} e^{-t^2+2tx} = \sum_{m,n=0}^{\infty} e^{-x^2} H_m(x)\, H_n(x) \frac{s^m t^n}{m!\,n!}. \tag{13.13}$$

[1] Rewrite the generating function as $g(x, t) = e^{x^2} e^{-(t-x)^2}$. Note that

$$\frac{\partial}{\partial t} e^{-(t-x)^2} = -\frac{\partial}{\partial x} e^{-(t-x)^2}.$$

When integrating over x from $-\infty$ to $+\infty$, the cross terms of the double sum drop out because of the orthogonality property[1]

$$\sum_{n=0}^{\infty} \frac{(st)^n}{n!n!} \int_{-\infty}^{\infty} e^{-x^2}[H_n(x)]^2\, dx = \int_{-\infty}^{\infty} e^{-x^2-s^2+2sx-t^2+2tx}\, dx$$

$$= \int_{-\infty}^{\infty} e^{-(x-s-t)^2} e^{2st}\, dx$$

$$= \pi^{1/2} e^{2st} = \pi^{1/2} \sum_{n=0}^{\infty} \frac{2^n (st)^n}{n!}. \tag{13.14}$$

By equating coefficients of like powers of st we obtain

$$\int_{-\infty}^{\infty} e^{-x^2}[H_n(x)]^2\, dx = 2^n \pi^{1/2} n!. \tag{13.15}$$

Quantum mechanical simple harmonic oscillator. As already indicated, the Hermite polynomials are used in analyzing the quantum mechanical simple harmonic oscillator. For a potential energy $V = \frac{1}{2}Kz^2$ (force $\mathbf{F} = -\nabla V = -Kz$), the Schrödinger wave equation is

$$-\frac{\hbar^2}{2m} \nabla^2 \Psi(z) + \frac{1}{2} Kz^2 \Psi(z) = E\, \Psi(z). \tag{13.16}$$

Our oscillating particle has mass m and total energy E. By use of the abbreviations

$$x = \alpha z \qquad \text{with} \qquad \alpha^4 = \frac{mK}{\hbar^2} = \frac{m^2\omega^2}{\hbar^2},$$

$$\lambda = \frac{2E}{\hbar}\left(\frac{m}{K}\right)^{1/2} = \frac{2E}{\hbar\omega}, \tag{13.17}$$

in which ω is the angular frequency of the corresponding classical oscillator, Eq. 13.16 becomes [with $\Psi(z) = \Psi(x/\alpha) = \psi(x)$]

$$\frac{d^2\psi(x)}{dx^2} + (\lambda - x^2)\, \psi(x) = 0. \tag{13.18}$$

This is Eq. 13.12 with $\lambda = 2n + 1$. Hence

$$\psi_n(x) = 2^{-n/2} \pi^{-1/4} (n!)^{-1/2} e^{-x^2/2} H_n(x) \qquad \text{(normalized)}. \tag{13.19}$$

The requirement that n be an integer is dictated by the boundary conditions of the quantum mechanical system,

$$\lim_{z \to \pm\infty} \Psi(z) = 0.$$

Specifically, if $n \to \nu$, not an integer, a power-series solution of Eq. 13.10 (Ex. 8.4.5) shows that $H_\nu(x)$ will behave as $x^\nu e^{x^2}$ for large x. The functions $\psi_\nu(x)$ and $\Psi_\nu(z)$ will therefore blow up at infinity, and it will be impossible to normalize the wave function $\Psi(z)$. With this requirement, energy E becomes

[1] The cross terms ($m \neq n$) may be left in, if desired. Then, when the coefficients of $s^\alpha t^\beta$ are equated, the orthogonality will be apparent.

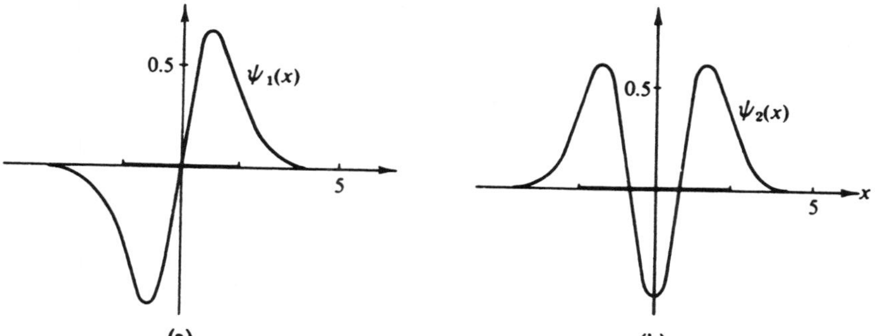

FIG. 13.2 Quantum mechanical oscillator wave functions: the heavy bar on the x-axis indicates the allowed range of the classical oscillator with the same total energy

$$E = (n + \tfrac{1}{2})\hbar\omega. \tag{13.20}$$

As n ranges over integral values ($n \geqslant 0$), we see that the energy is quantized and that there is a minimum or zero point energy

$$E_{\min} = \tfrac{1}{2}\hbar\omega. \tag{13.21}$$

This zero point energy is an aspect of the uncertainty principle, a purely quantum phenomenon.

An alternate treatment of the quantum mechanical oscillator found in many quantum mechanics texts employs raising and lowering operators:

$$\frac{1}{\sqrt{2}}\left(x + \frac{d}{dx}\right)\psi_n(x) = n^{1/2}\psi_{n-1}(x), \tag{13.22a}$$

$$\frac{1}{\sqrt{2}}\left(x - \frac{d}{dx}\right)\psi_n(x) = (n+1)^{1/2}\psi_{n+1}(x). \tag{13.22b}$$

The wave function ψ_n (actually given by Eq. 13.19) is unknown. The development is similar to the use of the raising and lowering operators presented in Section 12.7. The minimum energy or ground state wave function, ψ_0, satisfies the equation

$$\left(x + \frac{d}{dx}\right)\psi_0(x) = 0. \tag{13.23}$$

Normalized to unity,

$$\psi_0(x) = \pi^{-1/4}e^{-\frac{1}{2}x^2}, \tag{13.23a}$$

in agreement with Eq. 13.19. The excited state wave functions, ψ_1, ψ_2, etc., are then generated by the raising operator—Eq. 13.22*b*. The verification of these raising and lowering operators, Eqs. 13.22*a* and 13.22*b*, is left as Ex. 13.1.14.

In quantum mechanical problems, particularly in molecular spectroscopy, a number of integrals of the form

$$\int_{-\infty}^{\infty} x^r e^{-x^2} H_n(x)\, H_m(x)\, dx$$

are needed. Examples for $r = 1$ and $r = 2$ (with $n = m$) are included in the exercises at the end of this section. A large number of other examples are contained in Wilson, Decius, and Cross.[1]

The oscillator potential has also been employed extensively in calculations of nuclear structure (nuclear shell model).

There is a second independent solution of Eq. 13.10. This Hermite function of the second kind is an infinite series (Sections 8.4, 8.5) and of no physical interest, at least not yet.

EXERCISES

13.1.1 From the generating function show that

$$H_n(x) = \sum_{s=0}^{[n/2]} (-1)^s \frac{n!}{(n-2s)!s!} (2x)^{n-2s}.$$

13.1.2 From the generating function derive the recurrence relations

$$H_{n+1}(x) = 2x\, H_n(x) - 2n\, H_{n-1}(x),$$
$$H_n'(x) = 2n\, H_{n-1}(x).$$

13.1.3 Prove that

$$\left(2x - \frac{d}{dx}\right)^n 1 = H_n(x).$$

Hint. Check out the first couple of examples and then use mathematical induction.

13.1.4 Prove that

$$|H_n(x)| \le |H_n(ix)|.$$

13.1.5 Rewrite the series form of $H_n(x)$, Eq. 13.9, as an *ascending* power series.

Ans.

$$H_{2n}(x) = (-1)^n \sum_{s=0}^{n} (-1)^s (2x)^{2s} \frac{(2n)!}{(2s)!(n-s)!},$$

$$H_{2n+1}(x) = (-1)^n \sum_{s=0}^{n} (-1)^s (2x)^{2s+1} \frac{(2n+1)!}{(2s+1)!(n-s)!}.$$

13.1.6 (a) Expand x^{2r} in a series of even order Hermite polynomials.
(b) Expand x^{2r+1} in a series of odd order Hermite polynomials

Ans. (a) $$x^{2r} = \frac{(2r)!}{2^{2r}} \sum_{n=0}^{r} \frac{H_{2n}(x)}{(2n)!(r-n)!}$$

(b) $$x^{2r+1} = \frac{(2r+1)!}{2^{2r+1}} \sum_{n=0}^{r} \frac{H_{2n+1}(x)}{(2n+1)!(r-n)!}, \qquad r = 0, 1, 2, \ldots.$$

Hint: Use a Rodrigues representation of $H_{2n}(x)$ and integrate by parts.

[1] Wilson, E., B. Jr., J. C. Decius, and P. C. Cross, *Molecular Vibrations*. New York: McGraw-Hill (1955).

13.1.7 Show that

(a)
$$\int_{-\infty}^{\infty} H_n(x) \exp[-x^2/2]\, dx = \begin{cases} \sqrt{2\pi}\, n!/(n/2)!, & n \text{ even} \\ 0, & n \text{ odd.} \end{cases}$$

(b)
$$\int_{-\infty}^{\infty} x H_n(x) \exp[-x^2/2]\, dx = \begin{cases} 0, & n \text{ even} \\ \sqrt{2\pi}\, \dfrac{(n+1)!}{\left(\dfrac{n+1}{2}\right)!}, & n \text{ odd.} \end{cases}$$

13.1.8 Show that

$$\int_{-\infty}^{\infty} x^m e^{-x^2} H_n(x)\, dx = 0 \quad \text{for} \quad m \text{ an integer}, \qquad 0 \leqslant m \leqslant n-1.$$

13.1.9 The transition probability between two oscillator states, m and n, depends on

$$\int_{-\infty}^{\infty} x e^{-x^2} H_n(x)\, H_m(x)\, dx.$$

Show that this integral equals $\pi^{1/2} 2^{n-1} n!\, \delta_{m,n-1} + \pi^{1/2} 2^n (n+1)!\, \delta_{m,n+1}$. This result shows that such transitions can occur only between states of adjacent energy levels, $m = n \pm 1$. *Hint*. Multiply the generating function (Eq. 13.1) by itself using two different sets of variables (x, s) and (x, t).

13.1.10 Show that

$$\int_{-\infty}^{\infty} x^2 e^{-x^2} H_n(x)\, H_n(x)\, dx = \pi^{1/2} 2^n n! \left(n + \frac{1}{2}\right).$$

This integral occurs in the calculation of the mean-square displacement of our quantum oscillator. *Hint*. Use the recurrence relation Eq. 13.2 and the orthogonality integral.

13.1.11 Evaluate

$$\int_{-\infty}^{\infty} x^2 \exp[-x^2] H_n(x) H_m(x)\, dx$$

in terms of n and m and appropriate Kronecker delta functions.

Ans. $2^{n-1}\pi^{1/2}(2n+1)n!\delta_{n,m} + 2^n\pi^{1/2}(n+2)!\delta_{n+2,m} + 2^{n-2}\pi^{1/2}n!\delta_{n-2,m}$.

13.1.12 Show that

$$\int_{-\infty}^{\infty} x^r \exp[-x^2] H_n(x) H_{n+p}(x)\, dx = \begin{cases} 0, & p > r \\ 2^n \pi^{1/2} (n+r)!, & p = r. \end{cases}$$

n, p, and r are nonnegative integers.

13.1.13 (a) Using the Cauchy integral formula, develop an integral representation of $H_n(x)$ based on Eq. 13.1 with the contour enclosing the point $z = -x$.

Ans. $H_n(x) = \dfrac{n!}{2\pi i} e^{x^2} \oint \dfrac{e^{-z^2}}{(z+x)^{n+1}}\, dz.$

(b) Show by direct substitution that this result satisfies the Hermite equation.

13.1.14 With

$$\psi_n(x) = e^{-\frac{1}{2}x^2} H_n(x)/(2^n n! \pi^{1/2})^{1/2},$$

verify that

$$\frac{1}{\sqrt{2}}\left(x + \frac{d}{dx}\right)\psi_n(x) = n^{1/2}\psi_{n-1}(x),$$

$$\frac{1}{\sqrt{2}}\left(x - \frac{d}{dx}\right)\psi_n(x) = (n+1)^{1/2}\psi_{n+1}(x).$$

Note: The usual quantum mechanical operator approach establishes these raising and lowering properties *before* the form of $\psi_n(x)$ is known.

13.1.15 (a) Verify the operator identity

$$x - \frac{d}{dx} = -\exp[x^2/2]\frac{d}{dx}\exp[-x^2/2].$$

(b) The normalized simple harmonic oscillator wave function is

$$\psi_n(x) = (\pi^{1/2}2^n n!)^{-1/2}\exp[-x^2/2]H_n(x).$$

Show that this may be written

$$\psi_n(x) = (\pi^{1/2}2^n n!)^{-1/2}\left(x - \frac{d}{dx}\right)^n \exp[-x^2/2].$$

Note: This corresponds to an n-fold application of the raising operator of Ex. 13.1.14.

13.2 Laguerre Functions

Differential equation—Laguerre polynomials. If we start with the appropriate generating function, it is possible to develop the Laguerre polynomials in exact analogy with the Hermite polynomials. Alternatively, a series solution may be developed by the methods of Section 8.4. Instead, to illustrate a different technique, let us start with Laguerre's differential equation and obtain a solution in the form of a contour integral, as we did with the modified Bessel function $K_\nu(x)$ (Section 11.6). From this integral representation a generating function will be derived.

Laguerre's differential equation is

$$xy''(x) + (1 - x)\, y'(x) + n\, y(x) = 0. \tag{13.24}$$

We shall attempt to represent y, or rather y_n, since y will depend on n, by the contour integral

$$y_n(x) = \frac{1}{2\pi i}\oint \frac{e^{-xz/(1-z)}}{(1-z)z^{n+1}}\,dz. \tag{13.25a}$$

The contour includes the origin but does not enclose the point $z = 1$. From Section 6.4

$$y_n'(x) = -\frac{1}{2\pi i}\oint \frac{e^{-xz/(1-z)}}{(1-z)^2 z^n}\,dz, \tag{13.25b}$$

$$y_n''(x) = \frac{1}{2\pi i}\oint \frac{e^{-xz/(1-z)}}{(1-z)^3 z^{n-1}}\,dz. \tag{13.25c}$$

Substituting into the left-hand side of Eq. 13.24 we obtain

$$\frac{1}{2\pi i}\oint\left[\frac{x}{(1-z)^3 z^{n-1}} - \frac{1-x}{(1-z)^2 z^n} + \frac{n}{(1-z)z^{n+1}}\right]e^{-xz/(1-z)}\,dz,$$

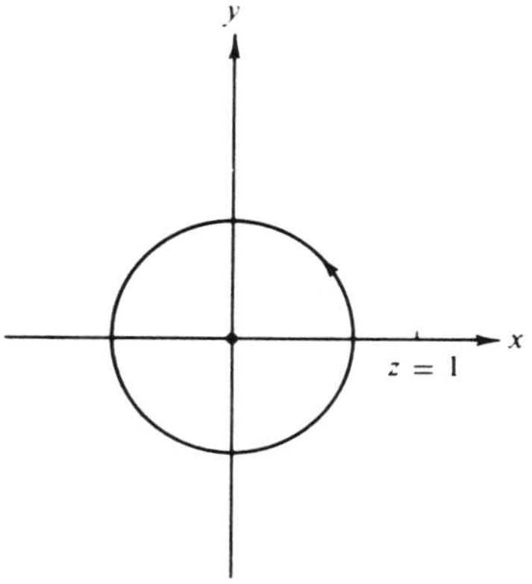

FIG. 13.3 Laguerre function contour

which is equal to

$$\frac{1}{2\pi i}\oint \frac{d}{dz}\left[\frac{e^{-xz/(1-z)}}{(1-z)z^n}\right] dz. \tag{13.26}$$

If we integrate our perfect differential around a contour chosen so that the final value equals the initial value (Fig. 13.3), the integral will vanish, thus verifying that $y_n(x)$ (Eq. 13.25*a*) is a solution of Laguerre's equation.

It has become customary to define $L_n(x)$, the Laguerre polynomial, by

$$L_n(x) = \frac{1}{2\pi i}\oint \frac{e^{-xz/(1-z)}}{(1-z)z^{n+1}} dz. \tag{13.27}$$

This is exactly what we would obtain from the series

$$g(x, z) = \frac{e^{-xz/(1-z)}}{1-z} = \sum_{n=0}^{\infty} L_n(x) z^n, \qquad |z| < 1 \tag{13.28}$$

if we multiplied by z^{-n-1} and integrated around the origin. As in the development of the calculus of residues (Section 7.2), only the z^{-1} term in the series survives. On this basis, we identify $g(x, z)$ as the generating function for the Laguerre polynomials.

With the transformation

$$\frac{xz}{1-z} = s - x \quad \text{or} \quad z = \frac{s-x}{s}, \tag{13.29}$$

$$L_n(x) = \frac{e^x}{2\pi i}\oint \frac{s^n e^{-s}}{(s-x)^{n+1}} ds, \tag{13.30}$$

the new contour enclosing the point $s = x$ in the s-plane. By Cauchy's integral formula (for derivatives)

$$L_n(x) = \frac{e^x}{n!}\frac{d^n}{dx^n}(x^n e^{-x}), \qquad \text{(integral } n\text{)}, \tag{13.31}$$

TABLE 13.2

Laguerre Polynomials

$L_0(x)$	$= 1$
$L_1(x)$	$= -x + 1$
$2!\,L_2(x)$	$= x^2 - 4x + 2$
$3!\,L_3(x)$	$= -x^3 + 9x^2 - 18x + 6$
$4!\,L_4(x)$	$= x^4 - 16x^3 + 72x^2 - 96x + 24$
$5!\,L_5(x)$	$= -x^5 + 25x^4 - 200x^3 + 600x^2 - 600x + 120$
$6!\,L_6(x)$	$= x^6 - 36x^5 + 450x^4 - 2400x^3 + 5400x^2 - 4320x + 720$

giving us Rodrigues' formula for Laguerre polynomials. From these representations of $L_n(x)$ we find the series form (for integral n):

$$\begin{aligned} L_n(x) &= \frac{(-1)^n}{n!}\left[x^n - \frac{n^2}{1!}x^{n-1} + \frac{n^2(n-1)^2}{2!}x^{n-2} - \cdots + (-1)^n n!\right], \\ &= \sum_{m=0}^{n}(-1)^m \frac{n!}{(n-m)!m!m!}x^m = \sum_{s=0}^{n}(-1)^{n-s}\frac{n!x^{n-s}}{(n-s)!(n-s)!s!} \end{aligned} \tag{13.32}$$

and the specific polynomials listed in Table 13.2.

By differentiating the generating function in Eq. 13.28 , with respect to x and z, we obtain the recurrence relations

$$(n+1)L_{n+1}(x) = (2n+1-x)L_n(x) - n\,L_{n-1}(x), \tag{13.33}$$

$$x\,L_n'(x) = n\,L_n(x) - n\,L_{n-1}(x). \tag{13.34}$$

Equation 13.33, modified to read

$$\begin{aligned} L_{n+1}(x) &= 2L_n(x) - L_{n-1}(x) \\ &\quad - [(1+x)L_n(x) - L_{n-1}(x)]/(n+1), \end{aligned} \tag{13.33a}$$

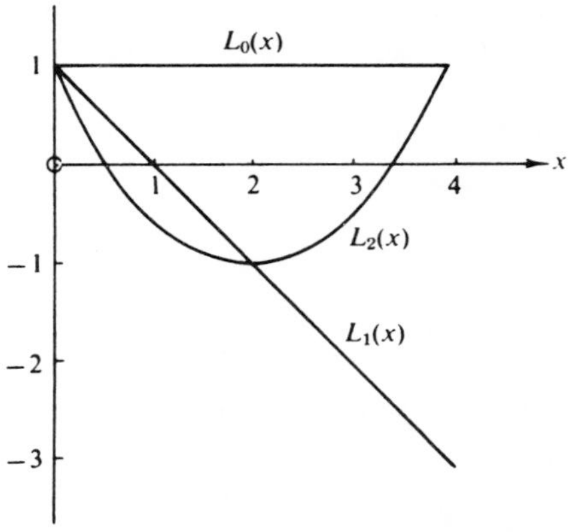

FIG. 13.4 Laguerre polynomials

for reasons of economy and numerical stability, is used for machine computation of numerical values of $L_n(x)$. The computing machine starts with known numerical values of $L_0(x)$ and $L_1(x)$, Table 13.2, and works up step by step—in milliseconds. This is the same technique discussed for computing Legendre polynomials, Section 12.2.

Also from Eq. 13.28 we find the special value

$$L_n(0) = 1. \tag{13.35}$$

As may be seen from the form of the generating function, the form of Laguerre's differential equation, or from Table 13.2, the Laguerre polynomials have neither odd nor even symmetry (parity).

The Laguerre differential equation is not self-adjoint and the Laguerre polynomials, $L_n(x)$, do not by themselves form an orthogonal set. However, the related set of functions[1]

$$\varphi_n(x) = e^{-x/2} L_n(x) \tag{13.36}$$

is orthonormal for the interval $(0 \leqslant x < \infty)$, that is,

$$\int_0^\infty e^{-x} L_m(x)\, L_n(x)\, dx = \delta_{m,n}. \tag{13.37}$$

Our new orthonormal function $\varphi_n(x)$ satisfies the differential equation

$$x\, \varphi_n''(x) + \varphi_n'(x) + \left(n + \frac{1}{2} - \frac{x}{4}\right) \varphi_n(x) = 0, \tag{13.38}$$

which is seen to have the Sturm-Liouville form (self-adjoint). Note that it is the boundary conditions in the Sturm-Liouville theory that fix our interval as $(0 \leqslant x < \infty)$. Equation 13.37 may be verified by using the generating function (Eq. 13.28).

Associated Laguerre polynomials. In many applications, particularly in quantum theory, we need the associated Laguerre polynomials defined by[2]

$$L_n^k(x) = (-1)^k \frac{d^k}{dx^k} [L_{n+k}(x)]. \tag{13.39}$$

From the series form of $L_n(x)$

$$\begin{aligned} L_0^k(x) &= 1 \\ L_1^k(x) &= -x + k + 1 \\ L_2^k(x) &= \frac{x^2}{2} - (k+2)x + \frac{(k+2)(k+1)}{2}. \end{aligned} \tag{13.40}$$

In general

[1] The factor $e^{-x/2}$ may be *derived* by the method of Section 9.1.

[2] Some authors use $\mathscr{L}_{n+k}^k(x) = (d^k/dx^k)[L_{n+k}(x)]$. Hence our $L_n^k(x) = (-1)^k \mathscr{L}_{n+k}^k(x)$.

$$L_n^k(x) = \sum_{m=0}^{n} (-1)^m \frac{(n+k)!}{(n-m)!(k+m)!m!} x^m, \quad (k > -1). \tag{13.41}$$

A generating function may be developed by differentiating the Laguerre generating function k times. Adjusting the index to L_{n+k}, we obtain

$$\frac{e^{-xz/(1-z)}}{(1-z)^{k+1}} = \sum_{n=0}^{\infty} L_n^k(x) z^n, \qquad |z| < 1. \tag{13.42}$$

From this

$$L_n^k(0) = \frac{(n+k)!}{n!k!}. \tag{13.43}$$

Recurrence relations can easily be derived from the generating function or by differentiating the Laguerre polynomial recurrence relations. Among the numerous possibilities are

$$(n+1)L_{n+1}^k(x) = (2n+k+1-x)\, L_n^k(x) - (n+k)\, L_{n-1}^k(x) \tag{13.44}$$

$$xL_n^{k\prime}(x) = n\, L_n^k(x) - (n+k)\, L_{n-1}^k(x). \tag{13.45}$$

From these or from differentiating Laguerre's differential equation k times we have the associated Laguerre equation

$$x\, L_n^{k\prime\prime}(x) + (k+1-x)\, L_n^{k\prime}(x) + n\, L_n^k(x) = 0. \tag{13.46}$$

A Rodrigues representation of the associated Laguerre polynomial is

$$L_n^k(x) = \frac{e^x x^{-k}}{n!} \frac{d^n}{dx^n} (e^{-x} x^{n+k}). \tag{13.47}$$

The reader will note that all of these formulas for $L_n^k(x)$ reduce to the corresponding expressions for $L_n(x)$ when $k = 0$.

The associated Laguerre equation (13.46) is not self-adjoint but it can be put in self-adjoint form by multiplying by $e^{-x}x^k$, which becomes the weighting function (Section 9.1). We obtain

$$\int_0^{\infty} e^{-x} x^k\, L_n^k(x)\, L_m^k(x)\, dx = \frac{(n+k)!}{n!} \delta_{m,n}. \tag{13.48}$$

By letting $\psi_n^k(x) = e^{-x/2} x^{k/2}\, L_n^k(x)$, $\psi_n^k(x)$ satisfies the self-adjoint equation

$$x\, \psi_n^{k\prime\prime}(x) + \psi_n^{k\prime}(x) + \left(-\frac{x}{4} + \frac{2n+k+1}{2} - \frac{k^2}{4x} \right) \psi_n^k(x) = 0. \tag{13.49}$$

A further useful form is given by defining[1]

$$\Phi_n^k(x) = e^{-x/2} x^{(k+1)/2}\, L_n^k(x). \tag{13.50}$$

Substitution into the associated Laguerre equation yields

$$\Phi_n^{k\prime\prime}(x) + \left(-\frac{1}{4} + \frac{2n+k+1}{2x} - \frac{k^2-1}{4x^2} \right) \Phi_n^k(x) = 0. \tag{13.51}$$

[1] This corresponds to modifying the function ψ in Eq. 13.49 to eliminate the first derivative (Cf. Exercise 8.5.7).

The corresponding normalization integral is

$$\int_0^\infty e^{-x}x^{k+1}\,L_n^k(x)\,L_n^k(x)\,dx = \frac{(n+k)!}{n!}(2n+k+1). \tag{13.52}$$

The reader may show that the $\Phi_n^k(x)$ do *not* form an orthogonal set (except with x^{-1} as a weighting function) because of the x^{-1} in the term $(2n+k+1)/2x$.

Example 13.2.1 The Hydrogen Atom

Perhaps the most important single application of the Laguerre polynomials is in the solution of the Schrödinger wave equation for the hydrogen atom. This equation is

$$-\frac{\hbar^2}{2m}\nabla^2\psi - \frac{Ze^2}{r}\psi = E\psi, \tag{13.53}$$

in which $Z = 1$ for hydrogen, 2 for singly ionized helium, and so on. Separating variables, we find that the angular dependence of ψ is $Y_L^M(\theta, \varphi)$. The radial part, $R(r)$, satisfies the equation

$$-\frac{\hbar^2}{2m}\frac{1}{r^2}\frac{d}{dr}\left(r^2\frac{dR}{dr}\right) - \frac{Ze^2}{r}R + \frac{\hbar^2}{2m}\frac{L(L+1)}{r^2}R = ER. \tag{13.54}$$

By use of the abbreviations

$$\begin{aligned} \rho &= \alpha r \quad \text{with } \alpha^2 = -\frac{8mE}{\hbar^2}, \qquad (E<0), \\ \lambda &= \frac{2mZe^2}{\alpha\hbar^2}, \end{aligned} \tag{13.55}$$

Eq. 13.54 becomes

$$\frac{1}{\rho^2}\frac{d}{d\rho}\left(\rho^2\frac{d\chi(\rho)}{d\rho}\right) + \left(\frac{\lambda}{\rho} - \frac{1}{4} - \frac{L(L+1)}{\rho^2}\right)\chi(\rho) = 0, \tag{13.56}$$

where $\chi(\rho) = R(\rho/\alpha)$. A comparison with Eq. 13.51 for $\Phi_n^k(x)$ shows that Eq. 13.56 is satisfied by

$$\rho\chi(\rho) = e^{-\rho/2}\rho^{L+1}L_{\lambda-L-1}^{2L+1}(\rho), \tag{13.57}$$

in which k is replaced by $2L+1$ and n by $\lambda - L - 1$.

We must restrict the parameter λ by requiring it to be an integer n, $n = 1, 2\ 3, \ldots$[1]

This is necessary because the Laguerre function of nonintegral n would diverge as $\rho^n e^\rho$, which is unacceptable for our physical problem in which

$$\lim_{r\to\infty} R(r) = 0.$$

[1] This is the conventional notation for λ. It is *not* the same n as the index n in $\Phi_n^k(x)$.

This restriction on λ, imposed by our boundary condition, has the effect of quantizing the energy

$$E_n = -\frac{Z^2 me^4}{2n^2\hbar^2}. \tag{13.58}$$

The negative sign enters because we are dealing here with bound states, $E = 0$ corresponding to an electron that is just able to escape to infinity. Using this result for E_n, we have

$$\alpha = 2\frac{me^2}{\hbar^2} \cdot \frac{Z}{n} = \frac{2Z}{na_0}$$

$$\rho = \frac{2Z}{na_0} r \tag{13.59}$$

with

$$a_0 = \frac{\hbar^2}{me^2} \quad \text{the Bohr radius.}$$

The final normalized hydrogen wave function may be written

$$\psi_{nLM}(r, \theta, \varphi) = \left[\left(\frac{2Z}{na_0}\right)^3 \frac{(n-L-1)!}{2n(n+L)!}\right]^{1/2} \times e^{-\alpha r/2}(\alpha r)^L L_{n-L-1}^{2L+1}(\alpha r)\, Y_L^M(\theta, \varphi). \tag{13.60}$$

EXERCISES

13.2.1 Show with the aid of the Leibnitz formula, that the series expansion of $L_n(x)$ (Eq. 13.32) follows from the Rodrigues representation (Eq. 13.31).

13.2.2 (a) Using the explicit series form (Eq. 13.32) show that

$$L_n'(0) = -n$$
$$L_n''(0) = \tfrac{1}{2}n(n-1).$$

(b) Repeat without using the explicit series form of $L_n(x)$.

13.2.3 From the generating function derive the Rodrigues representation

$$L_n^k(x) = \frac{e^x x^{-k}}{n!}\frac{d^n}{dx^n}(e^{-x}x^{n+k}).$$

13.2.4 Derive the normalization relation (Eq. 13.48) for the associated Laguerre polynomials.

13.2.5 Expand x^r in a series of associated Laguerre polynomials $L_n^k(x)$, k fixed and n ranging from 0 to r (or to ∞ if r is not an integer).
Hint: The Rodrigues form of $L_n^k(x)$ will be useful.

$$\textit{Ans.}\ x^r = (r+k)!r! \sum_{n=0}^{r} \frac{(-1)^n L_n^k(x)}{(n+k)!(r-n)!}, \qquad 0 \le x < \infty.$$

13.2.6 Expand e^{-ax} in a series of associated Laguerre polynomials $L_n^k(x)$, k fixed and n ranging from 0 to ∞.

(a) Evaluate directly the coefficients in your assumed expansion.

(b) Develop the desired expansion from the generating function.

$$\textit{Ans.}\ e^{-ax} = \frac{1}{(1+a)^{1+k}} \sum_{n=0}^{\infty} \left(\frac{a}{1+a}\right)^n L_n^k(x), \qquad 0 \leq x < \infty.$$

13.2.7 Show that

$$\int_0^\infty e^{-x} x^{k+1} L_n^k(x)\, L_n^k(x)\, dx = \frac{(n+k)!}{n!}(2n+k+1).$$

Hint: Note that

$$xL_n^k = (2n+k+1)\, L_n^k - (n+k)\, L_{n-1}^k - (n+1)\, L_{n+1}^k.$$

13.2.8 Assume that a particular problem in quantum mechanics has led to the differential equation

$$\frac{d^2y}{dx^2} - \left[\frac{k^2-1}{4x^2} - \frac{2n+k+1}{2x} + \frac{1}{4}\right] y = 0.$$

Write $y(x)$ as

$$y(x) = A(x)B(x)C(x)$$

with the requirement that

(a) $A(x)$ be a *negative* exponential giving the required asymptotic behavior of $y(x)$ and

(b) $B(x)$ be a *positive* power of x giving the behavior of $y(x)$ for $x \ll 1$.

Determine $A(x)$ and $B(x)$. Find the relation between $C(x)$ and the associated Laguerre polynomial.

$$\textit{Ans.}\ A(x) = e^{-x/2}$$
$$B(x) = x^{(k+1)/2},$$
$$C(x) = L_n^k(x).$$

13.2.9 From Eq. 13.60 the normalized radial part of the hydrogenic wave function is

$$R_{nL}(r) = \left[\alpha^3 \frac{(n-L-1)!}{2n(n+L)!}\right]^{1/2} e^{-\alpha r/2} (\alpha r)^L\, L_{n-L-1}^{2L+1}(\alpha r),$$

in which $\alpha = 2Z/na_0 = 2Zme^2/n\hbar^2$. Evaluate

(a) $\langle r \rangle = \int_0^\infty r\, R_{nL}(\alpha r)\, R_{nL}(\alpha r) r^2\, dr,$ (b) $\langle r^{-1} \rangle = \int_0^\infty r^{-1}\, R_{nL}(\alpha r)\, R_{nL}(\alpha r) r^2\, dr.$

The quantity $\langle r \rangle$ is the average displacement of the electron from the nucleus, whereas $\langle r^{-1} \rangle$ is the average of the reciprocal displacement.

$$\textit{Ans.}\ \langle r \rangle = \frac{a_0}{2}[3n^2 - L(L+1)]$$

$$\langle r^{-1} \rangle = \frac{1}{n^2 a_0}.$$

13.2.10 Derive the recurrence relation for the hydrogen wave function expectation values.

$$\frac{s+2}{n^2}\langle r^{s+1} \rangle - (2s+3)a_0 \langle r^s \rangle + \frac{s+1}{4}[(2L+1)^2 - (s+1)^2]a_0^2 \langle r^{s-1} \rangle = 0$$

with $s \geq -2L-1$. $\langle r^s \rangle \equiv \overline{r^s}$.

Hint: Transform Eq. 13.56 into a form analogous to Eq. 13.51. Multiply by $\rho^{s+2}u' - c\rho^{s+1}u$. Here $u = \rho\chi$. Adjust c to cancel terms that do not yield expectation values.

13.2.11 The hydrogen wave functions, Eq. 13.60, are mutually orthogonal as they should be, since they are eigenfunctions of the self-adjoint Schrödinger equation.

$$\int \psi^*_{n_1L_1M_1}\psi_{n_2L_2M_2}r^2\, dr\, d\Omega = \delta_{n_1n_2}\delta_{L_1L_2}\delta_{m_1m_2}$$

Yet the radial integral has the (misleading) form

$$\int_0^\infty e^{-\alpha r/2}(\alpha r)^L L^{2L+1}_{n_1-L-1}(\alpha r)e^{-\alpha r/2}(\alpha r)^L L^{2L+1}_{n_2-L-1}(\alpha r)^2\, dr$$

which *appears* to match Eq. 13.52 and not the associated Laguerre orthogonality relation, Eq. 13.48. How do you resolve this paradox?

Ans. The parameter α is dependent upon n. The first three α's above are $2Z/n_1a_0$. The last three are $2Z/n_2a_0$. For $n_1 = n_2$ Eq. 13.52 applies. For $n_1 \neq n_2$, neither Eq. 13.48 nor Eq. 13.52 is applicable.

13.2.12 A quantum mechanical analysis of the Stark effect (parabolic coordinate) leads to the differential equation

$$\frac{d}{d\xi}\left(\xi\frac{du}{d\xi}\right) + \left(\frac{1}{2}E\xi + L - \frac{m^2}{4\xi} - \frac{1}{4}F\xi^2\right)u = 0.$$

Here F is a measure of the perturbation energy introduced by an external electric field. Find the unperturbed wave functions ($F = 0$) in terms of associated Laguerre polynomials.

Ans. $u(\xi) = e^{-\varepsilon\xi/2}\xi^{m/2}L^m_p(\varepsilon\xi)$, with $\varepsilon = \sqrt{-2E} > 0$, $p = \alpha/\varepsilon - (m+1)/2$, a non-negative integer.

13.2.13 The wave equation for the three-dimensional harmonic oscillator is

$$-\frac{\hbar^2}{2M}\nabla^2\psi + \tfrac{1}{2}M\omega^2r^2\psi = E\psi.$$

Here ω is the angular frequency of the corresponding classical oscillator. Show that the radial part of ψ (in spherical polar coordinates) may be written in terms of associated Laguerre functions of argument (βr^2), where $\beta = M\omega/\hbar$.

Hint. As in Ex. 13.2.8, split off radial factors of r^l and $e^{-\beta r^2/2}$. The associated Laguerre function will have the form $L^{l+1/2}_{(n-l-1)/2}(\beta r^2)$.

13.3 Chebyshev (Tschebyscheff) Polynomials

Generating function. In Chapter 12 the generating function for the ultraspherical or Gegenbauer polynomials[1]

$$\frac{2^\beta}{(1 - 2xt + t^2)^{\beta+1/2}} = \frac{\pi^{1/2}}{(\beta - \frac{1}{2})!}\sum_{n=0}^\infty T^\beta_n(x)t^n, \qquad |t| < 1 \tag{13.61}$$

was mentioned, with $\beta = 0$ giving rise to the Legendre polynomials. In this chapter, by letting $\beta = \pm\frac{1}{2}$, we may generate two sets of polynomials known as the Chebyshev polynomials.

[1] AMS-55 denotes the ultraspherical polynomials by $C^{(\alpha)}_n(x)$:

$$C^{(\alpha)}_n(x) = \frac{\pi^{1/2}}{2^{\alpha-1/2}(\alpha-1)!}T^{\alpha-1/2}_n(x), \quad \frac{n}{2}C^{(0)}_n(x) = T_n(x), \quad C^{(1/2)}_n(x) = P_n(x), \quad C^{(1)}_n(x) = U_n(x).$$

Type II. With $\beta = +\frac{1}{2}$, Eq. 13.61 becomes

$$\frac{2^{1/2}}{(1 - 2xt + t^2)} = \pi^{1/2} \sum_{n=0}^{\infty} T_n^{1/2}(x)t^n. \tag{13.62}$$

It will be convenient to change the notation slightly and to use

$$\sqrt{\frac{\pi}{2}}\, T_n^{1/2}(x) \equiv U_n(x). \tag{13.63}$$

This gives us

$$\frac{1}{1 - 2xt + t^2} = \sum_{n=0}^{\infty} U_n(x)t^n. \tag{13.64}$$

The functions $U_n(x)$ generated by $(1 - 2xt + t^2)^{-1}$ are labeled Chebyshev polynomials type II.

Type I. With $\beta = -\frac{1}{2}$ there is trouble. The t- and x-dependence on the left disappears and the $(\beta - \frac{1}{2})!$ blows up. We may avoid this difficulty by first differentiating Eq. 13.61 with respect to t and then setting $\beta = -\frac{1}{2}$. This yields

$$\frac{x - t}{1 - 2xt + t^2} = \sqrt{\frac{\pi}{2}} \sum_{n=0}^{\infty} n\, T_n^{-1/2}(x)t^{n-1}. \tag{13.65}$$

By multiplying Eq. 13.65 by $2t$ and adding one we obtain

$$\frac{1 - t^2}{1 - 2tx + t^2} = 1 + \sqrt{\frac{\pi}{2}} \sum_{n=0}^{\infty} 2n\, T_n^{-1/2}(x)t^n. \tag{13.66}$$

For $n > 0$ we define $T_n(x)$ by

$$T_n(x) = \sqrt{\frac{\pi}{2}}\, n\, T_n^{-1/2}(x). \tag{13.67}$$

Then

$$\frac{1 - t^2}{1 - 2tx + t^2} = 1 + 2\sum_{n=1}^{\infty} T_n(x)t^n. \tag{13.68}$$

For $n = 0$ we define

$$T_0(x) = 1 \tag{13.69}$$

to preserve the recurrence relation (Eq. 13.70). We label $T_n(x)$ the Chebyshev polynomials type I. The reader should be warned that the notation for these functions differs from reference to reference. There is almost no general agreement. Here we follow the usage of AMS-55.

These Chebyshev polynomials (type I) combine useful features of (a) Fourier series and (b) orthogonal polynomials. They are of great interest in numerical computation. For example, a least squares approximation minimizes the average squared error. An approximation using Chebyshev polynomials allows a larger average squared error but may keep extreme errors down.

From the generating functions (Eqs. 13.64 and 13.68) we obtain recurrence relations

$$T_{n+1}(x) - 2x\, T_n(x) + T_{n-1}(x) = 0, \tag{13.70}$$

$$U_{n+1}(x) - 2x\, U_n(x) + U_{n-1}(x) = 0. \tag{13.71}$$

Then, using the generating functions for the first few values of n and these recurrence relations for the higher-order polynomials, we get Tables 13.3 and 13.4.

TABLE 13.3

Chebyshev Polynomials, Type I

$T_0 = 1$
$T_1 = x$
$T_2 = 2x^2 - 1$
$T_3 = 4x^3 - 3x$
$T_4 = 8x^4 - 8x^2 + 1$
$T_5 = 16x^5 - 20x^3 + 5x$
$T_6 = 32x^6 - 48x^4 + 18x^2 - 1$

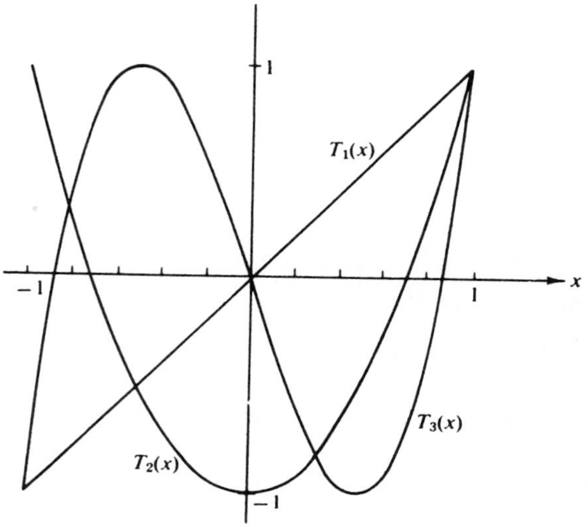

FIG. 13.5 Chebyshev polynomials, $T_n(x)$

Table 13.4

Chebyshev Polynomials, Type II

$U_0 = 1$
$U_1 = 2x$
$U_2 = 4x^2 - 1$
$U_3 = 8x^3 - 4x$
$U_4 = 16x^4 - 12x^2 + 1$
$U_5 = 32x^5 - 32x^3 + 6x$
$U_6 = 64x^6 - 80x^4 + 24x^2 - 1$

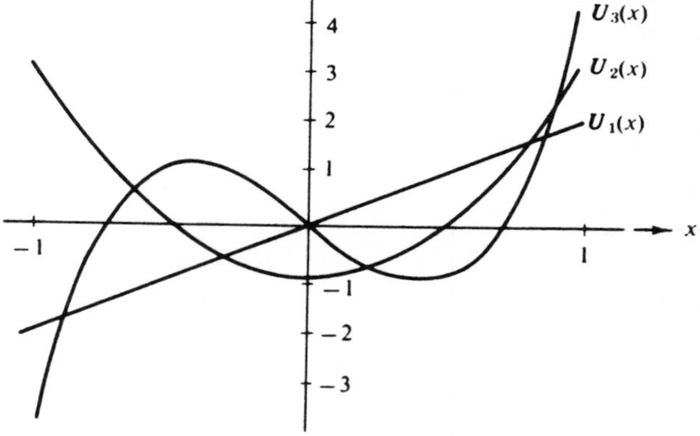

FIG. 13.6 Chebyshev polynomials, $U_n(x)$

As with the Hermite polynomials, Section 13.1, the recurrence relations, Eqs. 13.70 and 13.71, together with the known values of $T_0(x)$, $T_1(x)$, $U_0(x)$, and $U_1(x)$, provide a convenient means of getting the numerical value of any $T_n(x_0)$ or $U_n(x_0)$, with x_0 a given number: convenient, that is, for a high-speed electronic computer.

Again, from the generating functions, we have the special values

$$\begin{aligned} T_n(1) &= 1 \\ T_n(-1) &= (-1)^n \\ T_{2n}(0) &= (-1)^n \\ T_{2n+1}(0) &= 0 \end{aligned} \tag{13.72}$$

$$\begin{aligned} U_n(1) &= n+1 \\ U_n(-1) &= (-1)^n(n+1) \\ U_{2n}(0) &= (-1)^n \\ U_{2n+1}(0) &= 0. \end{aligned} \tag{13.73}$$

The parity relations for T_n and U_n are

$$\begin{aligned} T_n(x) &= (-1)^n\, T_n(-x) \\ U_n(x) &= (-1)^n\, U_n(-x). \end{aligned} \tag{13.74}$$

Rodrigues' representations of $T_n(x)$ and $U_n(x)$ are

$$T_n(x) = \frac{(-1)^n \pi^{1/2}(1-x^2)^{1/2}}{2^n(n-\frac{1}{2})!} \frac{d^n}{dx^n}[(1-x^2)^{n-1/2}], \tag{13.75}$$

and

$$U_n(x) = \frac{(-1)^n(n+1)\pi^{1/2}}{2^{n+1}(n+\frac{1}{2})!(1-x^2)^{1/2}} \frac{d^n}{dx^n}[(1-x^2)^{n+1/2}]. \tag{13.76}$$

Recurrence relations—derivatives. From the generating functions for $T_n(x)$ and $U_n(x)$, we can obtain a variety of recurrence relations involving derivatives. Among the more useful equations are

$$(1-x^2)\, T_n'(x) = -nx\, T_n(x) + n\, T_{n-1}(x), \tag{13.77}$$

and

$$(1-x^2)\, U_n'(x) = -nx\, U_n(x) + (n+1)\, U_{n-1}(x). \tag{13.78}$$

From Eqs. 13.70 and 13.77 $T_n(x)$, the Chebyshev polynomial type I, satisfies

$$(1-x^2)\, T_n''(x) - x\, T_n'(x) + n^2\, T_n(x) = 0, \tag{13.79}$$

whereas $U_n(x)$ the Chebyshev polynomial of type II satisfies

$$(1-x^2)\, U_n''(x) - 3x\, U_n'(x) + n(n+2)\, U_n(x) = 0. \tag{13.80}$$

Gegenbauer's equation,

$$(1-x^2)y'' - 2(1+\beta)xy' + n(n+2\beta+1)y = 0, \tag{13.81}$$

is a generalization of these equations reducing to Eq. 13.79 for $\beta = -\frac{1}{2}$, to Eq. 13.80 for $\beta = +\frac{1}{2}$, and to Legendre's equation for $\beta = 0$.

It is sometimes convenient to define a new function, $V_n(x)$, in terms of the type II polynomials by

$$V_{n+1}(x) = \sqrt{1 - x^2}\, U_n(x). \tag{13.82}$$

Note that the index has been increased from n to $n + 1$. It may be shown that $V_n(x)$ satisfies Eq. 13.79.

From the generating functions, or from the differential equations, power series representations are

$$\begin{aligned} T_n(x) &= \frac{n}{2} \sum_{m=0}^{[n/2]} (-1)^m \frac{(n - m - 1)!}{m!(n - 2m)!} (2x)^{n-2m} \\ &= x^n - \binom{n}{2} x^{n-2}(1 - x^2) + \binom{n}{4} x^{n-4}(1 - x^2)^2 - \cdots, \end{aligned} \tag{13.83}$$

$$U_n(x) = \sum_{m=0}^{[n/2]} (-1)^m \frac{(n - m)!}{m!(n - 2m)!} (2x)^{n-2m}. \tag{13.84}$$

Using Eq. 13.82, we obtain

$$\begin{aligned} V_n(x) = \sqrt{1 - x^2} \Bigg[\binom{n}{1} x^{n-1} - \binom{n}{3} x^{n-3}(1 - x^2) \\ + \binom{n}{5} x^{n-5}(1 - x^2)^2 - \cdots \Bigg]. \end{aligned} \tag{13.85}$$

A combination of Eqs. 13.83 and 13.85 produces

$$T_n(x) + i\, V_n(x) = [x + i(1 - x^2)^{1/2}]^n \qquad |x| \leqslant 1. \tag{13.86}$$

Orthogonality. If Eq. 13.79 is put into self-adjoint form (Section 9.1), we obtain $w(x) = (1 - x^2)^{-1/2}$ as a weighting factor. For Eq. 13.80 the corresponding weighting factor is $(1 - x^2)^{+1/2}$. The resulting orthogonality integrals are

$$\int_{-1}^{1} T_m(x)\, T_n(x)(1 - x^2)^{-1/2}\, dx = \begin{cases} 0 & m \neq n, \\ \dfrac{\pi}{2}, & m = n \neq 0, \\ \pi, & m = n = 0, \end{cases} \tag{13.87}$$

$$\int_{-1}^{1} V_m(x)\, V_n(x)(1 - x^2)^{-1/2}\, dx = \begin{cases} 0, & m \neq n, \\ \dfrac{\pi}{2}, & m = n \neq 0, \\ 0, & m = n = 0, \end{cases} \tag{13.88}$$

and

$$\int_{-1}^{1} U_m(x)\, U_n(x)(1 - x^2)^{1/2}\, dx = \frac{\pi}{2}\, \delta_{m,n}. \tag{13.89}$$

The $\pi/2$ factors may be obtained from the generating functions by using a technique similar to that in Section 12.3. Notice the change in pattern when $m = n = 0$. The orthogonal relations for sine and cosine needed for the development of Fourier series (Chapter 14) show this same behavior.

Numerical applications. The Chebyshev polynomials (T_n) are useful in numerical work for an interval $[-1, 1]$, because

(a) $|T_n(x)| \leqslant 1, \quad -1 \leqslant x \leqslant 1$ (cf. Ex. 13.3.4);

(b) the maxima and minima are of comparable magnitude;

(c) the maxima and minima are spread reasonably uniformly over the range $[-1, 1]$.

These properties follow from $T_n(x) = \cos(n \cos^{-1} x)$, Ex. 13.3.13.

One application is in the truncation and rearrangement of power series expansions. Suppose we want $\cosh x$ in the interval $[-1, 1]$ to ± 0.0001. We might start with

$$\begin{aligned}\cosh x &= 1.0 + \frac{x^2}{2!} + \frac{x^4}{4!} + \frac{x^6}{6!} + \frac{x^8}{8!} + \cdots \\ &= 1.000\,00 + 0.500\,00\, x^2 + 0.041\,67\, x^4 \\ &\quad + 0.001\,39\, x^6 + 0.000\,02\, x^8 + \cdots.\end{aligned} \tag{13.90}$$

At $x = \pm 1$, the last term contributes

$$1/8! = 0.000\,024\,8.$$

For four decimal accuracy (± 0.0001), the $x^8/8!$ may be dropped, but all the lower order terms must be retained—if we wish to use this power series expansion.

Rewriting $\cosh x$ as a series of Chebyshev polynomials

$$\begin{aligned}\cosh x = 1.266\,06\, T_0(x) &+ 0.271\,48\, T_2(x) \\ &+ 0.005\,47\, T_4(x) + 0.000\,04\, T_6(x).\end{aligned} \tag{13.90a}$$

Since $|T_6(x)| \leqslant 1$, this last term in the Chebyshev series may be dropped—to within our desired accuracy.

Now, returning to a power series form

$$\begin{aligned}\cosh x &= 1.266\,06\, T_0 + 0.271\,48\, T_2 + 0.005\,47\, T_4 \\ &= 1.000\,05 + 0.499\,20\, x^2 + 0.043\,76\, x^4.\end{aligned} \tag{13.90b}$$

This (truncated and rearranged) series will give us the *same accuracy* (to ± 0.0001) as the original series through x^6. The difference is that the error in the original form was concentrated near the ends of the range. In Eq. 13.90*b*, the error is distributed more uniformly across the entire range.

This is not the same as a least squares fit. In a least squares fit, one attempts to minimize the average of the square of the error. This does not preclude occasional large errors. Here, in contrast, we seek to minimize the *maximum* (absolute value) of the error, or at least hold the error within some predetermined limit. For a given number of powers, this can be done better with the Chebyshev adjusted coefficients (Eq. 13.90*b*) than with the original power series (Eq. 13.90, through x^6).

This type of Chebyshev truncation was used in the polynomial fits for the factorial function, $x!$, [0, 1], Section 10.3.

EXERCISES

13.3.1 Given

$$(1 - x^2)\, U_n''(x) - 3x\, U_n'(x) + n(n+2)\, U_n(x) = 0,$$

show that $V_n(x)$ satisfies

$$(1 - x^2)\, V_n''(x) - x\, V_n'(x) + n^2\, V_n(x) = 0,$$

which is Chebyshev's equation.

13.3.2 Show that the Wronskian of $T_n(x)$ and $V_n(x)$ is given by

$$T_n(x)\, V_n'(x) - T_n'(x)\, V_n(x) = -\frac{n}{(1 - x^2)^{1/2}}.$$

This verifies that T_n and V_n ($n \neq 0$) are independent solutions of Eq. 13.79. Conversely, for $n = 0$, we do not have linear independence. What happens at $n = 0$? Where is the "second" solution?

13.3.3 Another Chebyshev generating function is

$$\frac{1 - xt}{1 - 2xt + t^2} = \sum_{n=0}^{\infty} W_n(x) t^n, \qquad |t| < 1.$$

How is $W_n(x)$ related to our $T_n(x)$ and $U_n(x)$?

13.3.4 The curves plotted in Fig. 13.5 suggest that $|T_n(x)| \leqslant 1$ for $-1 \leqslant x \leqslant 1$ (real x). Prove this.

13.3.5 Establish the following bounds, $-1 \leq x \leq 1$:
(a) $|U_n(x)| \leq n + 1$,
(b) $\left| \dfrac{d}{dx} T_n(x) \right| \leq n^2$.

13.3.6 A number of equations relate the two types of Chebyshev polynomials. As examples show that

$$T_n(x) = U_n(x) - x\, U_{n-1}(x)$$

and

$$(1 - x^2)\, U_n(x) = x T_{n+1}(x) - T_{n+2}(x).$$

13.3.7 Show that

$$\frac{dV_n(x)}{dx} = -n \frac{T_n(x)}{\sqrt{1 - x^2}},$$

(a) using the trigonometric forms of V_n and T_n,
(b) using the Rodrigues representation.

13.3.8 Derive the Rodrigues representation of $T_n(x)$.

$$T_n(x) = \frac{(-1)^n \pi^{1/2}(1-x^2)^{1/2}}{2^n(n-\frac{1}{2})!} \frac{d^n}{dx^n} [(1-x^2)^{n-1/2}].$$

Hints. One possibility is to use the hypergeometric function relation

$${}_2F_1(a, b, c; z) = (1-z)^{-a}\, {}_2F_1\left(a, c-b, c; \frac{-z}{1-z}\right),$$

with $z = (1-x)/2$. An alternate approach is to develop a first-order differential equation for $y = (1-x^2)^{n-1/2}$. Repeated differentiation of this equation leads to the Chebyshev equation.

13.3.9 Show that $V_n(x)$ satisfies the same three-term recurrence relation as $T_n(x)$ (Eq. 13.70).

13.3.10 Verify the series solutions for $T_n(x)$ and $U_n(x)$ (Eqs. 13.83 and 13.84).

13.3.11 Transform the series form of $T_n(x)$, Eq. 13.83, into an *ascending* power series.

Ans.

$$T_{2n}(x) = (-1)^n n \sum_{m=0}^{n} (-1)^m \frac{(n+m-1)!}{(n-m)!(2m)!} (2x)^{2m},$$

$$T_{2n+1}(x) = (-1)^n \frac{2n+1}{2} \sum_{m=0}^{n} (-1)^m \frac{(n+m)!}{(n-m)!(2m+1)!} (2x)^{2m+1}.$$

13.3.12 Rewrite the series form of $U_n(x)$, Eq. 13.84, as an ascending power series.

Ans.

$$U_{2n}(x) = (-1)^n \sum_{m=0}^{n} (-1)^m \frac{(n+m)!}{(n-m)!(2m)!} (2x)^{2m},$$

$$U_{2n+1}(x) = (-1)^n \sum_{m=0}^{u} (-1)^m \frac{(n+m+1)!}{(n-m)!\,(2m+1)!} (2x)^{2m+1}.$$

13.3.13 From

$$T_n(x) + i\, V_n(x) = [x + i(1-x^2)^{1/2}]^n$$

let $x = \cos\theta$ and show that

(a) $T_n(x) = \cos n\theta = \cos(n \cos^{-1} x)$, (b) $V_n(x) = \sin n\theta = \sin(n\cos^{-1}x)$,

(c) $U_n(x) = \dfrac{\sin[(n+1)\theta]}{\sin\theta} = \dfrac{\sin[(n+1)\cos^{-1} x]}{\sin(\cos^{-1}x)}$.

The result in part (a) shows that the expansion of an even function in a series of $T_n(x)$ is equivalent to a Fourier cosine approximation (cf. Section 14.1).

13.3.14 Derive

(a) $T_{n+1}(x) + T_{n-1}(x) = 2x\, T_n(x)$, (b) $T_{m+n}(x) + T_{m-n}(x) = 2T_m(x)\, T_n(x)$,

from the "corresponding" cosine identities.

13.3.15 Starting with $x = \cos\theta$ and $T_n(\cos\theta) = \cos n\theta$, expand

$$x^k = \left(\frac{e^{i\theta} + e^{-i\theta}}{2}\right)^k$$

and show that

$$x^k = \frac{1}{2^{k-1}}\left[T_k(x) + \binom{k}{1} T_{k-2}(x) + \binom{k}{2} T_{k-4}(x) + \cdots\right],$$

the series in brackets terminating with $\binom{k}{m} T_1(x)$ for $k = 2m+1$ or $\frac{1}{2}\binom{k}{m} T_0$ for $k = 2m$.

13.3.16 (a) From the differential equation for T_n (in self-adjoint form) show that

$$\int_{-1}^{1} \frac{dT_m(x)}{dx}\frac{dT_n(x)}{dx}(1-x^2)^{1/2}\,dx = 0, \qquad m \neq n.$$

(b) Confirm the above result by showing that

$$\frac{dT_n(x)}{dx} = nU_{n-1}(x).$$

13.4 Hypergeometric Functions

In Chapter 8 the hypergeometric equation[1]

$$x(1-x)\,y''(x) + [c-(a+b+1)x]\,y'(x) - ab\,y(x) = 0 \qquad (13.91)$$

was introduced as a canonical form of a linear second-order differential equation with regular singularities at $x = 0$, 1, and ∞. One solution is

$$y(x) = {}_2F_1(a, b, c; x) \qquad (13.91a)$$
$$= 1 + \frac{a \cdot b}{c}\frac{x}{1!} + \frac{a(a+1)b(b+1)}{c(c+1)}\frac{x^2}{2!} + \cdots, \qquad c \neq 0, -1, -2, -3, \cdots,$$

which is known as the hypergeometric function or hypergeomeric series. The range of convergence $|x| < 1$ and $x = 1$, for $c > a + b$, and $x = -1$, for $c > a + b - 1$. In terms of the often-used Pochhammer symbol,

$$(a)_n = a(a+1)(a+2)\cdots(a+n-1) = \frac{(a+n-1)!}{(a-1)!}, \qquad (13.92)$$
$$(a)_0 = 1,$$

the hypergeometric function becomes

$${}_2F_1(a, b, c; x) = \sum_{n=0}^{\infty} \frac{(a)_n(b)_n}{(c)_n}\frac{x^n}{n!}. \qquad (13.93)$$

In this form, the subscripts 2 and 1 become clear. The leading subscript 2 indicates that two Pochhammer symbols appear in the numerator and the final subscript 1 indicates one Pochhammer symbol in the denominator.[2] The confluent hypergeometric function ${}_1F_1$ with one Pochhammer symbol in the numerator and one in the denominator appears in Section 13.5

[1] This is sometimes called Gauss's differential equation. The solutions then become Gauss functions.

[2] The Pochhammer symbol is often useful in other expressions involving factorials, for instance,

$$(1-z)^{-a} = \sum_{n=0}^{\infty}(a)_n\, z^n/n! \qquad |z| < 1.$$

From the form of Eq. 13.91a we see that the parameter c may not be zero or a negative integer. On the other hand, if a or b equals 0 or a negative integer, the series terminates and the hypergeometric function becomes a simple polynomial.

Many more or less elementary functions can be represented by the hypergeometric function.[1] We find

$$\ln(1+x) = x\,{}_2F_1(1, 1, 2; -x). \tag{13.94}$$

For the complete elliptic integrals K and E

$$\begin{aligned} K &= \int_0^{\pi/2} (1 - k^2 \sin^2\theta)^{-1/2}\, d\theta \\ &= \frac{\pi}{2}\,{}_2F_1\left(\frac{1}{2}, \frac{1}{2}, 1; k^2\right), \end{aligned} \tag{13.95}$$

$$\begin{aligned} E &= \int_0^{\pi/2} (1 - k^2 \sin^2\theta)^{1/2}\, d\theta \\ &= \frac{\pi}{2}\,{}_2F_1\left(\frac{1}{2}, -\frac{1}{2}, 1; k^2\right). \end{aligned} \tag{13.96}$$

The explicit series forms and other properties of the elliptic integrals are developed in Section 5.8.

The hypergeometric equation as a second-order linear differential equation has a second independent solution. The usual form is

$$y(x) = x^{1-c}\,{}_2F_1(a+1-c, b+1-c, 2-c; x), \qquad c \neq 2, 3, 4, \cdots. \tag{13.97}$$

The reader may show (Exercise 13.4.1) that if c is an integer either the two solutions coincide or (barring a rescue by integral a or integral b) one of the solutions will blow up. In such a case the second solution is expected to include a logarithmic term.

Alternate forms of the hypergeometric equation include

$$\begin{aligned} &(1 - z^2)\frac{d^2}{dz^2}\, y\left(\frac{1-z}{2}\right) - [(a+b+1)z - (a+b+1-2c)]\frac{d}{dz}\, y\left(\frac{1-z}{2}\right) \\ &\quad - ab\, y\left(\frac{1-z}{2}\right) = 0, \end{aligned} \tag{13.98}$$

$$\begin{aligned} &(1 - z^2)\frac{d^2}{dz^2}\, y(z^2) - \left[(2a+2b+1)z + \frac{1-2c}{z}\right]\frac{d}{dz}\, y(z^2) \\ &\quad - 4ab\, y(z^2) = 0. \end{aligned} \tag{13.99}$$

Contiguous function relations. The parameters a, b, and c enter in the same way as the parameter n of Bessel, Legendre, and other special functions. As we found with these functions, we expect recurrence relations involving unit changes in the parameters a, b, and c. The usual nomenclature for the hypergeometric

[1] With three parameters, a, b, and c, we can represent almost anything.

functions in which *one* parameter changes by + or -1 is "contiguous function." Generalizing this term to include simultaneous unit changes in more than one parameter, we find 26 functions contiguous to ${}_2F_1(a, b, c; x)$. Taking them two at a time, we can develop the formidable total of 325 equations among the contiguous functions. One typical example is

$$\begin{aligned}(a-b)\{c(a+b-1)+1-a^2-b^2+[(a-b)^2-1](1-x)\}\,{}_2F_1(a,b,c;x)\\ =(c-a)(a-b+1)b\,{}_2F_1(a-1,b+1,c;x)\qquad(13.100)\\ +(c-b)(a-b-1)a\,{}_2F_1(a+1,b-1,c;x).\end{aligned}$$

Another contiguous function relation appears in Ex. 13.4.10.

Hypergeometric representations. Since the Gegenbauer equation (13.81) in Section 13.3 is a special case of Eq. 13.91, we see that Gegenbauer functions (and Legendre and Chebyshev functions) may be expressed as hypergeometric functions. For the Gegenbauer function we obtain

$$T_n^\beta(x)=\frac{(n+2\beta)!}{2^\beta n!\,\beta!}\,{}_2F_1\left(-n,n+2\beta+1,1+\beta;\frac{1-x}{2}\right).\qquad(13.101)$$

For Legendre and associated Legendre functions

$$P_n(x)={}_2F_1\left(-n,n+1,1;\frac{1-x}{2}\right),\qquad(13.102)$$

$$P_n^m(x)=\frac{(n+m)!}{(n-m)!}\frac{(1-x^2)^{m/2}}{2^m m!}\,{}_2F_1\left(m-n,m+n+1,m+1;\frac{1-x}{2}\right).\qquad(13.103)$$

Alternate forms are

$$\begin{aligned}P_{2n}(x)&=(-1)^n\frac{(2n)!}{2^{2n}n!\,n!}\,{}_2F_1\left(-n,n+\frac{1}{2},\frac{1}{2};x^2\right)\\ &=(-1)^n\frac{(2n-1)!!}{(2n)!!}\,{}_2F_1\left(-n,n+\frac{1}{2},\frac{1}{2};x^2\right),\qquad(13.104)\end{aligned}$$

$$\begin{aligned}P_{2n+1}(x)&=(-1)^n\frac{(2n+1)!}{2^{2n}n!\,n!}\,x\,{}_2F_1\left(-n,n+\frac{3}{2},\frac{3}{2};x^2\right)\\ &=(-1)^n\frac{(2n+1)!!}{(2n)!!}\,x\,{}_2F_1\left(-n,n+\frac{3}{2},\frac{3}{2};x^2\right).\qquad(13.105)\end{aligned}$$

In terms of hypergeometric functions, the Chebyshev functions become

$$T_n(x)={}_2F_1\left(-n,n,\frac{1}{2};\frac{1-x}{2}\right),\qquad(13.106)$$

$$U_n(x)=(n+1)\,{}_2F_1\left(-n,n+2,\frac{3}{2};\frac{1-x}{2}\right),\qquad(13.107)$$

$$V_n(x) = \sqrt{1-x^2}\, n\, {}_2F_1\left(-n+1, n+1, \frac{3}{2}; \frac{1-x}{2}\right). \tag{13.108}$$

The leading factors are determined by direct comparison of complete power series, comparison of coefficients of particular powers of the variable, or evaluation at $x = 0$ or 1, etc.

EXERCISES

13.4.1 (a) For c, an integer, and a and b, nonintegral, show that

$${}_2F_1(a, b, c; x) \text{ and } x^{1-c}\, {}_2F_1(a+1-c, b+1-c, 2-c; x)$$

yield only one solution to the hypergeometric equation.

(b) What happens if a is an integer, say $a = -1$, and $c = -2$?

13.4.2 Find the Legendre, Chebyshev I, and Chebyshev II recurrence relations corresponding to the contiguous hypergeometric function equation (13.100).

13.4.3 Write the following polynomials as hypergeometric functions of argument x^2.

(a) $T_{2n}(x)$; (b) $x^{-1}T_{2n+1}(x)$; (c) $U_{2n}(x)$; (d) $x^{-1}U_{2n+1}(x)$.

Ans. (a) $T_{2n}(x) = (-1)^n\, {}_2F_1(-n, n, \frac{1}{2}; x^2)$.

(b) $x^{-1}T_{2n+1}(x) = (-1)^n (2n+1)\, {}_2F_1(-n, n+1, \frac{3}{2}; x^2)$.

(c) $U_{2n}(x) = (-1)^n\, {}_2F_1(-n, n+1, \frac{1}{2}; x^2)$.

(d) $x^{-1}U_{2n+1}(x) = (-1)^n (2n+2)\, {}_2F_1(-n, n+2, \frac{3}{2}; x^2)$.

13.4.4 Derive or verify the leading factor in the hypergeometric representations of the Chebyshev functions.

13.4.5 Verify that the Legendre function of the second kind, $Q_\nu(z)$, is given by

$$Q_\nu(z) = \frac{\pi^{1/2}\nu!}{(\nu+\frac{1}{2})!(2z)^{\nu+1}}\, {}_2F_1\left(\frac{\nu}{2}+\frac{1}{2}, \frac{\nu}{2}+1, \frac{\nu}{2}+\frac{3}{2}; z^{-2}\right)$$

$$|z| > 1, \quad |\arg z| < \pi, \qquad \nu \neq -1, -2, -3, \ldots .$$

13.4.6 Analogous to the incomplete gamma function one may define an incomplete beta function by

$$B_x(a, b) = \int_0^x t^{a-1}(1-t)^{b-1}\, dt.$$

Show that

$$B_x(a, b) = a^{-1}x^a\, {}_2F_1(a, 1-b, a+1; x).$$

13.4.7 Verify the integral representation

$${}_2F_1(a, b, c; z) = \frac{\Gamma(c)}{\Gamma(b)\Gamma(c-b)} \int_0^1 t^{b-1}(1-t)^{c-b-1}(1-tz)^{-a}\, dt.$$

What restrictions must you place on the parameters b and c, on the variable z?

Note: The restriction on $|z|$ can be dropped—analytic continuation. For nonintegral a, the real axis in the z-plane 1 to ∞ is a cut line.

Hint. The integral is suspiciously like a beta function and can be expanded into a series of beta functions.

Ans. $\mathcal{R}(c) > \mathcal{R}(b) > 0$, and $|z| < 1$.

13.4.8 Prove that

$$ {}_2F_1(a, b, c; 1) = \frac{\Gamma(c)\Gamma(c-a-b)}{\Gamma(c-a)\Gamma(c-b)}, \qquad c \neq 0, -1, -2, \ldots, \qquad c > a + b $$

Hint. Here is a chance to use the integral representation, Ex. 13.4.7.

13.4.9 Prove that

$$ {}_2F_1(a, b, c; x) = (1-x)^{-a}\, {}_2F_1\left(a, c-b, c; \frac{-x}{1-x}\right). $$

Hint. Try the integral representation, Ex. 13.4.7.
Note: This relation is useful in developing a Rodrigues representation of $T_n(x)$ (cf. Ex. 13.3.8).

13.4.10 Verify

$$ {}_2F_1(-n, b, c; 1) = \frac{(c-b)_n}{(c)_n}. $$

Hint. Here is a chance to use the contiguous function relation

$[2a - c + (b-a)x]\, F(a, b, c; x) = a(1-x)\, F(a+1, b, c; x) - (c-a)\, F(a-1, b, c; x)$
and mathematical induction.

13.5 Confluent Hypergeometric Functions

The confluent hypergeometric equation[1]

$$ x\, y''(x) + (c - x)\, y'(x) - a\, y(x) = 0 \tag{13.109} $$

may be obtained from the hypergeometric equation of Section 13.4 by merging two of its singularities. The resulting equation has a regular singularity at $x = 0$ and an irregular one at $x = \infty$. One solution of the confluent hypergeometric equation is

$$ \begin{aligned} y(x) &= {}_1F_1(a, c; x) = M(a, c; x) \\ &= 1 + \frac{a}{c}\frac{x}{1!} + \frac{a(a+1)}{c(c+1)}\frac{x^2}{2!} + \cdots, \qquad c \neq 0, -1, -2, \cdots. \end{aligned} \tag{13.110} $$

This solution is convergent for all finite x (or z). In terms of the Pochhammer symbols, we have

$$ M(a, c; x) = \sum_{n=0}^{\infty} \frac{(a)_n}{(c)_n}\frac{x^n}{n!}. \tag{13.111} $$

Clearly $M(a, c; x)$ becomes a polynomial if the parameter a is 0 or a negative integer. Numerous more or less elementary functions may be represented by the confluent hypergeometric function. Examples are the error function and the incomplete gamma function.

$$ \mathrm{erf}(x) = \frac{2}{\pi^{1/2}} \int_0^x e^{-t^2}\, dt = \frac{2}{\pi^{1/2}}\, x\, M\left(\frac{1}{2}, \frac{3}{2}; -x^2\right), \tag{13.112} $$

[1] This is often called Kummer's equation. The solutions, then, are Kummer functions.

$$\begin{aligned}\gamma(a, x) &= \int_0^x e^{-t} t^{a-1}\, dt \\ &= a^{-1} x^a M(a, a+1; -x), \qquad \mathscr{R}(a) > 0.\end{aligned} \tag{13.113}$$

The error function and the incomplete gamma function are discussed further in Section 10.5.

A second solution of Eq. 13.109 is given by

$$y(x) = x^{1-c} M(a+1-c, 2-c; x), \qquad c \neq 2, 3, 4, \ldots. \tag{13.114}$$

The standard form of the second solution of Eq. 13.109 is a linear combination of Eqs. 13.110 and 13.114.

$$U(a, c; x) = \frac{\pi}{\sin \pi c}\left[\frac{M(a, c; x)}{(a-c)!(c-1)!} - \frac{x^{1-c} M(a+1-c, 2-c; x)}{(a-1)!(1-c)!}\right]. \tag{13.115}$$

Note the resemblance to our definition of the Neumann function, Eq. 11.60.

An alternate form of the confluent hypergeometric equation that will be useful later is obtained by changing the independent variable from x to x^2.

$$\frac{d^2}{dx^2} y(x^2) + \left[\frac{2c-1}{x} - 2x\right]\frac{d}{dx} y(x^2) - 4a y(x^2) = 0. \tag{13.116}$$

As with the hypergeometric functions, contiguous functions exist in which the parameters a and c are changed by ± 1. Including the cases of simultaneous changes in the two parameters,[1] we have eight possibilities. Taking the original function and pairs of the contiguous functions, we can develop a total of 28 equations.[2]

Integral representations. It is frequently convenient to have the confluent hypergeometric functions in integral form. We find (Ex. 13.5.10)

$$M(a, c; x) = \frac{\Gamma(c)}{\Gamma(a)\Gamma(c-a)} \int_0^1 e^{xt} t^{a-1} (1-t)^{c-a-1}\, dt,$$
$$\mathscr{R}(c) > \mathscr{R}(a) > 0, \tag{13.117a}$$

$$U(a, c; x) = \frac{1}{\Gamma(a)} \int_0^\infty e^{-xt} t^{a-1} (1+t)^{c-a-1}\, dt,$$
$$\mathscr{R}(x) > 0, \mathscr{R}(a) > 0. \tag{13.117b}$$

Three important techniques for deriving or verifying integral representations are as follows:

1. Transformation of generating function expansions and Rodrigues representations: The Bessel and Legendre functions provide examples of this approach.
2. Direct integration to yield a series: This direct technique is useful for a Bessel

[1] Slater refers to these as associated functions.

[2] The recurrence relations for Bessel, Hermite, and Laguerre functions are special cases of these equations.

function representation (Ex. 11.1.18) and a hypergeometric integral (Ex. 13.4.7).

3. (a) Verification that the integral representation satisfies the differential equation. (b) Exclusion of the other solution. (c) Verification of normalization. This is the method used in Section 11.6 to establish an integral representation of the modified Bessel function, $K_\nu(z)$. It will work here to establish Eqs. 13.117*a* and 13.117*b*.

Bessel and modified Bessel functions. Kummer's first formula,

$$M(a, c; x) = e^x M(c - a, c; -x), \tag{13.118}$$

is useful in representing the Bessel and modified Bessel functions. The formula may be verified by series expansion or by use of an integral representation (cf. Ex. 13.5.10).

As expected from the form of the confluent hypergeometric equation and the character of its singularities, the confluent hypergeometric functions are useful in representing a number of the special functions of mathematical physics. For the Bessel functions

$$J_\nu(x) = \frac{e^{-ix}}{\nu!}\left(\frac{x}{2}\right)^\nu M\left(\nu + \frac{1}{2}, 2\nu + 1; 2ix\right), \tag{13.119a}$$

whereas for the modified Bessel functions of the first kind,

$$I_\nu(x) = \frac{e^{-x}}{\nu!}\left(\frac{x}{2}\right)^\nu M\left(\nu + \frac{1}{2}, 2\nu + 1; 2x\right). \tag{13.119b}$$

Hermite functions. The Hermite functions are given by

$$H_{2n}(x) = (-1)^n \frac{(2n)!}{n!} M\left(-n, \frac{1}{2}; x^2\right), \tag{13.120}$$

$$H_{2n+1}(x) = (-1)^n \frac{2(2n+1)!}{n!} x\, M\left(-n, \frac{3}{2}; x^2\right), \tag{13.121}$$

using Eq. 13.116.

Comparing the Laguerre differential equation with the confluent hypergeometric equation, we have

$$L_n(x) = M(-n, 1; x). \tag{13.122}$$

The constant is fixed as unity by noting Eq. 13.35 for $x = 0$. For the associated Laguerre functions

$$\begin{aligned} L_n^m(x) &= (-1)^m \frac{d^m}{dx^m} L_{n+m}(x) \\ &= \frac{(n+m)!}{n!\,m!} M(-n, m+1; x). \end{aligned} \tag{13.123}$$

Alternate verification is obtained by comparing Eq. 13.123 with the power series solution (Eq. 13.41 of Section 13.2). Note that in the hypergeometric form, as distinct from a Rodrigues representation, the indices n and m need not be integers and, if they are not integers, $L_n^m(x)$ will not be a polynomial.

Miscellaneous cases. There are certain advantages in expressing our special functions in terms of hypergeometric and confluent hypergeometric functions. If the general behavior of the latter functions is known, the behavior of the special functions we have investigated follows as a series of special cases. This may be useful in determining asymptotic behavior or evaluating normalization integrals. The asymptotic behavior of $M(a, c; x)$ and $U(a, c; x)$ may be conveniently obtained from integral representations of these functions, Eqs. 13.117*a* and 13.117*b*. The further advantage is that the relations between the special functions are clarified. For instance, an examination of Eqs. 13.120, 13.121, and 13.123 suggests that the Laguerre and Hermite functions are related.

The confluent hypergeometric equation (13.109) is clearly not self-adjoint. For this and other reasons it is convenient to define

$$M_{k\mu}(x) = e^{-x/2}x^{\mu+1/2}\,M(\mu - k + \tfrac{1}{2}, 2\mu + 1; x). \tag{13.124}$$

This new function $M_{k\mu}(x)$ is a Whittaker function which satisfies the self-adjoint equation

$$M''_{k\mu}(x) + \left(-\frac{1}{4} + \frac{k}{x} + \frac{\frac{1}{4} - \mu^2}{x^2}\right) M_{k\mu}(x) = 0. \tag{13.125}$$

The corresponding second solution is

$$W_{k\mu}(x) = e^{-x/2}x^{\mu+1/2}\,U(\mu - k + \tfrac{1}{2}, 2\mu + 1; x). \tag{13.126}$$

EXERCISES

13.5.1 Verify the confluent hypergeometric representation of the error function

$$\text{erf}\,(x) = \frac{2x}{\pi^{1/2}}\,M(\tfrac{1}{2}, \tfrac{3}{2}, -x^2).$$

13.5.2 Show that the Fresnel integrals $C(x)$ and $S(x)$ of Exercise 5.11.2 may be expressed in terms of the confluent hypergeometric function as

$$C(x) + iS(x) = xM\left(\tfrac{1}{2}, \tfrac{3}{2}; \frac{i\pi x^2}{2}\right).$$

13.5.3 By direct differentiation and substitution verify that

$$y = ax^{-a}\int_0^x e^{-t}t^{a-1}\,dt = ax^{-a}\gamma(a, x)$$

actually does satisfy

$$xy'' + (a + 1 + x)y' + ay = 0.$$

13.5.4 Show that the modified Bessel function of the second kind $K_\nu(x)$ is given by

$$K_\nu(x) = \pi^{1/2} e^{-x}(2x)^\nu\, U(\nu + \tfrac{1}{2}, 2\nu + 1; 2x).$$

13.5.5 Show that the cosine and sine integrals of Section 10.5 may be expressed in terms of confluent hypergeometric functions as

$$Ci(x) + i\, si(x) = -e^{ix}\, U(1, 1; -ix).$$

This relation is useful in numerical computation of $Ci(x)$ and $si(x)$ for large values of x.

13.5.6 Verify the confluent hypergeometric form of the Hermite polynomial $H_{2n+1}(x)$ (Eq. 13.121) by showing that

(a) $H_{2n+1}(x)/x$ satisfies the confluent hypergeometric equation with $a = -n$, $c = \frac{3}{2}$ and argument x^2,

(b) $\lim\limits_{x \to 0} \dfrac{H_{2n+1}(x)}{x} = (-1)^n \dfrac{2(2n+1)!}{n!}$.

13.5.7 Show that the contiguous confluent hypergeometric function equation,

$$(c - a)\, M(a - 1, c; x) + (2a - c + x)\, M(a, c; x) - a\, M(a + 1, c; x) = 0,$$

leads to the associated Laguerre function recurrence relation (Eq. 13.44).

13.5.8 Verify the Kummer transformations:

(a) $M(a, c; x) = e^x\, M(c - a, c; -x)$, (b) $U(a, c; x) = x^{1-c}\, U(a - c + 1, 2 - c; x)$.

13.5.9 Prove that

(a) $$\frac{d^n}{dx^n} M(a, c; x) = \frac{(a)_n}{(b)_n} M(a + n, b + n; x),$$

(b) $$\frac{d^n}{dx^n} U(a, c; x) = (-1)^n (a)_n U(a + n, c + n; x).$$

13.5.10 Verify the following integral representations:

(a) $$M(a, c; x) = \frac{\Gamma(c)}{\Gamma(a)\Gamma(c-a)} \int_0^1 e^{xt} t^{a-1} (1 - t)^{c-a-1}\, dt, \quad \mathscr{R}(c) > \mathscr{R}(a) > 0,$$

(b) $$U(a, c; x) = \frac{1}{\Gamma(a)} \int_0^\infty e^{-xt} t^{a-1} (1 + t)^{c-a-1}\, dt, \quad \mathscr{R}(x) > 0, \quad \mathscr{R}(a) > 0.$$

13.5.11 From the integral representation of $M(a,c;x)$, Ex. 13.5.10(a), show that

$$M(a, c; x) = e^x M(c - a, c; -x).$$

Hint. Replace the variable of integration t by $1 - s$ to release a factor e^x from the integral.

13.5.12 From the integral representation of $U(a, c; x)$, Ex. 13.5.10(b), show that the exponential integral is given by

$$E_1(x) = e^{-x} U(1, 1; x).$$

Hint. Replace the variable of integration t in $E_1(x)$ by $x(1 + s)$.

13.5.13 From the integral representations of $M(a, c; x)$ and $U(a, c; x)$ in Ex. 13.5.10 develop asymptotic expansions of

(a) $M(a, c; x)$, (b) $U(a, c; x)$.

Hint. You can use the technique that was employed with $K_\nu(z)$, Section 11.6.

Ans.

(a) $\dfrac{\Gamma(c)}{\Gamma(a)} \dfrac{e^x}{x^{c-a}} \left\{1 + \dfrac{(1-a)(c-a)}{x} + \cdots\right\}.$

(b) $\dfrac{1}{x^a} \left\{1 - \dfrac{a(1+a-c)}{x} + \cdots\right\}.$

13.5.14 Show that the Wronskian of the two confluent hypergeometric functions, $M(a, c; x)$ and $U(a, c; x)$ is given by

$$MU' - M'U = -\frac{(c-1)!}{(a-1)!} \frac{e^x}{x^c}.$$

What happens if a is 0 or a negative integer?

13.5.15 The Coulomb wave equation (radial part of the Schrödinger wave equation with Coulomb potential) is

$$\frac{d^2y}{d\rho^2} + \left[1 - \frac{2\eta}{\rho} - \frac{L(L+1)}{\rho^2}\right] y = 0.$$

Show that a regular solution, $y = F_L(\eta, \rho)$, is given by

$$F_L(\eta, \rho) = C_L(\eta)\rho^{L+1}e^{-i\rho}M(L + 1 - i\eta, 2L + 2; 2i\rho).$$

13.5.16 (a) Show that the radial part of the hydrogen wave function, Eq. 13.60, may be written

$$e^{-\alpha r/2}(\alpha r)^L L_{n-L-1}^{2L+1}(\alpha r) = \frac{(n+L)!}{(n-L-1)!\,(2L+1)!} e^{-\alpha r/2}(\alpha r)^L M(L + 1 - n, 2L + 2; \alpha r).$$

(b) It was assumed above that the total (kinetic + potential) energy E of the electron was negative. Rewrite the (unnormalized) radial wave function for the free electron $E > 0$.

Ans. $e^{+i\alpha r/2}(\alpha r)^L M(L + 1 - in, 2L + 2, -i\alpha r)$, outgoing wave. This representation provides a powerful alternative technique for the calculation of photoionization and recombination coefficients.

13.5.17 Show that the Laplace transform of $M(a, c; x)$ is

$$\mathscr{L}\{M(a, c; x)\} = \frac{1}{s}\, {}_2F_1\left(a, 1, c; \frac{1}{s}\right).$$

13.5.18 Evaluate

(a) $\displaystyle\int_0^\infty [M_{k\mu}(x)]^2\, dx$

(b) $\displaystyle\int_0^\infty [M_{k\mu}(x)]^2 \frac{dx}{x},$

where $2\mu = 0, 1, 2, \ldots, k - \mu - \frac{1}{2} = 0, 1, 2, \ldots, a > -2\mu - 1.$

Ans. (*a*) $(2\mu)!\,2k$.
(b) $(2\mu)!$.
(c) $(2\mu)!\,(2k)^a$.

REFERENCES

ABRAMOWITZ, M., and I. A. STEGUN, editors, *Handbook of Mathematical Functions*. Washington, D.C.: National Bureau of Standards, Applied Mathematics Series-55 (1964). Chapter 22 is a detailed summary of the properties and representations of orthogonal polynomials. Other chapters summarize properties of Bessel, Legendre, hypergeometric, and confluent hypergeometric functions and much more.

LEBEDEV, N. N., *Special Functions and their Applications*. Translated by R. A. Silverman. Englewood Cliffs, New Jersey: Prentice-Hall (1965).

LUKE, Y. L., *The Special Functions and Their Approximations*. Academic Press: New York (1969). Two volumes. Volume 1 is a thorough theoretical treatment of gamma functions, hypergeometric functions, confluent hypergeometric functions, and related functions. Volume 2 develops approximations and other techniques for numerical work.

MAGNUS, W., F. OBERHETTINGER, and R. P. SONI, *Formulas and Theorems for the Special Functions of Mathematical Physics*. Springer: New York (1966). A new and enlarged edition. An excellent summary of just what the title says, including the topics of Chapters 10–13.

RAINVILLE, E. D., *Special Functions*. New York: Macmillan (1960). A coherent, comprehensive account of almost all the special functions of mathematical physics that the reader is likely to encounter.

SANSONE, G. *Orthogonal Functions*. Translated by A. H. Diamond. New York: Interscience Publishers (1959).

SLATER, L. J., *Confluent Hypergeometric Functions*. Cambridge: Cambridge University Press (1960). A clear and detailed development of the properties of the confluent hypergeometric functions and of relations of the confluent hypergeometric equation to other differential equations of mathematical physics.

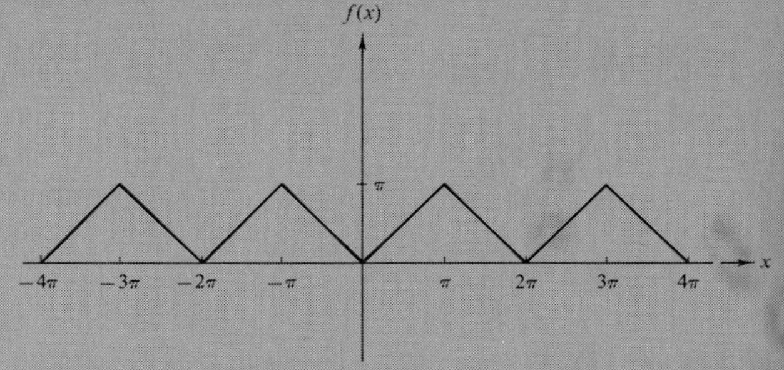

CHAPTER 14

FOURIER SERIES

14.1 General Properties

Fourier series. A Fourier series may be defined as an expansion of a function or representation of a function in a series of sines and cosines such as

$$f(x) = \frac{a_0}{2} + \sum_{n=1}^{\infty} a_n \cos nx + \sum_{n=1}^{\infty} b_n \sin nx. \tag{14.1}$$

The conditions imposed on $f(x)$ to make this equation valid are that $f(x)$ has only a finite number of finite discontinuities and only a finite number of extreme values, maxima, and minima.[1] Functions satisfying these conditions may be called piecewise regular. The conditions themselves are known as the Dirichlet conditions. Although there are some functions that do not obey these Dirichlet conditions, they may well be labeled pathological for purposes of Fourier expansions. In the vast majority of physical problems involving a Fourier series these conditions will be satisfied.

Expressing $\cos nx$ and $\sin nx$ in exponential form, we may rewrite Eq. 14.1 as

$$f(x) = \sum_{n=-\infty}^{\infty} c_n e^{inx} \tag{14.2}$$

in which

$$\begin{aligned} c_n &= \tfrac{1}{2}(a_n - ib_n), \\ c_{-n} &= \tfrac{1}{2}(a_n + ib_n), \qquad n > 0, \end{aligned} \tag{14.3}$$

and

$$c_0 = \tfrac{1}{2}a_0 .$$

[1] These conditions are *sufficient* but not *necessary*.

If we expand $f(z)$ in a Laurent series[1] (assuming $f(z)$ is analytic),

$$f(z) = \sum_{n=-\infty}^{\infty} d_n z^n. \tag{14.4}$$

On the unit circle $z = e^{i\theta}$ and

$$f(z) = f(e^{i\theta}) = \sum_{n=-\infty}^{\infty} d_n e^{in\theta}. \tag{14.5}$$

Completeness—Laurent series. The Laurent expansion on the unit circle (Eq. 14.5) has the same form as the complex Fourier series (Eq. 14.2), which shows the equivalence between the two expansions. Since the Laurent series as a power series has the property of completeness, we see that the Fourier functions, e^{inx}, form a complete set; that is the series (Eqs. 14.1 and 14.2) is able to represent any piecewise regular function in the range $(0, 2\pi)$. This range and the related question of periodicity are considered in the next section.

The Fourier expansion and the completeness property may be expected, for the functions $\sin nx$, $\cos nx$, e^{inx} are all eigenfunctions of a self-adjoint linear differential equation,

$$y'' + n^2 y = 0. \tag{14.6}$$

We obtain orthogonal eigenfunctions for different values of the eigenvalue n by choosing the interval $[0, p\pi]$, p an integer, to satisfy the boundary conditions in the Sturm-Liouville theory (Chapter 9). If we further choose $p = 2$, the different eigenfunctions for the same eigenvalue n may be orthogonal. We have

$$\int_0^{2\pi} \sin mx \sin nx \, dx = \begin{cases} \pi\delta_{m,n}, & m \neq 0, \\ 0, & m = 0, \end{cases} \tag{14.7}$$

$$\int_0^{2\pi} \cos mx \cos nx \, dx = \begin{cases} \pi\delta_{m,n}, & m \neq 0, \\ 2\pi, & m = n = 0, \end{cases} \tag{14.8}$$

$$\int_0^{2\pi} \sin mx \cos nx \, dx = 0 \quad \text{for all integral } m \text{ and } n. \tag{14.9}$$

Note carefully that any interval $x_0 \leqslant x \leqslant x_0 + 2\pi$ will be equally satisfactory. Frequently we shall use $x_0 = -\pi$ to obtain the interval $-\pi \leqslant x \leqslant \pi$. For the complex eigenfunctions $e^{\pm inx}$ orthogonality is usually *defined* in terms of the complex conjugate of one of the two factors,

$$\int_0^{2\pi} (e^{imx})^* e^{inx} \, dx = 2\pi\delta_{m,n}. \tag{14.10}$$

This agrees with the treatment of the spherical harmonics (Section 12.6).

Sturm-Liouville theory. The Sturm-Liouville theory guarantees the validity of Eq. 14.1 (for functions satisfying the Dirichlet conditions) and, by use of the

[1] Section 6.5.

orthogonality relations, allows us to compute the expansion coefficients

$$a_n = \frac{1}{\pi} \int_0^{2\pi} f(t) \cos nt \, dt, \tag{14.11}$$

$$b_n = \frac{1}{\pi} \int_0^{2\pi} f(t) \sin nt \, dt, \qquad n = 0, 1, 2, \ldots . \tag{14.12}$$

This, of course, is subject to the requirement that the integrals exist. They do if $f(t)$ is piecewise continuous. Substituting Eqs. 14.11 and 14.12 into Eq. 14.1, we write our Fourier expansion

$$f(x) = \frac{1}{2\pi} \int_0^{2\pi} f(t)\, dt + \frac{1}{\pi} \sum_{n=1}^{\infty} \left(\cos nx \int_0^{2\pi} f(t) \cos nt \, dt + \sin nx \int_0^{2\pi} f(t) \sin nt \, dt \right)$$

$$= \frac{1}{2\pi} \int_0^{2\pi} f(t)\, dt + \frac{1}{\pi} \sum_{n=1}^{\infty} \int_0^{2\pi} f(t) \cos n(t - x)\, dt, \tag{14.13}$$

the first (constant) term being the average value of $f(x)$ over the interval $[0, 2\pi]$.

Another way of describing what we are doing here is to say that $f(x)$ is part of an infinite-dimensional Hilbert space, with the orthogonal $\cos nx$ and $\sin nx$ as the basis. (They can always be renormalized to unity if desired.) The statement that $\cos nx$ and $\sin nx$ ($n = 0, 1, 2 \cdots$) span this Hilbert space is equivalent to saying that they form a complete set. Finally, the expansion coefficients a_n and b_n correspond to the projections of $f(x)$ with the integral inner products (Eqs. 14.11 and 14.12) playing the role of the dot product of Section 1.3.

Sawtooth wave. An idea of the convergence of a Fourier series and the error in using only a finite number of terms in the series may be obtained by considering the expansion of

$$f(x) = \begin{cases} x, & 0 \leqslant x < \pi, \\ x - 2\pi, & \pi < x \leqslant 2\pi. \end{cases} \tag{14.14}$$

This is a sawtooth wave, and for convenience we shall shift our interval from $[0, 2\pi]$ to $[-\pi, \pi]$. In this interval we have simply $f(x) = x$. Using Eqs. 14.11 and

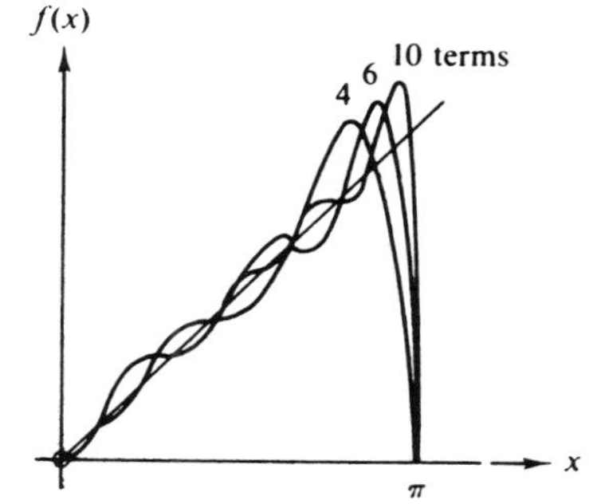

FIG. 14.1 Fourier representation of sawtooth wave

14.12 we may show the expansion to be

$$f(x) = x = 2\left[\sin x - \frac{\sin 2x}{2} + \frac{\sin 3x}{3} - \cdots + (-1)^{n+1}\frac{\sin nx}{n} + \cdots\right]. \qquad (14.15)$$

Figure 14.1 shows $f(x)$ for $0 \leqslant x < \pi$ for the sum of four, six, and ten terms of the series. Three features deserve comment.

1. There is a steady increase in the accuracy of the representation as the number of terms included is increased.
2. All the curves pass through the midpoint $y = 0$ at $x = \pi$
3. In the vicinity of $x = \pi$ there is an overshoot that persists and shows no sign of diminishing.

As a matter of incidental interest, setting $x = \pi/2$ in Eq. 14.15 provides an alternate derivation of Leibnitz' formula, Ex. 5.7.6.

Behavior of discontinuities. The behavior at $x = \pi$ is an example of a general rule that at a finite discontinuity the series converges to the arithmetic mean. For a discontinuity at $x = x_0$ the series yields

$$f(x_0) = \tfrac{1}{2}[f(x_0+) + f(x_0-)], \qquad (14.16)$$

the arithmetic mean of the right and left approaches to $x = x_0$. A general proof using partial sums, as in Section 14.5, is given by Jeffreys and by Carslaw. The proof may be simplified by the use of Dirac delta functions—Ex. 14.5.1.

The overshoot just before $x = \pi$ is an example of the Gibbs phenomenon, discussed in Section 14.5.

EXERCISES

14.1.1 A function $f(x)$ (quadratically integrable) is to be represented by a *finite* Fourier series. A convenient measure of the accuracy of the series is given by the integrated square of the deviation

$$\Delta_p = \int_0^{2\pi}\left[f(x) - \frac{a_0}{2} - \sum_{n=1}^{p}(a_n \cos nx + b_n \sin nx)\right]^2 dx.$$

Show that the requirement that Δ_p be minimized, that is,

$$\frac{\partial \Delta_p}{\partial a_n} = 0, \qquad \frac{\partial \Delta_p}{\partial b_n} = 0,$$

for all n, leads to choosing a_n and b_n, as given in Eqs. 14.11 and 14.12.

14.1.2 In the analysis of a complex wave form (ocean tides, earthquakes, musical tones, etc.) it might be more convenient to have the Fourier series written

$$f(x) = \frac{a_0}{2} + \sum_{n=1}^{\infty}\alpha_n \cos(nx - \theta_n).$$

Show that this is equivalent to Eq. 14.1 with

$$a_n = \alpha_n \cos\theta_n, \qquad \alpha_n^2 = a_n^2 + b_n^2,$$

$$b_n = \alpha_n \sin\theta_n, \qquad \tan\theta_n = b_n/a_n.$$

14.1.3 A function $f(x)$ is expanded in an exponential Fourier series

$$f(x) = \sum_{n=-\infty}^{\infty} c_n e^{inx}.$$

If $f(x)$ is real, $f(x) = f^*(x)$, what restriction is imposed on the coefficients c_n?

14.1.4 Assuming that $\int_{-\pi}^{\pi} f(x)\,dx$ and $\int_{-\pi}^{\pi} [f(x)]^2\,dx$ are finite, show that

$$\lim_{m\to\infty} a_m = 0, \qquad \lim_{m\to\infty} b_m = 0.$$

Hint. Integrate $[f(x) - s_n(x)]^2$, where $s_n(x)$ is the nth partial sum and use Bessel's inequality, Section 9.4.

14.2 Advantages, Uses of Fourier Series

Discontinuous function. One of the advantages of a Fourier representation over some other representation, such as a Taylor series, is that it may represent a discontinuous function. An example is the sawtooth wave in the preceding section. Other examples are considered in Section 14.3 and in the exercises.

Periodic functions. Related to this advantage is the usefulness of a Fourier series in representing a periodic function. If $f(x)$ has a period of 2π, perhaps it is only natural that we expand it in a series of functions with period $2\pi, 2\pi/2, 2\pi/3, \ldots$. This guarantees that if our periodic $f(x)$ is represented over one interval $[0, 2\pi]$ or $[-\pi, \pi]$ the representation holds for all finite x.

At this point we may conveniently consider the properties of symmetry. Using the interval $[-\pi, \pi]$, $\sin x$ is odd and $\cos x$ is an even function of x. Hence by Eqs. 14.11 and 14.12,[1] if $f(x)$ is odd, all $a_n = 0$ and if $f(x)$ is even all $b_n = 0$. In other words,

$$f(x) = \frac{a_0}{2} + \sum_{n=1}^{\infty} a_n \cos nx, \qquad f(x) \text{ even}, \tag{14.17}$$

$$f(x) = \sum_{n=1}^{\infty} b_n \sin nx, \qquad f(x) \text{ odd}. \tag{14.18}$$

Frequently these properties are helpful in expanding a given function.

We have noted that the Fourier series is periodic. This is important in considering whether Eq. 14.1 holds outside the initial interval. Suppose we are given

[1] With the range of integration $-\pi \leqslant x \leqslant \pi$.

only that

$$f(x) = x, \qquad 0 \leqslant x < \pi \tag{14.19}$$

and are asked to represent $f(x)$ by a series expansion. Let us take three of the infinite number of possible expansions.

1. If we assume a Taylor expansion, we have

$$f(x) = x, \tag{14.20}$$

a one-term series. This (one term) series is defined for all finite x.

2. Using the Fourier cosine series (Eq. 14.17), we predict that

$$\begin{aligned} f(x) &= -x, \qquad -\pi < x \leqslant 0, \\ f(x) &= 2\pi - x, \qquad \pi < x < 2\pi. \end{aligned} \tag{14.21}$$

3. Finally, from the Fourier sine series (Eq. 14.18), we have

$$\begin{aligned} f(x) &= x, \qquad -\pi < x \leqslant 0, \\ f(x) &= x - 2\pi, \qquad \pi < x < 2\pi. \end{aligned} \tag{14.22}$$

These three possibilities, Taylor series, Fourier cosine series, and Fourier sine series, are each perfectly valid in the original interval $[0, \pi]$. Outside, however, their behavior is strikingly different (cf. Fig. 14.2). Which of the three, then, is correct? This question has no answer, unless we are given more information about $f(x)$. It may be any of the three or none of them. Our Fourier expansions are valid over the basic interval. Unless the function $f(x)$ is known to be periodic with a period equal to our basic interval, or $(1/n)$th of our basic interval, there is no assurance whatever that the representation (Eq. 14.1) will have any meaning outside the basic interval.

In addition to the advantages of representing discontinuous and periodic

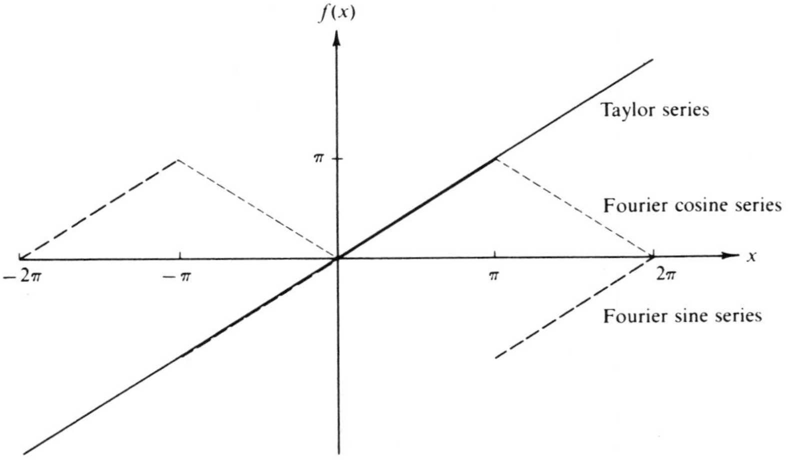

FIG. 14.2 Comparison of Fourier cosine series, Fourier sine series, and Taylor series

functions, there is a third very real advantage in using a Fourier series. Suppose that we are solving the equation of motion of an oscillating particle subject to a periodic driving force. The Fourier expansion of the driving force then gives us the fundamental term and a series of harmonics. The (linear) differential equation may be solved for each of these harmonics individually, a process that may be much easier than dealing with the original driving force. Then, as long as the differential equation is linear, all the solutions may be added together to obtain the final solution.[1] This is more than just a clever mathematical trick. It corresponds to finding the response of the system to the fundamental frequency and to each of the harmonic frequencies.

One question that is sometimes raised is, "Were the harmonics there all along or were they created by our Fourier analysis?" One answer compares the functional resolution into harmonics with the resolution of a vector into rectangular components. The components may have been present in the sense that they may be isolated and observed, but the resolution is certainly not unique. Hence many authorities prefer to say that the harmonics were created by our choice of expansion. Other expansions in other sets of orthogonal functions would give different results. For further discussion the reader should consult a series of notes and letters in the *American Journal of Physics*.[2]

Change of interval. So far attention has been restricted to an interval of length 2π. This restriction may easily be relaxed. If $f(x)$ is periodic with a period $2L$, we may write

$$f(x) = \frac{a_0}{2} + \sum_{n=1}^{\infty} \left[a_n \cos \frac{n\pi x}{L} + b_n \sin \frac{n\pi x}{L} \right] \tag{14.23}$$

with

$$a_n = \frac{1}{L} \int_{-L}^{L} f(t) \cos \frac{n\pi t}{L}\, dt, \qquad n = 0, 1, 2, 3, \dots, \tag{14.24}$$

$$b_n = \frac{1}{L} \int_{-L}^{L} f(t) \sin \frac{n\pi t}{L}\, dt, \qquad n = 1, 2, 3, \dots, \tag{14.25}$$

replacing x in Eq. 14.1 with $\pi x/L$ and t in Eqs. 14.11 and 14.12 with $\pi t/L$. (For convenience the interval in Eqs. 14.11 and 14.12 is shifted to $-\pi \leqslant t \leqslant \pi$.)

[1] One of the nastier features of nonlinear differential equations is that this principle of superposition is not valid.

[2] B. L. Robinson, *Am. J. Phys.* **21**, 391 (1953).
F. W. Van Name, Jr., *Am. J. Phys.* **22**, 94 (1954).

EXERCISES

14.2.1 The boundary conditions (such as $\psi(0) = \psi(l) = 0$) may suggest solutions of the form $\sin(n\pi x/l)$ and eliminate the corresponding cosines.

(a) Verify that the boundary conditions used in the Sturm-Liouville theory are satisfied for the interval $(0, l)$. Note that this is only half of the usual Fourier interval.

(b) Show that the set of functions $\varphi_n(x) = \sin(n\pi x/l)$, $n = 1, 2, 3, \ldots$ satisfies an orthogonality relation

$$\int_0^l \varphi_m(x)\varphi_n(x)\,dx = \frac{l}{2}\,\delta_{nm}, \qquad n > 0.$$

14.2.2 (a) Expand $f(x) = x$ in the interval $(0, 2L)$.

Ans. $$x = L - \frac{2L}{\pi}\sum_{n=1}^{\infty}\frac{1}{n}\sin\left(\frac{n\pi x}{L}\right).$$

(b) Expand $f(x) = x$ as a sine series in the *half* interval $(0, L)$.

Ans. $$x = \frac{2L}{\pi}\sum_{n=1}^{\infty}\frac{(-1)^{n+1}}{n}\sin\left(\frac{n\pi x}{L}\right).$$

14.3 Applications of Fourier Series

EXAMPLE 14.3.1. SQUARE WAVE—HIGH FREQUENCIES

One simple application of Fourier series, the analysis of a "square" wave in terms of its Fourier components, may occur in electronic circuits designed to handle sharply rising pulses. Suppose that our wave is defined by

$$\begin{aligned} f(x) &= 0, \qquad -\pi < x < 0, \\ f(x) &= h, \qquad 0 < x < \pi. \end{aligned} \tag{14.26}$$

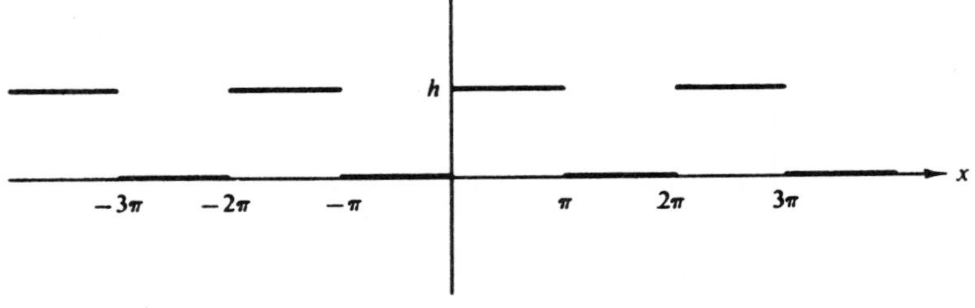

FIG. 14.3 Square wave

From Eqs. 14.11 and 14.12 we find

$$a_0 = \frac{1}{\pi}\int_0^\pi h\,dt = h, \tag{14.27}$$

$$a_n = \frac{1}{\pi}\int_0^\pi h\cos nt\,dt = 0, \qquad n = 1, 2, 3, \ldots, \tag{14.28}$$

$$b_n = \frac{1}{\pi}\int_0^\pi h\sin nt\,dt = \frac{h}{n\pi}(1 - \cos n\pi); \tag{14.29}$$

$$\therefore\quad b_n = \frac{2h}{n\pi}, \qquad n \text{ odd}, \tag{14.30}$$

$$b_n = 0, \qquad n \text{ even}. \tag{14.31}$$

The resulting series is

$$f(x) = \frac{h}{2} + \frac{2h}{\pi}\left(\frac{\sin x}{1} + \frac{\sin 3x}{3} + \frac{\sin 5x}{5} + \cdots\right). \tag{14.32}$$

Except for the first term which represents an average of $f(x)$ over the interval $[-\pi, \pi]$ all the cosine terms have vanished. Since $f(x) - h/2$ is odd, we have a Fourier sine series. Although only the odd terms in the sine series occur, they fall only as n^{-1}. This is similar to the convergence (or lack of convergence) of the harmonic series. Physically this means that our square wave contains a lot of high-frequency components. If the electronic apparatus will not pass these components, our square wave input will emerge more or less rounded off, perhaps as an amorphous blob.

Example 14.3.2. Full Wave Rectifier

As a second example, let us ask how well the output of a full wave rectifier approaches pure direct current. Our rectifier may be thought of as having passed the positive peaks of an incoming sine wave and inverting the negative peaks. This yields

$$\begin{aligned} f(t) &= \sin \omega t, & 0 < \omega t < \pi, \\ f(t) &= -\sin \omega t, & -\pi < \omega t < 0. \end{aligned} \tag{14.33}$$

Since $f(t)$ defined here is even, no terms of the form $\sin n\omega t$ will appear. Again, from Eqs. 14.11 and 14.12, we have

$$\begin{aligned} a_0 &= \frac{1}{\pi}\int_{-\pi}^0 -\sin \omega t\,d(\omega t) + \frac{1}{\pi}\int_0^\pi \sin \omega t\,d(\omega t) \\ &= \frac{2}{\pi}\int_0^\pi \sin \omega t\,d(\omega t) = \frac{4}{\pi}, \end{aligned} \tag{14.34}$$

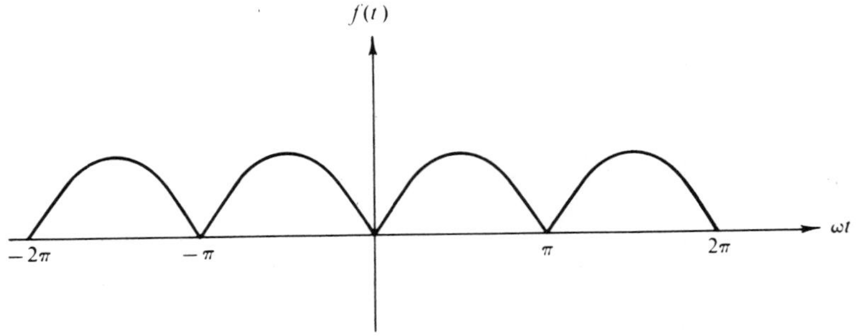

FIG. 14.4 Full wave rectifier

$$a_n = \frac{2}{\pi} \int_0^{\pi} \sin \omega t \cos n\omega t \, d(\omega t)$$

$$= -\frac{2}{\pi} \frac{2}{n^2 - 1}, \qquad n \text{ even}$$

$$= 0, \qquad n \text{ odd}. \tag{14.35}$$

Note carefully that $[0, \pi]$ is not an orthogonality interval and we do not get zero for even n. The resulting series is

$$f(t) = \frac{2}{\pi} - \frac{4}{\pi} \sum_{n=2,4,6,\cdots}^{\infty} \frac{\cos n\omega t}{n^2 - 1}. \tag{14.36}$$

The original frequency ω has been eliminated. The lowest frequency oscillation is 2ω. The high-frequency components fall off as n^{-2}, showing that the full wave rectifier does a fairly good job of approximating direct current. Whether this good approximation is adequate depends on the particular application. If the remaining ac components are objectionable, they may be further suppressed by appropriate filter circuits.

These two examples bring out two features characteristic of Fourier expansions.[1]

1. If $f(x)$ has discontinuities (as in the square wave in Example 14.3.1), we can expect the nth coefficient to be decreasing as $1/n$. Convergence is relatively slow.

2. If $f(x)$ is continuous (though possibly with discontinuous derivatives as in the full wave rectifier of Example 14.3.2), we can expect the nth coefficient to be decreasing as $1/n^2$.

Example 14.3.3. Infinite Series, Riemann Zeta Function

As a final example, we consider the purely mathematical problem of expanding x^2. Let

$$f(x) = x^2, \qquad -\pi < x < \pi. \tag{14.37}$$

[1] G. Raisbeck, "Order of Magnitude of Fourier Coefficients," *Am. Math. Monthly* **62**, 149–155 (1955).

By symmetry all $b_n = 0$. For the a_n's we have

$$a_0 = \frac{1}{\pi}\int_{-\pi}^{\pi} x^2\,dx = \frac{2\pi^2}{3}, \tag{14.38}$$

$$\begin{aligned} a_n &= \frac{2}{\pi}\int_0^{\pi} x^2 \cos nx\,dx \\ &= \frac{2}{\pi}\cdot(-1)^n \frac{2\pi}{n^2} \\ &= (-1)^n \frac{4}{n^2}. \end{aligned} \tag{14.39}$$

From this we obtain

$$x^2 = \frac{\pi^2}{3} + 4\sum_{n=1}^{\infty}(-1)^n \frac{\cos nx}{n^2}. \tag{14.40}$$

As it stands, Eq. 14.40 is of no particular importance, but if we set $x = \pi$ [1]

$$\cos n\pi = (-1)^n \tag{14.41}$$

and Eq. 14.40 becomes

$$\pi^2 = \frac{\pi^2}{3} + 4\sum_{n=1}^{\infty}\frac{1}{n^2} \tag{14.42}$$

or

$$\frac{\pi^2}{6} = \sum_{n=1}^{\infty}\frac{1}{n^2} \equiv \zeta(2), \tag{14.43}$$

thus yielding the Riemann zeta function, $\zeta(2)$, in closed form. From our expansion of x^2 and expansions of other powers of x numerous other infinite series can be evaluated. A few are included in the list of exercises below.

EXERCISES

14.3.1 Develop the Fourier series representation of

$$f(t) = \begin{cases} 0, & -\pi \leqslant \omega t \leqslant 0, \\ \sin \omega t, & 0 \leqslant \omega t \leqslant \pi. \end{cases}$$

This is the output of a simple half-wave rectifier. It is also an approximation of the solar thermal effect that produces "tides" in the atmosphere.

Ans. $f(t) = \dfrac{1}{\pi} + \dfrac{1}{2}\sin \omega t - \dfrac{2}{\pi}\displaystyle\sum_{\substack{n=2,4,6,\ldots \\ \text{even}}}^{\infty} \frac{\cos n\omega t}{n^2 - 1}.$

[1] Note that the point $x = \pi$ is *not* a point of discontinuity.

14.3.2 A sawtooth wave is given by

$$f(x) = x, \qquad -\pi < x < \pi.$$

Show that

$$f(x) = 2\sum_{n=1}^{\infty} \frac{(-1)^{n+1}}{n} \sin nx.$$

14.3.3 A triangular wave is represented by

$$f(x) = \begin{cases} x, & 0 < x < \pi \\ -x, & -\pi < x < 0. \end{cases}$$

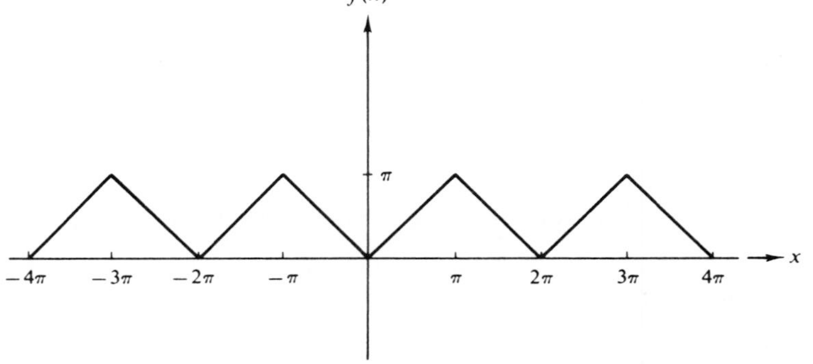

FIG. 14.5 Triangular wave

Represent $f(x)$ by a Fourier series.

$$Ans.\ f(x) = \frac{\pi}{2} - \frac{4}{\pi} \sum_{\substack{n=1,3,5,\dots \\ \text{odd}}} \frac{\cos nx}{n^2}.$$

14.3.4 A metal cylindrical tube of radius a is split lengthwise into two nontouching halves. The top half is maintained at a potential $+V$, the bottom half at a potential $-V$. Separate the variables in Laplace's equation and solve for the electrostatic potential for $r \leqslant a$. Observe the resemblance between your solution for $r = a$ and the Fourier series for a square wave.

Note: A closed form solution is obtained in Section 6.7 by complex variable analysis.

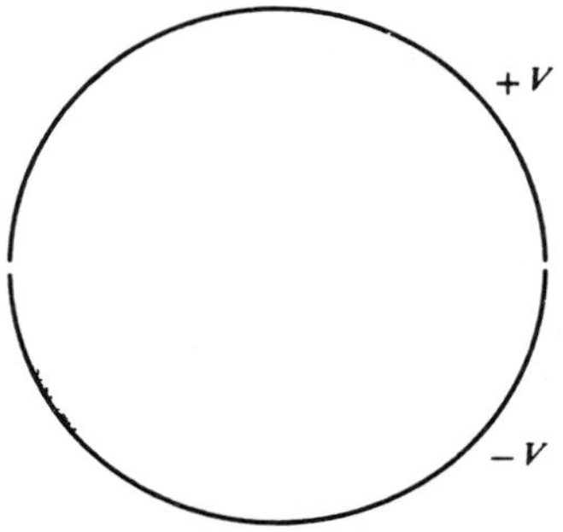

FIG. 14.6

14.3.5 A metal cylinder is placed in a (previously) uniform electric field, E_0, the axis of the cylinder perpendicular to that of the original field.
(a) Find the perturbed electrostatic potential.
(b) Find the induced surface charge on the cylinder as a function of angular position.

Note: Exercise 6.7.12 asks for a conformal mapping solution of this same problem.

14.3.6 Transform the Fourier expansion of a square wave, Eq. 14.32, into a power series. Show that the coefficients of x^1 form a *divergent* series. Repeat for the coefficients of x^3.

A power series cannot handle a discontinuity. These infinite coefficients are the result of attempting to beat this basic limitation on power series.

14.3.7 (a) Show that the Fourier expansion of $\cos ax$ is

$$\cos ax = \frac{2a \sin a\pi}{\pi} \left\{ \frac{1}{2a^2} - \frac{\cos x}{a^2 - 1^2} + \frac{\cos 2x}{a^2 - 2^2} - \cdots \right\},$$

$$a_n = (-1)^n \frac{2a \sin a\pi}{\pi(a^2 - n^2)}.$$

(b) From the above result show that

$$a\pi \cot a\pi = 1 - 2 \sum_{p=1}^{\infty} \zeta(2p) a^{2p}.$$

This provides an alternate derivation of the relation between the Riemann zeta function and the Bernoulli numbers, Eq. 5.151.

14.3.8 Derive the Fourier series expansion of the Dirac delta function in the interval $-\pi < x < \pi$.
(a) What significance can be attached to the constant term?
(b) In what region is this representation valid?

14.3.9 Verify that

$$\delta(\varphi_1 - \varphi_2) = \frac{1}{2\pi} \sum_{m=-\infty}^{\infty} e^{im(\varphi_1 - \varphi_2)}.$$

Note: The continuum analog of this expression is developed in Section 15.2. The most important application of this expression is in the determination of Green's functions, Section 16.6.

14.3.10 (a) Using

$$f(x) = x^2, \qquad -\pi < x < \pi,$$

show that

$$\sum_{n=1}^{\infty} \frac{(-1)^{n+1}}{n^2} = \frac{\pi^2}{12}.$$

(b) Using the Fourier series for a triangular wave developed in Exercise 14.3.3 show that

$$\sum_{n=1}^{\infty} \frac{1}{(2n-1)^2} = \frac{\pi^2}{8}.$$

(c) Using

$$f(x) = x^4, \qquad -\pi < x < \pi,$$

show that

$$\sum_{n=1}^{\infty} \frac{1}{n^4} = \frac{\pi^4}{90} = \zeta(4),$$

$$\sum_{n=1}^{\infty} \frac{(-1)^{n+1}}{n^4} = \frac{7\pi^4}{720}.$$

(d) Using

$$f(x) = \begin{cases} x(\pi - x), & 0 < x < \pi, \\ x(\pi + x), & -\pi < x < 0, \end{cases}$$

derive

$$f(x) = \frac{8}{\pi} \sum_{\substack{n=1,3,5,\cdots \\ \text{odd}}}^{\infty} \frac{\sin nx}{n^3}$$

and show that

$$\sum_{\substack{n=1,3,5,\cdots \\ \text{odd}}}^{\infty} (-1)^{(n-1)/2} n^{-3} = 1 - \frac{1}{3^3} + \frac{1}{5^3} - \frac{1}{7^3} + \cdots = \frac{\pi^3}{32}.$$

(e) Using the Fourier series for a square wave, show that

$$\sum_{\substack{n=1,3,5,\cdots \\ \text{odd}}}^{\infty} (-1)^{(n-1)/2} n^{-1} = 1 - \frac{1}{3} + \frac{1}{5} - \frac{1}{7} + \cdots = \frac{\pi}{4}.$$

This is Leibnitz' formula for π, obtained by a different technique in Ex. 5.7.6.

14.4 Properties of Fourier Series

Convergence. It might be noted, first, that our Fourier series should not be expected to be uniformly convergent if it represents a discontinuous function. A uniformly convergent series of continuous functions ($\sin nx$, $\cos nx$) always yields a continuous function (cf. Section 5.5). If, however, (a) $f(x)$ is continuous, $-\pi \leqslant x \leqslant \pi$, (b) $f(-\pi) = f(+\pi)$, and (c) $f'(x)$ is sectionally continuous, the Fourier series for $f(x)$ will converge uniformly. These restrictions do not demand that $f(x)$ be periodic, but they will be satisfied by continuous, differentiable, periodic functions (period of 2π). For a proof of uniform convergence the reader is referred to the literature.[1]

Integration. Term-by-term integration of the series

$$f(x) = \frac{a_0}{2} + \sum_{n=1}^{\infty} a_n \cos nx + \sum_{n=1}^{\infty} b_n \sin nx \tag{14.44}$$

yields

$$\int_{x_0}^{x} f(x)\,dx = \frac{a_0 x}{2}\bigg|_{x_0}^{x} + \sum_{n=1}^{\infty} \frac{a_n}{n} \sin nx\bigg|_{x_0}^{x} - \sum_{n=1}^{\infty} \frac{b_n}{n} \cos nx\bigg|_{x_0}^{x}. \tag{14.45}$$

Clearly, the effect of integration is to place an additional power of n in the denominator of each coefficient. This results in more rapid convergence than before. Consequently, a convergent Fourier series may always be integrated term by term, the resulting series converging uniformly to the integral of the original function. Indeed, term-by-term integration may be valid even if the original series (Eq. 14.44) is not itself convergent! A discussion will be found in Jeffreys and Jeffreys, Section 14.06.

Strictly speaking, Eq. 14.45 may not be a Fourier series; that is, if $a_0 \neq 0$, there

[1] See, for instance, R. V. Churchill, *Fourier Series and Boundary Value Problems*. New York: McGraw-Hill (1941), Section 38.

will be a term $\frac{1}{2}a_0 x$. However,

$$\int_{x_0}^{x} f(x)\,dx - \tfrac{1}{2}a_0 x \tag{14.46}$$

will still be a Fourier series.

Differentiation. The situation regarding differentiation is quite different from that of integration. Here, the word is *caution*. Consider the series for

$$f(x) = x, \qquad -\pi < x < \pi. \tag{14.47}$$

We readily find (cf. Ex. 14.3.2) that the Fourier series is

$$x = 2\sum_{n=1}^{\infty}(-1)^{n+1}\frac{\sin nx}{n}, \qquad -\pi < x < \pi. \tag{14.48}$$

Differentiating term by term, we obtain

$$1 = 2\sum_{n=1}^{\infty}(-1)^{n+1}\cos nx, \tag{14.49}$$

which is not convergent!

For a triangular wave (Exercise 14.3.3), in which the convergence is more rapid (and uniform),

$$f(x) = \frac{\pi}{2} - \frac{4}{\pi}\sum_{n=1,\text{odd}}^{\infty}\frac{\cos nx}{n^2}. \tag{14.50}$$

Differentiating term by term

$$f'(x) = \frac{4}{\pi}\sum_{n=1,\text{odd}}^{\infty}\frac{\sin nx}{n}, \tag{14.51}$$

which is the Fourier expansion of a square wave

$$f'(x) = \begin{cases} 1, & 0 < x < \pi, \\ -1, & -\pi < x < 0. \end{cases} \tag{14.52}$$

Inspection of Fig. 14.5 verifies that this is indeed the derivative of our triangular wave.

As the inverse of integration, the operation of differentiation has placed an additional factor n in the numerator of each term. This reduces the rate of convergence and may, as in the first case mentioned, render the differentiated series divergent.

In general, term-by-term differentiation is permissible under the same conditions listed for uniform convergence.

Lanczos convergence factors. The operation of differentiation is also troublesome in another way. Lanczos has pointed out the danger and a possible remedy. First we note that the use of an *infinite* number of terms in the Fourier expansion is an often unobtainable ideal. Instead, let

$$f(x) = f_m(x) + \eta_m(x), \tag{14.53}$$

with

$$f_m(x) = \sum_{n=-(m-1)}^{m-1} c_n e^{inx}, \tag{14.54}$$

reverting to the exponential form for convenience, and η_m is a remainder or error term given by

$$\begin{aligned}
\eta_m(x) &= \sum_{n=m}^{\infty} (c_n e^{inx} + c_{-n} e^{-inx}) \\
&= e^{imx} \sum_{n=0}^{\infty} c_{m+n} e^{inx} + e^{-imx} \sum_{n=0}^{\infty} c_{-m-n} e^{-inx} \\
&= e^{imx} \rho_m(x) + e^{-imx} \rho_{-m}(x).
\end{aligned} \tag{14.55}$$

In electronic terms $\eta_m(x)$ is a modulated carrier wave with a high-frequency carrier, e^{imx}, and a modulation, $\rho_m(x)$. Differentiation of this remainder term, $\eta_m(x)$, leads to

$$\frac{d\eta_m(x)}{dx} = ime^{imx} \rho_m(x) + e^{imx} \frac{d\rho_m(x)}{dx} + \cdots. \tag{14.56}$$

The trouble arises in the first term on the right which blows up as $m \to \infty$. Now this corresponds to differentiating the carrier frequency e^{imx}. Actually we want only to differentiate the modulation $\rho_m(x)$, and this part is well behaved.

Lanczos suggested that we could do what we wanted to do and avoid the extraneous infinity by defining a new difference operator, $\mathscr{D}_m$, by the equation

$$\mathscr{D}_m f(x) = \frac{f(x + \pi/m) - f(x - \pi/m)}{2\pi/m}. \tag{14.57}$$

This represents a sort of average derivative. Clearly

$$\lim_{m \to \infty} \mathscr{D}_m = \frac{d}{dx}. \tag{14.58}$$

By applying $\mathscr{D}_m$ to the remainder, $\eta_m(x)$, we obtain

$$\begin{aligned}
\mathscr{D}_m \eta_m(x) &= \frac{e^{im(x+\pi/m)} \rho_m(x + \pi/m) - e^{im(x-\pi/m)} \rho_m(x - \pi/m)}{2\pi/m} \\
&\quad + \frac{e^{-im(x+\pi/m)} \rho_{-m}(x + \pi/m) - e^{-im(x-\pi/m)} \rho_{-m}(x - \pi/m)}{2\pi/m},
\end{aligned} \tag{14.59}$$

$$\mathscr{D}_m \eta_m(x) = -e^{imx} \mathscr{D}_m \rho_m(x) - e^{-imx} \mathscr{D}_m \rho_{-m}(x),$$

showing that $\mathscr{D}_m$ effectively operates on the modulation ρ_m and *not* on the carrier e^{imx}. The diverging terms of the forms $ime^{imx} \rho_m(x)$ have been eliminated. (The minus signs appearing in this case are the result of picking points $x \pm \pi/m$ which are shifted one half cycle of e^{imx}.)

Applied to a complex Fourier series,

$$\mathscr{D}_m e^{inx} = \frac{e^{in(x+\pi/m)} - e^{+in(x-\pi/m)}}{2\pi/m}$$

$$= ine^{inx} \frac{\sin(n\pi/m)}{n\pi/m}. \tag{14.60}$$

Hence

$$\mathscr{D}_m e^{inx} = \left[\frac{\sin(n\pi/m)}{n\pi/m}\right] \frac{d}{dx} e^{inx}. \tag{14.61}$$

The additional factor $\sin(n\pi/m)/(n\pi/m)$ is called a Lanczos smoothing factor, or Lanczos convergence factor, σ_n; that is

$$\sigma_n = \frac{\sin(n\pi/m)}{n\pi/m}. \tag{14.62}$$

The inclusion of these σ_n's in a differentiated series is sometimes helpful to convergence. The *finite* Fourier series

$$f(x) = \frac{a_0}{2} + \sum_{n=1}^{m-1} (a_n \cos nx + b_n \sin nx) \tag{14.63}$$

is replaced with

$$f(x) = \frac{a_0}{2} + \sum_{n=1}^{m-1} \frac{\sin(n\pi/m)}{n\pi/m} (a_n \cos nx + b_n \sin nx). \tag{14.64}$$

As will be seen in Section 14.5, the Lanczos σ_n's can almost eliminate the Gibbs phenomenon.

EXERCISES

14.4.1 Show that integration of the Fourier expansion of $f(x) = x$, $-\pi < x < \pi$, leads to

$$\frac{\pi^2}{12} = \sum_{n=1}^{\infty} (-1)^{n+1} n^{-2}$$
$$= 1 - \tfrac{1}{4} + \tfrac{1}{9} - \tfrac{1}{16} + \cdots.$$

14.4.2. Parseval's identity.

(a) Assuming that the Fourier expansion of $f(x)$ is uniformly convergent, show that

$$\frac{1}{\pi}\int_{-\pi}^{\pi} [f(x)]^2\, dx = \frac{a_0^2}{2} + \sum_{n=1}^{\infty} (a_n^2 + b_n^2).$$

This is Parseval's identity. It is actually a special case of the completeness relation, Eq. 9.72.

(b) Given

$$x^2 = \frac{\pi^2}{3} + 4\sum_{n=1}^{\infty} \frac{(-1)^n \cos nx}{n^2}, \qquad -\pi \leqslant x \leqslant \pi,$$

apply Parseval's identity to obtain $\zeta(4)$ in closed form.

14.4.3 Show that integrating the Fourier expansion of the Dirac delta function (Ex. 14.3.8) leads to the Fourier representation of the square wave, Eq. 14.32, with $h = 1$.
Note: Integrating the constant term $(1/2\pi)$ leads to a term $x/2\pi$. What are you going to do with this?

14.4.4 In the interval $(-\pi, \pi)$,

$$\delta_n(x) = n, \quad \text{for } |x| < \frac{1}{2n},$$
$$0, \quad \text{for } |x| > \frac{1}{2n}.$$

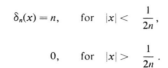

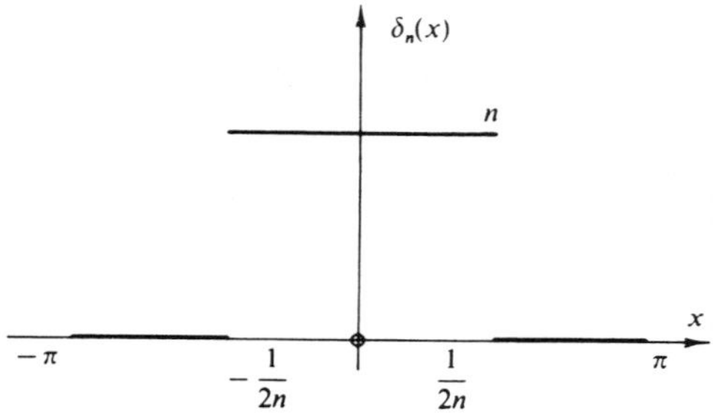

FIG. 14.7 Rectangular pulse

(a) Expand $\delta_n(x)$ as a Fourier cosine series.
(b) Show that your Fourier series agrees with a Fourier expansion of $\delta(x)$ in the limit as $n \to \infty$.

14.4.5 Confirm the delta function nature of your Fourier series of Ex. 14.4.4 by showing that for any $f(x)$ that is finite in the interval $[-\pi, \pi]$ and continuous at $x = 0$,

$$\int_{-\pi}^{\pi} f(x) \,[\text{Fourier expansion of } \delta_\infty(x)]\, dx = f(0).$$

14.4.6 (a) Show that the Dirac delta function $\delta(x - a)$, expanded in a Fourier sine series in the half interval $(0, L)$, $(0 < a < L)$, is given by

$$\delta(x - a) = \frac{2}{L} \sum_{n=1}^{\infty} \sin\left(\frac{n\pi a}{L}\right) \sin\left(\frac{n\pi x}{L}\right)$$

Note that this series actually describes

$$-\delta(x + a) + \delta(x - a) \text{ in the interval } (-L, L).$$

(b) By integrating both sides of the above equation from 0 to x show that the cosine expansion of the square wave

$$f(x) = \begin{cases} 0, & 0 \le x < a \\ 1, & a < x < L, \end{cases}$$

is

$$f(x) = \frac{2}{\pi} \sum_{n=1}^{\infty} \frac{1}{n} \sin\left(\frac{n\pi a}{L}\right) - \frac{2}{\pi} \sum_{n=1}^{\infty} \frac{1}{n} \sin\left(\frac{n\pi a}{L}\right) \cos\left(\frac{n\pi x}{L}\right), \quad 0 \le x < L.$$

(c) Verify that the term $\frac{2}{\pi}\sum_{n=1}^{\infty}\frac{1}{n}\sin\left(\frac{n\pi a}{L}\right)$ is $<f(x)>$.

14.4.7 Verify the Fourier cosine expansion of the square wave, Ex. 14.4.6(b), by direct calculation of the Fourier coefficients.

14.4.8 (a) A string is clamped at both ends $x = 0$ and $x = L$. Assuming small amplitude vibrations the amplitude $y(x,t)$ satisfies the wave equation

$$\frac{\partial^2 y}{\partial x^2} = \frac{1}{v^2}\frac{\partial^2 y}{\partial t^2}.$$

Here v is the wave velocity. The string is set in vibration by a sharp blow at $x = a$. Hence, we have

$$y(x, 0) = 0$$

$$\frac{\partial y(x, t)}{\partial t} = Lv_0\,\delta(x - a) \qquad \text{at } t = 0.$$

The constant L is included to compensate for the dimensions (inverse length) of $\delta(x - a)$. With $\delta(x - a)$ given by Exercise 14.4.6 (a) solve the wave equation subject to these initial conditions.

$$\textit{Ans. } y(x, t) = \frac{2v_0 L}{\pi v}\sum_{n=1}^{\infty}\frac{1}{n}\sin\frac{n\pi a}{L}\sin\frac{n\pi x}{L}\sin\frac{n\pi vt}{L}.$$

(b) Show that the transverse velocity of the string $\frac{\partial y(x, t)}{\partial t}$ is given by

$$\frac{\partial y(x, t)}{\partial t} = 2v_0\sum_{n=1}^{\infty}\sin\frac{n\pi a}{L}\sin\frac{n\pi x}{L}\cos\frac{n\pi vt}{L}.$$

14.4.9 A string, clamped at $x = 0$ and at $x = l$, is vibrating freely. Its motion is described by the wave equation

$$\frac{\partial^2 u(x, t)}{\partial t^2} = v^2\frac{\partial^2 u(x, t)}{\partial x^2}.$$

Assume a Fourier expansion of the form

$$u(x, t) = \sum_{n=1}^{\infty} b_n(t)\sin\frac{n\pi x}{l}$$

and determine the coefficients $b_n(t)$. The initial conditions are

$$u(x, 0) = f(x) \qquad \text{and} \qquad \frac{\partial}{\partial t}u(x, 0) = g(x).$$

Note: This is only half of the conventional Fourier orthogonality integral interval. However as long as only the sines are included here the Sturm-Liouville boundary conditions are still satisfied and the functions are orthogonal.

$$\textit{Ans. } b_n(t) = A_n\cos\frac{n\pi vt}{l} + B_n\sin\frac{n\pi vt}{l},$$

$$A_n = \frac{2}{l}\int_0^l f(x)\sin\frac{n\pi x}{l}\,dx,\ B_n = \frac{2}{n\pi v}\int_0^l g(x)\sin\frac{n\pi x}{l}\,dx.$$

14.4.10 (a) Continuing the vibrating string problem, Exercise 14.4.9, the presence of a resisting medium will damp the vibrations according to the equation

$$\frac{\partial^2 u(x,t)}{\partial t^2} = v^2 \frac{\partial^2 u(x,t)}{\partial x^2} - k \frac{\partial u(x,t)}{\partial t}.$$

Assume a Fourier expansion

$$u(x,t) = \sum_{n=1}^{\infty} b_n(t) \sin \frac{n\pi x}{l}$$

and again determine the coefficients $b_n(t)$. Take the initial and boundary conditions to be the same as in Exercise 14.4.9. Assume the damping to be small.

(b) Repeat but assume the damping to be large.

Ans. (a) $b_n(t) = e^{-kt/2}\{A_n \cos \omega_n t + B_n \sin \omega_n t\}$,

$$A_n = \frac{2}{l}\int_0^l f(x) \sin \frac{n\pi x}{l}\, dx,$$

$$B_n = \frac{2}{\omega_n l}\int_0^l g(x) \sin \frac{n\pi x}{l}\, dx + \frac{k}{2\omega_n} A_n, \quad \omega_n^2 = \left(\frac{n\pi v}{l}\right)^2 - \left(\frac{k}{2}\right)^2 > 0.$$

(b) $b_n(t) = e^{-kt/2}\{A_n \cosh \sigma_n t + B_n \sinh \sigma_n t\}$,

$$A_n = \frac{2}{l}\int_0^l f(x) \sin \frac{n\pi x}{l}\, dx,$$

$$B_n = \frac{2}{\sigma_n l}\int_0^l g(x) \sin \frac{n\pi x}{l}\, dx + \frac{k}{2\sigma_n} A_n, \quad \sigma_n^2 = \left(\frac{k}{2}\right)^2 - \left(\frac{n\pi v}{l}\right)^2 > 0.$$

14.4.11 Find the charge distribution over the interior surfaces of the semicircles of Ex. 14.3.4. *Note:* You obtain a divergent series and this Fourier approach fails. From Section 6.7 $\sigma \sim \csc\theta$. Does $\csc\theta$ have a Fourier expansion?

14.5 Gibbs Phenomenon

This is a peculiarity of the Fourier series at a simple discontinuity. An example is seen in Fig. 14.1. For convenience of numerical calculation we consider the behavior of the Fourier series which represents the periodic square wave

$$f(x) = \begin{cases} \dfrac{h}{2}, & 0 < x < \pi, \\ -\dfrac{h}{2}, & -\pi < x < 0. \end{cases} \tag{14.65}$$

This is essentially the square wave used in Section 14.3, and we see immediately that the solution is

$$f(x) = \frac{2h}{\pi}\left(\frac{\sin x}{1} + \frac{\sin 3x}{3} + \frac{\sin 5x}{5} + \cdots\right) \tag{14.66}$$

Summation of series.[1] In Section 14.1 the sum of the first several terms of the Fourier series for a sawtooth wave was plotted (Fig. 14.1). Now we develop an analytic method of summing the first r terms of our Fourier series (Eq. 14.66).

From Eq. 14.13

$$a_n \cos nx + b_n \sin nx = \frac{1}{\pi}\int_{-\pi}^{\pi} f(t) \cos n(t - x)\, dt. \tag{14.67}$$

Then the rth partial sum becomes

$$\begin{aligned} s_r(x) &= \sum_{n=0}^{r} (a_n \cos nx + b_n \sin nx) \\ &= \mathcal{R}\ \frac{1}{\pi}\int_{-\pi}^{\pi} f(t)\left[\frac{1}{2} + \sum_{n=1}^{r} e^{-i(t-x)n}\right] dt. \end{aligned} \tag{14.68}$$

Summing the finite series of exponentials (geometric progression),[2] we obtain

$$s_r(x) = \frac{1}{2\pi}\int_{-\pi}^{\pi} f(t)\frac{\sin[(r + \frac{1}{2})(t - x)]}{\sin \frac{1}{2}(t - x)}\, dt. \tag{14.69}$$

This is convergent at all points, including $t = x$.

Square wave. Applying this result to our square wave (Eq. 14.66), we have the sum of the first r terms (plus $\frac{1}{2}a_0$, which is zero here).

$$\begin{aligned} s_r(x) &= \frac{h}{4\pi}\int_0^{\pi} \frac{\sin(r + \frac{1}{2})(t - x)}{\sin \frac{1}{2}(t - x)}\, dt - \frac{h}{4\pi}\int_{-\pi}^{0} \frac{\sin(r + \frac{1}{2})(t - x)}{\sin \frac{1}{2}(t - x)}\, dt \\ &= \frac{h}{4\pi}\int_0^{\pi} \frac{\sin(r + \frac{1}{2})(t - x)}{\sin \frac{1}{2}(t - x)}\, dt - \frac{h}{4\pi}\int_0^{\pi} \frac{\sin(r + \frac{1}{2})(t + x)}{\sin \frac{1}{2}(t + x)}\, dt. \end{aligned} \tag{14.70}$$

This last result follows from the transformation

$$t = -t \qquad \text{in the second integral.} \tag{14.71}$$

[1] It is of some interest to note that this series also occurs in the analysis of the diffraction grating (r slits).

[2] Cf. Ex. 6.1.5 with initial value $n = 1$.

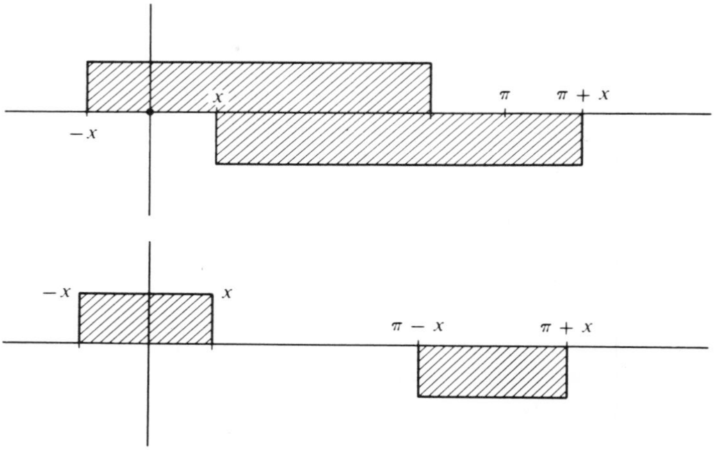

FIG. 14.8 Intervals of integration—Eq. 14.72

Replacing $t - x$ in the first term with s and $t + x$ in the second term with s we obtain

$$s_r(x) = \frac{h}{4\pi}\int_{-x}^{\pi-x} \frac{\sin(r+\frac{1}{2})s}{\sin\frac{1}{2}s}\,ds - \frac{h}{4\pi}\int_{x}^{\pi+x} \frac{\sin(r+\frac{1}{2})s}{\sin\frac{1}{2}s}\,ds$$

$$= \frac{h}{4\pi}\int_{-x}^{x} \frac{\sin(r+\frac{1}{2})s}{\sin\frac{1}{2}s}\,ds - \frac{h}{4\pi}\int_{\pi-x}^{\pi+x} \frac{\sin(r+\frac{1}{2})s}{\sin\frac{1}{2}s}\,ds. \qquad (14.72)$$

Consider the partial sum in the vicinity of the discontinuity at $x = 0$. As $x \to 0$, the second integral becomes negligible, and we associate the first integral with the discontinuity at $x = 0$. Using $(r + \frac{1}{2}) = p$ and $ps = \xi$, we obtain

$$s_r(x) = \frac{h}{2\pi}\int_0^{px} \frac{\sin\xi}{\sin(\xi/2p)} \cdot \frac{d\xi}{p}. \qquad (14.73)$$

Calculation of overshoot. Our partial sum, $s_r(x)$, starts at zero when $x = 0$ (in agreement with Eq. 14.16) and increases until $px = \pi$, at which point the numerator, $\sin\xi$, goes negative. For large r, and therefore large p, our denominator remains positive.

The maximum value of the partial sum is then

$$s_r(x)_{\max} = \frac{h}{2}\cdot\frac{1}{\pi}\int_0^{\pi} \frac{\sin\xi\,d\xi}{\sin(\xi/2p)p}$$

$$\approx \frac{h}{2}\cdot\frac{2}{\pi}\int_0^{\pi} \frac{\sin\xi}{\xi}\,d\xi. \qquad (14.74)$$

In terms of the sine integral, $si(x)$ of Section 10.5,

$$\int_0^{\pi} \frac{\sin\xi}{\xi}\,d\xi = \frac{\pi}{2} + si(\pi). \qquad (14.74a)$$

The integral is clearly greater than $\pi/2$, since it can be written

$$\left(\int_0^{\infty} - \int_{\pi}^{3\pi} - \int_{3\pi}^{5\pi} - \cdots\right)\frac{\sin\xi}{\xi}\,d\xi = \int_0^{\pi}\frac{\sin\xi}{\xi}\,d\xi. \tag{14.75}$$

We saw in Section 7.2 that the integral from 0 to ∞ is $\pi/2$. From this integral we are subtracting a series of *negative* terms. A power series expansion and term-by-term integration yields

$$\frac{2}{\pi}\int_0^{\pi}\frac{\sin\xi}{\xi}\,d\xi = 1.1789798\ldots, \tag{14.76}$$

which means that the Fourier series tends to overshoot the positive corner by some 18 per cent and to undershoot the negative corner by the same amount, as suggested in Fig. 14.9. The inclusion of more terms (increasing r) does nothing to remove this overshoot but merely moves it closer to the point of discontinuity. The overshoot is the Gibbs phenomenon, and because of it the Fourier series representation may be highly unreliable for precise numerical work, especially in the vicinity of a discontinuity.

As suggested in Section 14.4, the Gibbs phenomenon may be suppressed drastically by the use of Lanczos convergence factors. Considering our square wave to be the derivative of a triangular wave, we replace Eq. 14.66 with $h = 2$ with

$$f(x) = \frac{4}{\pi}\sum_{n=1}^{m}\frac{\sin\{[(2n-1)\pi]/2m\}}{[(2n-1)\pi]/2m}\cdot\frac{\sin(2n-1)x}{(2n-1)}. \tag{14.77}$$

The results, with $m = 100$, are plotted in Fig. 14.10. The 18 per cent overshoot is sharply reduced, but the reader will note that a price has been paid for this smoothing: the rate of rise has been cut roughly in half.

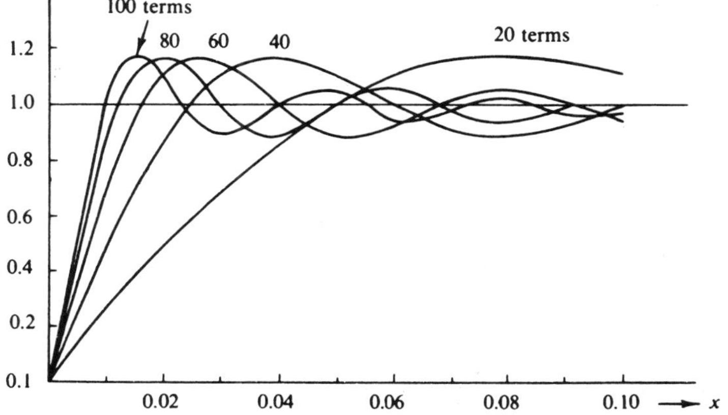

FIG. 14.9 Square wave—Gibbs phenomenon

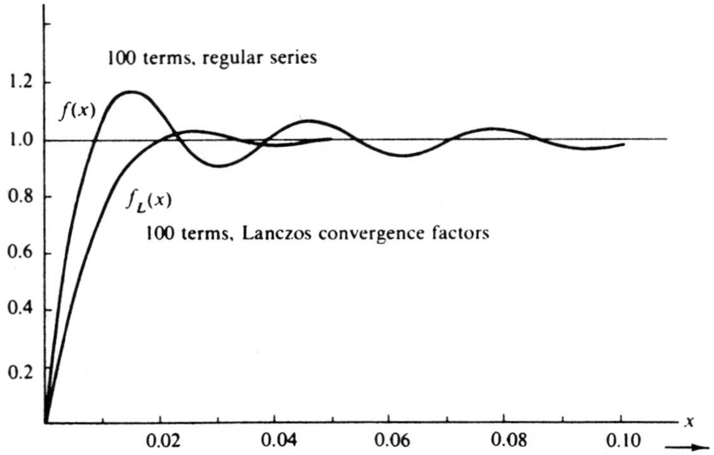

FIG. 14.10 Square wave—smoothing effect of Lanczos convergence factors

EXERCISES

14.5.1 With the partial sum summation techniques of this section, show that at a discontinuity in $f(x)$ the Fourier series for $f(x)$ takes on the arithmetic mean of the right and left hand limits:

$$f(x_0) = \tfrac{1}{2}[f(x_0+) + f(x_0-)].$$

In evaluating $\lim_{r\to\infty} s_r(x_0)$ you may find it convenient to identify part of the integrand as a Dirac delta function.

14.5.2 Determine the partial sum, s_n, of the series in Eq. 14.66 by using

(a) $$\frac{\sin mx}{m} = \int_0^x \cos my\, dy$$

and

(b) $$\sum_{p=1}^{n} \cos(2p-1)y = \frac{\sin 2ny}{2\sin y}.$$

Do you agree with the result given in Eq. 14.73?

REFERENCES

CARSLAW, H. S., *Introduction to the Theory of Fourier's Series and Integrals*. 2nd Ed. London: Macmillan (1921). A detailed and classic work. Includes a considerable discussion of Gibbs phenomenon in Chapter IX.

JEFFREYS, H. and B. S. JEFFREYS, *Methods of Mathematical Physics*. 3rd Ed. Cambridge: Cambridge University Press (1966).

LANCZOS, C., *Applied Analysis*. Englewood Cliffs, New Jersey: Prentice-Hall (1956). A well-written presentation of the Lanczos convergence technique for Fourier series. This and several other topics are presented from the point of view of a mathematician who wants useful numerical results and not just abstract existence theorems.

ZYGMUND, A., *Trigonometric Series*, Vols. I and II. Cambridge: Cambridge University Press (1959). An extremely complete exposition, including relatively recent results in the realm of pure mathematics.

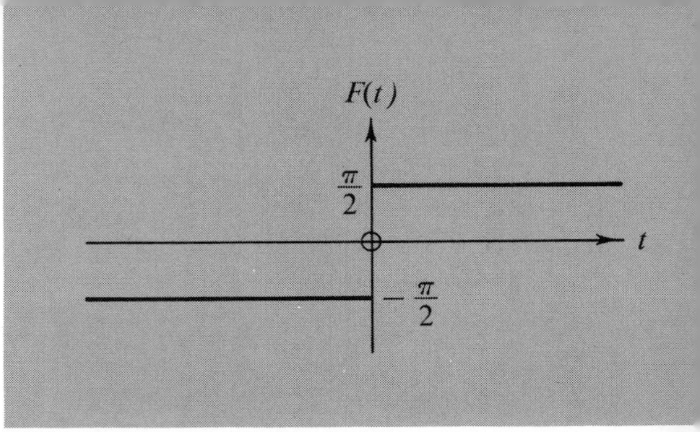

CHAPTER 15

INTEGRAL TRANSFORMS

15.1 Integral Transforms

Frequently in mathematical physics we encounter pairs of functions related by an expression of the following form:

$$g(\alpha) = \int_a^b f(t)\, K(\alpha, t)\, dt. \tag{15.1}$$

The function $g(\alpha)$ is called the (integral) transform of $f(t)$ by the kernel $K(\alpha, t)$.

Fourier transform. One of the most useful of the infinite number of possible transforms is the Fourier transform given by

$$g(\alpha) = \frac{1}{\sqrt{2\pi}} \int_{-\infty}^{\infty} f(t) e^{i\alpha t}\, dt. \tag{15.2}$$

Two modifications of this form, developed in Section 15.3, are the Fourier cosine and Fourier sine transforms:

$$g_c(\alpha) = \sqrt{\frac{2}{\pi}} \int_0^{\infty} f(t) \cos \alpha t\, dt, \tag{15.3}$$

$$g_s(\alpha) = \sqrt{\frac{2}{\pi}} \int_0^{\infty} f(t) \sin \alpha t\, dt. \tag{15.4}$$

The Fourier transform is based on the kernel $e^{i\alpha t}$ and its real and imaginary parts taken separately, $\cos \alpha t$ and $\sin \alpha t$. Because these kernels are the functions used to describe waves, Fourier transforms appear frequently in studies of waves and of the extraction of information from waves, particularly when phase information is involved. The output of a stellar interferometer, for instance, involves a Fourier transform of the brightness across a stellar disk. The electron distribution in an atom may be obtained from a Fourier transform of the amplitude of scattered X-rays.

Laplace, Mellin, and Hankel transforms. Three other useful kernels are

$$e^{-\alpha t}, \qquad t\,J_n(\alpha t), \qquad t^{\alpha-1}.$$

These give rise to the following transforms

$$g(\alpha) = \int_0^\infty f(t)\,e^{-\alpha t}\,dt, \qquad \text{Laplace transform} \tag{15.5}$$

$$g(\alpha) = \int_0^\infty f(t)t\,J_n(\alpha t)\,dt, \qquad \text{Hankel transform (Fourier-Bessel)} \tag{15.6}$$

$$g(\alpha) = \int_0^\infty f(t)t^{\alpha-1}\,dt, \qquad \text{Mellin transform.} \tag{15.7}$$

Clearly the possible types are unlimited. These transforms have been useful in mathematical analysis and in physical applications. We have actually used the Mellin transform without calling it by name; that is, $g(\alpha) = (\alpha - 1)!$ is the Mellin transform of $f(t) = e^{-t}$. Of course, we could just as well say $g(\alpha) = n!/\alpha^{n+1}$ is the Laplace transform of $f(t) = t^n$. Of the three, the Laplace transform is by far the most used. It is discussed at length in Sections 15.7–15.11. The Hankel transform, a Fourier transform for a Bessel function expansion, represents a limiting case of a Fourier-Bessel series. It occurs in potential problems in cylindrical coordinates.

Linearity. All of these integral transforms are linear; that is

$$\int_a^b [c_1 f_1(t) + c_2 f_2(t)]\,K(\alpha, t)\,dt = \int_a^b c_1 f_1(t)\,K(\alpha, t)\,dt + \int_a^b c_2 f_2(t)\,K(\alpha, t)\,dt, \tag{15.8}$$

$$\int_a^b c\,f(t)\,K(\alpha, t)\,dt = c\int_a^b f(t)\,K(\alpha, t)\,dt, \tag{15.9}$$

where c_1 and c_2 are constants and $f_1(t)$ and $f_2(t)$ are functions for which the transform operation is defined.

Representing our linear integral transform by the operator $\mathscr{L}$, we obtain

$$g(\alpha) = \mathscr{L}f(t). \tag{15.10}$$

We expect an inverse operator $\mathscr{L}^{-1}$ exists such that[1]

$$f(t) = \mathscr{L}^{-1} g(\alpha). \tag{15.11}$$

For our three Fourier transforms $\mathscr{L}^{-1}$ is given in Section 15.3. In general, the determination of the inverse transform is the main problem in using integral transforms. The inverse Laplace transform is discussed in Section 15.11. For details of the inverse Hankel and inverse Mellin transforms the reader is referred to the references at the end of the chapter.

EXERCISES

15.1.1 The Fourier transforms for a function of two variables are

$$F(u, v) = \frac{1}{2\pi} \iint_{-\infty}^{\infty} f(x, y) e^{i(ux+vy)}\, dx\, dy,$$

$$f(x, y) = \frac{1}{2\pi} \iint_{-\infty}^{\infty} F(u, v) e^{-i(ux+vy)}\, du\, dv.$$

Using $f(x, y) = f([x^2 + y^2]^{1/2})$, show that the *zero*-order Hankel transforms

$$F(\rho) = \int_0^\infty r f(r) J_0(\rho r)\, dr,$$

$$f(r) = \int_0^\infty \rho F(\rho) J_0(\rho r)\, d\rho,$$

are a special case of the Fourier transforms.

This technique may be generalized to derive the nth order Hankel transforms (cf. Sneddon, *Fourier Transforms*, Chapter 2). It might also be noted that the Hankel transforms of nonintegral order $\nu = \pm\frac{1}{2}$ reduce to Fourier sine and cosine transforms.

15.1.2 Assuming the validity of the Hankel transform–inverse transform pair of equations

$$g(\alpha) = \int_0^\infty f(t) J_n(\alpha t)\, t\, dt$$

$$f(t) = \int_0^\infty g(\alpha) J_n(\alpha t)\, \alpha\, d\alpha,$$

show that the Dirac delta function has a Bessel integral representation

$$\delta(t - t') = t \int_0^\infty J_n(\alpha t) J_n(\alpha t')\, \alpha\, d\alpha.$$

This expression is useful in developing Green's functions in cylindrical coordinates–where the eigenfunctions are Bessel functions.

[1] Expectation is not proof, and here proof of existence is complicated because we are actually in an *infinite*-dimensional Hilbert space. We shall prove existence in the special cases of interest by actual construction.

15.1.3 From the Fourier transforms, Eqs. 15.22 and 15.23, show that the transformation

$$t \to \ln x$$
$$i\omega \to \alpha - \gamma$$

leads to

$$G(\alpha) = \int_0^\infty F(x) x^{\alpha-1}\, dx$$

and

$$F(x) = \frac{1}{2\pi i} \int_{\gamma - i\infty}^{\gamma + i\infty} G(\alpha) x^{-\alpha}\, d\alpha.$$

These are the Mellin transforms. A similar change of variables is employed in Section 15.11 to derive the inverse Laplace transform.

15.1.4 Verify the following Mellin transforms:

(a) $\displaystyle\int_0^\infty x^{\alpha-1} \sin(kx)\, dx = k^{-\alpha}(\alpha - 1)! \sin\frac{\pi\alpha}{2}, \qquad -1 < \alpha < 1.$

(b) $\displaystyle\int_0^\infty x^{\alpha-1} \cos(kx)\, dx = k^{-\alpha}(\alpha - 1)! \cos\frac{\pi\alpha}{2}, \qquad 0 < \alpha < 1.$

Hint: You can force the integrals into a tractable form by inserting a convergence factor e^{-bx} and (after integrating) letting $b \to 0$.

15.2 Development of the Fourier Integral

In Chapter 14 it was shown that Fourier series are useful in representing certain functions (a) over a limited range $[0, 2\pi]$, $[-L, L]$, etc., or (b) for the infinite interval $(-\infty, \infty)$, *if the function is periodic.* We now turn our attention to the problem of representing a nonperiodic function over the infinite range. Physically this means resolving a single pulse or wave packet into sinusoidal waves.

We have seen (Section 14.2) that for the interval $[-L, L]$ the coefficients a_n and b_n could be written

$$a_n = \frac{1}{L} \int_{-L}^{L} f(t) \cos\frac{n\pi t}{L}\, dt \tag{15.12}$$

$$b_n = \frac{1}{L} \int_{-L}^{L} f(t) \sin\frac{n\pi t}{L}\, dt. \tag{15.13}$$

The resulting Fourier series is

$$f(x) = \frac{1}{2L} \int_{-L}^{L} f(t)\, dt + \frac{1}{L} \sum_{n=1}^{\infty} \cos\frac{n\pi x}{L} \int_{-L}^{L} f(t) \cos\frac{n\pi t}{L}\, dt$$
$$+ \frac{1}{L} \sum_{n=1}^{\infty} \sin\frac{n\pi x}{L} \int_{-L}^{L} f(t) \sin\frac{n\pi t}{L}\, dt \tag{15.14}$$

or

$$f(x)=\frac{1}{2L}\int_{-L}^{L} f(t)\,dt+\frac{1}{L}\sum_{n=1}^{\infty}\int_{-L}^{L} f(t)\cos\frac{n\pi}{L}(t-x)\,dt. \tag{15.15}$$

We now let the parameter L approach infinity, transforming the finite interval $[-L, L]$ into the infinite interval $(-\infty, \infty)$. We set

$$\frac{n\pi}{L}=\omega, \qquad \frac{\pi}{L}=\Delta\omega, \qquad \text{with } L\to\infty.$$

Then we have

$$f(x)\to\frac{1}{\pi}\sum_{n=1}^{\infty}\Delta\omega\int_{-\infty}^{\infty} f(t)\cos\omega(t-x)\,dt \tag{15.16}$$

$$f(x)=\frac{1}{\pi}\int_{0}^{\infty} d\omega\int_{-\infty}^{\infty} f(t)\cos\omega(t-x)\,dt, \tag{15.17}$$

replacing the infinite sum by the integral over ω. The first term (corresponding to a_0) has vanished, assuming that $\int_{-\infty}^{\infty} f(t)\,dt$ exists.

It must be emphasized that this result (Eq. 15.17) is purely formal. It is not intended as a rigorous derivation, but it can be made rigorous (cf. I. N. Sneddon, *Fourier Transforms*, Section 3.2). However, for those interested in contour integration, a completely different derivation, based on the Cauchy integral formula, is given in our Section 6.4. We shall take Eq. 15.17 as the Fourier integral. It is subject to the restrictions that $f(x)$ satisfy the Dirichlet conditions (Chapter 14) and that $\int_{-\infty}^{\infty}|f(t)|\,dt$ be convergent.

Fourier integral—exponential form. Our Fourier integral (Eq. 15.17) may be put into exponential form by noting that

$$f(x)=\frac{1}{2\pi}\int_{-\infty}^{\infty} d\omega\int_{-\infty}^{\infty} f(t)\cos\omega(t-x)\,dt, \tag{15.18}$$

whereas

$$\frac{1}{2\pi}\int_{-\infty}^{\infty} d\omega\int_{-\infty}^{\infty} f(t)\sin\omega(t-x)\,dt=0; \tag{15.19}$$

$\cos\omega(t-x)$ is an even function of ω and $\sin\omega(t-x)$ is an odd function of ω. Adding Eqs. 15.18 and 15.19 (with a factor i), we obtain

$$f(x)=\frac{1}{2\pi}\int_{-\infty}^{\infty} e^{-i\omega x}\,d\omega\int_{-\infty}^{\infty} f(t)e^{i\omega t}\,dt. \tag{15.20}$$

The variable ω introduced here is an arbitrary mathematical variable. In many physical problems, however, it corresponds to the angular frequency ω. We may then interpret Eq. 15.18 or 15.20 as a representation of $f(x)$ in terms of a distribution of infinitely long sinusoidal wave trains of angular frequency ω in which this frequency is a *continuous* variable.

Dirac delta function derivation. If the order of integration of Eq. 15.20 may be reversed, we may rewrite it as

$$f(x) = \int_{-\infty}^{\infty} f(t)\left\{\frac{1}{2\pi}\int_{-\infty}^{\infty} e^{i\omega(t-x)}\,d\omega\right\} dt \tag{15.20a}$$

Apparently the quantity in curly brackets behaves as a delta function—$\delta(t - x)$. We might take Eq. 15.20*a* as presenting us with a representation of the Dirac delta function. Alternatively, we take it as a clue to a new derivation of the Fourier integral theorem.

From Eq. 8.85*a* (shifting the singularity from $t = 0$ to $t = x$)

$$f(x) = \lim_{n\to\infty} \int_{-\infty}^{\infty} f(t)\,\delta_n(t - x)\,dt, \tag{15.21a}$$

where $\delta_n(t - x)$ is a sequence defining the distribution $\delta(t - x)$. Note that Eq. 15.21*a* assumes that $f(t)$ is continuous at $t = x$.

We take $\delta_n(t - x)$ to be

$$\delta_n(t - x) = \frac{\sin n(t - x)}{\pi(t - x)} = \frac{1}{2\pi}\int_{-n}^{n} e^{i\omega(t-x)}\,d\omega, \tag{15.21b}$$

using Eq. 8.83*d*. Substituting into Eq. 15.21*a*, we have

$$f(x) = \lim_{n\to\infty} \frac{1}{2\pi}\int_{-\infty}^{\infty} f(t)\int_{-n}^{n} e^{i\omega(t-x)}\,d\omega\,dt. \tag{15.21c}$$

Interchanging the order of integration and then taking the limit as $n \to \infty$, we have Eq. 15.20, the Fourier integral theorem.

With the understanding that it belongs under an integral sign as in Eq. 15.21*a*, the identification

$$\delta(t - x) = \frac{1}{2\pi}\int_{-\infty}^{\infty} e^{i\omega(t-x)}\,d\omega, \tag{15.21d}$$

provides a very useful representation of the delta function. It is used to great advantage in Sections 15.5 and 15.6.

15.3 Fourier Transforms—Inversion Theorem

Let us *define* $g(\omega)$, the Fourier transform of the function $f(t)$, by

$$g(\omega) \equiv \frac{1}{\sqrt{2\pi}}\int_{-\infty}^{\infty} f(t)e^{i\omega t}\,dt. \tag{15.22}$$

Exponential transform. Then from Eq. 15.20 (or Section 6.4) we have the inverse relation

$$f(x) = \frac{1}{\sqrt{2\pi}}\int_{-\infty}^{\infty} g(\omega)e^{-i\omega x}\,d\omega. \tag{15.23}$$

It will be noted that Eqs. 15.22 and 15.23 are almost but not quite symmetrical, differing in the sign of i.

Cosine transform. If $f(x)$ is odd or even, these transforms may be expressed in a somewhat different form. Consider, first, $f(x) = f(-x)$, even. Writing the exponential of Eq. 15.22 in trigonometric form, we have

$$g_c(\omega) = \frac{1}{\sqrt{2\pi}} \int_{-\infty}^{\infty} f_c(t)(\cos \omega t + i \sin \omega t)\, dt$$

$$= \sqrt{\frac{2}{\pi}} \int_0^{\infty} f_c(t) \cos \omega t\, dt, \tag{15.24}$$

the $\sin \omega t$ dependence vanishing on integration over the symmetric interval $(-\infty, \infty)$. Similarly, Eq. 15.23 transforms to

$$f_c(x) = \sqrt{\frac{2}{\pi}} \int_0^{\infty} g_c(\omega) \cos \omega x\, d\omega. \tag{15.25}$$

Equations 15.24 and 15.25 are known as Fourier cosine transforms.

Sine transform. The corresponding pair of Fourier sine transforms is obtained by assuming that $f(x) = -f(-x)$, odd, and applying the same symmetry arguments. The equations are

$$g_s(\omega) = \sqrt{\frac{2}{\pi}} \int_0^{\infty} f_s(t) \sin \omega t\, dt,^1 \tag{15.26}$$

$$f_s(x) = \sqrt{\frac{2}{\pi}} \int_0^{\infty} g_s(\omega) \sin \omega x\, d\omega. \tag{15.27}$$

From the last equation, we may develop the physical interpretation that $f(x)$ is being described by a continuum of sine waves. The amplitude of $\sin \omega x$ is given by $\sqrt{2/\pi}\, g_s(\omega)$, in which $g_s(\omega)$ is the Fourier sine transform of $f_s(x)$. It will be seen that Eq. 15.27 is the integral analog of the summation (Eq. 14.18). Similar interpretations hold for the cosine and exponential cases.

If we take Eqs. 15.22, 15.24, and 15.26 as the direct integral transforms, described by $\mathscr{L}$ in Eq. 15.10 (Section 15.1), the corresponding inverse transforms, $\mathscr{L}^{-1}$ of Eq. 15.11, are given by Eqs. 15.23, 15.25, and 15.27.

Example 15.3.1. Finite Wave Train

An important application of the Fourier transform is the resolution of a finite pulse into sinusoidal waves. Imagine that an infinite wave train $\sin \omega_0 t$ is clipped by Kerr cell shutters so that we have

$$f(t) = \begin{cases} \sin \omega_0 t, & |t| < \dfrac{N\pi}{\omega_0}, \\ 0, & |t| > \dfrac{N\pi}{\omega_0}. \end{cases} \tag{15.28}$$

[1] Note that a factor $-i$ has been absorbed into this $g(\omega)$.

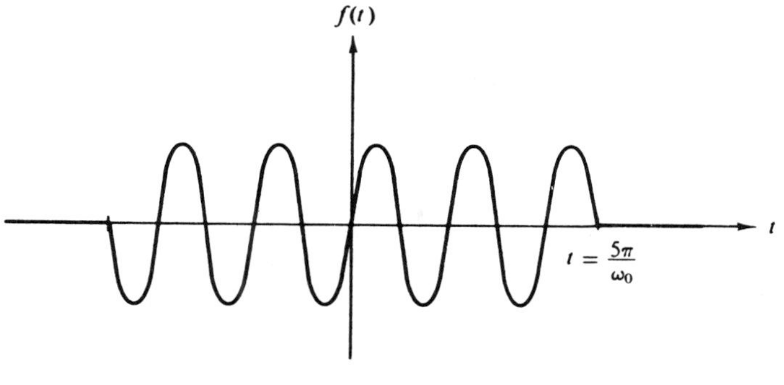

FIG. 15.1 Finite wave train

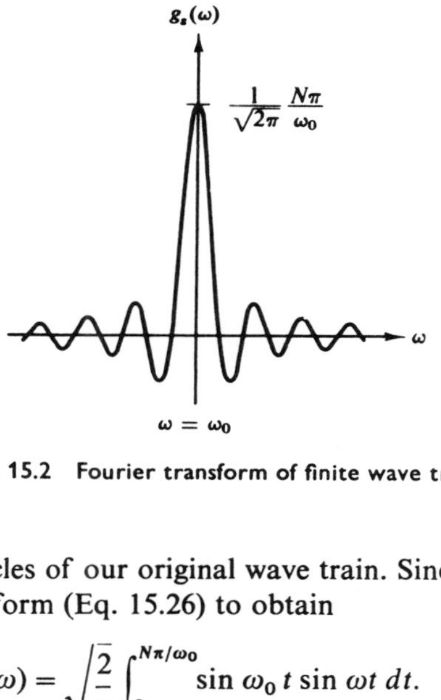

FIG. 15.2 Fourier transform of finite wave train

This corresponds to N cycles of our original wave train. Since $f(t)$ is odd, we may use the Fourier sine transform (Eq. 15.26) to obtain

$$g_s(\omega) = \sqrt{\frac{2}{\pi}} \int_0^{N\pi/\omega_0} \sin \omega_0 t \sin \omega t \, dt. \tag{15.29}$$

Integrating, we find our amplitude function

$$g_s(\omega) = \sqrt{\frac{2}{\pi}} \left[\frac{\sin[(\omega_0 - \omega)(N\pi/\omega_0)]}{2(\omega_0 - \omega)} - \frac{\sin[(\omega_0 + \omega)(N\pi/\omega_0)]}{2(\omega_0 + \omega)} \right] \tag{15.30}$$

It is of some considerable interest to see how $g_s(\omega)$ depends on frequency. For large ω_0 and $\omega \approx \omega_0$ only the first term will be of any importance. It is plotted in Fig. 15.2. This is the amplitude curve for the single slit diffraction pattern. There are zeroes at

$$\frac{\omega_0 - \omega}{\omega_0} = \frac{\Delta\omega}{\omega_0} = \pm\frac{1}{N}, \qquad \pm\frac{2}{N}, \text{ etc.} \tag{15.31}$$

Since the contributions outside the central maximum are small, we may take

$$\Delta\omega = \frac{\omega_0}{N} \tag{15.32}$$

as a good measure of the spread in frequency of our wave pulse. Clearly, if N is large (a long pulse), the frequency spread will be small. On the other hand, if our pulse is clipped short, N small, the frequency distribution will be wider.

Uncertainty principle. Here is a classical analog of the famous uncertainty principle of quantum mechanics. If we are dealing with electromagnetic waves,

$$\begin{aligned} \frac{h\omega}{2\pi} &= E, \qquad \text{energy (of our wave pulse or photon)} \\ \frac{h\,\Delta\omega}{2\pi} &= \Delta E, \end{aligned} \tag{15.33}$$

h being Planck's constant, which represents an uncertainty in the energy of our pulse. There is also an uncertainty in the time, for our wave of N cycles requires $2N\pi/\omega_0$ seconds to pass. Taking

$$\Delta t = \frac{2N\pi}{\omega_0}, \tag{15.34}$$

we have the product of these two uncertainties:

$$\begin{aligned} \Delta E \cdot \Delta t &= \frac{h\,\Delta\omega}{2\pi} \cdot \frac{2\pi N}{\omega_0} \\ &= h\,\frac{\omega_0}{2\pi N} \cdot \frac{2\pi N}{\omega_0} = h. \end{aligned} \tag{15.35}$$

The Heisenberg uncertainty principle actually states

$$\Delta E \cdot \Delta t \geqslant \frac{h}{4\pi} \tag{15.36}$$

and this is clearly satisfied in our example.

EXERCISES

15.3.1 (a) Show that $g(-\omega) = g^*(\omega)$ is a necessary and sufficient condition for $f(x)$ to be real.
(b) Show that $g(-\omega) = -g^*(\omega)$ is a necessary and sufficient condition for $f(x)$ to be pure imaginary.

15.3.2 The function

$$f(x) = \begin{cases} 1, & |x| < 1 \\ 0, & |x| > 1 \end{cases}$$

is a symmetrical finite step function.

(a) Find the $g_c(\omega)$, Fourier cosine transform of $f(x)$.

(b) Taking the inverse cosine transform show that

$$f(x) = \frac{2}{\pi} \int_0^\infty \frac{\sin \omega \cos \omega x}{\omega} \, d\omega.$$

(c) From part (b) show that

$$\int_0^\infty \frac{\sin \omega \cos \omega x}{\omega} \, d\omega = \begin{cases} 0, & |x| > 1, \\ \dfrac{\pi}{4}, & |x| = 1, \\ \dfrac{\pi}{2}, & |x| < 1. \end{cases}$$

Ans. (a) $g_c(\omega) = \sqrt{\dfrac{2}{\pi}} \dfrac{\sin \omega}{\omega}$.

15.3.3 (a) Show that the Fourier sine and cosine transforms of e^{-at} are

$$g_s(\omega) = \sqrt{\frac{2}{\pi}} \frac{\omega}{\omega^2 + a^2}$$

$$g_c(\omega) = \sqrt{\frac{2}{\pi}} \frac{a}{\omega^2 + a^2}.$$

Hint: Each of the transforms can be related to the other by integration by parts.

(b) Show that

$$\int_0^\infty \frac{\omega \sin \omega x}{\omega^2 + a^2} \, d\omega = \frac{\pi}{2} e^{-ax}, \quad x > 0,$$

$$\int_0^\infty \frac{\cos \omega x}{\omega^2 + a^2} \, d\omega = \frac{\pi}{2a} e^{-ax}, \quad x \geq 0.$$

These results may also be obtained by contour integration (Exercise 7.2.10).

15.3.4 Find the Fourier transform of the triangular pulse

$$f(x) = \begin{cases} h(1 - a|x|) & |x| < 1/a, \\ 0, & |x| > 1/a. \end{cases}$$

Note: This function provides another delta sequence with $h = a$ and $a \to \infty$.

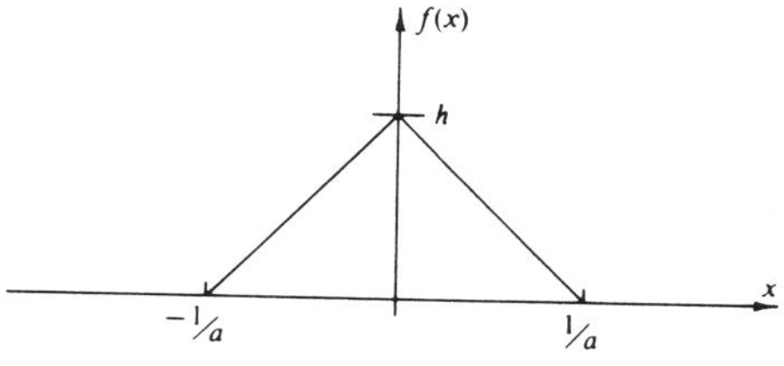

15.3.5 In a resonant cavity an electromagnetic oscillation of frequency ω_0 dies out as

$$A(t) = A_0 e^{-\omega_0 t/2Q} e^{-i\omega_0 t}, \qquad t > 0.$$

(Take $A(t) = 0$ for $t < 0$.)

The parameter Q is a measure of the ratio of stored energy to energy loss per cycle. Calculate the frequency distribution of the oscillation, $a^*(\omega)a(\omega)$, where $a(\omega)$ is the Fourier transform of $A(t)$.
Note: The larger Q is, the sharper your resonance line.

$$\textit{Ans.}\quad a^*(\omega)a(\omega) = \frac{{A_0}^2}{2\pi}\,\frac{1}{(\omega-\omega_0)^2 + (\omega_0/2Q)^2}.$$

15.3.6 Verify that the following are Fourier integral transforms of one another:

(a) $$\left.\begin{array}{ll} \sqrt{\dfrac{2}{\pi}}\cdot\dfrac{1}{\sqrt{a^2-x^2}}, & |x| < a \\ 0, & |x| > a \end{array}\right\} \text{ and } J_0(ay),$$

(b) $$\left.\begin{array}{ll} 0, & |x| < a \\ -\sqrt{\dfrac{2}{\pi}}\,\dfrac{1}{\sqrt{x^2-a^2}}, & |x| > a \end{array}\right\} \text{ and } N_0(a|y|),$$

(c) $$\sqrt{\frac{\pi}{2}}\cdot\frac{1}{\sqrt{x^2+a^2}} \quad \text{and} \quad K_0(a|y|).$$

(d) Can you suggest why $I_0(ay)$ is not included in this list?
Hint. J_0, N_0, and K_0 may be transformed most easily by using an exponential representation, reversing the order of integration, and employing the Dirac delta function exponential representation (Section 15.2). These cases can be treated equally well as Fourier cosine transforms.

15.3.7 A calculation of the magnetic field of a circular current loop in circular cylindrical coordinates leads to the integral

$$\int_0^\infty \cos kz\, k\, K_1(ka)\, dk.$$

Show that this integral is equal to

$$\frac{\pi a}{2(z^2+a^2)^{3/2}}.$$

Hint. Try differentiating Ex. 15.3.6(c).

15.3.8 We may define a sequence

$$\delta_n(x) = \begin{cases} n & |x| < 1/2n. \\ 0 & |x| > 1/2n. \end{cases}$$

(This is Eq. 8.83a.) Express $\delta_n(x)$ as a Fourier integral (via the Fourier integral theorem, inverse transform, etc.). Finally show that we may write

$$\delta(x) = \lim_{n\to\infty} \delta_n(x) = \frac{1}{2\pi}\int_{-\infty}^{\infty} e^{-ikx}\, dk.$$

15.3.9 Using the sequence

$$\delta_n(x) = \frac{n}{\sqrt{\pi}} \exp(-n^2x^2),$$

show that $$\delta(x) = \frac{1}{2\pi}\int_{-\infty}^{\infty} e^{-ikx}\,dk.$$

Note. Remember that $\delta(x)$ is defined in terms of its behavior as part of an integrand—Section 8.6, especially Eqs. 8.85*a* and 8.85*b*.

15.3.10 The Fourier integral, Eq. 15.18, has been held meaningless for $f(t) = \cos \alpha t$. Show that the Fourier integral can be extended to cover $f(t) = \cos \alpha t$ by use of the Dirac delta function.

15.3.11 Show that

$$\int_0^{\infty} \sin ka\, J_0(k\rho)\,dk = \begin{cases}(a^2-\rho^2)^{-1/2}, & \rho < a,\\ 0, & \rho > a.\end{cases}$$

Here a and ρ are positive. The equation comes from the determination of the distribution of charge on an isolated conducting disk, radius a. Cf. Exercises 2.11.2 and 11.1.28.

15.4 Fourier Transform of Derivatives

Using the exponential form, the Fourier transform of $f(x)$ is

$$g(\omega) = \frac{1}{\sqrt{2\pi}}\int_{-\infty}^{\infty} f(x)e^{i\omega x}\,dx \tag{15.37}$$

and for $df(x)/dx$

$$g_1(\omega) = \frac{1}{\sqrt{2\pi}}\int_{-\infty}^{\infty} \frac{d\,f(x)}{dx}\,e^{i\omega x}\,dx. \tag{15.38}$$

Integrating Eq. 15.38 by parts, we obtain

$$g_1(\omega) = \frac{e^{i\omega x}}{\sqrt{2\pi}}f(x)\Big|_{-\infty}^{\infty} - \frac{i\omega}{\sqrt{2\pi}}\int_{-\infty}^{\infty} f(x)e^{i\omega x}\,dx. \tag{15.39}$$

If $f(x)$ vanishes as $x \to \pm\infty$, we obtain

$$g_1(\omega) = -i\omega\, g(\omega); \tag{15.40}$$

that is, the transform of the derivative is $(-i\omega)$ times the transform of the original function. This may readily be generalized to the nth derivative to yield

$$g_n(\omega) = (-i\omega)^n\, g(\omega), \tag{15.41}$$

provided all the integrated parts vanish as $x \to \pm\infty$.

Example 15.4.1. Wave Equation

This technique may be used to advantage in handling partial differential equations. To illustrate the technique let us derive a familiar expression of elementary physics. An infinitely long string is vibrating freely. The amplitude y of the (small) vibrations satisfies the wave equation

$$\frac{\partial^2 y}{\partial x^2} = \frac{1}{v^2}\frac{\partial^2 y}{\partial t^2}. \tag{15.42}$$

We shall assume that

$$y = f(x) \tag{15.43}$$

at $t = 0$.

Applying our Fourier transform, which means multiplying by $e^{i\alpha x}$ and integrating over x, we obtain

$$\int_{-\infty}^{\infty} \frac{\partial^2 y(x, t)}{\partial x^2} e^{i\alpha x}\, dx = \frac{1}{v^2} \int_{-\infty}^{\infty} \frac{\partial^2 y(x, t)}{\partial t^2} e^{i\alpha x}\, dx \tag{15.44}$$

or

$$(-i\alpha)^2\, Y(\alpha, t) = \frac{1}{v^2} \frac{\partial^2 Y(\alpha, t)}{\partial t^2}. \tag{15.45}$$

Here we have used

$$Y(\alpha, t) = \frac{1}{\sqrt{2\pi}} \int_{-\infty}^{\infty} y(x, t) e^{i\alpha x}\, dx \tag{15.46}$$

and Eq. 15.41 for the second derivative. Note that the integrated part of Eq. 15.39 vanishes. The wave has not yet got out to ∞. Since no derivatives with respect to α appear, Eq. 15.45 is actually an ordinary differential equation—in fact the linear oscillator equation. This transformation, from a partial to an ordinary differential equation, is a significant achievement. We solve Eq. 15.45 subject to the appropriate initial conditions. At $t = 0$, applying Eq. 15.43, Eq. 15.46 reduces to

$$\begin{aligned} Y(\alpha, 0) &= \frac{1}{\sqrt{2\pi}} \int_{-\infty}^{\infty} f(x) e^{i\alpha x}\, dx \\ &= F(\alpha). \end{aligned} \tag{15.47}$$

The general solution of Eq. 15.45 in exponential form is

$$Y(\alpha, t) = F(\alpha) e^{\pm iv\alpha t}. \tag{15.48}$$

Using the inversion formula (Eq. 15.23), we have

$$y(x, t) = \frac{1}{\sqrt{2\pi}} \int_{-\infty}^{\infty} Y(\alpha, t) e^{-i\alpha x}\, d\alpha \tag{15.49}$$

and, by Eq. 15.48,

$$y(x, t) = \frac{1}{\sqrt{2\pi}} \int_{-\infty}^{\infty} F(\alpha) e^{-i\alpha(x \mp vt)}\, d\alpha. \tag{15.50}$$

Since $f(x)$ is the Fourier inverse transform of $F(\alpha)$,

$$y(x, t) = f(x \mp vt), \tag{15.51}$$

corresponding to waves advancing in the $+x$- and $-x$-directions, respectively.

The particular linear combinations of waves is given by the boundary condition of Eq. 15.43 and some other boundary condition such as a restriction on $\partial y/\partial t$.

EXERCISES

15.4.1 The one-dimensional Fermi age equation for the diffusion of neutrons slowing down in some medium (such as graphite) is

$$\frac{\partial^2 q(x, \tau)}{\partial x^2} = \frac{\partial q(x, \tau)}{\partial \tau}.$$

Here q is the number of neutrons that "slow down," falling below some given energy per second per unit volume. The Fermi age, τ, is a measure of the energy loss.

If $q(x, 0) = S\,\delta(x)$, corresponding to a plane source of neutrons at $x = 0$, emitting S neutrons per unit area per second, derive the solution

$$q = S\frac{e^{-x^2/4\tau}}{\sqrt{4\pi\tau}}$$

Hint: Replace $q(x, \tau)$ with

$$p(k, \tau) = \frac{1}{\sqrt{2\pi}} \int_{-\infty}^{\infty} q(x, \tau) e^{ikx}\, dx.$$

This is analogous to the diffusion of heat in an infinite medium.

15.4.2 Equation 15.41 yields

$$g_2(\omega) = -\omega^2 g(\omega)$$

for the Fourier transform of the second derivative of $f(x)$. The condition $f(x) \to 0$ for $x \to \pm\infty$ may be relaxed slightly. Find the *least* restrictive condition for the above equation for $g_2(\omega)$ to hold.

$$\textit{Ans.} \quad \left[\frac{df(x)}{dx} - i\omega f(x)\right] e^{i\omega x}\Big|_{-\infty}^{\infty} = 0.$$

15.5 Convolution Theorem

We consider two functions $f(x)$ and $g(x)$ with Fourier transforms $F(t)$ and $G(t)$, respectively. We define the operation

$$f * g \equiv \frac{1}{\sqrt{2\pi}} \int_{-\infty}^{\infty} g(y) f(x - y)\, dy \tag{15.52}$$

as the *convolution* of the two functions f and g over the interval $(-\infty, \infty)$. In other works this is sometimes referred to as the *Faltung*, to use the German term for "folding."[1] We now transform the integral in Eq. 15.52 by introducing the Fourier transforms.

[1] For $f(y) = e^{-y}$, $f(y)$ and $f(x - y)$ are plotted in Fig. 15.3. Clearly, $f(y)$ and $f(x - y)$ are mirror images of each other in relation to the vertical line $y = x/2$, that is, we could generate $f(x - y)$ by "folding" over $f(y)$ on the line $y = x/2$.

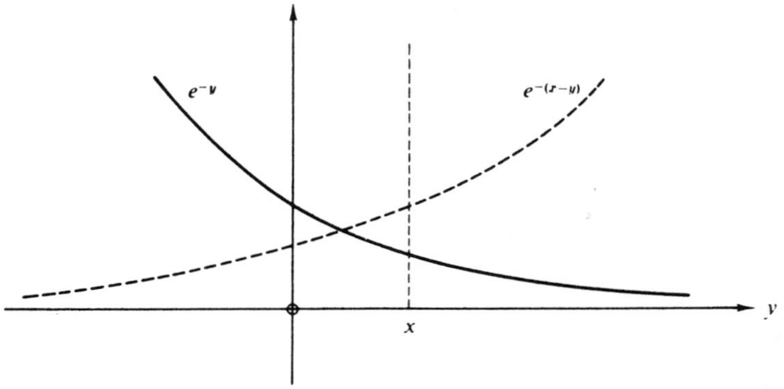

FIG. 15.3

$$\begin{aligned}\int_{-\infty}^{\infty} g(y) f(x-y)\,dy &= \frac{1}{\sqrt{2\pi}} \int_{-\infty}^{\infty} g(y) \int_{-\infty}^{\infty} F(t) e^{-it(x-y)} dt\,dy \\ &= \frac{1}{\sqrt{2\pi}} \int_{-\infty}^{\infty} F(t) e^{-itx} dt \int_{-\infty}^{\infty} g(y) e^{ity}\,dy \\ &= \int_{-\infty}^{\infty} F(t)\,G(t) e^{-itx}\,dt, \end{aligned} \tag{15.53}$$

interchanging the order of integration and transforming $g(y)$. This result may be interpreted as follows: the Fourier inverse transform of a *product* of Fourier transforms is the convolution of the original functions, $f * g$.

For the special case $x = 0$ we have

$$\int_{-\infty}^{\infty} F(t)\,G(t)\,dt = \int_{-\infty}^{\infty} f(-y)\,g(y)\,dy. \tag{15.54}$$

Parseval's relation. Results analogous to Eqs. 15.53 and 15.54 may be derived for the Fourier sine and cosine transforms (Exercises 15.5.1, 2). Equation 15.54 and the corresponding sine and cosine convolutions are often labeled "Parseval's relations" by analogy with Parseval's theorem for Fourier series (Chapter 14, Exercise 14.4.2).

The Parseval relation[1]

$$\int_{-\infty}^{\infty} F(\omega)\,G^*(\omega)\,d\omega = \int_{-\infty}^{\infty} f(t)\,g^*(t)\,dt, \tag{15.55}$$

may be derived very beautifully using the Dirac delta function representation, Eq. 15.21*d*. We have

$$\int_{-\infty}^{\infty} f(t)\,g^*(t)\,dt = \int_{-\infty}^{\infty} \frac{1}{\sqrt{2\pi}} \int_{-\infty}^{\infty} F(\omega)\,e^{-i\omega t}\,d\omega \cdot \frac{1}{\sqrt{2\pi}} \int_{-\infty}^{\infty} G^*(x)\,e^{ixt}\,dx\,dt, \tag{15.56}$$

[1] Note that all arguments are positive in contrast to Eq. 15.54.

with attention to the complex conjugation in the $G^*(x)$ to $g^*(t)$ transform. Integrating over t first, and using Eq. 15.21d,

$$\int_{-\infty}^{\infty} f(t)\, g^*(t)\, dt = \int_{-\infty}^{\infty} F(\omega) \int_{-\infty}^{\infty} G^*(x)\, \delta(x-\omega)\, dx\, d\omega$$

$$= \int_{-\infty}^{\infty} F(\omega)\, G^*(\omega)\, d\omega, \tag{15.57}$$

our desired Parseval relation. An extremely important physical interpretation of this Parseval relation is presented in the next section.

Parseval's relations may be developed independently of the inverse Fourier transform and then used rigorously to derive the inverse transform. Details are given by Morse and Feshbach,[1] Section 4.8 (see also Exercise 15.5.3).

EXERCISES

15.5.1 Work out the convolution equation corresponding to Eq. 15.53 for
(a) Fourier sine transforms

$$\frac{1}{2}\int_{-\infty}^{\infty} g(y) f(x-y)\, dy = -\int_0^{\infty} F_s(s)\, G_s(s) \cos sx\, ds,$$

where f and g are odd functions.
(b) Fourier cosine transforms

$$\frac{1}{2}\int_{-\infty}^{\infty} g(y) f(x-y)\, dy = \int_0^{\infty} F_c(s)\, G_c(s) \cos sx\, ds,$$

where f and g are even functions.

15.5.2 Show that for both Fourier sine and Fourier cosine transforms Parseval's relation has the form

$$\int_0^{\infty} F(t)\, G(t)\, dt = \int_0^{\infty} f(y)\, g(y)\, dy.$$

15.5.3 Starting from Parseval's relation (Eq. 15.54), let $g(y) = 1$, $0 \leqslant y \leqslant \alpha$, and zero elsewhere. From this derive the Fourier inverse transform, (Eq. 15.23).
Hint. Differentiate with respect to α.

15.6 Momentum Representation

In advanced dynamics and in quantum mechanics linear momentum and spacial position occur on an equal footing. In this section we shall start with the usual space distribution and derive the corresponding momentum distribution. For the

[1] Morse, P. M. and H. Feshbach, *Methods of Theoretical Physics*. New York: McGraw-Hill (1953).

one-dimensional case our wave function $\psi(x)$, a solution of the Schrödinger wave equation, has the following properties:

1. $\psi^*(x)\,\psi(x)\,dx$ is the probability of finding the quantum particle between x and $x + dx$ and

2. $$\int_{-\infty}^{\infty} \psi^*(x)\,\psi(x)\,dx = 1, \tag{15.58}$$

corresponding to *one* particle (along the x-axis).

In addition, we have

3. $$\langle x \rangle = \int_{-\infty}^{\infty} \psi^*(x) x\, \psi(x)\,dx \tag{15.59}$$

for the *average* position of the particle along the x-axis. This is often called an expectation value.

We want a function $g(p)$ that will give the same information about the momentum.

1. $g^*(p)\,g(p)\,dp$ is the probability that our quantum particle has a momentum between p and $p + dp$.

2. $$\int_{-\infty}^{\infty} g^*(p)\,g(p)\,dp = 1. \tag{15.60}$$

3. $$\langle p \rangle = \int_{-\infty}^{\infty} g^*(p) p\, g(p) dp \tag{15.61}$$

As shown below, such a function is given by the Fourier transform of our space function $\psi(x)$. Specifically,[1]

$$g(p) = \frac{1}{\sqrt{2\pi\hbar}} \int_{-\infty}^{\infty} \psi(x) e^{-ipx/\hbar}\,dx \tag{15.62}$$

$$g^*(p) = \frac{1}{\sqrt{2\pi\hbar}} \int_{-\infty}^{\infty} \psi^*(x) e^{ipx/\hbar}\,dx. \tag{15.63}$$

To verify this, let us check on properties 2 and 3.

Property 2, the normalization, is automatically satisfied as a Parseval relation, Eq. 15.55. If the space function $\psi(x)$ is normalized to unity, the momentum function $g(p)$ is also.

To check on property 3 we must show that

$$\langle p \rangle = \int_{-\infty}^{\infty} g^*(p) p\, g(p)\,dp = \int_{-\infty}^{\infty} \psi^*(x) \frac{\hbar}{i} \frac{d}{dx} \psi(x)\,dx, \tag{15.64}$$

[1] The $\hbar$ may be avoided by using the wave number k, $p = k\hbar$ (and $\mathbf{p} = \mathbf{k}\hbar$), so that

$$\varphi(k) = \frac{1}{(2\pi)^{1/2}} \int \psi(x)\, e^{-ikx}\,dx.$$

An example of this notation appears in Section 16.1.

where $(\hbar/i)(d/dx)$ is the momentum operator in the space representation. We replace the momentum functions and the first integral becomes

$$\frac{1}{2\pi\hbar}\iiint_{-\infty}^{\infty} pe^{-ip(x-x')/\hbar}\,\psi^*(x')\,\psi(x)\,dp\,dx'\,dx. \tag{15.65}$$

Now

$$pe^{-ip(x-x')/\hbar} = \frac{d}{dx}\left[-\frac{\hbar}{i}\,e^{-ip(x-x')/\hbar}\right]. \tag{15.66}$$

Substituting into Eq. 15.65 and integrating by parts, holding x' and p constant, we obtain

$$\langle p\rangle = \iint_{-\infty}^{\infty}\left[\frac{1}{2\pi\hbar}\int_{-\infty}^{\infty} e^{-ip(x-x')/\hbar}\,dp\right]\cdot\psi^*(x')\frac{\hbar}{i}\frac{d}{dx}\psi(x)\,dx'\,dx. \tag{15.67}$$

Here we assume $\psi(x)$ vanishes as $x \to \pm\infty$, eliminating the integrated part. Again using the Dirac delta function, Eq. 15.21c, Eq. 15.67 reduces to Eq. 15.64 to verify our momentum representation. The reader will note that technically we have employed the inverse Fourier transform in Eq. 15.62. This was chosen deliberately to yield the proper sign in Eq. 15.67.

EXAMPLE 15.6.1. HYDROGEN ATOM

The hydrogen atom ground state may be described by the spacial wave function

$$\psi(\mathbf{r}) = \left(\frac{1}{\pi a_0^3}\right)^{1/2} e^{-r/a_0}, \tag{15.68}$$

a_0 being the Bohr radius, $\hbar^2/me^2$. We now have a three-dimensional wave function. The transform corresponding to Eq. 15.62 is

$$g(\mathbf{p}) = \frac{1}{(2\pi\hbar)^{3/2}}\int \psi(\mathbf{r})e^{-i\mathbf{p}\cdot\mathbf{r}/\hbar}\,d^3r. \tag{15.69}$$

Substituting Eq. 15.68 into Eq. 15.69 and using

$$\int e^{-ar+i\mathbf{b}\cdot\mathbf{r}}\,d^3r = \frac{8\pi a}{(a^2+b^2)^2}, \tag{15.70}$$

we obtain the hydrogenic momentum wave function

$$g(\mathbf{p}) = \frac{2^{3/2}}{\pi}\,\frac{a_0^{3/2}\hbar^{5/2}}{(a_0^2p^2+\hbar^2)^2}. \tag{15.71}$$

Such momentum functions have been found useful in problems like Compton scattering from atomic electrons, the wavelength distribution of the scattered radiation depending on the momentum distribution of the target electrons.

The relation between the ordinary space representation and the momentum representation may be clarified by considering the basic commutation relations of quantum mechanics. We can go from a classical Hamiltonian to the Schrödinger wave equation by requiring that momentum p and position x *not* commute. Instead, we require that

$$[p, x] \equiv (px - xp) = -i\hbar. \tag{15.72}$$

For the multidimensional case Eq. 15.72 is replaced by

$$[p_i, x_j] = -i\hbar\delta_{ij}. \tag{15.73}$$

The Schrödinger (space) representation is obtained by using

$$\begin{aligned} x_j &\to x_j, \\ p_i &\to -i\hbar\frac{\partial}{\partial x_i}, \end{aligned} \tag{x}$$

replacing the momentum by a partial space derivative. The reader will easily see that

$$[p, x]\psi(x) = -i\hbar\,\psi(x). \tag{15.74}$$

However, Eq. 15.72 can equally well be satisfied by using

$$\begin{aligned} x_j &\to i\hbar\frac{\partial}{\partial p_j}, \\ p_i &\to p_i. \end{aligned} \tag{p}$$

This is the momentum representation. Then

$$[p, x]\,g(p) = -i\hbar\,g(p). \tag{15.75}$$

Hence the representation (x) is not unique; (p) is an alternate possibility.

In general the Schrödinger representation (x) leading to the Schrödinger wave equation is more convenient because the potential energy V is generally given as a function of position $V(x, y, z)$. The momentum representation (p) usually leads to an integral equation (cf. Chapter 16 for the pros and cons of the integral equations). For an exception, consider the harmonic oscillator.

Example 15.6.2. Harmonic Oscillator

The classical Hamiltonian (kinetic energy + potential energy = total energy) is

$$H(p, x) = \frac{p^2}{2m} + \frac{1}{2}kx^2 = E. \tag{15.76}$$

In the Schrödinger representation we obtain

$$-\frac{\hbar^2}{2m}\frac{d^2\psi(x)}{dx^2} + \frac{1}{2}kx^2\,\psi(x) = E\,\psi(x). \tag{15.77}$$

For total energy E equal to $\sqrt{(k/m)}\hbar/2$ there is a solution

$$\psi(x) = c^{-(\sqrt{mk}/2\hbar)x^2} \tag{15.78}$$

The momentum representation leads to

$$\frac{p^2}{2m}g(p) - \frac{\hbar^2 k}{2}\frac{d^2 g(p)}{dp^2} = E\,g(p). \tag{15.79}$$

Again for

$$E = \sqrt{\frac{k}{m}}\frac{\hbar}{2} \tag{15.80}$$

the momentum wave equation (15.79) is satisfied by

$$g(p) = e^{-p^2/(2\hbar\sqrt{mk})}. \tag{15.81}$$

Either representation, space or momentum (and an infinite number of other possibilities), may be used, depending on which is more convenient for the particular problem under attack.

The demonstration that $g(p)$ is the momentum wave function corresponding to Eq. 15.78, that it is the Fourier inverse transform of Eq. 15.78, is left as Exercise 15.6.3.

EXERCISES

15.6.1 The function $e^{i\mathbf{k}\cdot\mathbf{r}}$ describes a plane wave of momentum $\mathbf{p} = \hbar\mathbf{k}$ normalized to unit density. (Time dependence of $e^{-i\omega t}$ is assumed.) Show that these plane wave functions satisfy an orthogonality relation

$$\int (e^{i\mathbf{k}\cdot\mathbf{r}})^* e^{i\mathbf{k}'\cdot\mathbf{r}}\, dx\, dy\, dz = (2\pi)^3\,\delta(\mathbf{k} - \mathbf{k}').$$

15.6.2 An infinite plane wave in quantum mechanics may be represented by the function

$$\psi(x) = e^{ip'x/\hbar}.$$

Find the corresponding momentum distribution function. Note that it has an infinity and that $\psi(x)$ is not normalized.

15.6.3 A linear quantum oscillator in its ground state has a wave function

$$\psi(x) = a^{-1/2}\pi^{-1/4}e^{-x^2/2a^2}.$$

Show that the corresponding momentum function is

$$g(p) = a^{1/2}\pi^{-1/4}\hbar^{-1/2}e^{-a^2p^2/2\hbar^2}.$$

15.6.4 A free particle in quantum mechanics is described by a plane wave

$$\psi_k(x, t) = e^{i[kx - (\hbar k^2/2m)t]}.$$

Combining waves of adjacent momentum with an amplitude weighting factor $\varphi(k)$ we form a wave packet

$$\Psi(x, t) = \int_{-\infty}^{\infty} \varphi(k) e^{i[kx - (\hbar k^2/2m)t]}\, dk.$$

(a) Solve for $\varphi(k)$ given that

$$\Psi(x, 0) = e^{-x^2/2a^2}.$$

(b) Using the known value of $\varphi(k)$ integrate to get the explicit form of $\Psi(x, t)$. Note that this wave packet diffuses or spreads out with time.

$$\textit{Ans.} \quad \Psi(x, t) = \frac{e^{-\{x^2/2[(a^2+(i\hbar/m)t]\}}}{[1 + (i\hbar t/ma^2)]^{1/2}}.$$

15.6.5 The deuteron, Example 9.1.2, may be described reasonably well with a Hulthen wave function

$$\psi(\mathbf{r}) = A[e^{-\alpha r} - e^{-\beta r}]/r$$

with A, α, and β constants. Find $g(\mathbf{p})$ the corresponding momentum function.
Note: The Fourier transform may be rewritten as Fourier sine and cosine transforms or as a Laplace transform, Section 15.7.

15.6.6 The nuclear form factor $F(k)$ and the charge distribution $\rho(r)$ are three-dimensional Fourier transforms of each other:

$$F(k) = \frac{1}{(2\pi)^{3/2}} \int \rho(r) e^{i\mathbf{k}\cdot\mathbf{r}} d^3r.$$

If the measured form factor is

$$F(k) = (2\pi)^{-3/2}\left(1 + \frac{k^2}{a^2}\right)^{-1},$$

find the corresponding charge distribution.

$$\textit{Ans.} \quad \rho(r) = \frac{a^2}{4\pi}\frac{e^{-ar}}{r}.$$

15.7 Elementary Laplace Transforms

Definition. The Laplace transform $f(s)$ or $\mathscr{L}$ of a function $F(t)$ is defined by[1]

$$f(s) = \mathscr{L}\{F(t)\} = \lim_{a\to\infty} \int_0^a e^{-st} F(t)\, dt$$

$$= \int_0^\infty e^{-st} F(t)\, dt. \tag{15.82}$$

A few comments on the existence of the integral might be in order. The infinite integral of $F(t)$,

$$\int_0^\infty F(t)\, dt,$$

need not exist. For instance, $F(t)$ may diverge exponentially for large t. However, if there is some *constant* s_0 such that

$$|e^{-s_0 t} F(t)| \leqslant M, \tag{15.83}$$

a positive constant for sufficiently large t, $t > t_0$, the Laplace transform (Eq. 15.82), will exist for $s > s_0$; $F(t)$ is said to be of exponential order. As a counterexample, $F(t) = e^{t^2}$ does not satisfy the condition given by Eq. 15.83 and is *not* of exponential

[1] This is sometimes called a one-sided Laplace transform; the integral from $-\infty$ to $+\infty$ is referred to as a two-sided Laplace transform. Some authors introduce an additional factor of s. This extra s appears to have little advantage and continually gets in the way (cf. Jeffreys and Jeffreys, Section 14.13 for additional comments). Generally we take s to be real and positive. It is possible to have s complex provided $\mathscr{R}(s) > 0$.

order. $\mathscr{L}\{e^{t^2}\}$ does *not* exist.

The Laplace transform may also fail to exist because of a sufficiently strong singularity in the function $F(t)$ as $t \to 0$; that is,

$$\int_0^\infty e^{-st} t^n \, dt$$

diverges at the origin for $n \leqslant -1$. The Laplace transform $\mathscr{L}\{t^n\}$ does not exist for $n \leqslant -1$.

Since, for two functions $F(t)$ and $G(t)$, for which the integrals exist

$$\mathscr{L}\{a\,F(t) + b\,G(t)\} = a\mathscr{L}\{F(t)\} + b\mathscr{L}\{G(t)\}, \tag{15.84}$$

the operation denoted by $\mathscr{L}$ is *linear*.

Elementary functions. To introduce the Laplace transform, let us apply the operation to some of the elementary functions. In all cases we assume that $F(t) = 0$ for $t < 0$.

$$F(t) = 1 \qquad (t > 0).$$

Then

$$\mathscr{L}\{1\} = \int_0^\infty e^{-st}\,dt = \frac{1}{s}, \qquad \text{for } s > 0. \tag{15.85}$$

Again, let

$$F(t) = e^{kt} \qquad (t > 0).$$

The Laplace transform becomes

$$\mathscr{L}\{e^{kt}\} = \int_0^\infty e^{-st} e^{kt}\,dt = \frac{1}{s-k} \qquad \text{for } s > k. \tag{15.86}$$

Using this relation we may easily obtain the Laplace transform of certain other functions. Since

$$\begin{aligned} \cosh kt &= \tfrac{1}{2}(e^{kt} + e^{-kt}), \\ \sinh kt &= \tfrac{1}{2}(e^{kt} - e^{-kt}), \end{aligned} \tag{15.87}$$

we have

$$\begin{aligned} \mathscr{L}\{\cosh kt\} &= \frac{1}{2}\left(\frac{1}{s-k} + \frac{1}{s+k}\right) = \frac{s}{s^2 - k^2}, \\ \mathscr{L}\{\sinh kt\} &= \frac{1}{2}\left(\frac{1}{s-k} - \frac{1}{s+k}\right) = \frac{k}{s^2 - k^2}, \end{aligned} \tag{15.88}$$

both valid for $s > k$. By the relations

$$\begin{aligned} \cos kt &= \cosh ikt, \\ \sin kt &= -i \sinh ikt, \end{aligned} \tag{15.89}$$

the Laplace transforms are

$$\begin{aligned} \mathscr{L}\{\cos kt\} &= \frac{s}{s^2 + k^2}, \\ \mathscr{L}\{\sin kt\} &= \frac{k}{s^2 + k^2}, \end{aligned} \tag{15.90}$$

both valid for $s > 0$. Another derivation of this last transform is given in the next section.

Finally, for $F(t) = t^n$, we have

$$\mathscr{L}\{t^n\} = \int_0^\infty e^{-st} t^n \, dt,$$

which is just the factorial function. Hence

$$\mathscr{L}\{t^n\} = \frac{n!}{s^{n+1}}, \qquad s > 0, \quad n > -1. \tag{15.91}$$

Inverse transform. There is little point to these operations unless we can carry out the inverse transform, as in Fourier transforms, that is, with

$$\mathscr{L}\{F(t)\} = f(s),$$

then

$$\mathscr{L}^{-1}\{f(s)\} = F(t). \tag{15.92}$$

Taken literally, this inverse transform is *not* unique; that is, two functions $F_1(t)$ and $F_2(t)$ may have the same transform, $f(s)$. However, in this case

$$F_1(t) - F_2(t) = N(t)$$

where $N(t)$ is a null function meaning that

$$\int_0^{t_0} N(t) \, dt = 0.$$

for all positive t_0. This result is known as Lerch's theorem. Therefore to the physicist and engineer $N(t)$ may almost always be taken as zero and the inverse operation becomes unique.

The determination of the inverse transform can be done in two ways. (1) a general technique for $\mathscr{L}^{-1}$ will be developed in Section 15.11 by using the calculus of residues. (2) A table of transforms can be built up and used to carry out the inverse transforms exactly as we can use a table of logarithms to look up antilogarithms. These transforms constitute the embryonic beginnings of such a table.

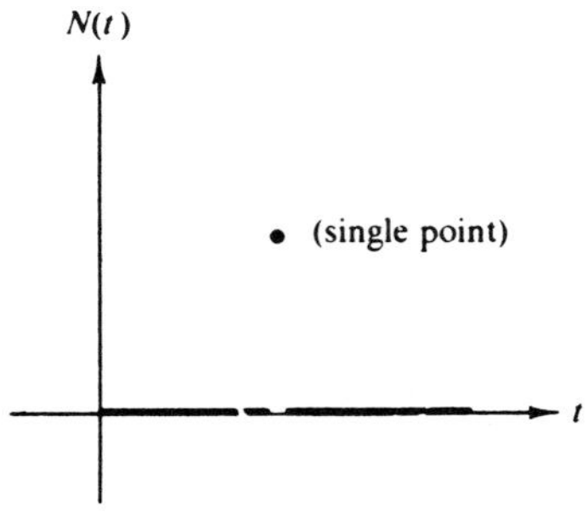

FIG. 15.4 A possible null function

For a more complete set, see Table 15.2 page 723 or AMS-55, Chap. 29.

Partial fraction expansion. Utilization of such a table of transforms (or inverse transforms) is facilitated by expanding $f(s)$ in *partial fractions*.

Frequently $f(s)$, our transform, occurs in the form $g(s)/h(s)$, where $g(s)$ and $h(s)$ are polynomials with no common factors, $g(s)$ being of lower degree than $h(s)$. If the factors of $h(s)$ are all linear and distinct, then by the theory of partial fractions we may write

$$f(s) = \frac{c_1}{s - a_1} + \frac{c_2}{s - a_2} + \cdots + \frac{c_n}{s - a_n}, \tag{15.93}$$

where the c_i's are independent of s. If any one of the roots, say a_1, is multiple (occurring m times), then $f(s)$ has the form

$$f(s) = \frac{c_{1,m}}{(s - a_1)^m} + \frac{c_{1,m-1}}{(s - a_1)^{m-1}} + \cdots + \frac{c_{1,1}}{s - a_1} + \sum_{i=2}^{n} \frac{c_i}{s - a_i}. \tag{15.94}$$

Finally, if one of the factors is quadratic, $(s^2 + ps + q)$, the numerator, instead of being a simple constant, will have the form

$$\frac{as + b}{s^2 + ps + q}.$$

There are various ways of determining the constants introduced. For instance, in Eq. 15.93 we may multiply through by $(s - a_i)$ and obtain

$$c_i = \lim_{s \to a_i} (s - a_i) f(s). \tag{15.95}$$

In elementary cases a direct solution is often the easiest.

Example 15.7.1 Partial Fraction Expansion

Let

$$f(s) = \frac{k^2}{s(s^2 + k^2)} = \frac{c}{s} + \frac{as + b}{s^2 + k^2} \tag{15.96}$$

Putting the right side of the equation over a common denominator and equating like powers of s in the numerator,

$$\frac{k^2}{s(s^2 + k^2)} = \frac{c(s^2 + k^2) + s(as + b)}{s(s^2 + k^2)}, \tag{15.97}$$

$$c + a = 0, \qquad s^2,$$

$$b = 0, \qquad s^1,$$

and

$$ck^2 = k^2, \qquad s^0.$$

Solving these ($s \neq 0$),

$$c = 1,$$

$$b = 0,$$

$$a = -1,$$

giving

$$f(s) = \frac{1}{s} - \frac{s}{s^2 + k^2}, \tag{15.98}$$

and

$$\mathscr{L}^{-1}\{f(s)\} = 1 - \cos kt \tag{15.99}$$

by Eqs. 15.85 and 15.90.

EXAMPLE 15.7.2

As one application of Laplace transforms, consider the evaluation of

$$F(t) = \int_0^\infty \frac{\sin tx}{x}\, dx. \tag{15.100}$$

Suppose we take the Laplace transform of this definite (and improper) integral:

$$\mathscr{L}\left\{\int_0^\infty \frac{\sin tx}{x}\, dx\right\} = \int_0^\infty e^{-st} \int_0^\infty \frac{\sin tx}{x}\, dx\, dt. \tag{15.101}$$

Now, interchanging the order of integration (which must be justified!),[1] we get

$$\int_0^\infty \frac{1}{x}\left[\int_0^\infty e^{-st} \sin tx\, dt\right] dx = \int_0^\infty \frac{dx}{s^2 + x^2}, \tag{15.102}$$

since the factor in square brackets is just the Laplace transform of sin tx. From the integral tables

$$\int_0^\infty \frac{dx}{s^2 + x^2} = \frac{1}{s} \tan^{-1}\left(\frac{x}{s}\right)\Bigg|_0^\infty = \frac{\pi}{2s} = f(s). \tag{15.103}$$

By Eq. 15.85 we carry out the inverse transformation to obtain

$$F(t) = \frac{\pi}{2}, \qquad t > 0, \tag{15.104}$$

in agreement with an evaluation by the calculus of residues (Section 7.2). It has been assumed that $t > 0$ in $F(t)$. For $F(-t)$ we need note only that $\sin(-tx) = -\sin tx$, giving $F(-t) = -F(t)$. Finally, if $t = 0$, $F(0)$ is clearly zero. Therefore

$$\int_0^\infty \frac{\sin tx}{x}\, dx = \begin{cases} \dfrac{\pi}{2}, & t > 0 \\ 0, & t = 0 \\ -\dfrac{\pi}{2}, & t < 0. \end{cases} \tag{15.105}$$

Note that $\int_0^\infty (\sin tx/x)\, dx$, taken as a function of t, describes a step function, a step of height π at $t = 0$. This is consistent with Eq. 8.83d.

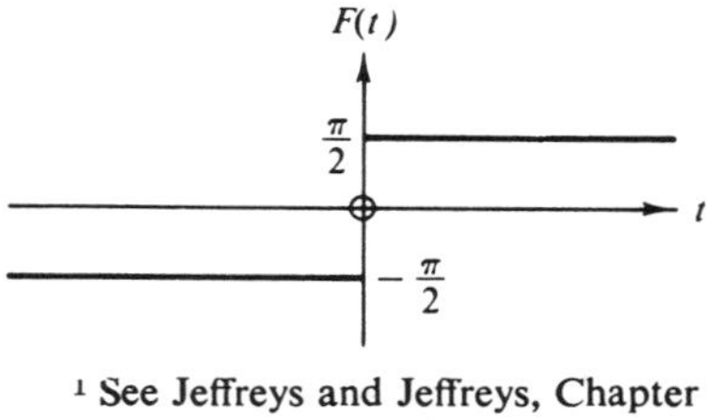

FIG. 15.5 $F(t) = \int_0^\infty \frac{\sin tx}{x}\, dx$, a step function

[1] See Jeffreys and Jeffreys, Chapter 1.

EXERCISES

15.7.1 Prove that

$$\lim_{s \to \infty} s f(s) = \lim_{t \to +0} F(t).$$

Hint. Assume that $F(t)$ can be expressed as $F(t) = \sum_{n=0}^{\infty} a_n t^n$.

15.7.2 Show that

$$\frac{1}{\pi} \lim_{s \to 0} \mathscr{L}\{\cos xt\} = \delta(x).$$

15.7.3 Verify that

$$\mathscr{L}\left\{\frac{\cos at - \cos bt}{b^2 - a^2}\right\} = \frac{s}{(s^2 + a^2)(s^2 + b^2)}, \qquad a^2 \neq b^2.$$

15.7.4 Show that

(a)

$$\mathscr{L}^{-1}\left\{\frac{1}{(s+a)(s+b)}\right\} = \frac{e^{-at} - e^{-bt}}{b - a}, \qquad a \neq b.$$

(b)

$$\mathscr{L}^{-1}\left\{\frac{s}{(s+a)(s+b)}\right\} = \frac{ae^{-at} - be^{-bt}}{a - b}, \qquad a \neq b.$$

15.7.5 The electrostatic potential of a charged conducting disk is known to have the general form (circular cylindrical coordinates)

$$\Phi(\rho, z) = \int_0^{\infty} e^{-k|z|} J_0(k\rho) f(k)\, dk$$

with $f(k)$ unknown. At large distances ($z \to \infty$), the potential must approach the Coulomb potential $Q/4\pi\varepsilon_0 z$. Show that

$$\lim_{k \to 0} f(k) = \frac{Q}{4\pi\varepsilon_0}.$$

Hint: You may set $\rho = 0$ and assume a Maclaurin expansion of $f(k)$ or, using e^{-i} construct a delta sequence.

15.8 Laplace Transform of Derivatives

Perhaps the main application of Laplace transforms is in converting differential equations into simpler forms that may be solved more easily. It will be seen, for instance, that coupled differential equations with constant coefficients transform to simultaneous linear algebraic equations.

Let us transform the first derivative of $F(t)$.

$$\begin{aligned} \mathscr{L}\{F'(t)\} &= \int_0^{\infty} e^{-st} \frac{dF(t)}{dt}\, dt \\ &= e^{-st} F(t) \Big|_0^{\infty} + s \int_0^{\infty} e^{-st} F(t)\, dt \\ &= s\mathscr{L}\{F(t)\} - F(0). \end{aligned} \tag{15.106}$$

Strictly speaking $F(0) = F(+0)$[1] and dF/dt is required to be at least piecewise continuous for $0 \leqslant t < \infty$. Naturally both $F(t)$ and its derivative must be such that the integrals do not diverge. Incidentally Eq. 15.106 provides another proof of Exercise 15.7.1.

An extension gives

$$\mathscr{L}\{F^{(2)}(t)\} = s^2\mathscr{L}\{F(t)\} - sF(+0) - F'(+0), \tag{15.107}$$

$$\mathscr{L}\{F^{(n)}(t)\} = s^n\mathscr{L}\{F(t)\} - s^{n-1}F(+0) - s^{n-2}F'(+0) - \cdots - F^{(n-1)}(+0). \tag{15.108}$$

Note carefully how the initial conditions, $F(+0)$, $F'(+0)$, and so on, are incorporated into the transform. Equation 15.107 may be used to derive $\mathscr{L}\{\sin kt\}$. We use the identity

$$-k^2 \sin kt = \frac{d^2}{dt^2}\sin kt. \tag{15.109}$$

Then, applying the Laplace transform operation,

$$\begin{aligned} -k^2\mathscr{L}\{\sin kt\} &= \mathscr{L}\left\{\frac{d^2}{dt^2}\sin kt\right\} \\ &= s^2\mathscr{L}\{\sin kt\} - s\sin(0) - \frac{d}{dt}\sin kt\,|_{t=0}. \end{aligned} \tag{15.110}$$

Since $\sin(0) = 0$ and $d/dt \sin kt|_{t=0} = k$,

$$\mathscr{L}\{\sin kt\} = \frac{k}{s^2 + k^2}, \tag{15.111}$$

verifying Eq. 15.90.

Example 15.8.1. Simple Harmonic Oscillator

As a simple but reasonably physical example, consider a mass m oscillating under the influence of an ideal spring, spring constant k. As usual, friction is neglected. Then Newton's second law becomes

$$m\frac{d^2X(t)}{dt^2} + kX(t) = 0; \tag{15.112}$$

also

$$X(0) = X_0,$$
$$X'(0) = 0.$$

Applying the Laplace transform, we obtain

$$m\mathscr{L}\left\{\frac{d^2X}{dt^2}\right\} + k\mathscr{L}\{X(t)\} = 0, \tag{15.113}$$

and by use of Eq. 15.107 this becomes

[1] Zero is approached from the *positive* side.

$$ms^2\, x(s) - msX_0 + k\, x(s) = 0, \tag{15.114}$$

$$x(s) = X_0 \frac{s}{s^2 + \omega_0^2}, \qquad \text{with } \omega_0^2 \equiv \frac{k}{m}. \tag{15.115}$$

From Eq. 15.90 this is seen to be the transform of $\cos \omega_0 t$, which gives

$$X(t) = X_0 \cos \omega_0 t \tag{15.116}$$

as expected.

EXAMPLE 15.8.2. EARTH'S NUTATION

A somewhat more involved example is provided by the nutation of the earth's poles (force-free precession). Treating the earth as a rigid (oblate) spheroid, the Euler equations of motion reduce to

$$\begin{aligned} \frac{dX}{dt} &= -aY \\ \frac{dY}{dt} &= +aX \end{aligned} \tag{15.117}$$

where $a \equiv [(I_z - I_x)/I_z]\omega_z$,

$X = \omega_x$,

$Y = \omega_y$ with angular velocity vector $\boldsymbol{\omega} = (\omega_x, \omega_y, \omega_z)$,

I_z = moment of inertia about the z-axis and $I_y = I_x$ moment of inertia about x (or y)-axis.

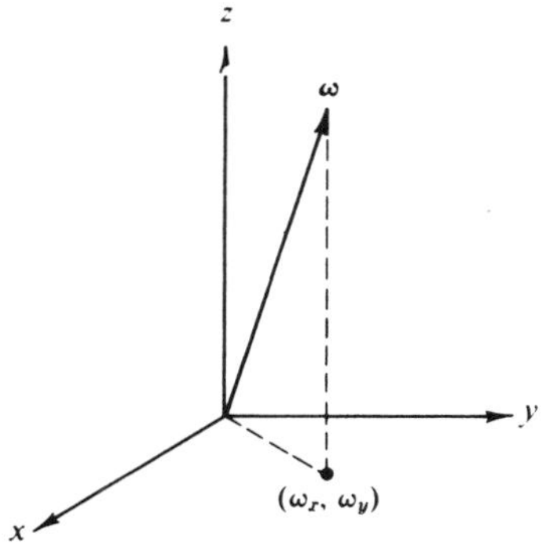

FIG. 15.6

The z-axis coincides with the axis of symmetry of the earth. It differs from the axis for the earth's daily rotation, $\boldsymbol{\omega}$, by some 15 meters, measured at the poles. Transformation of these coupled differential equations yields

$$\begin{aligned} s\,x(s) - X(0) &= -a y(s), \\ s\,y(s) - Y(0) &= a x(s). \end{aligned} \tag{15.118}$$

Combining to eliminate $y(s)$, we have

$$s^2 x(s) - s\,X(0) + a Y(0) = -a^2 x(s)$$

or

$$x(s) = X(0)\frac{s}{s^2 + a^2} - Y(0)\frac{a}{s^2 + a^2}. \tag{15.119}$$

Hence

$$X(t) = X(0) \cos at - Y(0) \sin at. \tag{15.120}$$

Similarly,

$$Y(t) = X(0) \sin at + Y(0) \cos at. \tag{15.121}$$

This is seen to be a rotation of the vector (X, Y) counterclockwise (for $a > 0$) about the z-axis with angle $\theta = at$ and angular velocity a.

A direct interpretation may be found by choosing the time axis so that $Y(0) = 0$. Then

$$\begin{aligned} X(t) &= X(0) \cos at, \\ Y(t) &= X(0) \sin at, \end{aligned} \tag{15.122}$$

which are the parametric equations for rotation of (X, Y) in a circular orbit of radius $X(0)$ with angular velocity a in the counterclockwise sense.

In the case of the earth's angular velocity vector $X(0)$ is about 15 meters, whereas a, as defined here, corresponds to a period $(2\pi/a)$ of some 300 days. Actually, because of departures from the idealized rigid body assumed in setting up Euler's equations, the period is about 427 days.[1]

If in Eq. 15.117 we set

$$X(t) = L_x,$$

$$Y(t) = L_y,$$

where L_x and L_y = the x- and y-components of the angular momentum $\mathbf{L}$,
$a = -g_L B_z$,
g_L = gyromagnetic ratio,
B_z = magnetic field (along the z-axis),
Eq. 15.117 describes the Larmor precession of charged bodies in a uniform magnetic field, B_z.

Dirac delta function. For use with differential equations one further transform is helpful—that of the Dirac delta function:[2]

$$\mathscr{L}\{\delta(t - t_0)\} = \int_0^\infty e^{-st}\delta(t - t_0)\,dt = e^{-st_0}, \qquad \text{for } t_0 \geqslant 0, \tag{15.123}$$

and for $t_0 = 0$

[1] Menzel, D., ed., *Fundamental Formulas of Physics*, p. 695. New York: Prentice Hall (1955).

[2] Strictly speaking, the Dirac delta function is undefined. However, the integral over it is well defined. This approach is developed in Section 8.6 using delta sequences.

$$\mathscr{L}\{\delta(t)\} = 1, \tag{15.124}$$

where it is assumed that we are using a representation of the delta function such that

$$\int_0^\infty \delta(t)\,dt = 1, \qquad \delta(t) = 0 \quad \text{for} \quad t > 0. \tag{15.125}$$

As an alternate method, $\delta(t)$ may be considered the limit as $\varepsilon \to 0$ of $F(t)$, where

$$F(t) = \begin{cases} 0, & t < 0, \\ \varepsilon^{-1}, & 0 < t < \varepsilon, \\ 0, & t > \varepsilon. \end{cases} \tag{15.126}$$

By direct calculation

$$\mathscr{L}\{F(t)\} = \frac{1 - e^{-\varepsilon s}}{\varepsilon s}. \tag{15.127}$$

Taking the limit of the integral (instead of the integral of the limit),

$$\lim_{\varepsilon \to 0} \mathscr{L}\{F(t)\} = 1$$

or Eq. 15.124

$$\mathscr{L}\{\delta(t)\} = 1.$$

This delta function is frequently called the impulse function because it is so useful in describing impulsive forces, that is, forces lasting only a short time.

EXAMPLE 15.8.3. IMPULSIVE FORCE

Newton's second law for impulsive force acting on a particle of mass m becomes

$$m\,\frac{d^2X}{dt^2} = P\,\delta(t) \tag{15.128}$$

where P is a constant.

Transforming, we obtain

$$ms^2\,x(s) - ms\,X(0) - m\,X'(0) = P. \tag{15.129}$$

For a particle starting from rest, $X'(0) = 0$.[1] We shall also take $X(0) = 0$. Then

$$x(s) = \frac{P}{ms^2}, \tag{15.130}$$

and

$$X(t) = \frac{P}{m}\,t, \tag{15.131}$$

$$\frac{dX(t)}{dt} = \frac{P}{m}, \qquad \text{a constant.} \tag{15.132}$$

[1] This really should be $X'(+0)$. To include the effect of the impulse consider that the impulse will occur at $t = \varepsilon$ and let $\varepsilon \to 0$.

The effect of the impulse $P\,\delta(t)$ is to transfer (instantaneously) P units of linear momentum to the particle.

A similar analysis applies to the ballistic galvanometer. The torque on the galvanometer is given initially by ki, in which i is a pulse of current and k is a proportionality constant. Since i is of short duration, we set

$$ki = kq\,\delta(t) \tag{15.133}$$

where q is the total charge carried by the current i. Then, with I the moment of inertia,

$$I\frac{d^2\theta}{dt^2} = kq\,\delta(t), \tag{15.134}$$

and transforming as above, we find that the effect of the current pulse is a transfer of kq units of *angular* momentum to the galvanometer.

EXERCISES

15.8.1 Use the expression for the transform of a second derivative to obtain the transform of $\cos kt$.

15.8.2 If $F(t)$ can be expanded in a power series (Taylor or Laurent), that is,

$$F(t) = \sum_n a_n t^n,$$

and

$$\int_0^\infty e^{-st} \sum_n a_n t^n \, dt$$

or

$$\sum_n a_n \int_0^\infty e^{-st}\, t^n \, dt$$

exists, show that $f(s)$, the Laplace transform of $F(t)$, contains no powers of s greater than s^{-1}. Check your result by calculating $\mathscr{L}\{\delta(t)\}$ and comment intelligently on this fiasco.

15.8.3 Radioactive nuclei decay according to the law

$$\frac{dN}{dt} = -\lambda N,$$

N being the concentration of a given nuclide and λ, the particular decay constant. This equation may be interpreted as stating that the rate of decay is proportional to the number of these radioactive nuclei present. They all decay independently.

In a radioactive series of n different nuclides, starting with N_1,

$$\frac{dN_1}{dt} = -\lambda_1 N_1,$$

$$\frac{dN_2}{dt} = \lambda_1 N_1 - \lambda_2 N_2, \text{ etc.,}$$

$$\frac{dN_n}{dt} = \lambda_{n-1} N_{n-1}, \qquad \text{stable.}$$

Find $N_1(t)$, $N_2(t)$, and $N_3(t)$, $n = 3$, with $N_1(0) = N_0$, $N_2(0) = N_3(0) = 0$.

Ans. $N_1(t) = N_0 e^{-\lambda_1 t}$,

$$N_2(t) = N_0 \frac{\lambda_1}{\lambda_2 - \lambda_1}(e^{-\lambda_1 t} - e^{-\lambda_2 t}),$$

$$N_3(t) = N_0\left(1 - \frac{\lambda_2}{\lambda_2 - \lambda_1} e^{-\lambda_1 t} + \frac{\lambda_1}{\lambda_2 - \lambda_1} e^{-\lambda_2 t}\right).$$

Find an approximate expression for N_2 and N_3, valid for small t when $\lambda_1 \approx \lambda_2$.

Ans. $N_2 \approx N_0 \lambda_1 t$

$$N_3 \approx \frac{N_0}{2}\lambda_1\lambda_2 t^2.$$

Find approximate expressions for N_2 and N_3, valid for large t, when
(a) $\lambda_1 \gg \lambda_2$, (b) $\lambda_1 \ll \lambda_2$.

Ans. (a) $N_2 \approx N_0 e^{-\lambda_2 t}$
$N_3 \approx N_0(1 - e^{-\lambda_2 t})$, $\lambda_1 t \gg 1$.

(b) $N_2 \approx N_0 \frac{\lambda_1}{\lambda_2} e^{-\lambda_1 t}$,
$N_3 \approx N_0(1 - e^{-\lambda_1 t})$, $\lambda_2 t \gg 1$.

15.8.4 The formation of an isotope in a nuclear reactor is given by

$$\frac{dN_2}{dt} = nv\sigma_1 N_{10} - \lambda_2 N_2(t) - nv\sigma_2 N_2(t).$$

Here the product nv is the neutron flux, neutrons per cubic centimeter, times centimeters per second mean velocity; σ_1 and σ_2 (cm^2) are measures of the probability of neutron absorption by the original isotope, concentration N_{10}, which is assumed constant and the newly formed isotope, concentration N_2, respectively. The radioactive decay constant for the isotope is λ_2.

(a) Find the concentration N_2 of the new isotope as a function of time.

(b) If the original element is Eu^{153}, $\sigma_1 = 400$ barns $= 400 \times 10^{-24}$ cm^2, $\sigma_2 = 1000$ barns $= 1000 \times 10^{-24}$ cm^2, and $\lambda_2 = 1.4 \times 10^{-9}$ sec^{-1}. If $N_{10} = 10^{20}$ and $(nv) = 10^9$ cm^{-2} sec^{-1}, find N_2, the concentration of Eu^{154} after one year of continuous irradiation. Is the assumption that N_1 is constant justified?

15.8.5 In a nuclear reactor Xe^{135} is formed both as a direct fission product and as a decay product of I^{135}, half-life, 6.7 hours. The half-life of Xe^{135} is 9.2 hours. As Xe^{135} strongly absorbs thermal neutrons thereby "poisoning" the nuclear reactor, its concentration is a matter of great interest. The relevant equations are

$$\frac{dN_I}{dt} = \gamma_I \varphi \sigma_f N_U - \lambda_I N_I,$$

$$\frac{dN_X}{dt} = \lambda_I N_I + \gamma_X \varphi \sigma_f N_U - \lambda_X N_X - \varphi \sigma_X N_X.$$

Here N_I = concentration of I^{135} (Xe^{135}, U^{235}), Assume N_U = constant.
γ_I = yield of I^{135} per fission = 0.060,
γ_X = yield of Xe^{135} direct from fission = 0.003,
λ_I = I^{135} (Xe^{135}) decay constant $= \dfrac{\ln 2}{t_{1/2}} = \dfrac{0.693}{t_{1/2}}$
σ_f = thermal neutron fission cross section for U^{235},
σ_X = thermal neutron absorption cross section for Xe^{135} = 3.5×10^6 barns,
= 3.5×10^{-18} cm^2.

(σ_I, the absorption cross section of I^{135} is negligible.)
φ = neutron flux = neutrons/cm^3 × mean velocity (cm/sec)

1. Find $N_X(t)$ in terms of neutron flux φ and the product $\sigma_f N_U$.

2. Find $N_X(t \to \infty)$.
3. After N_X has reached equilibrium, the reactor is shut down, $\varphi = 0$. Find $N_X(t)$ following shut down. Notice the increase in N_X which may for a few hours interfere with starting the reactor up again.

15.9 Other Properties

Substitution. If we replace the parameter s by $s - a$ in the definition of the Laplace transform (Eq. 15.82), we have

$$\begin{aligned} f(s-a) &= \int_0^\infty e^{-(s-a)t}\, F(t)\, dt = \int_0^\infty e^{-st} e^{at}\, F(t)\, dt \\ &= \mathscr{L}\{e^{at}\, F(t)\}. \end{aligned} \tag{15.135}$$

Hence the replacement of s with $s - a$ corresponds to multiplying $F(t)$ by e^{at} and conversely. This result can be used to good advantage in extending our table of transforms. From Eq. 15.90 we find immediately that

$$\mathscr{L}\{e^{at} \sin kt\} = \frac{k}{(s-a)^2 + k^2}; \tag{15.136}$$

also

$$\mathscr{L}\{e^{at} \cos kt\} = \frac{s-a}{(s-a)^2 + k^2}, \qquad s > a.$$

EXAMPLE 15.9.1. DAMPED OSCILLATOR

These expressions are useful when we consider an oscillating mass with damping proportional to the velocity. Equation 15.112, with such damping added, becomes

$$m\, X''(t) + b\, X'(t) + k\, X(t) = 0, \tag{15.137}$$

in which b is a proportionality constant. Let us assume that the particle starts from rest at $X(0) = X_0$, $X'(0) = 0$. The transformed equation is

$$m[s^2\, x(s) - sX_0] + b[s\, x(s) - X_0] + k\, x(s) = 0 \tag{15.138}$$

and

$$x(s) = X_0 \frac{ms + b}{ms^2 + bs + k}. \tag{15.139}$$

This may be handled by completing the square of the denominator,

$$s^2 + \frac{b}{m} s + \frac{k}{m} = \left(s + \frac{b}{2m}\right)^2 + \left(\frac{k}{m} - \frac{b^2}{4m^2}\right). \tag{15.140}$$

If the damping is small, $b^2 < 4$ km, the last term is positive and will be denoted by ω_1^2.

$$\begin{aligned} x(s) &= X_0 \frac{s + b/m}{(s + b/2m)^2 + \omega_1^2} \\ &= X_0 \frac{s + b/2m}{(s + b/2m)^2 + \omega_1^2} + X_0 \frac{(b/2m\omega_1)\omega_1}{(s + b/2m)^2 + \omega_1^2}. \end{aligned} \tag{15.141}$$

By Eqs. 15.136

$$X(t) = X_0 e^{-(b/2m)t}\left(\cos \omega_1 t + \frac{b}{2m\omega_1} \sin \omega_1 t\right)$$

$$= X_0 \frac{\omega_0}{\omega_1} e^{-(b/2m)t} \cos(\omega_1 t - \varphi) \tag{15.142}$$

where

$$\tan \varphi = \frac{b}{2m\omega_1},$$

$$\omega_0^2 = \frac{k}{m}.$$

Of course, as $b \to 0$, this solution goes over to the undamped solution, (Section 15.8).

RLC analog. It is worth noting the similarity between this damped simple harmonic oscillation of a mass on a spring and an *RLC* circuit (resistance, inductance, and capacitance). At any instant the sum of the potential differences around the loop must be zero (Kirchhoff's law, conservation of energy). This gives

$$L\frac{dI}{dt} + RI + \frac{1}{C}\int^t I\,dt = 0. \tag{15.143}$$

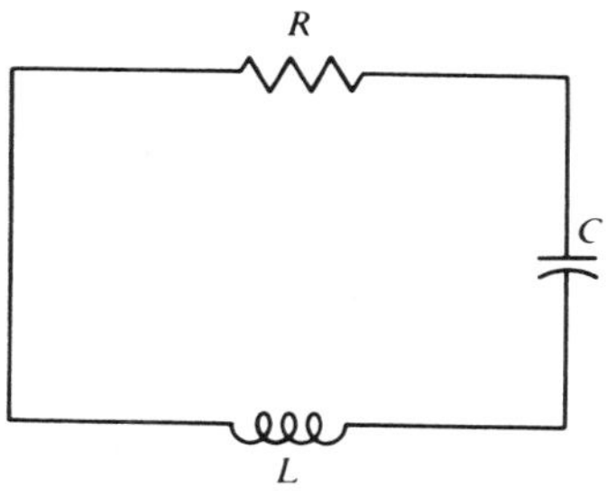

FIG. 15.7 *RLC circuit*

Differentiating the current I with respect to time (to eliminate the integral), we have

$$L\frac{d^2I}{dt^2} + R\frac{dI}{dt} + \frac{1}{C}I = 0. \tag{15.144}$$

If we replace

$I(t)$ with $X(t)$,

L with m,

R with b,

C^{-1} with k,

this is identical with the mechanical problem. It is but one example of the unification of diverse branches of physics by mathematics. A more complete discussion will be found in Olson's book.[1]

Translation. This time let $f(s)$ be multiplied by e^{-bs}, $b > 0$.

$$e^{-bs}f(s) = e^{-bs}\int_0^\infty e^{-st}F(t)\,dt$$

$$= \int_0^\infty e^{-s(t+b)}\,F(t)\,dt. \tag{15.145}$$

Now let $t + b = \tau$. Eq. 15.145 becomes

$$e^{-bs}f(s) = \int_b^\infty e^{-s\tau}\,F(\tau - b)\,d\tau. \tag{15.146}$$

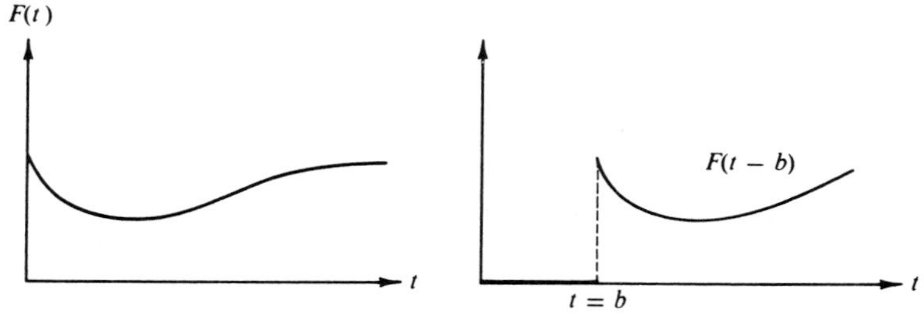

FIG. 15.8 Translation

Since $F(t)$ is assumed to be equal to zero for $t < 0$, $F(\tau - b) = 0$ for $0 \leqslant \tau < b$. Therefore we can extend the lower limit to zero without changing the value of the integral. Then, noting that τ is only a variable of integration,

$$e^{-bs}f(s) = \mathscr{L}\{F(t - b)\}. \tag{15.147}$$

Example 15.9.2. Electromagnetic Waves

The electromagnetic wave equation with $E = E_y$ or E_z, a transverse wave propagating along the x-axis, is

$$\frac{\partial^2 E(x, t)}{\partial x^2} - \frac{1}{v^2}\frac{\partial^2 E(x, t)}{\partial t^2} = 0. \tag{15.148}$$

Transforming this equation with respect to t,

$$\frac{\partial^2}{\partial x^2}\mathscr{L}\{E(x, t)\} - \frac{s^2}{v^2}\mathscr{L}\{E(x, t)\} + \frac{s}{v^2}E(x, 0) + \frac{1}{v^2}\left.\frac{\partial E(x, t)}{\partial t}\right|_{t=0} = 0. \tag{15.149}$$

If we have the initial condition $E(x, 0) = 0$ and

$$\left.\frac{\partial E(x, t)}{\partial t}\right|_{t=0} = 0,$$

[1] Olsen, H. F., *Dynamical Analogies*. New York: Van Nostrand (1943).

then

$$\frac{\partial^2}{\partial x^2}\mathscr{L}\{E(x,t)\} = \frac{s^2}{v^2}\mathscr{L}\{E(x,t)\}. \tag{15.150}$$

The solution (of this *ordinary* differential equation) is

$$\mathscr{L}\{E(x,t)\} = c_1 e^{-(s/v)x} + c_2 e^{+(s/v)x}; \tag{15.151}$$

c_1 and c_2 are obtained by additional boundary conditions. If our wave remains finite as $x \to \infty$, $\mathscr{L}\{E(x,t)\}$ will also. Hence $c_2 = 0$.

If $E(0, t)$ is denoted by $F(t)$, then $c_1 = f(s)$ and

$$\mathscr{L}\{E(x,t)\} = e^{-(s/v)x} f(s). \tag{15.152}$$

From the translation property (Eq. 15.147) we find immediately

$$E(x,t) = \begin{cases} F\left(t - \dfrac{x}{v}\right), & t \geqslant \dfrac{x}{v}, \\ 0, & t < \dfrac{x}{v}. \end{cases} \tag{15.153}$$

Differentiation and substitution into Eq. 15.148 verifies Eq. 15.153. Our solution represents a wave (or pulse) moving in the positive x-direction with velocity v. Note that for $x > vt$ the region remains undisturbed; the pulse has not had time to get there. If we had wanted a signal propagated along the negative x-axis, c_1 would have been set equal to 0 and we would have obtained

$$E(x,t) = \begin{cases} F\left(t + \dfrac{x}{v}\right), & t \geqslant -\dfrac{x}{v}, \\ 0, & t < -\dfrac{x}{v}, \end{cases} \tag{15.154}$$

a wave along the negative x-axis.

Derivative of a transform. When $F(t)$, which is at least piecewise continuous, and s are chosen so that $e^{-st} F(t)$ converges exponentially for large s, the integral

$$\int_0^\infty e^{-st} F(t)\, dt$$

is uniformly convergent and may be differentiated (under the integral sign) with respect to s. Then

$$f'(s) = \int_0^\infty (-t) e^{-st} F(t)\, dt = \mathscr{L}\{-t\, F(t)\}. \tag{15.155}$$

Continuing this process,

$$f^{(n)}(s) = \mathscr{L}\{(-t)^n F(t)\}. \tag{15.156}$$

All the integrals so obtained will be uniformly convergent because of the decreasing exponential behavior of $e^{-st} F(t)$.

This same technique may be applied to generate more transforms. For example

$$\mathcal{L}\{e^{kt}\} = \int_0^\infty e^{-st} e^{kt}\, dt$$

$$= \frac{1}{s-k}, \qquad (s > k). \tag{15.157}$$

Differentiating with respect to s (or with respect to k), we obtain

$$\mathcal{L}\{te^{kt}\} = \frac{1}{(s-k)^2}, \qquad s > k. \tag{15.158}$$

EXAMPLE 15.9.3. BESSEL'S EQUATION

An interesting application of a differentiated Laplace transform appears in the solution of Bessel's equation with $n = 0$. From Chapter 11 we have

$$x^2\, y''(x) + x\, y'(x) + x^2\, y(x) = 0. \tag{15.159}$$

Dividing by x and substituting $t = x$ and $F(t) = y(x)$ to agree with the present notation, the Bessel equation becomes

$$t\, F''(t) + F'(t) + t\, F(t) = 0. \tag{15.160}$$

We need a regular solution, in particular $F(0) = 1$. Also we assume that our unknown $F(t)$ has a transform. Then, transforming and using Eqs. 15.107 and 15.155, we have

$$-\frac{d}{ds}[s^2 f(s) - s] + s f(s) - 1 - \frac{d}{ds} f(s) = 0. \tag{15.161}$$

Since $F(0)$ is finite, substitution into Eq. 15.160 indicates that $F'(0) = 0$. Rearranging Eq. 15.161, we obtain

$$(s^2 + 1) f'(s) + s f(s) = 0 \tag{15.162}$$

or

$$\frac{df}{f} = -\frac{s\, ds}{s^2 + 1}. \tag{15.163}$$

By integration,

$$\ln f(s) = -\tfrac{1}{2} \ln (s^2 + 1) + \ln C, \tag{15.164}$$

which may be rewritten

$$f(s) = \frac{C}{\sqrt{s^2 + 1}}. \tag{15.165}$$

To make use of Eq. 15.91, $f(s)$ is expanded in a series of negative powers of s, convergent for $s > 1$.

$$f(s) = \frac{C}{s}\left(1 + \frac{1}{s^2}\right)^{-1/2}$$

$$= \frac{C}{s}\left[1 - \frac{1}{2s^2} + \frac{1 \cdot 3}{2^2 \cdot 2! s^4} - \cdots + \frac{(-1)^n (2n)!}{(2^n n!)^2 s^{2n}} + \cdots\right]. \quad (15.166)$$

Inverting, term by term, we obtain

$$F(t) = C \sum_{n=0}^{\infty} \frac{(-1)^n t^{2n}}{(2^n n!)^2}. \quad (15.167)$$

When C is set equal to 1, as required by the initial condition $F(0) = 1$, $F(t)$ is just $J_0(t)$, our familiar Bessel function of order zero. Hence

$$\mathscr{L}\{J_0(t)\} = \frac{1}{\sqrt{s^2 + 1}}. \quad (15.168)$$

Note that we assumed $s > 1$. The proof for $s > 0$ is left as a problem.

It is perhaps worth noting that this application was successful and relatively easy because we took $n = 0$ in Bessel's equation. This made it possible to divide out a factor of x (or t). If this had not been done, the terms of the form $t^2 F(t)$ would have introduced a second derivative of $f(s)$. The resulting equation would have been no easier to solve than the original one.

When we go beyond linear differential equations with constant coefficients, the Laplace transform may still be applied, but there is no guarantee that it will be helpful.

The application to Bessel's equation, $n \neq 0$, will be found in the references. Alternatively, we can show that

$$\mathscr{L}\{J_n(at)\} = \frac{a^{-n}(\sqrt{s^2 + a^2} - s)^n}{\sqrt{s^2 + a^2}} \quad (15.169)$$

by expressing $J_n(t)$ as an infinite series and transforming term by term.

Integration of transforms. Again with $F(t)$ at least piecewise continuous and x large enough so that $e^{-xt} F(t)$ decreases exponentially (as $x \to \infty$), the integral

$$f(x) = \int_0^\infty e^{-xt} F(t)\, dt \quad (15.170)$$

is uniformly convergent with respect to x. This justifies reversing the order of integration in the following equation:

$$\int_s^b f(x)\, dx = \int_s^b \int_0^\infty e^{-xt} F(t)\, dt\, dx$$

$$= \int_0^\infty \frac{F(t)}{t} (e^{-st} - e^{-bt})\, dt, \quad (15.171)$$

on integrating with respect to x. The lower limit s is chosen large enough so that $f(s)$ is within the region of uniform convergence. Now, letting $b \to \infty$, we have

$$\int_s^\infty f(x)\,dx = \int_0^\infty \frac{F(t)}{t} e^{-st}\,dt$$

$$= \mathscr{L}\left\{\frac{F(t)}{t}\right\}, \tag{15.172}$$

provided that $F(t)/t$ is finite at $t = 0$ or diverges less strongly than t^{-1} (so that $\mathscr{L}\{F(t)/t\}$ will exist).

EXERCISES

15.9.1 Solve Eq. 15.137, which describes a damped simple harmonic oscillator for $X(0) = X_0$, $X'(0) = 0$, and

(a) $b^2 = 4\,km$ (critically damped), (b) $b^2 > 4\,km$ (overdamped).

Ans. (a) $X(t) = X_0 e^{-(b/2m)t}\left(1 + \dfrac{b}{2m}t\right)$.

15.9.2 Solve Eq. 15.137, which describes a damped simple harmonic oscillator for $X(0) = 0$, $X'(0) = v_0$, and

(a) $b^2 < 4\,km$ (underdamped), (b) $b^2 = 4\,km$ (critically damped),

Ans. (a) $X(t) = \dfrac{v_0}{\omega_1} e^{-(b/2m)t} \sin \omega_1 t$,

(b) $X(t) = v_0 t\, e^{-(b/2m)t}$.

(c) $b^2 > 4\,km$ (overdamped).

15.9.3 The motion of a body falling in a resisting medium may be described by

$$m\frac{d^2 X(t)}{dt^2} = mg - b\frac{d\,X(t)}{dt}$$

when the retarding force is proportional to the velocity. Find $X(t)$ and $dX(t)/dt$ for the initial conditions

$$X(0) = \left.\frac{dX}{dt}\right|_{t=0} = 0.$$

15.9.4 Ringing circuit. In certain electronic circuits resistance, inductance, and capacitance are placed in the plate circuit in parallel (Fig. 15.9). A constant voltage is maintained across the parallel elements, keeping the capacitor charged. At time $t = 0$ the circuit is disconnected from the voltage source. Find the voltages across the parallel elements R, L, and C as a function of time. Assume R to be *large*. *Hint.* By Kirchhoff's laws

$$I_R + I_C + I_L = 0 \quad \text{and} \quad E_R = E_C = E_L$$

where

$$E_R = I_R R,$$

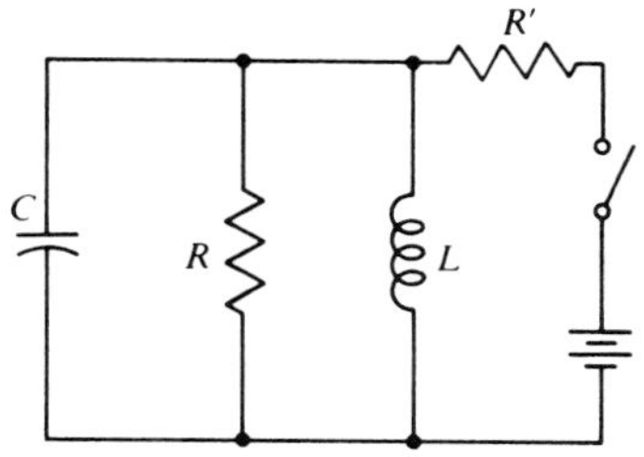

FIG. 15.9 Ringing circuit

$$E_C = \frac{q_0}{C} + \frac{1}{C}\int_0^t I_C\, dt,$$

$$E_L = L\frac{dI_L}{dt},$$

$q_0 =$ initial charge of capacitor.

With the DC impedance of $L = 0$, let $I_L(0) = I_0$, $E_L(0) = 0$. This means $q_0 = 0$.

15.9.5 **With $J_0(t)$ expressed as a contour integral, apply the Laplace transform operation, reverse the order of integration, and thus show that**

$$\mathscr{L}\{J_0(t)\} = (s^2 + 1)^{-1/2}, \qquad \text{for } s > 0.$$

15.9.6 Evaluate $\int_0^\infty xJ_0(x)\, dx$.

Hint: Direct application of Eq. 11.15 will lead you into infinite oscillations. Try something else—like a Laplace transform with $s \to 0$.

15.9.7 **Develop the Laplace transform of $J_n(t)$ from $\mathscr{L}\{J_0(t)\}$ by using the Bessel function recurrence relations.**

Hint. Here is a chance to use mathematical induction.

15.9.8 A calculation of the magnetic field of a circular current loop in circular cylindrical coordinates leads to the integral

$$\int_0^\infty e^{-kz}kJ_1(ka)\, dk, \qquad \mathscr{R}(z) \geq 0.$$

Show that this integral is equal to $a/(z^2 + a^2)^{3/2}$.

15.9.9 The electrostatic potential of a point charge q at the origin in circular cylindrical coordinates is

$$\frac{q}{4\pi\varepsilon_0}\int_0^\infty e^{-kz}J_0(k\rho)\, dk = \frac{q}{4\pi\varepsilon_0} \cdot \frac{1}{(\rho^2 + z^2)^{1/2}}, \qquad \mathscr{R}(z) \geq 0.$$

From this relation show that the Fourier cosine and sine transforms of $J_0(k\rho)$ are

(a) $$\sqrt{\frac{\pi}{2}}\, F_c\{J_0(k\rho)\} = \int_0^\infty J_0(k\rho) \cos k\zeta\, dk = \begin{cases} (\rho^2 - \zeta^2)^{-1/2}, & \rho > \zeta, \\ 0, & \rho < \zeta. \end{cases}$$

(b) $$\sqrt{\frac{\pi}{2}}\, F_s\{J_0(k\rho)\} = \int_0^\infty J_0(k\rho) \sin k\zeta\, dk = \begin{cases} 0, & \rho > \zeta, \\ (\zeta^2 - \rho^2)^{-1/2}, & \rho < \zeta. \end{cases}$$

15.9.10 Show that

$$\mathcal{L}\{I_0(at)\} = (s^2 - a^2)^{-1/2}, \qquad s > a.$$

15.9.11 Verify the following Laplace transforms:

(a) $\mathcal{L}\{j_0(at)\} = \mathcal{L}\left\{\frac{\sin at}{at}\right\} = \frac{1}{a}\cot^{-1}\left(\frac{s}{a}\right),$ (b) $\mathcal{L}\{n_0(at)\}$ does not exist

(c) $\mathcal{L}\{i_0(at)\} = \mathcal{L}\left\{\frac{\sinh at}{at}\right\} = \frac{1}{2a}\ln\frac{s+a}{s-a}$

$$= \frac{1}{a}\coth^{-1}\left(\frac{s}{a}\right),$$

(d) $\mathcal{L}\{k_0(at)\}$ does not exist.

15.9.12 Develop a Laplace transform solution of Laguerre's equation

$$t\,F''(t) + (1-t)\,F'(t) + n\,F(t) = 0.$$

Note that you need a derivative of a transform and a transform of derivatives. Go as far as you can with $n = n$; then (and only then) set $n = 0$.

15.9.13 Show that the Laplace transform of the Laguerre polynomial $L_n(at)$ is given by

$$\mathcal{L}\{L_n(at)\} = \frac{(s-a)^n}{s^{n+1}}, \qquad s > 0.$$

15.9.14 Show that

$$\mathcal{L}\{E_1(t)\} = \frac{1}{s}\ln(s+1), \qquad s > 0,$$

where

$$E_1(t) = \int_t^\infty \frac{e^{-\tau}d\tau}{\tau} = \int_1^\infty \frac{e^{-xt}}{x}\,dx.$$

$E_1(t)$ is the exponential-integral function.

15.9.15 (a) From Eq. 15.172 show that

$$\int_0^\infty f(x)\,dx = \int_0^\infty \frac{F(t)}{t}\,dt$$

provided the integrals exist.

(b) From the above result show that

$$\int_0^\infty \frac{\sin t}{t}\,dt = \frac{\pi}{2}$$

in agreement with Eqs. 15.105 and 7.41.

15.9.16 (a) Show that

$$\mathcal{L}\left\{\frac{\sin kt}{t}\right\} = \cot^{-1}\left(\frac{s}{k}\right).$$

(b) Using this result (with $k = 1$), prove that

$$\mathcal{L}\{si(t)\} = -\frac{1}{s}\tan^{-1} s,$$

where

$$si(t) = -\int_t^\infty \frac{\sin x}{x}\,dx, \qquad \text{the sine integral.}$$

15.9.17 If $F(t)$ is periodic with a period a so that $F(t+a) = F(t)$ for all $t \geqslant 0$, show that

$$\mathscr{L}\{F(t)\} = \frac{\int_0^a e^{-st} F(t)\, dt}{1 - e^{-as}}$$

with the integration now over only the *first period* of $F(t)$.

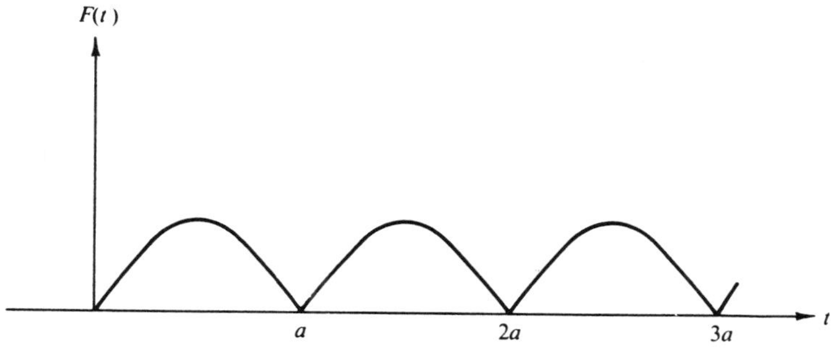

FIG. 15.10 Periodic function

15.9.18 Find the Laplace transform of the square wave (period a) defined by

$$F(t) = \begin{cases} 1, & 0 < t < a/2 \\ 0, & a/2 < t < a. \end{cases}$$

Ans. $f(s) = \dfrac{1}{s} \cdot \dfrac{1 - e^{-as/2}}{1 - e^{-as}}$.

15.9.19 Show that

(a) $\mathscr{L}\{\cosh at \cos at\} = \dfrac{s^3}{s^4 + 4a^4}$,

(b) $\mathscr{L}\{\cosh at \sin at\} = \dfrac{as^2 + 2a^3}{s^4 + 4a^4}$,

(c) $\mathscr{L}\{\sinh at \cos at\} = \dfrac{as^2 - 2a^3}{s^4 + 4a^4}$,

(d) $\mathscr{L}\{\sinh at \sin at\} = \dfrac{2a^2 s}{s^4 + 4a^4}$.

15.10 Convolution or Faltung Theorem

One of the most important properties of the Laplace transform is that given by the convolution or faltung theorem.[1] We take two transforms

$$f_1(s) = \mathscr{L}\{F_1(t)\} \quad \text{and} \quad f_2(s) = \mathscr{L}\{F_2(t)\} \tag{15.173}$$

[1] An alternate derivation employs the Bromwich integral (Section 15.11). This is Ex. 15.11.3.

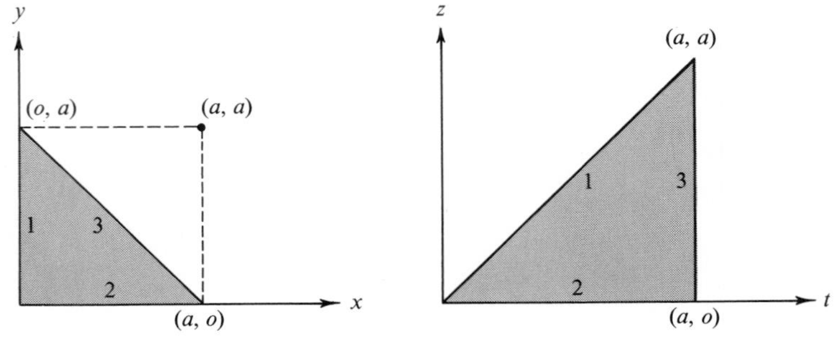

(a) (b)
FIG. 15.11 Change of variables, (a) *xy*-plane (b) *zt*-plane

and multiply them together. To avoid complications when changing variables, the upper limits are held finite.

$$f_1(s) \cdot f_2(s) = \lim_{a\to\infty} \int_0^a e^{-sx} F_1(x)\, dx \int_0^{a-x} e^{-sy} F_2(y)\, dy. \qquad (15.174)$$

The upper limits are chosen so that the area of integration, shown in Fig. 15.11*a*, is the shaded triangle, not the square. This modification is permissible because the two integrands are assumed to decrease exponentially. In the limit $a \to \infty$ the integral over the unshaded triangle will give zero contribution. Substituting $x = t - z$, $y = z$ the region of integration is mapped into the triangle shown in Fig. 15.11*b*. Using Jacobians to transform the element of area, we have

$$dx\, dy = \begin{vmatrix} \dfrac{\partial x}{\partial t} & \dfrac{\partial y}{\partial t} \\ \dfrac{\partial x}{\partial z} & \dfrac{\partial y}{\partial z} \end{vmatrix} dt\, dz = \begin{vmatrix} 1 & 0 \\ -1 & 1 \end{vmatrix} dt\, dz \qquad (15.175)$$

or $dx\, dy = dt\, dz$. With this substitution Eq. 15.174 becomes

$$\begin{aligned} f_1(s) \cdot f_2(s) &= \lim_{a\to\infty} \int_0^a e^{-st} \int_0^t F_1(t - z)\, F_2(z)\, dz\, dt \\ &= \mathscr{L}\left\{\int_0^t F_1(t - z)\, F_2(z)\, dz\right\}. \end{aligned} \qquad (15.176)$$

For convenience this integral is represented by the symbol

$$\int_0^t F_1(t - z)\, F_2(z)\, dz \equiv F_1 * F_2 \qquad (15.177)$$

and referred to as the convolution, closely analogous to the Fourier convolution (Section 15.5). If we substitute $w = t - z$, we find

$$F_1 * F_2 = F_2 * F_1, \qquad (15.178)$$

showing that the relation is symmetric.

Carrying out the inverse transform, we also find

$$\mathscr{L}^{-1}\{f_1(s) \cdot f_2(s)\} = \int_0^t F_1(t - z)\, F_2(z)\, dz. \tag{15.179}$$

This can be useful in the development of new transforms or as an alternative to a partial fraction expansion. One immediate application is in the solution of integral equations (Section 16.2).

EXAMPLE 15.10.1. DRIVEN OSCILLATOR WITH DAMPING

As one illustration of the use of the convolution theorem; let us return to the mass m on a spring, with damping and a driving force $F(t)$. The equation of motion (15.112) now becomes

$$m\, X''(t) + b\, X'(t) + k\, X(t) = F(t). \tag{15.180}$$

Initial conditions $X(0) = 0$, $X'(0) = 0$ are used to simplify this illustration, and the transformed equation is

$$ms^2\, x(s) + bs\, x(s) + k\, x(s) = f(s) \tag{15.181}$$

or

$$x(s) = \frac{f(s)}{m} \times \frac{1}{(s + b/2m)^2 + \omega_1^2}, \tag{15.182}$$

where $\omega_1^2 \equiv k/m - b^2/4m^2$, as before.

By the convolution theorem (Eq. 15.176 or 15.179),

$$X(t) = \frac{1}{m\omega_1} \int_0^t F(t - z) e^{-(b/2m)z} \sin \omega_1 z\, dz. \tag{15.183}$$

If the force is impulsive, $F(t) = P\, \delta(t)$,[1]

$$X(t) = \frac{P}{m\omega_1} e^{-(b/2m)t} \sin \omega_1 t. \tag{15.184}$$

P represents the momentum transferred by the impulse and the constant P/m takes the place of an initial velocity $X'(0)$.

If $F(t) = F_0 \sin \omega t$, Eq. 15.183 may be used, but a partial fraction expansion is perhaps more convenient. With

$$f(s) = \frac{F_0 \omega}{s^2 + \omega^2}$$

Eq. 15.182 becomes

$$x(s) = \frac{F_0 \omega}{m} \times \frac{1}{s^2 + \omega^2} \times \frac{1}{(s + b/2m)^2 + \omega_1^2}$$

[1] Note that $\delta(t)$ lies *inside* the interval $[0, t]$.

$$= \frac{F_0\omega}{m}\left[\frac{a's + b'}{s^2 + \omega^2} + \frac{c's + d'}{(s + b/2m)^2 + \omega_1^2}\right]. \tag{15.185}$$

The coefficients a', b', d', and d' are independent of s. Direct calculation shows

$$a' = \frac{b}{m}\omega^2 + \frac{m}{b}(\omega_0^2 - \omega^2)^2,$$

$$b' = -\frac{m}{b}(\omega_0^2 - \omega^2)\left[\frac{b}{m}\omega^2 + \frac{m}{b}(\omega_0^2 - \omega^2)^2\right].$$

Since c' and d' will lead to exponentially decreasing terms (transients), they will be discarded here. Carrying out the inverse operation, we find for the steady-state solution

$$X(t) = \frac{F_0}{[b^2\omega^2 + m^2(\omega_0^2 - \omega^2)^2]^{1/2}} \sin(\omega t - \varphi), \tag{15.186}$$

where

$$\tan\varphi = \frac{b\omega}{m(\omega_0^2 - \omega^2)}.$$

Differentiating the denominator, we find that the amplitude has a maximum when

$$\omega^2 = \omega_0^2 - \frac{b^2}{2m^2} = \omega_1^2 - \frac{b^2}{4m^2}. \tag{15.187}$$

This is the resonance condition. At resonance the amplitude becomes $F_0/b\omega_1$, showing that the mass m goes into infinite oscillation at resonance if damping is neglected ($b = 0$). It is worth noting that we have had three different characteristic frequencies:

$$\omega_2^2 = \omega_0^2 - \frac{b^2}{2m^2},$$

resonance for forced oscillations, with damping,

$$\omega_1^2 = \omega_0^2 - \frac{b^2}{4m^2},$$

free oscillation frequency, with damping,

$$\omega_0^2 = \frac{k}{m},$$

free oscillation frequency, no damping. They coincide *only* if the damping is zero.

Returning to Eqs. 15.180 and 15.182, Eq. 15.180 is our differential equation for the response of a dynamical system to an arbitrary driving force. The final response clearly depends on both the driving force and the characteristics of our system. This dual dependence is separated in the transform space. In Eq. 15.182, the transform of the response (output) appears as the product of two factors, one describing the driving force (input) and the other describing the dynamical system. This latter part, which modifies the input and yields the output, is often called a *transfer function*. Specifically $[(s + b/2m)^2 + \omega_1^2]^{-1}$ is the transfer function corresponding to this damped oscillator. The concept of a transfer function is of great use in the field of servomechanisms. Often the characteristics of a particular servomechanism are described by giving its transfer function. The convolution theorem then yields the output signal for a particular input signal.

EXAMPLE 15.10.2. TAUTACHRONE

Another illustration of the convolution theorem is provided by the tautachrone, or equal time, problem, which is related to the famous brachistochrone problem (calculus of variations, Section 17.2). The problem is to determine a curve passing through the origin for which the time required for a particle to slide down the curve to the origin (with no friction) is *independent* of the starting point. The particle starts from rest.

By conservation of energy, kinetic energy gained equals potential energy lost,

$$\frac{1}{2}m\left(\frac{d\lambda}{dt}\right)^2 = mg(y_0 - y), \qquad \left(\frac{d\lambda}{dt} < 0\right), \tag{15.188}$$

where λ is the distance along the curve from the origin, m is the mass of the particle,

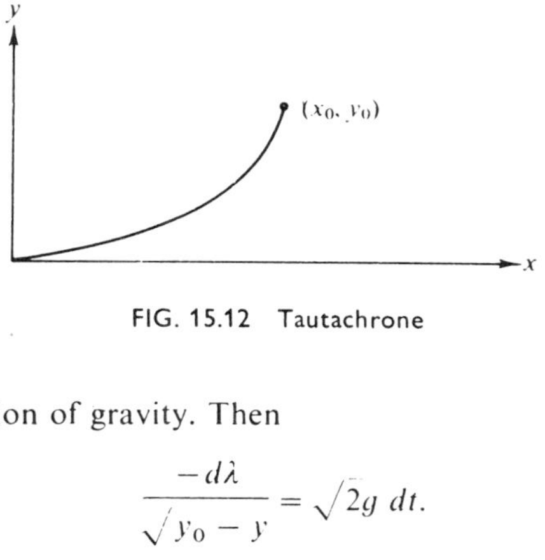

FIG. 15.12 Tautachrone

and g is the acceleration of gravity. Then

$$\frac{-d\lambda}{\sqrt{y_0 - y}} = \sqrt{2g}\, dt. \tag{15.189}$$

Integrating from top to bottom (time $t = 0$ to time $t = T$), we have

$$\sqrt{2g}T = -\int_{y=y_0}^{y=0} \frac{d\lambda}{\sqrt{y_0 - y}} = \int_{y=0}^{y=y_0} \frac{d\lambda}{\sqrt{y_0 - y}}. \tag{15.190}$$

The time of descent, T, is to be constant, independent of y_0. Now our path length λ is some function of the height, say $\lambda = F(y)$ and

$$\frac{d\lambda}{dy} = F'(y) = \sqrt{1 + \left(\frac{dx}{dy}\right)^2}. \tag{15.191}$$

Therefore

$$\sqrt{2g}T = \int_0^{y_0} (y_0 - y)^{-1/2} F'(y)\, dy$$
$$= y^{-1/2} * F'(y), \tag{15.192}$$

by the definition of convolution. Since the transform of the convolution of two functions is the product of the transforms,

$$\sqrt{2g}T\,\frac{1}{s} = \mathscr{L}\{F'(y)\} \cdot \mathscr{L}\{y^{-1/2}\}$$
$$= \mathscr{L}\{F'(y)\} \cdot \sqrt{\frac{\pi}{s}} \tag{15.193}$$

by Eq. 15.91. Note that $(-\frac{1}{2})! = \sqrt{\pi}$. Hence

$$\mathscr{L}\{F'(y)\} = \frac{\sqrt{2g}T}{\pi}\sqrt{\frac{\pi}{s}}, \tag{15.194}$$

which may be inverted to give

$$F'(y) = \frac{\sqrt{2g}T}{\pi}\, y^{-1/2}. \tag{15.195}$$

Squaring Eq. 15.191, we obtain

$$1 + \left(\frac{dx}{dy}\right)^2 = \frac{c}{y}, \tag{15.196}$$

where

$$c = \frac{2gT^2}{\pi^2}.$$

Separation of variables leads to

$$dx = \sqrt{\frac{c - y}{y}}\, dy, \tag{15.197}$$

which is satisfied by

$$x = \frac{c}{2}(\theta + \sin\theta),$$
$$y = \frac{c}{2}(1 - \cos\theta), \tag{15.198}$$

the parametric equations for a cycloid passing through the origin.

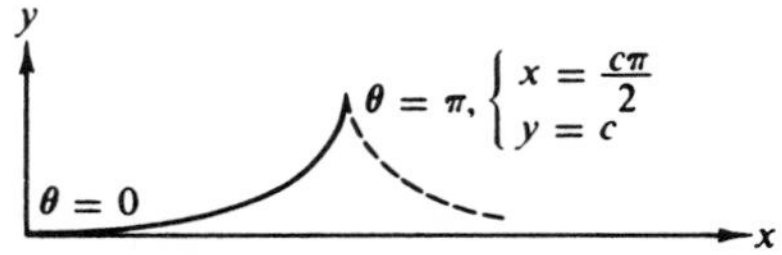

FIG. 15.13 Tautachrone, cycloidal path

Figure 15.13 is our tautachrone. This analysis demonstrates that a particle sliding (with no friction) down the curve in the first quadrant, up the reflected curve in the second quadrant, sliding back and forth, would have a period independent of its amplitude. This amplitude independence would also apply to a pendulum whose bob was constrained to follow such a cycloidal path. Some of the earliest pendulum clocks with large amplitude swings included just such cycloidal constraints.

EXERCISES

15.10.1 From the convolution theorem show that

$$\frac{1}{s}f(s) = \mathscr{L}\left\{\int_0^t F(x)\,dx\right\}$$

where $f(s) = \mathscr{L}\{F(t)\}$.

15.10.2 If $F(t) = t^a$ and $G(t) = t^b$,

(a) show that the convolution

$$F * G = t^{a+b+1}\int_0^1 y^a(1-y)^b\,dy.$$

(b) By using the convolution theorem show that

$$\int_0^1 y^a(1-y)^b\,dy = \frac{a!\,b!}{(a+b+1)!}.$$

If we replace a by $a-1$ and b by $b-1$, this is the Euler formula for the beta function (Eq. 10.60).

15.10.3 Using the convolution integral calculate

$$\mathscr{L}^{-1}\left\{\frac{s}{(s^2+a^2)(s^2+b^2)}\right\}, \qquad a^2 \neq b^2.$$

15.10.4 An undamped oscillator is driven by a force $F_0 \sin \omega t$. Find the displacement as a function of time. Notice that it is a linear combination of two simple harmonic motions, one with the frequency of the driving force and one with the frequency ω_0 of the free oscillator. (Assume $X(0) = X'(0) = 0$).

Ans. $$X(t) = \frac{F_0/m}{\omega^2 - \omega_0^2}\left(\frac{\omega}{\omega_0}\sin \omega_0 t - \sin \omega t\right).$$

15.11 Inverse Laplace Transformation

Bromwich integral. We shall now develop an expression for the inverse Laplace transform, $\mathscr{L}^{-1}$, appearing in the equation

$$F(t) = \mathscr{L}^{-1}\{f(s)\}. \qquad (15.199)$$

One approach lies in the Fourier transform for which we know the inverse relation. There is a difficulty, however. Our Fourier transformable function had to satisfy the Dirichlet conditions. In particular, we required that

$$\lim_{\omega \to \infty} G(\omega) = 0 \tag{15.200}$$

so that the infinite integral would be well defined.[1] Now we wish to treat functions, $F(t)$, that may diverge exponentially. To surmount this difficulty we extract an exponential factor, $e^{\gamma t}$ from our (possibly) divergent Laplace function and write

$$F(t) = e^{\gamma t}\, G(t). \tag{15.201}$$

If $F(t)$ diverges as $e^{\alpha t}$, we require γ to be greater than α *so that* $G(t)$ *will be convergent*. Now, with $G(t) = 0$ for $t < 0$ and otherwise suitably restricted so that it may be represented by a Fourier integral (Eq. 15.20),

$$G(t) = \frac{1}{2\pi} \int_{-\infty}^{\infty} e^{iut}\, du \int_0^{\infty} G(v) e^{-iuv}\, dv. \tag{15.202}$$

Using Eq. 15.201, we may rewrite (15.202)

$$F(t) = \frac{e^{\gamma t}}{2\pi} \int_{-\infty}^{\infty} e^{iut}\, du \int_0^{\infty} F(v) e^{-\gamma v} e^{-iuv}\, dv. \tag{15.203}$$

Now with the change of variable,

$$s = \gamma + iu, \tag{15.204}$$

the integral over v is thrown into the form of a Laplace transform

$$\int_0^{\infty} F(v) e^{-sv}\, dv = f(s); \tag{15.205}$$

s is now a complex variable and $\mathcal{R}(s) \geqslant \gamma$ to guarantee convergence. Notice that the Laplace transform has mapped a function specified on the positive real axis onto the complex plane, $\mathcal{R}(s) \geqslant \gamma$.

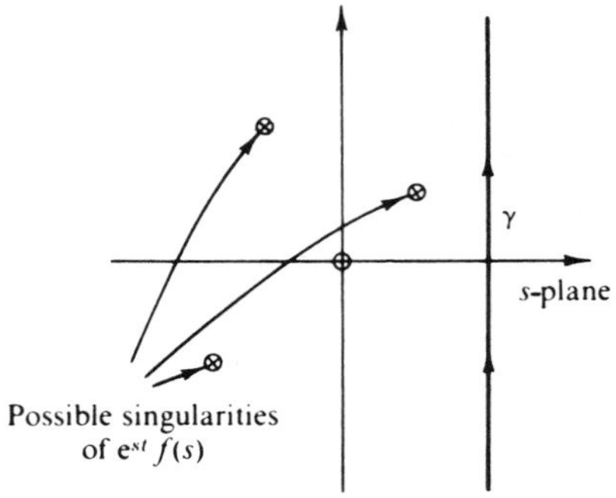

FIG. 15.14 Singularities of $e^{st} f(s)$

[1] If delta functions are included, $G(\omega)$ may be a cosine. While this does not satisfy Eq. 15.200, $G(\omega)$ is still bounded.

With γ as a constant, $ds = i\,du$. Substituting Eq. 15.205 into Eq. 15.203, we obtain

$$F(t) = \frac{1}{2\pi i}\int_{\gamma - i\infty}^{\gamma + i\infty} e^{st} f(s)\,ds. \tag{15.206}$$

Here is our inverse transform. We have rotated the line of integration through 90° (by using $ds = i\,du$). The path has become an infinite *vertical* line in the complex plane, the constant γ having been chosen so that all the singularities of $f(s)$ are on the left-hand side.

Equation 15.206, our inverse transformation, is usually known as the Bromwich integral, although sometimes it is referred to as the Fourier-Mellin theorem or Fourier-Mellin integral. This integral may now be evaluated by the regular methods of contour integration (Chapter 7). If $t > 0$, the contour may be closed by an infinite semicircle in the left half plane. Then by the residue theorem (Section 7.2)

$$F(t) = \sum (\text{residues included for } \mathscr{R}(s) < \gamma). \tag{15.207}$$

Possibly this means of evaluation with $\mathscr{R}(s)$ ranging through negative values seems paradoxical in view of our previous requirement that $\mathscr{R}(s) \geqslant \gamma$. The paradox disappears when we recall that the requirement $\mathscr{R}(s) \geqslant \gamma$ was imposed to guarantee convergence of the Laplace transform integral that defined $f(s)$. Once $f(s)$ is obtained we may then proceed to exploit its properties as an analytical function in the complex plane wherever we choose. In effect, we are employing analytical continuation to get $\mathscr{L}\{F(t)\}$ in the left half plane exactly as the recurrence relation for the factorial function was used to extend the Euler integral definition (Eq. 10.5) to the left half plane.

Perhaps a pair of examples may clarify the evaluation of Eq. 15.206.

EXAMPLE 15.11.1

If $f(s) = a/(s^2 - a^2)$, then

$$e^{st} f(s) = \frac{ae^{st}}{s^2 - a^2} = \frac{ae^{st}}{(s+a)(s-a)}. \tag{15.208}$$

There is a simple pole at $s = a$, residue $= \frac{1}{2}e^{at}$ and a second pole at $s = -a$, residue $= -\frac{1}{2}e^{-at}$,

$$\sum \text{residues} = \tfrac{1}{2}(e^{at} - e^{-at}) = \sinh at \tag{15.209}$$

in agreement with Eq. 15.88.

EXAMPLE 15.11.2

If

$$f(s) = \frac{1 - e^{-as}}{s},$$

then we have

$$e^{st}f(s) = \frac{e^{st}}{s} - e^{-as}\left(\frac{e^{st}}{s}\right). \tag{15.210}$$

The first term on the right has a simple pole at $s = 0$, residue $= 1$. Then by Eq. 15.207

$$F_1(t) = \begin{cases} 1, & t > 0, \\ 0, & t < 0. \end{cases} \tag{15.211}$$

Neglecting the minus sign and the e^{-as}, we find that the second term on the right also has a simple pole at $s = 0$, residue $= 1$. Noting the translation property (Eq. 15.147), we have

$$F_2(t) = \begin{cases} 1, & t - a > 0, \\ 0, & t - a < 0. \end{cases} \tag{15.212}$$

Therefore

$$F(t) = F_1(t) - F_2(t) = \begin{cases} 0, & t < 0, \\ 1, & 0 < t < a, \\ 0, & t > a, \end{cases} \tag{15.213}$$

a step function of unit height and length a.

Two general comments may be in order. First, these two examples hardly begin to show the usefulness and power of the Bromwich integral. It is always available for inverting a complicated transform when the tables prove inadequate.

Second, this derivation is not presented as a rigorous one. Rather, it is given more

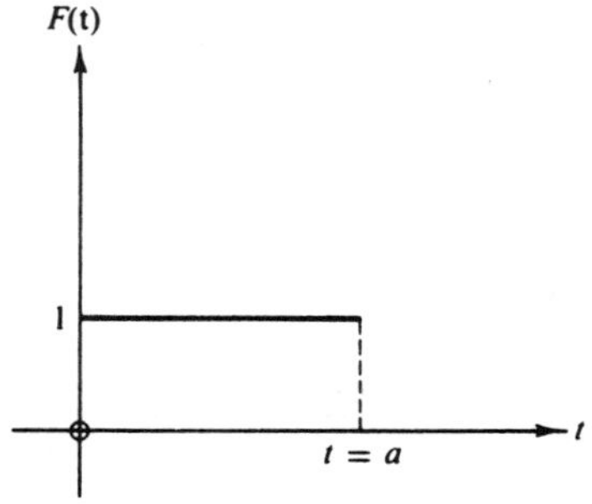

FIG. 15.15 Finite-length step function

as a plausibility argument, though it can be made rigorous. The determination of the inverse transform is somewhat similar to the solution of a differential equation. It makes little difference how you get the solution. Guess at it if you want to. The solution can always be checked by substitution back into the original differential equation. Similarly, $F(t)$ can (and, to check on careless errors, should) be checked by determining whether by Eq. 15.82

$$\mathcal{L}\{F(t)\} = f(s).$$

As a final illustration of the use of the Laplace inverse transform we have some results from the work of Brillouin and Sommerfeld (1914) in electromagnetic theory.

EXAMPLE 15.11.3. VELOCITY OF ELECTROMAGNETIC WAVES IN A DISPERSIVE MEDIUM

The group velocity u of traveling waves is related to the phase velocity v by the equation

$$u = v - \lambda \frac{dv}{d\lambda}. \tag{15.214}$$

Here λ is the wavelength. In the vicinity of an absorption line (resonance) $dv/d\lambda$ may

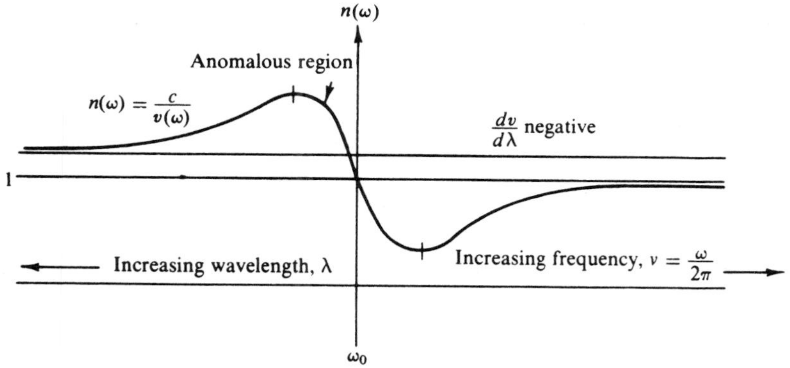

FIG. 15.16 Optical dispersion

be sufficiently negative so that $u > c$. The question immediately arises whether a signal can be transmitted faster than c, the velocity of light in vacuum. This question, which assumes that such a group velocity is meaningful, is of fundamental importance to the theory of special relativity.

We need a solution to the wave equation

$$\frac{\partial^2 \psi}{\partial x^2} = \frac{1}{v^2} \frac{\partial^2 \psi}{\partial t^2} \tag{15.215}$$

corresponding to a harmonic vibration starting at the origin at time zero. Since our medium is dispersive, v is a function of the angular frequency. Imagine, for instance,

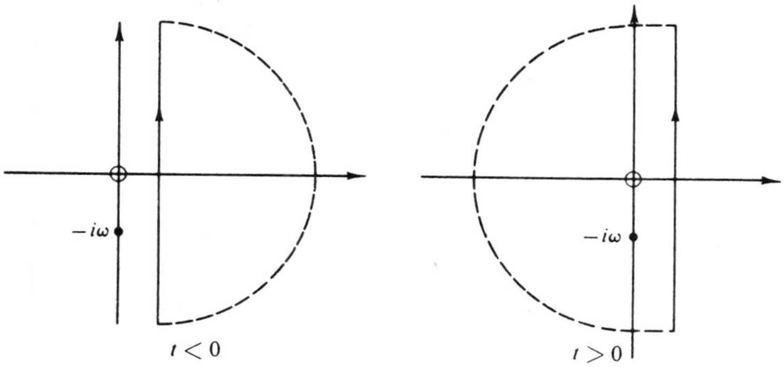

FIG. 15.17 Possible closed contours

a plane wave, angular frequency ω, incident on a shutter at the origin. At $t = 0$ the shutter is (instantaneously) opened, and the wave is permitted to advance along the positive x-axis.

Let us then build up a solution starting at $x = 0$. It is convenient to use the Cauchy integral

$$\psi(0, t) \quad \frac{1}{2\pi i}\int_{\gamma-i\infty}^{\gamma+i\infty} \frac{e^{st}}{s+i\omega}\, ds = \begin{cases} 0, & t < 0, \\ e^{-i\omega t}, & t > 0. \end{cases} \tag{15.216}$$

That this is just the Bromwich integral may be verified by noting that

$$F(t) = \begin{cases} 0, & t < 0, \\ e^{-i\omega t}, & t > 0, \end{cases} \tag{15.217}$$

and applying the Laplace transform. The transformed function $f(s)$ becomes

$$f(s) = \frac{1}{s + i\omega}. \tag{15.218}$$

Our Cauchy-Bromwich integral provides us with the time-dependence of a signal leaving the origin at $t = 0$. To include the space-dependence we note that

$$e^{s(t - x/v)}$$

satisfies the wave equation. With this as a clue, we replace t by $t - x/v$ and write a solution

$$\psi(x, t) = \frac{1}{2\pi i}\int_{\gamma-i\infty}^{\gamma+i\infty} \frac{e^{s(t-x/v)}}{s+i\omega}\, ds. \tag{15.219}$$

It was seen in the derivation of the Bromwich integral that our variable s replaces the ω of the Fourier transformation. Hence the wave velocity v becomes a function of s, that is, $v(s)$. Its particular form need not concern us here. We need only the property

$$\lim_{|s|\to\infty} v(s) = \text{constant}, c. \tag{15.220}$$

This is suggested by the asymptotic behavior of the curve on the right side of Fig. 15.16.[1]

Evaluating Eq. 15.219 by the calculus of residues, we may close the path of integration by a semicircle in the *right* half plane, provided

$$t - \frac{x}{c} < 0.$$

Hence

$$\psi(x, t) = 0, \qquad t - \frac{x}{c} < 0, \tag{15.221}$$

which means that the velocity of our signal cannot exceed the velocity of light in vacuum c. This simple but very significant result was extended by Sommerfeld and Brillouin to show just how the wave advanced in the dispersive medium.

[1] Equation 15.220 follows rigorously from the theory of anomalous dispersion.

EXERCISES

15.11.1 Derive the Bromwich integral from Cauchy's integral formula. *Hint.* Apply the inverse transform $\mathcal{L}^{-1}$ to

$$f(s) = \frac{1}{2\pi i} \lim_{\alpha \to \infty} \int_{\gamma - i\alpha}^{\gamma + i\alpha} \frac{f(z)}{s - z}\, dz,$$

where $f(z)$ is analytic for $\mathcal{R}(z) \geqslant \gamma$.

15.11.2 Starting with

$$\frac{1}{2\pi i} \int_{\gamma - i\infty}^{\gamma + i\infty} e^{st} f(s)\, ds,$$

show that by introducing

$$f(s) = \int_0^\infty e^{-sz} F(z)\, dz$$

we can convert one integral into the Fourier representation of a Dirac delta function. From this derive the inverse Laplace transform.

15.11.3 Derive the Laplace transformation convolution theorem by use of the Bromwich integral.

15.11.4 Find

$$\mathcal{L}^{-1}\left\{\frac{s}{s^2 - k^2}\right\}$$

(a) by a partial fraction expansion,
(b) repeat, using the Bromwich integral.

15.11.5 Find

$$\mathcal{L}^{-1}\left\{\frac{k^2}{s(s^2 + k^2)}\right\}$$

(a) by using a partial fraction expansion,
(b) repeat using the convolution theorem,
(c) repeat using the Bromwich integral.

Ans. $F(t) = 1 - \cos kt$.

15.11.6 Use the Bromwich integral to find the function whose transform is $f(s) = s^{-1/2}$. Note that $f(s)$ has a branch point at $s = 0$. The negative x-axis may be taken as a cut line.

Ans. $F(t) = (\pi t)^{-1/2}$.

15.11.7 Show that

$$\mathcal{L}^{-1}\{(s^2 + 1)^{-1/2}\} = J_0(t)$$

by evaluation of the Bromwich integral.

Hint. Convert your Bromwich integral into an integral representation of $J_0(t)$.

15.11.8 Evaluate the inverse Laplace transform

(a) $\mathcal{L}^{-1}\{(s^2 - a^2)^{-1/2}\} = I_0(at)$

(b) $\mathcal{L}^{-1}\left\{\dfrac{(s - \sqrt{s^2 - a^2})^\nu}{\sqrt{s^2 - a^2}}\right\} = a^\nu I_\nu(at), \qquad \nu > -1.$

15.11.9 Show that

$$\mathcal{L}^{-1}\left\{\frac{\ln s}{s}\right\} = -\ln t - \gamma,$$

where $\gamma = 0.5772 \cdots$, the Euler-Mascheroni constant.

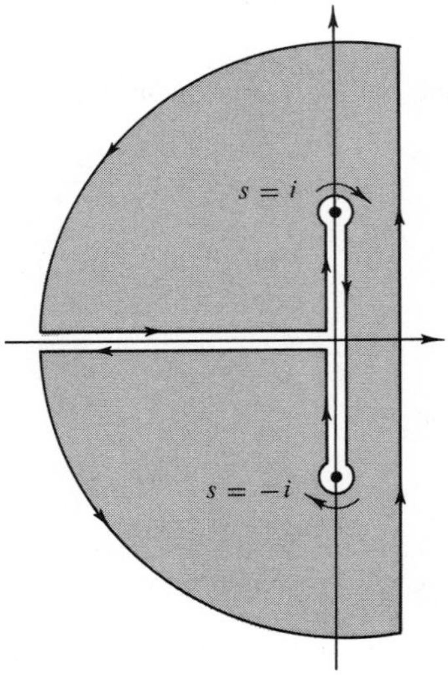

FIG. 15.18 A possible contour for the inversion of $J_0(t)$

TABLE 15.1

Laplace Transform Operations

		Operations	Equation
1.	Laplace transform	$f(s) = \mathscr{L}\{F(t)\} = \int_0^\infty e^{-st} F(t)\, dt$	(15.82)
2.	Transform of derivative	$sf(s) - F(+0) = \mathscr{L}\{F'(t)\}$ $s^2 f(s) - s\,F(+0) - F'(+0) = \mathscr{L}\{F''(t)\}$	(15.106)
3.	Transform of integral	$\frac{1}{s} f(s) = \mathscr{L}\left\{\int_0^t F(x)\, dx\right\}$	(Exercise 15.10.1)
4.	Substitution	$f(s-a) = \mathscr{L}\{e^{at} F(t)\}$	(15.135)
5.	Translation	$e^{-bs} f(s) = \mathscr{L}\{F(t-b)\}$	(15.147)
6.	Derivative of transform	$f^{(n)}(s) = \mathscr{L}\{(-t)^n F(t)\}$	(15.156)
7.	Integral of transform	$\int_s^\infty f(x)\, dx = \mathscr{L}\left\{\frac{F(t)}{t}\right\}$	(15.172)
8.	Convolution	$f_1(s) \cdot f_2(s) = \mathscr{L}\left\{\int_0^t F_1(t-z)\, F_2(z)\, dz\right\}$	(15.176)
9.	Inverse transform, Bromwich integral	$\frac{1}{2\pi i} \int_{\gamma - i\infty}^{\gamma + i\infty} e^{st} f(s)\, ds = F(t)$	(15.206)

TABLE 15.2

Laplace Transforms

	$f(s)$	$F(t)$	Limitation	Equation
1.	1	$\delta(t)$	Singularity at $+0$	(15.124)
2.	$\frac{1}{s}$	1	$s > 0$	(15.85)
3.	$\frac{n!}{s^{n+1}}$	t^n	$s > 0$ $n > -1$	(15.91)
4.	$\frac{1}{s-k}$	e^{kt}	$s > k$	(15.86)
5.	$\frac{1}{(s-k)^2}$	te^{kt}	$s > k$	(15.158)
6.	$\frac{s}{s^2-k^2}$	$\cosh kt$	$s > k$	(15.88)
7.	$\frac{k}{s^2-k^2}$	$\sinh kt$	$s > k$	(15.88)
8.	$\frac{s}{s^2+k^2}$	$\cos kt$	$s > 0$	(15.90)
9.	$\frac{k}{s^2+k^2}$	$\sin kt$	$s > 0$	(15.90)
10.	$\frac{s-a}{(s-a)^2+k^2}$	$e^{at}\cos kt$	$s > a$	(15.136)
11.	$\frac{k}{(s-a)^2+k^2}$	$e^{at}\sin kt$	$s > a$	(15.136)
12.	$(s^2+a^2)^{-1/2}$	$J_0(at)$	$s > 0$	(15.168)
13.	$(s^2-a^2)^{-1/2}$	$I_0(at)$	$s > a$	(Exercise 15.9.10)
14.	$\frac{1}{a}\cot^{-1}\left(\frac{s}{a}\right)$	$j_0(at)$	$s > 0$	(Exercise 15.9.11)
15.	$\left.\begin{array}{l}\frac{1}{2a}\ln\frac{s+a}{s-a}\\ \frac{1}{a}\coth^{-1}\left(\frac{s}{a}\right)\end{array}\right\}$	$i_0(at)$	$s > a$	(Exercise 15.9.11)
16.	$\frac{(s-a)^n}{s^{n+1}}$	$L_n(at)$	$s > 0$	(Exercise 15.9.13)
17.	$\frac{1}{s}\ln(s+1)$	$E_1(x) = -Ei(-x)$	$s > 0$	(Exercise 15.9.14)
18.	$\frac{\ln s}{s}$	$-\ln t - C$	$s > 0$	(Exercise 15.11.9)

A more extensive table of Laplace transforms appears in Chapter 29 of AMS-55.

REFERENCES

Carslaw H. S., and J. C. Jaeger, *Operational Methods in Applied Mathematics*. 2nd Ed. Oxford: Oxford University Press (1947). Here the Laplace transform is developed and applied to a wide variety of fields in physics and engineering.

Erdelyi, A., Ed., *Tables of Integral Transforms*, Volumes I and II. Bateman Manuscript Project. New York: McGraw-Hill. Volume I: *Fourier, Laplace, Mellin Transforms*. Volume II: *Hankel Transforms and Special Functions*. An encyclopedic compilation of transforms, special functions, and their properties. Useful primarily as a reference.

Jeffreys, H., and B. S. Jeffreys, *Methods of Mathematical Physics*. 3rd Ed. Cambridge: Cambridge University Press (1966).

LePage, W. R., *Complex Variables and the Laplace Transform for Engineers*. New York: McGraw-Hill (1961). Complex variable analysis is carefully developed and then applied to Fourier and Laplace transforms. Written to be read by students, but intended for the serious student.

Papoulis, A., *The Fourier Integral and Its Applications*. New York: McGraw-Hill (1962). A rigorous development of Fourier and Laplace transforms. Extensive applications in science and engineering.

Scott, E. J., *Transform Calculus with an Introduction to Complex Variables*. New York: Harper (1955). The primary emphasis is on the Laplace transform. Complex variable contour integration is developed and used extensively.

Sneddon, I. N., *Fourier Transforms*. New York: McGraw-Hill (1951). A detailed comprehensive treatment, loaded with applications to a wide variety of fields of modern and classical physics.

Titchmarsh, E. C., *Introduction to the Theory of Fourier Integrals*. Oxford: Oxford University Press (1948). Comprehensive treatment of Fourier integrals and related integral transforms.

Tranter, C. J., *Integral Transforms in Mathematical Physics*. 2nd Ed. London: Methuen, New York: Wiley (1951). A short, concise, but useful work covering Laplace, Fourier, Hankel, and Mellin transforms and their inverses.

Van der Pol, B., and H. Bremmer, *Operational Calculus Based on the Two-sided Laplace Integral*. 2nd Ed. Cambridge: Cambridge University Press (1955). Here is a development based on the integral range $-\infty$ to $+\infty$, rather than the usual 0 to $+\infty$. Chapter V contains a detailed study of the Dirac delta function (impulse function).

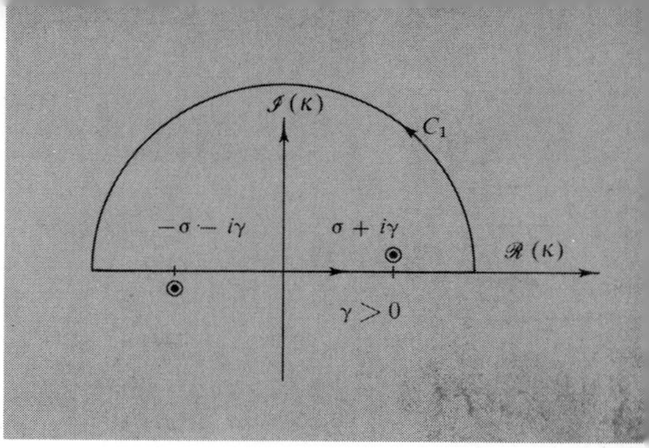

CHAPTER 16

INTEGRAL EQUATIONS

16.1 Introduction

With the exception of the integral transforms of the last chapter, we have been considering equations with relations between the unknown function $\varphi(x)$ and one or more of its derivatives. We now proceed to investigate equations containing the unknown function within an integral. These equations are classified in two ways:

1. If the limits of integration are fixed, we shall call the equation a Fredholm equation; if one limit is variable, it is a Volterra equation.
2. If the unknown function appears *only* under the integral sign, we shall label it " first kind." If it appears both inside and outside the integral, it will be labeled, " second kind."

Definitions. Symbolically we have
Fredholm equation of the first kind:

$$f(x) = \int_a^b K(x, t)\, \varphi(t)\, dt. \tag{16.1}$$

Fredholm equation of the second kind:

$$\varphi(x) = f(x) + \lambda \int_a^b K(x, t)\, \varphi(t)\, dt. \tag{16.2}$$

Volterra equation of the first kind:

$$f(x) = \int_a^x K(x, t)\, \varphi(t)\, dt. \tag{16.3}$$

Volterra equation of the second kind:

$$\varphi(x) = f(x) + \int_a^x K(x, t)\, \varphi(t)\, dt. \tag{16.4}$$

In all four cases $\varphi(t)$ is the unknown function. $K(x, t)$, which we call the kernel, and $f(x)$ are assumed to be known. When $f(x) = 0$, the equation is said to be homogeneous.

The reader may wonder, with some justification, why we bother about integral equations. After all, the differential equations have done a rather good job of describing our physical world so far. There are two reasons for introducing integral equations here.

First, we have placed considerable emphasis on the solution of differential equations *subject to particular boundary conditions.* For instance, the boundary condition at $r = 0$ determines whether the Neumann function $N_n(r)$ is present when Bessel's equation is solved. The boundary condition for $r \to \infty$ determines whether the $I_n(r)$ is present in our solution of the modified Bessel equation. The integral equation relates the unknown function not only to its values at neighboring points (derivatives) but also to its values throughout a region, including the boundary. In a very real sense the boundary conditions are built into the integral equation rather than imposed at the final stage of the solution. It will be seen later, when we construct kernels (Section 16.5), that the form of the kernel depends on the values on the boundary. The integral equation, then, is compact and may turn out to be a more convenient or more powerful form than the differential equation. Second, whether we like it or not, there are some problems, such as some diffusion and transport phenomena, that cannot be represented by differential equations. If we wish to solve such problems, we are forced to handle integral equations. A most important example of this sort of physical situation follows.

EXAMPLE 16.1.1. NEUTRON TRANSPORT THEORY—BOLTZMANN EQUATION

The fundamental equation of neutron transport theory is an expression of the equation of continuity for neutrons:

$$\text{production} = \text{losses} + \text{leakage}.$$

Under production we have sources

$$S(v, \mathbf{\Omega}, \mathbf{r})dv\, d\mathbf{\Omega}$$

representing the introduction of S neutrons per cubic centimeter per second with *speeds* between v and $v + dv$ and direction of motion $\mathbf{\Omega}$ within a solid angle $d\mathbf{\Omega}$.

An additional source is provided by scattering collisions that *scatter* neutrons into the ranges just listed. The rate of scattering is given by

$$\sum_s (v, v', \mathbf{\Omega}, \mathbf{\Omega}')\, \varphi(v', \mathbf{\Omega}', \mathbf{r}),$$

where $\sum_s$ is the (macroscopic) probability that a neutron of speed v', direction $\mathbf{\Omega}'$, will be scattered with resultant speed v, direction $\mathbf{\Omega}$. The quantity $\varphi(v', \mathbf{\Omega}', \mathbf{r})$ is the neutron flux. Expressed as a vector, $\boldsymbol{\varphi} = \mathbf{\Omega}\varphi$ has the direction of the neutron velocity and a magnitude equal to the number of neutrons per second of speed v crossing a unit area at position $\mathbf{r}$ and in a direction $\mathbf{\Omega}$ (Fig. 16.1).

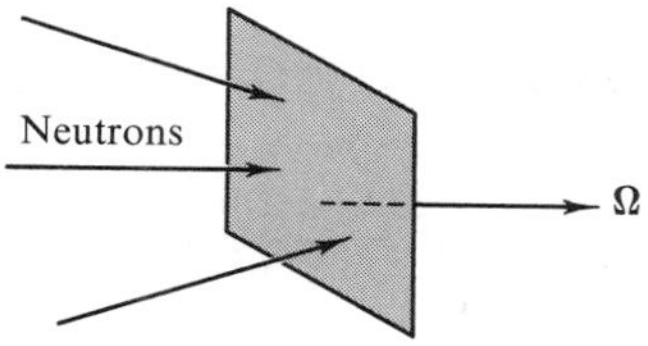

FIG. 16.1 Neutron flux

Integrating over available initial speeds (v') and over all directions ($\boldsymbol{\Omega}'$), we obtain

$$\iint \sum_s (v, v', \boldsymbol{\Omega}, \boldsymbol{\Omega}')\, \varphi(v', \boldsymbol{\Omega}', \mathbf{r})\, dv'\, d\boldsymbol{\Omega}'$$

for the second production term.

Losses come from leakage given by

$$\nabla \cdot \boldsymbol{\varphi}(v, \boldsymbol{\Omega}, \mathbf{r})$$

and from absorption and scattering into another (lower) velocity range. These are

$$\left[\sum_a (v) + \sum_s (v)\right] \varphi(v, \boldsymbol{\Omega}, \mathbf{r}).$$

If the medium is not homogeneous and isotropic, the $\sum$'s may have position- and direction-dependence in addition to the indicated speed or energy dependence.

Our equation of continuity finally becomes

$$\iint \sum_a (v, v', \boldsymbol{\Omega}, \boldsymbol{\Omega}')\, \varphi(v', \boldsymbol{\Omega}', \mathbf{r})\, dv'\, d\boldsymbol{\Omega}' + S(v, \boldsymbol{\Omega}, \mathbf{r})$$

$$= \nabla \cdot \boldsymbol{\varphi}(v, \boldsymbol{\Omega}, \mathbf{r}) + \left[\sum_a (v) + \sum_s (v)\right] \varphi(v, \boldsymbol{\Omega}, \mathbf{r}). \tag{16.5}$$

This is the steady-state Boltzmann equation, an integro-differential equation. In this form the Boltzmann equation is almost impossible to handle. Most of neutron transport theory is a development of methods that are compromises between physical accuracy and mathematical feasibility.[1]

An integral equation may also appear as a matter of deliberate choice based on convenience or the need for the mathematical power of an integral equation formulation.

Example 16.1.2. Momentum Representation in Quantum Mechanics

The Schrödinger equation (in ordinary space representation) is

$$-\frac{\hbar^2}{2m} \nabla^2 \psi(\mathbf{r}) + V(\mathbf{r})\, \psi(\mathbf{r}) = E\, \psi(\mathbf{r}) \tag{16.6}$$

[1] Cf. *Reactor Handbook*, 2nd Ed., Vol. III, Part A, *Physics*. H. Soodak, Ed. New York: Interscience Publishers (1962). Cf. Chapter 3.

or

$$(-\nabla^2 + a^2)\,\psi(\mathbf{r}) = v(\mathbf{r})\,\psi(\mathbf{r}) \tag{16.7}$$

where

$$a^2 = -\frac{2m}{\hbar^2}E,$$

$$v(\mathbf{r}) = -\frac{2m}{\hbar^2}V(\mathbf{r}).$$

We may generalize Eq. 16.7 to

$$(-\nabla^2 + a^2)\,\psi(\mathbf{r}) = \int v(\mathbf{r}, \mathbf{r}')\,\psi(\mathbf{r}')\,d^3x'. \tag{16.8}$$

For the special case of

$$v(\mathbf{r}, \mathbf{r}') = v(\mathbf{r}')\,\delta(\mathbf{r} - \mathbf{r}'), \tag{16.9}$$

which represents local interaction Eq. 16.8 reduces to Eq. 16.7. Equation 16.8 is now subjected to the Fourier transform (cf. Section 15.6).

$$\Phi(\mathbf{k}) = \frac{1}{(2\pi)^{3/2}}\int \psi(\mathbf{r})e^{-i\mathbf{k}\cdot\mathbf{r}}\,d^3x. \tag{16.10}$$

$$\psi(\mathbf{r}) = \frac{1}{(2\pi)^{3/2}}\int \Phi(\mathbf{k})e^{i\mathbf{k}\cdot\mathbf{r}}\,d^3k.$$

Here, the abbreviation

$$\frac{\mathbf{p}}{\hbar} = \mathbf{k} \quad (2\pi \times \text{wave number}) \tag{16.11}$$

has been introduced. Developing Eq. 16.10, we obtain

$$\int (-\nabla^2 + a^2)\,\psi(\mathbf{r})e^{-i\mathbf{k}\cdot\mathbf{r}}\,d^3x = \iint v(\mathbf{r}, \mathbf{r}')\,\psi(\mathbf{r}')e^{-i\mathbf{k}\cdot\mathbf{r}}\,d^3x'\,d^3x. \tag{16.12}$$

Note that the ∇^2 on the left operates only on the $\psi(\mathbf{r})$. Integrating the left-hand side by parts and substituting Eq. 16.10 for $\psi(\mathbf{r}')$ on the right,

$$\begin{aligned}\int (k^2 + a^2)\,\psi(\mathbf{r})e^{-i\mathbf{k}\cdot\mathbf{r}}\,d^3x &= (2\pi)^{3/2}(k^2 + a^2)\,\Phi(\mathbf{k})\\ &= \frac{1}{(2\pi)^{3/2}}\iiint v(\mathbf{r}, \mathbf{r}')\,\Phi(\mathbf{k}')e^{-i(\mathbf{k}\cdot\mathbf{r} - \mathbf{k}'\cdot\mathbf{r}')}\,d^3x'\,d^3x\,d^3k'.\end{aligned} \tag{16.13}$$

If we use

$$f(\mathbf{k}, \mathbf{k}') = \frac{1}{(2\pi)^3}\iint v(\mathbf{r}, \mathbf{r}')e^{-i(\mathbf{k}\cdot\mathbf{r} - \mathbf{k}'\cdot\mathbf{r}')}\,d^3x'\,d^3x, \tag{16.14}$$

Eq. 16.13 becomes

$$(k^2 + a^2)\,\Phi(\mathbf{k}) = \int f(\mathbf{k}, \mathbf{k}')\,\Phi(\mathbf{k}')\,d^3k', \tag{16.15}$$

a homogeneous Fredholm equation of the second kind in which the parameter a^2 corresponds to the eigenvalue.

For our special but important case of local interaction, application of Eq. 16.9 leads to

$$f(\mathbf{k}, \mathbf{k}') = f(\mathbf{k} - \mathbf{k}'). \tag{16.16}$$

This is our momentum representation equivalent to an ordinary static interaction potential in ordinary space. Our momentum function $\Phi(\mathbf{k})$ satisfies the integral equation (Eq. 16.15). It must be emphasized that all through here we have assumed that the required Fourier integrals exist. For a linear oscillator potential, $V(\mathbf{r}) = r^2$, the required integrals would not exist. Equation 16.10 would lead to divergent oscillations and we would have no Eq. 16.15.

Often we find that we have a choice. The physical problem may be represented by a differential or an integral equation. Let us assume that we have the differential equation and wish to transform it into an integral equation. Starting with a *linear* second-order differential equation

$$y'' + A(x)y' + B(x)y = g(x) \tag{16.17}$$

with initial conditions

$$y(a) = y_0,$$

$$y'(a) = y_0',$$

we integrate to obtain

$$y' = -\int_a^x Ay'\,dx - \int_a^x By\,dx + \int_a^x g\,dx + y_0'. \tag{16.18}$$

Integrating the first integral on the right by parts yields

$$y' = -Ay - \int_a^x (B - A')y\,dx + \int_a^x g\,dx + A(a)y_0 + y_0'. \tag{16.19}$$

Notice how the initial conditions are being absorbed into our new version. Integrating a second time, we obtain

$$\begin{aligned} y = -\int_a^x Ay\,dx - \int_a^x\int_a^x [B(t) - A'(t)]y(t)\,dt\,dx \\ + \int_a^x\int_a^x g(t)\,dt\,dx + [A(a)y_0 + y_0'](x - a) + y_0. \end{aligned} \tag{16.20}$$

To transform this equation into a neater form, we use the relation

$$\int_a^x\int_a^x f(t)\,dt\,dx = \int_a^x (x - t)f(t)\,dt. \tag{16.21}$$

This may be verified by differentiating both sides. Since the derivatives are equal, the original expressions can differ only by a constant. Letting $x \to a$, the constant vanishes and Eq. 16.21 is established. Applying it to Eq. 16.20

$$\begin{aligned} y(x) = -\int_a^x \{A(t) + (x - t)[B(t) - A'(t)]\}\,y(t)\,dt \\ + \int_a^x (x - t)\,g(t)\,dt + [A(a)y_0 + y_0'](x - a) + y_0. \end{aligned} \tag{16.22}$$

If we now introduce the abbreviations

$$K(x, t) = (t - x)[B(t) - A'(t)] - A(t),$$
$$f(x) = \int_a^x (x - t)\, g(t)\, dt + [A(a)\, y_0 + y_0'](x - a) + y_0, \tag{16.23}$$

Eq. 16.22 becomes

$$y(x) = f(x) + \int_a^x K(x, t)\, y(t)\, dt, \tag{16.24}$$

which is a Volterra equation of the second kind.

EXAMPLE 16.1.3

As a simple illustration consider the linear oscillator equation

$$y'' + \omega^2 y = 0 \tag{16.25}$$

with

$$y(0) = 0$$
$$y'(0) = 1.$$

This yields

$$A(x) = 0,$$
$$B(x) = \omega^2,$$
$$g(x) = 0,$$

and our integral equation becomes

$$y(x) = x + \omega^2 \int_0^x (t - x) y(t)\, dt. \tag{16.26}$$

The reader may show that this is indeed satisfied by $y(x) = (1/\omega) \sin \omega x$.

Let us reconsider the linear oscillator equation (16.25) but now with the boundary conditions

$$y(0) = 0,$$
$$y(b) = 0.$$

Since $y'(0)$ is not given, we must modify the procedure. The first integration gives

$$y' = -\omega^2 \int_0^x y\, dx + y'(0). \tag{16.27}$$

Integrating a second time and again using Eq. 16.21, we have

$$y = -\omega^2 \int_0^x (x - t) y(t)\, dt + y'(0)x. \tag{16.28}$$

We now impose the condition $y(b) = 0$. This gives

$$\omega^2 \int_0^b (b - t)\, y(t)\, dt = by'(0). \tag{16.29}$$

Substituting this back into Eq. 16.28, we obtain

$$y(x) = -\omega^2 \int_0^x (x - t)\, y(t)\, dt + \omega^2 \frac{x}{b} \int_0^b (b - t) y(t)\, dt. \tag{16.30}$$

Now let us break the interval $[0, b]$ into two intervals $[0, x]$ and $[x, b]$. Since

$$\frac{x}{b}(b - t) - (x - t) = \frac{t}{b}(b - x), \tag{16.31}$$

we find

$$y(x) = \omega^2 \int_0^x \frac{t}{b}(b - x)\, y(t)\, dt + \omega^2 \int_x^b \frac{x}{b}(b - t)\, y(t)\, dt. \tag{16.32}$$

Finally, if we define a kernel

$$K(x, t) = \begin{cases} \dfrac{t}{b}(b - x), & t < x, \\ \dfrac{x}{b}(b - t), & x < t, \end{cases} \tag{16.33}$$

we have

$$y(x) = \omega^2 \int_0^b K(x, t)\, y(t)\, dt, \tag{16.34}$$

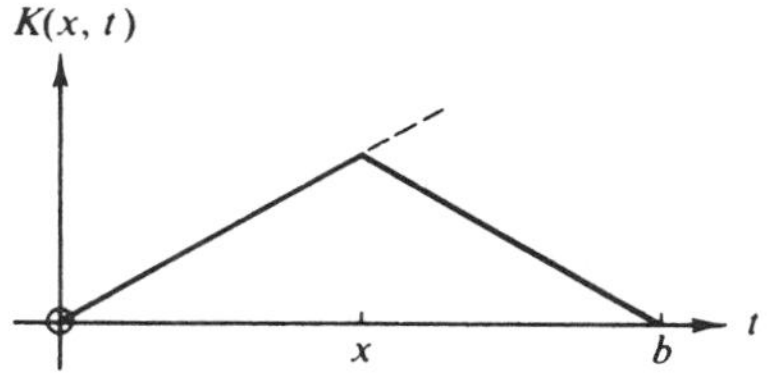

FIG. 16.2

a homogeneous Fredholm equation of the second kind.

Our new kernel, $K(x, t)$, has some interesting properties.

1. It is symmetric, $K(x, t) = K(t, x)$.
2. It is continuous in the sense that

$$\left.\frac{t}{b}(b - x)\right|_{t=x} = \left.\frac{x}{b}(b - t)\right|_{t=x}$$

3. Its derivative with respect to t is *discontinuous*. As t increases through the point $t = x$, there is a discontinuity of -1 in $\partial K(x, t)/\partial t$.

We shall return to these properties in Section 16.5 in which we identify $K(x, t)$ as a Green's function.

EXERCISES

16.1.1 Derive the Volterra integral equation corresponding to

(a) $y''(x) - y(x) = 0; \quad y(0) = 0, \quad y'(0) = 1.$

$$Ans.\ y = \int_0^x (x-t)y(t)dt + x.$$

(b) $y''(x) - y(x) = 0; \quad y(0) = 1, \quad y'(0) = -1.$

$$Ans.\ y = \int_0^x (x-t)y(t)dt - x + 1.$$

16.1.2 Derive a Fredholm integral equation corresponding to

$$y''(x) - y(x) = 0; \quad y(1) = 1, \quad y(-1) = 1,$$

(a) by integrating twice, (b) by forming the Green's function.

$$Ans.\quad y(x) = 1 - \int_{-1}^{1} K(x,t)y(t)dt,$$

$$K(x,t) = \begin{cases} \frac{1}{2}(1-x)(t+1), & x > t, \\ \frac{1}{2}(1-t)(x+1), & x < t. \end{cases}$$

16.1.3 (a) Starting with the given answers of Exercise 16.1.1, differentiate and recover the original differential equations *and the boundary conditions.*
(b) Repeat for Exercise 16.1.2.

16.1.4 The general second-order linear differential equation with constant coefficients is

$$y''(x) + a_1 y'(x) + a_2 y(x) = 0.$$

Given the boundary conditions

$$y(0) = y(1) = 0,$$

integrate twice and develop the integral equation

$$y(x) = \int_0^1 K(x,t)\, y(t)\, dt$$

with

$$K(x,t) = \begin{cases} a_2 t(1-x) + a_1(x-1), & t < x, \\ a_2 x(1-t) + a_1 x, & x < t. \end{cases}$$

Note that $K(x, t)$ is symmetric and continuous if $a_1 = 0$. How is this related to self-adjointness of the differential equation?

16.2 Integral Transforms, Generating Functions

Although there is no general method available for inverting integral equations, certain special cases may be treated with our *integral transforms* (Chapter 15.) For convenience, these are listed here. If

$$\psi(x) = \frac{1}{\sqrt{2\pi}} \int_{-\infty}^{\infty} e^{ixt}\, \varphi(t)\, dt,$$

then

$$\varphi(x) = \frac{1}{\sqrt{2\pi}} \int_{-\infty}^{\infty} e^{-ixt}\, \psi(t)\, dt \qquad \text{(Fourier).} \tag{16.35}$$

If

$$\psi(x) = \int_0^\infty e^{-xt}\, \varphi(t)\, dt,$$

then

$$\varphi(x) = \frac{1}{2\pi i} \int_{\gamma - i\infty}^{\gamma + i\infty} e^{xt}\, \psi(t)\, dt \qquad \text{(Laplace).} \tag{16.36}$$

If

$$\psi(x) = \int_0^\infty t^{x-1}\, \varphi(t)\, dt,$$

then

$$\varphi(x) = \frac{1}{2\pi i} \int_{\gamma - i\infty}^{\gamma + i\infty} x^{-t}\, \psi(t)\, dt \qquad \text{(Mellin).} \tag{16.37}$$

If

$$\psi(x) = \int_0^\infty t\, \varphi(t)\, J_\nu(xt)\, dt,$$

then

$$\varphi(x) = \int_0^\infty t\, \psi(t)\, J_\nu(xt)\, dt \qquad \text{(Hankel).} \tag{16.38}$$

Actually, the usefulness of the integral transform technique extends a bit beyond these four rather specialized forms.

Example 16.2.1. Fourier Transform Solution

Let us consider a Fredholm equation of the first kind with a kernel of the general type $k(x - t)$:

$$f(x) = \int_{-\infty}^{\infty} k(x - t)\, \varphi(t)\, dt, \tag{16.39}$$

in which $\varphi(t)$ is our unknown function. *Assuming that the needed transforms exist*, we apply the Fourier convolution theorem (Section 15.5) to obtain

$$f(x) = \int_{-\infty}^{\infty} K(t)\, \Phi(t) e^{-ixt}\, dt. \tag{16.40}$$

The functions $K(t)$ and $\Phi(t)$ are the Fourier transforms of $k(x)$ and $\varphi(x)$, respectively. Inverting, by Eq. 16.35, we have

$$K(t)\, \Phi(t) = \frac{1}{2\pi} \int_{-\infty}^{\infty} f(x) e^{ixt}\, dx = \frac{F(t)}{\sqrt{2\pi}}. \tag{16.41}$$

Then

$$\Phi(t) = \frac{1}{\sqrt{2\pi}} \cdot \frac{F(t)}{K(t)}, \tag{16.42}$$

and again inverting we have

$$\varphi(x) = \frac{1}{2\pi} \int_{-\infty}^{\infty} \frac{F(t)}{K(t)} e^{-ixt}\, dt. \tag{16.43}$$

For a rigorous justification of this result the reader is invited to follow Morse and

Feshbach across complex planes. An extension of this transformation solution appears as Exercise 16.2.1.

EXAMPLE 16.2.2. GENERALIZED ABEL EQUATION, CONVOLUTION THEOREM

In Section 15.10 Laplace transforms were used to solve the integral equation of the tautachrone problem. Generalizing slightly, we write the integral equation as

$$f(x) = \int_0^x \frac{\varphi(t)}{(x-t)^\alpha}\,dt, \qquad 0 < \alpha < 1, \quad \text{with} \begin{cases} f(x) & \text{known,} \\ \varphi(t) & \text{unknown.} \end{cases} \tag{16.44}$$

Taking the Laplace transform of both sides of this equation, we obtain

$$\begin{aligned} \mathscr{L}\{f(x)\} &= \mathscr{L}\left\{\int_0^x \frac{\varphi(t)}{(x-t)^\alpha}\,dt\right\} \\ &= \mathscr{L}\{x^{-\alpha}\}\mathscr{L}\{\varphi(x)\}, \end{aligned} \tag{16.45}$$

the last step following by the Laplace convolution theorem (Section 15.10). Then

$$\mathscr{L}\{\varphi(x)\} = \frac{s^{1-\alpha}\mathscr{L}\{f(x)\}}{(-\alpha)!}. \tag{16.46}$$

Dividing by s, we obtain

$$\begin{aligned} \frac{1}{s}\mathscr{L}\{\varphi(x)\} &= \frac{s^{-\alpha}\mathscr{L}\{f(x)\}}{(-\alpha)!} \\ &= \frac{\mathscr{L}\{x^{\alpha-1}\}\mathscr{L}\{f(x)\}}{(\alpha-1)!(-\alpha)!}. \end{aligned} \tag{16.47}$$

Combining the factorials (Eq. 10.32), and applying the Laplace convolution theorem again, we discover that

$$\frac{1}{s}\mathscr{L}\{\varphi(x)\} = \frac{\sin \pi\alpha}{\pi}\mathscr{L}\left\{\int_0^x \frac{f(t)}{(x-t)^{1-\alpha}}\,dt\right\}. \tag{16.48}$$

Inverting with the aid of Exercise 15.10.1, we get

$$\int_0^x \varphi(t)\,dt = \frac{\sin \pi\alpha}{\pi}\int_0^x \frac{f(t)}{(x-t)^{1-\alpha}}\,dt, \tag{16.49}$$

and finally, by differentiating,

$$\varphi(x) = \frac{\sin \pi\alpha}{\pi}\frac{d}{dx}\int_0^x \frac{f(t)}{(x-t)^{1-\alpha}}\,dt. \tag{16.50}$$

Our solution of Abel's equation in Section 15.10 is a special case ($\alpha = \frac{1}{2}$).

Generating functions. Occasionally, the reader may encounter integral equations that involve generating functions. Suppose we have the admittedly special case

$$f(x) = \int_{-1}^1 \frac{\varphi(t)}{(1-2xt+x^2)^{1/2}}\,dt, \qquad -1 \leqslant x \leqslant 1. \tag{16.51}$$

We notice two important features:

1. $(1 - 2xt + x^2)^{-1/2}$ generates the Legendre polynomials.
2. $[-1, 1]$ is the orthogonal interval for the Legendre polynomials.

If we now expand the denominator (property 1) and assume that our unknown $\varphi(t)$ may be written as a series of these same Legendre polynomials,

$$f(x) = \int_{-1}^{1} \sum_{n=0}^{\infty} a_n P_n(t) \sum_{r=0}^{\infty} P_r(t) x^r \, dt \tag{16.52}$$

by utilizing the orthogonality of the Legendre polynomials (property 2),

$$f(x) = \sum_{r=0}^{\infty} \frac{2a_r}{2r+1} x^r. \tag{16.53}$$

We may identify the a_n's by differentiating n times and then setting $x = 0$.

$$f^{(n)}(0) = n! \frac{2}{2n+1} a_n. \tag{16.54}$$

Hence,

$$\varphi(t) = \sum_{n=0}^{\infty} \frac{2n+1}{2} \frac{f^{(n)}(0)}{n!} P_n(t). \tag{16.55}$$

Similar results may be obtained with the other generating functions (cf. Exercise 16.2.5). Actually, the technique of expanding in a series of special functions is always available. It is worth a try whenever the expansion is possible (and convenient) and the interval is appropriate.

EXERCISES

16.2.1 The kernel of a Fredholm equation of the second kind,

$$\varphi(x) = f(x) + \lambda \int_{-\infty}^{\infty} K(x, t) \varphi(t) \, dt,$$

is of the form $k(x - t)$.[1] Assuming that the required transforms exist, show that

$$\varphi(x) = \frac{1}{\sqrt{2\pi}} \int_{-\infty}^{\infty} \frac{F(t) e^{-ixt} \, dt}{1 - \sqrt{2\pi} \lambda K(t)}.$$

$F(t)$ and $K(t)$ are the Fourier transforms of $f(x)$ and $k(x)$, respectively.

16.2.2 The kernel of a Volterra equation of the first kind,

$$f(x) = \int_0^x K(x, t) \varphi(t) \, dt,$$

has the form $k(x - t)$. Assuming that the required transforms exist, show that

$$\varphi(x) = \frac{1}{2\pi i} \int_{\gamma - i\infty}^{\gamma + i\infty} \frac{F(s)}{K(s)} e^{xs} \, ds.$$

$F(s)$ and $K(s)$ are the Laplace transforms of $f(x)$ and $k(x)$, respectively.

[1] This kernel and a range $0 \leqslant x < \infty$ are the characteristics of integral equations of the Wiener-Hopf type. Details will be found in Chapter 8 of Morse and Feshbach.

16.2.3 The kernel of a Volterra equation of the second kind,

$$\varphi(x) = f(x) + \lambda \int_0^x K(x, t)\, \varphi(t)\, dt$$

has the form $k(x - t)$. Assuming that the required transforms exist, show that

$$\varphi(x) = \frac{1}{2\pi i} \int_{\gamma - i\infty}^{\gamma + i\infty} \frac{F(s)}{1 - \lambda K(s)} e^{xs}\, ds.$$

16.2.4 Using the Laplace transform solution (Exercise 16.2.3), solve

(a) $\varphi(x) = x + \int_0^x (t - x)\, \varphi(t)\, dt,$ *Ans.* $\varphi(x) = \sin x,$

(b) $\varphi(x) = x - \int_0^x (t - x)\, \varphi(t)\, dt.$ $\varphi(x) = \sinh x.$

Check your results by substituting back into the original integral equations.

16.2.5 A Fredholm equation of the first kind has a kernel $e^{-(x-t)^2}$.

$$f(x) = \int_{-\infty}^{\infty} e^{-(x-t)^2} \varphi(t)\, dt.$$

Show that the solution is

$$\varphi(x) = \frac{1}{\sqrt{\pi}} \sum_{n=0}^{\infty} \frac{f^{(n)}(0)}{2^n n!} H_n(x),$$

in which $H_n(x)$ is an nth-order Hermite polynomial.

16.2.6 Solve Abel's equation,

$$f(x) = \int_0^x \frac{\varphi(t)}{(x - t)^\alpha}\, dt, \qquad 0 < \alpha < 1;$$

by the following method:

(a) Multiply both sides by $(z - x)^{\alpha - 1}$ and integrate with respect to x over the range $0 \leqslant x \leqslant z$.

(b) Reverse the order of integration and evaluate the integral on the right-hand side (with respect to x) by the beta function.

Note:

$$\begin{aligned}\int_t^z \frac{dx}{(z - x)^{1-\alpha}(x - t)^\alpha} &= B(1 - \alpha, \alpha) \\ &= (-\alpha)!(\alpha - 1)! \\ &= \frac{\pi}{\sin \pi\alpha}.\end{aligned}$$

16.2.7 Solve the integral equation

$$f(x) = \int_{-1}^{1} \frac{\varphi(t)}{(1 - 2xt + x^2)^{1/2}}\, dt, \qquad -1 \leq x \leq 1,$$

for the unknown function $\varphi(t)$ if (a) $f(x) = x^{2s}$, (b) $f(x) = x^{2s+1}$.

Ans. (a) $\varphi(t) = \dfrac{4s + 1}{2} P_{2s}(t),$

(b) $\varphi(t) = \dfrac{4s + 3}{2} P_{2s+1}(t).$

16.2.8 A Kirchhoff diffraction theory analysis of a laser leads to the integral equation

$$v(\mathbf{r}_2) = \gamma \iint K(\mathbf{r}_1, \mathbf{r}_2)\, v(\mathbf{r}_1)\, dA.$$

The unknown $v(\mathbf{r}_1)$ gives the geometric distribution of the radiation field over one mirror surface; the range of integration is over the surface of that mirror. For square confocal spherical mirrors the integral equation becomes

$$v(x_2, y_2) = \frac{-i\gamma e^{ikb}}{\lambda b} \int_{-a}^{a} \int_{-a}^{a} e^{-(ik/b)(x_1x_2+y_1y_2)}\, v(x_1, y_1)\, dx_1\, dy_1,$$

in which b is the centerline distance between the laser mirrors. This can be put in a somewhat simpler form by the substitutions

$$\frac{kx_i^2}{b} = \xi_i^2, \qquad \frac{ky_i^2}{b} = \eta_i^2 \quad \text{and} \quad \frac{ka^2}{b} = \frac{2\pi a^2}{\lambda b} = \alpha^2.$$

(a) Show that the variables separate and we get two integral equations.
(b) Show that the new limits $\pm\alpha$ may be approximated by $\pm\infty$ for a mirror dimension $a \gg \lambda$.
(c) Solve the resulting integral equations.

16.3 Neumann Series, Separable (Degenerate) Kernels

Many and probably most integral equations cannot be solved by the specialized integral transform techniques of the preceding section. Here we develop two rather general techniques for solving integral equations. The first, due largely to Neumann, Liouville, and Volterra, develops the unknown function $\varphi(x)$ as a power series in λ, where λ is a given constant. The method is applicable whenever the series converges.

The second method is somewhat restricted because it requires that the two variables appearing in the kernel $K(x, t)$ be separable. However, there are two major rewards: (1) the relation between an integral equation and a set of simultaneous linear algebraic equations is shown explicitly, and (2) the method leads to eigenvalues and eigenfunctions—in close analogy to Section 4.6.

Neumann series. We solve a linear integral equation of the second kind by successive approximations; our integral equation is the Fredholm equation

$$\varphi(x) = f(x) + \lambda \int_a^b K(x, t)\, \varphi(t)\, dt \tag{16.56}$$

in which $f(x) \neq 0$. If the upper limit of the integral is a variable (Volterra equation), the following development will still hold, but with minor modifications. Let us try (there is no *guarantee* that it will work) to approximate our unknown function by

$$\varphi(x) \approx \varphi_0(x) = f(x). \tag{16.57}$$

This choice is not mandatory. If you can make a better guess, go ahead and guess. The choice here is equivalent to saying that the integral or the constant λ is small. To improve this first crude approximation we feed $\varphi_0(x)$ back into the integral, getting

$$\varphi_1(x) = f(x) + \lambda \int_a^b K(x, t)\, f(t)\, dt. \tag{16.58}$$

Repeating this process of substituting the new $\varphi_n(x)$ back into Eq. 16.56, we develop the sequence

$$\varphi_2(x) = f(x) + \lambda \int_a^b K(x, t_1) f(t_1)\, dt_1$$
$$+ \lambda^2 \int_a^b \int_a^b K(x, t_1)\, K(t_1, t_2) f(t_2)\, dt_2\, dt_1 \tag{16.59}$$

and

$$\varphi_n(x) = \sum_{i=0}^{n} \lambda^i u_i(x), \tag{16.60}$$

where

$$\begin{aligned} u_0(x) &= f(x) \\ u_1(x) &= \int_a^b K(x, t_1) f(t_1)\, dt_1 \\ u_2(x) &= \int_a^b \int_a^b K(x, t_1)\, K(t_1, t_2) f(t_2)\, dt_2\, dt_1 \\ u_n(x) &= \iiint K(x, t_1)\, K(t_1, t_2) \cdots K(t_{n-1}, t_n) \cdot f(t_n)\, dt_n \cdots dt_1. \end{aligned} \tag{16.61}$$

We expect that our solution $\varphi(x)$ will be

$$\varphi(x) = \lim_{n \to \infty} \varphi_n(x) = \lim_{n \to \infty} \sum_{i=0}^{n} \lambda^i u_i(x), \tag{16.62}$$

provided that our infinite series converges.

We may conveniently check the convergence by the Cauchy ratio test, Section 5.2, noting that

$$|\lambda^n u_n(x)| \leqslant |\lambda|^n |f|_{\max} |K|_{\max}^n |b - a|^n, \tag{16.63}$$

using $|f|_{\max}$ to represent the *maximum* value of $|f(x)|$ in the interval $[a, b]$ and $|K|_{\max}$ to represent the maximum value of $|K(x, t)|$ in its domain in the x, t-plane. We have convergence if

$$|\lambda|\, |K|_{\max} |b - a| < 1. \tag{16.64}$$

Note that $\lambda\, u_n(\max)$ is being used as a *comparison* series. If it converges, our actual series must converge. If this condition is not satisfied, we may have convergence or we may not. A more sensitive test is required. Of course, even if the Neumann series diverges, there still may be a solution obtainable by another method.

EXAMPLE 16.3.1

To illustrate the Neumann method we consider the integral equation

$$\varphi(x) = x + \frac{1}{2} \int_{-1}^{1} (t - x)\, \varphi(t)\, dt. \tag{16.65}$$

The Neumann series yields

$$\varphi_0(x) = x$$

$$\varphi_1(x) = x + \tfrac{1}{3} \tag{16.66}$$

$$\varphi_2(x) = x + \frac{1}{3} - \frac{x}{3}$$

$$\varphi_3(x) = x + \frac{1}{3} - \frac{x}{3} - \frac{1}{3^2},$$

and by induction

$$\varphi_{2n}(x) = x + \sum_{s=1}^{n}(-1)^{s-1}3^{-s} - x\sum_{s=1}^{n}(-1)^{s-1}3^{-s} \tag{16.67}$$

Letting $n \to \infty$,

$$\varphi(x) = \tfrac{3}{4}x + \tfrac{1}{4}. \tag{16.68}$$

It is interesting to note that our series converged easily even though Eq. 16.64 is *not* satisfied in this particular case. Actually Eq. 16.64 is a rather crude upper bound on λ. It can be shown that a necessary and sufficient condition for the convergence of our series solution is that $|\lambda| < |\lambda_e|$, where λ_e is the eigenvalue of smallest magnitude of the corresponding homogeneous equation $[f(x) = 0)]$. For this particular example $\lambda_e = \sqrt{3}/2$ which is clearly greater than $\frac{1}{2}$.

One approach to the calculation of time dependent perturbations in quantum mechanics starts with the integral equation for the evolution operator

$$U(t, t_0) = 1 - \frac{i}{\hbar}\int_{t_0}^{t} V(t_1)\, U(t_1, t_0)\, dt_1. \tag{16.69a}$$

Iteration leads to

$$U(t, t_0) = 1 - \frac{i}{\hbar}\int_{t_0}^{t} V(t_1)\, dt_1 + \left(\frac{i}{\hbar}\right)^2 \int_{t_0}^{t}\int_{t_0}^{t_1} V(t_1)\, V(t_2)\, dt_2\, dt_1 + \cdots \tag{16.69b}$$

The evolution operator is obtained as a series of multiple integrals of the perturbing potential $V(t)$, closely analogous to the Neumann series, Eq. 16.60. A second and similar relationship between the Neumann series and quantum mechanics appears when the Schrödinger wave equation for scattering is reformulated as an integral equation. The first term in a Neumann series solution is the incident (unperturbed) wave. The second term is the Born approximation, Eq. 16.191 of Section 16.6 (cf. Ex. 16.6.13).

The Fredholm method of solving Fredholm equations consists of subdividing the integration interval and essentially replacing the integral by a sum. The single integral equation is replaced by a large number (approaching infinity) of simultaneous linear algebraic equations. For details the reader is again referred to the references listed at the end of this chapter.

Separable kernel. The technique of replacing our integral equation by simultaneous algebraic equations may also be used whenever our kernel $K(x, t)$ is separable in the sense that

$$K(x, t) = \sum_{j=1}^{n} M_j(x) N_j(t), \tag{16.70}$$

where n, the upper limit of the sum, is *finite*. Such kernels are sometimes called degenerate. Our class of separable kernels includes all polynomials and many of the elementary transcendental functions, that is,

$$\cos(t - x) = \cos t \cos x + \sin t \sin x. \tag{16.70a}$$

If Eq. 16.70 is satisfied, substitution into the Fredholm equation of second kind yields

$$\varphi(x) = f(x) + \lambda \sum_{j=1}^{n} M_j(x) \int_a^b N_j(t)\, \varphi(t)\, dt, \tag{16.71}$$

interchanging integration and summation. Now the integral with respect to t is a constant,

$$\int_a^b N_j(t)\, \varphi(t)\, dt = c_j. \tag{16.72}$$

Hence Eq. 16.71 becomes

$$\varphi(x) = f(x) + \lambda \sum_{j=1}^{n} c_j M_j(x). \tag{16.73}$$

This gives us $\varphi(x)$, our solution, once the constants c_i have been determined.

We may find c_i by multiplying Eq. 16.73 by $N_i(x)$ and integrating, to eliminate the x-dependence. Use of Eq. 16.72 yields

$$c_i = b_i + \lambda \sum_{j=1}^{n} a_{ij} c_j, \tag{16.74}$$

where

$$b_i = \int_a^b N_i(x) f(x)\, dx, \tag{16.75}$$

$$a_{ij} = \int_a^b N_i(x)\, M_j(x)\, dx.$$

It is perhaps helpful to write Eq. 16.74 in matrix form, with $\mathsf{A} = (a_{ij})$.

$$\mathbf{c} - \lambda \mathsf{A}\mathbf{c} = \mathbf{b}$$

$$= (1 - \lambda \mathsf{A})\mathbf{c},$$

or

$$\mathbf{c} = (1 - \lambda \mathsf{A})^{-1}\mathbf{b}. \tag{16.76}$$

This is equivalent to a set of simultaneous linear algebraic equations

$$\begin{aligned} (1 - \lambda a_{11})c_1 - \lambda a_{12}c_2 - \lambda a_{13}c_3 - \cdots &= b_1, \\ -\lambda a_{21}c_1 + (1 - \lambda a_{22})c_2 - \lambda a_{23}c_3 - \cdots &= b_2, \\ -\lambda a_{31}c_1 - \lambda a_{32}c_2 + (1 - \lambda a_{33})c_3 - \cdots &= b_3, \quad \text{etc.} \end{aligned} \tag{16.77}$$

If our integral equation is homogeneous, $[f(x) = 0]$, then $\mathbf{b} = 0$. To get a solution we set the determinant of the coefficients of c_i equal to zero,

$$|1 - \lambda \mathsf{A}| = 0, \tag{16.78}$$

exactly as in Section 4.6. The roots of Eq. 16.78 yield our eigenvalues. Substituting into Eq. 16.76, we find the c_i's and then Eq. 16.73 gives our solution.

EXAMPLE 16.3.2

To illustrate this technique for determining eigenvalues and eigenfunctions of the homogeneous Fredholm equation, we consider the simple case

$$\varphi(x) = \lambda \int_{-1}^{1} (t + x)\, \varphi(t)\, dt. \tag{16.79}$$

Here

$$M_1 = 1, \qquad M_2(x) = x,$$

$$N_1(t) = t, \qquad N_2 = 1.$$

Equation 16.75 yields

$$a_{11} = a_{22} = 0,$$

$$a_{12} = \tfrac{2}{3},$$

$$a_{21} = 2.$$

Equation 16.78, our secular equation, becomes

$$\begin{vmatrix} 1 & -\dfrac{2\lambda}{3} \\ -2\lambda & 1 \end{vmatrix} = 0. \tag{16.80}$$

Expanding, we obtain

$$1 - \frac{4\lambda^2}{3} = 0,$$

$$\lambda = \pm \frac{\sqrt{3}}{2}. \tag{16.81}$$

Substituting the eigenvalues $\lambda = \pm\sqrt{3}/2$ into Eq. 16.76, we have

$$c_1 \mp \frac{c_2}{\sqrt{3}} = 0. \tag{16.82}$$

Finally, with a choice of $c_1 = 1$, Eq. 16.73 gives

$$\varphi_1(x) = \frac{\sqrt{3}}{2}(1 + \sqrt{3}x), \qquad \lambda = \frac{\sqrt{3}}{2}, \tag{16.83}$$

$$\varphi_2(x) = -\frac{\sqrt{3}}{2}(1 - \sqrt{3}x), \qquad \lambda = -\frac{\sqrt{3}}{2}. \tag{16.84}$$

Since our equation is homogeneous, the normalization of $\varphi(x)$ is arbitrary.

EXERCISES

16.3.1 Using the Neumann series, solve

(a) $\varphi(x) = 1 - 2\int_0^x t\,\varphi(t)\,dt$, *Ans.* (a) $\varphi(x) = e^{-x^2}$. (b) $\varphi(x) = x + \int_0^x (t - x)\,\varphi(t)\,dt$,

(c) $\varphi(x) = x - \int_0^x (t - x)\,\varphi(t)\,dt$.

16.3.2 Solve the equation

$$\varphi(x) = x + \tfrac{1}{2}\int_{-1}^{1} (t + x)\,\varphi(t)\,dt$$

by the separable kernel method. Compare with the Neumann method solution of Section 16.3.

Ans. $\varphi(x) = \frac{1}{2}(3x + 1)$.

16.3.3 Find the eigenvalues and eigenfunctions of

$$\varphi(x) = \lambda\int_{-1}^{1} (t - x)\,\varphi(t)\,dt.$$

16.3.4 Find the eigenvalues and eigenfunctions of

$$\varphi(x) = \lambda\int_0^{2\pi} \cos(x - t)\,\varphi(t)\,dt.$$

Ans. $\lambda_1 = \lambda_2 = \dfrac{1}{\pi}$,

$\varphi(x) = A\cos x + B\sin x$.

16.3.5 Find the eigenvalues and eigenfunctions of

$$y(x) = \lambda\int_{-1}^{1} (x - t)^2\,y(t)\,dt.$$

Hint. This problem may be treated by the separable kernel method or by a Legendre expansion.

16.3.6 If the separable kernel technique of this section is applied to a Fredholm equation of the first kind, show that Eq. 16.76 is replaced by

$$\mathbf{c} = \mathbf{A}^{-1}\mathbf{b}.$$

In general the solution for the unknown $\varphi(t)$ is *not* unique.

16.3.7 Solve

$$\psi(x) = x + \int_0^1 (1 + xt)\,\psi(t)\,dt$$

by each of the following methods:

(a) the Neumann series technique, (b) the separable kernel technique, and (c) educated guessing.

16.3.8 Use the separable kernel technique to show that

$$\psi(x) = \lambda\int_0^{\pi} \cos x \sin t\,\psi(t)\,dt$$

has *no* solution (apart from the trivial $\psi = 0$). Explain this result in terms of separability and symmetry.

16.4 Hilbert-Schmidt Theory

Symmetrization of kernels. This is the development of the properties of linear integral equations (Fredholm type) with symmetric kernels.

$$K(x, t) = K(t, x). \tag{16.85}$$

Before plunging into the theory, we note that some important nonsymmetric kernels can be symmetrized. If we have the equation

$$\varphi(x) = f(x) + \lambda \int_a^b K(x, t)\, \rho(t)\, \varphi(t)\, dt, \tag{16.86}$$

the total kernel is actually $K(x, t)\,\rho(t)$, clearly not symmetric if $K(x, t)$ alone is symmetric. However, if we multiply Eq. 16.86 by $\sqrt{\rho(x)}$ and substitute

$$\sqrt{\rho(x)}\, \varphi(x) = \psi(x), \tag{16.87}$$

we obtain

$$\psi(x) = \sqrt{\rho(x)}\, f(x) + \lambda \int_a^b [K(x, t)\sqrt{\rho(x)\, \rho(t)}]\, \psi(t)\, dt \tag{16.88}$$

with a symmetric total kernel, $K(x, t)\sqrt{\rho(x)\, \rho(t)}$. We shall meet $\rho(x)$ later as a weighting factor in this integral equation Sturm-Liouville theory.

Orthogonal eigenfunctions. We now focus our attention on the homogeneous Fredholm equation of the second kind:

$$\varphi(x) = \lambda \int_a^b K(x, t)\, \varphi(t)\, dt. \tag{16.89}$$

We shall assume that the kernel $K(x, t)$ is symmetric and real. Perhaps one of the first questions the mathematician might ask about the equation is, "Does it make sense?" or more precisely, "Does an eigenvalue λ satisfying this equation exist?" With the aid of the Schwarz and Bessel inequalities, Courant and Hilbert (Chapter III, Section 4) show that if $K(x, t)$ is continuous there is at least one such eigenvalue and possibly an infinite number of them.

We shall show that the eigenvalues, λ, are real and that the corresponding eigenfunctions, $\varphi_i(x)$, are orthogonal. Let λ_i, λ_j be two *different* eigenvalues and $\varphi_i(x)$, $\varphi_j(x)$, the corresponding eigenfunctions. Equation 16.89 then becomes

$$\varphi_i(x) = \lambda_i \int_a^b K(x, t)\, \varphi_i(t)\, dt, \tag{16.90a}$$

$$\varphi_j(x) = \lambda_j \int_a^b K(x, t)\, \varphi_j(t)\, dt. \tag{16.90b}$$

If we multiply Eq. 16.90*a* by $\lambda_j \varphi_j(x)$, Eq. 16.90*b* by $\lambda_i\, \varphi_i(x)$, and then integrate with respect to x, the two equations become[1]

[1] We assume that the necessary integrals exist. For an example of a simple pathological case, see Exercise 16.4.3.

$$\lambda_j \int_a^b \varphi_i(x)\, \varphi_j(x)\, dx = \lambda_i \lambda_j \int_a^b \int_a^b K(x, t)\, \varphi_i(t)\, \varphi_j(x)\, dt\, dx, \tag{16.91a}$$

$$\lambda_i \int_a^b \varphi_i(x)\, \varphi_j(x)\, dx = \lambda_i \lambda_j \int_a^b \int_a^b K(x, t)\, \varphi_j(t)\, \varphi_i(x)\, dt\, dx. \tag{16.91b}$$

Since we have demanded that $K(x, t)$ be symmetric Eq. 16.91*b* may be rewritten

$$\lambda_i \int_a^b \varphi_i(x)\, \varphi_j(x)\, dx = \lambda_i \lambda_j \int_a^b \int_a^b K(x, t)\, \varphi_i(t)\, \varphi_j(x)\, dt\, dx. \tag{16.92}$$

Subtracting Eq. 16.92 from Eq. 16.91*a*, we obtain

$$(\lambda_j - \lambda_i) \int_a^b \varphi_i(x)\, \varphi_j(x)\, dx = 0. \tag{16.93}$$

Since $\lambda_i \neq \lambda_j$,

$$\int_a^b \varphi_i(x)\, \varphi_j(x)\, dx = 0, \qquad i \neq j, \tag{16.94}$$

proving orthogonality. Note that with a symmetric kernel no complex conjugates are involved in Eq. 16.94. For the self-adjoint or Hermitian kernel see Exercise 16.4.1.

If the eigenvalue λ_i is degenerate,[1] the eigenfunctions for that particular eigenvalue may be orthogonalized by the Schmidt method (Section 9.3). Our orthogonal eigenfunctions may, of course, be normalized, and we will assume that this has been done. The result is

$$\int_a^b \varphi_i(x)\, \varphi_j(x)\, dx = \delta_{ij}. \tag{16.95}$$

To demonstrate that the λ_i are real it is necessary to get into complex conjugates. Taking the complex conjugate of Eq. 16.90*a*, we have

$$\varphi_i^*(x) = \lambda_i^* \int_a^b K(x, t) \varphi_i^*(t)\, dt, \tag{16.96}$$

provided the kernel $K(x, t)$ is real. Now, using Eq. 16.96 instead of Eq. 16.90*b* the analysis leads to

$$(\lambda_i^* - \lambda_i) \int_a^b \varphi_i^*(x)\, \varphi_i(x)\, dx = 0. \tag{16.97}$$

This time the integral cannot vanish (unless we have the trivial solution, $\varphi_i(x) = 0$) and

$$\lambda_i^* = \lambda_i \tag{16.98}$$

or λ_i, our eigenvalue, is real.

If the reader feels that somehow this state of affairs is vaguely familiar, he is right. This is the *third* time we have passed this way, first with Hermitian matrices, then with Sturm-Liouville (self-adjoint) equations, and now with Hilbert-Schmidt

[1] If more than one distinct eigenfunction corresponds to the same eigenvalue (satisfying Eq. 16.89), that eigenvalue is said to be degenerate.

integral equations. The correspondence between the Hermitian matrices and the self-adjoint differential equations shows up in modern physics as the two outstanding formulations of quantum mechanics—the Heisenberg matrix approach and the Schrödinger differential operator approach. In Section 16.5 we shall explore further the correspondence between the Hilbert-Schmidt symmetric kernel integral equations and the Sturm-Liouville self-adjoint differential equations.

The eigenfunctions of our integral equation form a complete set[1] in the sense that any function $g(x)$ that can be generated by the integral

$$g(x) = \int K(x, t)\, h(t)\, dt, \tag{16.99}$$

in which $h(t)$ is any piecewise continuous function, can be represented by a series of eigenfunctions,

$$g(x) = \sum_{n=1}^{\infty} a_n\, \varphi_n(x). \tag{16.100}$$

The series converges uniformly and absolutely.

Let us extend this to the kernel, $K(x, t)$, by asserting that

$$K(x, t) = \sum_{n=1}^{\infty} a_n\, \varphi_n(t), \tag{16.101}$$

and $a_n = a_n(x)$. Substituting into the original integral equation (Eq. 16.89) and using the orthogonality integral, we obtain

$$\varphi_i(x) = \lambda_i\, a_i(x). \tag{16.102}$$

Therefore for our homogeneous Fredholm equation of the second kind the kernel may be expressed in terms of the eigenfunctions and eigenvalues by

$$K(x, t) = \sum_{n=1}^{\infty} \frac{\varphi_n(x)\, \varphi_n(t)}{\lambda_n}, \qquad \text{(zero not an eigenvalue)}. \tag{16.103}$$

It is possible that the expansion given by Eq. 16.101 may not exist. As an illustration of the sort of pathological behavior that may occur, the reader is invited to apply this analysis to

$$\varphi(x) = \lambda \int_0^{\infty} e^{-xt}\, \varphi(t)\, dt$$

(cf. Exercise 16.4.3).

It should be emphasized that this Hilbert-Schmidt theory is concerned with the establishment of properties of the eigenvalues (real) and eigenfunctions (orthogonality, completeness), properties that may be of great interest and value. The Hilbert-Schmidt theory does *not* solve the homogeneous integral equation for us any more than the Sturm-Liouville theory of Chapter 9 solved the differential equations. The solutions of the integral equation come from Sections 16.2 and 16.3 (or from numerical analysis).

[1] For a proof of this statement see Courant and Hilbert, Chapter III, Section 5.

Nonhomogeneous integral equation. We need a solution of the nonhomogeneous equation

$$\varphi(x) = f(x) + \lambda \int_a^b K(x, t)\, \varphi(t)\, dt. \tag{16.104}$$

Let us assume that the solutions of the corresponding homogeneous integral equation are known.

$$\varphi_n(x) = \lambda_n \int_a^b K(x, t)\, \varphi_n(t)\, dt, \tag{16.105}$$

the solution $\varphi_n(x)$ corresponding to the eigenvalue λ_n. We shall expand both $\varphi(x)$ and $f(x)$ in terms of this set of eigenfunctions.

$$\varphi(x) = \sum_{n=1}^{\infty} a_n\, \varphi_n(x), \tag{16.106}$$

$$f(x) = \sum_{n=1}^{\infty} b_n\, \varphi_n(x). \tag{16.107}$$

Substituting into Eq. 16.104, we obtain

$$\sum_{n=1}^{\infty} a_n\, \varphi_n(x) = \sum_{n=1}^{\infty} b_n\, \varphi_n(x) + \lambda \int_a^b K(x, t) \sum_{n=1}^{\infty} a_n\, \varphi_n(t)\, dt. \tag{16.108}$$

By interchanging the order of integration and summation the integral may be evaluated by Eq. 16.105, and we get

$$\sum_{n=1}^{\infty} a_n\, \varphi_n(x) = \sum_{n=1}^{\infty} b_n\, \varphi_n(x) + \lambda \sum_{n=1}^{\infty} \frac{a_n\, \varphi_n(x)}{\lambda_n}. \tag{16.109}$$

If we multiply by $\varphi_i(x)$ and integrate from $x = a$ to $x = b$, the orthogonality of our eigenfunctions leads to

$$a_i = b_i + \lambda \frac{a_i}{\lambda_i}. \tag{16.110}$$

This can be rewritten as

$$a_i = b_i + \frac{\lambda}{\lambda_i - \lambda}\, b_i \tag{16.111}$$

which brings us to our solution

$$\varphi(x) = f(x) + \lambda \sum_{i=1}^{\infty} \frac{\int_a^b f(t)\, \varphi_i(t)\, dt}{\lambda_i - \lambda}\, \varphi_i(x). \tag{16.112}$$

Here it is assumed that the eigenfunctions, $\varphi_i(x)$, are normalized to unity. *Note that if $f(x) = 0$ there is no solution unless $\lambda = \lambda_i$.* This means that our homogeneous equation has no solution (except the trivial $\varphi(x) = 0$) unless λ is an eigenvalue, λ_i.

In the event that λ for the nonhomogeneous equation (16.104) is equal to one of the eigenvalues, λ_p, of the homogeneous equation, our solution (Eq. 16.112) blows up. To repair the damage we return to Eq. 16.110 and give the value

$$a_p = b_p + \lambda_p \frac{a_p}{\lambda_p} = b_p + a_p \tag{16.113}$$

special attention. Clearly, a_p drops out and is no longer determined by b_p, whereas $b_p = 0$. This implies that $\int f(x)\, \varphi_p(x)\, dx = 0$, that is, that $f(x)$ is orthogonal to the eigenfunction $\varphi_p(x)$. If this is *not* the case, we have no solution.

Equation 16.111 still holds for $i \neq p$, so we multiply by $\varphi_i(x)$ and sum over $i\ (i \neq p)$ to obtain

$$\varphi(x) = f(x) + a_p\varphi_p + \lambda_p \sum_{\substack{i=1 \\ i\neq p}}^{\infty}{}' \frac{\int_a^b f(t)\, \varphi_i(t)\, dt}{\lambda_i - \lambda_p}\, \varphi_i(x); \tag{16.114}$$

the prime emphasizes that the value $i = p$ is omitted. In this solution the a_p remains as an undetermined constant.[1]

EXERCISES

16.4.1 In the Fredholm equation

$$\varphi(x) = \lambda \int_a^b K(x, t)\, \varphi(t)\, dt$$

the kernel $K(x, t)$ is self-adjoint or Hermitian.

$$K(x, t) = K^*(t, x).$$

Show that

(a) the eigenfunctions are orthogonal in the sense

$$\int_a^b \varphi_m^*(x)\, \varphi_n(x)\, dx = 0, \qquad m \neq n\ (\lambda_m \neq \lambda_n),$$

(b) the eigenvalues are real.

16.4.2 Solve the integral equation

$$\varphi(x) = x + \frac{1}{2}\int_{-1}^{1} (t + x)\, \varphi(t)\, dt$$

(cf. Exercise 16.3.2) by the Hilbert-Schmidt method.

The application of the Hilbert-Schmidt technique here is somewhat like using a shotgun to kill a mosquito, especially when the equation can be solved in about fifteen seconds by expanding in Legendre polynomials.

16.4.3 Solve the Fredholm integral equation

$$\varphi(x) = \lambda \int_0^\infty e^{-xt}\, \varphi(t)\, dt.$$

[1] This is like the inhomogeneous linear differential equation. To its solution we may add any constant times a solution of the corresponding homogeneous differential equation.

Note. A series expansion of the kernel e^{-xt} would permit a separable kernel-type solution (Section 16.3), except that the series is infinite. This suggests an infinite number of eigenvalues and eigenfunctions. If you stop with

$$\varphi(x) = x^{-1/2},$$
$$\lambda = \pi^{-1/2},$$

you will have missed most of the solutions! Show that the normalization integrals of the eigenfunctions do *not* exist. A basic reason for this anomalous behavior is that the range of integration is infinite, making this a "singular" integral equation.

16.4.4 Given

$$y(x) = x + \lambda \int_0^1 xt\, y(t)\, dt.$$

(a) Determine $y(x)$ as a Neumann series.

(b) Find the range of λ for which your Neumann series solution is convergent. Compare with the value obtained from

$$|\lambda|\, |K|_{\max} < 1.$$

(c) Find the eigenvalue and the eigenfunction of the corresponding homogeneous integral equation.

(d) By the separable kernel method show that the solution is

$$y(x) = \frac{3x}{3-\lambda}.$$

(e) Find $y(x)$ by the Hilbert-Schmidt method.

16.4.5 In Exercise 16.3.4

$$K(x, t) = \cos(x - t).$$

The (unnormalized) eigenfunctions are $\cos x$ and $\sin x$.

(a) Show that there is a function $h(t)$ such that $K(x, s)$, considered as a function of s alone, may be written

$$K(x, s) = \int_0^{2\pi} K(s, t)\, h(t)\, dt.$$

(b) Show that $K(x, t)$ may be expanded as

$$K(x, t) = \sum_{n=1}^{2} \frac{\varphi_n(x)\, \varphi_n(t)}{\lambda_n}.$$

16.5 Green's Functions—One Dimension

As part of the investigation of differential operators in Section 8.6 we see that Poisson's equation of electrostatics

$$\nabla^2 \varphi(\mathbf{r}) = -\frac{\rho(\mathbf{r})}{\varepsilon_0} \tag{16.115}$$

has a solution

$$\varphi(\mathbf{r}_1) = \frac{1}{4\pi\varepsilon_0} \int \frac{\rho(\mathbf{r}_2)}{|\mathbf{r}_1 - \mathbf{r}_2|}\, d\tau_2. \tag{16.116}$$

Here we have the infinite case in which the range of integration covers all space. If desired, the potential $\varphi(\mathbf{r}_1)$ may be developed for a finite case by using appropriate charge and dipole layer distributions on the boundaries.[1]

[1] Cf. Stratton, J. A., *Electromagnetic Theory*. New York: McGraw-Hill (1941).

Equation 16.116 may be given two interpretations.

1. If the potential function $\varphi(\mathbf{r}_1)$ is known and we seek the charge distribution $\rho(\mathbf{r}_2)$, which produces the given potential, Eq. 16.116 is an integral equation for $\rho(\mathbf{r}_2)$.
2. If the charge distribution $\rho(\mathbf{r}_2)$ is known, Eq. 16.116 yields the electrostatic potential $\varphi(\mathbf{r}_1)$ as a definite integral.

Following up this second (and more frequently encountered) situation, we may use the physicists' customary cause and effect vocabulary. We might label $\rho(\mathbf{r}_2)$ the "cause" that gives rise to the "effect" $\varphi(\mathbf{r}_1)$; that is, the charge distribution produces a potential field. However, the *effectiveness* of the charge in producing this potential depends on the distance between the element of charge $\rho(\mathbf{r}_2)\,d\tau_2$ and the point of interest given by $\mathbf{r}_1$. This effectiveness or, let us say, the influence of the element of charge is given by the function $(4\pi|\mathbf{r}_1 - \mathbf{r}_2|)^{-1}$.

For this reason $(4\pi|\mathbf{r}_1 - \mathbf{r}_2|)^{-1}$ is often called an influence function. Although we shall relabel it a Green's function, the physical basis for the term influence function remains important and may well be helpful in determining the form of other Green's functions.

Also in Section 8.6, the Green's function (for the operator ∇^2) is described as satisfying the point source equation

$$\nabla^2 G(\mathbf{r}_1, \mathbf{r}_2) = -\delta(\mathbf{r}_1 - \mathbf{r}_2). \tag{8.91}$$

A detailed discussion of the Dirac delta function in terms of sequences is included. Using Eq. 8.91 and Green's theorem, Section 1.11, the Green's function is shown to be symmetric:

$$G(\mathbf{r}_1, \mathbf{r}_2) = G(\mathbf{r}_2, \mathbf{r}_1). \tag{8.105}$$

In Section 9.4, the Dirac delta and Green's functions are expanded in series of eigenfunctions. These expansions make the symmetry properties explicit.

Further examples of Green's functions appear in connection with the Neumann function (Ex. 11.5.12) and the modified Bessel function $K_0(z)$. Exercise 11.5.13 points up the relations between point, line, and plane sources.

Moving into this chapter, in Section 16.1 it is seen that the integral equation corresponding to a differential equation *and certain boundary conditions* may lead to a peculiar kernel. This kernel is our Green's function.

The development of Green's functions from Eq. 8.91 for two- and three-dimensional systems is the topic of Section 16.6. Here, for simplicity, we restrict ourselves to one-dimensional cases and follow a somewhat different approach.[1]

Defining properties. In our one-dimensional analysis we consider first the nonhomogeneous Sturm-Liouville equation (Chapter 9)

$$\mathcal{L}\, y(x) + f(x) = 0, \tag{16.117}$$

in which $\mathcal{L}$ is the self-adjoint differential operator

$$\mathcal{L} = \frac{d}{dx}\left(p(x)\frac{d}{dx}\right) + q(x). \tag{16.118}$$

As in Section 9.1, $y(x)$ is required to satisfy certain boundary conditions at the end

[1] Equation 8.91 can be used for one-dimensional systems. The relationship between these two different approaches to Green's functions is shown at the end of this section.

points a and b of our interval $[a, b]$. Indeed, the interval may well be chosen so that appropriate boundary conditions can be satisfied. We now proceed to define a rather strange and arbitrary function G over the interval $[a, b]$. At this stage the most that can be said in defense of G is that the defining properties are legitimate, or mathematically acceptable.[1] Later, it is hoped, G may appear reasonable if not obvious.

1. The interval $a \leqslant x \leqslant b$ is divided by a parameter t. We shall label $G(x) = G_1(x)$ for $a \leqslant x < t$ and $G(x) = G_2(x)$ for $t < x \leqslant b$.

2. The functions $G_1(x)$ and $G_2(x)$ each satisfy the homogeneous[2] Sturm-Liouville equation: that is,

$$\begin{aligned} \mathscr{L}\, G_1(x) &= 0, \qquad a \leqslant x < t, \\ \mathscr{L}\, G_2(x) &= 0, \qquad t < x \leqslant b. \end{aligned} \tag{16.119}$$

3. At $x = a$, $G_1(x)$ satisfies the boundary conditions we impose on $y(x)$. At $x = b$, $G_2(x)$ satisfies the boundary conditions imposed on $y(x)$ at this end point of the interval. For convenience in renormalizing the boundary conditions are taken to be homogeneous; that is, at $x = a$

$$y(a) = 0,$$

$$y'(a) = 0,$$

or

$$\alpha\, y(a) + \beta\, y'(a) = 0$$

and similarly for $x = b$.

4. We demand that $G(x)$ be *continuous*,[3]

$$G_1(t) = G_2(t). \tag{16.120}$$

5. We require that $G'(x)$ be *discontinuous*, specifically that[3]

$$\frac{d}{dx} G_2(x)\bigg|_t - \frac{d}{dx} G_1(x)\bigg|_t = -\frac{1}{p(t)}. \tag{16.121}$$

where $p(t)$ comes from the self-adjoint operator, Eq. 16.118. Note that with the first derivative discontinuous the second derivative does not exist.

These requirements, in effect, make G a function of two variables, $G(x, t)$. Also we note that $G(x, t)$ depends on both the form of the differential operator $\mathscr{L}$ *and* the boundary conditions that $y(x)$ must satisfy.

Now, assuming that we can find a function $G(x, t)$ which has these properties, we shall label it a Green's function and proceed to show that a solution of Eq. 16.117 is

$$y(x) = \int_a^b G(x, t) f(t)\, dt. \tag{16.122}$$

[1] Note, however, that these properties are just those of the kernel of the Fredholm equation which had been derived from a self-adjoint differential equation, Example 16.1.3.

[2] Homogeneous with respect to the unknown function. The function $f(x)$ in Eq. 16.117 is set equal to zero.

[3] Strictly speaking, this is the limit as $x \to t$.

To do this we shall first construct the Green's function, $G(x, t)$. Let $u(x)$ be a solution of the homogeneous Sturm-Liouville equation which satisfies the boundary conditions at $x = a$ and $v(x)$ is a solution which satisfies the boundary conditions at $x = b$. Then we may take[1]

$$G(x, t) = \begin{cases} c_1\, u(x), & a \leqslant x < t, \\ c_2\, v(x), & t < x \leqslant b. \end{cases} \tag{16.123}$$

Continuity at $x = t$ (Eq. 16.120) requires

$$c_2\, v(t) - c_1\, u(t) = 0. \tag{16.124}$$

Finally, the discontinuity in the first derivative (Eq. 16.121) becomes

$$c_2\, v'(t) - c_1\, u'(t) = -\frac{1}{p(t)}. \tag{16.125}$$

There will be a unique solution for our unknown coefficients c_1 and c_2 if the Wronskian determinant

$$\begin{vmatrix} u(t) & v(t) \\ u'(t) & v'(t) \end{vmatrix} = u(t)\, v'(t) - v(t)\, u'(t)$$

does not vanish. We have seen in Section 8.5 that the nonvanishing of this determinant is a necessary condition for linear independence. Let us consider $u(x)$ and $v(x)$ to be independent. The contrary, which occurs when $u(x)$ satisfies the boundary conditions at both end points, requires a generalized Green's function. Strictly speaking, no Green's function exists when $u(x)$ and $v(x)$ are linearly dependent. This is also true when $\lambda = 0$ is an eigenvalue of the homogeneous equation. However, a "generalized Green's function" may be defined. This situation, which occurs with Legendre's equation, is discussed in Courant and Hilbert and other references. For independent $u(x)$ and $v(x)$ we have the Wronskian (again from Section 8.5)

$$u(t)\, v'(t) - v(t)\, u'(t) = \frac{A}{p(t)}, \tag{16.126}$$

in which A is a constant. Equation 16.126 is sometimes called Abel's formula. Numerous examples have appeared in connection with Bessel and Legendre functions. Now, from Eq. 16.125, we identify

$$\begin{aligned} c_1 &= -\frac{v(t)}{A}, \\ c_2 &= -\frac{u(t)}{A}. \end{aligned} \tag{16.127}$$

Equation 16.124 is clearly satisfied. Substitution into Eq. 16.123 yields our Green's function.

[1]The "constants" c_1 and c_2 are independent of x, but they may (and do) depend on the other variable, t.

$$G(x, t) = \begin{cases} -\dfrac{1}{A} u(x)\, v(t), & a \leqslant x < t, \\ \\ -\dfrac{1}{A} u(t)\, v(x), & t < x \leqslant b. \end{cases} \tag{16.128}$$

Note carefully that $G(x, t) = G(t, x)$. This is the symmetry property that was proved earlier in Section 8.6. Its physical interpretation is given by the reciprocity principle (via our influence function)—a cause at t yields the same effect at x as a cause at x produces at t. In terms of our electrostatic analogy this is obvious, the influence function depending only on the magnitude of the distance between the two points

$$|\mathbf{r}_1 - \mathbf{r}_2| = |\mathbf{r}_2 - \mathbf{r}_1|.$$

We have constructed $G(x, t)$, but there still remains the task of showing that the integral (Eq. 16.122) with our new Green's function is indeed a solution of the original differential equation (16.117). This we do by direct substitution. With $G(x, t)$ given by Eq. 16.128,[1] Eq. 16.122 becomes

$$y(x) = -\frac{1}{A}\int_a^x v(x)\, u(t) f(t)\, dt - \frac{1}{A}\int_x^b u(x)\, v(t) f(t)\, dt. \tag{16.129}$$

Differentiating, we obtain

$$y'(x) = -\frac{1}{A}\int_a^x v'(x)\, u(t) f(t)\, dt - \frac{1}{A}\int_x^b u'(x)\, v(t) f(t)\, dt, \tag{16.130}$$

the derivatives of the limits cancelling. A second differentiation yields

$$\begin{aligned} y''(x) = &-\frac{1}{A}\int_a^x v''(x)\, u(t) f(t)\, dt - \frac{1}{A}\int_x^b u''(x)\, v(t) f(t)\, dt \\ &-\frac{1}{A}[u(x)\, v'(x) - v(x)\, u'(x)] f(x). \end{aligned} \tag{16.131}$$

By Eqs. 16.125 and 16.127 this may be rewritten

$$y''(x) = -\frac{v''(x)}{A}\int_a^x u(t) f(t)\, dt - \frac{u''(x)}{A}\int_x^b v(t) f(t)\, dt - \frac{f(x)}{p(x)}. \tag{16.132}$$

Now, by substituting into Eq. 16.118, we have

$$\mathscr{L}\, y(x) = -\frac{[\mathscr{L}\, v(x)]}{A}\int_a^x u(t) f(t)\, dt - \frac{[\mathscr{L}\, u(x)]}{A}\int_x^b v(t) f(t)\, dt - f(x). \tag{16.133}$$

Since $u(x)$ and $v(x)$ were chosen to satisfy the homogeneous Sturm-Liouville equation, the factors in brackets are zero and the integral terms vanish. Transposing $f(x)$, we see that Eq. 16.117 is satisfied.

We must also check that $y(x)$ satisfies the required boundary conditions. At

[1] In the first integral $a \leqslant t \leqslant x$. Hence $G(x, t) = G_2(x, t) = -(1/A)u(t)\, v(x)$. Similarly, the second integral requires $G = G_1$.

point $x = a$

$$y(a) = -\frac{u(a)}{A}\int_a^b v(t) f(t)\, dt = c\, u(a), \tag{16.134}$$

$$y'(a) = -\frac{u'(a)}{A}\int_a^b v(t) f(t)\, dt = c\, u'(a), \tag{16.135}$$

since the definite integral is a constant. We chose $u(x)$ to satisfy

$$\alpha\, u(a) + \beta\, u'(a) = 0. \tag{16.136}$$

Multiplying by the constant c, we verify that $y(x)$ also satisfies Eq. 16.136. This illustrates the utility of the *homogeneous* boundary conditions: the normalization does not matter. In quantum mechanical problems the boundary condition on the wave function is often expressed in terms of the ratio

$$\frac{\psi'(x)}{\psi(x)} = \frac{d}{dx} \ln \psi(x),$$

equivalent to Eq. 16.136. The advantage is that the wave function need not be normalized.

Summarizing, we have Eq. 16.122

$$y(x) = \int_a^b G(x, t) f(t)\, dt$$

which satisfies the differential equation (Eq. 16.117)

$$\mathscr{L}\, y(x) + f(x) = 0$$

and the boundary conditions, these boundary conditions having been built into the Green's function $G(x, t)$.

Basically, what we have done is to use the solutions of the homogeneous Sturm-Liouville equation to construct a solution of the nonhomogeneous equation. Again Poisson's equation is an illustration. The solution (Eq. 16.116) represents a weighted $[\rho(\mathbf{r}_2)]$ combination of solutions of the corresponding homogeneous Laplace's equation.

The preceding analysis placed no special restrictions on our $f(x)$. Let us now assume that $f(x) = \lambda\, \rho(x)\, y(x)$. Then we have

$$y(x) = \lambda \int_a^b G(x, t)\, \rho(t)\, y(t)\, dt \tag{16.137}$$

as a solution of

$$\mathscr{L}\, y(x) + \lambda\, \rho(x)\, y(x) = 0 \tag{16.138}$$

and its boundary conditions. Equation 16.137 is a homogeneous Fredholm equation of the second kind and Eq. 16.138 is the Sturm-Liouville eigenvalue equation of Chapter 9.

Notice the change from Eqs. 16.117 and 16.122 to 16.137 and 16.138. There is a corresponding change in the interpretation of our Green's function. It started as an importance or influence function, a weighting function giving the importance of the charge $\rho(\mathbf{r}_2)$ in producing the potential $\varphi(\mathbf{r}_1)$. The charge ρ was the nonhomogeneous term in the nonhomogeneous differential equation 16.117. Now, the differential equation and the integral equation are both *homogeneous.* $G(x, t)$ has become a link relating the two equations, differential and integral.

To complete the discussion of this differential equation—integral equation equivalence—let us now show that Eq 16.138 implies Eq. 16.137; that is, a solution of our differential equation (16.138) with its boundary conditions satisfies the integral equation (16.137). We multiply Eq. 16.138 by $G(x, t)$, the appropriate Green's function, and integrate from $x = a$ to $x = b$ to obtain

$$\int_a^b G(x, t)\mathscr{L}\, y(x)\, dx + \lambda \int_a^b G(x, t)\, \rho(x)\, y(x)\, dx = 0. \qquad (16.139)$$

The first integral is split in two ($x < t$, $x > t$), according to the construction of our Green's function, giving

$$-\int_a^t G_1(x, t)\mathscr{L}\, y(x)\, dx - \int_t^b G_2(x, t)\mathscr{L}\, y(x)\, dx = \lambda \int_a^b G(x, t)\, \rho(x)\, y(x)\, dx. \qquad (16.140)$$

Applying Green's theorem to the left-hand side or, equivalently, integrating by parts, we obtain

$$-\int_a^t G_1(x, t)\left[\frac{d}{dx}\left(p(x)\frac{d}{dx}y(x)\right) + q(x)\, y(x)\right]dx$$
$$= -\left|G_1(x, t)\, p(x)\, y'(x)\right|_a^t + \int_a^t G_1'(x, t)\, p(x)\, y'(x)\, dx - \int_a^t G_1(x, t)\, q(x)\, y(x)\, dx, \qquad (16.141)$$

with an equivalent expression for the second integral. A second integration by parts yields

$$-\int_a^t G_1(x, t)\mathscr{L}\, y(x)\, dx = -\int_a^t y(x)\, \mathscr{L}\, G_1(x, t)\, dx$$
$$- \left|G_1(x, t)\, p(x)\, y'(x)\right|_a^t + \left|G_1'(x, t)\, p(x)\, y(x)\right|_a^t. \qquad (16.142)$$

The integral on the right vanishes because $\mathscr{L}G_1 = 0$. By combining the integrated terms with those from integrating G_2, we have

$$-p(t)[G_1(t, t)\, y'(t) - G_1'(t, t)\, y(t) - G_2(t, t)\, y'(t) + G_2'(t, t)\, y(t)]$$
$$+ p(a)[G_1(a, t)\, y'(a) - G_1'(a, t)\, y(a)]$$
$$- p(b)[G_2(b, t)\, y'(b) - G_2'(b, t)\, y(b)]. \qquad (16.143)$$

Each of the last two expressions vanishes, for $G(x, t)$ and $y(x)$ satisfy the same boundary conditions. The first expression, with the help of Eqs. 16.120 and 16.121, reduces to $y(t)$. Substituting into Eq. 16.140, we have Eq. 16.137, thus completing the demonstration of the equivalence of the integral equation and the differential equation plus boundary conditions.

EXAMPLE 16.5.1. LINEAR OSCILLATOR

As a simple example, consider the linear oscillator equation (for a vibrating string)

$$y''(x) + \lambda\, y(x) = 0. \tag{16.144}$$

We impose the conditions $y(0) = y(1) = 0$, which correspond to a string clamped at both ends. Now, to construct our Green's function, we need solutions of the homogeneous Sturm-Liouville equation, $\mathscr{L}y(x) = 0$, which is $y''(x) = 0$. To satisfy the boundary conditions one solution must vanish at $x = 0$, the other at $x = 1$. Such solutions (unnormalized) are

$$\begin{aligned} u(x) &= x, \\ v(x) &= 1 - x. \end{aligned} \tag{16.145}$$

We find that

$$uv' - vu' = -1 \tag{16.146}$$

or, by Eq. 16.126 with $p(x) = 1$, $A = -1$. Our Green's function becomes

$$G(x, t) = \begin{cases} x(1-t), & 0 \leqslant x < t, \\ t(1-x), & t < x \leqslant 1. \end{cases} \tag{16.147}$$

Hence by Eq. 16.137 our clamped vibrating string satisfies

$$y(x) = \lambda \int_0^1 G(x, t)\, y(t)\, dt. \tag{16.148}$$

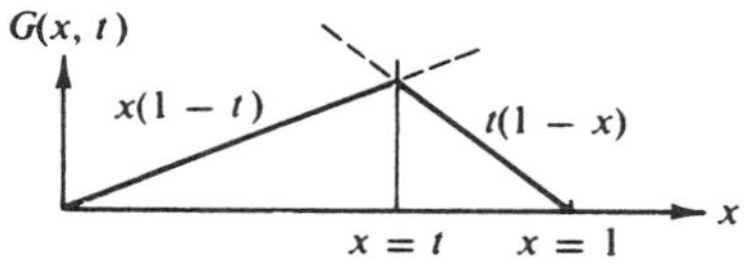

FIG. 16.3 A linear oscillator Green's function

The reader may show that the known solutions of Eq. 16.144.

$$y = \sin n\pi x \qquad (\lambda = n^2\pi^2),$$

do indeed satisfy Eq. 16.148. Note that our eigenvalue λ is *not* the wavelength.

One more approach to the Green's function may shed additional light on our formulation and particularly on its relation to physical problems. Let us refer once more to Poisson's equation, this time for a point charge

$$\nabla^2\varphi(\mathbf{r}) = -\frac{\rho_{\text{point}}}{\varepsilon_0}. \tag{16.149}$$

The Green's function solution of this equation was developed in Section 8.6. This time let us take a one-dimensional analog

$$\mathscr{L}\, y(x) + f(x)_{\text{point}} = 0. \tag{16.150}$$

Here $f(x)_{\text{point}}$ refers to a unit point "charge" or a point force. We may represent it by a number of forms, but perhaps the most convenient is

$$f(x)_{\text{point}} = \begin{cases} \dfrac{1}{2\varepsilon}, & t - \varepsilon < x < t + \varepsilon, \\ 0, & \text{elsewhere}, \end{cases} \tag{16.151}$$

which is essentially the same as Eq. 8.83*a*. Then, integrating Eq. 16.150, we have

$$\begin{aligned} \int_{t-\varepsilon}^{t+\varepsilon} \mathscr{L}\, y(x)\, dx &= -\int_{t-\varepsilon}^{t+\varepsilon} f(x)_{\text{point}}\, dx \\ &= -1 \end{aligned} \tag{16.152}$$

from the definition of $f(x)$. Let us examine $\mathscr{L}\, y(x)$ more closely. We have

$$\begin{aligned} &\int_{t-\varepsilon}^{t+\varepsilon} \frac{d}{dx}[p(x)\, y'(x)]\, dx + \int_{t-\varepsilon}^{t+\varepsilon} q(x)\, y(x)\, dx \\ &\qquad = p(x)\, y'(x)\Big|_{t-\varepsilon}^{t+\varepsilon} + \int_{t-\varepsilon}^{t+\varepsilon} q(x)\, y(x)\, dx = -1. \end{aligned} \tag{16.153}$$

In the limit $\varepsilon \to 0$ we may satisfy this relation by *permitting* $y'(x)$ to have a discontinuity of $-1/p(x)$ at $x = t$, $y(x)$ itself remaining continuous.[1] These, however, are just the properties used to define our Green's function, $G(x, t)$. In addition, we note that in the limit $\varepsilon \to 0$

$$f(x)_{\text{point}} = \delta(x - t), \tag{16.154}$$

in which $\delta(x - t)$ is our Dirac delta function, defined in this manner in Section 8.6. Hence Eq. 16.150 has become

$$\mathscr{L}\, G(x, t) = -\delta(x - t). \tag{16.155}$$

This is Eq. 8.91, which we exploit for the development of Green's functions in two and three dimensions—Section 16.6. It will be recalled that we used this relation in Sections 8.6 and 11.5 to determine our Green's functions.

EXERCISES

16.5.1 Show that $G(x, t) = \begin{cases} x, & 0 \leqslant x < t, \\ t, & t < x \leqslant 1. \end{cases}$

is the Green's function for the operator $\mathscr{L} = d^2/dx^2$ and the boundary conditions

$$y(0) = 0,$$
$$y'(1) = 0.$$

[1] The functions $p(x)$ and $q(x)$ appearing in the operator $\mathscr{L}$ are continuous functions. With $y(x)$ remaining continuous, $\int q(x)\, y(x)\, dx$ is certainly continuous. Hence this integral over an interval 2ε (Eq. 16.153) vanishes as ε vanishes.

16.5.2 Find the Green's function for

(a) $\mathscr{L}\, y(x) = \dfrac{d^2y(x)}{dx^2} + y(x), \qquad \begin{cases} y(0) = 0, \\ y'(1) = 0, \end{cases}$

(b) $\mathscr{L}\, y(x) = \dfrac{d^2y(x)}{dx^2} - y(x), \qquad y(x)$ finite for $-\infty < x < \infty$.

16.5.3 Find the Green's function for the Bessel operators

(a) $\mathscr{L}\, y(x) = \dfrac{d}{dx}\left(x\dfrac{dy(x)}{dx}\right), \qquad$ *Ans.* (a) $G(x, t) = \begin{cases} -\ln t, & 0 \leqslant x < t, \\ -\ln x, & t < x \leqslant 1, \end{cases}$

(b) $\mathscr{L}\, y(x) = \dfrac{d}{dx}\left(x\dfrac{dy(x)}{dx}\right) - \dfrac{n^2}{x}\, y(x),$

with $y(0)$ finite and $y(1) = 0$.

(b) $G(x, t) = \begin{cases} \dfrac{1}{2n}\left[\left(\dfrac{x}{t}\right)^n - (xt)^n\right], & 0 \leqslant x < t, \\ \dfrac{1}{2n}\left[\left(\dfrac{t}{x}\right)^n - (xt)^n\right], & t < x \leqslant 1. \end{cases}$

16.5.4 Construct the Green's function for

$$x^2\frac{d^2y}{dx^2} + x\frac{dy}{dx} + (k^2x^2 - 1)y = 0,$$

subject to the boundary conditions

$$y(0) = 0,$$
$$y(1) = 0.$$

16.5.5 Given

(1) $$\mathscr{L} = (1 - x^2)\frac{d^2}{dx^2} - 2x\frac{d}{dx}$$

and

(2) $G(\pm 1, t)$ remains finite.

Show that no Green's function can be constructed by the techniques of this section. ($u(x)$ and $v(x)$ are linearly dependent.)

16.5.6 Construct the infinite one-dimensional Green's function for the Helmholtz equation

$$(\nabla^2 + k^2)\psi(x) = g(x).$$

The boundary conditions are those for a wave advancing in the positive x direction—assuming a time dependence $e^{-i\omega t}$.

Ans. $G(x_1, x_2) = \dfrac{i}{2k}\exp(ik|x_1 - x_2|)$.

16.5.7 Construct the infinite one-dimensional Green's function for the modified Helmholtz equation

$$(\nabla^2 - k^2)\psi(x) = f(x).$$

The boundary conditions are that the Green's function must vanish for $x \to \infty$ and for $x \to -\infty$.

Ans. $G(x_1, x_2) = \dfrac{1}{2k}\exp(-k|x_1 - x_2|)$.

16.5.8 From the eigenfunction expansion of the Green's function show that

(a) $$\frac{2}{\pi^2}\sum_{n=1}^{\infty}\frac{\sin n\pi x \sin n\pi t}{n^2} = \begin{cases} x(1-t), & 0 \leqslant x < t, \\ t(1-x), & t < x \leqslant 1. \end{cases}$$

(b) $$\frac{2}{\pi^2}\sum_{n=0}^{\infty}\frac{\sin (n+\frac{1}{2})\pi x \sin (n+\frac{1}{2})\pi t}{(n+\frac{1}{2})^2} = \begin{cases} x, & 0 \leqslant x < t, \\ t, & t < x \leqslant 1. \end{cases}$$

Note : In Section 9.4, the Green's function of $\mathscr{L} + \lambda$ is expanded in eigenfunctions. The λ there is an adjustable parameter, not an eigenvalue.

16.5.9 In the Fredholm equation,

$$f(x) = \lambda^2 \int_a^b G(x, t)\, \varphi(t)\, dt,$$

$G(x, t)$ is a Green's function given by

$$G(x, t) = \sum_{n=1}^{\infty} \frac{\varphi_n(x)\, \varphi_n(t)}{\lambda_n^2 - \lambda^2}.$$

Show that the solution is

$$\varphi(x) = \sum_{n=1}^{\infty} \frac{\lambda_n^2 - \lambda^2}{\lambda^2}\, \varphi_n(x) \int_a^b f(t)\, \varphi_n(t)\, dt.$$

16.6 Green's Functions—Two and Three Dimensions

As in the preceding section (and in Section 8.6), we consider a nonhomogeneous differential equation

$$\mathscr{L} y(\mathbf{r}_1) = -f(\mathbf{r}_1). \tag{16.156}$$

We seek a solution which might be represented by

$$y(\mathbf{r}_1) = -\mathscr{L}^{-1} f(\mathbf{r}_1). \tag{16.156a}$$

It might be expected that with $\mathscr{L}$ a differential operator, the inverse operator $\mathscr{L}^{-1}$ will involve integration. To proceed further, we define the Green's function corresponding to the differential operator $\mathscr{L}$ as a solution of the *point source* nonhomogeneous equation[1]

$$\mathscr{L}_1\, G(\mathbf{r}_1, \mathbf{r}_2) = -\delta(\mathbf{r}_1 - \mathbf{r}_2), \tag{16.156b}$$

which satisfies the required boundary conditions. Here the subscript 1 on $\mathscr{L}$ emphasizes that $\mathscr{L}$ operates on $\mathbf{r}_1$.

Let us assume that $\mathscr{L}_1$ is a self-adjoint differential operator of the general form[2]

$$\mathscr{L}_1 = \nabla_1 \cdot [p(\mathbf{r}_1)\, \nabla_1] + q(\mathbf{r}_1). \tag{16.156c}$$

Then, as a simple generalization of Green's theorem, Eq. 1.97, we have

$$\int (v \mathscr{L}_2 u - u \mathscr{L}_2 v)\, d\tau_2 = \int p(v \nabla_2 u - u \nabla_2 v) \cdot d\boldsymbol{\sigma}_2, \tag{16.156d}$$

in which all quantities have $\mathbf{r}_2$ as their argument. We let $u(\mathbf{r}_2) = y(\mathbf{r}_2)$ so that

[1] As stressed in Section 8.6, the delta function will be part of an integrand.

[2] $\mathscr{L}_1$ may be in 1, 2, or 3 dimensions (with appropriate interpretation of ∇_1).